PROMPT®
PUBLICATIONS

REVISED EDITION, 1992

PROMPT® Publications is an imprint of Sams Technical Publishing, 5436 W. 78th St., Indianapolis, IN 46268.

International Standard Book Number: 0-7906-1041-8

Acquisitions Editor: Alice J. Tripp
Cover Design: Sarah Wright
Illustrations and Other Materials: Unless credited otherwise, courtesy author, Van Valekenburgh, Nooger & Neville, Inc., or Sams Technical Publishing.

PRINTED IN THE UNITED STATES OF AMERICA

15 14 13 12 11

Preface to Revision of BASIC ELECTRICITY

The COMMON-CORE® Program – *Basic Electricity, Basic Electronics, Basic Synchros and Servomechanisms,* etc. – was designed and developed during the years 1952–1954. On the basis of a job task analysis of a broad spectrum of U.S. Navy electrical/electronics equipment of that era, there was established a "common-core" of prerequisite knowledge and skills. This "common-core" prerequisite was then programed into a teaching/learning system which had as its primary instructional objective the effective training of U.S. Navy electrical/electronics technicians who could understand and apply such understanding in meaningful job problem situations.

Since that time, over 100,000 U.S. Navy technicians have been efficiently trained by this performance-based system. Civilian students and technicians have accounted for hundreds of thousands more. The military and civilian education and training programs in South America, Europe, the Middle East, Asia, Australia, and Africa have also recognized its usefulness with some 12 foreign-language editions presently in print.

Now the foundation of the COMMON-CORE Program, *Basic Electricity,* is being updated and improved. Its equipment job task base has been enlarged to cover the understanding and skills needed for the spectrum of present-day electrical/electronic equipment – modern industrial machines, controls, instrumentation, computers, communications, radar, lasers, etc. Its technological components/circuits/functions base has been revised and broadened to incorporate the generations of development in electrical/electronics technology – namely, from (1) vacuum tubes to (2) transistors and semiconductors to (3) integrated circuits, large scale integration, and microminiaturization.

Educationally, considerable effort has been given to incorporating individualized learning/testing features and techniques within the texts themselves, and in the accompanying interactive student mastery tests.

Notwithstanding the passage of time, the original innovative, basic text-format, system-design elements of the COMMON-CORE Program still stand – this solid framework of proved effectiveness that has been the stimulus for many of the improvements in vocational/technical education.

VAN VALKENBURGH, NOOGER & NEVILLE, INC.

New York, N.Y.

CONTENTS

Volume 1

Ohm's Law

Resistance

DC Series Circuits

DC Parallel Circuits

DC Series-Parallel Circuits

Electric Power

Volume 3

AC Meters

Resistance in AC Circuits

Inductance in DC and AC Circuits

Capacitance in DC and AC Circuits

Troubleshooting Simple AC Circuits

Volume 4

AC Parallel Circuits

AC Complex Circuits

Transformers

Introduction to AC Power Distribution

Troubleshooting AC Circuits

Appendix

Volume 5

Introduction to Generators and Motors

Introduction to Generators

The Elementary Generator

The DC Generator

The DC Motor

DC Systems and Controls

DC Machinery Maintenance and Troubleshooting

AC Generators—Alternators

AC Motors

AC Systems and Controls

AC Systems Troubleshooting

Basic Electricity

REVISED EDITION

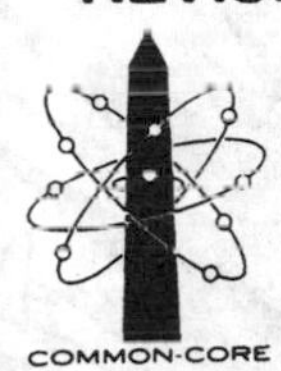

VAN VALKENBURGH,
NOOGER & NEVILLE, INC.

VOL. 1

SPACE
INDUSTRY
ELECTRICITY
IS THE KEY TO
OPPORTUNITY
IN...
SCIENCE
DEFENSE

The Importance of the Study of Electricity

It is hard to imagine a world without electricity. It touches and influences our daily lives in hundreds of ways. We see the use of electricity directly in our homes for lighting, the operation of appliances, telephone, television, radio, stereo, heating, etc. We see the use of electricity in transportation. Electricity has been used in the manufacture of most of the things we use either directly or to operate machines that make or process the products that we need. Without electricity, most of the things we use and enjoy today would not be possible.

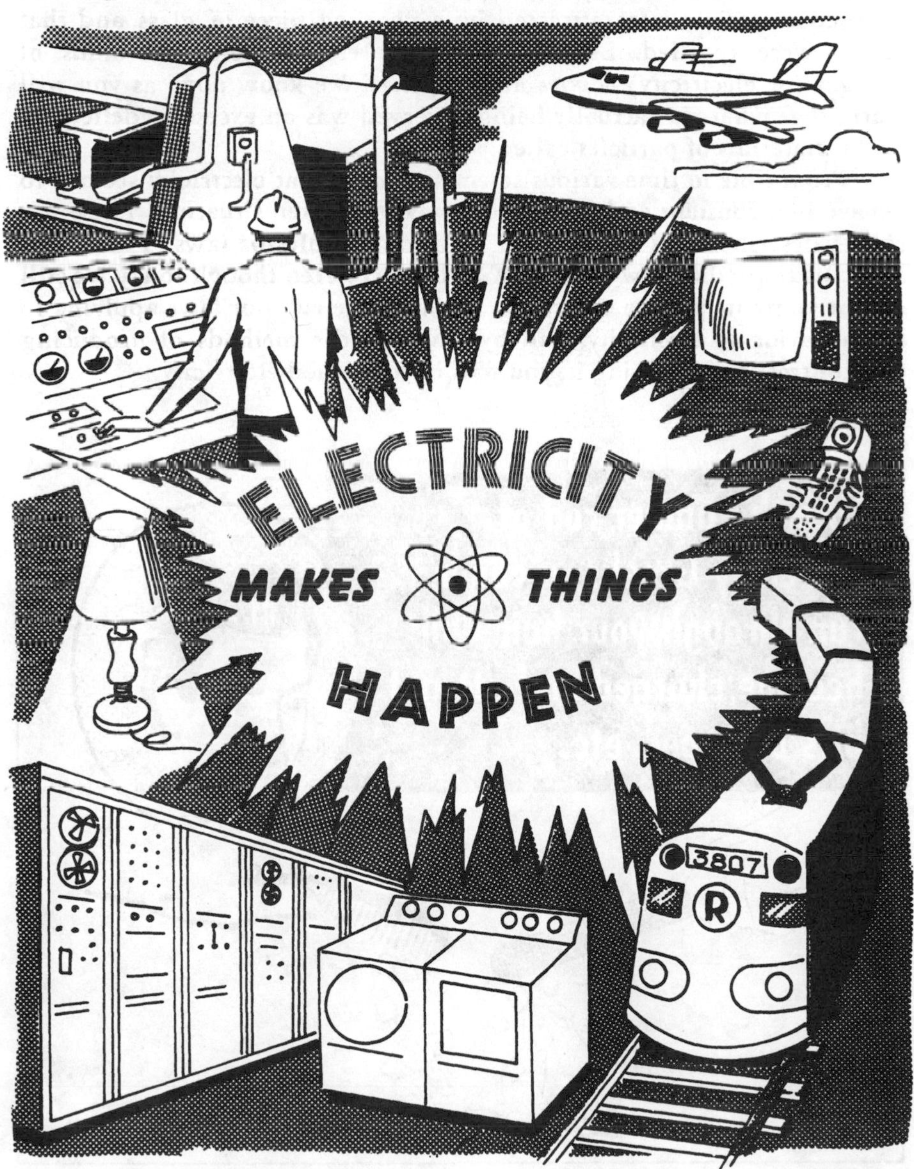

Early History

The word electricity comes from the ancient Greek word for amber —*elektron*. The early Greeks observed that when amber (a fossilized resin) was rubbed with a cloth, it would attract bits of material such as dried leaves. Later, scientists showed that this property of attraction occurred in other materials such as rubber and glass but did not occur with materials such as copper or iron. The materials that had this property of attraction when rubbed with a cloth were described as being charged with an *electric force*; and it was noticed that some of these charged materials were attracted by a charged piece of glass and that others were repelled. Benjamin Franklin called these two kinds of charges (or electricity) *positive* and *negative*. We know now, as you will learn, that what was actually being observed was an excess or deficiency in the materials of particles called *electrons*.

From time to time various scientists found that electricity seemed to behave in a constant and predictable way in a given situation. These scientists described this behavior in the form of rules or laws. These laws allow us to predict how electricity will behave even though today we still do not know its precise nature. By learning the rules or laws applying to the behavior of electricity, and by learning the methods of producing and controlling and using it, you will have learned electricity.

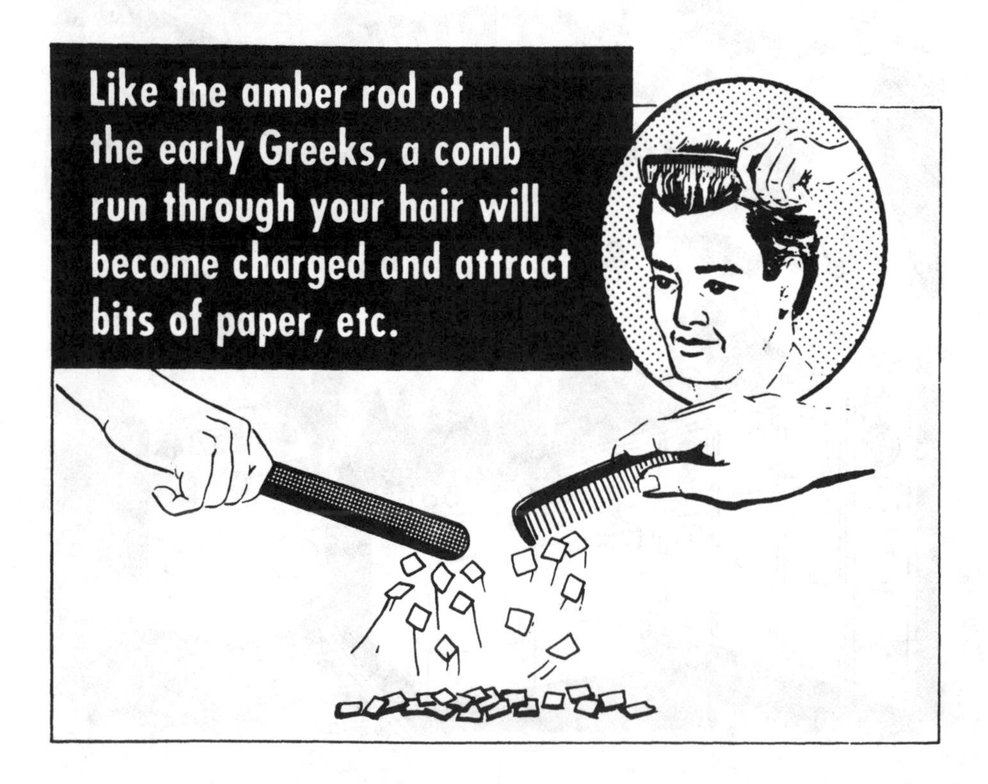

Like the amber rod of the early Greeks, a comb run through your hair will become charged and attract bits of paper, etc.

The Electron Theory

All the effects of electricity take place because of the existence of a tiny particle called the *electron*. Since no one has actually seen an electron, but only the effects it produces, we call the laws governing its behavior the *electron theory*. The electron theory is not only the basis for the design for all electrical and electronic equipment, it also explains physical and chemical action and helps scientists to probe into the very nature of the universe and life itself.

The Electron Theory (continued)

Since assuming that the electron exists has led to so many important discoveries in electricity, electronics, chemistry, and atomic physics, we can safely assume that the electron really exists. All electrical and electronic equipment has been designed using this theory. Since the electron theory has always worked for everyone, it will always work for you.

Your entire study of electricity will be based upon the electron theory, which assumes that all electrical and electronic effects are due to the movement of electrons from place to place or that there are too many or too few electrons in a particular place.

According to the electron theory, all electrical and electronic effects are caused either by the movement of electrons from place to place or because there exist too many or too few electrons in a particular place at a particular time.

Before you can usefully begin to consider the forces that make electrons move or accumulate, you must first find out what an electron is.

All matter is composed of atoms of many different sizes, degrees of structural complexity, and weight. But all atoms are alike in consisting of a nucleus—which differs from atom to atom of the 100-odd chemical elements that either exist in Nature or have been made by man—and of a varying number of electrons which move about that nucleus.

You will get an idea of what the atom is essentially like by looking at the picture below.

THE ELECTRON IS ELECTRICITY

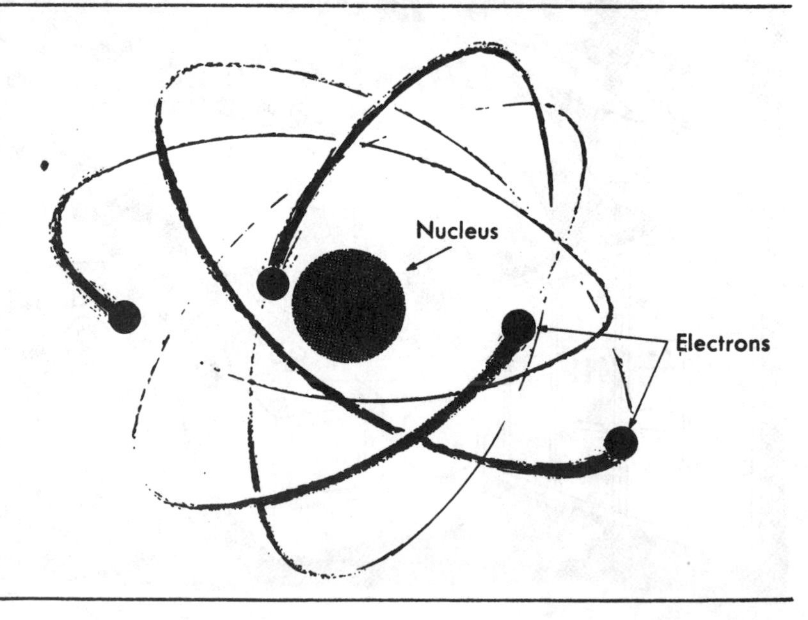

The Breakdown of Matter

A good way of understanding more about what the electron is like is to examine closely the composition of a drop of ordinary water.

If you take this drop of water and divide it into two drops, divide one of these two drops into two smaller drops, and repeat this process thousands of times, you will have a very tiny drop of water. This tiny drop will be so small that you will need the best microscope made today in order to see it.

This tiny drop of water will still have all the chemical characteristics of water. If examined by a chemist, he will not be able to find any chemical difference between this microscopic drop and an ordinary glass of water.

The Breakdown of Matter (continued)

Now if you take this tiny drop of water and try to divide it in half any further, you will not be able to see it in your microscope. Imagine then that you have available a super microscope that will magnify many times more than any microscope presently existing. This microscope can give you any magnification you want, so you can put your tiny drop of water under it and divide it into smaller and smaller droplets.

As the droplet of water is divided into smaller and smaller droplets, these tiny droplets will still have all the chemical characteristics of water. However, you eventually will have a droplet so small that any further division will cause it to lose the chemical characteristics of water. This last bit of water is called a *molecule.* Thus, a molecule is the smallest unit into which a substance can be divided and still be identified as that substance.

The Structure of the Molecule

When you increase the magnifying power of the microscope, you will see that the water molecule is made up of two tiny structures that are the same and a larger structure that is different from the two. These structures are called *atoms*. The two smaller atoms which are the same are hydrogen atoms and the larger, different one is an oxygen atom. When two atoms of hydrogen combine with one atom of oxygen, you have a molecule of water.

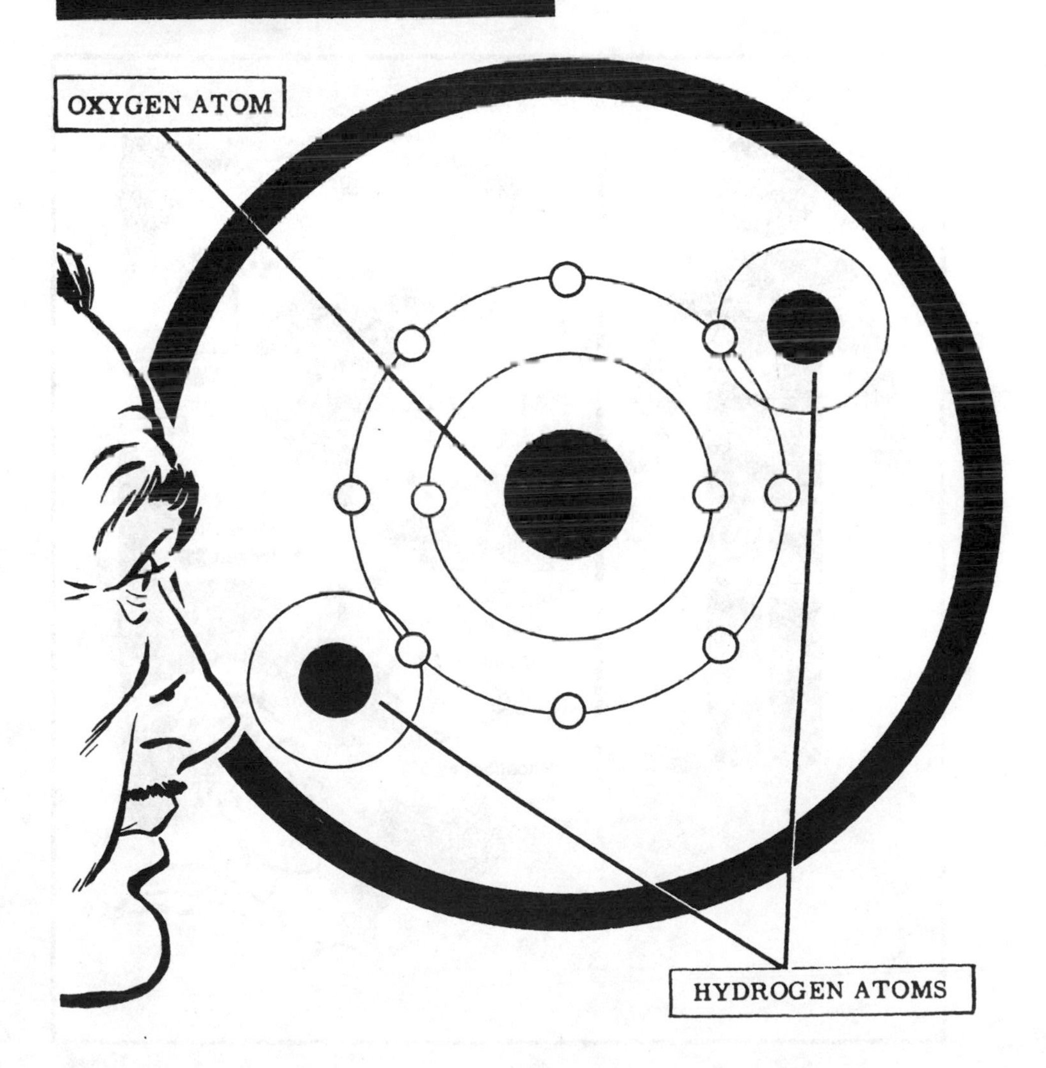

The Structure of the Molecule (continued)

While water is made up of only two kinds of atoms—oxygen and hydrogen—the molecules of many materials are more complex in structure. Cellulose molecules, the basic molecules of which wood is made, consist of three different kinds of atoms—carbon, hydrogen, and oxygen. All materials are made up of different combinations of atoms to form molecules of the materials. There are only about 100 different kinds of atoms and these are known as elements: oxygen, carbon, hydrogen, iron, gold, and nitrogen are all elements. The human body with all its complex tissues, bones, teeth, etc., is made up mainly of only 15 elements, and only six of these are found in quantity. (See Table of Elements at back of book.)

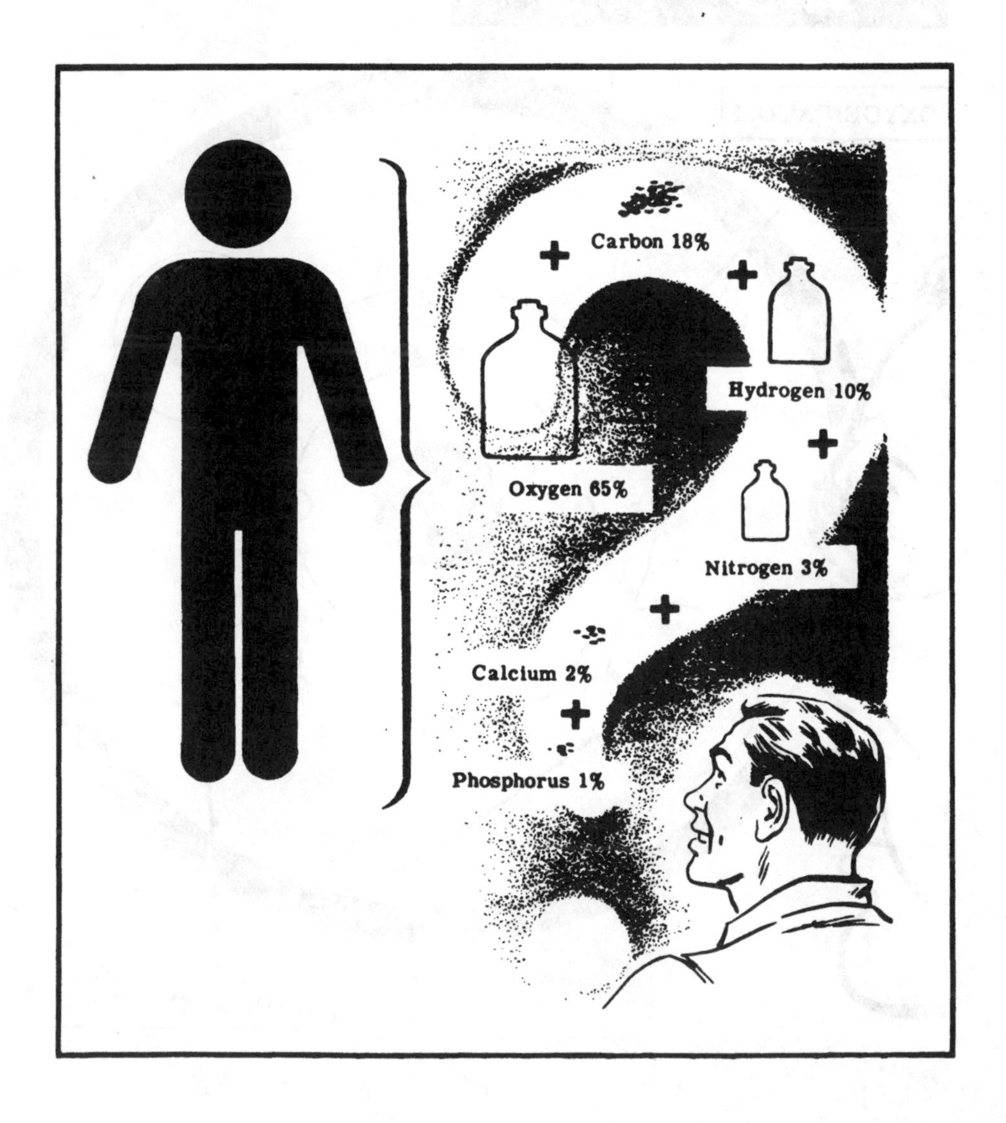

The Structure of the Atom

Now that you know that all materials are made up of molecules which consist of various combinations of about 100 different types of atoms, you will want to know what all this has to do with electricity. Increase the magnification of your imaginary super microscope still further and examine the atoms in the water molecule. Pick out the smallest atom you can see—the hydrogen atom—and examine it closely.

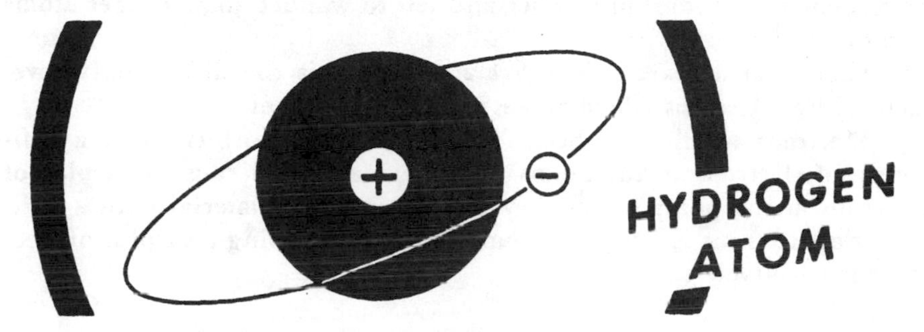

You see that the hydrogen atom is like a sun with one planet spinning around it. The planet is known as an electron and the sun is known as the nucleus. The electron has a *negative* (—) charge of electricity and the nucleus has a *positive* (+) charge of electricity.

In an atom, the total number of negatively charged electrons circling around the nucleus equals exactly the number of positive charges in the nucleus. The positive charges are called *protons.* Besides the protons, the nucleus also contains electrically neutral particles called *neutrons,* which are like a proton and an electron bonded together. Atoms of different elements contain different numbers of neutrons within the nucleus, but the number of electrons spinning about the nucleus always equals the number of protons (or positive charges) within the nucleus.

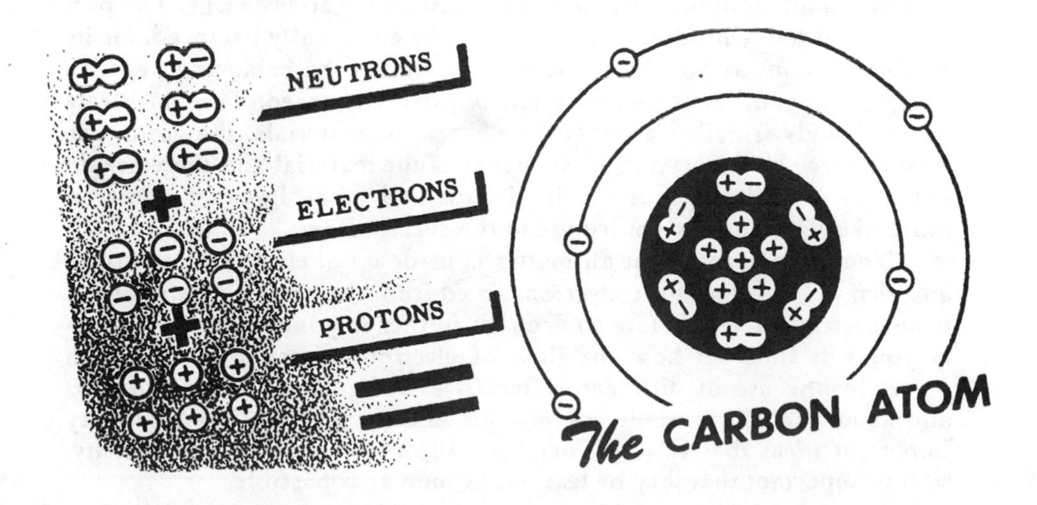

Electric Current and Electric Charge

All atoms are bound together by powerful forces of attraction between the nucleus and its electrons. Electrons in the outer orbits of an atom, however, are attracted to their nucleus less powerfully than are electrons whose orbits are nearer the nucleus.

In certain materials (they are known as electrical conductors), these outer electrons are so weakly bound to their nucleus that they can easily be forced away from it altogether and left to wander among other atoms at random.

Such electrons are called *free electrons*. It is the directional movement of free electrons which makes an electric current.

Electrons which have been forced out of their orbits create a deficiency of electrons in the atoms they leave and will cause a surplus of electrons at the point where they come to rest. A material with a deficiency of electrons is positively charged; one possessing a surplus of electrons is negatively charged.

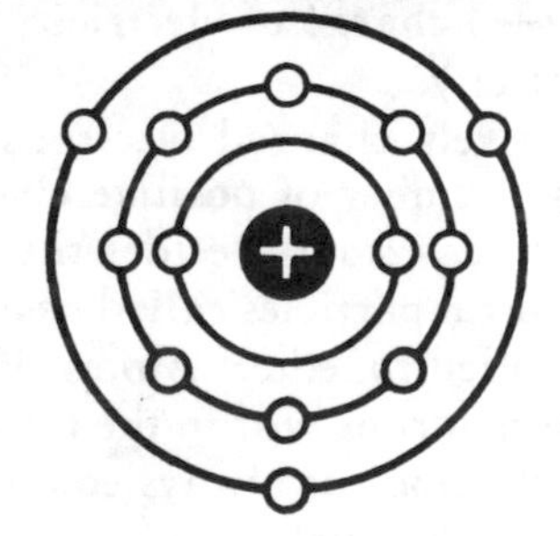

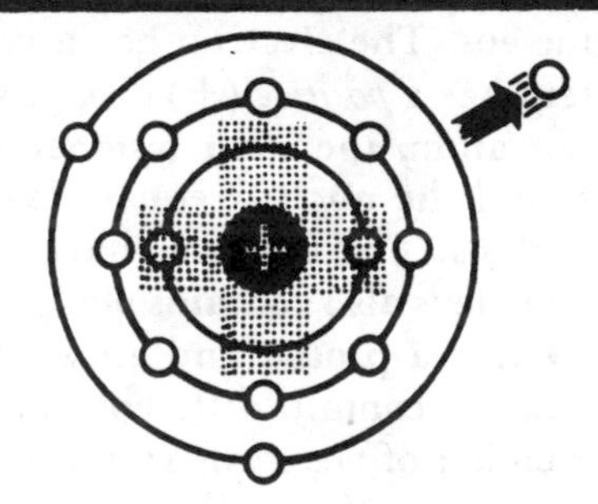

When an atom loses an electron, it loses a negative charge. The part of the atom left behind therefore ceases to be electrically balanced, for its nucleus remains as positive as before, but one of the balancing negative charges has gone. It is therefore left positively charged. This positively charged body is called a *positive ion*. In solid materials, the atoms are held in place by the crystalline structure of the material and therefore do not move as free electrons do. In liquids and gases, however, ions can move like electrons and contribute to the current flow.

You have learned that all matter is made up of electronic structures and that the motion of the electrons freed from the outer orbits of atoms is an electric current. Before you can go further in your study of electricity you will find out how the flow of electrons is confined to certain places by the use of different materials called *conductors* and *insulators* and about the nature of electric charges and magnetism. These are very important ideas that you will need for all of your studies in electricity, so it is important that they be learned as soon as is possible.

Review of Electricity—What It Is

Now stop and review what you have found out about electricity and the electron theory. Then you will be ready to learn about conductors, insulators, semiconductors, electric charges, etc.

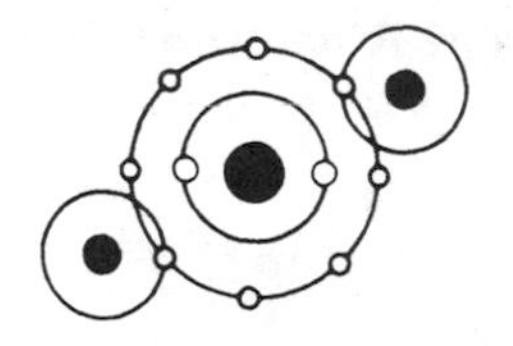

1. MOLECULE—The combination of two or more atoms. The smallest unit into which a substance—such as water—can be divided and still be identified as that substance.

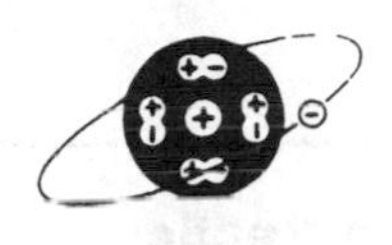

2. ATOM—The smallest particle into which an element—such as oxygen—can be divided and still retain its original properties.

3. NUCLEUS—The heavy positively charged central part of the atom.

4. NEUTRON—The heavy neutral particles in the nucleus that behave like a combination of a proton and an electron.

5. PROTON—The heavy positively charged particles in the nucleus.

6. ELECTRON—The very small negatively charged particles which are practically weightless and circle the nucleus in orbits.

7. BOUND ELECTRONS—Electrons in orbit in an atom.

8. FREE ELECTRONS—Electrons that have left their orbit in an atom and are wandering freely through a material.

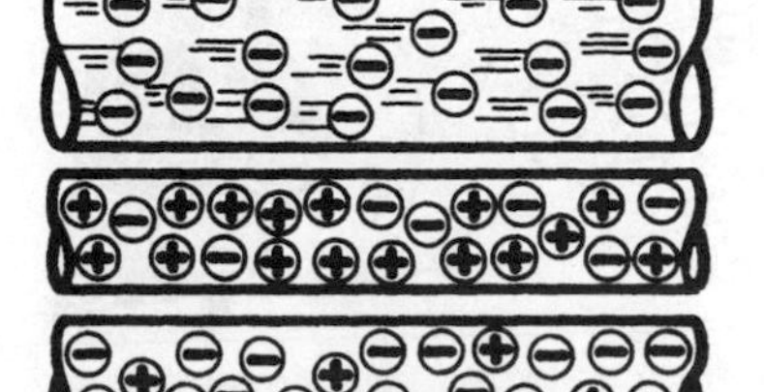

9. ELECTRIC CURRENT—The movement of free electrons.

10. POSITIVE CHARGE—A deficiency of electrons.

11. NEGATIVE CHARGE—A surplus of electrons.

Self-Test—Review Questions

1. What is the basic point of the electron theory?
2. Why is the electron theory still called a theory?
3. What is a molecule? An element?
4. Is the nucleus of an atom positively or negatively charged?
5. What is the charge on an electron? A proton? A neutron?
6. What are electrons?
7. What are free electrons?
8. Define positive and negative charges.
9. Define an electric current.
10. What is the difference between an electric charge and an electric current?

Learning Objectives—Next Section

Overview—You have learned that electricity is the flow of electrons. In the next section, you will learn about conductors, insulators, and semiconductors. The proper use of these makes current flow where we want it to.

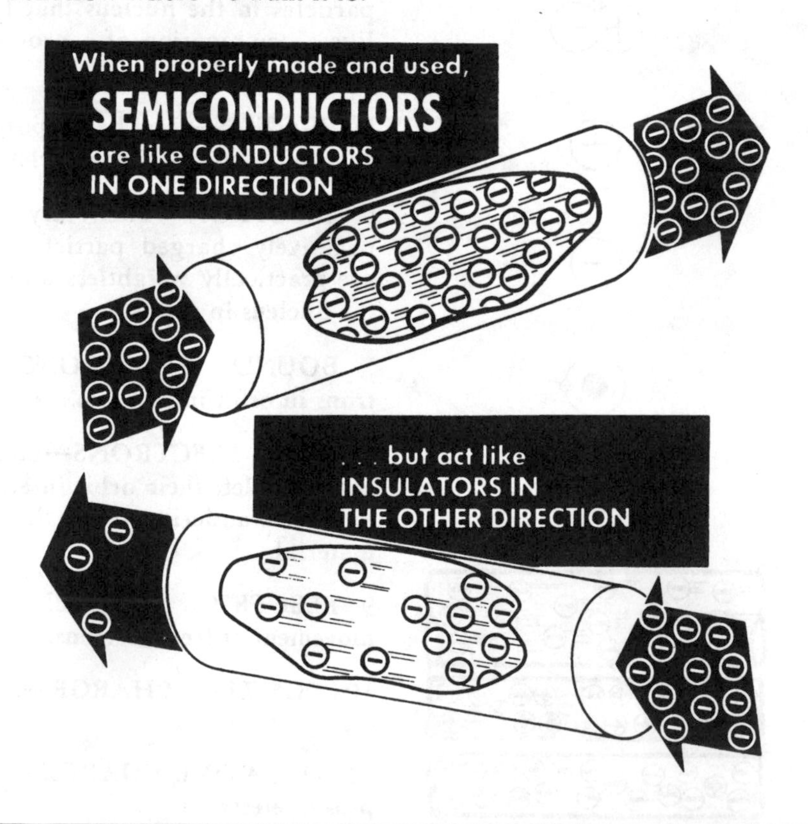
When properly made and used,
SEMICONDUCTORS
are like CONDUCTORS
IN ONE DIRECTION
. . . but act like
INSULATORS IN
THE OTHER DIRECTION

What a Conductor Is

You have learned that an electric current is the flow of electrons. Materials that permit the free motion of electrons are called *conductors*. Copper wire is considered a good conductor because it has many free electrons. The atoms of copper are held in place by the structure that copper forms when it is a solid. The electrons in the outer orbit of the copper atom are not very strongly bound and can be readily freed from the atom.

Electrical energy is transferred through conductors by means of the movement of free electrons that migrate from atom to atom inside the conductor. Each electron moves a very short distance to a neighboring atom where it replaces one or more of its electrons by forcing them out of its outer orbit. The replaced electrons repeat the process in other nearby atoms until the movement of electrons has been transmitted throughout the entire conductor. The more electrons that can be made to move in a material for a given applied force, the better conductor you have. Silver is the best conductor but we usually use copper, the next best, because it is cheaper. Recently, we have begun to use aluminum; when properly used it is almost as good a conductor as copper but has become much cheaper. Zinc, brass, and iron come next. In fact, most common metals are relatively good conductors. Salt water and similar solutions of salts or acids are also good conductors of electricity. Carbon is a good conductor, too.

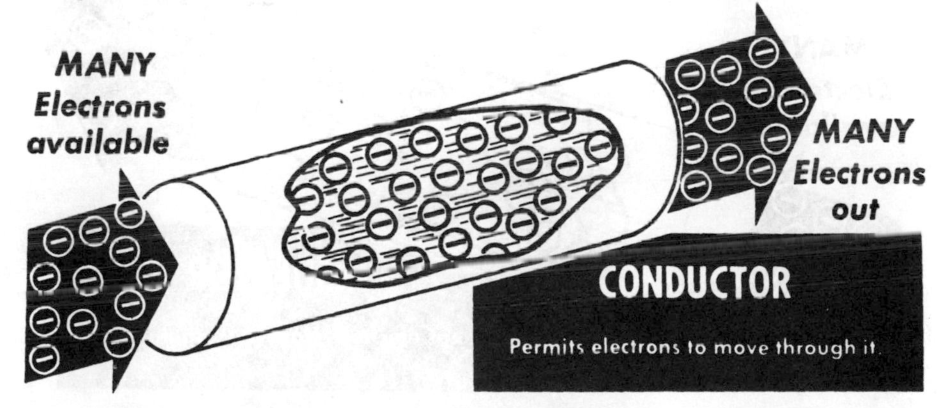

When you have learned more about electricity, you will learn how important it is to choose the right conductor of the right size to do a particular job. You will also learn that certain metals and alloys (mixtures of metals) are only fair conductors, although these materials are very useful, too.

When some metals are cooled to about —270 degrees Celsius (centigrade scale), they exhibit *superconductivity*. Under such conditions, these metals have essentially *no* resistance to the flow of electrons. Practical use is being made of superconductivity in cryogenic (supercold) electric motors and in strong electromagnets used in nuclear fusion work.

What an Insulator Is

Materials that have very few free electrons are *insulators*. In these materials, a lot of energy is needed to get the electrons out of the orbit of the atom. Even then only a few can be released at a time. Actually, there is no such thing as a perfect insulator. As a result, there is no sharp division between conductors and insulators; insulators can be thought of as poor conductors. Materials such as glass, mica, rubber, plastics, ceramics, and slate are considered to be among the best insulators. Dry air is also a good insulator. Another name for insulator is *dielectric*.

It may surprise you to know that insulators are just as important as conductors, because without them it would not be possible to keep electrons flowing in the places that we want and to keep electrons from flowing in places where we do not want them.

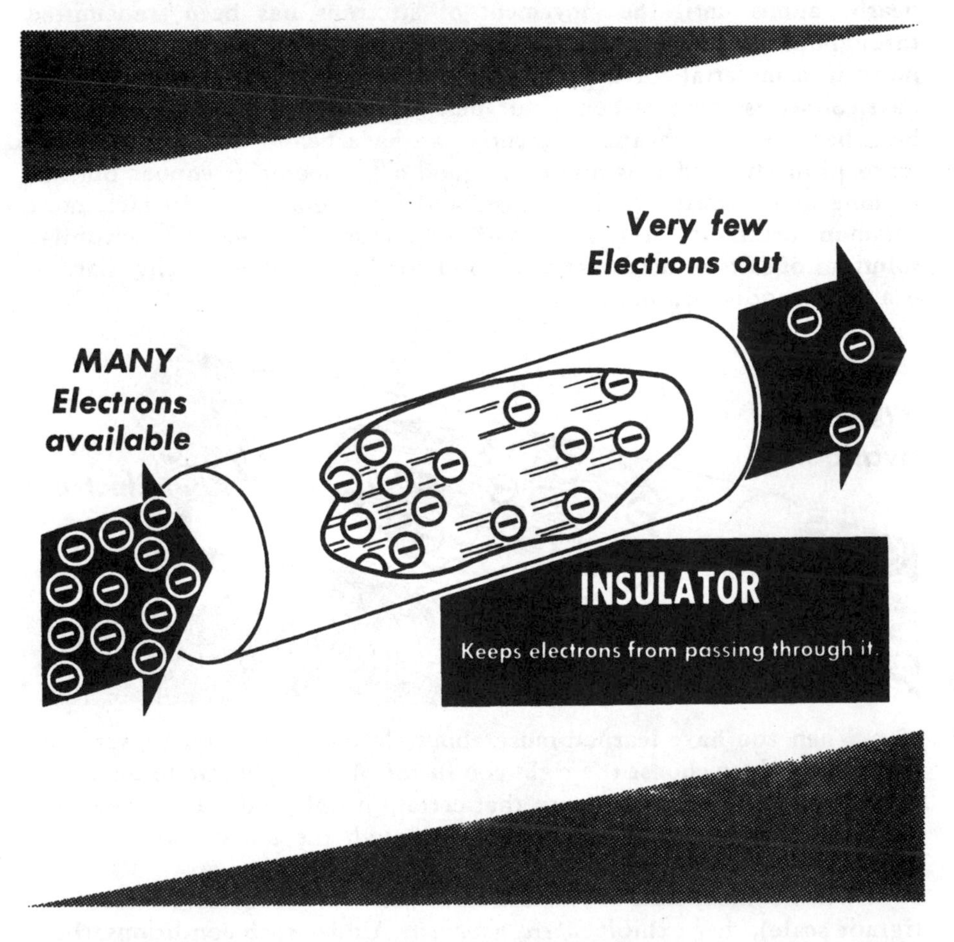

When you have learned more about electricity you will see how important it is to choose the right insulator of the right size to do a particular job.

What a Semiconductor Is

As the name implies, a *semiconductor* is a material that has some characteristics of both insulators and conductors. In recent years, these semiconductor materials have become extremely important as the basis for transistors, diodes, and other solid-state devices that you have probably heard about. Semiconductors are usually made from germanium or silicon, but selenium and copper oxide, as well as other materials, are also used. To make these materials into semiconductors, carefully controlled impurities are added to them during manufacture. The important thing about semiconductors is not that they are midway between insulators and conductors. It is that when properly made, they will conduct electricity in one direction better than they will in the other direction. As you will see later, this is an extremely valuable property that you can take advantage of in a number of ways. You do not know enough at the present time to be able to learn more about semiconductors. However, you will learn more about them when you have learned more facts about electricity.

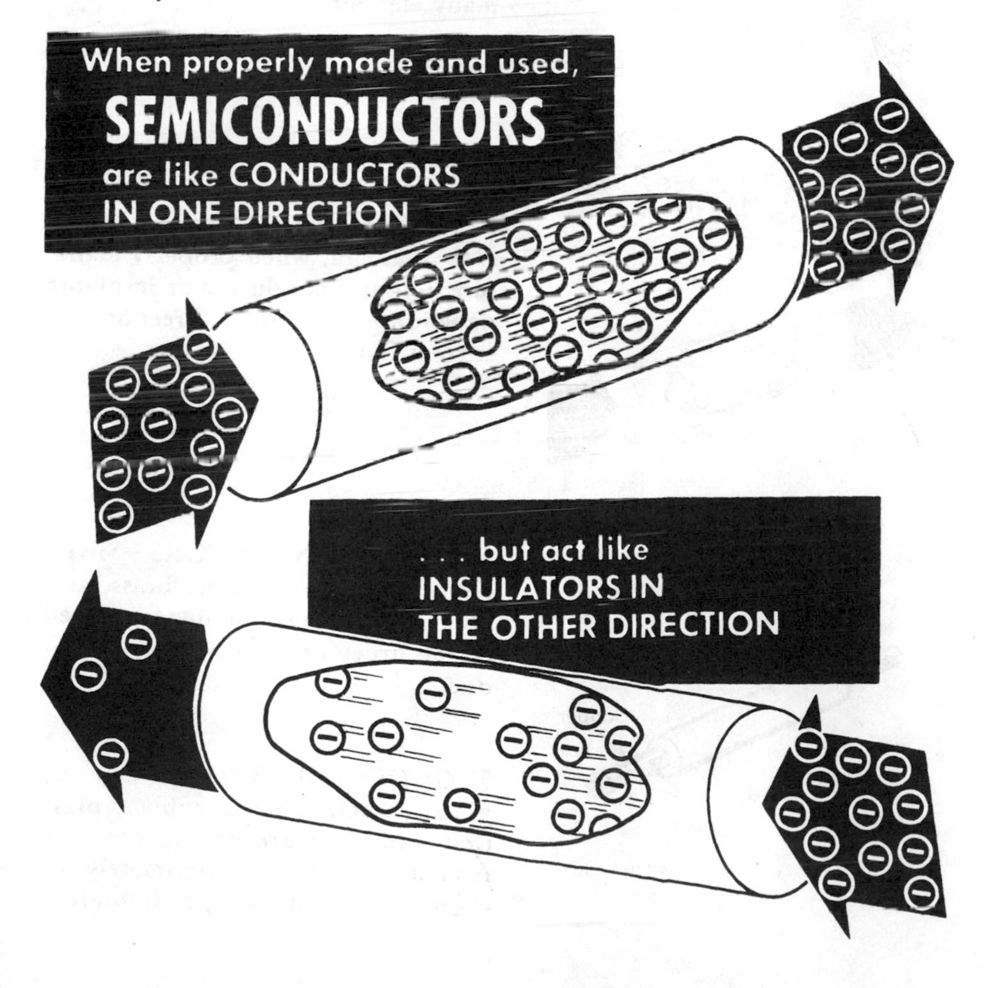
When properly made and used,
SEMICONDUCTORS
are like CONDUCTORS
IN ONE DIRECTION
. . . but act like
INSULATORS IN
THE OTHER DIRECTION

Review of Conductors, Insulators, Semiconductors

All materials can be classified as conductors, insulators, or semiconductors. There are no firm dividing lines. Also, there is no perfect conductor or perfect insulator. We use conductors and insulators in the right places to make electricity go where we want it and to keep it out of places where it should not be.

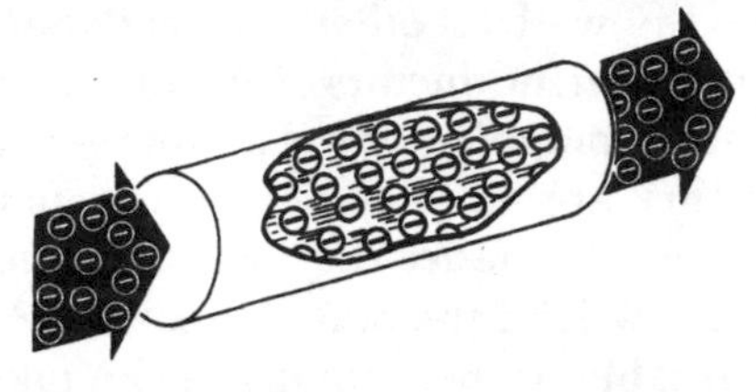

1. CONDUCTORS—Materials that permit the free movement of many electrons.

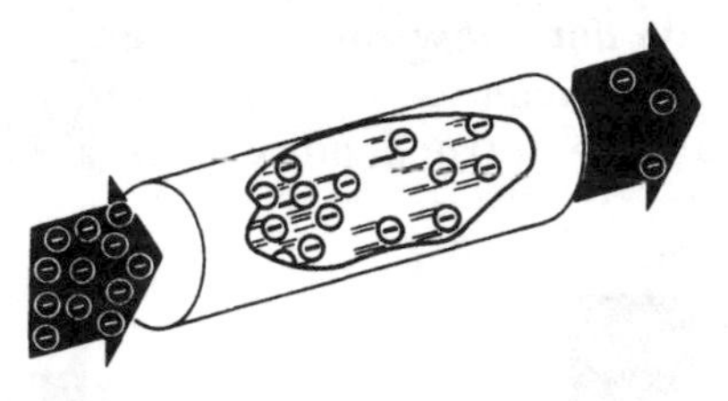

2. INSULATORS—Materials that do not permit the free movement of many electrons.

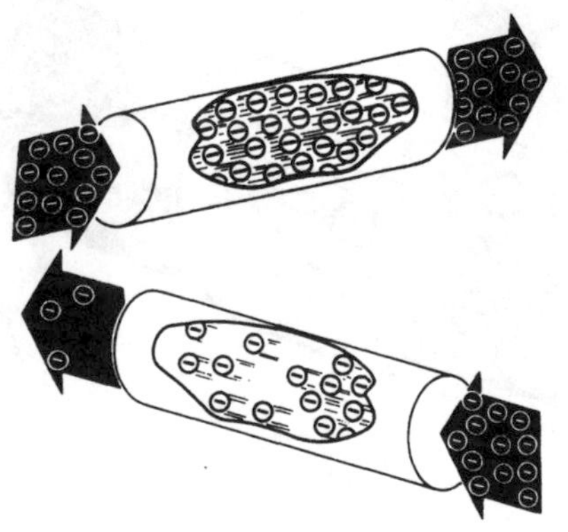

3. SEMICONDUCTORS — Materials that can, when properly made, function as a conductor or insulator depending on the direction of current flow.

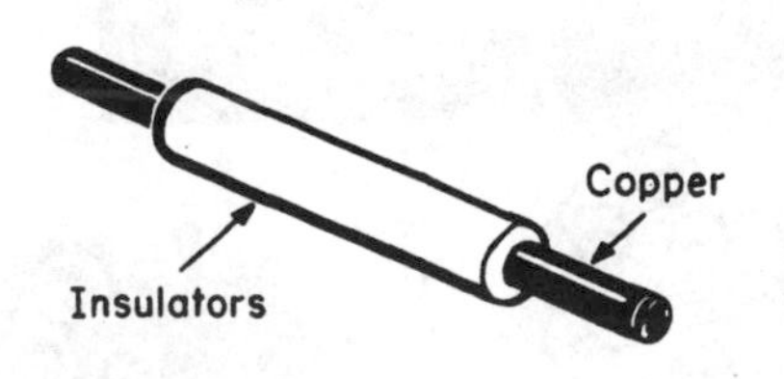

4. GOOD CONDUCTORS—Silver, copper, aluminum, zinc, brass, and iron are the best conductors, listed in the order of their ability to conduct.

5. GOOD INSULATORS—Dry air, glass, ceramics, mica, rubber, plastics, and slate are among the best insulators, listed approximately in the order of their ability to insulate.

Self-Test—Review Questions

1. What makes a good conductor?
2. Can materials other than metals be conductors?
3. Why is copper used as a conductor?
4. What makes a good insulator?
5. Describe some common insulators that you have seen.
6. Glass is a better insulator than rubber. Why then, do you find that rubber is very commonly used as an insulator?
7. What are the most important properties of semiconductors?
8. Compare semiconductors to conductors and insulators.
9. Are insulators as important in electricity as conductors? Why?
10. Choose a common electrical device with which you are familiar. Describe how the conductors and insulators are used. Why were the particular materials that were used chosen?

Learning Objectives—Next Section

Overview—Now that you know about conductors and insulators, you can learn about electric charges and static electricity. You will learn, in the next section how static charges can be generated and moved and how electric fields exist around a charged body.

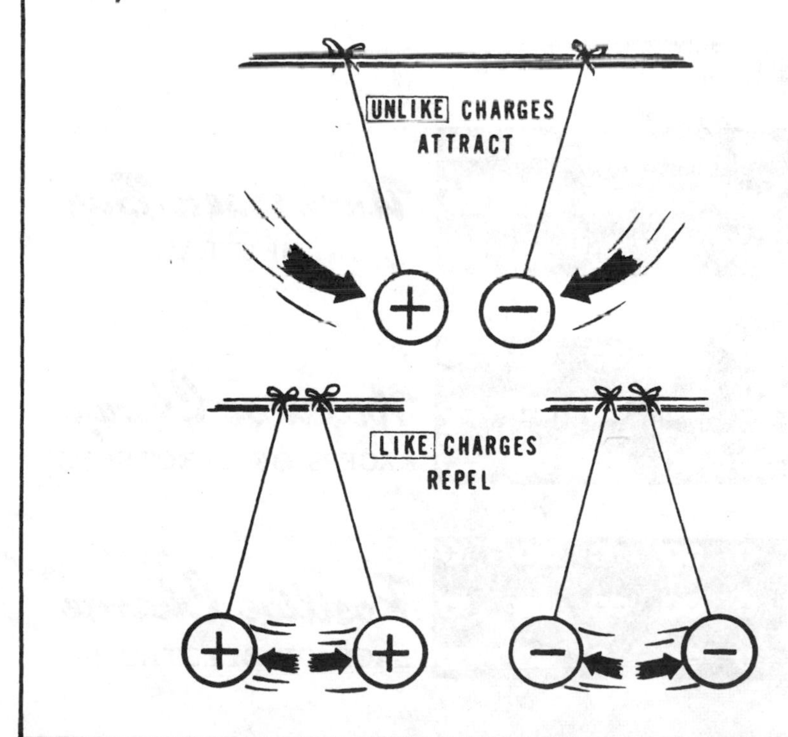

What Electric Charges Are

You have learned that electrons travel around the nucleus of an atom and are held in orbits by the attraction of the positive charge in the nucleus. When you force an electron out of its orbit, then the electron's action becomes what is known as electricity.

Electrons which are forced out of their orbits in some way will cause a lack of electrons in the material which they leave and will cause an excess of electrons at the point where they come to rest. This excess of electrons is called a *negative* charge and the lack of electrons is called a *positive* charge. When these charges exist and are not in motion, you have what is called *static electricity*.

To cause either a positive or negative charge, the electron must move while the positive charges in the nucleus do not move. Any material which has a positive charge will have its normal number of positive charges in the nucleus but will have electrons missing or lacking. However, a material which is negatively charged actually has an excess of electrons. Static electricity usually involves nonconductors since, if the materials were conductors, then the free electrons or negative charges could easily flow back toward the positive charges and the material would be neutral or uncharged.

You are now ready to find out how friction can produce this excess or lack of electrons to cause static electricity.

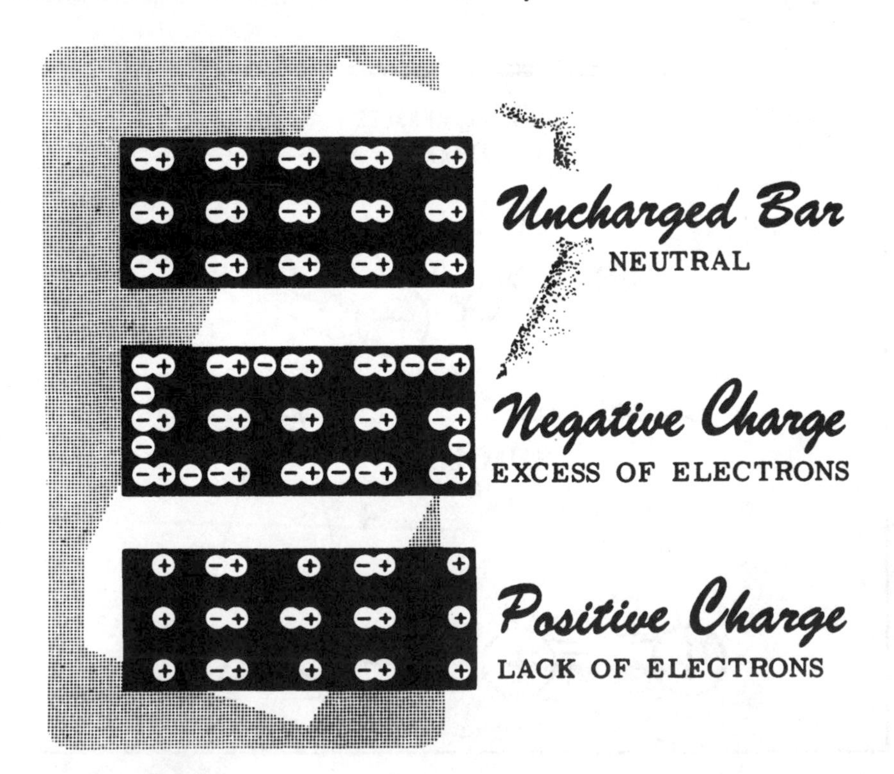

Static Charges from Friction

You have studied the electron and the meaning of positive and negative charges, so that you are now ready to find out how these charges are produced. The main source of static electricity is *friction*. If you rub two different materials together, electrons may be forced out of their orbits in one material and captured in the other. The material which captured electrons would then have a negative charge and the material which lost electrons would have a positive charge. If the materials are conductors, the electrons will move freely and the charges will be quickly neutralized. If the materials are insulators, however, then the charges will stay separated in the two materials.

When two materials are rubbed together, due to friction contact, some electron orbits at the surface of the materials will cross each other and one material may give up electrons to the other. If this happens, static charges are built up in the two materials, and friction has thus been a source of an electric charge. The charge could be either positive or negative depending on which material gives up electrons more freely.

Some materials which easily build up static electricity are glass, amber, hard rubber, wax, flannel, silk, rayon, and nylon. When hard rubber is rubbed with fur, the fur loses electrons to the rod—the rod becomes negatively charged and the fur positively charged. When glass is rubbed with silk, the glass rod loses electrons—the rod becomes positively charged and the silk, negatively charged. You will find out that a static charge may transfer from one material to another without friction, but the original source of these static charges is friction.

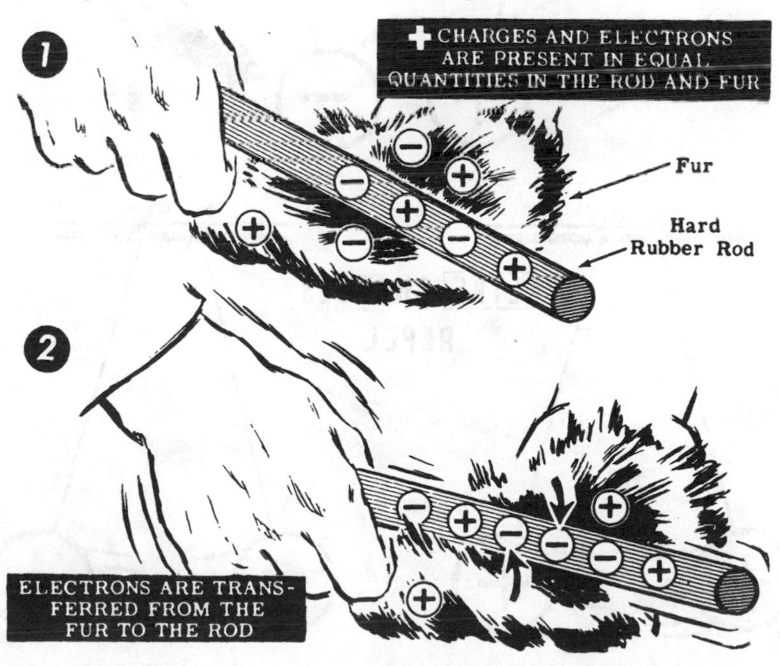
1
+ CHARGES AND ELECTRONS ARE PRESENT IN EQUAL QUANTITIES IN THE ROD AND FUR
Fur
Hard Rubber Rod
2
ELECTRONS ARE TRANSFERRED FROM THE FUR TO THE ROD

Attraction and Repulsion of Electric Charges

When materials are charged with static electricity, they behave in a different manner. For instance, if you place a positively charged ball near one which is negatively charged, the balls will attract each other. If the charges are great enough and the balls are light and free enough to move, they will come into contact. Whether they are free to move or not, a force of attraction always exists between unlike charges.

If you bring two materials of opposite charges together, the excess electrons of the negative charge will transfer to the material having a lack of electrons. This transfer or crossing over of electrons from a negative to a positive charge is called *discharge*, and by definition represents a current flow.

Using two balls with the same type of charge, either positive or negative, you will find that they repel each other.

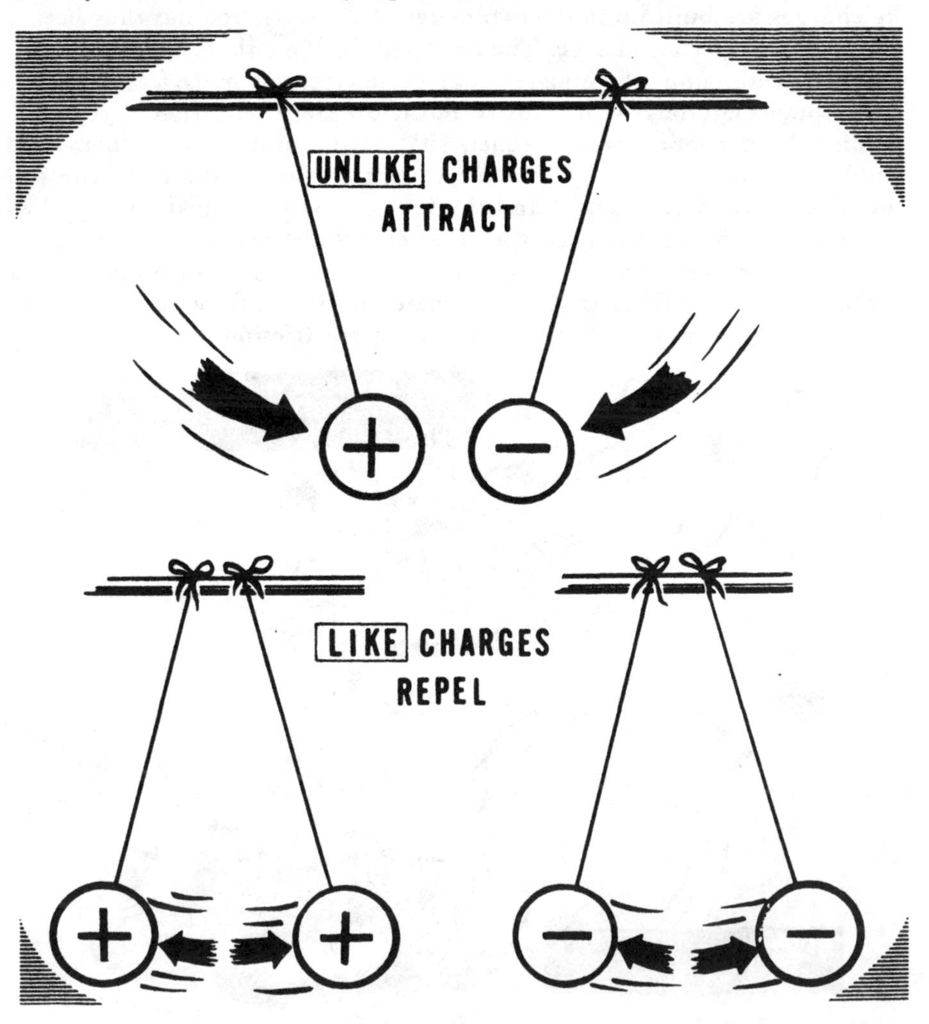

Electric Fields

You have learned that like charges repel each other and unlike charges attract. Since this happens when the charged bodies are separated, it must mean that there is a field of force that surrounds the charges and the effect of attraction or repulsion is due to this field. This field of force is called an *electric field of force*. It is also sometimes called an *electrostatic field* or a *dielectric field* since it can exist in air, glass, paper, a vacuum, or in any other dielectric or insulating material. Charles A. Coulomb, a French scientist, studied these fields in the 18th century and found that they behave in a predictable way according to what we now call *Coulomb's Law*. His law states that the force of attraction or repulsion between two charged bodies is proportional to the amount of charge present on both bodies divided by the square of the distance between them. Thus, the bigger we make the charge on our materials, the greater will be the attraction or repulsion between them; the further away we move the charged bodies, the less influence they will have on each other.

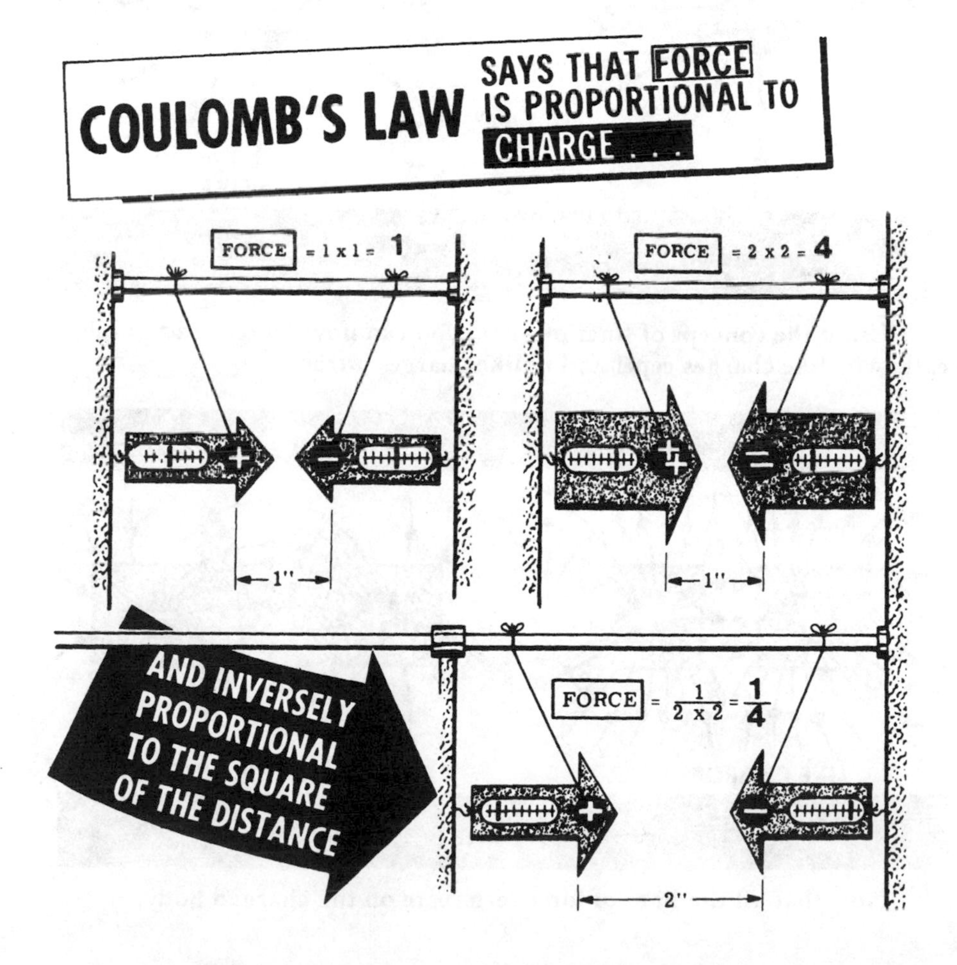

Electric Fields (continued)

The electric field around a charged body is usually represented by lines that are referred to as *electrostatic lines of force*. These lines are imaginary and are used to show the direction and strength of the field. Thus, they help us to understand what happens when these fields interact. To avoid confusion, the lines of force of a positive charge are always shown leaving the charge and the lines of force of a negative charge are always shown entering the charge.

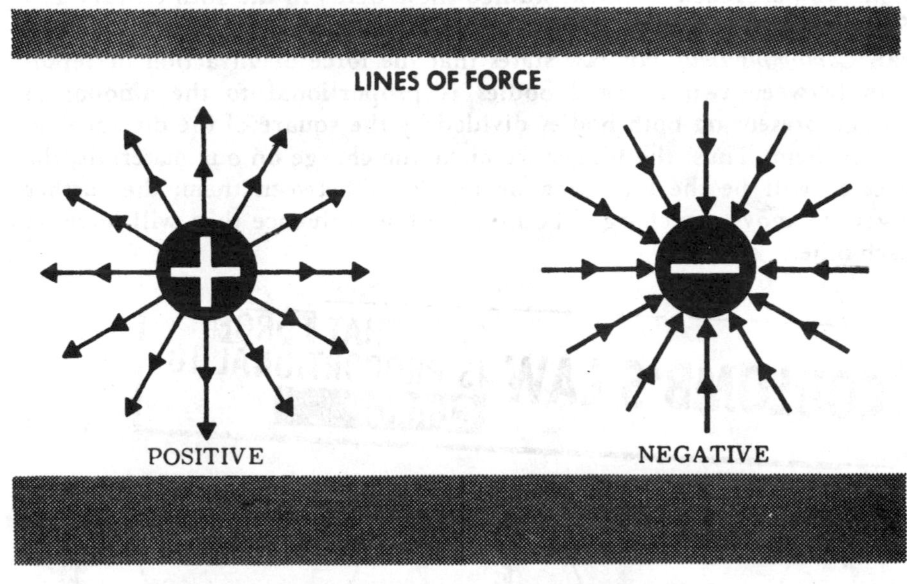

Using the concept of lines of force, you can now understand graphically why like charges repel and unlike charges attract.

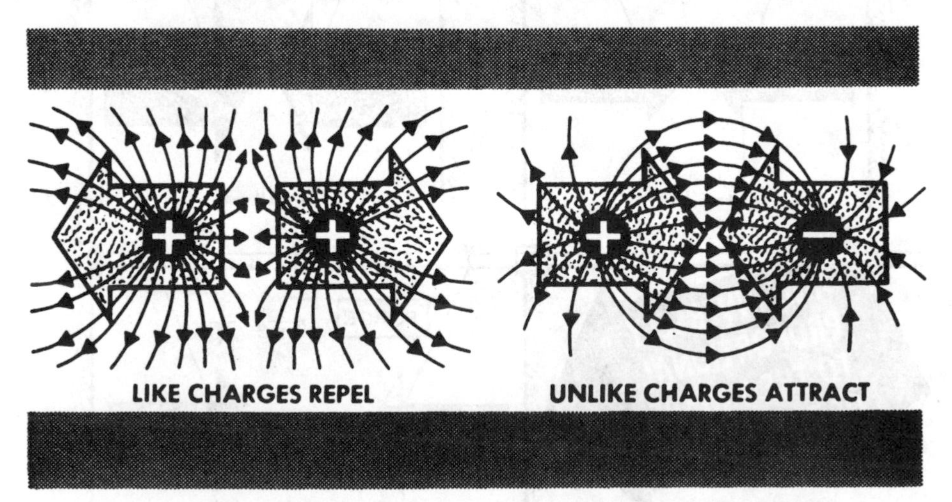

Note that all the lines of force terminate on the charged body.

Transfer of Electric Charges through Contact

Most electrostatic charges are due to friction. If an object has a static charge, it will influence other nearby objects. This influence may be exerted through contact or induction.

Positive charges mean a lack of electrons and always attract electrons, while negative charges mean an excess of electrons and always repel electrons.

If you should touch a positively charged rod to an uncharged metal bar that is supported on an insulator, it will attract electrons in the bar to the point of contact. Some of these electrons will leave the bar and enter the rod, causing the bar to become positively charged and decreasing the positive charge of the rod. When a charged object touches an uncharged object, it loses some of its charge.

In a similar way, the reverse happens when you start with a negatively charged bar.

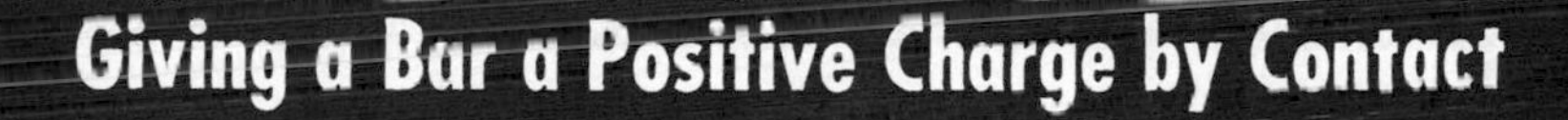

POSITIVELY CHARGED ROD ALMOST TOUCHING UNCHARGED BAR

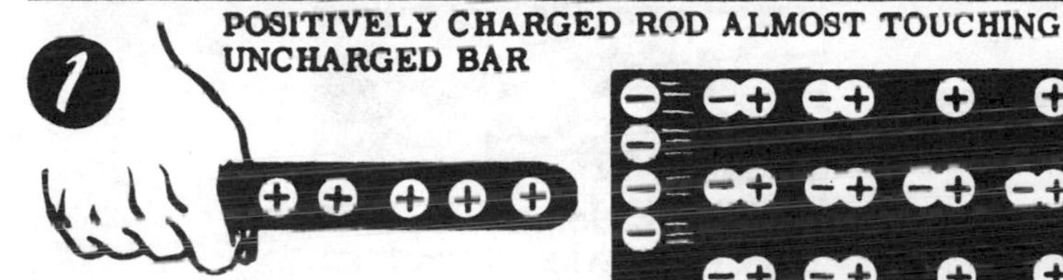

ELECTRONS ARE ATTRACTED BY POSITIVE CHARGE

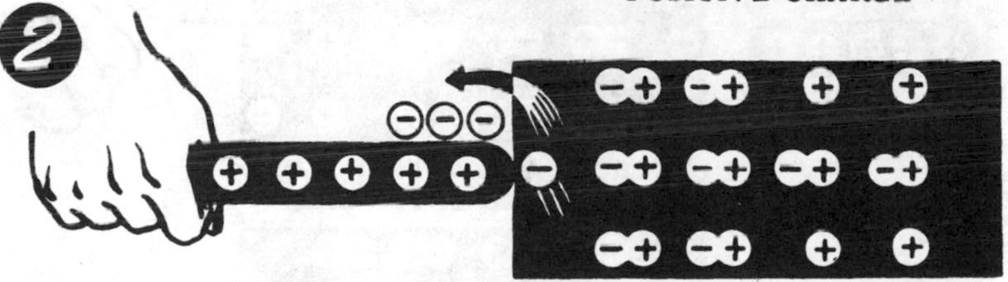

WHEN ROD TOUCHES BAR, ELECTRONS ENTER ROD

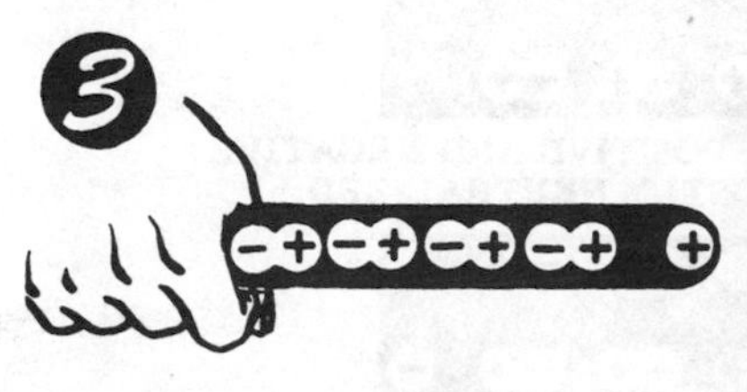

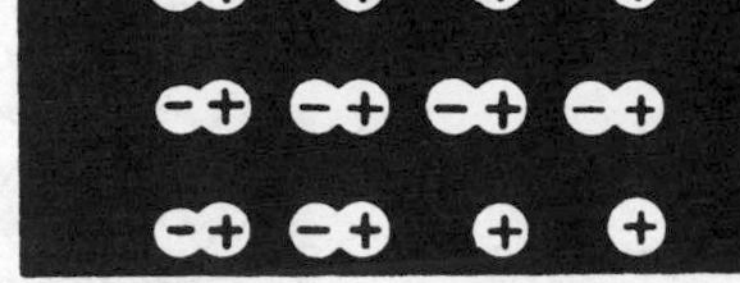

THE ROD IS NOW LESS POSITIVELY CHARGED

METAL BAR NOW HAS POSITIVE CHARGE

Transfer of Electric Charges through Induction

You have seen what happens when you touch a metal bar with a positively charged rod. Some of the charge on the rod is transferred and the bar becomes charged. Suppose that instead of touching the bar with the rod, you only bring the positively charged rod near to the bar. In that case, electrons in the bar would be attracted to the point nearest the rod, causing a negative charge to be induced at that point. The opposite side of the bar would again lack electrons and be positively charged. Three charges would then exist, the positive charge in the rod, the negative charge in the bar at the point nearest the rod, and a positive charge in the bar on the side opposite the rod. By allowing electrons from an outside source (your finger, for instance) to enter the positive end of the bar, you can give the bar a negative charge. This method of charge transfer is called *induction* because the charge distribution is induced by the presence of the charged rod rather than by actual contact.

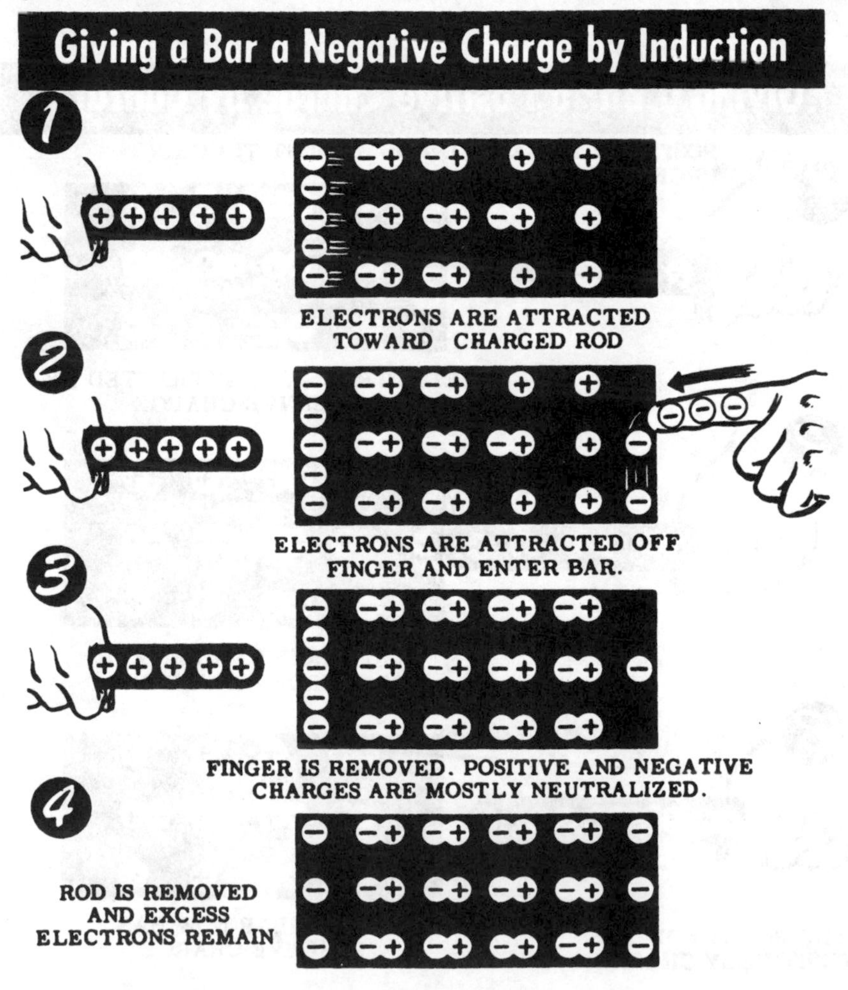
Giving a Bar a Negative Charge by Induction
1
ELECTRONS ARE ATTRACTED TOWARD CHARGED ROD
2
ELECTRONS ARE ATTRACTED OFF FINGER AND ENTER BAR.
3
FINGER IS REMOVED. POSITIVE AND NEGATIVE CHARGES ARE MOSTLY NEUTRALIZED.
4
ROD IS REMOVED AND EXCESS ELECTRONS REMAIN

Discharge of Electric Charges

Whenever two materials are charged with opposite charges and placed near one another, the excess electrons on the negatively charged material will be pulled toward the positively charged material. By connecting a wire (conductor) from one material to the other, you would provide a path for the electrons of the negative charge to cross over to the positive charge, and the charges would thereby neutralize. Instead of connecting the materials with a wire, you might touch them together (contact), and again the charges would disappear.

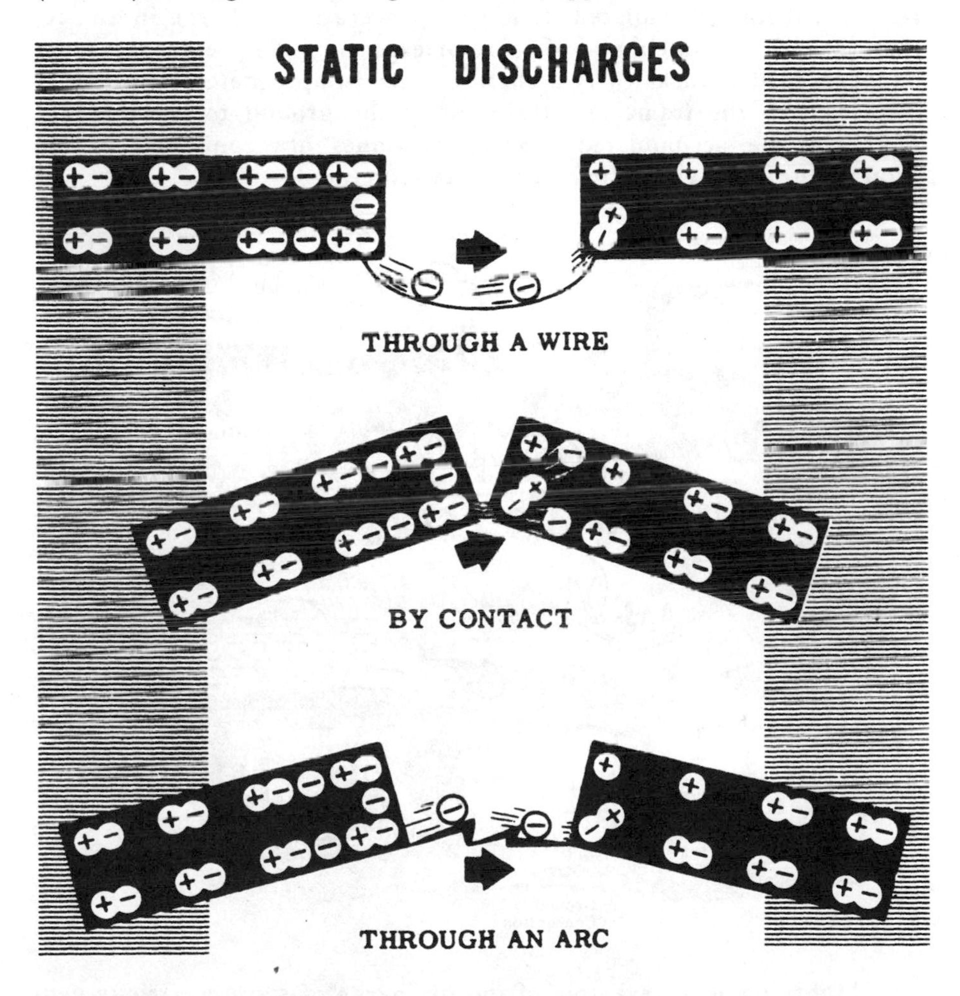

If you use materials with strong charges, the electrons may jump from the negative charge to the positive charge before the two materials are in contact. In that case, you would actually see the discharge in the form of an *arc*. With very strong charges, static electricity can discharge across large gaps, causing arcs many feet in length.

Discharge of Electric Charges (continued)

Although static electricity has limited practical use, its presence can be unpleasant and even dangerous if it discharges through an arc. You have probably had the experience of accumulating a static charge on a dry day and getting an unpleasant shock when you touched a metal object. Automobiles and trucks can pick up static charges from the friction of their tires on the road. Airplanes also can pick up static charges from the friction of their motion through the air. When a vehicle or truck is carrying an inflammable liquid such as gasoline, or an airplane is being refueled, if the accumulated static charge were to discharge in an arc, there would be a likelihood of a fire or explosion. To prevent this from happening, fuel trucks carry a chain or metal-impregnated strip that is connected to the frame and trails along the ground to continuously discharge the accumulated charge. Airplanes are connected to the ground through a grounding jack before refueling.

Lightning is an example of the discharge of static electricity generated from the friction between a cloud and the surrounding air. As you probably know, the energy in a stroke of lightning is enormous. Stationary objects, such as houses, can be protected from the effects of lightning by a lightning rod that minimizes the attracting (+) charge in the vicinity of the house.

Review of Electric Charges

1. NEGATIVE CHARGE—An excess of electrons.

2. POSITIVE CHARGE—A lack of electrons.

3. REPULSION OF CHARGES—Like charges repel each other.

4. ATTRACTION OF CHARGES—Unlike charges attract each other.

5. STATIC ELECTRICITY—Electric charges at rest.

6. FRICTION CHARGE—A charge caused by rubbing one material against another.

7. ELECTRIC FIELD—A field of force that surrounds a charged body.

8. CONTACT CHARGE—Transfer of a charge from one material to another by direct contact.

9. INDUCTION CHARGE—Transfer of a charge from one material to another without actual contact.

10. CONTACT DISCHARGE — Electrons crossing over from a negative charge to positive through contact.

11. ARC DISCHARGE—Electrons crossing over from a negative charge to positive through an arc.

12. COULOMB'S LAW—The force of attraction or repulsion is proportional to the amount of charge on each body and inversely proportional to the square of the distance between them.

Self-Test—Review Questions

1. Define a negative charge. A positive charge.
2. What are the rules for attraction and repulsion of charges?
3. According to Coulomb's Law, what happens to a force of attraction or repulsion when the distance is cut in half?
4. What concept do we use to account for the force between two charged bodies?
5. Describe, using diagrams, what happens when a negatively charged rod is used to charge a metal bar by contact.
6. Describe, using diagrams, what happens when a negatively charged rod is used to charge a metal bar by induction.
7. Describe two ways that a pair of charged bodies can be discharged.
8. Assume that you have two bodies, one with a negative charge and the other with a positive charge that is twice that of the negative body. What is the charge on each body after they have been discharged by each other?
9. What is the purpose of the grounding strap on gasoline trucks?
10. What is lightning?

Learning Objectives—Next Section

Overview—One of the most important effects of electricity is the generation of magnetic fields. To know about electricity, you must know about magnetism. In the next section, you are going to learn about the properties of magnets and how magnetic fields behave.

Natural Magnets

In ancient times, the Greeks discovered that a certain kind of rock, which they originally found near the city of Magnesia in Asia Minor, had the power to attract and pick up bits of iron. This rock was actually a type of iron ore called *magnetite*, and its power of attraction is called *magnetism*. Rocks containing ore that has this power of attraction are called *natural magnets*.

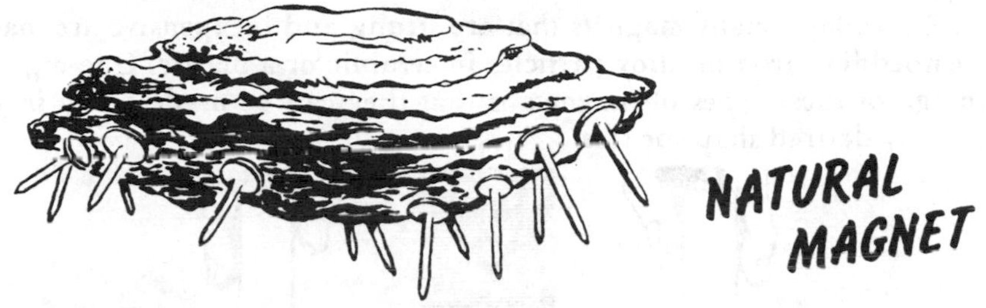

Natural magnets were seldom used until it was discovered that a magnet mounted so that it could turn freely would always turn so that one side would point to the north. Bits of magnetite suspended on a string were called *lodestones*, meaning a leading stone, and were used as crude compasses for desert travel by the Chinese more than 2,000 years ago. Crude mariner's compasses constructed of natural magnets were used by sailors in the early voyages of exploration.

The Earth itself is a large natural magnet, and the action of a natural magnet in turning toward the north is caused by the magnetism of the Earth.

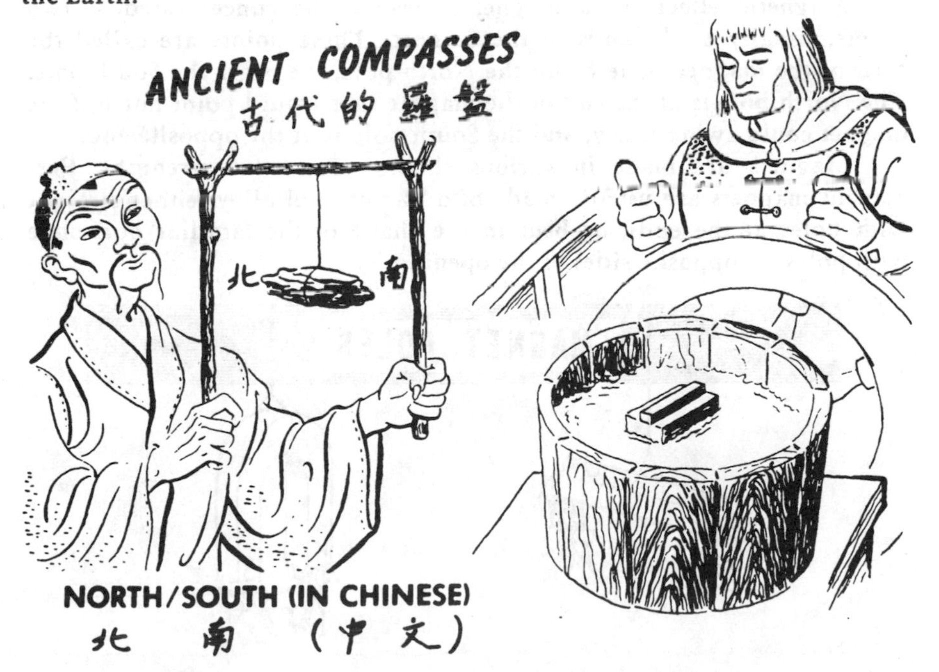

Permanent Magnets

In using natural magnets, it was found that a piece of iron stroked with a natural magnet became magnetized to form an *artificial magnet.* Artificial magnets may also be made electrically and materials other than iron may be used to form stronger magnets. Alloys containing nickel and cobalt make the best magnets and are usually used in strong magnets.

Nowadays, many magnets that are strong and inexpensive are made by embedding iron or alloy particles in ceramic or a plastic. One big advantage of these types of magnets is that they can be made easily in almost any desired shape or size.

Iron becomes magnetized more easily than other materials, but it also loses its magnetism easily so that magnets of soft iron are called *temporary magnets.* Magnets made of steel alloys hold their magnetism for a long period of time and are called *permanent magnets.*

Magnetic effects in a magnet appear to be concentrated at two points, usually at the ends of the magnet. These points are called the *poles* of the magnet—one being the North pole, the other the South pole. The North pole is at the end of the magnet that would point north if the magnet could swing freely, and the South pole is at the opposite end.

Magnets are made in various shapes, sizes, and strengths. Permanent magnets are usually made of a bar of steel alloy, either straight with poles at the ends, or bent in the shape of the familiar horseshoe with poles on opposite sides of the opening.

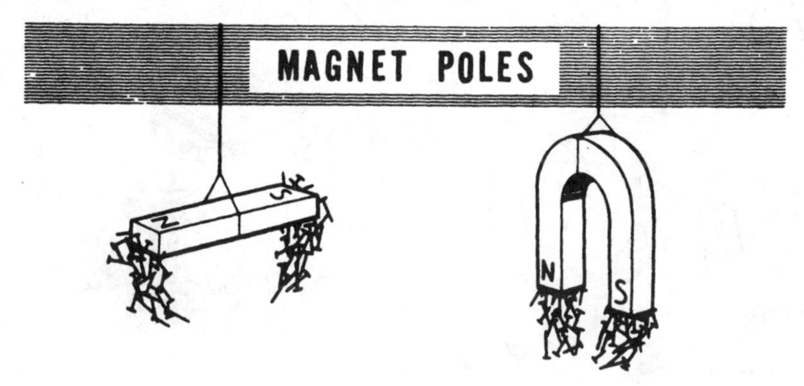

The Nature of Magnetic Materials

Magnetism is a property shown by only a few types of materials, for example iron, cobalt, and nickel, and alloys containing these materials. Two questions that you might ask are (1) why only a few materials show magnetic properties, and (2) why these materials must be magnetized to become magnets? You can get an answer to these questions by looking at what happens when you take a bar magnet and break it into pieces.

If you did this, you would find that each of the pieces was a magnet, but of course, it would be much weaker. If you did the same thing with an unmagnetized bar of the same material, you would get small pieces of unmagnetized material. If, however, you could break the bar into very small pieces consisting of only a few million, million atoms, you would find that these very small pieces for both the magnetized and unmagnetized bar had magnetic properties.

Physicists tell us that the electrons that orbit the nucleus of an atom create a magnetic field in all atoms. In most materials, the electrons go in different directions and their fields cancel so the individual atoms have no net (resultant) magnetic field. Even most atoms with an odd number of electrons are nonmagnetic since these atoms are arranged in groups of about a million, million atoms called *domains*; and these atoms are arranged at random so there is no net (resultant) magnetic field (even though an individual atom with an odd number of electrons might be magnetic). In magnetic materials, the atoms do not all oppose each other in their orbit, in fact they add, so that each domain is strongly magnetic. In unmagnetized magnetic material, these domains are oriented randomly so that the magnetic fields from each are in all directions and there is no net (resultant) field. When, however, we stroke the material with another magnet or by other means, we align all the domains in one direction, the magnetic fields add together, and the bar becomes magnetic.

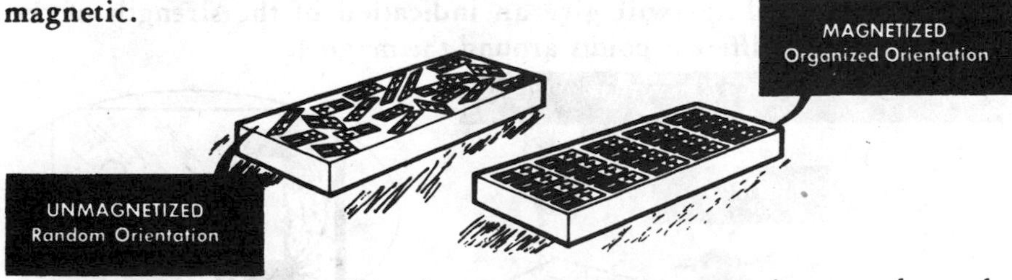

We can see from this that a *permanent magnet* is one where the domains remain aligned when they have been aligned once; and a *temporary magnet* is one where the domains go back to the original random alignment when the aligning source has been removed.

Magnetic Fields

Magnetic fields and forces, just like electrostatic fields and forces, are invisible and can be observed only in terms of the effects they produce. In spite of this, the interaction of magnetic fields with each other and with conductors moving through them are among the most important things in electricity, since these effects are used to generate most of the electricity that we use and provide the power that can be derived from electricity.

The magnetic field about a magnet can best be explained as invisible lines of force leaving the magnet at one point and entering it at another. These invisible lines of force are referred to as *flux lines* and the shape of the area they occupy is called the *flux pattern*. The number of flux lines per unit area is called the *flux density*. When the flux density is measured in lines per square centimeter, we use a unit of flux density called the *gauss*, named for an 18th century physicist who investigated magnetism. The points at which most of the flux lines leave or enter the magnet are called the *poles*. The magnetic circuit is the path taken by the magnetic lines of force.

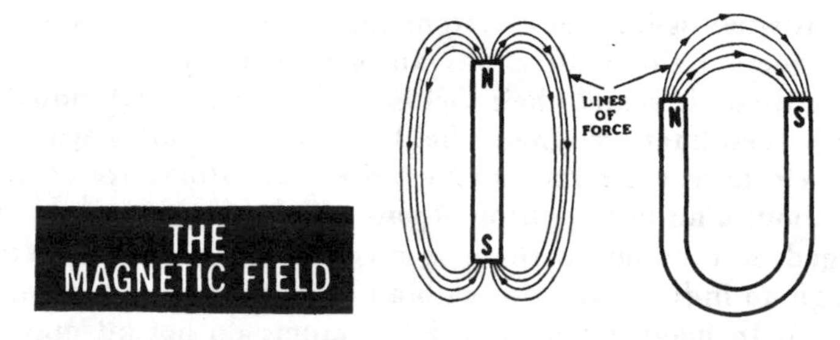

You can visualize the field around a magnet by using iron filings since the filings will become magnetized in the field of the magnet and then align themselves along the lines of force. If you place a sheet of paper or plastic over a magnet and then sprinkle the paper with iron filings, you will find that the filings will arrange themselves in a series of lines that do not cross and that terminate at the poles of the magnet. The concentration of filings will give an indication of the strength of the magnetic field at different points around the magnet.

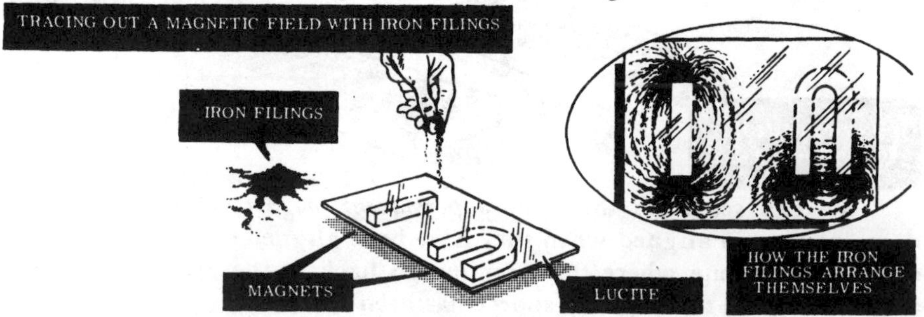

Magnetic Fields (continued)

If you were to bring two magnets together with the North poles facing each other, you would feel a force of repulsion between the poles. Bringing the South poles together would also result in repulsion, but if a North pole is brought near a South pole, a force of attraction exists. In this respect, magnetic poles are very much like static charges. Like charges or poles repel each other and unlike charges or poles attract. The laws of repulsion and attraction are like those for electric charges—that is, the force of repulsion or attraction is proportional to the strengths of the poles and inversely proportional to the distance between them.

The action of the magnetic poles in attracting and repelling each other is due to the magnetic field around the magnet. As has already been explained, the invisible magnetic field is represented by lines of force which leave a magnet at the North pole and enter it at the South pole. Inside the magnet the lines travel from the South pole to the North pole so that a line of force is continuous and unbroken.

One characteristic of magnetic lines of force is that they repel each other, never crossing or uniting. If two magnetic fields are placed near each other, as illustrated by the placement of the two magnets below, the magnetic fields will not combine but will reform in a distorted flux pattern. *Note that the flux lines do not cross each other.*

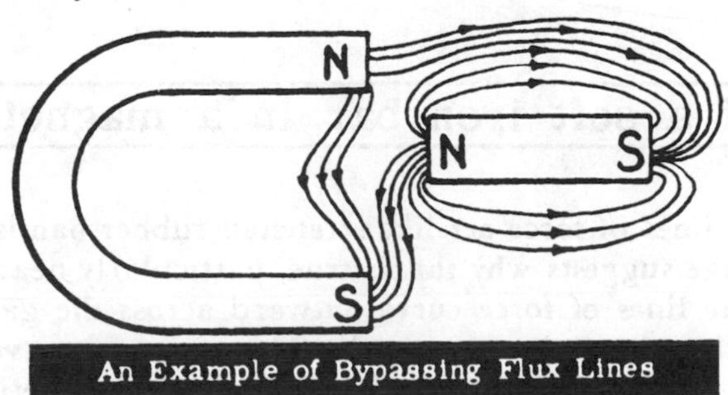

An Example of Bypassing Flux Lines

Magnetic Fields (continued)

There is no known insulator for magnetic lines of force. It has been found that flux lines will pass through all materials. Thus, most materials, except for magnetic materials, have no effect on magnetic fields. Conductors, insulators, air, or even a vacuum do not affect magnetic fields. However, they will go through some materials more easily than others. This fact makes it possible to concentrate flux lines where they are used, or to bypass them around an area or instrument.

On the previous page you were told that magnetic lines of force will go through some materials more easily than others. Those materials which will not pass flux lines so readily, or which seem to hinder the passage of the lines are said to have a comparatively *high reluctance* to magnetic fields. Materials which pass or do not hinder the flow of flux lines are said to have a comparatively *low reluctance* to magnetic fields of force.

Magnetic lines of force take the path of least reluctance; for example, they travel more easily through iron than through air. Since air has a greater reluctance than iron, the concentration of the magnetic field becomes greater in the iron (as compared to air) because the reluctance is decreased. In other words, the addition of iron to a magnetic circuit concentrates the magnetic field which is in use.

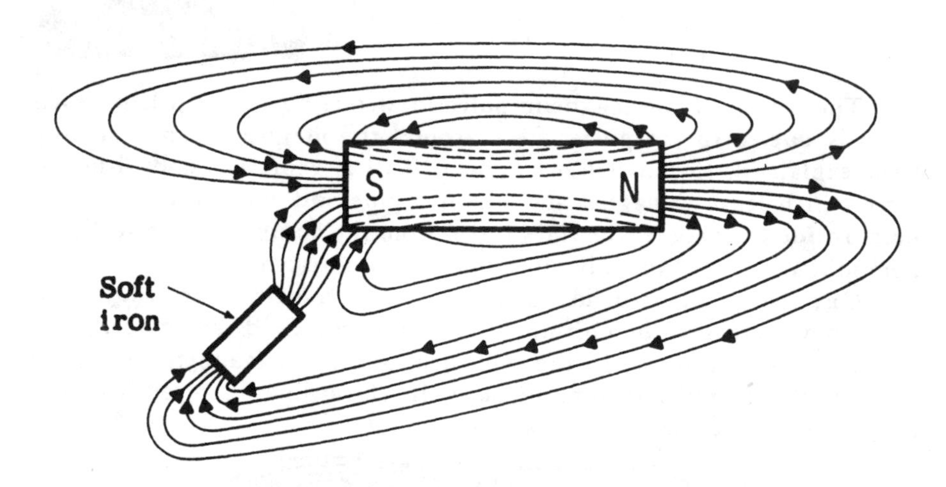

Effect of a soft iron bar in a magnetic field

Magnetic lines of force act like stretched rubber bands. The figure on the next page suggests why this is true, particularly near the air gap. Note that some lines of force curve outward across the gap in moving from the North pole to the South pole. This outward curve, or stretching effect, is caused by the repulsion of each magnetic line from

Magnetic Fields (continued)

its neighbor. However, the lines of force tend to resist the stretching effect and therefore resemble rubber bands under tension.

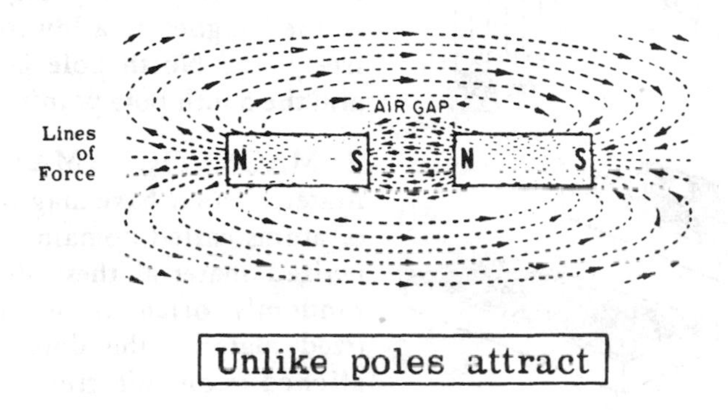

Unlike poles attract

As has already been mentioned, magnetic lines of force tend to repel each other. By tracing the flux pattern of the two magnets with like poles together in the diagram below, it can be seen why this characteristic exists.

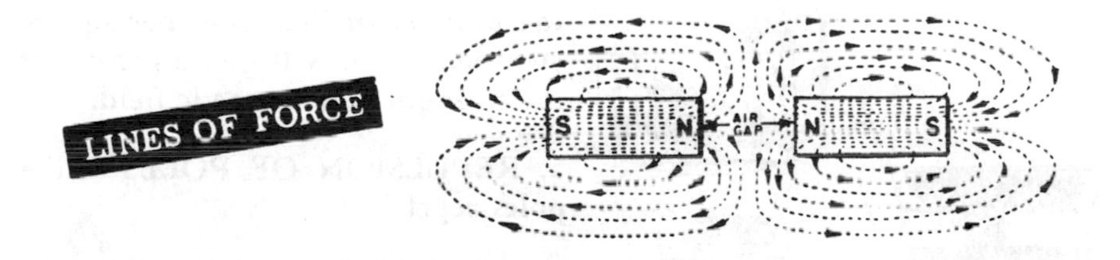

The reaction between the fields of the two magnets is caused by the fact that lines of force cannot cross each other. The lines, therefore, turn aside and travel in the same direction between the pole faces of the two magnets. Since lines of force which are directed in such a manner tend to push each other apart, the magnets mutually repel each other.

Only a certain number of magnetic lines can be crowded into a piece of material. This varies with each type of material. When the maximum number has been attained, the material is said to be *saturated.* This phenomenon is made use of in some pieces of electrical equipment but in most electrical equipment it is a very undesirable effect, since it limits the strength of a particular magnet made of a particular material. Thus, for equipment requiring very strong magnetic fields, a material with a very high magnetic saturation would be required, or it would be necessary to increase the amount of iron or other magnetic material.

Review of Magnetism

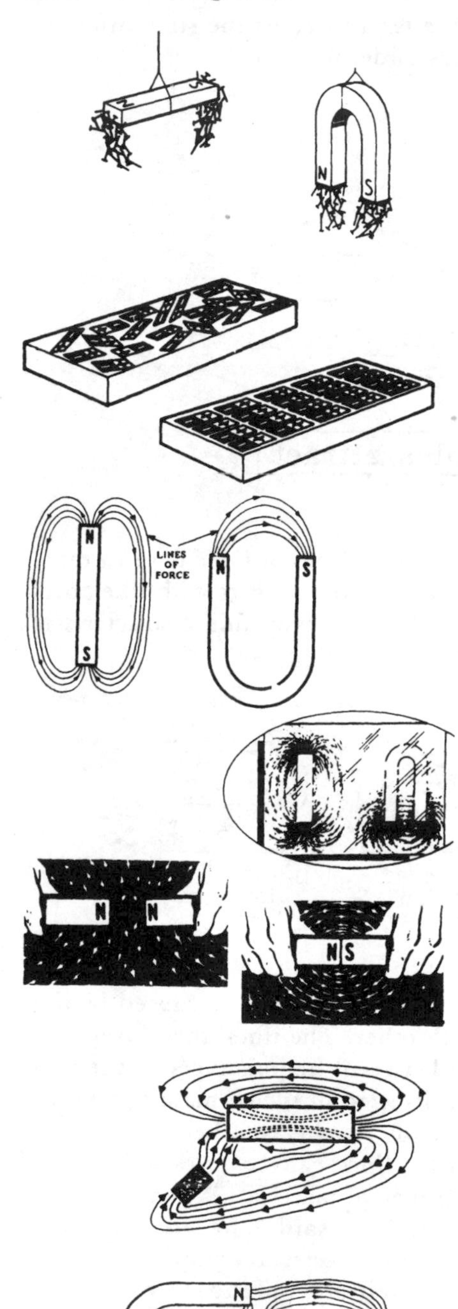

1. MAGNETIC POLES—Points on a magnet where there is a strong concentration of the magnetic field. If the magnet is allowed to swing freely, the North pole points north and the South pole points south.

2. MAGNETIC MATERIALS—Materials that have magnetic groups of atoms called domains. In unmagnetized material these domains are randomly oriented but in magnetized material the domains are all aligned in one direction.

3. MAGNETIC FIELD—Invisible lines of force that leave the magnet at the North pole and enter at the South pole. These lines are often called flux lines.

4. FLUX DENSITY—A measure of the number of flux lines per square centimeter that will give a picture of the strength of a magnetic field.

5. REPULSION OF POLES—Like poles repel.

6. ATTRACTION OF POLES—Unlike poles attract.

7. RELUCTANCE—A measure of the ease with which a material concentrates lines of force or flux lines. Materials with low reluctance tend strongly to concentrate flux lines. Magnetic lines of force take the path of least reluctance.

8. FLUX LINES DO NOT CROSS—Since flux lines do not cross because of the repulsion between them, they lie in parallel lines in a magnetic field.

Self-Test—Review Questions

1. What are permanent magnets? Temporary magnets?
2. Define a North Pole. A South Pole.
3. Based on what you know about the way magnetic materials work, do you think that it is necessary that all magnets have both a North and a South Pole? Why?
4. Draw a representation of the magnetic field around a bar magnet. A horseshoe magnet.
5. Of what use is the concept of flux density? Define it.
6. State the rules governing the interaction between magnetic fields or poles. Draw the fields for each case of like and unlike poles.
7. Why do you think saturation is a problem in some equipment that uses magnetism?
8. What is reluctance? Define high and low reluctance.
9. Can magnetic lines of force cross each other? Explain.
10. What happens to a magnetic field when another magnet is introduced? When a soft iron bar is introduced? Explain the effects in terms of what you have learned about magnetism.

Learning Objectives—Next Section

Overview—The study of electricity is mainly the study of current flow and the effects of current flow. You know that current flow is electrons in motion, now you will learn more about current flow.

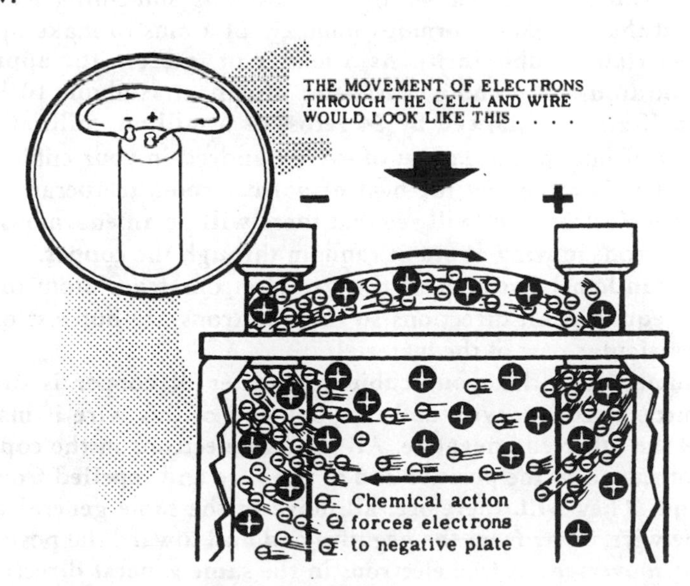
THE MOVEMENT OF ELECTRONS THROUGH THE CELL AND WIRE WOULD LOOK LIKE THIS . . . .
-
+
Chemical action forces electrons to negative plate

Electrons in Motion

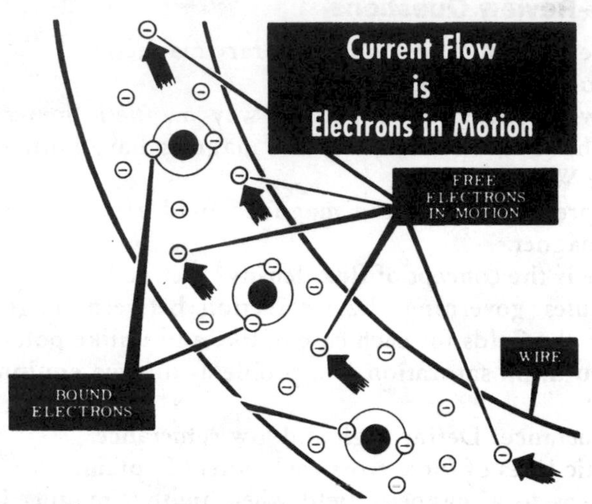

You already know that electrons in the outer orbits of an atom, being bound to the nucleus less tightly than electrons whose orbits are nearer the nucleus, can easily be forced from their orbits. You also know that in certain materials called *conductors* (they are generally, but not always, metals), very little energy is needed to liberate the outer electrons in this way.

In practice, the heat of normal room temperature is quite enough to liberate outer electrons in materials that are good conductors. The result is that a large number of electrons are normally *free* in these materials.

Now you must remember that an atom is something exceedingly small, and that it takes enormous numbers of atoms to make up a cubic centimeter (0.061 cubic inch). As a matter of interest, the approximate number of atoms in a cubic centimeter of copper is about 10^{24}—which means the figure 1 followed by 24 zeros, or a million, million, million, million! So if only one atom out of every hundred in your cubic centimeter of copper is forced by the heat of normal room temperature to give up a single electron, you will see that there will be an enormous number of free electrons moving about at random through the copper.

The random movement of the *free* electrons from atom to atom is normally equal in all directions so that electrons are not lost or gained by any particular part of the material.

Assume, now, that your cubic centimeter of copper is drawn out into a piece of copper wire and that one end of this wire is made positive, and the other end negative. All the free electrons in the copper wire will be attracted to the positive end of the wire and repelled from the negative end. They will, therefore, all move in the same general direction along the wire, away from the negative end and toward the positive end.

This movement of free electrons in the same general direction along the wire is called *current flow.*

Electrons in Motion (continued)

All electrons (being negative) are attracted by positive charges and are repelled by negative ones. They will always be attracted *from* a point having an excess of electrons, *toward* a point having a deficiency of them.

An electric cell or battery has exactly this property of having an excess of electrons at its negative terminal and a shortage of electrons at its positive terminal. This unbalance is maintained by chemical action as you will learn later.

Let's look at what happens when we connect a wire across the terminals of a cell. Instantly, a force will be exerted on all the free electrons in the wire, drawing some of them out of the end of the wire connected to the positive terminal of the cell. At the same time, the negative terminal of the cell will be pushing a lot more free electrons into the other end of the wire.

When electrons are drawn from one end of a piece of wire, there will result a *lack* of electrons (and therefore a positive charge) at that end. Similarly, when a lot of electrons are pushed by some outside source into the other end of the wire, there will be an *excess* of electrons (and, therefore, a negative charge) at that end. All these excess electrons will not only be repelled by one another, but also (and much more importantly) they will be attracted to the positive charge at the other end of the wire.

In this way, a continuous movement of electrons will take place *from the negatively charged* end of the wire *to the positively charged end*, for as long as electrons are furnished to one end of the wire and removed from it at the other end.

A battery (being a series of electric cells connected together) is a good way of maintaining current flow because, as you will learn, it is capable of furnishing a constant flow of electrons to its negative terminal, and of removing a constant stream of electrons from its positive terminal—and of continuing this for a long time.

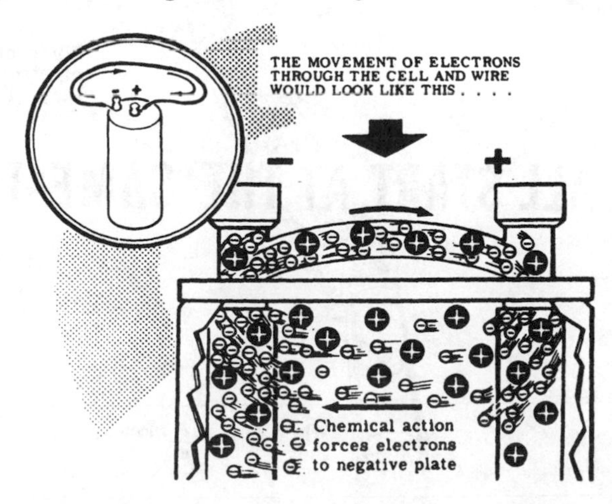
THE MOVEMENT OF ELECTRONS THROUGH THE CELL AND WIRE WOULD LOOK LIKE THIS . . . .
−
+
Chemical action forces electrons to negative plate

Electrons in Motion (continued)

By the way, do not think of electrons coming into direct physical contact with one another. You know that like charges repel, and that all electrons are negative. So, when a moving electron comes close to another electron, the second electron will be pushed away by the electric field of the first, without the two electrons themselves ever coming into contact.

When current flow starts in a wire, electrons start to move throughout the wire *at the same time*, just as the cars of a long train start and stop together.

If one car of a train moves, it causes all the cars of the train to move by the same amount, and free electrons in a wire act in the same manner. Free electrons are always present throughout the wire, and as each electron moves slightly it exerts a force on the next electron, causing it to move slightly and, in turn, to exert a force on the next electron. This effect continues throughout the wire.

When electrons move away from one end of a wire it becomes positively charged, causing all the free electrons in the wire to move in that direction. This movement, taking place throughout the wire simultaneously, moves electrons away from the other end of the wire and allows more electrons to enter the wire at that point.

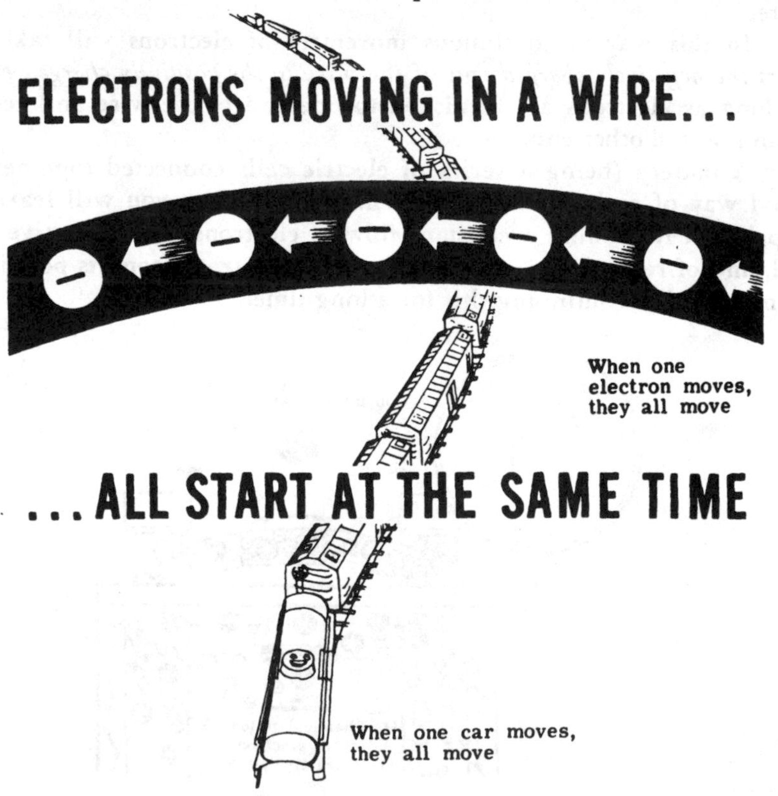

ELECTRONS MOVING IN A WIRE...
When one electron moves, they all move
...ALL START AT THE SAME TIME
When one car moves, they all move

Direction of Current Flow

According to the electron theory, current flow is always from a negative (—) charge to a positive (+) charge. Thus, if a wire is connected between the terminals of a battery, current will flow from the (−) terminal to the (+) terminal.

Before the electron theory of matter had been worked out, electricity was in use to operate lights, motors, etc. Electricity had been harnessed but no one knew exactly why it worked. It was believed that an electric fluid moved in the wire from (+) to (—). This conception of current flow is called *conventional current flow*. You will find the conventional flow of (+) to (—) is often used in working with electrical and electronic equipment. Actually, it doesn't matter which direction you choose as long as you are consistent in working out any particular problem.

In all of our studies in electricity, we will be consistent and always use the direction of electron flow as the direction of current flow: that is, current flow is from *negative* to *positive*.

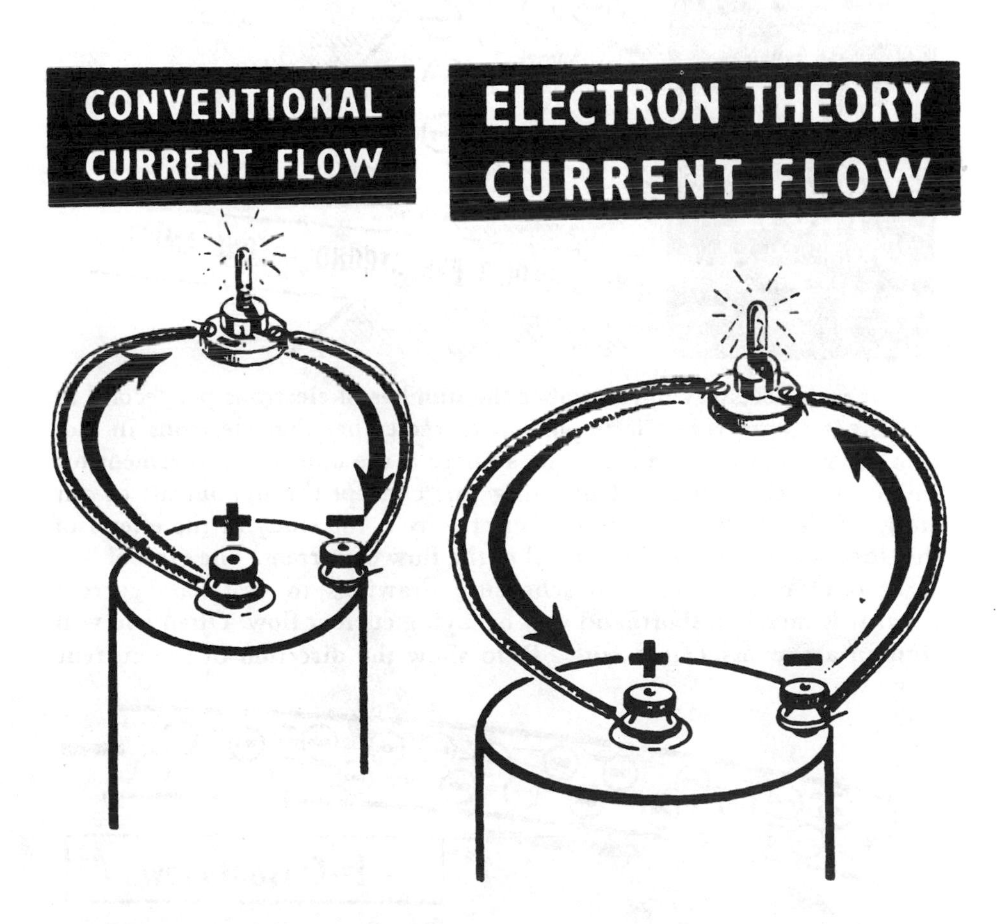

Units of Current Flow

Current flow, as you have learned, is the movement of electrons through a material. We measure the flow of current by measuring the number of electrons that flow past a given point in a given period of time. Since the coulomb is a measure of the number of electrons present, we can use it as the basis for the measurement of current flow. A coulomb is defined as about six and a quarter million, million, million electrons (or exactly 6.289×10^{18} electrons in mathematical terms). The unit of current flow is the ampere, which is defined as 1 coulomb flowing in 1 second. Thus, 1 ampere is a current flow of 1 coulomb per second and 2 amperes is a current flow of 2 coulombs per second, etc.

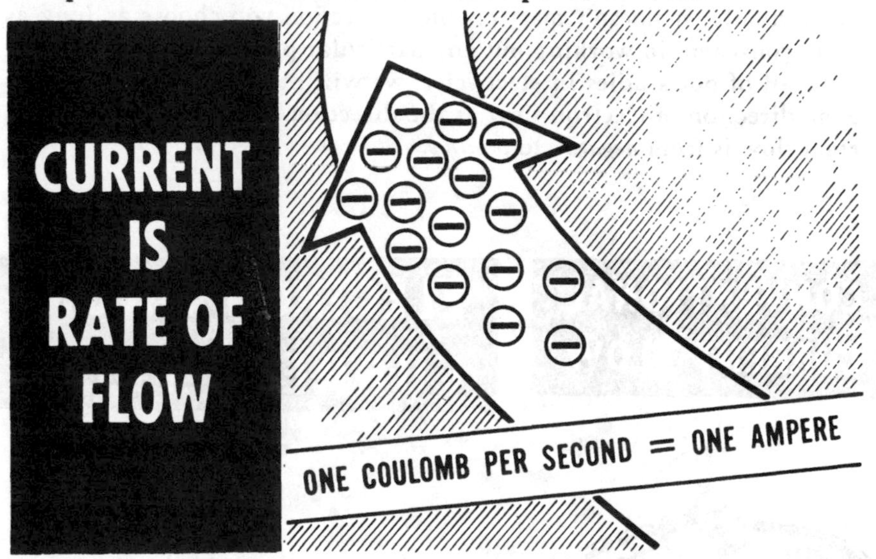

It isn't necessary to remember the number of electrons per second in an ampere; however, it is important to remember that electrons in motion are current flow and that the ampere is the unit of measurement for this current flow. We will be using this concept throughout all of our study of electricity. The study of electricity is the study of the effects of the flow of current and the control of the flow of current. The symbol "I" is used in calculations and schematic drawings to designate current flow. It is merely a shorthand way of saying current flow. Often you will find an arrow associated with "I" to show the direction of the current flow.

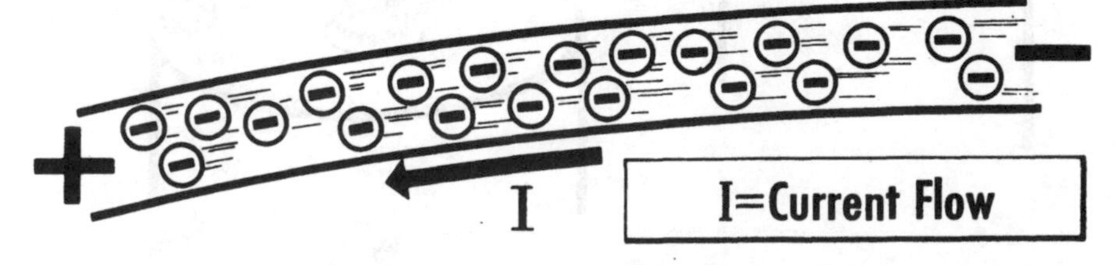

Review of Current Flow

Current flow does all the work involved in the operation of electrical equipment, whether it be a simple light bulb or some complicated electronic equipment such as a radio receiver or transmitter. For current to flow, a continuous path must be provided between the two terminals of a source of electric charges. Now suppose you review what you have found out about current flow.

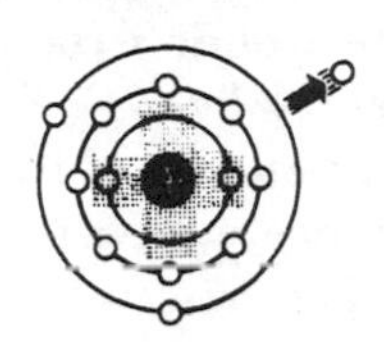

1. FREE ELECTRONS—Electrons in the outer orbits of an atom which can easily be forced out of their orbits.

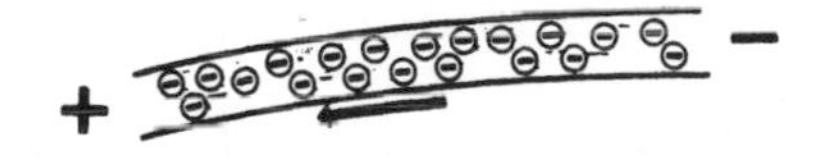

2. CURRENT FLOW—Movement of *free* electrons in the same direction in a material.

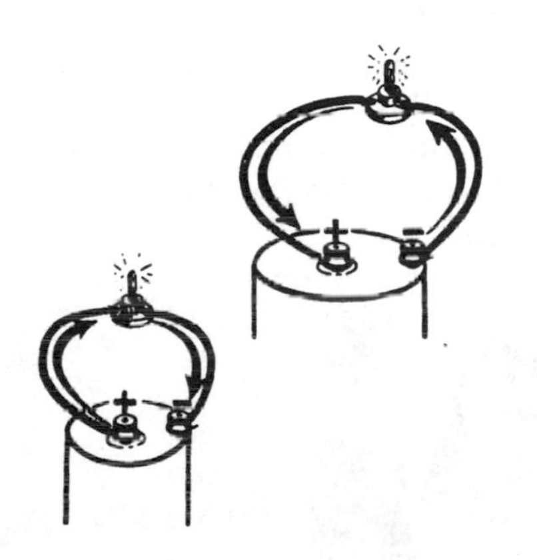

3. ELECTRON CURRENT—Current flow from a *negative* charge to a *positive* charge.

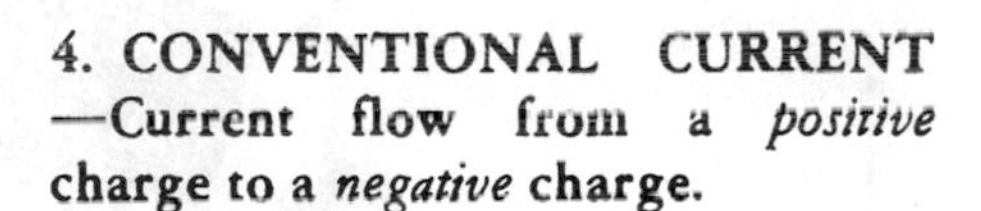

4. CONVENTIONAL CURRENT —Current flow from a *positive* charge to a *negative* charge.

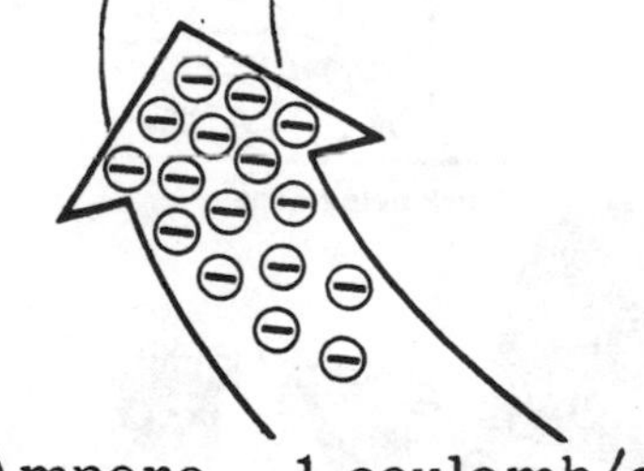

1 Ampere = 1 coulomb/sec

5. AMPERE—The unit of measure of current flow. It is equal to 1 coulomb per second.

I=Current

6. "I"–The symbol used to designate current in schematic drawings and formulas.

Self-Test—Review Questions

1. What are free electrons?
2. Conductors have many free electrons. Why is it that they don't have an electric charge?
3. What is current flow?
4. Describe what happens when a piece of wire is connected across the terminals of a battery or a cell.
5. Describe current flow in a wire in terms of the atomic structure.
6. Do all of the electrons in a wire move together? Why?
7. What is the difference between electron theory current flow and conventional current flow? Does it make any difference which is used? Which will we use throughout the rest of our study of electricity?
8. What is the unit of measure of current flow?
9. How is it defined?
10. What is the symbol for current flow?

Learning Objectives—Next Section

Overview—Now that you know about current flow, you will learn about the *force* that makes current flow. This force called *electromotive force (or EMF)* must be maintained if current flow is to be maintained and it takes work (or energy) to do this, as you will see.

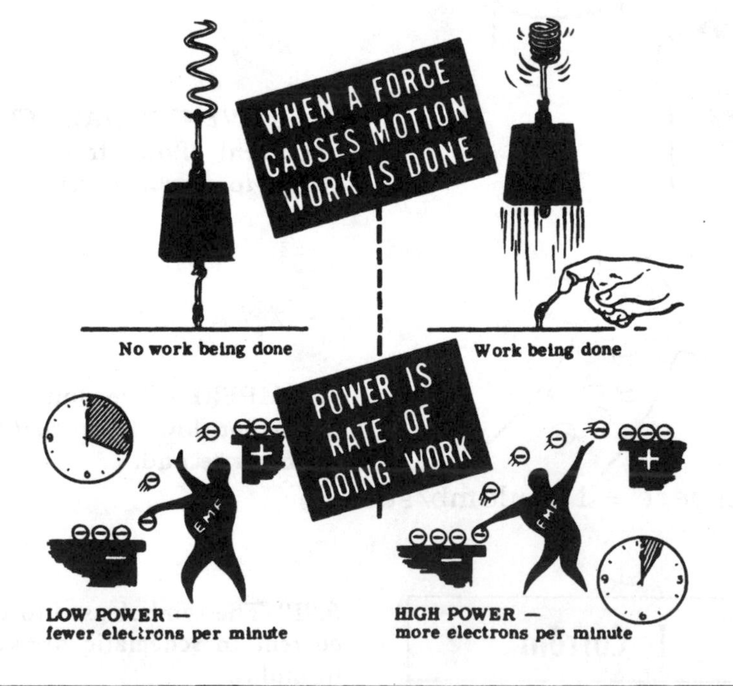
WHEN A FORCE CAUSES MOTION WORK IS DONE
No work being done
Work being done
POWER IS RATE OF DOING WORK
EMF
EMF
LOW POWER — fewer electrons per minute
HIGH POWER — more electrons per minute

What Work Is

You have learned that the flow of electrons is an electric current and that its unit of measure is the ampere, which is equal to a flow of 1 coulomb per second. Energy is required to obtain or maintain a *charge difference* between two points. When a conductor is connected between these two points, a current will flow. The expenditure of energy is called *work* and like most things has a unit of measure. In the English system of measure based on pounds and feet, the unit of work is the *foot pound*; that is, the energy required to raise 1 pound a distance of 1 foot. On the other hand, if we had a 1-pound weight that already was raised off the ground 1 foot, we could connect it to a mechanism and get that amount of work back if we let it fall the 1 foot. When the weight is poised to do work, we say that it has *potential energy*, that is, the potential to do work. The meter and the gram are the units of distance and weight used by scientists in the metric system, which is being adopted worldwide. In the metric system the unit of work has been given a special name—the *joule*. A joule is equal to about ¾ of a foot pound (to put it exactly, 1 joule = 0.7376 foot pound).

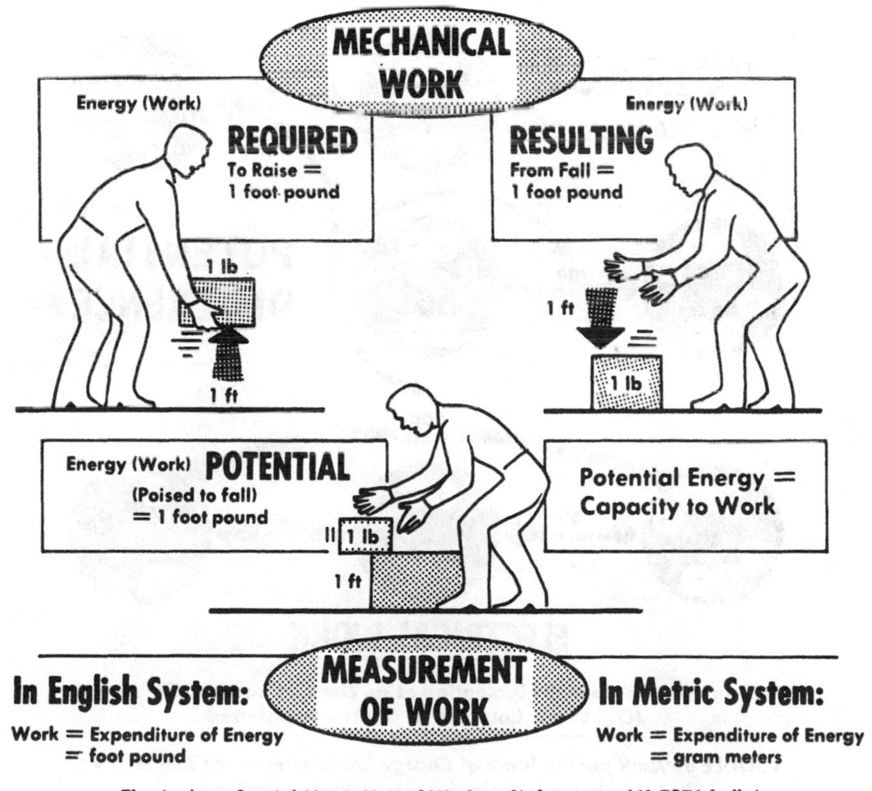

What EMF (Electromotive Force) Is

You may ask what all this discussion about work has to do with electricity, so let's see what the connection is. Since it is known that work is required to separate electric charges, then it follows that work results when these charges come back together. The separated charges, or potential difference, represent the *potential* (capacity) to do work just like our weight poised off the ground. As you shall see in the next section, several kinds of energy can be used to generate electricity; that is, to maintain potential difference. It is this potential difference that makes electrons move and thus do work or be used to generate other forms of energy.

The unit of potential difference is the *volt*. A volt is defined as the potential difference necessary to obtain 1 joule of work when one coulomb of charge flows. Thus, we have an *electromotive force* (EMF)—electron moving force—or potential difference of one volt when this happens.

We measure potential difference or emf in volts and we call the measured difference *voltage*. (And because emf is measured in volts, emf is often called, or is equated with, *voltage*.) The symbol that we use for voltage is "E" or "V."

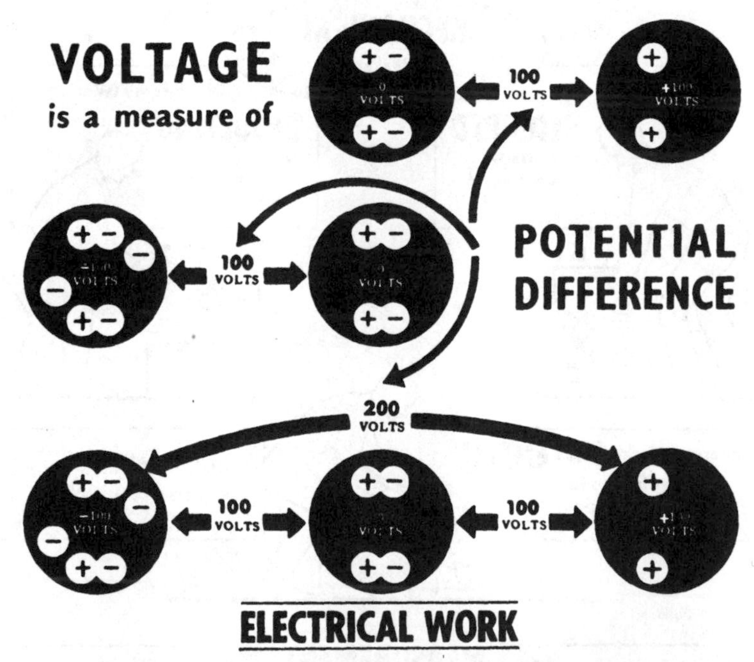

Measure For The Capacity (potential) of an Electric Charge To Do Work = JOULES per Coulomb of Charge Transferred

1 JOULE of Work per Coulomb of Charge Transferred = an EMF of 1 *Volt*

VOLTAGE = MEASURED POTENTIAL DIFFERENCE = *E* or *V*

What Electric Power Is

As you have just learned, whenever a force of any kind causes motion, *work* is done. When a mechanical force, for instance, is used to lift a weight, work is done. A force which is exerted *without* causing motion, however, such as the force of a spring held under tension between two objects which do not move, does *not* cause work to be done.

You know that a difference in potential between any two points in an electric circuit gives rise to a *voltage* which (when the two points are connected) causes electrons to move and current to flow. Here is an obvious case of a *force* causing motion, and causing *work* to be done. *Whenever a voltage causes electrons to move, therefore, work is done in moving them.*

The *rate* at which the work is done when moving electrons from point to point is called *electric power.* (It is represented by the symbol "P.")

The basic unit in which electric power is measured is the *watt.* It can be defined as the rate at which work is being done in a circuit in which a current of 1 ampere is flowing when the emf applied is 1 volt.

WHEN A FORCE CAUSES MOTION WORK IS DONE
No work being done
Work being done
POWER IS RATE OF DOING WORK
EMF
EMF
LOW POWER —
fewer electrons per minute
HIGH POWER —
more electrons per minute

How EMF Is Maintained

In order to cause continuous current flow, electric charges must be maintained so that difference of potential, or voltage, exists all all times. At the terminals of a battery, for example, this difference is caused by the chemical action within the battery, and as electrons flow from the (—) terminal to the (+) terminal, the chemical action maintains this difference. The electric generator in your car acts in the same manner, with the action of a wire moving through a magnetic field maintaining a charge difference at the generator terminals. The energy to move the wire through the magnetic field of the generator comes from the engine. The voltage difference across the generator or battery terminals remains constant, and the charges on the terminals never become equal as long as the chemical action continues in the battery, and as long as the generator wire continues to move through the magnetic field.

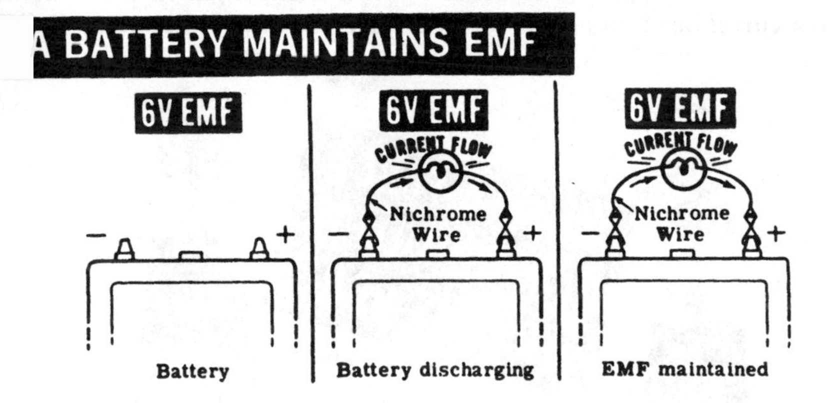

If the charge difference were not maintained at the terminals, as in the case of the two charged bars shown below, current flow would cause the two charges to become equal as the excess electrons of the (—) bar moved to the (+) bar. The voltage between the terminals would then fall to 0 volts, and current flow would no longer take place.

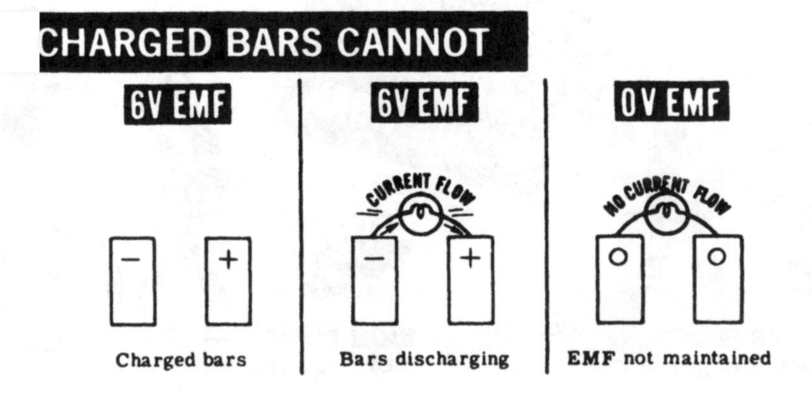

Voltage and Current Flow

Whenever two points of unequal potential or voltage are connected, current flows. The greater the emf or voltage, the greater the amount of current flow. Electrical equipment is designed to operate with a certain amount of current flow, and when this amount is exceeded, the equipment may be damaged. You have seen all kinds of equipment such as electric lamps, motors, radios, etc., with the voltage rating indicated. The voltage will differ on certain types of equipment, but it is usually 120 volts here in North America. This rating on a lamp, for example, means that 120 volts will cause the correct current flow. Using a higher voltage will result in a greater current flow and *burn out* the lamp, while a lower voltage will not cause enough current flow to make the lamp light up normally. While current flow makes equipment work, it takes emf or voltage to cause the current to flow, and the value of the voltage determines how much current will flow.

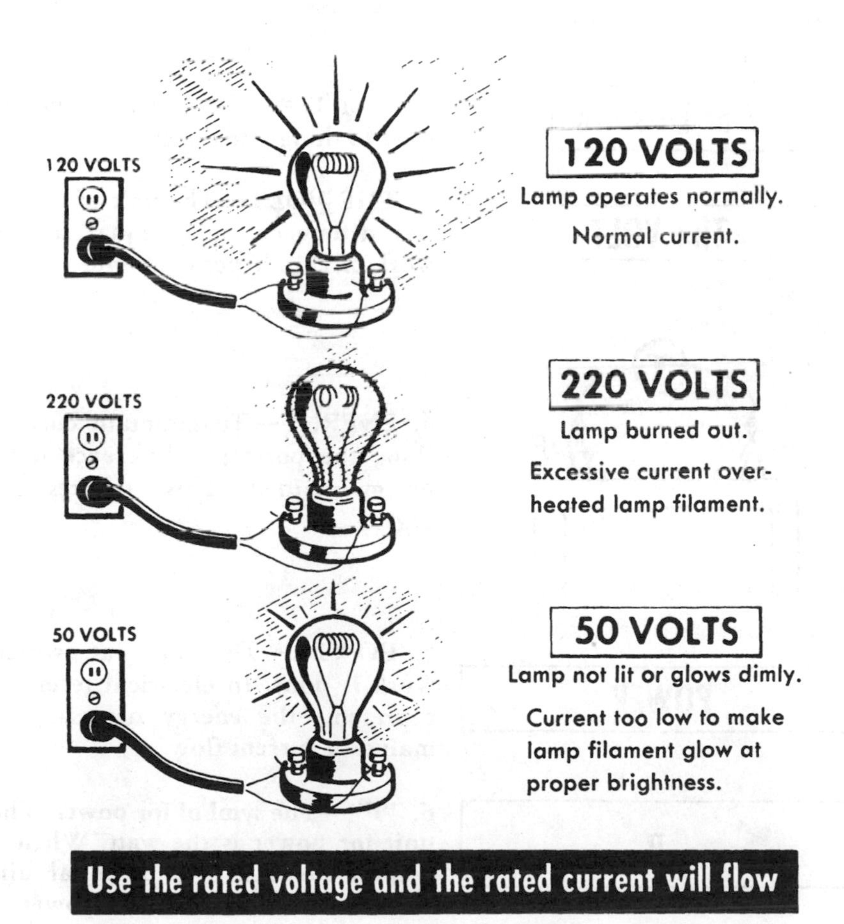
120 VOLTS
120 VOLTS
Lamp operates normally.
Normal current.
220 VOLTS
220 VOLTS
Lamp burned out.
Excessive current over-heated lamp filament.
50 VOLTS
50 VOLTS
Lamp not lit or glows dimly.
Current too low to make lamp filament glow at proper brightness.

Use the rated voltage and the rated current will flow

Review of EMF or Voltage

To make current flow, a potential difference must be maintained between the terminals. When current flows, energy is required to maintain this difference and work must be done. Thus, the generation of electricity is the conversion of other forms of energy into this potential difference. At the device where the electricity is to be used, the potential difference makes the current flow and this is used to convert the electrical energy to some other form of energy or work. The rate at which this work is done is called *power*. We will say more about this in the next section.

EMF=Electromotive Force

1. EMF = ELECTROMOTIVE FORCE—The force that makes the current flow. Potential difference between terminals.

V or E=Voltage

2. V or E = VOLTAGE—Symbols used to designate the emf.

The VOLT

3. THE VOLT—The unit of potential difference. It is equal to the work of 1 joule per coulomb.

4. ENERGY—To maintain current flow, the potential difference must be maintained. This requires energy.

POWER

5. POWER—The rate at which work is done. In electrical terms, it represents the energy necessary to maintain current flow.

P

6. "P"—The symbol for power. The unit for power is the watt. When 1 ampere flows with a potential difference of 1 volt, 1 watt of power is generated.

Self-Test—Review Questions

1. What is emf?
2. What quantity is used to designate the magnitude of emf?
3. What are the symbols used to designate emf?
4. Define the unit of potential difference.
5. What is the essential difference between current flow between two charged bars and that across the terminals of a battery?
6. What is needed to maintain a potential difference between terminals where current is flowing?
7. What happens if the potential difference is not maintained?
8. What happens in a circuit where the voltage is too low? Why?
9. What happens in a circuit where the voltage is too high? Why?
10. What happens when the rated voltage is applied? Explain.
11. What is electrical power?
12. What is the symbol for power?
13. Define the watt as the unit of power.
14. Name some sources for energy that can be used to generate power.

Learning Objectives—Next Section

Overview—Now that you know something about what electricity is, you will study how electricity is produced and the many uses we have for electricity.

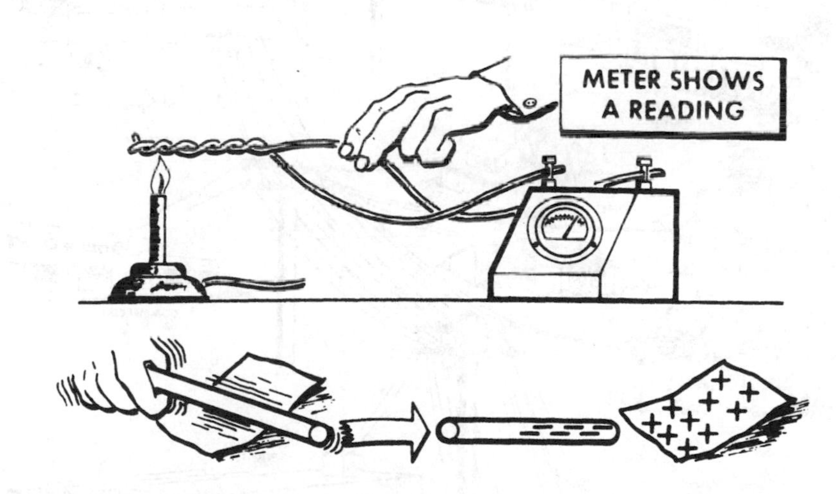

Electricity Is the Means for Transporting Power

You have learned about the electronic nature of matter and how electricity is the flow of electrons from place to place or the accumulation of electrons on a charged body. It is apparent that most materials are electrically balanced. What is needed is a source of *external* energy so that excesses and/or deficiencies of electrons can be maintained when current flows. By supplying external energy in the right way to the right device, we can generate electricity. On the other hand, by supplying electricity to the right devices, we can convert the energy of electricity to other useful forms such as mechanical power from motors, heat from appliances, light from lamps, etc. Thus, *electricity can be considered as the means for the convenient transportation and distribution of power*. For example, the energy from a waterfall can be harnessed to a generator to make electric power that is transported by transmission lines for hundreds of miles to a city where it is used to provide mechanical power, light, heat, cooling, and other necessities. In essence, then, we are using the power of the falling water when we use this electricity.

How Electricity Is Produced

The most common source of electricity and, in fact, the source for almost all the electrical energy that we use is electricity obtained from the interaction of conductors with magnetic fields. The second most common source involves chemical action and the example that you are most likely to see is the battery. Other energy sources for generating electricity in decreasing order of importance are light, heat, pressure, and friction. Remember, the generation of electricity is the conversion of other forms of energy into the potential difference required to make current flow.

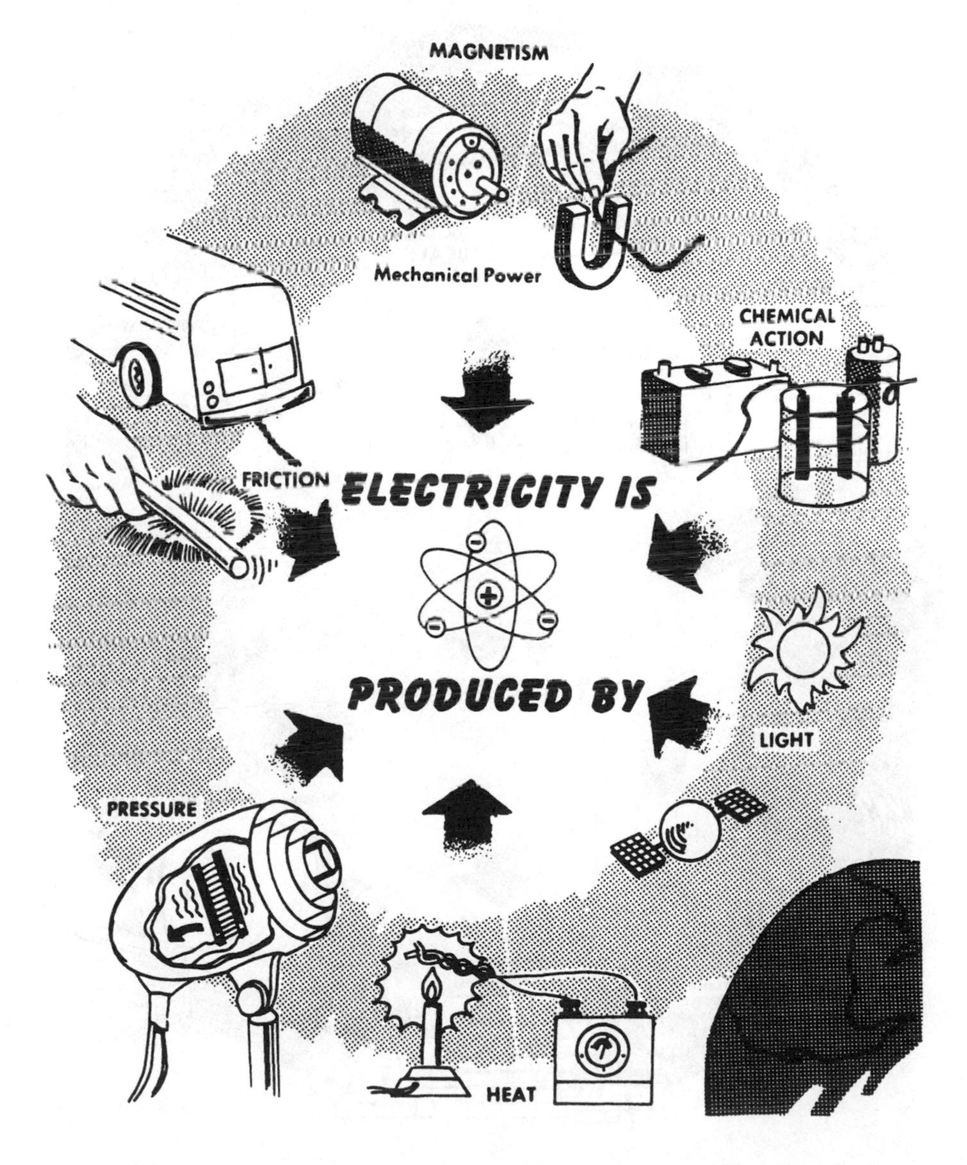

Uses of Electricity

Electricity, in turn, can be used to produce the very same effects that originally were utilized to produce that electricity with the exception of friction. These common uses of electricity (in reverse order) are mechanical power from motors, chemical action, light, heat, and pressure; and to operate electronic devices which will not be considered here.

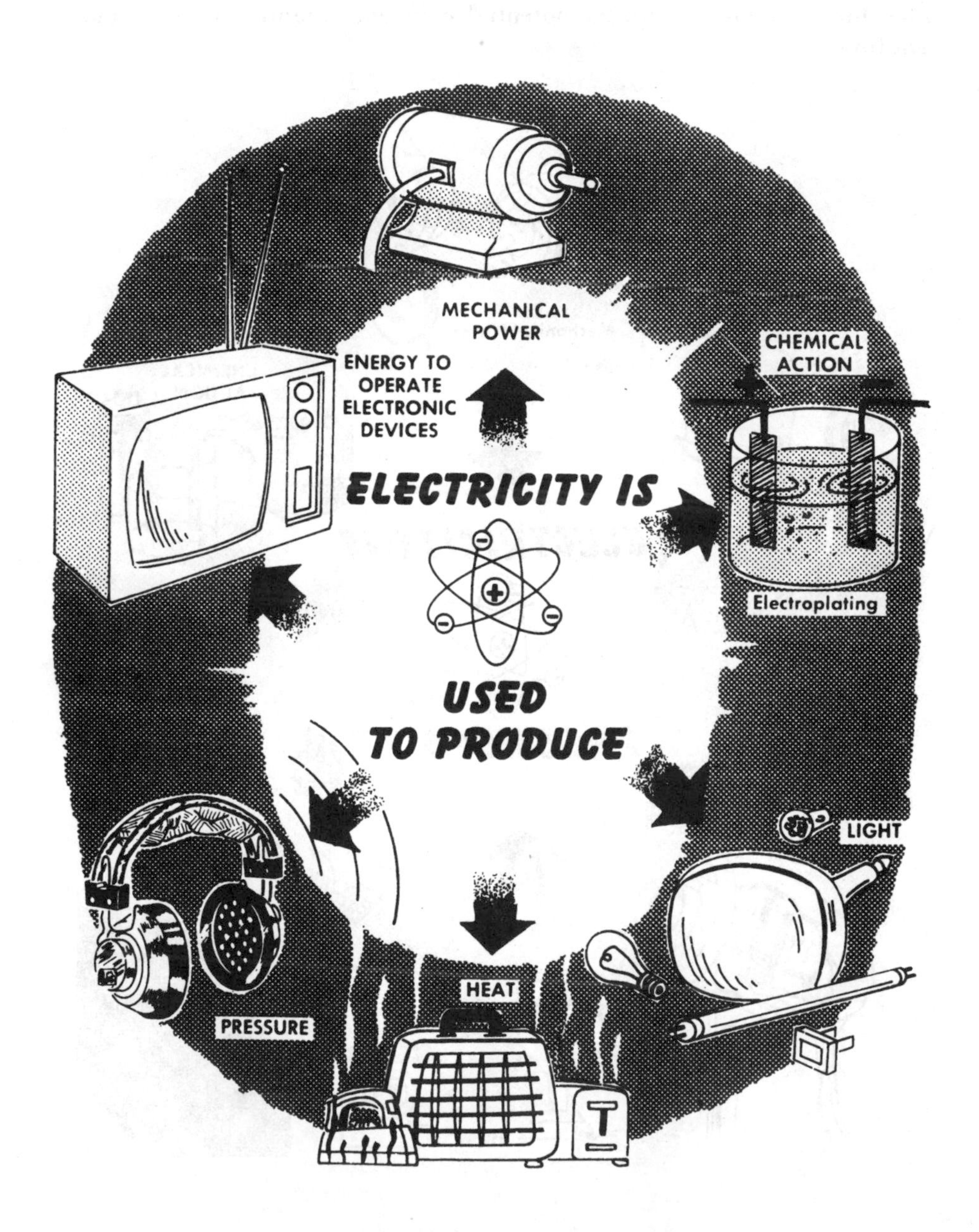

Electricity Produced from Frictional Energy (Static Electricity)

Although frictional electricity is the least important of all methods for the production of electricity, it is of value to study since it does have some useful applications and is important in understanding electric charges. You learned earlier that static (or frictional) electricity can be produced by rubbing certain dissimilar materials together. The source of energy in this case is from the muscles in your arm that, in turn, causes the separation of the charges. One of the applications of frictional or static electricity is in a device used in atomic research called the Van de Graf generator that will generate miniature bolts of lightning. Frictional electricity is usually a nuisance as mentioned earlier.

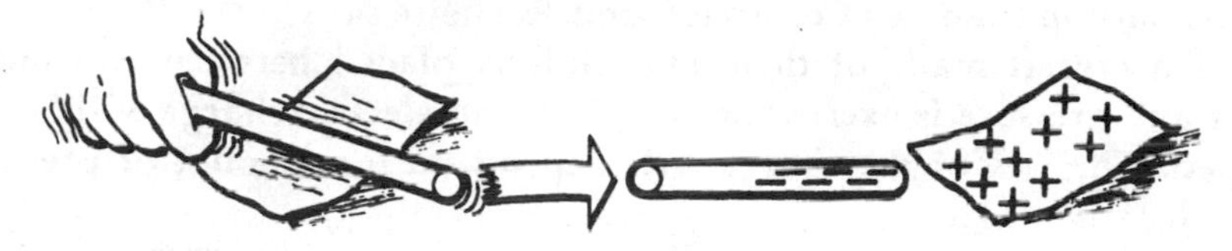

Although now we do not usually generate static electricity for use by friction, static electricity has some important applications. A most important application is the use of static electricity in electrostatic precipitators to remove carbon, fly ash, and other particles from the gases leaving a smokestack. This is done by giving the particles a charge of one polarity as they start to pass up the stack and then attracting them to collecting surfaces with a charge of opposite polarity further up the stack. In this way, most of the solid material present in the smoke can be removed. Techniques like this are most important for control of air pollution from industrial plants.

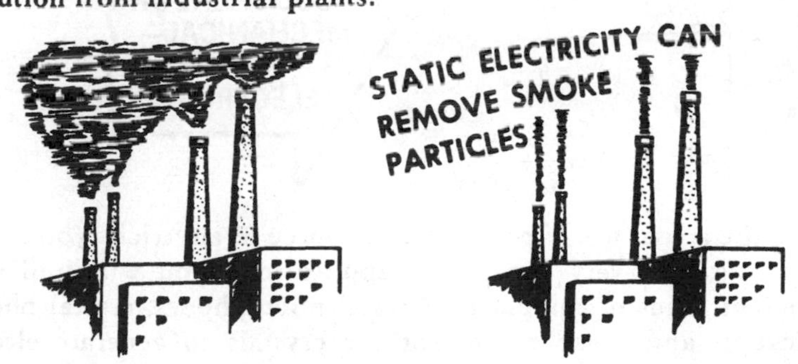
STATIC ELECTRICITY CAN
REMOVE SMOKE
PARTICLES

Another important application of static electricity is its use in the development of the xerographic copier—Xerox, IBM, etc. Electrostatic effects are utilized in the techniques of printing copies of the originals being duplicated.

Another useful application of static electricity is in electrostatic painting procedures (developed by the Ransburg Corporation) to paint on a high-volume, assembly-line basis even such irregularly shaped objects as automobile bodies, refrigerator cabinets, etc.

Electricity Produced from Pressure/Pressure Produced from Electricity

Electricity produced from pressure is called *piezoelectricity,* which is produced by certain crystalline materials. Crystals are orderly arrays of atoms in contrast to noncrystalline materials that have their atoms in a random pattern. Many crystalline materials exist in nature and many more can be made in the laboratory. Crystalline materials can be either pure elements or can be a compound. Two examples of naturally occurring crystals are quartz, which is the major component of common beach sand, and the diamond, which is the crystalline form of carbon. The most common materials used for the production of piezoelectricity are quartz, barium titanate (a ceramic), and Rochelle salts.

If a crystal made of these materials is placed between two metal plates and pressure is exerted on the plates, an electric charge will be developed. The size of the charge will depend on the amount of pressure exerted.

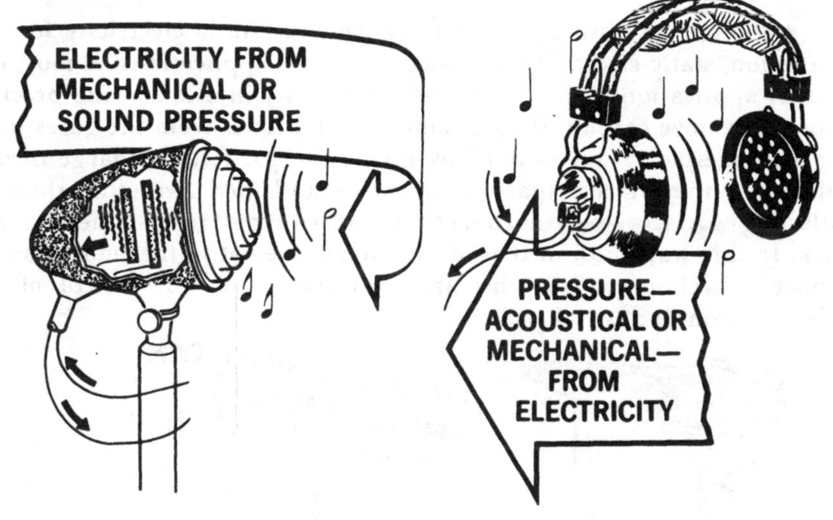

While the actual use of pressure as a source of electricity (piezoelectricity) is limited to very low power applications, you will find it in many different kinds of equipment. Crystal microphones, crystal phonograph pickups, and sonar equipment use crystals to generate electric charges from pressure. In these applications the mechanical energy comes from sound pressure or acoustical energy that moves a diaphragm that is mechanically coupled to the crystal; or, in the case of the phonograph pickup, the motion of the needle is coupled to the crystal. *Conversely,* if an electric charge is placed across the metal plates, the crystal will distort or physically change its shape, generating acoustical or mechanical energy. This is the principle used in crystal headphones. These sources and uses are entirely involved with electronic equipment and will not be further discussed in our study of electricity.

Electricity Produced from Heat (and Cold)

If a length of metal, such as copper, is heated at one end, electrons tend to move away from the hot end toward the cooler end. While this is true of most materials, some, such as iron, work the other way; that is, the electrons tend to flow toward the hot end. Thus, if an iron wire and a copper wire are twisted together to form a junction and the junction is heated, the flow of electrons will result in a charge difference between the free ends of the wires. (It should also be mentioned that if *cold* is applied, electrons will flow, but in the *opposite* direction.) In the illustration, the energy is supplied as heat from the burner.

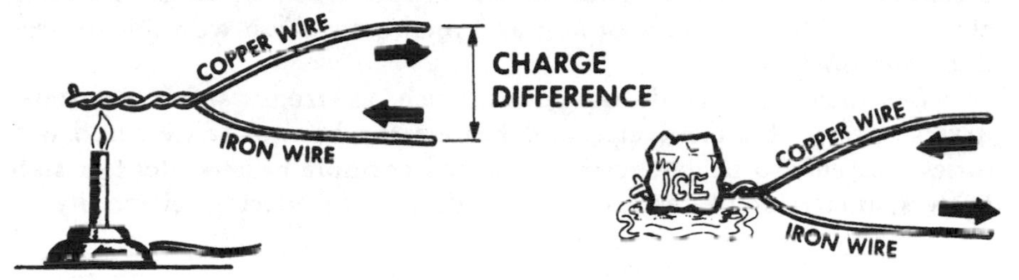

The amount of charge produced depends on the difference in temperature between the junction and the opposite ends of the two wires. A greater temperature difference results in a greater charge.

A junction of this type is called a *thermocouple* and will produce electricity as long as heat is applied. While twisted wires may form a thermocouple, more efficient thermocouples are constructed of two pieces of dissimilar metal riveted or welded together.

Since the current flow is proportional to the temperature of the junction, a thermocouple can be used to measure temperature when connected to a suitable indicating device. It is often used for this purpose, for example in an automobile engine temperature indicator. Many thermocouples can be connected together to form a *thermopile* or *thermoelectric generator*. Thermopiles are used in measuring heat inside high-temperature furnaces; also, in the fail-safe pilot flame thermal device in home gas furnaces to shut off the gas valve in case the pilot goes out. These generators using semiconductor materials are becoming more common as a replacement for batteries, particularly for military application. Their major advantage is that they will deliver power as long as the heat source is turned on. Their major disadvantages are that they are not capable of delivering very much power even in the larger sizes and are not very efficient.

Heat Produced from Electricity

Some heat is always produced when an electric current flows through a wire that is not a perfect conductor. This is because some energy is lost or used up—in the form of heat—in causing the electric current to flow. Good conductors produce less heat—although always some—because it is easy to cause current to flow through them. Poor conductors—as, for example, nichrome, an alloy of nickel and chromium used to make heating elements—produce a great deal of heat when current flows through them. Copper, for example, is about 60 times as good a conductor as nichrome. Thus, we use copper when we want to deliver electricity with a *minimum* of *loss*; and nichrome, when we want to produce *heat efficiently*.

Everyday household appliances, such as irons, stoves, toasters, dryers, electric blankets, etc., and heaters for houses, offices, and factories—baseboard heaters, ceiling heaters, portable heaters, floor or slab heaters, immersion heaters, etc—all use the heating effects of electricity.

Electricity Produced from Light

Electricity may be produced (or controlled) by using light as a source of energy. This occurs because materials like potassium, sodium, germanium, cadmium, cesium, selenium, and silicon release electrons when excited by light under the right conditions. This release of electrons is called the *photoelectric effect*.

The photoelectric effect is used in three ways. Light causes *photoemission*; the incident beam of light causes a surface to emit electrons that are collected to form an electric current. Light causes a change in how well a material conducts electricity. This is called the *photoconductive effect*. The third effect is the *photovoltaic effect*; the energy from the incident beam of light is converted directly into a flow of electrons. Although the photoconductive and photoemissive effects are very useful, particularly in electronic systems, the only source for significant amounts of electricity from light involves the principle of the photovoltaic effect. The solar cells that are used to power space vehicles and satellites are of this type. Let's take a look at how the photovoltaic cell is made and how it works.

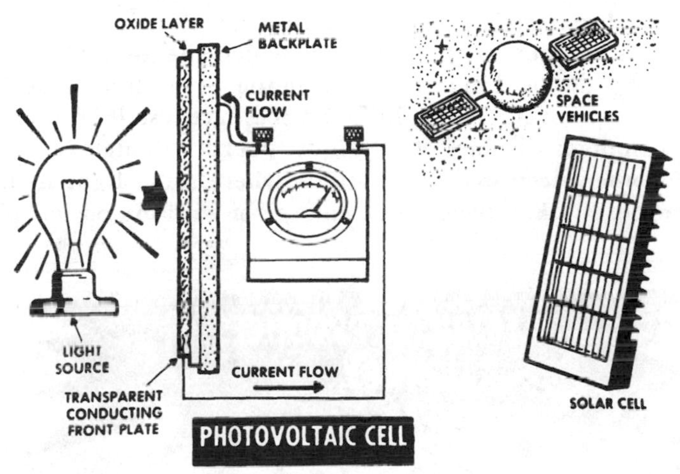

PHOTOVOLTAIC CELL

Most photovoltaic cells are made with selenium or silicon as the basic material. The cell is made up as a three-layer sandwich consisting of a pure material backing plate (copper) covered with a layer of oxide (selenium or silicon dioxide) that forms the center of the sandwich and a very thin transparent or translucent conducting layer as the other side of the sandwich. Light releases electrons from the junction of the oxide and the front plate. These electrons flow through the external circuit and back to the back plate. The electrical energy from each cell is very small, but if the cells are made larger to intercept more light, or many cells are connected together, then a significant amount of power can be developed. At least, there can be enough generated to power electronic equipment in satellites and similar space vehicles.

Light Produced from Electricity

Many of the poorer conductor materials, such as tungsten, glow red and even white hot when they become heated from conducting electric current. This radiant glow—incandescence—gives off *light* as well as heat. The incandescent lamp works this way to produce much of the light that we use.

Light is also produced with less electricity and without much heat by *fluorescence, phosphorescence* and *electroluminescence.* The fluorescent lights used in houses, offices, factories, etc. contain a gas (argon, mercury vapor, etc.) which, when forced to conduct an electric current, becomes ionized and produces ultraviolet and some visible radiation. When this radiation strikes a fluorescent coating inside the fluorescent light tube, it gives off a colored light. By using a mixture of fluorescent materials of different colors, white light can be produced.

Television picture tubes work on the principle of *phosphorescence* in which an electron beam strikes a surface coated with phosphorescent material which, in turn, gives off light.

Some light sources work on the principle of *electroluminescence* in which a solid material, when conducting an electric current, gives off light. This is the general principle by which the electroluminescent or light-emitting diode, LED (typically made of gallium arsenide phosphide wafers) works. These LED devices are used in digital wristwatches, computer displays of all kinds, pocket calculators, security systems, and other electronic applications where some light is needed for special purposes. LEDs have the advantage of working on low power.

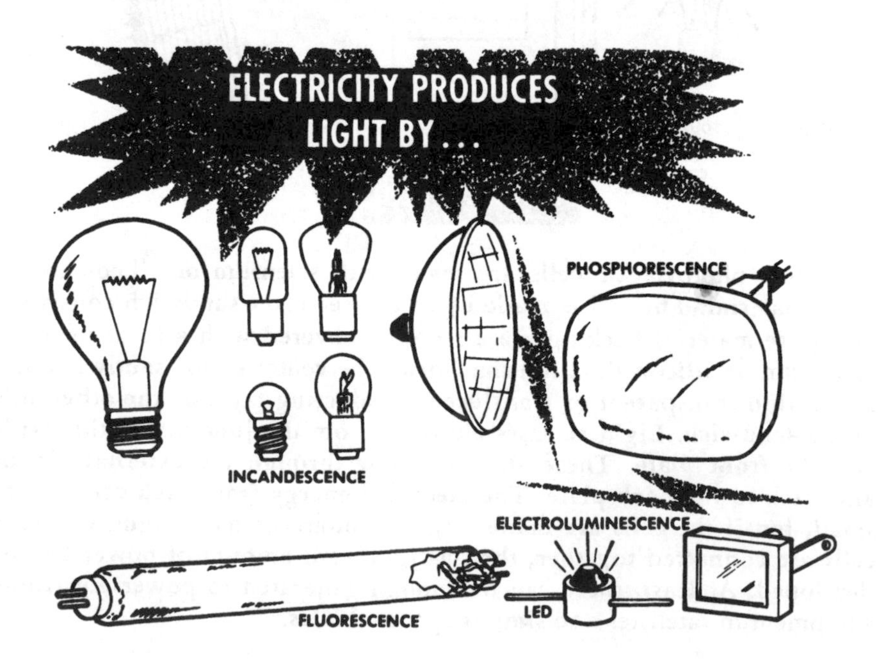

Electricity Produced from Chemical Action

Batteries are the most common source of electricity from chemical action. A *fuel cell* is another device used to generate electricity by chemical action. Fuel cells have the advantage of being light and capable of long life but are extremely expensive and, therefore, are only used for military and spacecraft applications at the present time. In the fuel cell, gases like hydrogen and oxygen are made to combine directly to form water and the energy released by this reaction is used directly to generate electricity. For most applications requiring a portable or emergency power source, batteries using chemical energy to generate electricity are used. We will discuss batteries only briefly here, but you will study them in much more detail later when you have learned more about electricity.

A battery is often made up of a number of identical cells connected together in a common container. We hook up cells this way so we can get more energy than we could from a single cell. Thus, the cell is the fundamental unit of the battery. When you learn about cells, you will know about batteries. Although materials may differ for different batteries, all batteries consist of two dissimilar metal plates called *electrodes* immersed in an *electrolyte* that may be either a paste or a liquid.

You will encounter batteries in many electric and electronic circuits. When reading diagrams showing how these circuits are hooked up, you will find that the various circuit elements are represented by a special representation called the *schematic symbol*. The schematic symbols shown above are used for all batteries. It is not necessary to show all of the cells in the schematic representation if we write adjacent to the battery the battery output potential, such as 6 volts.

Electricity Produced from Chemical Action (continued)

All cells consist of two electrodes and an electrolyte. Electricity is produced by the chemical reactions that take place between the electrodes and the electrolyte. The simplest cell—a *wet cell*—consists of strips of zinc metal and copper metal for the electrodes and an acid solution, for example sulfuric acid and water, as the electrolyte. In a *dry cell*, such as a flashlight battery, the electrolyte is in the form of a paste, rather than a fluid.

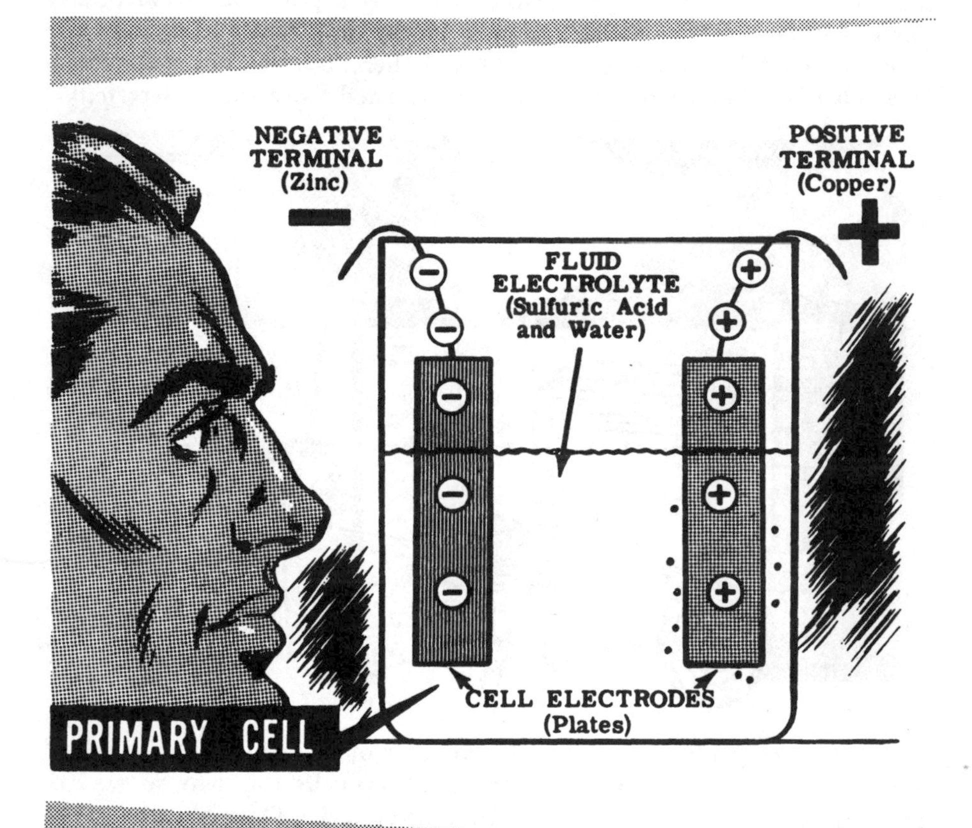

We will discuss the details of how electricity is obtained by chemical action later. In the cell above, the zinc is very slowly dissolved in the electrolyte. As the zinc atoms go into the solution they leave electrons on the undissolved zinc electrode. Thus, the zinc electrode develops a negative charge. By a similar process, electrons leave the copper electrode to unite with hydrogen atoms from the sulfuric acid electrolyte to form neutral hydrogen atoms and a positive charge develops on the electrode. The copper electrode is not dissolved since it supplies only electrons.

Electricity Produced from Chemical Action (continued)

With nothing connected to the cell terminals, you would see that electrons are pushed onto the negative plate until there is room for no more. The electrolyte would take enough electrons from the positive plate to make up for those it pushed onto the negative plate. Both plates would then be fully charged and no electrons would be moving between the plates.

Now suppose you connected a wire between the negative and positive terminals of the cell. You would see the electrons on the negative terminal leave the terminal and travel through the wire to the positive terminal. Since there would now be more room on the negative terminal, the electrolyte would carry more electrons across from the positive plate to the negative plate. As long as electrons leave the negative terminal and travel to the positive terminal outside the cell, the electrolyte will carry electrons from the positive plate to the negative plate inside the cell.

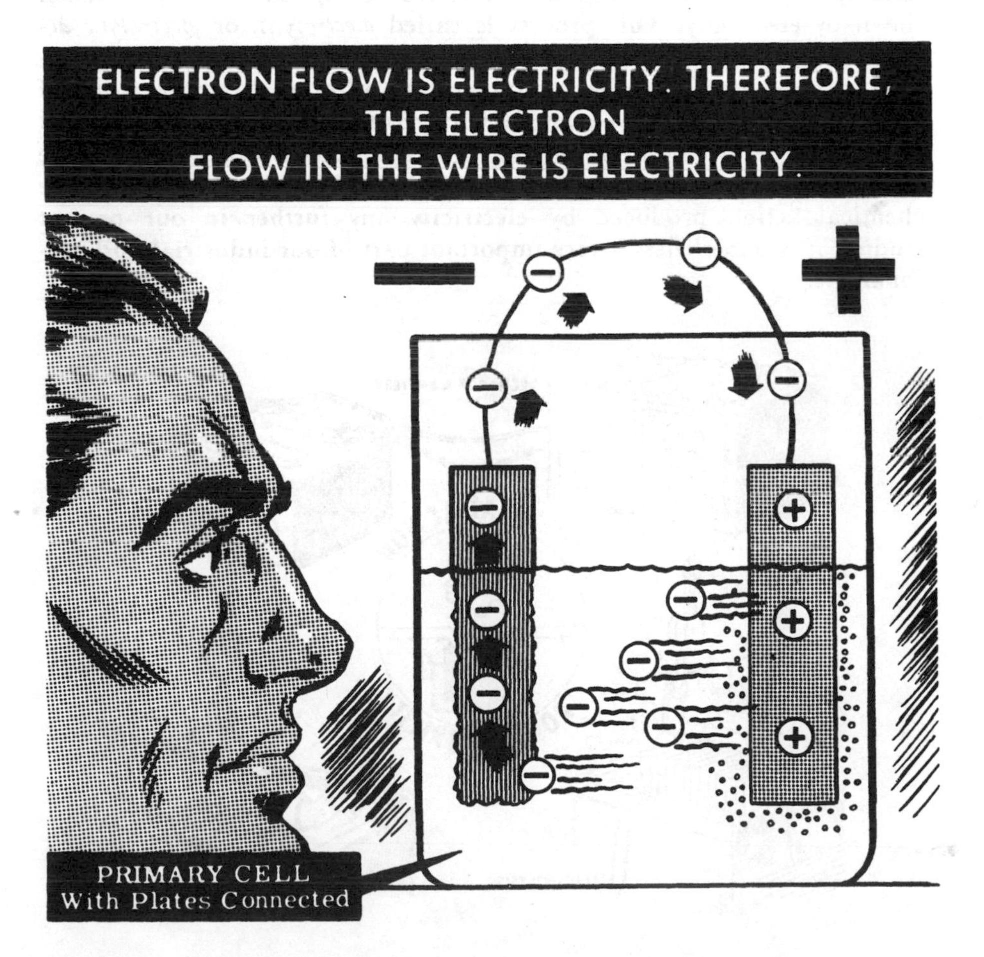

ELECTRON FLOW IS ELECTRICITY. THEREFORE, THE ELECTRON
FLOW IN THE WIRE IS ELECTRICITY.
PRIMARY CELL
With Plates Connected

Chemical Action Produced from Electricity

Probably the most common example of chemical action produced from electricity is the recharging of the ordinary automobile storage battery. When the cells of the storage battery are being used to generate electricity, a chemical reaction takes place. If a current is sent through the cells in the opposite direction, the reaction runs in the other direction and the battery is recharged. Cells that do this are called *secondary cells.* Most secondary cells used in storage batteries are of the lead-acid type. In this cell, the electrolyte is sulfuric acid, the positive plate is lead peroxide, and the negative plate is lead. During discharge of the cell, the acid becomes weaker and both plates change chemically to lead sulfate. Recharging reconverts the lead sulfate to pure lead on one plate and lead peroxide on the other, and the strength of the sulfuric acid electrolyte increases. Other types of secondary cells use nickel and iron, nickel and cadmium, or silver and zinc in a potassium hydroxide electrolyte.

Since the basic force that holds compounds together is electrical in nature, it is not surprising that chemical compounds can be broken down by electricity. This process is called *electrolysis*, or *electrolytic action*, and is very important in the manufacture of many metals (aluminum, copper, etc.) and other substances. An additional important use of chemical action produced from electricity is in *electroplating.* Here, metal ions are made to migrate to an electrode and adhere to it when they are changed from ions to the metal. Although we will not look into the chemical action produced by electricity any further in our present studies, it is nonetheless a very important part of our industrial and personal life.

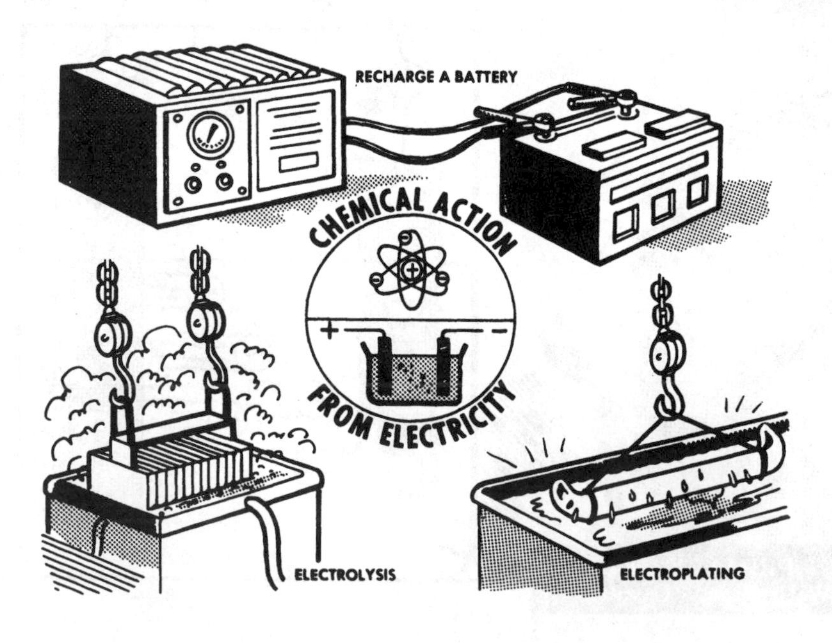

Electricity Produced from Magnetism

The most common method of producing electricity for electric power is by the use of magnetism. The source of electricity must be able to maintain a large potential difference because the charge is being used to furnish electric power.

Almost all of the electric power used, except for emergency and portable equipment operated from batteries, originally comes from a generator in a power plant. The generator may be driven by water power, a steam turbine with its steam heated by coal, oil, gas, or atomic power, or an internal combustion engine. No matter how the generator is driven, the electric power it produces is the result of the action between the wires and the magnets inside the generator. Remember, electricity does not by itself produce power, it transports it.

When wires move past a magnet or a magnet moves past wires, electricity is produced in the wires. Now you will find out how magnetism is used to produce electricity.

Electricity Produced from Magnetism (continued)

One method by which magnetism produces electricity is through the movement of a magnet past a stationary wire. If you connect a very sensitive meter (a device for indicating current flow) across the ends of a stationary wire and then move a magnet past the wire, the meter needle will deflect. This deflection indicates that electricity is produced in the wire. Repeating the movement and observing the meter closely, you will see that the meter moves only while the magnet is passing *near* the wire.

Placing the magnet near the wire and holding it at rest, you will observe no deflection of the meter. Moving the magnet from this position, however, does cause the meter to deflect and shows that, alone, the magnet and wire are not able to produce electricity. In order to deflect the needle, *movement* of the magnet past the wire is necessary.

Movement is necessary because the magnetic field, or flux lines, around a magnet produces an electric current in a wire only when the magnetic field is moved across the wire. When the magnet and its field are stationary, the field is not moving across the wire and will not produce a movement of electrons in it.

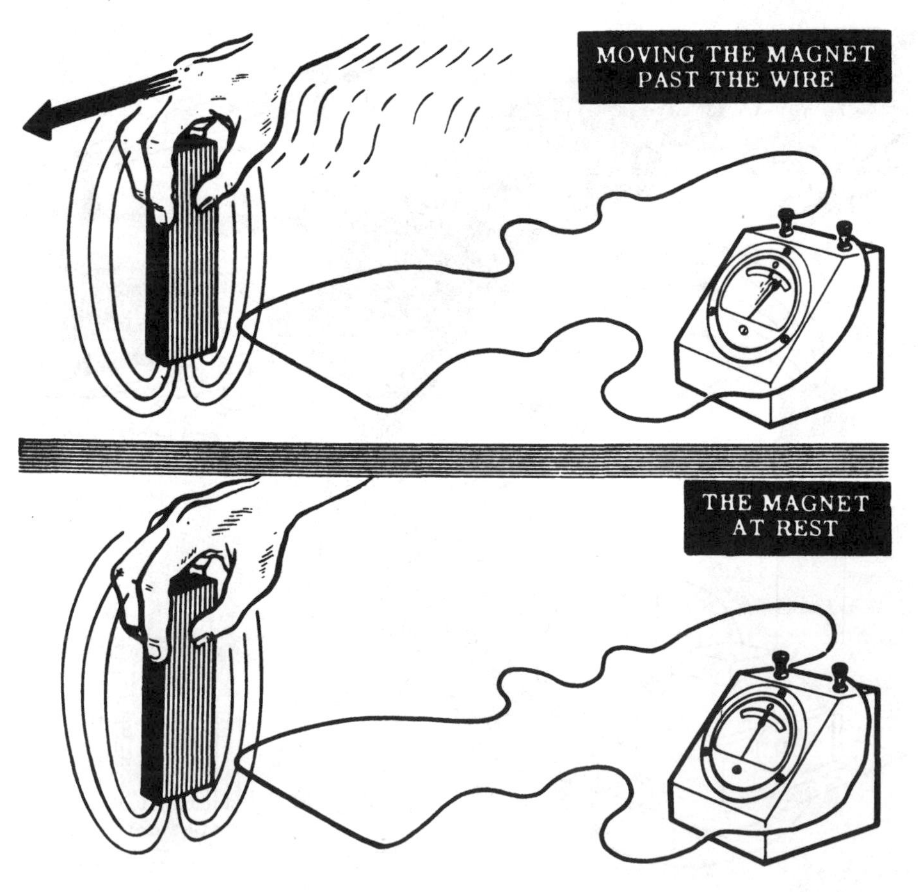

Electricity Produced from Magnetism (continued)

In studying the effects of moving a magnet past a wire, you discovered that electricity was produced only while the magnet and its field were actually moving. If you move the wire past a stationary magnet, you again will notice a deflection of the meter. This deflection will occur only while the wire is moving across the magnetic field.

To use magnetism to produce electricity, you may either move a magnetic field across a wire, or move a wire across a magnetic field. In either case, it is the wire cutting across the lines of force or flux lines that produces electricity. For a continuous source of electricity, however, you need to maintain a *continuous* motion of either the wire or the magnetic field.

To provide a continuous motion, the wire or the magnet would need to move back and forth constantly. A more practical way is to cause the wire to travel in a circle through the magnetic field.

This method of producing electricity—that of the wire traveling in a circle past the magnets—is the principle of the *electric generator* (dynamo) and is the source of most electricity used for electric power.

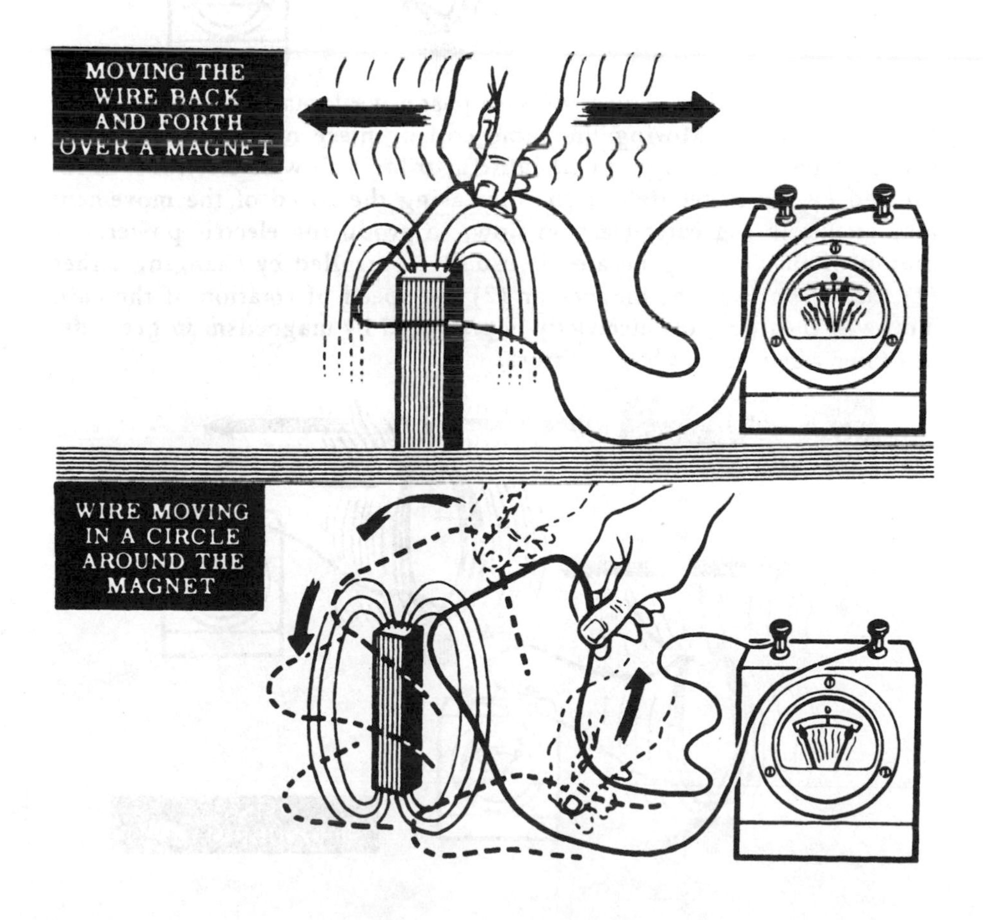

Electricity Produced from Magnetism (continued)

Since the electricity is produced by the wire cutting past the flux lines, you can change the *amount* of electricity produced by changing the strength of the magnetic field or by cutting more flux lines with the wire in a shorter length of time.

To increase the amount of electricity produced by moving a wire past a magnet, you might increase the length of the wire that passes through the magnetic field, use a stronger magnet, or move the wire faster. The length of the wire can be increased by winding it in several turns to form a *coil*. Moving the coil past the magnet will result in a much greater deflection of the meter than resulted with a single wire. Each additional coil turn will add an amount equal to that of one wire.

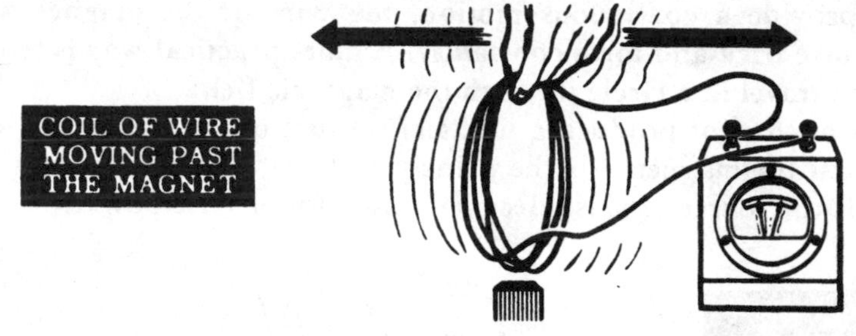

Moving a coil or a piece of wire past a weak magnet causes a weak flow of electrons. Moving the same coil or piece of wire at the same speed past a strong magnet will cause a stronger flow of electrons, as indicated by the meter deflection. Increasing the speed of the movement also results in a greater electron flow. In producing electric power, the output of an electric generator is usually controlled by changing either (1) the strength of the magnet or (2) the speed of rotation of the coil. You will consider how electricity is produced by magnetism in great detail later when you study generators.

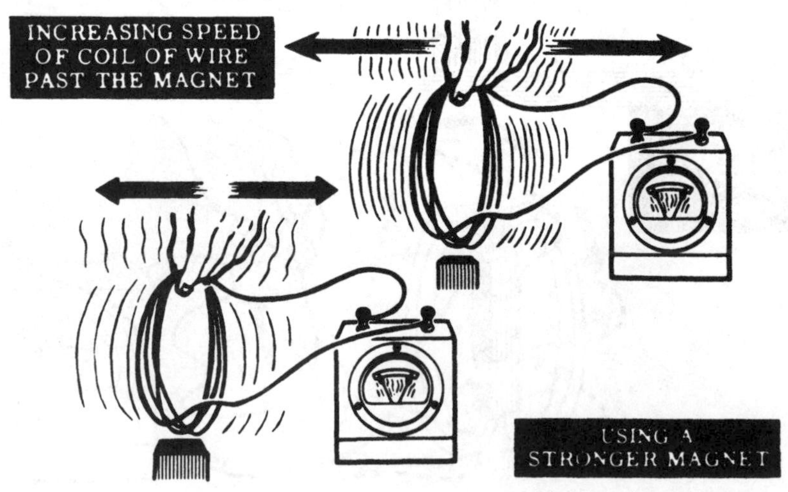

Magnetism Produced from Electricity

Magnetic fields can be created conversely, by electricity, just as you might suspect, because you know that you can produce electricity from magnetism. Any conductor that carries a current will act like and is, in fact, a magnet. If the wire is wound into a coil, the magnet will be stronger. The magnet will be stronger also if the current is increased. Since the magnetic field depends on the flow of current, there is no magnetic field if the current is removed. Temporary magnets of this type are called *electromagnets* and the effect is called *electromagnetism*. Electromagnetism is so important in your study of electricity that the entire next section is devoted to that subject.

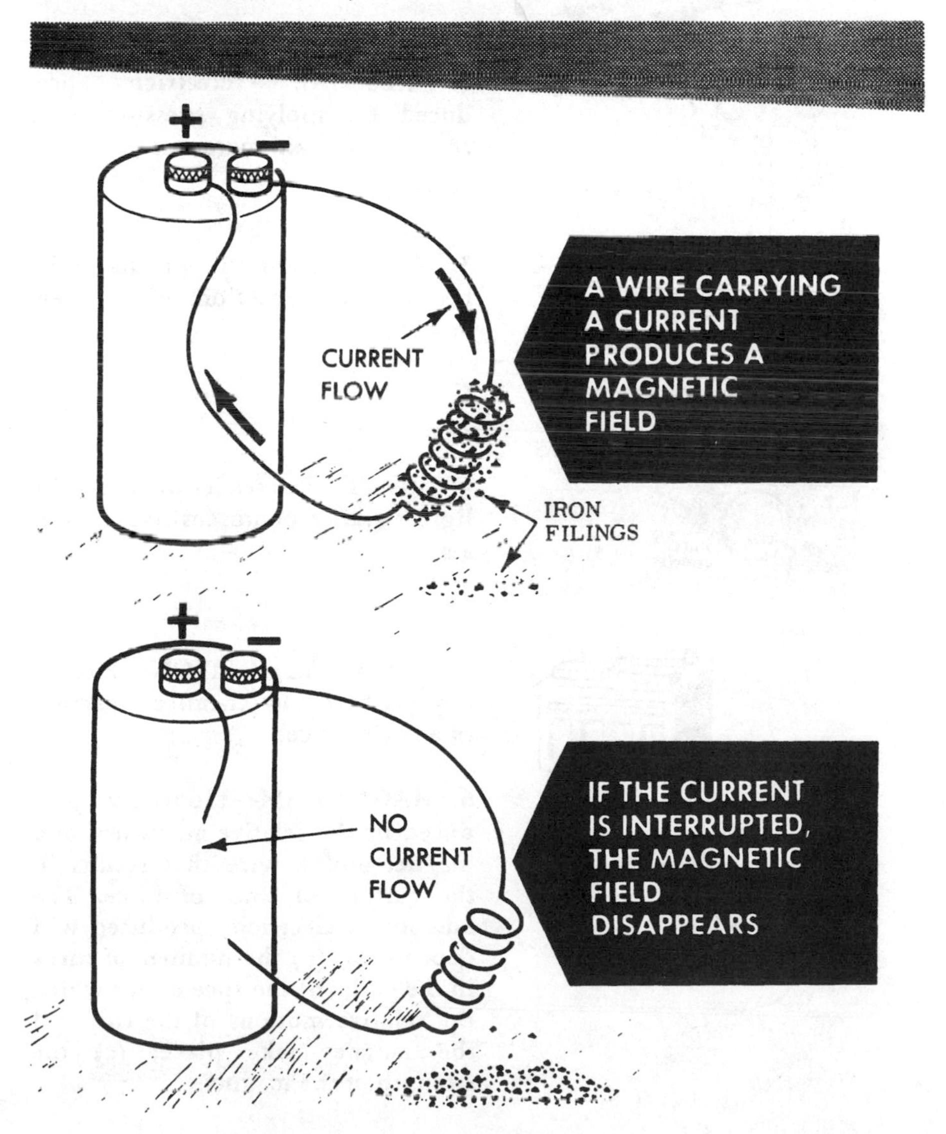
+
−
CURRENT FLOW
A WIRE CARRYING A CURRENT PRODUCES A MAGNETIC FIELD
IRON FILINGS
+
−
NO CURRENT FLOW
IF THE CURRENT IS INTERRUPTED, THE MAGNETIC FIELD DISAPPEARS

Review of How Electricity Is Produced

Electricity is the action of electrons which have been forced from their normal orbits around the nucleus of an atom. To force electrons out of their orbits so they can become a source of electricity, some kind of energy is required.

Six kinds of energy can be used:

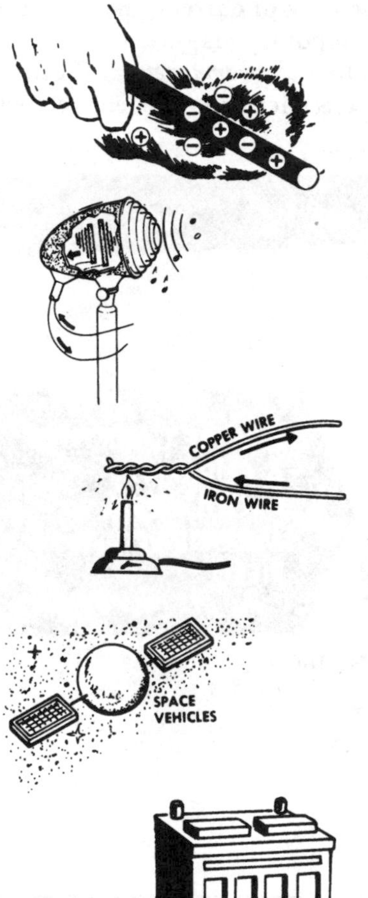

1. FRICTION — Electricity produced by rubbing two materials together.

2. PRESSURE — Electricity produced by applying pressure to a crystal of certain materials.

3. HEAT—Electricity produced by heating the junction of a thermocouple.

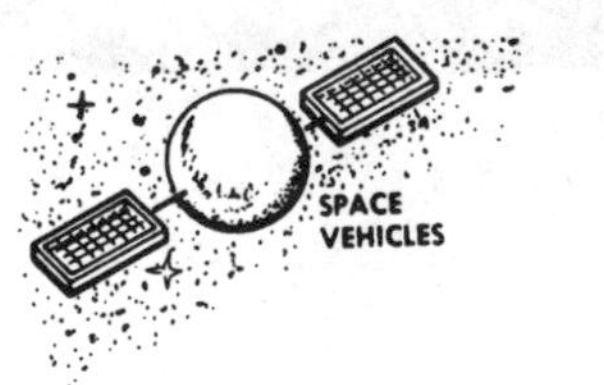

4. LIGHT—Electricity produced by light striking photosensitive materials.

5. CHEMICAL ACTION—Electricity produced by chemical reaction in an electric cell.

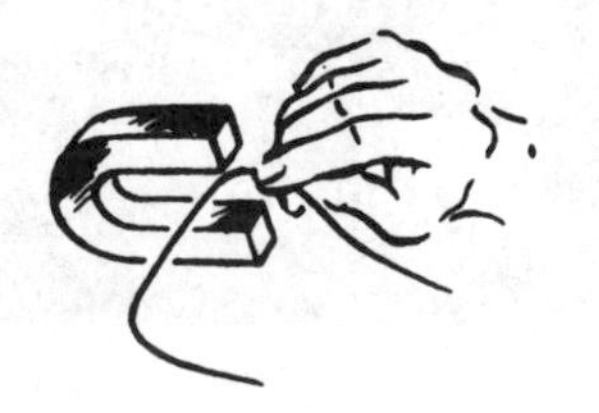

6. MAGNETISM—Electricity produced by the relative movement of a magnet and a wire that results in the cutting of lines of force. The amount of electricity produced will depend on: (a) the number of turns in the coil; (b) the speed with which the relative motions of the coil and the magnet take place; (c) the strength of the magnet.

Review of How Electricity Is Used

Electricity, or the flow of current, is used by all of us to do a number of tasks in all aspects of our everyday life, as well as in some ways that are not so obvious. Usually when electricity is used, it is converted into some useful form of energy.

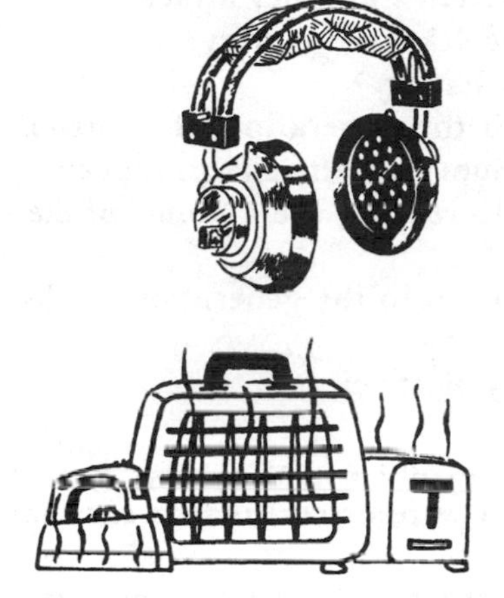

1. PRESSURE—If a potential difference is applied across the faces of certain kinds of crystals, such as Rochelle salt, the crystal will distort and produce pressure or mechanical movement. This is the principle used in crystal headphones.

2. HEAT—When current flows through an imperfect conductor, some of the energy is used up in getting the electrons through. This energy appears as heat. Poor conductors that do not melt easily, such as nichrome wire, are used as heating elements.

3. LIGHT—When enough current is passed through a wire, it can become white hot or incandescent. This is the way our ordinary light bulb works, and it produces light as well as heat. To keep the filament (heated wire) from burning up, the filament is enclosed in a bulb with an inert gas. Electricity can also produce light by *electroluminescence*, *phosphorescence*, and *fluorescence*.

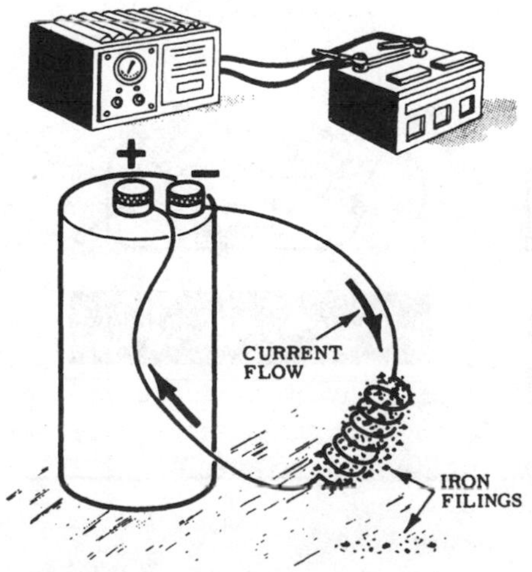

4. CHEMICAL ACTION—Electriccity can cause the decomposition of chemical compounds. This is the principle behind the secondary cells that are used in storage batteries. It is also the basis for electroplating and electrolytic action.

5. ELECTROMAGNETISM—Current passing through a wire produces a magnetic field around the wire as long as the current is flowing. This effect is *electromagnetism*.

Self-Test—Review Questions

1. What are the six common sources of electricity?
2. Which of these sources is most important? Least important? Why?
3. Describe how a thermocouple is made and how it operates.
4. What are the three necessary components for each cell of a battery?
5. How do primary and secondary cells differ? How are they alike?
6. What is the circuit schematic symbol for a cell? For a battery?
7. What is the difference between a cell and a battery?
8. Describe the basic principle involved with the generation of electricity from magnetism. What determines the amount of electricity produced?
9. How would you go about increasing or decreasing the amount of electricity produced by magnetism?
10. What is the common principle that is involved in the generation of electricity from any source?
11. What are the five major effects produced by electricity?
12. How does the incandescent lamp work?
13. Why does the flow of current in conductors produce heat?
14. What happens to iron filings placed near a wire or coil that is carrying current? Why?
15. What happens to the iron filings of Question 14 when the current is removed? Why?

Learning Objectives—Next Section

Overview—Earlier you learned about magnetism and some of the effects of magnetism. Now you will learn about a very important type of magnet called the *electromagnet*. Electromagnetism is one of the most important effects of electricity as you will see as you study further.

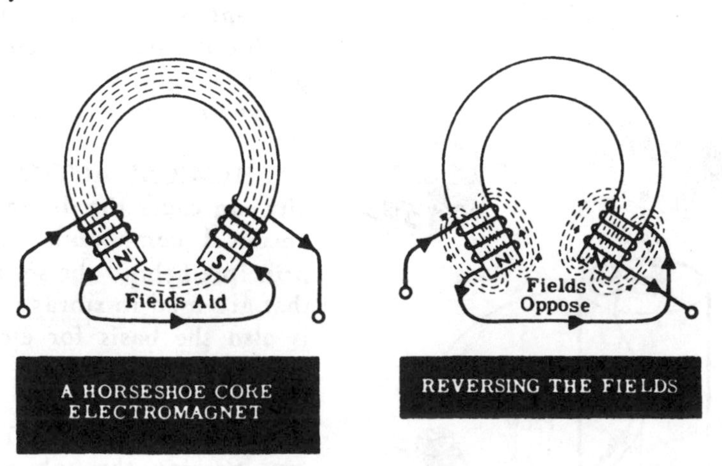

A HORSESHOE CORE ELECTROMAGNET

REVERSING THE FIELDS

Electromagnetism

Earlier you learned the very important fact that an electric current will flow when you move a coil of wire so that it cuts through a magnetic field. You also learned that this is the most widespread manner in which electricity is generated for the home, industry, aboard ship, etc. You learned too that electricity can generate magnetism. In this section you will see for yourself exactly how this is done.

Earlier you made use of permanent magnets to cause an electric current to flow. You saw that more current could be generated as you increased the number of turns of wire, the speed of motion of the coil, and the strength of the magnetic field. It is a simple matter to accomplish the first two of these in a practical electric generator, but it is very difficult to increase the strength of a permanent magnet beyond certain limits. In order to generate large amounts of electricity, a much stronger magnetic field must be used. This is accomplished by means of an *electromagnet*. Electromagnets work on the simple principle that a magnetic field is generated by passing an electric current through a coil of wire. As you learned, electromagnets differ from permanent magnets in that they are magnetic only when an electric current is supplied.

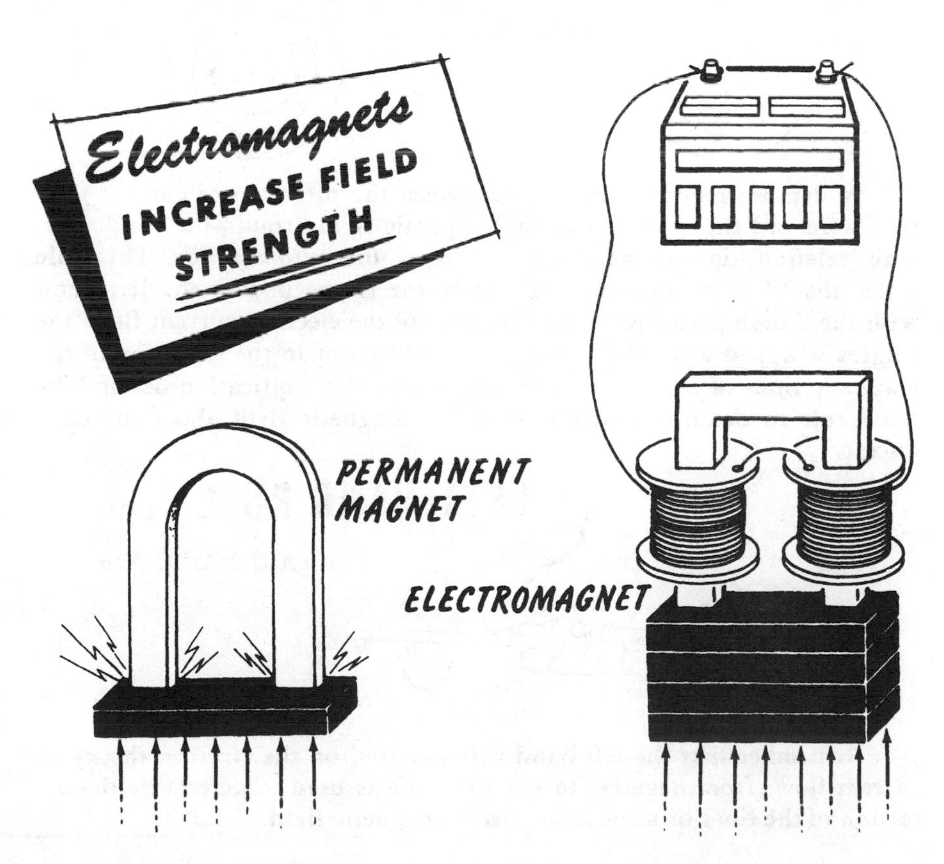

Magnetic Fields around a Conductor

An electromagnetic field is a magnetic field caused by the current flow in a wire. Whenever electric current flows, a magnetic field exists around the conductor, and the direction of this magnetic field depends upon the direction of current flow. The illustration shows conductors carrying current in different directions. The direction of the magnetic field is counterclockwise when current flows from left to right. If the direction of current flow *reverses*, the direction of the magnetic field also *reverses*, as shown. In the cross-sectional view of the magnetic field around the conductors, the dot in the center of the circle represents the head of the arrow indicating the current flowing out of the paper *toward* you; the cross represents the tail of the arrow indicating the current flowing into the paper *away* from you.

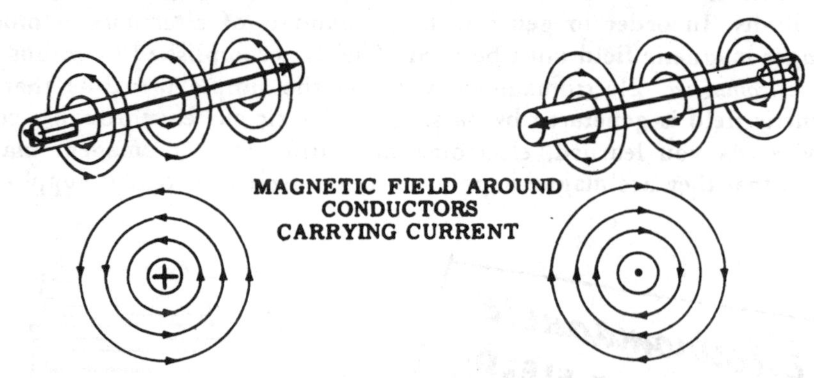

A definite relationship exists between the direction of current flow in a wire and the direction of the magnetic field around the conductor. This relationship can be shown by using the *left-hand rule*. This rule states that if a current-carrying conductor is grasped in the left hand with the thumb pointing in the direction of the electron current flow, the fingers wrapped around the conductor will point in the direction of the magnetic lines of force. The drawing shows the application of the left-hand rule to determine direction of the magnetic field about the conductor.

Remember that the left-hand rule is based on the electron theory of current flow (from negative to positive) and is used to determine the direction of the lines of force in an electromagnetic field.

Magnetic Fields around a Conductor (continued)

It is easy to demonstrate in an experiment that a magnetic field exists around a current-carrying conductor. Connect a heavy copper wire in series with a switch and a dry cell battery. The copper wire is bent to support itself vertically and then inserted through a hole in a lucite plastic sheet, which is held in a horizontal position. When the switch is closed, iron filings—which have the property of aligning themselves along the lines of force in a magnetic field—are sprinkled on the lucite. The lucite is tapped lightly to make it easier for the iron filings to fall into position.

If you did this experiment, you would see that the filings arrange themselves in concentric circles, showing that the magnetic lines of force form a circular pattern around the conductor. To show that the circular pattern is actually the result of the magnetic field, you could open the switch and spread the filings evenly over the lucite, then repeat the experiment. Each time the circuit current flows, the filings arrange themselves to show the magnetic field.

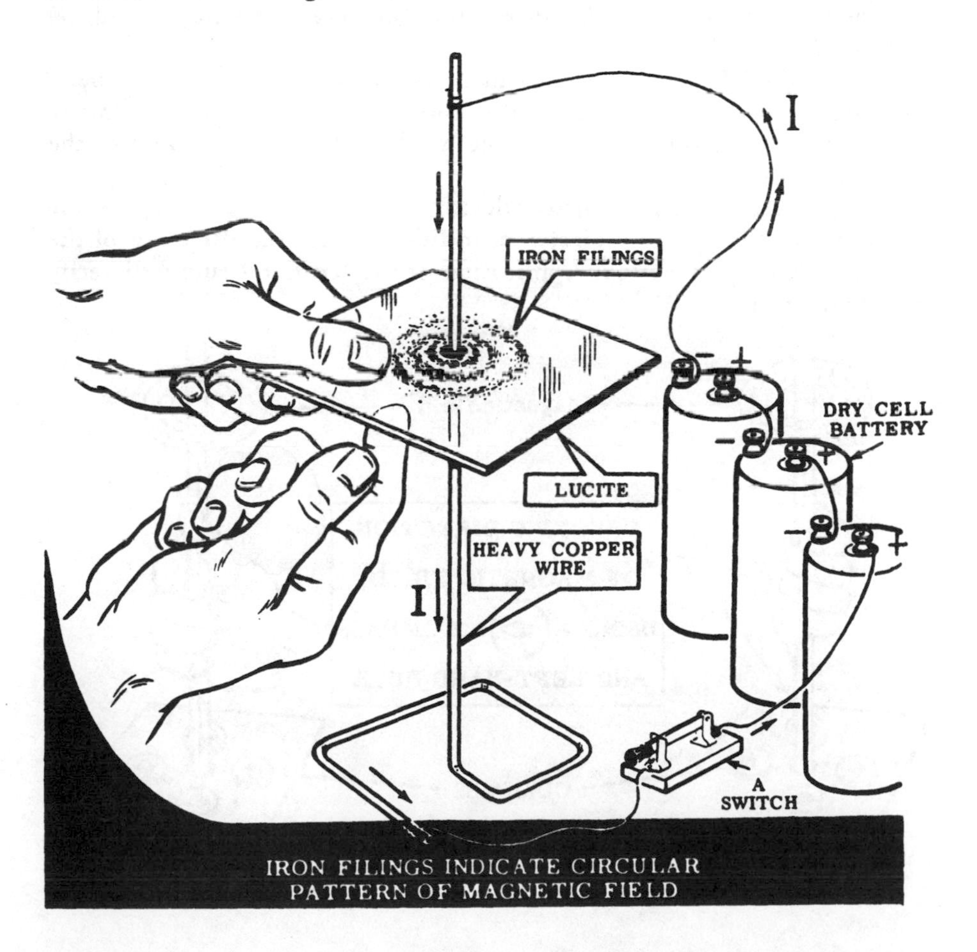

IRON FILINGS INDICATE CIRCULAR PATTERN OF MAGNETIC FIELD

Magnetic Fields around a Conductor (continued)

To demonstrate experimentally the direction of the magnetic field around the current-carrying conductor, a compass needle can be used instead of iron filings.

A compass needle is nothing more than a small bar magnet that will line itself up with the lines of force in a magnetic field. You know from the previous experiment that the magnetic field is circular. Therefore, the compass needle always will be positioned at right angles to the current-carrying conductor.

If you remove the iron filings from the lucite sheet, and the compass is placed on the lucite about 2 inches, or approximately 5 centimeters, away from the conductor, you can trace the direction of the magnetic field around the conductor. With no current flowing, the North pole end of the compass needle will point to the Earth's magnetic North pole. When current flows through the conductor, the compass needle lines itself up at right angles to a radius drawn from the conductor. If the compass needle is moved around the conductor, the needle always maintains itself at right angles to it. This proves that the magnetic field around the conductor is circular.

Using the left-hand rule you can check the direction of the magnetic field indicated by the compass needle. The direction in which the fingers go around the conductor is the same as that of the North pole of the compass needle.

If the current through the conductor is reversed, the compass needle will point in the opposite direction, indicating that the direction of the magnetic field has reversed. Application of the left-hand rule will verify this observation.

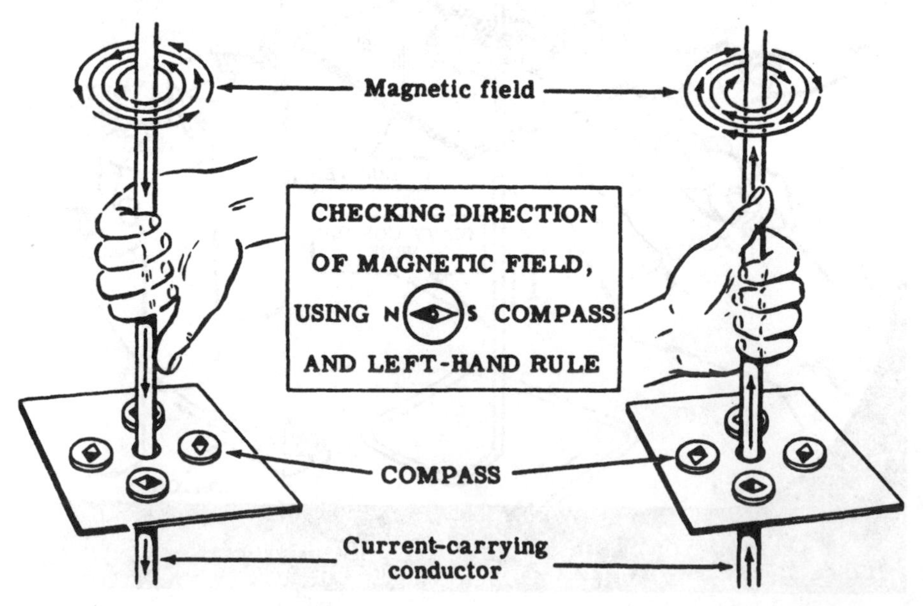

Magnetic Fields around a Coil

Magnetic fields around a coil of wire are extremely important in many pieces of electrical equipment. A coil of wire carrying a current acts as a magnet. If a length of wire carrying a current is bent to form a loop, the lines of force around the conductor all leave at one side of the loop and enter at the other side. Thus, the loop of wire carrying a current will act as a weak magnet having a North pole and a South pole. The North pole is on the side where the lines of force leave the loop and the South pole is on the side where they enter the loop.

If you desire to make the magnetic field of the loop stronger, you can form the wire into a coil of many loops as shown. Now the individual fields of each loop add together and form one strong magnetic field inside and outside the loop. In the spaces between the turns, the lines of force are in opposition and cancel each other out. The coil acts as a strong magnet with the North pole being the end where the lines of force leave the loop.

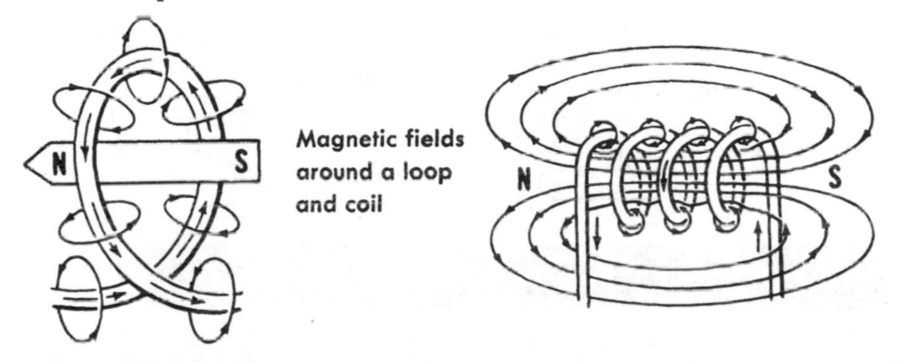

Magnetic fields around a loop and coil

A left-hand rule also exists for coils to determine the direction of the magnetic field. If the fingers of the left hand are wrapped around the coil in the direction of the current flow, the thumb will point toward the North pole end of the coil.

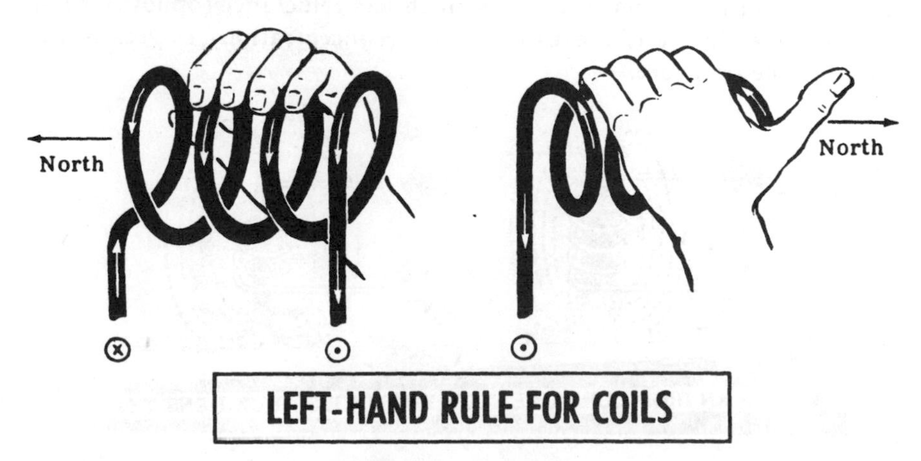

LEFT-HAND RULE FOR COILS

Magnetic Fields around a Coil (continued)

Adding more turns to a current-carrying coil increases the number of lines of force, causing it to act as a stronger magnet. An increase in current also strengthens the magnetic field. Strong electromagnets have coils of many turns and carry as large a current as the wire size permits.

In comparing coils using the same core or similar cores, a unit called the *ampere-turn* is used. This unit is the product of the current in amperes and the number of turns on the wire.

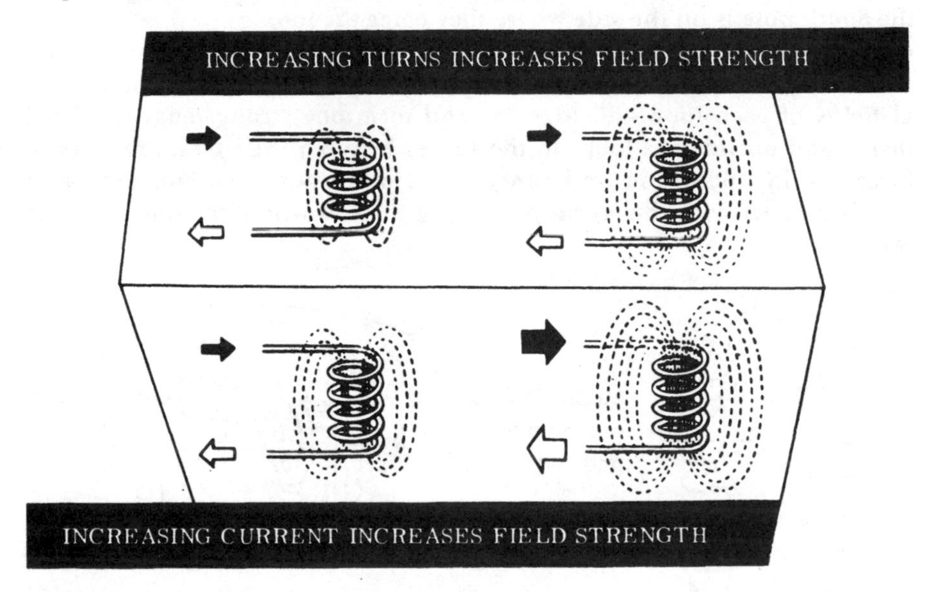
INCREASING TURNS INCREASES FIELD STRENGTH
INCREASING CURRENT INCREASES FIELD STRENGTH

Although the field strength of an electromagnet is increased by using both a large current flow and many turns to form the coil, these factors do not usually concentrate the field enough for use in a practical device. To further increase the flux density, an iron core is inserted in the coil. Because the iron core offers much less reluctance (opposition) to lines of force than air, the flux density (concentration) is greatly increased in the iron core.

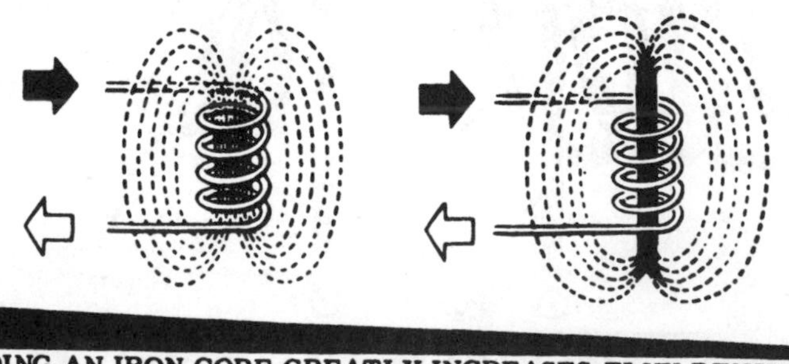

ADDING AN IRON CORE GREATLY INCREASES FLUX DENSITY

Magnetic Fields around a Coil (continued)

You can show what the field around a coil carrying current is like by taking a piece of wire and forming it into a coil that is threaded through some holes in a piece of lucite plastic.

The rest of the circuit is the same as those showing the fields around a conductor. When iron filings are sprinkled on the lucite and current is passed through the coil, tapping the lucite will cause the iron filings to line up parallel to the lines of force. If you did this, you would observe that the iron filings have formed the same pattern of a magnetic field that exists around a bar magnet.

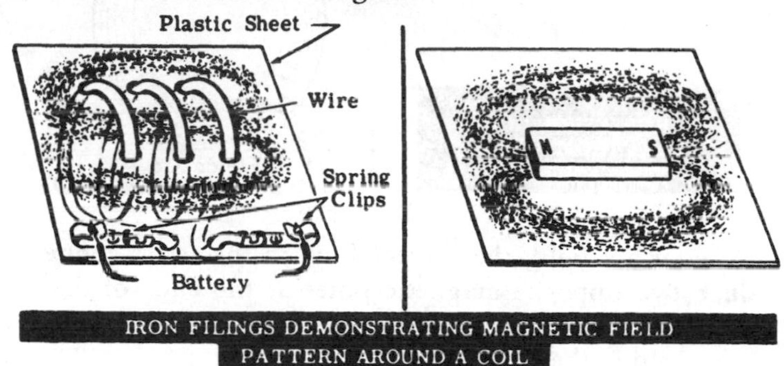

IRON FILINGS DEMONSTRATING MAGNETIC FIELD PATTERN AROUND A COIL

If the filings are removed, and a compass is placed inside the coil, the needle will line up along the axis of the coil with the North pole end of the compass pointing to the North pole end of the coil. Remember that the lines of force inside a magnet or coil flow from the South pole to the North pole. The North pole end of the coil can be verified by using the left-hand rule for coils. If the compass is placed outside the coil and moved from the North pole to the South pole, the compass needle will follow the direction of a line of force as it moves from the North pole to the South pole. When the current through the coil is reversed, the compass needle will also reverse its direction.

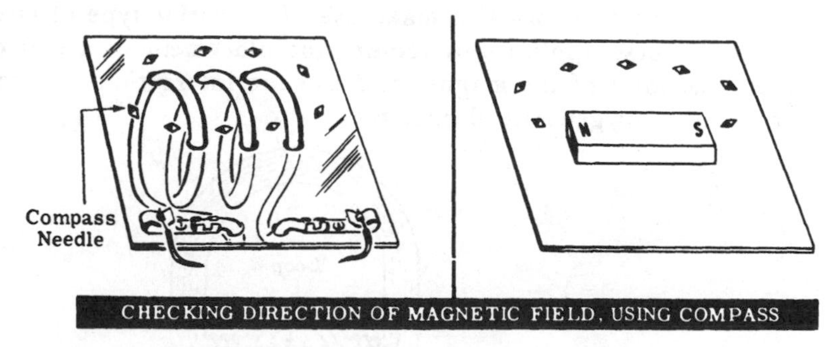

CHECKING DIRECTION OF MAGNETIC FIELD, USING COMPASS

If you placed a soft iron core inside the coil and tested the fields, you would observe that there is a strong concentration of the magnetic field in the iron core, as you would expect from your study of the effects of iron on magnetic fields.

Electromagnets

If the iron core is bent to form a horseshoe and two coils are used, one on each leg of the horseshoe-shaped core as illustrated, the lines of force will travel around the horseshoe and across the air gap, causing a very concentrated field to exist across the air gap. The shorter the air gap, the greater the flux density between the poles.

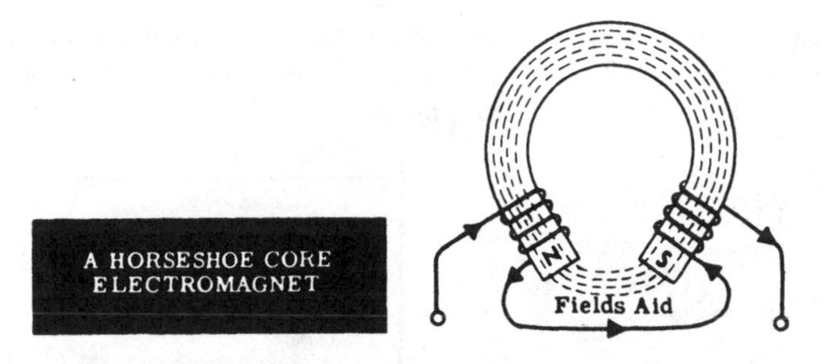

A HORSESHOE CORE ELECTROMAGNET

To cause such a field, the current flow in the series-connected coils must produce two opposite magnetic poles at the ends of the core. Reversing either coil would cause the two fields to oppose each other, canceling out the field in the air gap.

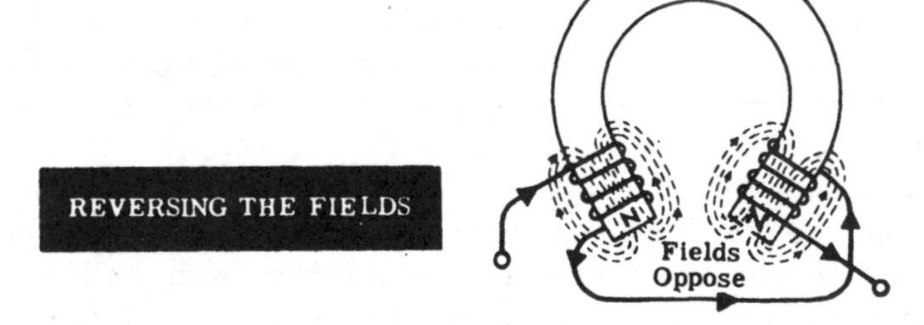

REVERSING THE FIELDS

Electric meters make use of horseshoe-type permanent magnets. Electric motors and generators also make use of a similar type of electromagnet. All of these applications require the placement of a coil of wire between the poles of the magnet and use the interaction between them either to utilize or generate electricity.

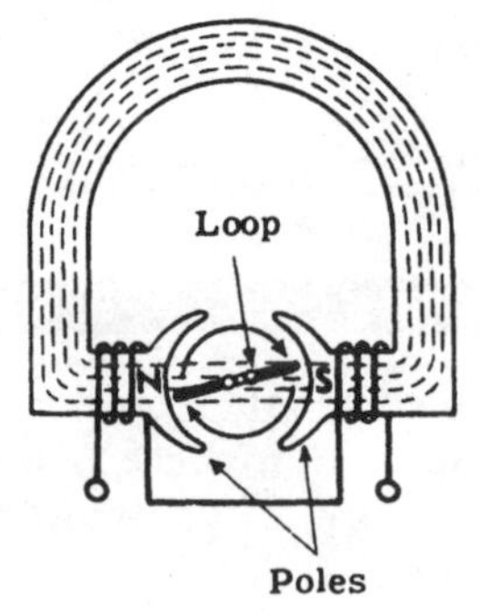

ELECTROMAGNET POLES

Review of Electromagnetism

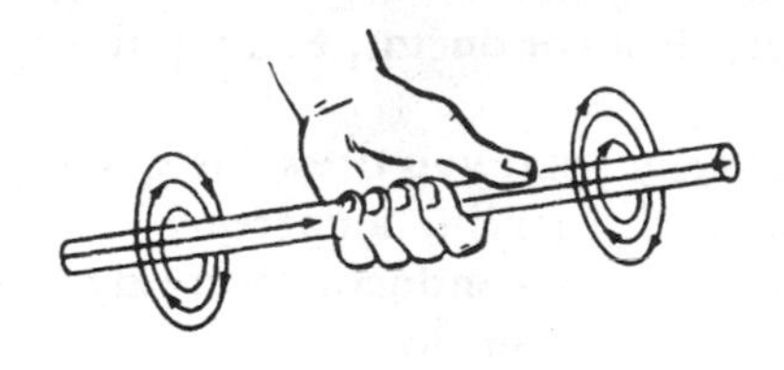

1. ELECTROMAGNETIC FIELD—Current flowing through a wire generates a magnetic field whose direction is determined by the direction of the current flow. The direction of the generated magnetic field is found by using the left-hand rule for a current-carrying conductor.

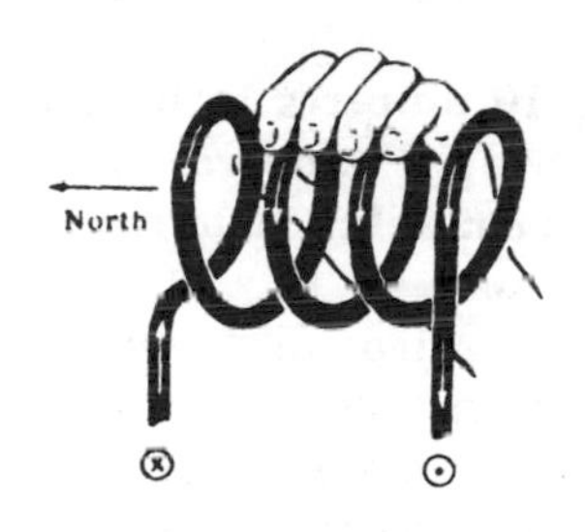

2. MAGNETIC FIELD OF A LOOP OR COIL—A loop generates a magnetic field exactly the same as a bar magnet. If many loops are added in series forming a coil, a stronger magnetic field is generated. The left-hand rule for a coil is used to determine the coil's magnetic polarity.

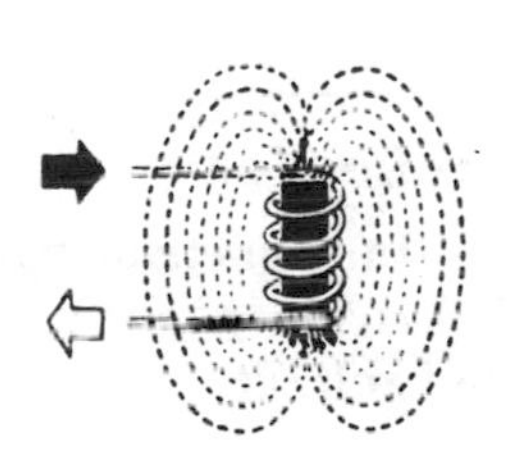

3. FIELD STRENGTH—Increasing the number of turns of a coil increases the field strength and increasing the coil current also increases the field strength. An iron core may be inserted to concentrate the field greatly (increase flux density) at the ends of the coil. The ampere-turn is the unit used in comparing the strength of electromagnetic fields.

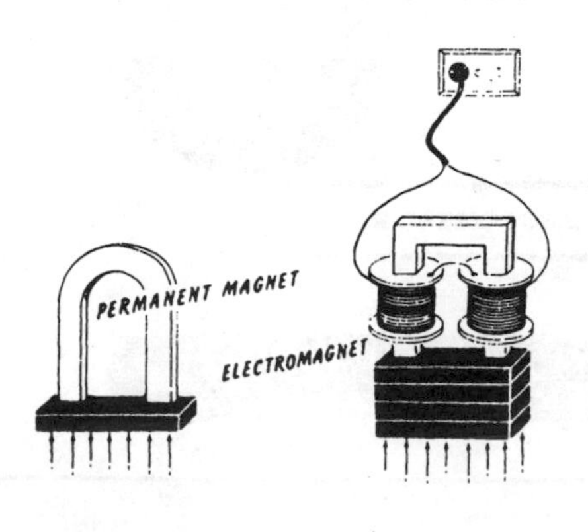

4. PERMANENT-MAGNETS and ELECTROMAGNETIC FIELDS—Electromagnetic fields are much stronger than the permanent magnet type, and are used in most practical electrical machinery. When electromagnets are used, the field strength can be varied by varying the amount of current flow through the field coils.

Self-Test—Review Questions

1. What are electromagnets and why are they used?
2. Draw the magnetic field that surrounds a conductor. Show polarity and direction of current and field.
3. Describe an experiment that would verify what you drew for Question 2.
4. What is the left-hand rule for a conductor carrying a current?
5. Invent a rule for the relationship between a conductor carrying current and its field for the case of conventional current flow.
6. We have stated that the magnetic field around a coil will be greater than for a wire (under the same current conditions). Why?
7. What is the rule for determining the direction of the field around a coil?
8. What are the factors that determine the strength of a magnetic field around a coil carrying current?
9. Would you expect a 2-turn coil carrying 10 amperes to have the same field strength as a coil of 20 turns carrying 1 ampere? Why? What would be the relative field strength of the 2 coils if the current in the second coil was changed to 2 amperes? To 0.5 ampere?
10. If you were to take a horseshoe shaped piece of iron and wound coils on each of the free ends and then connected them together and to a source of electricity, how would you explain the following observations? (a) Current flowing, but no magnetic field or very weak magnetic field. (b) The polarity of the poles of the electromagnet reverses when the current is reversed. (c) The polarity of the poles reverses when the coils are wound in the opposite direction.

Learning Objectives—Next Section

Overview—Now that you know about electromagnetism, you can learn about an important application of electromagnetism as it is used in meters to measure current flow and voltage.

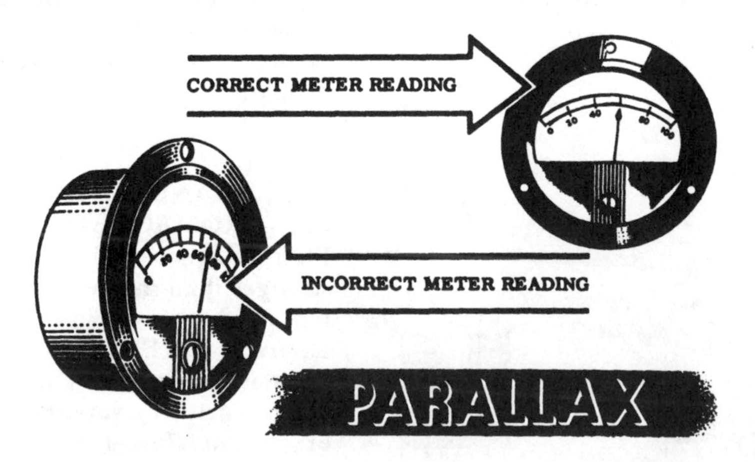

The Basic Meter Movement

You now have learned enough facts about electricity and magnetism to study a very important practical application in meters—the devices that measure current flow.

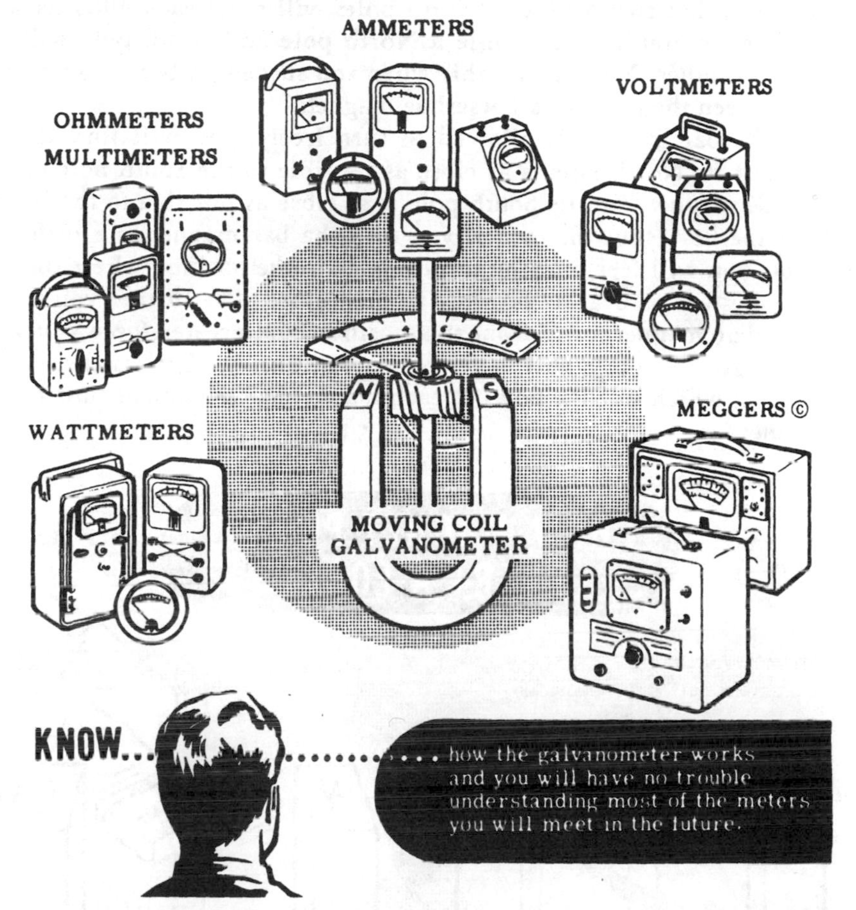

KNOW... . . . how the galvanometer works and you will have no trouble understanding most of the meters you will meet in the future.

You have probably used meters to show you whether or not an electric current was flowing and how much current was flowing. As you proceed further with your work in electricity, you will find yourself using meters more and more often. Meters are the right hand of anyone working in electricity or electronics, so now is the time for you to find out how they operate.

All the meters you have used and nearly all the meters you will use are made with the same type of meter *works* or movement. This meter movement is based on the principles of an electric current measuring device called the *moving-coil galvanometer*. Nearly all modern meters use the moving-coil galvanometer as a basic meter movement, so once you know how it works, you will have no trouble understanding most of the meters you will be using in the future.

The Basic Meter Movement (continued)

The galvanometer works on the principle of magnetic attraction and repulsion. According to this principle, which you have already learned, like poles repel each other and unlike poles attract each other. This means that two magnetic North poles will repel each other as will two magnetic South poles, while a North pole and South pole will attract one another. You can see this when you suspend a bar magnet on a pivot between the poles of a horseshoe magnet.

If the bar magnet is allowed to turn freely, you will find that it turns until its North pole is as close as possible to the South pole of the horseshoe magnet, and its South pole is as close as possible to the North pole of the horseshoe magnet. If you turn the bar magnet to a different position, you will feel it trying to turn back to the position where the opposite poles are as near as possible to each other. The further you try to turn the bar magnet away from this position, the greater force you will feel. The greatest force will be felt when you turn the bar magnet to the position in which the like poles of each magnet are as close as possible to each other.

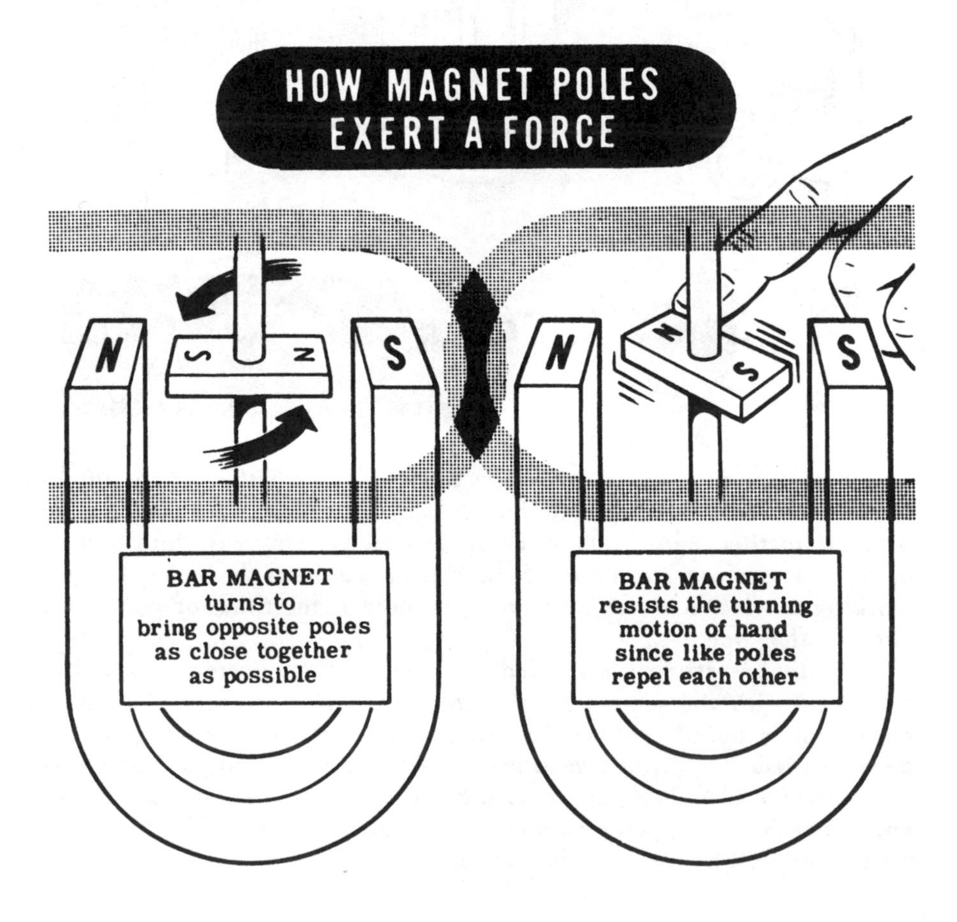

Basic Meter Movement (continued)

The forces of attraction and repulsion between magnetic poles become greater when stronger magnets are used. You can see this when you attach a spring to the bar magnet in such a way that the spring will have no tension when the North poles of the two magnets are as close as possible to each other. With the magnets in this position, the bar magnet would normally turn freely to a position which would bring its North pole as close as possible to the South pole of the horseshoe magnet. With a spring attached, it will turn only part way, to a position where its turning force is balanced by the force of the spring. If you were to replace the bar magnet with a stronger magnet, the force of repulsion between the like poles would be greater and the bar magnet would turn further against the force of the spring.

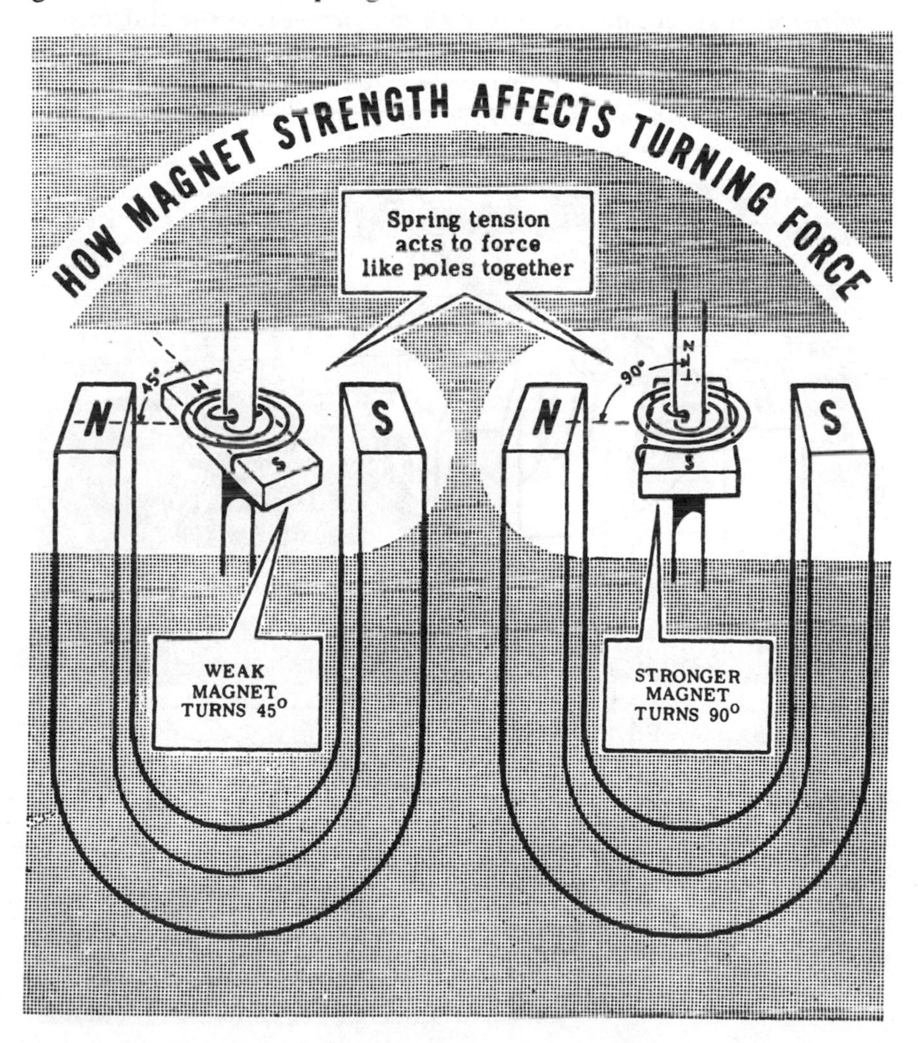
HOW MAGNET STRENGTH AFFECTS TURNING FORCE
Spring tension acts to force like poles together
45°
N
S
N
S
90°
N
S
N
S
WEAK MAGNET TURNS 45°
STRONGER MAGNET TURNS 90°

Basic Meter Movement (continued)

If you remove the bar magnet and replace it with a coil of wire, you have a *galvanometer*. Whenever an electric current flows through this coil of wire, it acts as a magnet. The strength of this wire-coil magnet depends on the size, shape, and number of turns in the coil and the amount of electric current flowing through the coil. If the coil itself is not changed in any way, the magnetic strength of the coil will depend on the amount of current flowing through the coil. The greater the current flow in the coil, the greater the magnetic strength of the wire-coil magnet.

If there is no current flow in the coil, it will have no magnetic strength and the coil will turn to a position where there will be no tension on the spring. If you cause a small electric current to flow through the coil, the coil becomes a magnet and the magnetic forces—between the wire-coil magnet and the horseshoe magnet—cause the coil to turn until the magnetic turning force is balanced by the force due to tension in the spring. When a larger current is made to flow through the coil, the magnetic strength of the coil is increased and the wire coil turns further against the spring tension.

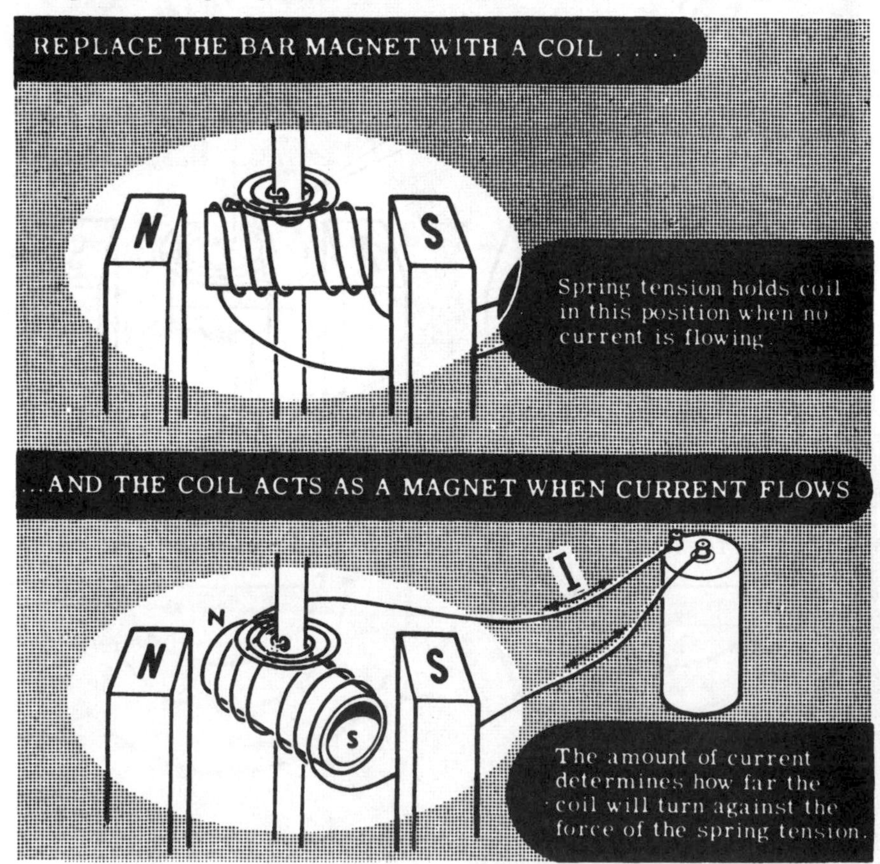

REPLACE THE BAR MAGNET WITH A COIL . . . .
N
S
Spring tension holds coil in this position when no current is flowing.
...AND THE COIL ACTS AS A MAGNET WHEN CURRENT FLOWS
I
N
S
N
S
The amount of current determines how far the coil will turn against the force of the spring tension.

Basic Meter Movement (continued)

When you want to find out how much current is flowing in a circuit, all you need to do is to connect the coil into the circuit and measure the angle through which the coil turns away from its position at rest. It is very difficult to measure this angle, and to calculate the amount of electric current which causes the coil to turn through this angle. However, by connecting a pointer to the coil and adding a scale for the pointer to travel across, you can read the amount of current directly from the scale.

Now that you have added a scale and a pointer, you have a basic dc meter, known as the *D'Arsonval-type* movement, which depends upon the operation of magnets and their magnetic fields. Actually, there are two magnets in this type of meter: one, a stationary permanent horseshoe magnet; the other, an electromagnet. The electromagnet consists of turns of wire wound on a frame, and the frame is mounted on a shaft fitted between two permanently mounted jewel bearings. A lightweight pointer is attached to the coil, and turns with it to indicate the amount of current flow. Current passing through the coil causes it to act as a magnet with poles being attracted and repelled by those of the horseshoe magnet. The strength of the magnetic field about the coil depends upon the amount of current flow. A greater current produces a stronger field, resulting in greater forces of attraction and repulsion between the coil sides and the magnet's poles.

Basic Meter Movement (continued)

The magnetic forces of attraction and repulsion cause the coil to turn so that the unlike poles of the coil and magnet will be brought together. As the coil current increases, the coil becomes a stronger magnet and turns further because of the greater magnetic forces between the coil and magnet poles. Since the amount by which the coil turns depends upon the amount of coil current, the meter indicates the current flow directly.

Although all meters that you will encounter work on the principle of two interacting magnetic fields, there are some modifications that you may encounter. For example, you may ask why the permanent magnet is not associated with the moving part of the meter and the electromagnet on the fixed part. This would eliminate the need to carry current through the springs. Actually there are meters built and used like those described above. These meters, however, are much less sensitive because the strength of the permanent magnet is very limited as a result of its small size; and also, they are less accurate because it is difficult to control the magnetic fields involved. Meters of this type are very inexpensive and are found in less expensive equipment.

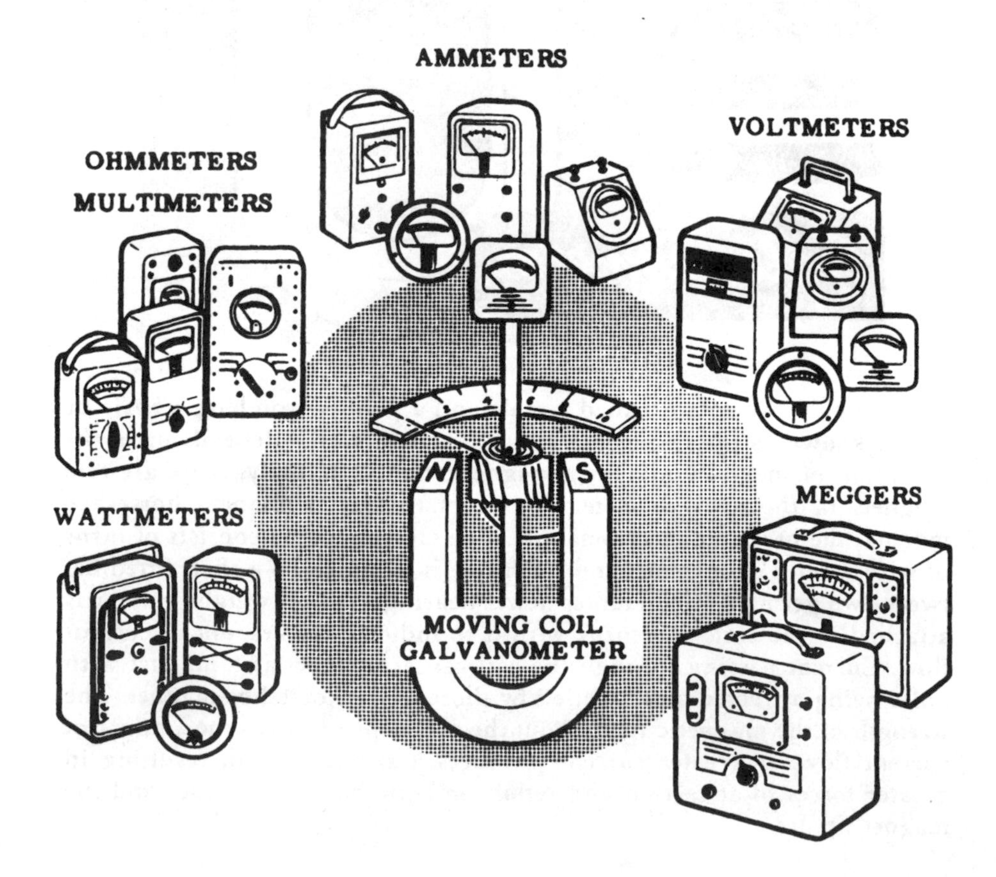

Meter Movement Considerations

While galvanometers are useful in laboratory measurements of extremely small currents, they are not portable, compact, or rugged enough for use in the field. A modern meter movement uses the principles of the galvanometer but is portable, compact, rugged, and easy to read. The coil is mounted on a shaft fitted between two permanently mounted jewel bearings. A lightweight pointer is attached to the coil and turns with the coil to indicate the amount of current flow.

Balance springs on each end of the shaft exert opposite turning forces on the coil and, by adjusting the tension of one spring, the meter pointer may be adjusted to read zero on the meter scale. Since temperature change affects both coil springs equally, the turning effect of the springs on the meter coil is canceled out. As the meter coil turns, one spring tightens to provide a retarding force, while the other spring releases its tension. In addition to providing tension, the springs are used to carry current from the meter terminals through the moving coil.

In order that the turning force will increase uniformly as the current increases, the horseshoe magnet poles are shaped to form semicircles. This brings the coil as near as possible to the North and South poles of the permanent magnet. The amount of current required to turn the meter pointer to full-scale deflection depends upon the magnet strength and the number of turns of wire in the moving coil.

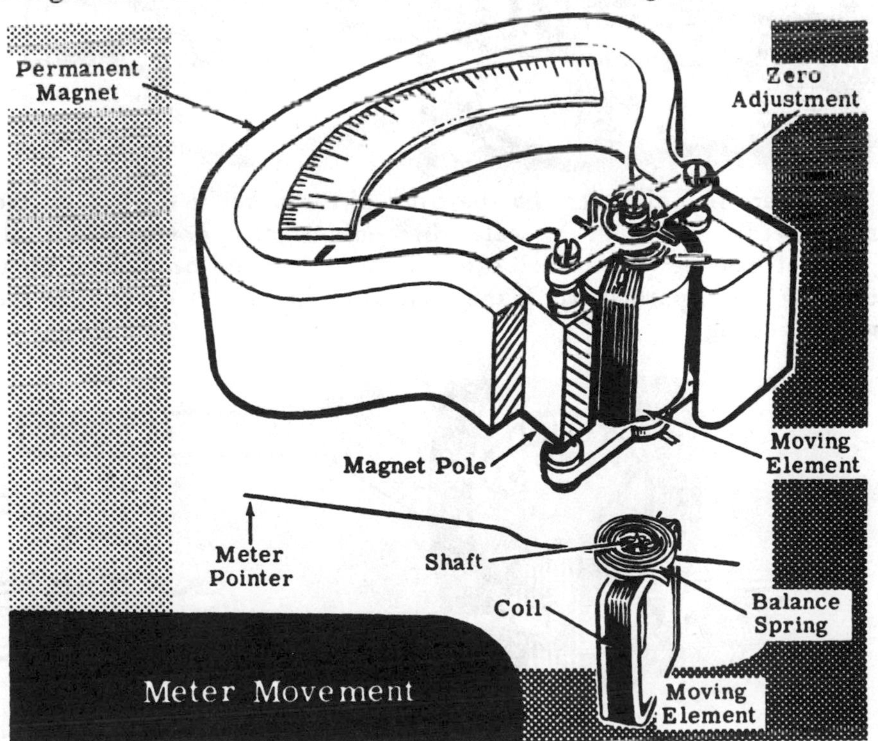

How Meter Scales Are Read

When you work with electricity, it is necessary that you take accurate meter readings to determine whether equipment is working properly, and to discover what is wrong with equipment that is not operating correctly. Many factors can cause meter readings to be inaccurate. It is necessary to keep them in mind whenever you use a meter. You will find the usable range of a meter scale does not include the extreme ends of the scale. For nearly all meters, the most accurate readings are those taken near the center of the scale. When current is measured with an *ammeter*, *milliammeter*, or *microammeter*, the range of the meter used should be chosen to give a reading near mid-scale.

All meters cannot be used in both horizontal and vertical positions. Due to the mechanical construction of many meters, the accuracy will vary considerably with the position of the meter. Normally, panel-mounted meters are calibrated and adjusted for use in a vertical position. Meters used in many test sets and in some electrical equipment are made for use in a horizontal position.

A zero-set adjustment on the front of the meter is used to set the meter needle at zero on the scale when no current is flowing. This adjustment is made with a small screwdriver and should be checked when using a meter, particularly if the vertical or horizontal position of the meter is changed.

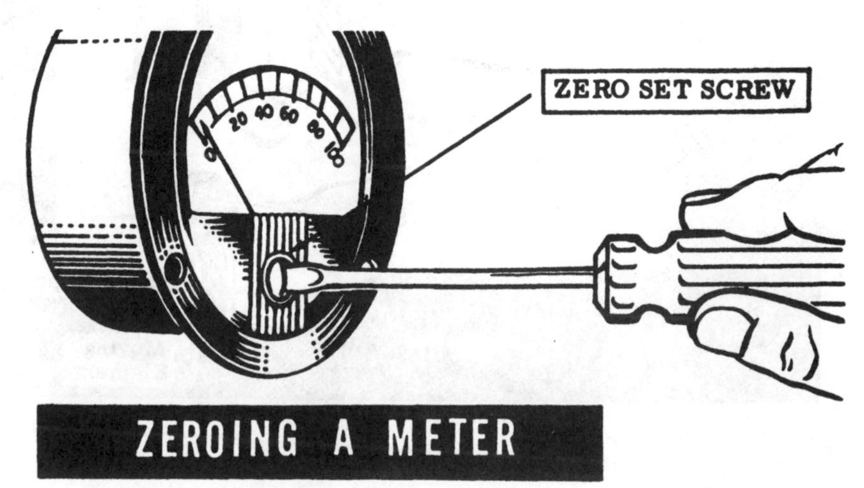

ZEROING A METER

How Meter Scales Are Read (continued)

Meter scales are usually divided into equal divisions, ordinarily with a total of between 30 and 50 divisions. The meter should always be read from a position at right angles to the meter face. Since the meter divisions are small and the meter pointer is raised above the scale, reading the position *from an angle* will result in an inaccurate reading—often as much as an entire scale division. This type of incorrect reading is called *parallax*. Most meters are slightly inaccurate due to the meter construction, and additional error from a parallax reading may result in a very inaccurate reading. Some precision meters have a mirror built into the scale. To read these meters you line up the needle and its image in the mirror with your eye before you read the meter. This eliminates the parallax problem.

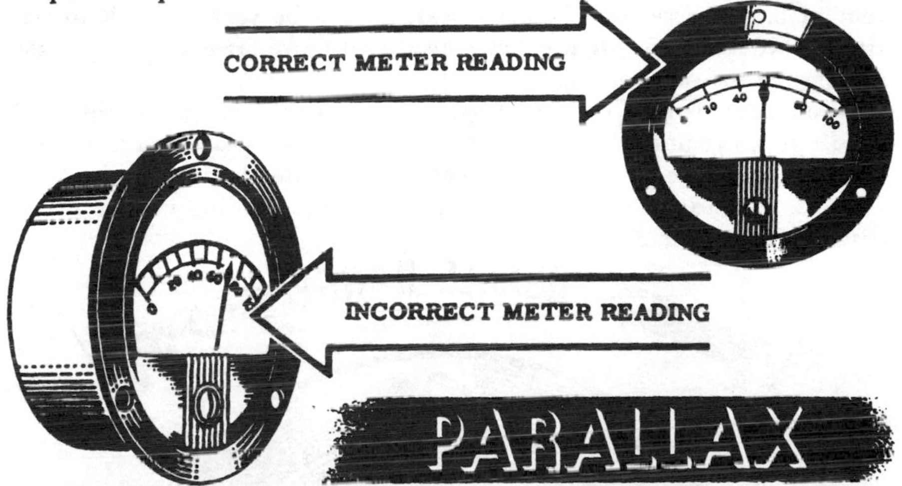

When the meter pointer reads a value of current between two divisions of the scale, usually the nearest division is used as the meter reading. However, if a more accurate reading is desired, the position of the pointer between the divisions is estimated, and the deflection between the scale divisions is added to the lower scale division. Estimating the pointer position is called *interpolation*, and you will use this process in many other ways in working with electricity.

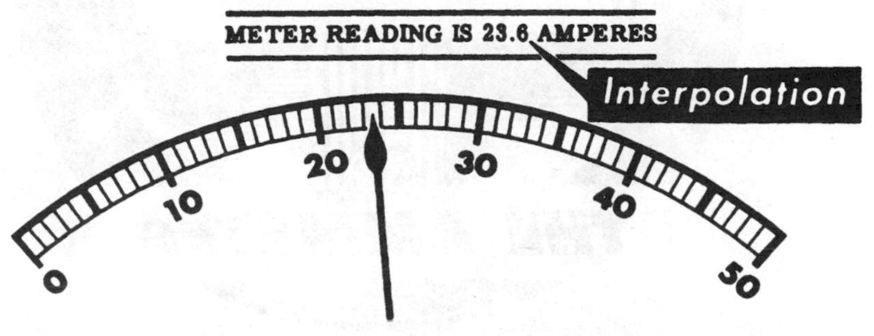

Usable Meter Range

The range of an ammeter (an ammeter measures current in amperes) indicates the maximum current which can be measured with the meter. Current in excess of this value can cause serious damage to the meter. If an ammeter has a range of 0-15 amperes, it will measure any current flow which does not exceed 15 amperes; but a current greater than 15 amperes can damage the meter.

While the meter scale may have a range of 0-15 amperes, its useful range for purposes of measurement will be from about 1 ampere to 14 amperes. When this meter scale indicates a current of 15 amperes, the actual current may be much greater but the meter can only indicate to its maximum range. For this reason, the useful maximum range of any meter is slightly less than the maximum range of the meter scale. A current of 0.1 ampere on this meter scale would be very difficult to read since it would not cause the meter needle to move far enough from zero to obtain a definite reading.

Smaller currents such as 0.001 ampere would not cause the meter needle to move and, thus, could not be measured at all with this meter. The useful minimum range of a meter never extends down to zero, but extends, instead, only to the point at which the reading can be readily distinguished from zero.

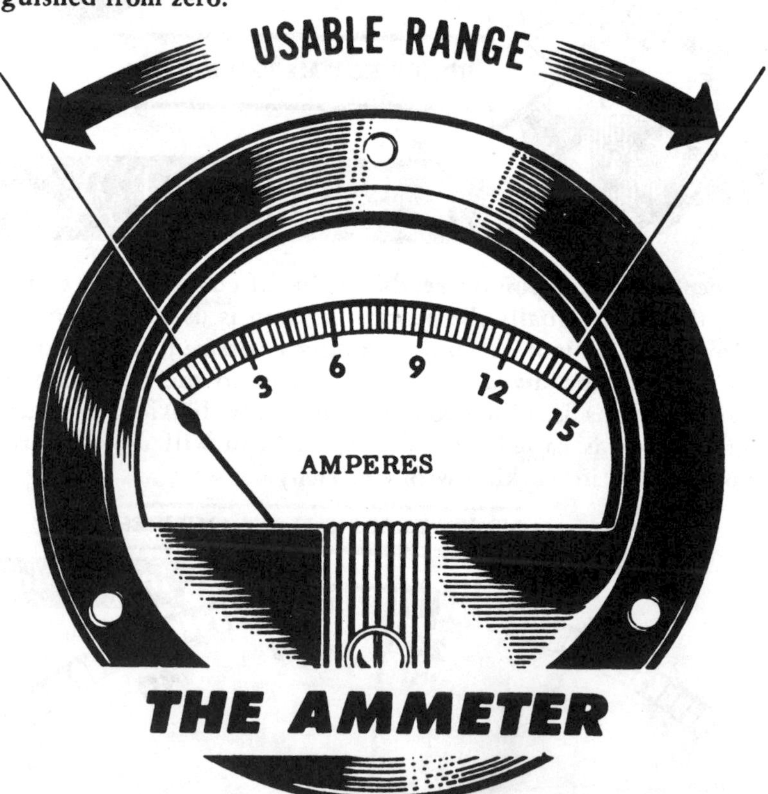

Review of How a Meter Works

Although you may have little reason to open up a meter to repair it, it is very important that you understand the principle of operation so that you can know how properly to use and take care of meters. As you might suspect, meters are delicate and must be handled with care. Meters are used everywhere electricity is used or generated to tell us what is going on in electrical systems. Therefore, it is imperative that you thoroughly understand how to read meters properly.

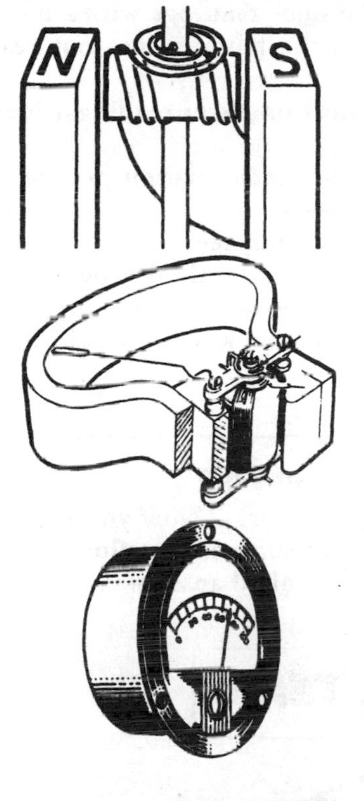

1. METER COIL—Moving coil which acts as a magnet when current flows in the coil.

2. METER MOVEMENT — Current-measuring instrument consisting of a moving coil suspended between the poles of a horseshoe magnet. Current in the coil causes the coil to turn.

3. PARALLAX — Meter reading error due to taking a reading from an angle.

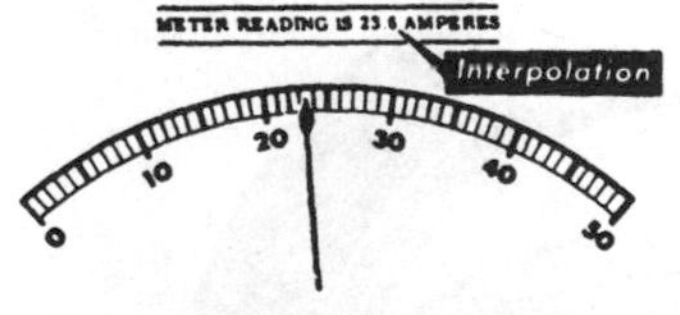

4. INTERPOLATION—Estimating the meter reading between two scale divisions.

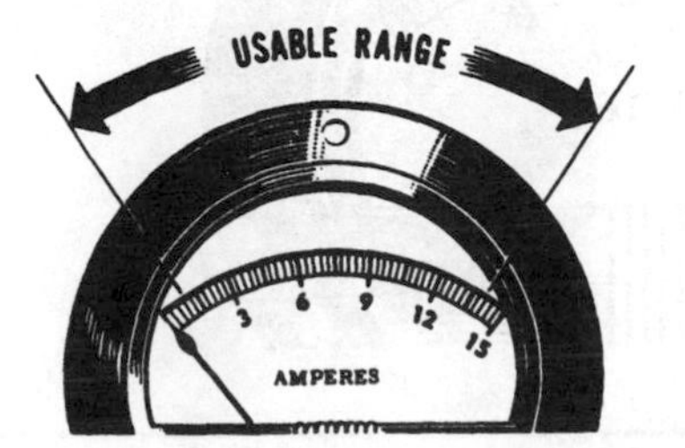

5. USEFUL METER RANGE

Self-Test—Review Questions

1. Why are meters important electrical devices?
2. What is the principle of operation of the moving-coil meter?
3. Why do you think it is necessary to handle meters with care?
4. What is the result of putting a current through a meter that is slightly in excess of the maximum scale reading? What is the result if the current is much greater?
5. Can meters be used in any position? Explain.
6. What should you do if you are ready to take some readings with a meter and you find that with no current flowing, the pointer does not read zero?
7. What is parallax? What is the effect of parallax on meter readings? How do you avoid it?
8. What is interpolation of meter readings? Does interpolation give more accurate results? Is interpolation always necessary?
9. What is normally considered the useful range of a meter?
10. In the meters we have been considering, what do you think would happen if you reversed the leads going to the meter in a circuit that had given a normal reading earlier? Would you say that a meter would have to be hooked up in a specific way?

Learning Objectives—Next Section

Overview—You have learned how a meter works. Now you can learn how it is connected and used to measure current flow. A meter connected to measure current flow is called an *ammeter*.

THE AMMETER

Measuring the Units of Current Flow

You have learned that the unit of current flow is the ampere and that this corresponds to the movement of charge (electrons) at the rate of 1 coulomb per second and that 1 coulomb is equal to 6.28 million, million, million electrons. You have also learned how current flow produces magnetic fields and how this principle is used in the construction of the meter. We will now examine how these ideas are put to practical use for the measurement of current.

The device which is used to measure the rate of current flow through a conducting material, and to display this information in such a way that you can use it, is called an *ammeter*. An ammeter indicates, in amperes, the number of electrons passing a given point in the material (which is, in practice, almost always a wire).

To be able to do this, the ammeter must somehow be connected into the wire in such a way that it is able to count *all* the electrons passing without letting any of them slip past uncounted. The only way to do this is to break the wire, or *open the line* as it is called, and to insert the ammeter physically in it.

When an ammeter is inserted in this way into a wire being used to carry current to an electric lamp, the ammeter is said to be *in series* with the lamp. The schematic symbol for the ammeter is a circle with the letter "A" or "I" in it.

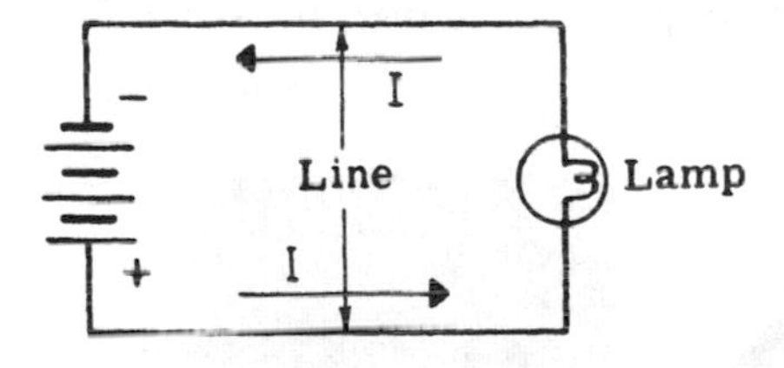

Without ammeter

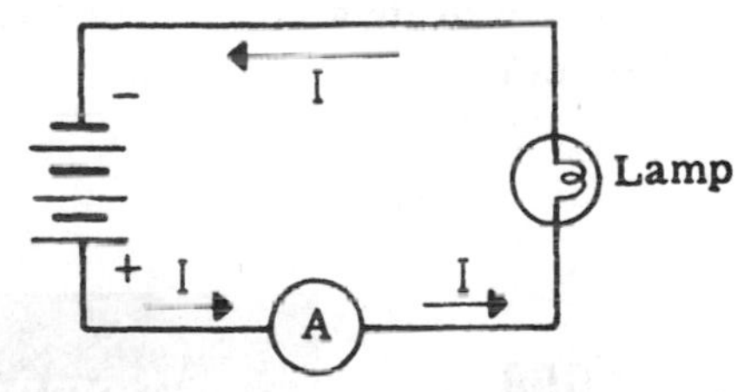

Ammeter connected in series with line to measure lamp current.

THE AMMETER

Whenever you use an ammeter, the pointer indicates on the meter scale the number of amperes of current flowing, which is also the number of coulombs passing per second.

How Small Currents Are Measured

While the ampere is the basic unit of measurement for current flow, it is not always a convenient unit to use. Current flows seldom exceed 1,000 amperes but may often be as little as 1/1,000 of an ampere. For measuring currents of less than 1 ampere, some other unit is needed. A cup of water is not measured in gallons, nor is the flow of water from a fire hydrant measured in cups. In any kind of measurement, a *usable* unit of measurement is needed. Since current flow seldom exceeds 1,000 amperes, the ampere can be used satisfactorily as the unit for currents in excess of 1 ampere. However, it is not convenient as the unit for currents of *less* than 1 ampere.

If the current flow is between 1/1,000 of an ampere and 1 ampere, the unit of measure used is the *milliampere* (abbreviated mA), which is equal to 1/1,000 ampere. For current flow of less than 1/1,000 ampere, the unit used is the *microampere* (abbreviated μA), which is equal to 1/1,000,000 ampere. Meters used for measuring milliamperes of current are called *milliammeters*, while meters used for measuring microamperes of current are called *microammeters*. Units of measurement are subdivided in such a way that a quantity expressed in one unit may be readily changed to another unit, either larger or smaller.

Fractions such as halves, quarters, thirds, etc., are seldom used in electrical work; decimals being generally preferred. A meter would therefore indicate a reading of half-an-ampere (½) either as "0.5 A" or as "500 mA."

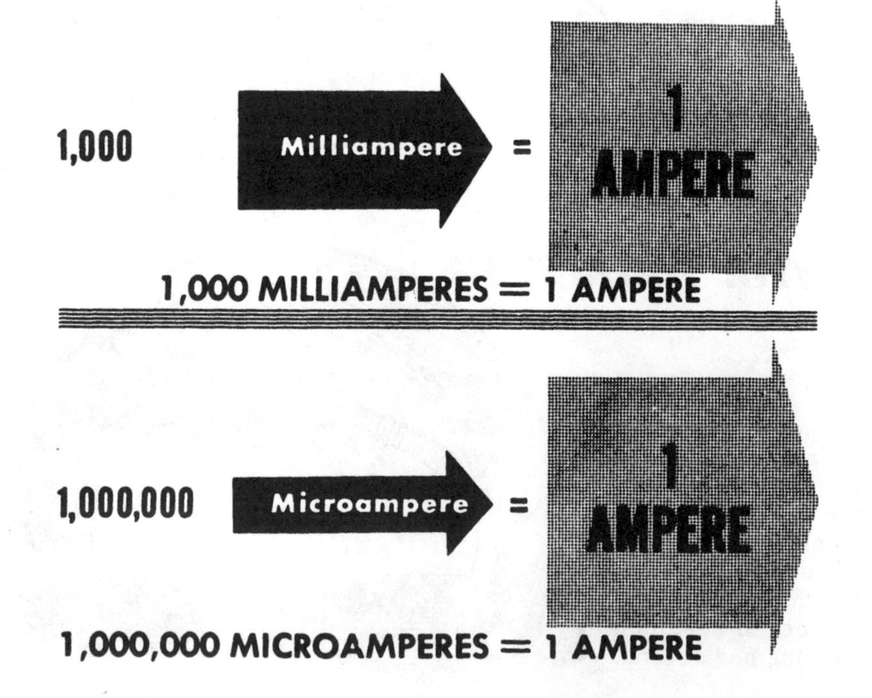

How Units of Current Are Converted

In order to work with electricity, you must be able to convert from one unit of current to another. Since a milliampere (mA) is 1/1,000 of an ampere, milliamperes can be converted to amperes by moving the decimal point three places to the *left*. For example, 35 milliamperes is equal to 0.035 ampere. There are two steps required in order to arrive at the correct answer. First, the original position of the decimal point must be located. The decimal is then moved three places to the left, thereby converting the unit from milliamperes to amperes. If no decimal point is given with the number, it is always understood to follow the last number in the quantity. In the example given, the reference decimal point is after the number 5, and to convert from milliamperes to amperes, it must be moved three places to the left. Since there are only two whole numbers to the left of the decimal point, a zero must be added to the left of the number to provide for a third place, as shown.

CONVERTING MILLIAMPERES TO AMPERES

35 milliamperes = ? ampere

Move decimal point three places to the left.

When converting amperes to milliamperes, you move the decimal point three places to the *right* instead of the left. For example, 0.125 ampere equals 125 milliamperes, and 16 amperes equals 16,000 milliamperes. In these examples, the decimal point is moved three places to the right of its reference position, with three zeros added in the second example to provide the necessary decimal places.

CONVERTING AMPERES TO MILLIAMPERES

.125 ampere = ? milliamperes

Move decimal point three places to the right.

How Units of Current Are Converted (continued)

Suppose that you are working with a current of 125 microamperes and you need to express this current in amperes. If you are converting from a larger unit to a smaller unit, the decimal point is moved to the *right*; while to change from a smaller unit to a larger unit, the decimal point is moved to the *left*. Since a microampere is 1/1,000,000 ampere, the ampere is the larger unit. Then converting microamperes to amperes is a conversion from small to large units and the decimal point should be moved to the left. In order to convert millionths to units, the decimal point must be moved six decimal places to the left; thus 125 microamperes equals 0.000125 ampere. The reference point in 125 microamperes is after the 5, and in order to move the decimal point six places to the left, you must add three zeros ahead of the number 125. When converting microamperes to milliamperes, the decimal point is moved only three places to the left; thus 125 microamperes equals 0.125 milliampere.

If your original current is in amperes and you want to express it in microamperes, the decimal point should be moved six places to the right. For example, 3 amperes equals 3,000,000 microamperes, because the reference decimal point after the 3 is moved six places to the right with the six zeros added to provide the necessary places. To convert milliamperes to microamperes, the decimal point should be moved three places to the right. For example, 125 milliamperes equals 125,000 microamperes, with the three zeros added to provide the necessary decimal places.

CONVERTING UNITS OF CURRENT

MICROAMPERES TO AMPERES

Move Decimal Point Six Places to the Left.

125. microamperes = .000125 ampere

MICROAMPERES TO MILLIAMPERES

Move Decimal Point Three Places to the Left.

125. microamperes = .125 milliampere

AMPERES TO MICROAMPERES

Move Decimal Point Six Places to the Right.

3. amperes = 3,000,000. microamperes

MILLIAMPERES TO MICROAMPERES

Move Decimal Point Three Places to the Right.

125. milliamperes = 125,000. microamperes

Milliammeters and Microammeters

An ammeter having a meter scale range of 0-1 ampere is actually a milliammeter with a range of 0-1,000 milliamperes. As fractions are seldom used in electricity, a meter reading of ½ ampere on the 0-1 ampere range is given as 0.5 ampere or 500 milliamperes (mA). For ranges less than 1 ampere, milliammeters and microammeters are used to measure current.

If you are using currents between 1 milliampere and 1,000 milliamperes, milliammeters are used to measure the amount of current. For currents of less than 1 milliampere, microammeters of the correct range are used. Very small currents of 1 microampere or less are measured on special laboratory-type instruments called *galvanometers.* You will not normally use the galvanometer, since the currents used in electrical equipment are usually between 1 microampere and 100 amperes and, thus, can be measured with a microammeter, milliammeter, or ammeter of the correct range. Meter scale ranges for milliammeters and microammeters, like ammeters, are in multiples of 5 or 10 since these multiples are easily converted to other units.

In using a meter to measure current, the maximum reading of the meter range should always be higher than the maximum current to be measured. A safe method of current measurement is to start with a meter having a range *much greater* than you expect to measure, in order to determine the correct meter to use.

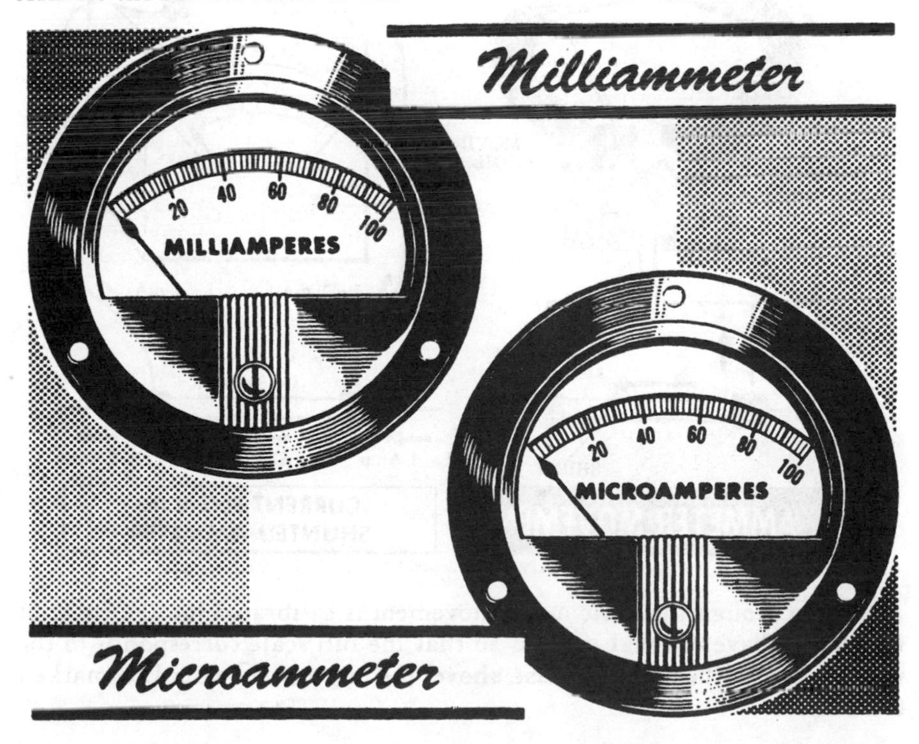

How Ammeter Ranges Are Converted

Meter ranges could be converted by using magnets of different strength or by changing the number of turns in the coil, since either of these changes would alter the amount of current needed for full-scale deflection. However, the wire used in the coil must always be large enough to carry the maximum current of the range the meter is intended for; therefore, changing the wire size would only be practical in the small current ranges, since large wire cannot be used as a moving coil. To keep the wire size and the coil small, basic meter movements are normally limited to a range of 1 milliampere or less. Also, for using a meter for more than one range, it is impractical to change the magnet or the coil each time the range is changed.

For measuring large currents, a low range meter is used with a *shunt*, which is a heavy wire connected across the meter terminals to carry most of the current. This shunt allows only a small part of the current to actually flow through the meter coil. Usually a 0-1 milliampere meter is used, with the proper-sized shunt connected across its terminals to achieve the desired range. The 0-1 milliammeter is a basic meter movement which you will find in many types of meters you will use. Other common basic current ranges are 0-100 μA and 0-50 μA.

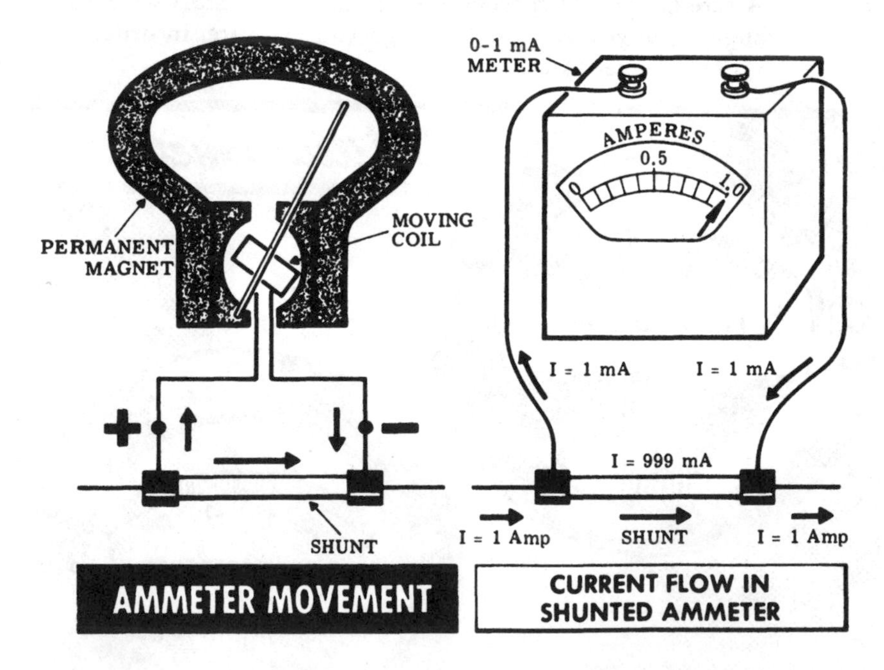

AMMETER MOVEMENT

CURRENT FLOW IN SHUNTED AMMETER

Even though the basic meter movement is calibrated for 0-1 mA, it is usual to have the dial marked so that the full scale corresponds to the value with the shunt. In the case above, the meter scale would be marked 0-1 ampere.

Multirange Ammeters

You have seen that you can change the range of an ammeter by the use of shunts. The range will vary according to the value of the shunt. Some ammeters are built with a number of internal shunts and a switching arrangement that is used to parallel different shunts across the meter movement to measure different currents. Thus, a single meter movement can be used as a *multirange* ammeter. A scale for each range is painted on the meter face. The diagram below shows a multirange ammeter with a 0-3, 0-30, 0-300 ampere range. Note the three scales on the meter face.

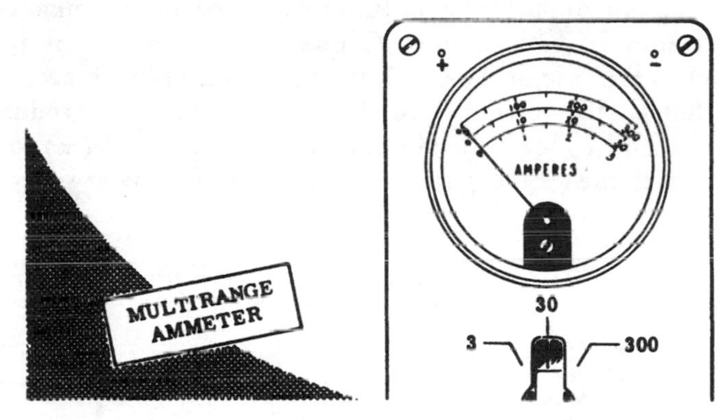

When a multirange ammeter is used to measure an unknown current, the *highest* range is always used first, then the next highest range, and so on, until the needle is positioned about midscale. In this way you can be assured that the current is not excessive for the meter range, and you will never have the unfortunate experience of burning out a meter movement, or of wrapping the needle around the stop-peg.

Some multimeters use external shunts and do away with internal shunts and the switching arrangement. Changing range for such a meter involves shunting it with the appropriate shunt. In the diagram, the ammeter is calibrated to read 30 amperes full-scale by shunting it with the 30-ampere shunt.

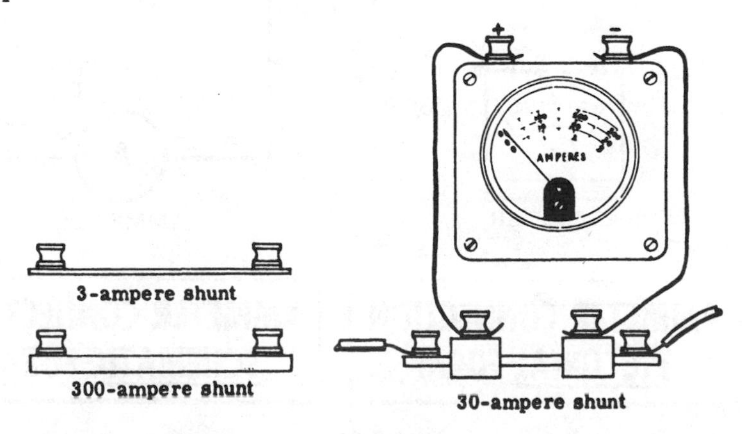

How Ammeters Are Connected into Circuits

All of the ammeters that have been described are called *direct-current meters*; that is, they are designed for circuits where the direction of the current flow is *constant*. When you see a meter of this type, you will notice that the meter terminals are marked with (+) and (—). These markings tell you how the meter should be connected into the electric circuit. You will remember that for the meter to read properly, it is necessary that the magnetic fields have the proper polarity. This means that the current must pass through the meter in the correct direction. The rule for connecting an ammeter, a milliammeter, or microammeter in the proper direction is very simple. Connect the terminal of the meter marked (—) to the side of the break in the circuit made to accommodate the meter that is still connected to the (—) or negative terminal of the power source; and connect the side of the meter marked (+) to the side still connected to the (+) or positive terminal of the power source.

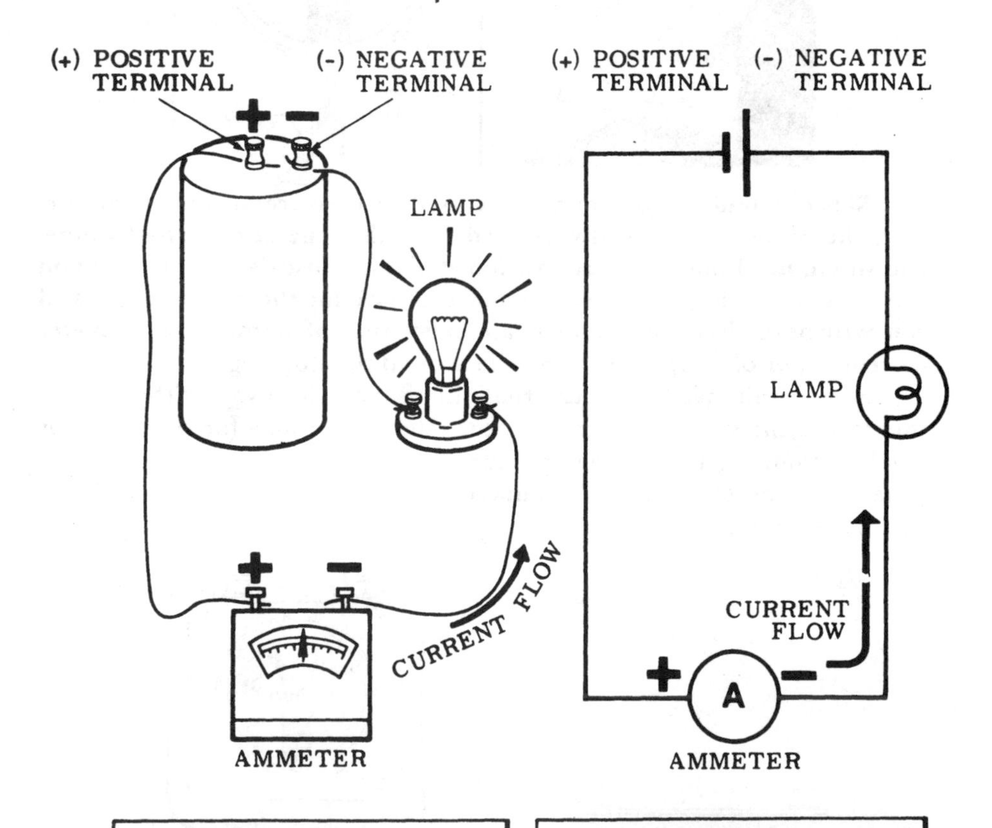

AMMETER CONNECTION PICTORIAL FORM

AMMETER CONNECTION SCHEMATIC FORM

Review of How Current Is Measured

To review what you have discovered about how current is measured, consider some of the important facts you have studied.

1. AMPERE—Unit of rate of flow of electrons, equal to 1 coulomb per second.

$$1\,\text{ma} = \frac{1}{1000}\,\text{amp}$$

2. MILLIAMPERE—A unit of current equal to 1/1,000 ampere.

$$1\,\mu\text{a} = \frac{1}{1{,}000{,}000}\,\text{amp}$$

3. MICROAMPERE—A unit of current equal to 1/1,000,000 ampere.

4. AMMETER—A meter used to measure currents of 1 ampere and greater.

5. MILLIAMMETER—A meter used to measure currents between 1/1,000 ampere and 1 ampere.

6. MICROAMMETER—A meter used to measure currents between 1/1,000,000 ampere and 1/1,000 ampere.

Review of How Current is Measured (continued)

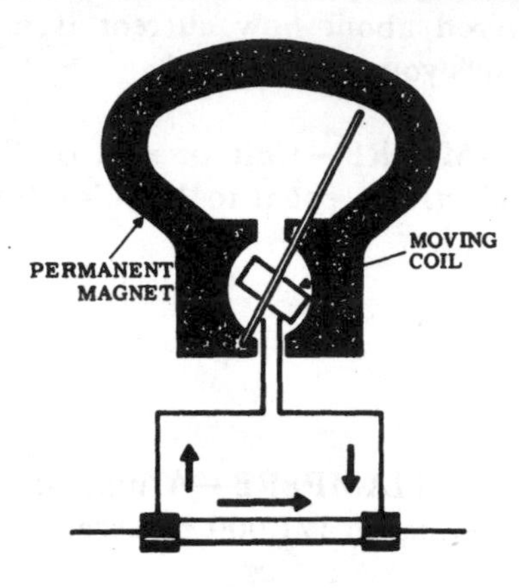

7. BASIC AMMETER MOVEMENT—0-1 mA (milliammeter) movement with shunt wire across the meter terminals to increase the meter scale range.

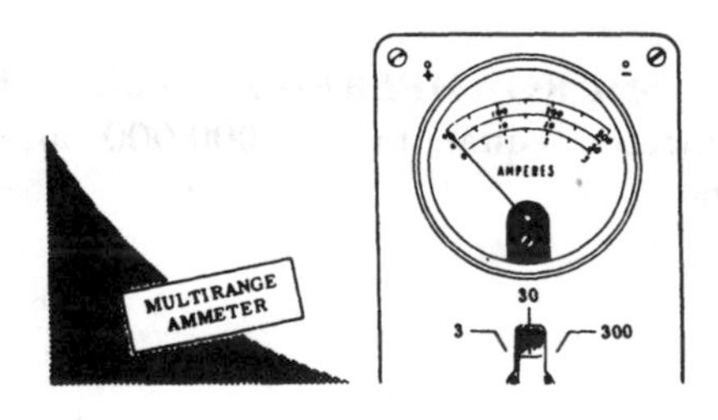

8. MULTIRANGE AMMETER—A single meter movement used for measuring different current ranges. Each range requires a different shunt. The shunts may be inside the meter movement and controlled by a switching arrangement, or they may be external, in which case they are connected in parallel with the meter binding posts.

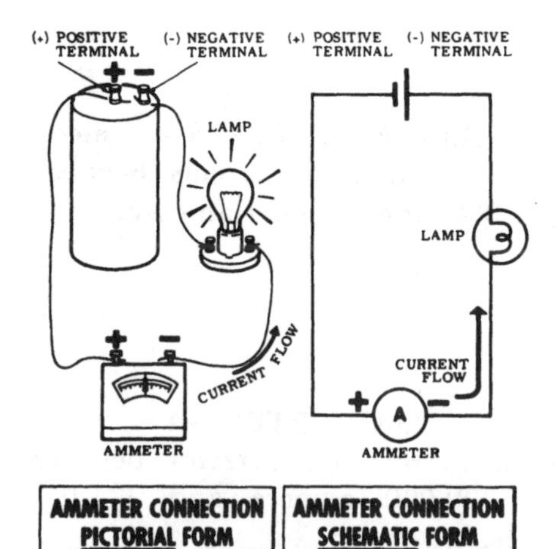

9. AMMETER CONNECTION—Ammeters are always connected in series with the circuit so that all of the circuit current flows through the ammeter. The connections are made by breaking the circuit and connecting the (—) terminal of the meter to the wire still connected to the (—) terminal of the power source; and similarly, for the (+) terminal of the ammeter.

Self-Test—Review Questions

1. Define the basic unit of current and name the instrument used to measure it.
2. Define the milliampere and the microampere. Why and when are these terms used?
3. Calculate the following conversions:

Convert to amperes	Convert to milliamperes	Convert to microamperes
10,000 mA	0.250 A	0.35 mA
10,000 μA	0.525 A	0.022 mA
2,500 mA	1.330 A	1.000 mA
1,000 mA	0.002 A	13.435 mA
33,500 mA	0.055 A	1.000 A
	1,000 μA	0.035 A
	13,200 μA	

4. Draw an electric circuit showing the way that an ammeter is connected into the system. Use a battery and a lamp as the other circuit elements. Indicate the direction of current flow and the proper polarity of the meter.
5. What are microammeters and milliammeters? How do they differ from an ammeter?
6. Why is a shunt used?
7. Show the way that a shunt is connected to make an ammeter.
8. What are the usual ranges of basic meter movement?
9. Assume that you have a basic meter movement of 0-250 μA, and it is desired to make this into a 0-10 ampere meter. Show how the shunt is connected and how the currents would distribute between the shunt and the meter movement with a current flow of 5 amperes.
10. How do multirange meters differ from single range meters? If you are not sure of the approximate amount of current flowing in the circuit, which range of a multirange meter should you start with if you are going to measure the current in the circuit?

Learning Objectives—Next Section

Overview—Just as an ammeter measures current, a voltmeter measures potential difference or voltage. In the next section, you will find out how a meter is used to measure voltage.

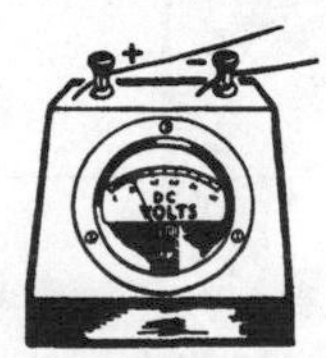

Units of Voltage

The electromotive force between two unequal charges is usually expressed in volts; but, when the difference in potential is only a fraction of a volt, or is more than 1,000 volts, other units are used. For voltages of less than 1 volt, *millivolts* and *microvolts* are used, just as milliamperes and microamperes are used to express currents less than 1 ampere. While current seldom exceeds 1,000 amperes, voltage often exceeds 1,000 volts, so that the kilovolt (abbreviated kV)—equal to 1,000 volts—is used as the unit of measurement. When the potential difference between two charges is between 1/1,000 volt and 1 volt, the unit of measure is the *millivolt* (abbreviated mV). When it is between 1/1,000,000 volt and 1/1,000 volt, the unit is the *microvolt* (abbreviated μV).

Meters for measuring voltage have scale ranges in microvolts, millivolts, volts, and kilovolts, depending on the units of voltage to be measured. Ordinarily, you will work with voltages between 1 and 500 volts and use the volt as a unit. Voltages of less than 1 volt, and more than 500 volts, are not used except in special applications of electrical and electronic equipment.

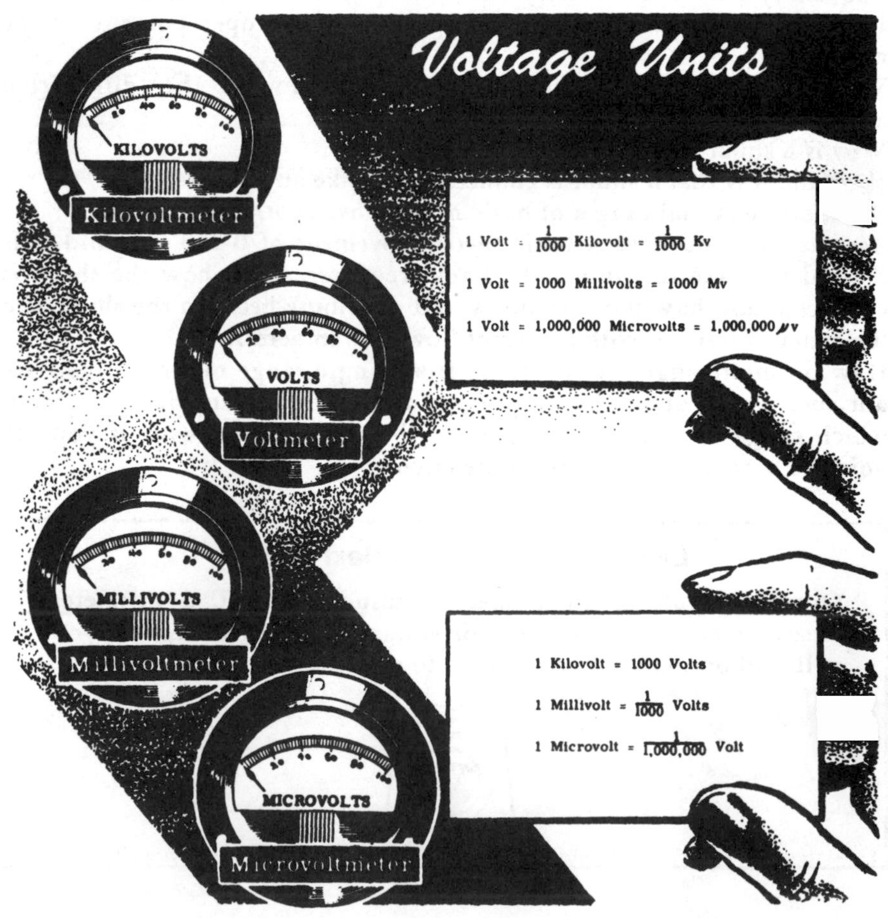

Converting Units of Voltage

Units of voltage measurement are converted in the same way that units of current are converted. In order to convert millivolts to volts, the decimal point is moved three places to the *left*; and to convert volts to millivolts, the decimal point is moved three places to the *right*. Similarly, in converting microvolts to volts, the decimal point is moved six places to the left; and in converting volts to microvolts, the decimal point is moved six places to the right. These examples show that in converting units, the same rules of moving the decimal point apply to *both* voltage and current.

Kilo (meaning one thousand) is not used to express current, but since it is used to express voltage, you may need to know how to convert *kilovolts* to volts, and the reverse. To convert kilovolts to volts, the decimal point is moved three places to the right; and to convert volts to kilovolts, it is moved three places to the left. For example, 5 kilovolts equals 5,000 volts. Since the decimal point is after the 5, three zeros are added to provide the necessary places. Also, 450 volts equals 0.45 kilovolt as the decimal point is moved three places to the left.

CONVERTING VOLTAGE UNITS

VOLTS TO KILOVOLTS

Move the decimal point
3 places to the left.

450 volts = .45 kilovolt

KILOVOLTS TO VOLTS

Move the decimal point
3 places to the right.

5 kilovolts = 5000 volts

VOLTS TO MILLIVOLTS

Move the decimal point
3 places to the right.

15 volts = 15,000 millivolts

MILLIVOLTS TO VOLTS

Move the decimal point
3 places to the left.

500 millivolts =.5 volt

VOLTS TO MICROVOLTS

Move the decimal point
6 places to the right.

15 volts = 15,000,000 microvolts

MICROVOLTS TO VOLTS

Move the decimal point
6 places to the left.

3505 microvolts =.003505 volt

How a Voltmeter Works

As you know, an ammeter measures the rate at which charges move through a material; and that the rate of current flow through a given material varies directly with the voltage difference. That is, the greater the voltage difference, the greater the current flow. Voltage is measured by a meter that is called a *voltmeter.* The voltmeter consists of an ammeter in series with a special piece of material called a *resistor* that limits the current flow. The measurement of voltage is made by measuring the current that flows in the meter circuit. When used in a voltmeter, the resistor (which you will learn about a little later) is called a *multiplier resistor.* For a given ammeter and multiplier resistor, a large current will flow when the voltage is high; a small current will flow when the voltage is low. The meter scale can be marked or calibrated in volts and read directly. Since it is desirable to keep the current in the voltmeter circuit as low as possible so that connection of the voltmeter will not disturb other measurements, the meter used is always a milliammeter or a microammeter.

The multiplier resistor determines the scale range of a voltmeter. Since the multiplier is built into most of the voltmeters you will use, you can measure voltage by making very simple connections. Whenever the (+) meter terminal is connected to the (+) terminal of the voltage source, and the (—) meter terminal to the (—) terminal of the voltage source, with nothing else connected in series, the meter reads voltage directly. When using a voltmeter, it is important to observe the correct meter polarity and to use a meter with a maximum scale range *greater* than the maximum voltage you expect to read.

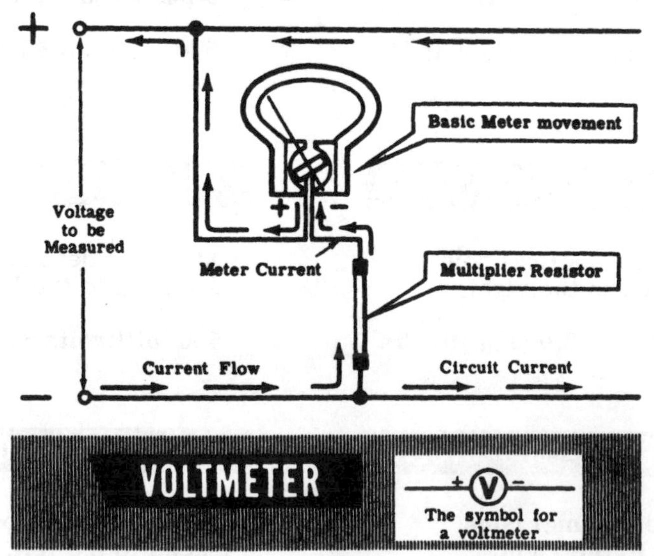

A voltmeter is always connected *across* the voltage source to be measured. When connected this way, the voltmeter is in a *parallel* circuit.

How a Voltmeter Is Used

A voltmeter is used to measure voltage anywhere in a circuit. If it is to measure a source of voltage, such as a battery, the negative (—) side of the voltmeter is always connected to the negative (—) side of the battery; the positive (+) side of the voltmeter is always connected to the positive (+) side of the battery. If connections are *reversed*, meter needle will move to the left of zero mark, and a reading will not be obtained.

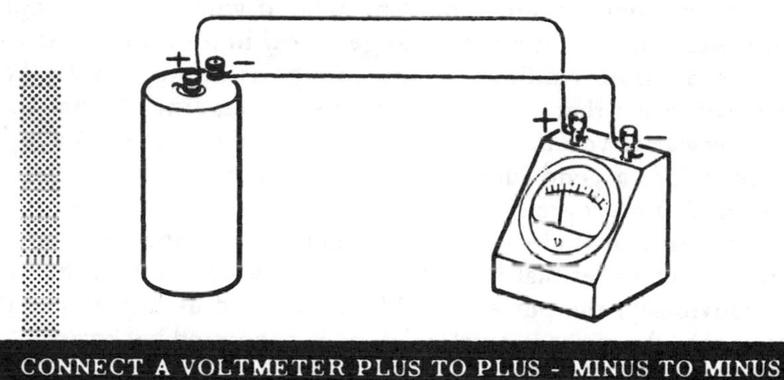

CONNECT A VOLTMETER PLUS TO PLUS - MINUS TO MINUS

The electric circuit through which the current flows is usually called the *load* and may consist of a single item such as a lamp, or may be extremely complex as in the case of all of the devices that are connected to the output of a large generator.

When the voltmeter is to measure the voltage drop across a load, the negative (—) lead is connected to the side of the load where the electrons *enter* (the (—) side); and the positive (+) lead is connected to the side of the load from which the electrons *emerge* (the (+) side).

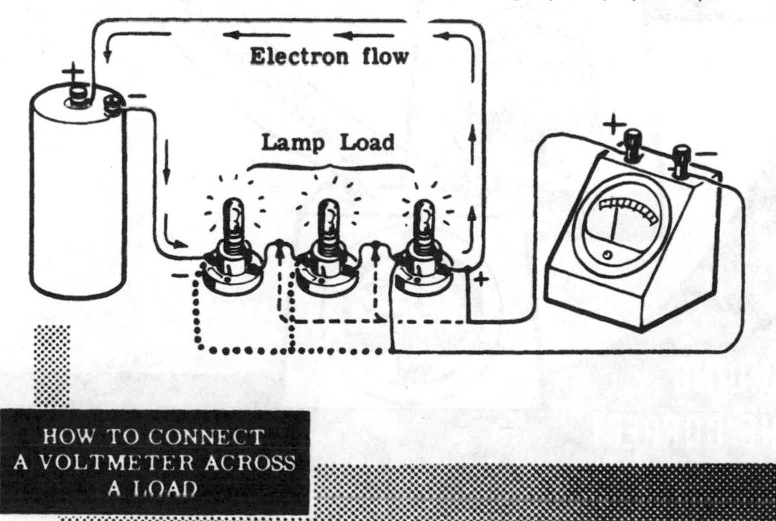

HOW TO CONNECT A VOLTMETER ACROSS A LOAD

Voltmeter Ranges

Just as you learned with the ammeter, it is necessary with a voltmeter to connect it into the circuit with the *proper polarity*; and to choose a voltmeter with the *proper range* for the measurement that is to be made. With direct current voltmeters such as those already discussed, it is a good idea to choose a voltmeter that reads between 10% and 90% of the scale. For example, a voltmeter with a full scale reading of 100 volts would be used to measure voltages between 10 and 90 volts. All meters are delicate instruments and should be treated with care. Improper use of a voltmeter can burn it out or change its calibration so that it will no longer give accurate readings. If the voltage is approximately known, choose a voltmeter that will give a mid-scale reading so that there is plenty of leeway if you have estimated incorrectly. If the voltage is not known, then it is a good idea to start with the *highest* range meter that you have and use progressively lower range meters until a satisfactory range is obtained. It is always a good idea to look at the meter when you throw the switch to see that it is properly connected and that the range is correct. Obviously, the power should be removed as fast as possible if you notice that the meter is overloaded or is connected backwards.

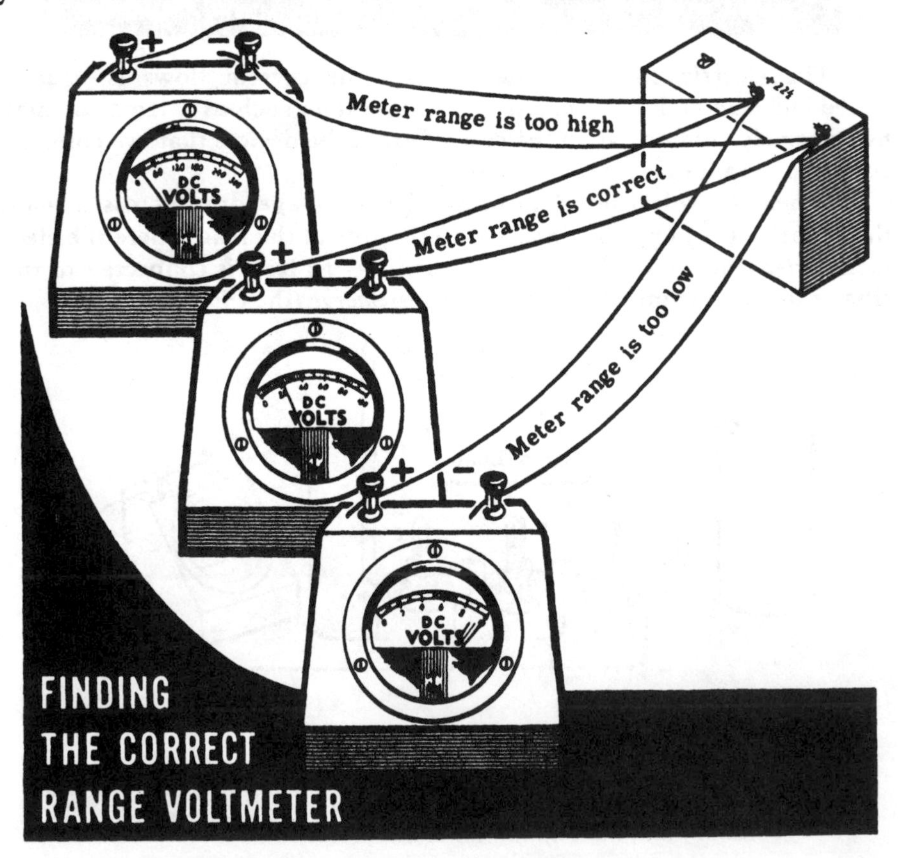

Multirange Voltmeters

As described, the range of any voltmeter can be increased by the addition of a multiplier to the voltmeter circuit in series with the basic meter movement. The multiplier causes reduction of the deflection of the pointer on the meter; by using multipliers of known values, the deflection can be reduced as much as desired.

Multirange voltmeters, like multirange ammeters, are instruments which you will frequently use. They are physically very similar to ammeters, and their multipliers are usually located inside the meter with suitable switches, or sets of terminals, on the outside for selecting range. Proper range is selected by starting with the *highest* range and working downward, until the needle reads about midscale.

Because they are lightweight, portable, and can be set up for different voltage ranges by the flick of a switch, multirange voltmeters are extremely useful.

The simplified drawing below shows a three-range, multirange voltmeter.

TYPICAL 3-RANGE MULTIRANGE VOLTMETER

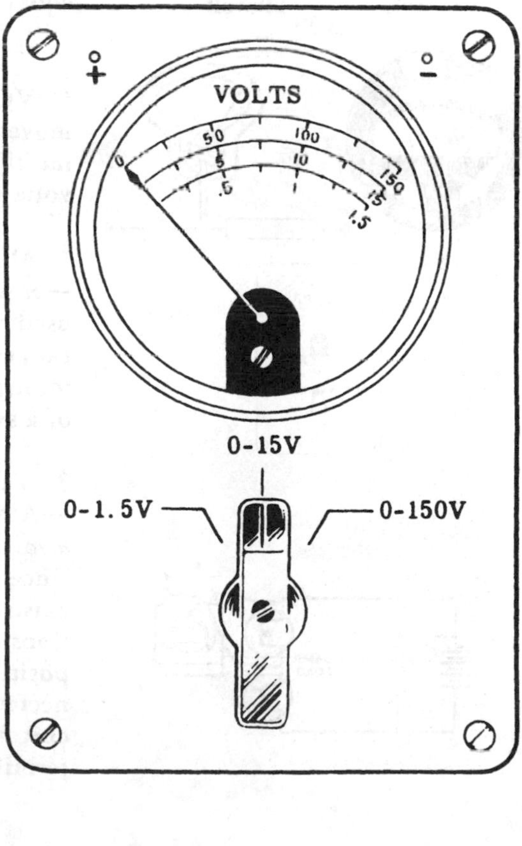

Review of Voltage Units and Measurement

V = Voltage (EMF)

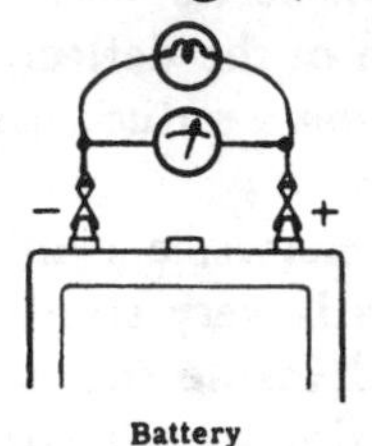

1. VOLT—The unit of potential difference. It is equal to work of 1 joule per coulomb.

2. MAINTENANCE OF EMF—EMF is maintained by having a source of energy that is converted into potential difference to keep the emf constant, regardless of load.

$$1 \text{ millivolt } (1 \text{ mV}) = \frac{1}{1000} \text{ volt}$$

$$1 \text{ volt} = 1000 \text{ millivolts}$$

3. MILLIVOLT—A unit of voltage equal to 1/1,000 volt.

$$1 \text{ microvolt } (1\ \mu\text{V}) = \frac{1}{1{,}000{,}000} \text{ volt}$$

$$1 \text{ volt} = 1{,}000{,}000 \text{ microvolts}$$

4. MICROVOLT—A unit of voltage equal to 1/1,000,000 volt.

$$1 \text{ kilovolt } (1 \text{ kV}) = 1000 \text{ volts}$$

$$1 \text{ volt} = \frac{1}{1000} \text{ kilovolt}$$

5. KILOVOLT—A unit of voltage equal to 1,000 volts.

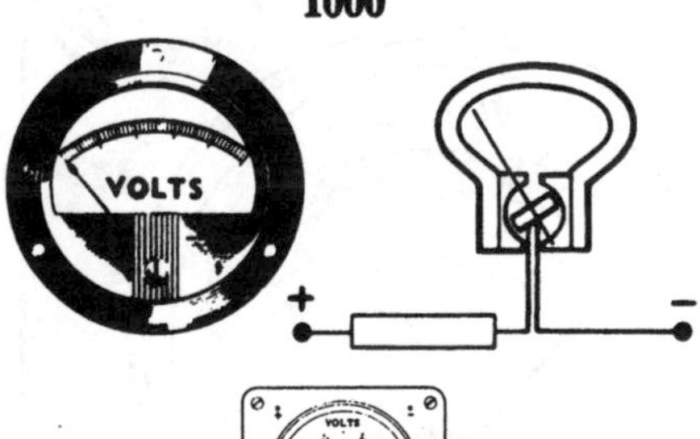

6. VOLTMETER — Basic meter movement with a series-connected multiplier, calibrated to measure voltage.

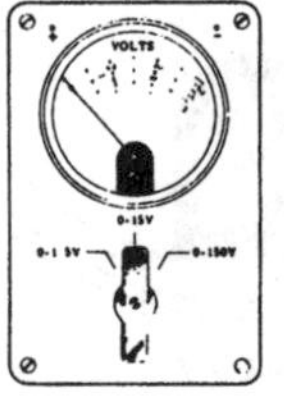

7. MULTIRANGE VOLTMETER—A single meter movement that is used to measure different voltage ranges. Each range uses a different multiplier that is selected by means of a switch.

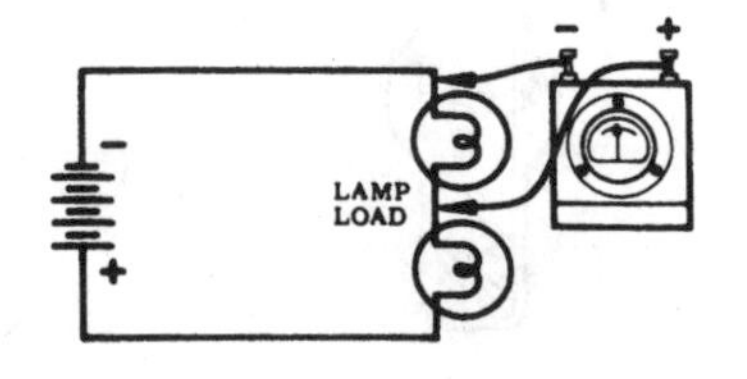

8. VOLTMETER CONNECTION—A voltmeter is always connected *across* the circuit to be measured since potential difference or voltage exists between two points. Connections are always made so that the positive (+) side of the meter is connected closest to the part of the circuit that goes to the positive (+) terminal of the power source.

Self-Test—Review Questions

1. What is the unit of potential difference? How is it defined?
2. What symbols are used to designate voltage?
3. Assume that you have three terminals available (A, B, and C) and that you are measuring the following potential differences with voltmeters of suitable range. What are the potential differences between the end (A to C) in each case? The polarity indications in parenthesis ( ) indicate the way the meter was connected to give a proper upscale reading.

   (a) A(−) to B(+) = 45 volts; B(−) to C(+) = 45 volts
   (b) A(−) to B(+) = 115 volts; B(−) to C(+) = 23.5 volts
   (c) A(+) to B(−) = 7.5 volts; B(−) to C(+) = 20 volts
   (d) A(−) to B(+) = 6 volts; B(+) to C(−) = 6 volts

4. What is the essential factor that allows a battery, or a generator to maintain an emf, while a pair of charged bars cannot?
5. Describe the components that make up a voltmeter and tell how each component functions.
6. Define the following and give the appropriate abbreviation:
   (a) millivolt (b) microvolt (c) kilovolt
7. Calculate the following conversions:

Convert to volts		Convert to millivolts	
0.01	kV	0.01	V
500	mV	1	V
25,000	μV	10	V
1,500	mV	10	μV
0.001	kV	1,000	μV
10	kV	320	μV
10	mV	3,200	μV

Convert to kilovolts		Convert to microvolts	
100	V	10	mV
17,500	V	0.001	mV
1,500,300	mV	1,450	mV
1,350	V	1	V
100,000	V	0.001	V
1	V	3.25	mV

8. What is the symbol for the voltmeter?
9. How is a voltmeter connected into a circuit? How does this differ from the connection for an ammeter?
10. How would you go about selecting the proper voltmeter to measure an unknown voltage? What sort of meter range would you select if you knew the voltage was approximately 80 volts?

Learning Objectives—Next Section

Overview—The flow of current is controlled by resistance. Devices called *resistors* can be used to do this. They are of basic importance in the study of electricity. In fact almost all of Volume 2 is devoted to the study of resistances and how they are used in electrical circuits.

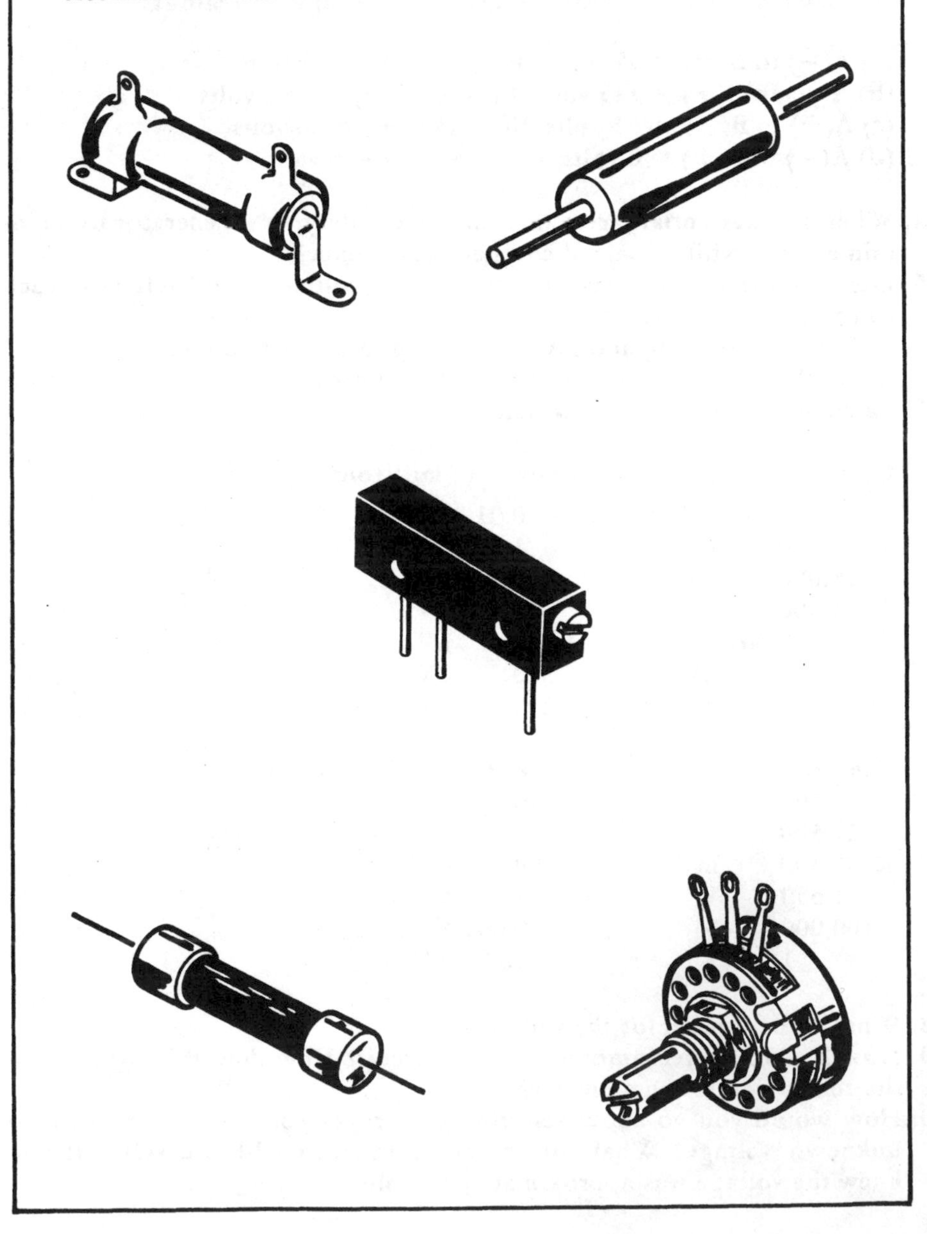

What Resistance Is

You know that an electric current is the movement of *free* electrons in a material, and that an electric current does not begin flowing all by itself because it needs a source of electric force to move the *free electrons* through the material. You have also found out that an electric current will not continue to flow if the source of electrical energy is removed. You can see from all this that there is something in a material that *resists* the flow of electric current—something that *holds onto* the *free* electrons and will not release them until force is applied.

As you learned earlier, the opposition to current flow is not the same for all materials. Current flow is the movement of *free* electrons through a material, and the *number* of *free* electrons in a material determines its *opposition* to current flow. Atoms of some materials give up their outer electrons easily. Such materials, called *conductors*, offer little opposition to current flow. Other materials hold onto their outer electrons. These materials, called *insulators*, offer considerable opposition to current flow. Every material has some opposition to current flow, whether large or small, and this opposition is called *resistance*.

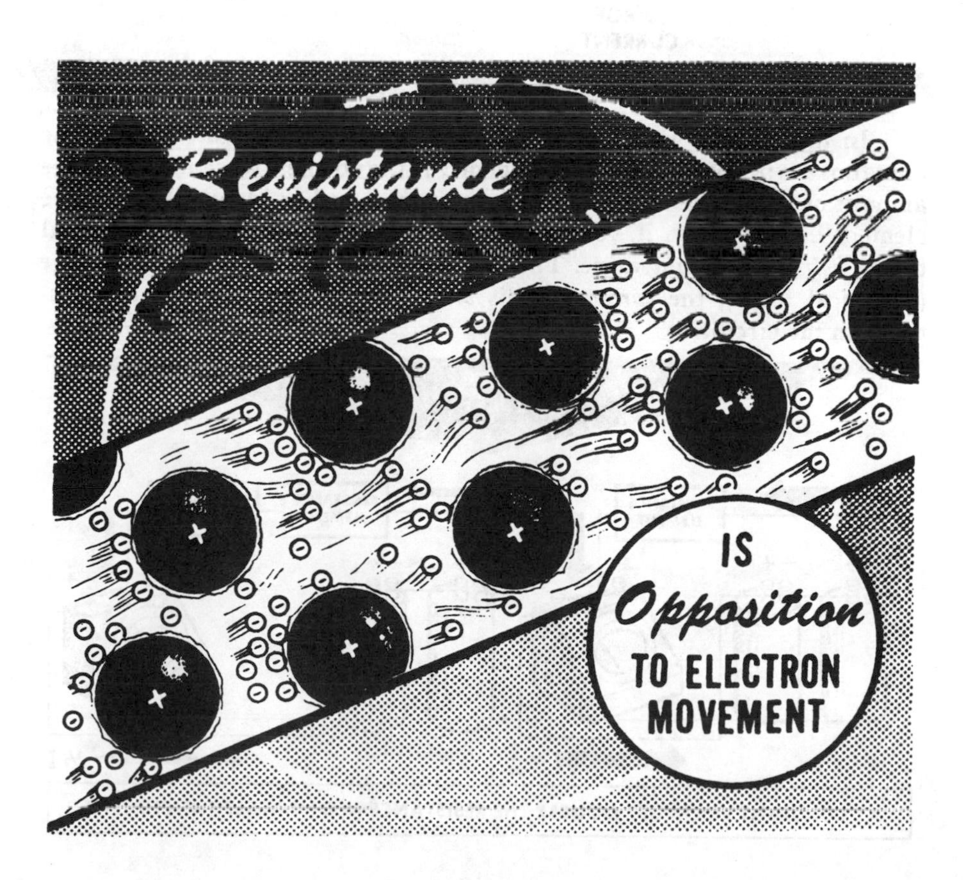

What Resistance Is (continued)

Let's assume that we have a source of constant electric force (voltage). If this is the case, then the more opposition that we have to current flow (resistance), the smaller will be the number of electrons flowing (current) through the material. Also, the *converse* is true. That is, for a constant voltage, the *smaller* the resistance (less opposition), the *greater* the current flow (more electrons).

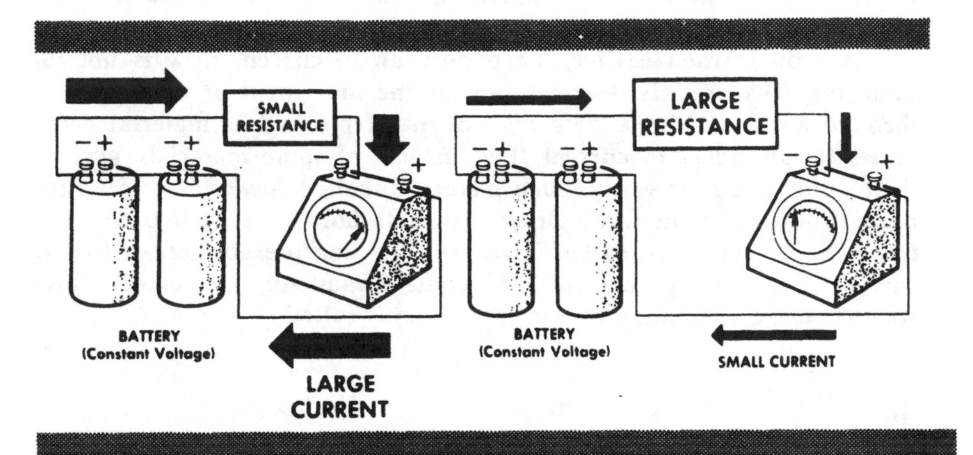

In fact, if you were to measure the current in a circuit such as the one above with one resistance element in it, and then were to *double* the amount of resistance in the circuit by putting another identical resistance element in series with it, you would find that the current would be equal to *one half* the original value. Thus, under conditions of constant voltage (electrical force), the *current flow* is *proportional* to the amount of resistance in the circuit.

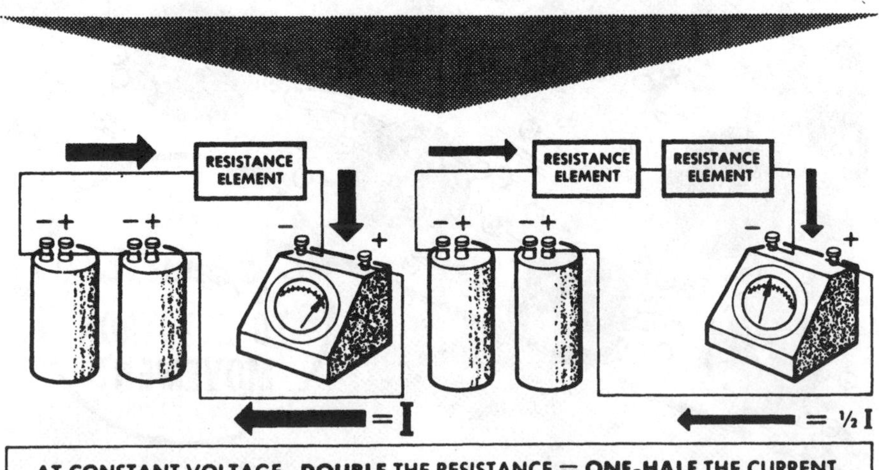

AT CONSTANT VOLTAGE, DOUBLE THE RESISTANCE = ONE-HALF THE CURRENT

What Resistance Is (continued)

Let's see now what happens when you hold the resistance constant and *vary* the voltage (electric force). You would expect that *increasing* the voltage (increased electric force) would permit the flow of *more* electrons (current) against the opposing effect of the resistance. This is exactly what happens.

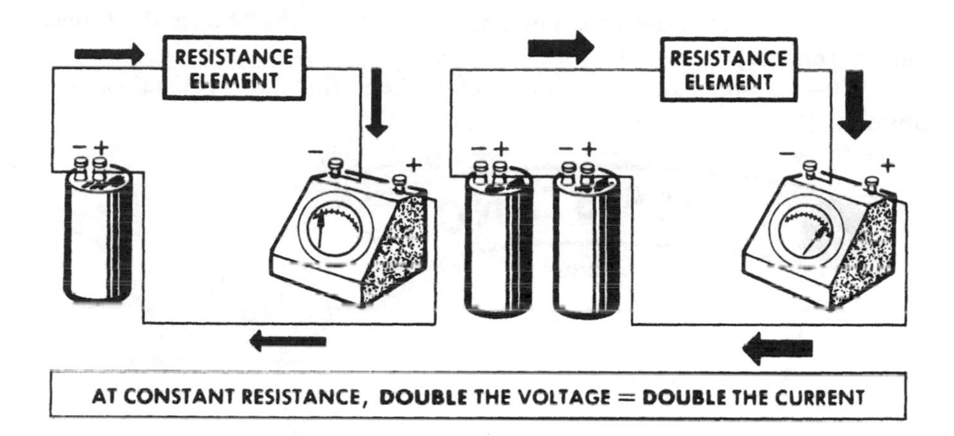

The above principles relating current flow to the amount of resistance in a circuit, or the voltage in a circuit, are of extreme importance and basic to the study of electricity. We will study these principles in much more detail later and use them throughout our study of electricity.

Although all conductors have resistance, you will have many occasions when you want to put in a *specific* amount of resistance in a circuit. Devices having known values of resistance are called *resistors*, designated with the letter "R," and are shown in circuit diagrams by the schematic symbol below.

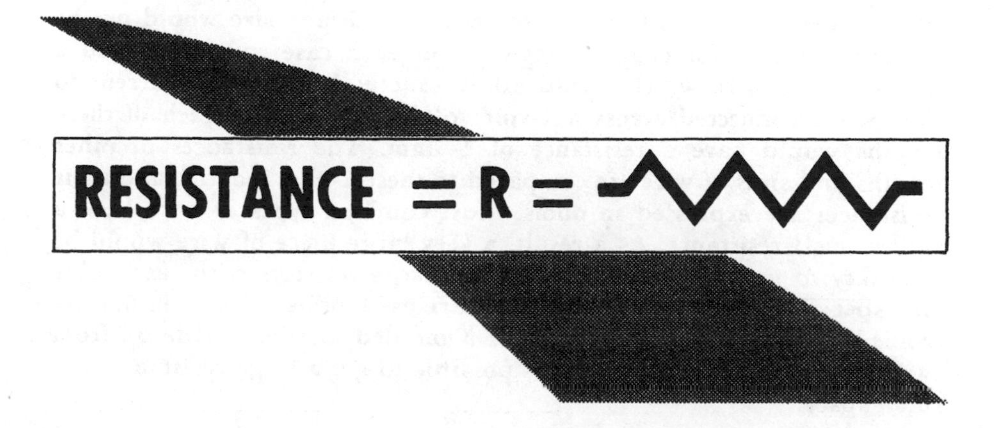

Units of Resistance

To measure current, the ampere is used as a unit of measure. To measure voltage, the volt is used. These units are necessary in order to compare different currents and different voltages. In the same manner, a unit of measure is needed to compare the resistance of different conductors. The basic unit of resistance is the *ohm*, equal to that resistance which will allow exactly *1 ampere* of *current* to flow when *1 volt* of *emf* is applied across the resistance. The abbreviation for the ohm is the Greek letter Ω (omega).

When 1 volt causes 1 ampere of current flow, the resistance is 1 ohm.

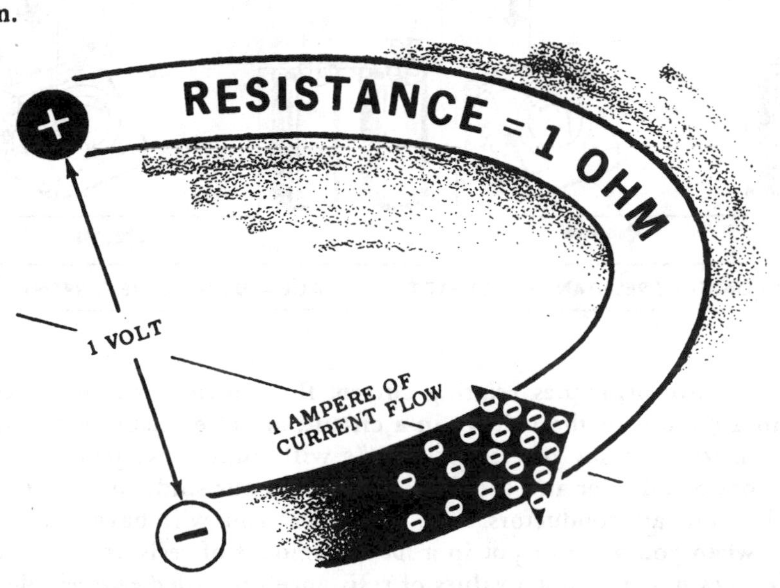

Suppose you connect a copper wire across a voltage source of 1 volt and adjust the length of the wire until the current flow through the wire is exactly 1 ampere. The resistance of the length of copper wire then is exactly 1 ohm. If you were to use wire of any other materials—iron, silver, etc.—you would find that the wire length and size would not be the same as that for copper. However, in each case you could find a length of the wire which would allow exactly 1 ampere of current to flow when connected across a 1-volt voltage source, and each of these lengths would have a resistance of 1 ohm. The resistances of other lengths and sizes of wire are compared to these 1-ohm lengths, and their resistances are expressed in ohms. Most common types of wire have a rather small resistance. As a result, a very large piece of wire would be necessary to get a large resistor. To get large resistors with reasonable size, special wires called *resistance wires* are used, or as is common in electronic circuits, resistors are made of a molded material made up from carbon and clay. In this way, it is possible to get a large resistance in a small space.

Units of Resistance (continued)

Although you will often find that resistance values are given in ohms, you will also find that in many cases large values of resistance will be used or indicated. In addition, on some occasions you will find it necessary to use small fractional values of the ohm. The previously learned prefixes—*micro*, *milli*, and *kilo*—that you have used with voltage and current are also used in the *same* way with resistance. In addition, we use another prefix *meg*, which, when put in front of *ohm*, is equal to 1,000,000 ohms; that is, 1 megohm is equal to 1,000,000 ohms.

Units of resistance are converted in the same manner as units of current or voltage. You will have to learn some new abbreviations, however, since K is often used to indicate *kilohms* and M or *meg* is often used to indicate *megohms*. Thus, 10 kilohms would be shown as 10 K and 3.3 megohms would be shown as 3.3 M or 3.3 meg.

CONVERTING RESISTANCE UNITS

MICROHMS TO OHMS

Move the decimal point 6 places to the left

35,000 microhms = 0.035 ohm

OHMS TO MICROHMS

Move the decimal point 6 places to the right

3.6 ohms = 3,600,000 microhms

MILLIOHMS TO OHMS

Move the decimal point 3 places to the left

2,700 milliohms = 2.7 ohms

OHMS TO MILLIOHMS

Move the decimal point 3 places to the right

0.68 ohms = 680 milliohms

KILOHMS TO OHMS

Move the decimal point 3 places to the right

6.2K = 6,200 ohms

OHMS TO KILOHMS

Move the decimal point 3 places to the left

47,000 ohms = 47 K

MEGOHMS TO OHMS

Move the decimal place 6 places to the right

2.7 Meg = 2,700,000 ohms

OHMS TO MEGOHMS

Move the decimal point 6 places to the left

620,000 ohms = 0.62 Meg

Factors Controlling Resistance

All materials have some resistance. In some cases this is *desirable*, as, for example, where you want to limit current flow deliberately and, thus, use components called *resistors* made up of materials chosen for their resistance properties. In other cases, resistance is an *undesirable* property and you want to keep it at a *minimum* as, for example, in the case where you want to deliver a large current to a load and do not want the current limited by the conductors. You will learn more about the undesirable, or desirable, resistance of conductors later when you study electric power. Now we will look at the factors that *control* resistance in a material.

The resistance of any object, such as a wire conductor, depends on four factors—the *material* from which it is made, the *length* of that material, the *cross-sectional area* of the material, and, finally, the *temperature* of the material.

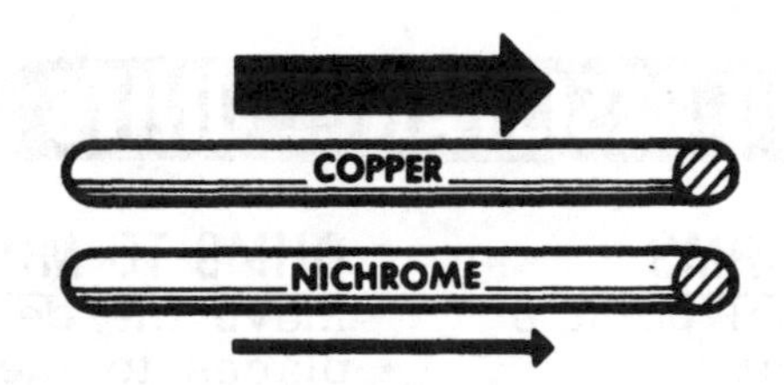

1. THE MATERIAL—Different materials have different resistances. Some, such as silver and copper, have a low resistance, while others, such as iron or nichrome (a special alloy of nickel, chromium and iron), have a higher resistance. Many resistors, such as those used in electronic circuits, are made of a molded mixture of carbon and clay.

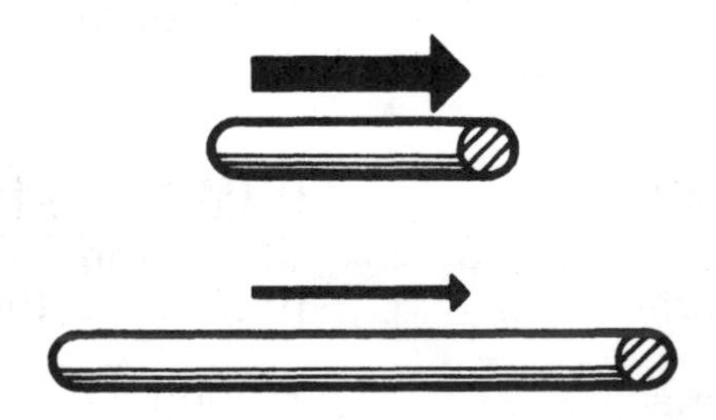

2. THE LENGTH—For a given material that has a constant cross-sectional area, the total resistance is *proportional* to the length. If a given length of the material has a resistance of 3 ohms, then twice that length will have a resistance of 6 ohms, a length of 3 times will have a resistance of 9 ohms, etc.

Factors in Controlling Resistance (continued)

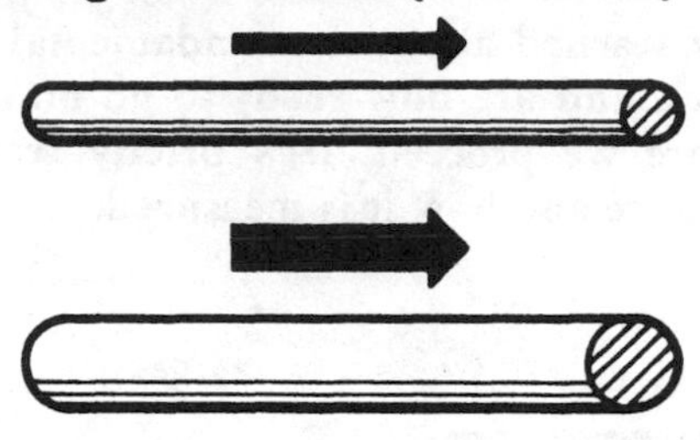

3. THE CROSS-SECTIONAL AREA—Current flow can be compared to the flow of water in a pipe. We know that if we make a pipe bigger (an increase in cross-sectional area), more water will flow even though the pressure is the same. There is a similar situation with respect to a conductor, in that the resistance *decreases* as the cross section *increases*. If we *double* the cross section of a material at constant length, the resistance will be *halved*. If we make the cross section one half, the resistance will be doubled.

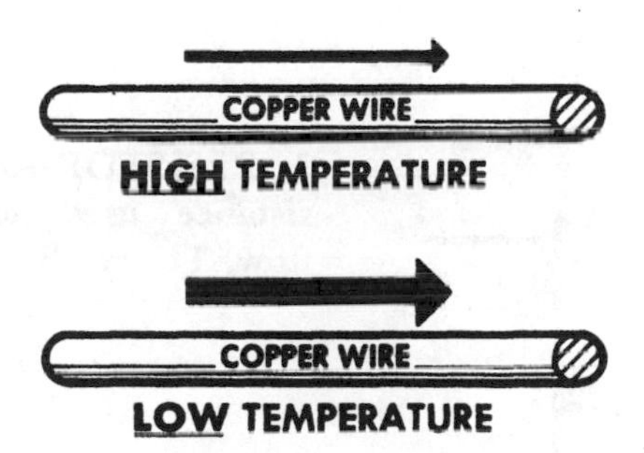

4. THE TEMPERATURE—Although temperature effects are generally small compared to the effects of material, length, and cross section, they can be important, particularly when we want to keep a resistance at a fixed value and the temperature is not constant. Generally, in metals, the resistance increases as the temperature increases. This is basically caused by the fact that the heat energy makes the free electrons in the material bounce around readily; and it is more difficult to get these electrons to flow along from atom to atom in an orderly way that we call current flow. In a few materials, such as carbon, the resistance decreases as the temperature increases.

Review of Resistance

You have now learned about the fundamental qualities of *voltage*, *current*, and *resistance*, and are now ready to go on and see how electric circuits work. Before we proceed, let's briefly review what you have learned about resistance and how it is measured.

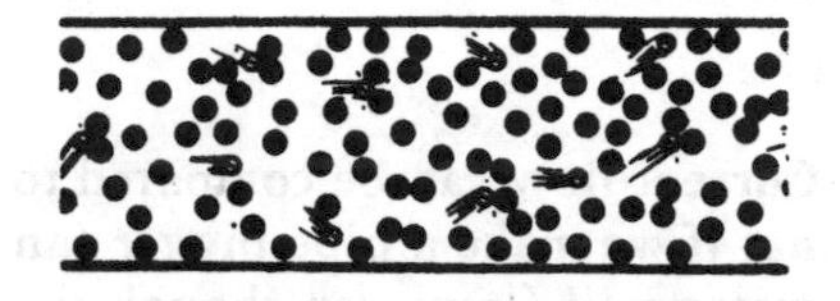

1. RESISTANCE—Opposition offered by a material to the flow of current.

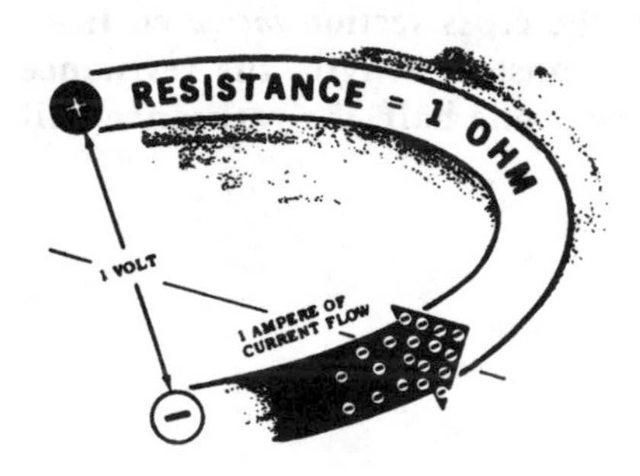

2. OHM—Basic unit of resistance measure equal to that resistance which allows 1 ampere of current to flow when an emf of 1 volt is applied across the resistance. The symbol for the ohm is Ω.

3. RESISTOR—Device having resistance used to control current flow. The symbol for a resistor is *R*.

4. OHMMETER—Meter used to measure resistance directly.

1 K = 1000Ω

5. KILOHM—One kilohm equals 1,000 ohms.

1 MegΩ = 1,000,000Ω

6. MEGOHM—One megohm is equal to 1,000,000 ohms.

Self-Test—Review Questions

1. Define what resistance is. What is a resistor? What is the symbol used to designate a resistor?
2. In a circuit with constant voltage, what happens to the current when the resistance is doubled? Halved? Tripled?
3. In a circuit with constant resistance, what happens to the current when the voltage is doubled? Halved? Quadrupled? Tripled?
4. Define the unit of resistance. What symbol is used to designate it?
5. What factors determine the resistance of a resistor? Give examples of their effect.

   Calculate the following conversions using appropriate symbols where applicable:
6. Convert to ohms

   6.2 K
   6.2 M
   270 milliohms
   3.3 K
   9.1 kilohms
   4.7 megohms
7. Convert to kilohms

   4,700 ohms
   8.2 megohms
   100,000 ohms
   0.1 megohms
   0.39 megohms
   24,000 ohms
8. Convert to megohms

   1,000 kilohms
   120,000 ohms
   82,000 ohms
   68 K
   470,000 ohms
   330 K
9. Draw a schematic diagram of a resistor. What is the symbol used to designate resistance?
10. What are the four factors that affect resistance in a conductor? How does the resistance vary as these change?

Review of Current (I), Voltage (E), and Resistance (R)

As a conclusion to your study of electricity in action, you should consider again what you have found out about current, voltage, and resistance.

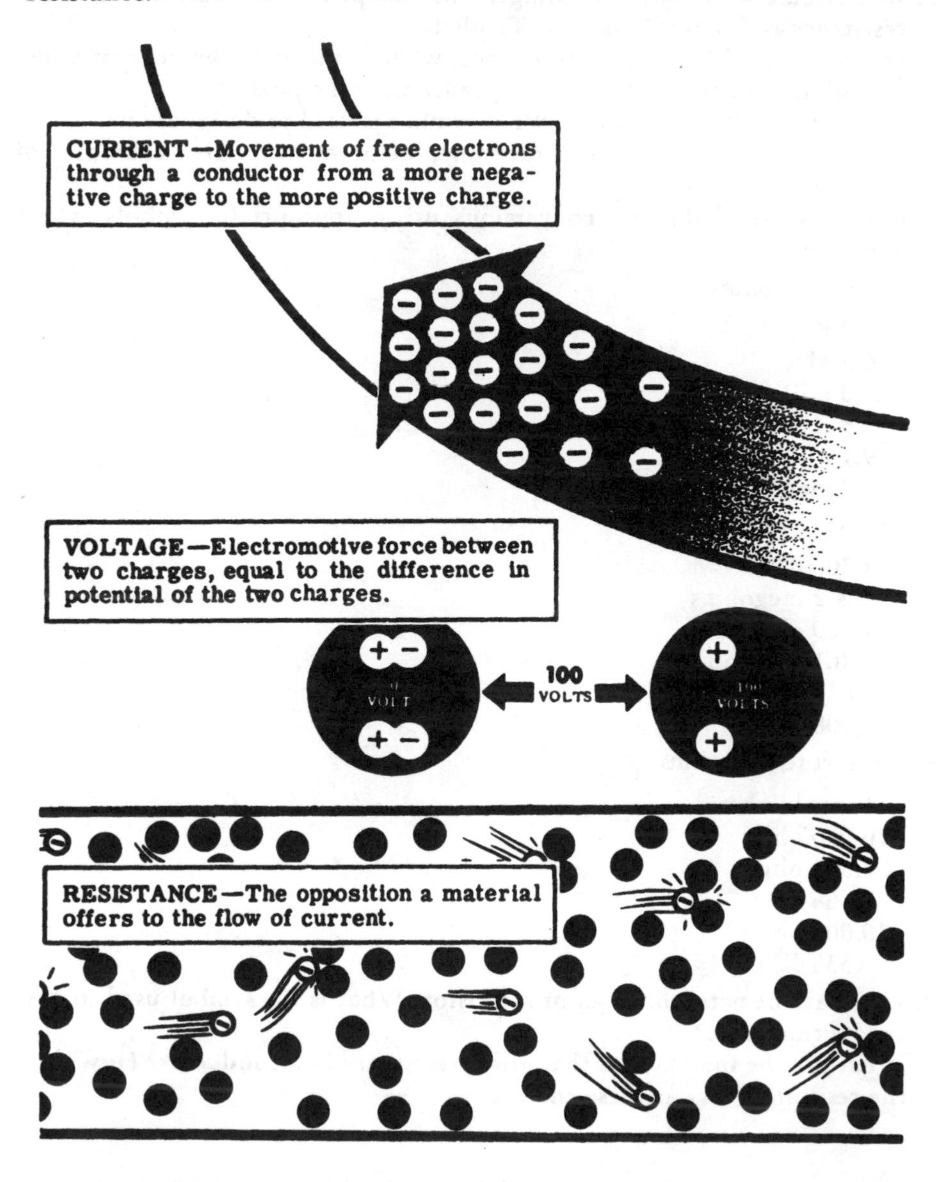

CURRENT—Movement of free electrons through a conductor from a more negative charge to the more positive charge.
VOLTAGE—Electromotive force between two charges, equal to the difference in potential of the two charges.
100 VOLTS
RESISTANCE—The opposition a material offers to the flow of current.

Particularly, you should recall the relationships between *current*, *voltage*, and *resistance*. Current flow is caused by the voltage between two points and is limited by the resistance between the points. In continuing your study of electricity, you will next find out about electric circuits and how they use current, voltage, and resistance.

The Relationship of Current, Voltage, and Resistance

As we said initially, the study of electricity is the study of the *effects* of the flow of current and the *control* of the flow of current.

Voltage, as you know, is the amount of electromotive force (emf) that is applied across a load (resistance) in order to make an electron current flow through the resistance. As you have learned, the *greater* the voltage you apply across a resistance, the *greater* the current flow. Similarly, the *lower* the voltage you apply, the *smaller* will be the current flow.

Resistance, as you also know, is the effect that impedes the flow of electrons. If you *increase* the resistance of the load across which a constant voltage is applied, *less* current will flow. Similarly, the *lower* you make the resistance, the *greater* will be the current flow.

This relationship between voltage, resistance, and current was studied by the German mathematician George Simon Ohm. His description, now know as *Ohm's Law*, says that current varies directly with the voltage and inversely with the resistance. The mathematical analysis of the law is of no concern to you at present, but you will learn about it when you get into Volume 2. Ohm's Law is a *basic tool* for all who work with electric circuits in any shape or form.

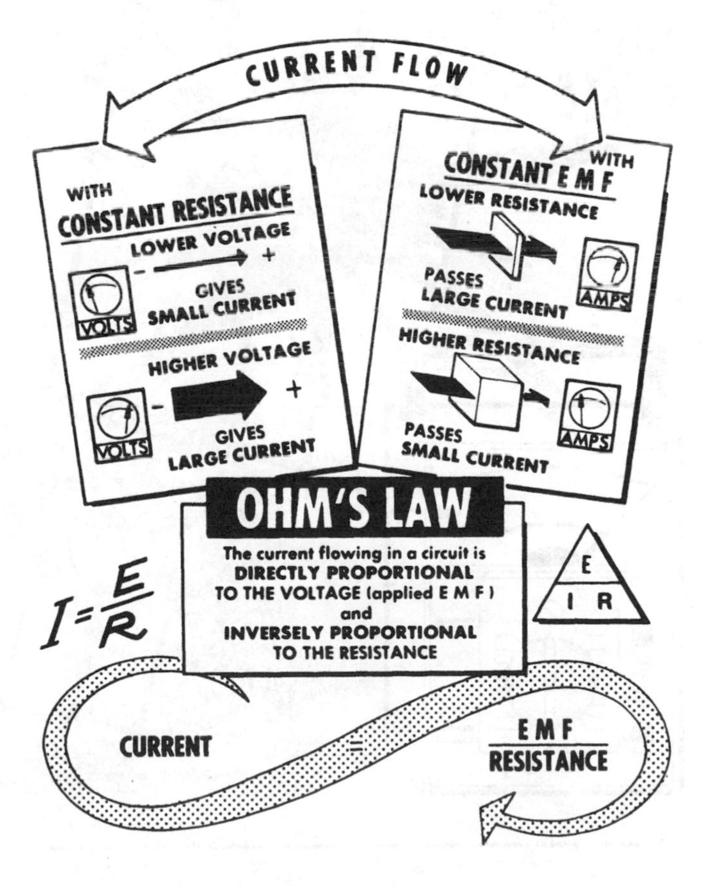

Learning Objectives—Next Volume

Overview—Now that you have learned some fundamentals of electricity, and know about current, voltage and resistance, you are ready to study the electric circuits in Volume 2. Sources and loads must be connected together to function and when connected together, are called *electric circuits*. What electric circuits are and how they are connected and behave is the most important part of the study of electricity.

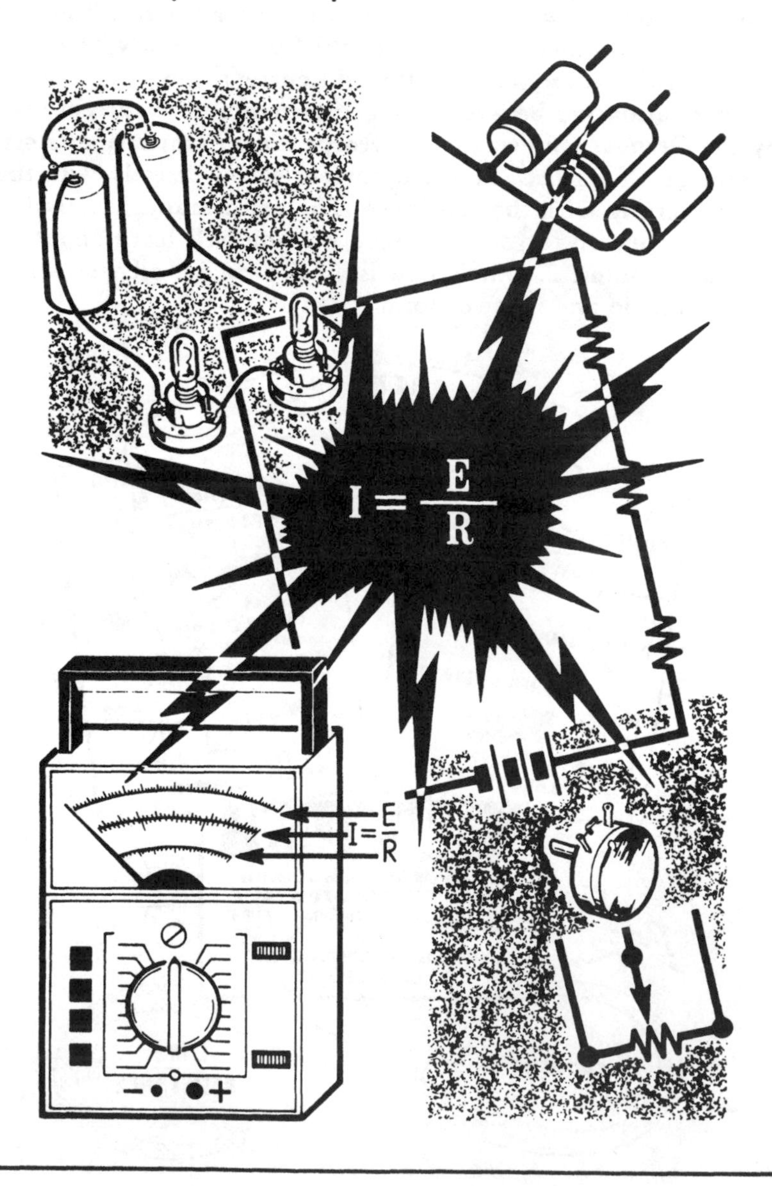

TABLE OF THE ELEMENTS

Atomic Number	Name of Element	Symbol of Element	Atomic Weight	Atomic Number	Name of Element	Symbol of Element	Atomic Weight
1	Hydrogen	H	1	53	Iodine	I	127
2	Helium	He	4	54	Xenon	Xe	131
3	Lithium	Li	7	55	Cesium	Cs	133
4	Beryllium	Be	9	56	Barium	Ba	137
5	Boron	B	11	57	Lanthanum	La	139
6	Carbon	C	12	58	Cerium	Ce	140
7	Nitrogen	N	14	59	Praseodymium	Pr	141
8	Oxygen	O	16	60	Neodymium	Nd	144
9	Fluorine	F	19	61	Promethium	Pm	147
10	Neon	Ne	20	62	Samarium	Sm	150
11	Sodium	Na	22	63	Europium	Eu	152
12	Magnesium	Mg	24	64	Gadolinium	Gd	157
13	Aluminum	Al	27	65	Terbium	Tb	159
14	Silicon	Si	28	66	Dysprosium	Dy	162
15	Phosphorus	P	31	67	Hilmium	Ho	165
16	Sulfur	S	32	68	Erbium	Er	167
17	Chlorine	Cl	35	69	Thulium	Tm	169
18	Argon	A	39	70	Ytterbium	Yb	173
19	Potassium	K	39	71	Lutecium	Lu	175
20	Calcium	Ca	40	72	Hafnium	Hf	179
21	Scandium	Sc	45	73	Tantalum	Ta	181
22	Titanium	Ti	48	74	Tungsten	W	184
23	Vanadium	V	51	75	Rhenium	Re	186
24	Chromium	Cr	52	76	Osmium	Os	190
25	Manganese	Mn	55	77	Iridium	Ir	193
26	Iron	Fe	56	78	Platinum	Pt	195
27	Cobalt	Co	59	79	Gold	Au	197
28	Nickel	Ni	59	80	Mercury	Hg	201
29	Copper	Cu	64	81	Thallium	Tl	204
30	Zinc	Zn	65	82	Lead	Pb	207
31	Gallium	Ga	70	83	Bismuth	Bi	209
32	Germanium	Ge	73	84	Polonium	Po	210
33	Arsenic	As	75	85	Astatine	At	211
34	Selenium	Se	79	86	Radon	Rn	222
35	Bromine	Br	80	87	Francium	Fr	223
36	Krypton	Kr	84	88	Radium	Ra	226
37	Rubidium	Rb	85	89	Actinium	Ac	227
38	Strontium	Sr	88	90	Thorium	Th	232
39	Yttrium	Y	89	91	Protactinium	Pa	231
40	Zirconium	Zr	91	92	Uranium	U	238
41	Columbium	Cb	93	93	Neptunium	Np	239
42	Molybdenum	Mo	96	94	Plutonium	Pu	239
43	Technetium	Tc	99	95	Americium	Am	241
44	Ruthenium	Ru	102	96	Curium	Cm	242
45	Rhodium	Rh	103	97	Berkelium	Bk	245
46	Palladium	Pd	107	98	Californium	Cf	246
47	Silver	Ag	108	99	Einsteinium	E	253
48	Cadmium	Cd	112	100	Fermium	Fm	256
49	Indium	In	115	101	Mendelevium	Mv	256
50	Tin	Sn	119	102	Nobelium	No	254
51	Antimony	Sb	122	103	Lawrencium	Lw	257
52	Tellurium	Te	128	104 & 105	Under Study		

Note: Elements 1 through 92 occur normally in nature. Elements 93 and above are those discovered by man as a result of transmutation.

Basic Electricity

REVISED EDITION

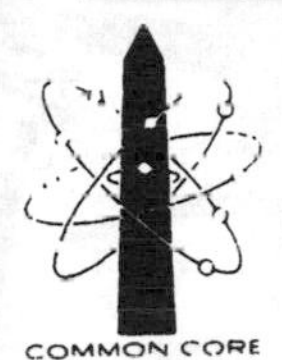

VAN VALKENBURGH,
NOOGER & NEVILLE, INC.

VOL. 2

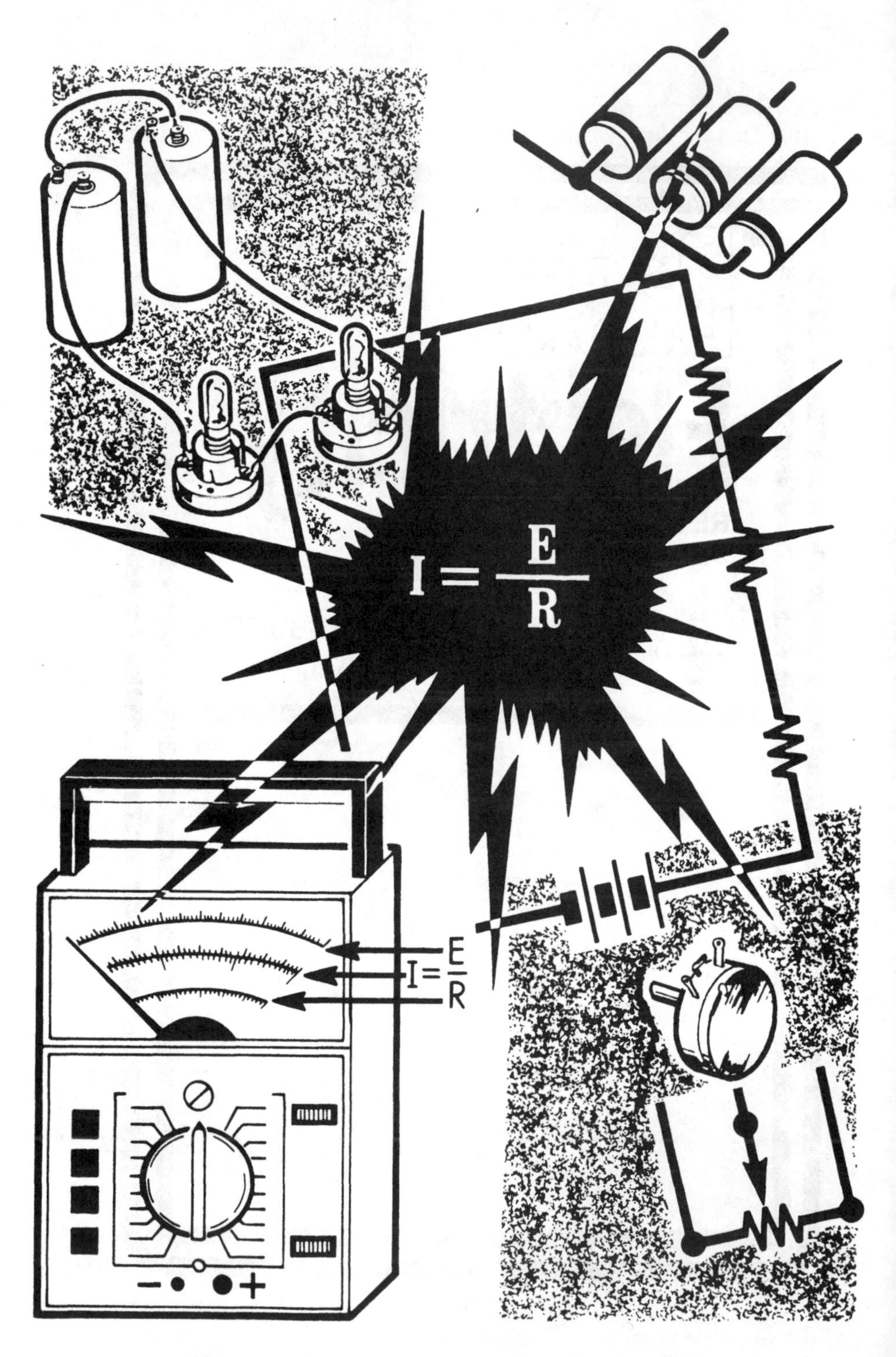

Direct current circuits

What a Circuit Is

It is hardly an exaggeration to say that the second half of the twentieth century runs on the flow of electric current. It is, therefore, essential that you should have an accurate picture of what electric current is, and how it behaves in a circuit.

Recall for a moment what you learned about current flow in *Basic Electricity*, Volume 1. You learned that if you connect a length of wire (a conductor) across the positive and negative terminals of a source of electromotive force (emf), say, a battery, the potential difference (voltage) makes the current flow; and also, that electrical energy is needed to keep the current flowing. Additionally, you know that for a battery, the electrical energy is produced from chemical action within the battery.

Many millions of free electrons that have already been separated from the outer orbits of their respective atoms by the heat of room temperature, and which have been wandering aimlessly in all directions through the wire, now come under a *common controlling force*. They are repelled by the *more negative* (or *less positive*) charge which has been set up at one end of the wire, and strongly attracted by the *less negative* (or *more positive*) charge which has been set up at the other end. Their aimless wanderings are converted into a disciplined current flow from more negative to more positive, and electric current flows.

Remember, these electrons are negative charges of electricity and have practically no weight at all. This means that when a potential difference is applied to the wire, they respond to it *immediately*. Similarly, when the potential difference is removed, the electrons stop their disciplined flow in a single direction at once and resume their random wanderings through the conductor material.

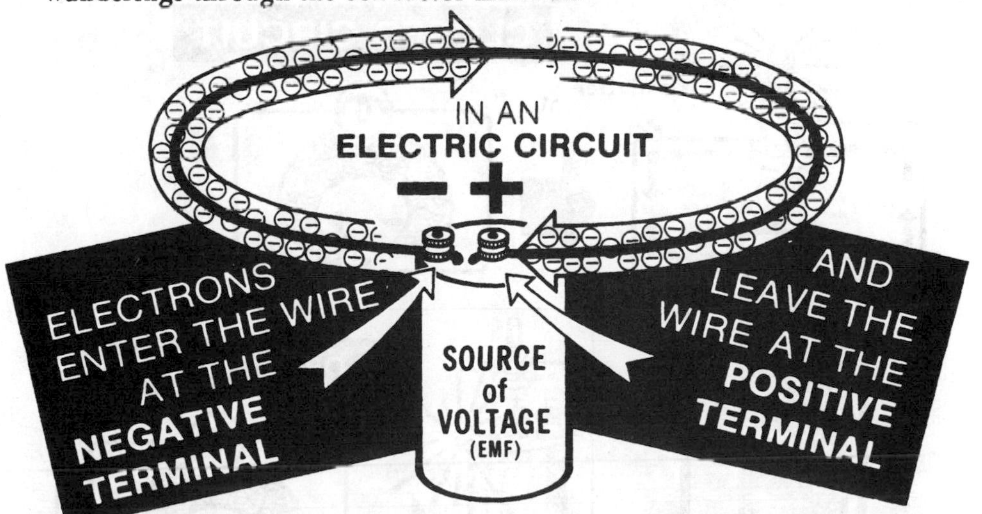

Any combination of a conductor and a source of electricity connected together to permit electrons to travel around in a continuous stream is called an *electric circuit*.

What a Circuit Is (continued)

The conditions required to set up and maintain the flow of an electric current in a circuit are as follows:

1. There must be a *source* of potential difference or voltage to provide the energy which forces electrons to move in a disciplined way in a specific direction.
2. There must be a *continuous (complete) external path* for the electrons to flow from the negative terminal to the positive terminal of the source of voltage.

This external path is usually made up of two parts: the *conductors*, or wires, and the *load* to which the electric power is to be delivered to accomplish some useful purpose or effect. In the illustration below, the load is the lamp.

An electric circuit is thus a completed electrical pathway, consisting not only of a conductor in which the current will flow from negative to positive, but also of a path through a source of potential difference (in this case, the battery) from the positive back to the negative.

A lamp connected across a dry cell battery is an example of a simple electric circuit. Current flows from the negative (−) terminal of the cell, through the lamp (the load), to the positive (+) terminal. The action of the cell is such that it provides a *regenerative* path for the flow of electrons to be maintained.

As long as this electrical pathway remains unbroken at any point, it is a *closed* circuit and current flows. But if the pathway is broken, it becomes an *open* circuit and no current flows.

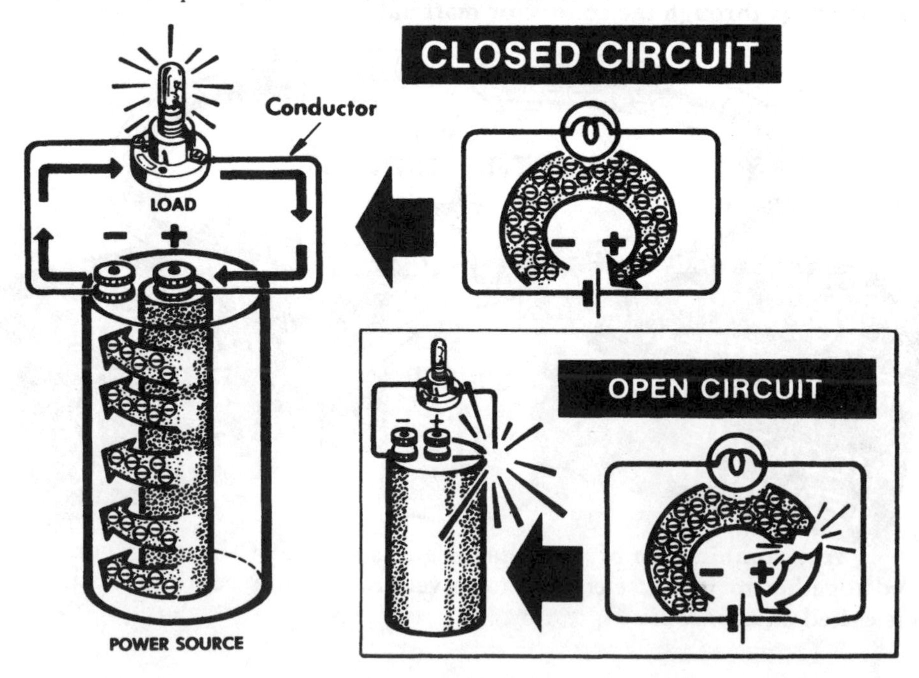

DC and AC Circuits

In electricity we deal with both *direct current* (abbreviated dc) and *alternating current* (abbreviated ac). In dc circuits, the current always flows in the *same* (one) direction. In ac circuits, the direction of current flow *reverses periodically*—in one instant, it will flow in one direction and in the next instant, in the opposite direction. This flow reversal in ac current is usually done regularly so that when we talk about 60-Hz ac power, we mean that the direction of flow reverses 60 times (or cycles) per second.

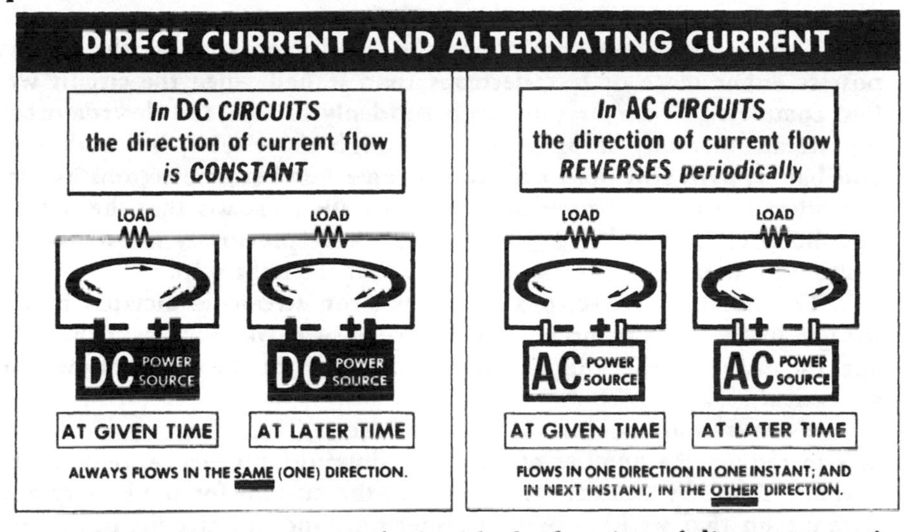
DIRECT CURRENT AND ALTERNATING CURRENT
In DC CIRCUITS the direction of current flow is CONSTANT
LOAD
LOAD
DC POWER SOURCE
DC POWER SOURCE
AT GIVEN TIME
AT LATER TIME
ALWAYS FLOWS IN THE SAME (ONE) DIRECTION.
In AC CIRCUITS the direction of current flow REVERSES periodically
LOAD
LOAD
AC POWER SOURCE
AC POWER SOURCE
AT GIVEN TIME
AT LATER TIME
FLOWS IN ONE DIRECTION IN ONE INSTANT; AND IN NEXT INSTANT, IN THE OTHER DIRECTION.

In this volume, we will deal with the function of direct current in circuits containing only resistance (resistive circuits) and we will use Ohm's law and Kirchhoff's laws as the tools for analysis and understanding the relationships of current, voltage and resistance. However, it is important to remember that what you learn here will be *directly applicable* to the ac circuits that you will study in Volumes 3 and 4. By proper interpretation of the concept of current, voltage and resistance, what you study and learn in this volume on dc circuits will be used *again* and *again* for understanding the operation of ac circuits. Therefore, it is *very* important that you completely understand the concepts in dc circuits since they are the foundation for your future understanding of ac circuitry.

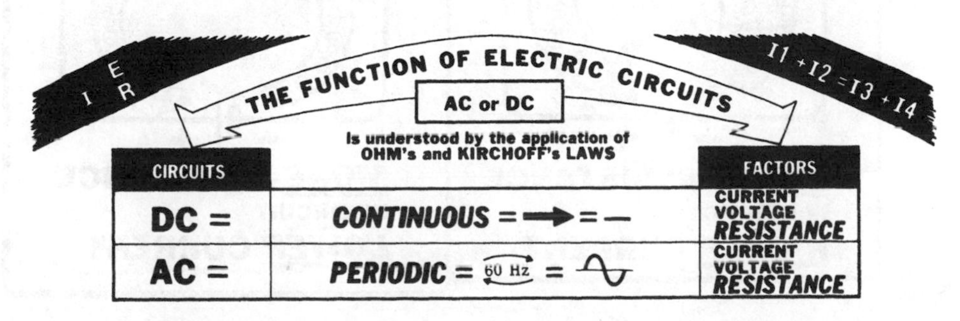
E
I R
THE FUNCTION OF ELECTRIC CIRCUITS
AC or DC
Is understood by the application of OHM's and KIRCHOFF's LAWS
I1 + I2 = I3 + I4
CIRCUITS
FACTORS
DC =
CONTINUOUS =
CURRENT VOLTAGE RESISTANCE
AC =
PERIODIC =
60 Hz
CURRENT VOLTAGE RESISTANCE

The Electric Circuit

It may help you grasp the concept of an electric current flowing through a closed circuit to imagine that the electrons, which make up the current, form a moving stream which revolves through the completed circuit.

This moving stream of electrons maintains a constant density throughout its entire length. The number of electrons entering the positive terminal of a battery from a wire is always exactly balanced by the number of electrons which the battery forces to move onto its own negative terminal, and, hence, out into the wire.

Thus, at no time does either the conductor wire or the battery possess either more or less electrons than it had when the circuit was first completed. If the circuit loop is suddenly broken, the electron orbiting stream instantly stops *revolving* through the circuit; but both wire and battery will still hold exactly the same number of electrons as they did when the circuit was made. The only difference is that the wire is now holding some of the electrons which were previously in the battery, while the battery has taken an equal number from the wire.

The number of electrons in the electron stream is dictated by the *strength* of the voltage forcing the electrons to move. The *lower* the voltage, the *weaker*—other things being equal—will be the current flow, and vice versa.

When a resistance of any kind is inserted into the circuit loop, it acts to restrict the number of electrons flowing and, hence, reduces the current. You may wonder what restricts the current for the battery and wire circuit that we have been considering. Since all circuits have some resistance, the flow of current is restricted by this resistance.

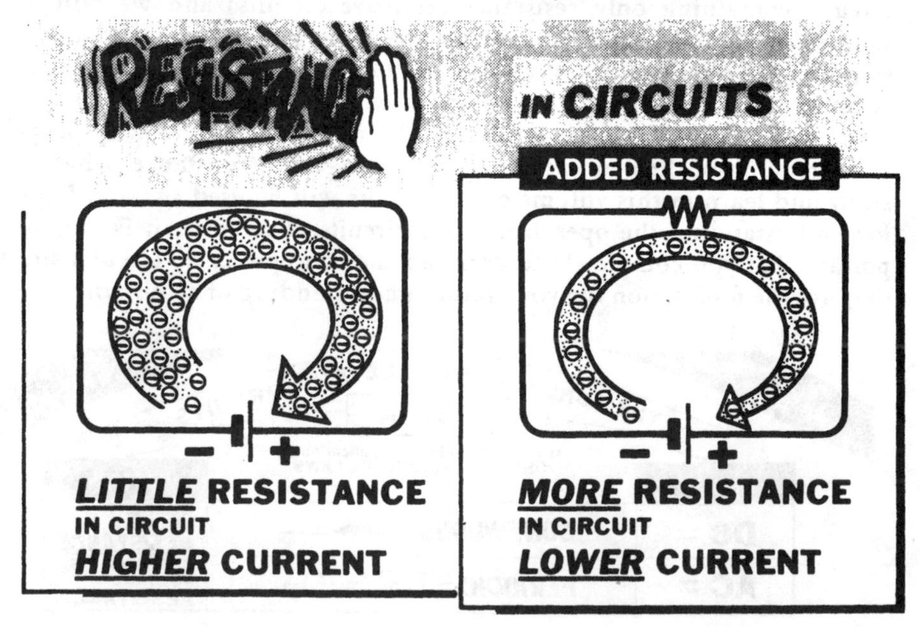

The Electric Circuit (continued)

A closed loop of wire is not always an electric circuit. Only if a source of emf is part of the loop do you have an electric circuit. Current, voltage, and resistance are present in any electric circuit where electrons move around a closed loop. The pathway for current flow is actually the circuit, and its resistance controls the amount of current flow around the circuit.

Direct-current circuits consist of a source of dc voltage, such as batteries, plus the combined resistance of the electrical load connected across this voltage. While working with dc circuits, you will find out how the total load of a circuit can be changed by using various combinations of resistances, and how these combinations of resistances control the circuit current and affect the voltage.

As you will see shortly, there are *two* basic types of circuits: *series circuits* and *parallel circuits*. No matter how complex a circuit you may work with, it can always be broken down into either a series circuit connection or a parallel circuit connection.

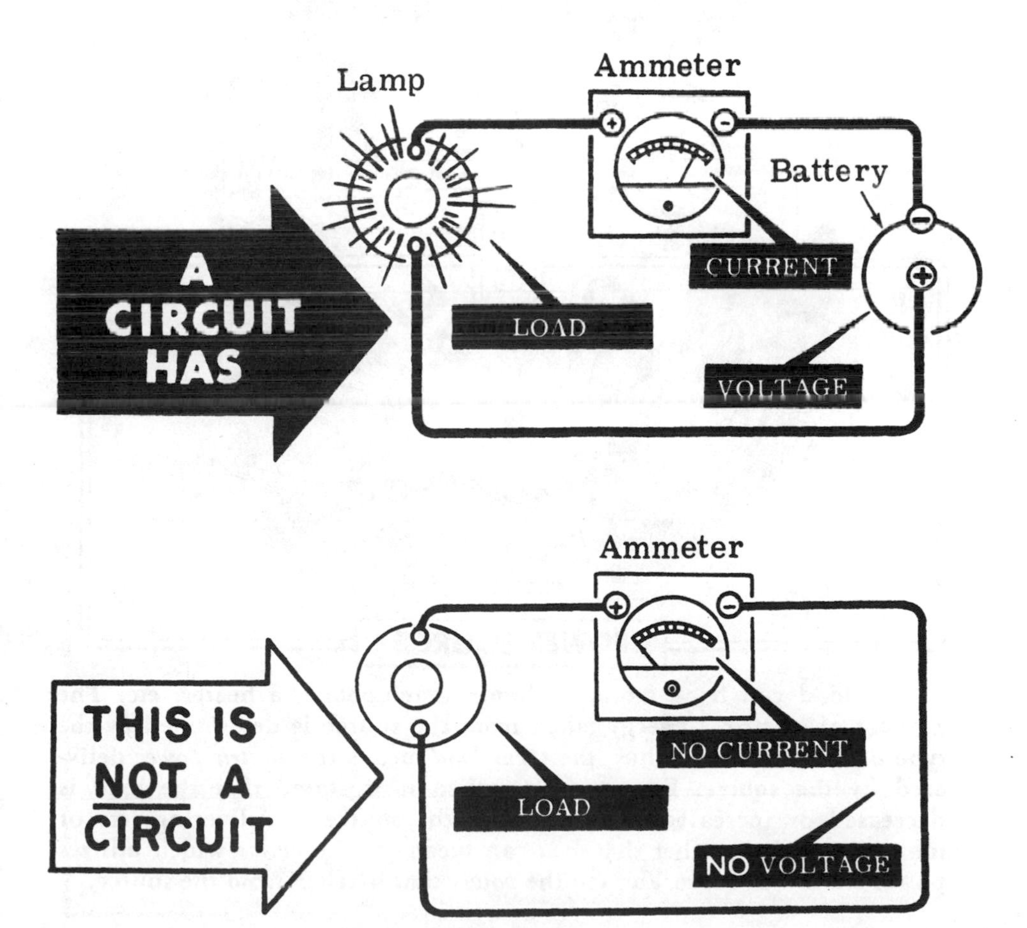

The Load

You learned in Volume 1 that electricity is used to produce pressure (sound), heat, light, chemical action, and magnetism (for mechanical power). In a basic electric circuit, the *device* that transforms the electrical energy from the source of power (emf) into some useful function—such as heat, light, mechanical power, etc.—is called the *load*. The load, besides transforming the electrical energy into one of the foregoing useful purposes, can also be utilized to change or control the *amount* of energy being delivered from the source.

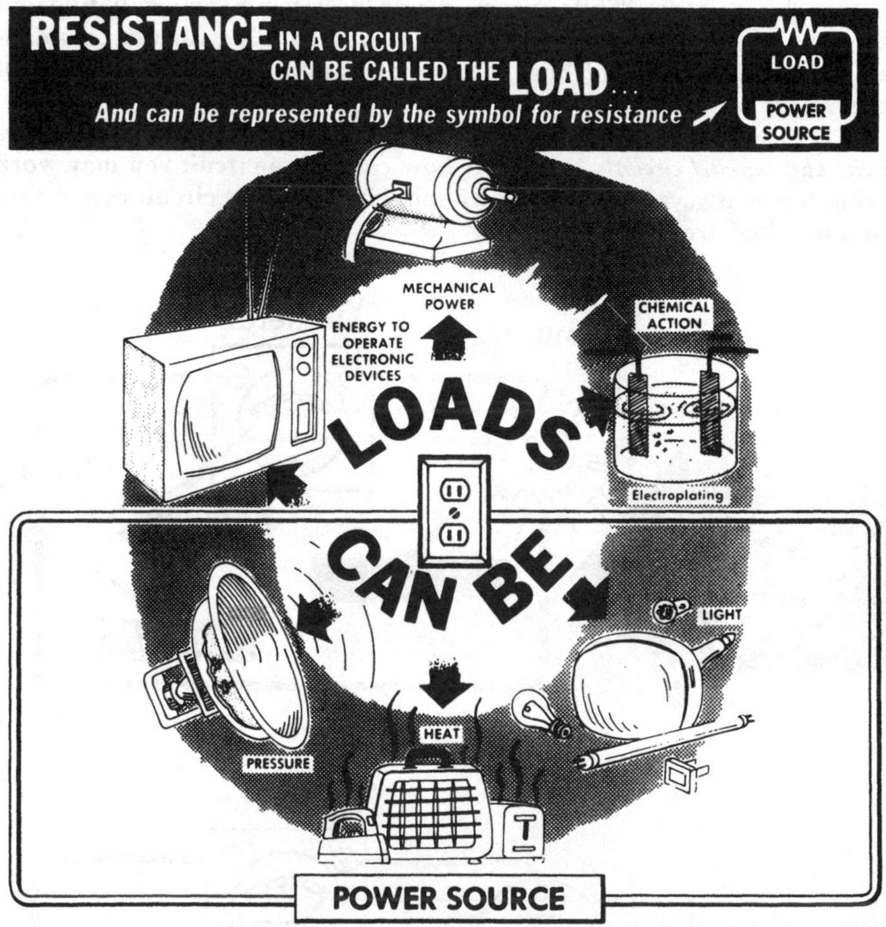

A load can be a motor, a lamp, a telephone, a heater, etc. The amount of electrical energy taken from the source is determined by the type or kind of load. Thus, the term *load* means the *electric power* delivered by the source. For example, when it is stated that the load is decreased or increased, it means that the source is delivering less or more power. Remember that *load* can mean (a) the *device* which utilizes power from the source, and (b) the *power* that is taken from the source.

Switches

A switch is a device used to *open* and *close* a circuit, or part of a circuit, when desired. You have been using switches all your life—in lamps, flashlights, radios, car ignitions, etc. You will meet many other kinds of switches while working with equipment.

You will encounter and use many different switches in your study of electricity. You will also need to know how they are shown symbolically on schematic diagrams. The simplest switch is the *single-pole, single-throw* switch, sometimes abbreviated SPST. More complicated switches can switch many circuits at the same time. These are *multipole, single-throw* switches. A switch that switches *two* circuits is called a *double-pole, single-throw* switch, sometimes abbreviated DPST. In some cases, a circuit is connected to one part of another circuit in one position of a switch, and to another part of the circuit in the other switch position. These are called *double-throw* (DPDT) switches. There are special symbols for these switches in circuit diagrams, as shown in the illustration.

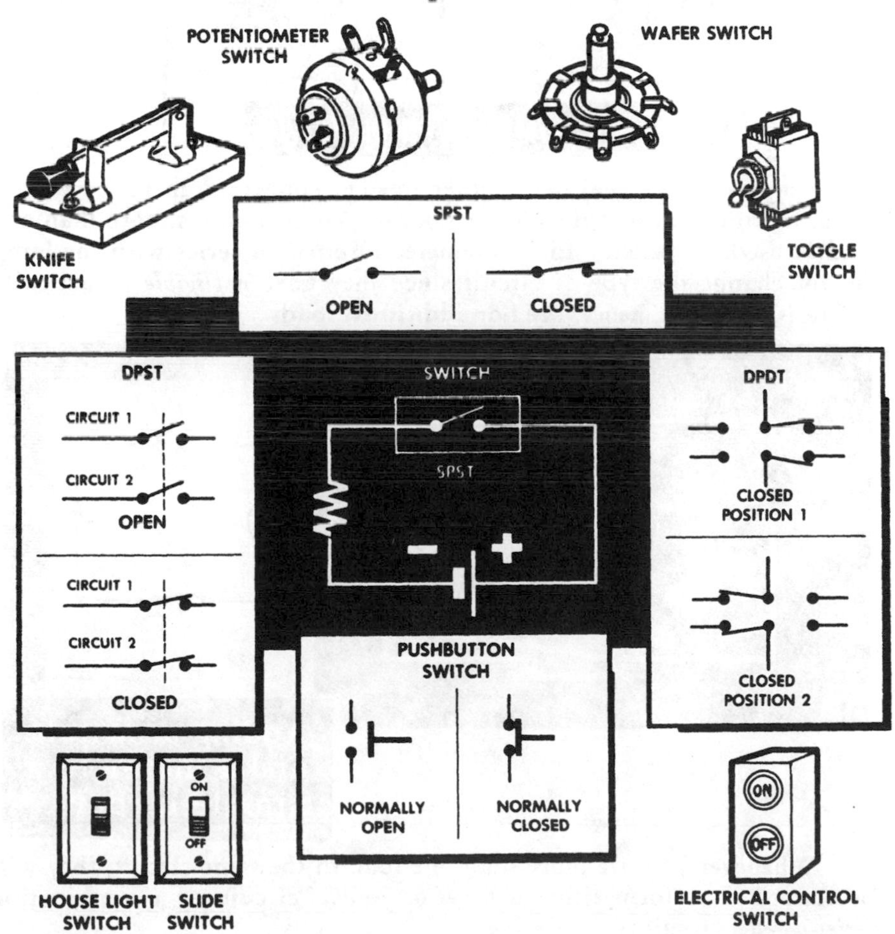

Simple Circuit Connections

Only the loads in the *external* circuit loop, between the terminals of the voltage source, are used to determine the type of circuit. When you have a circuit consisting of only one device, a voltage source, and the connecting wires, it is called a *simple* circuit. For example, a lamp connected directly across the terminals of a dry cell forms a simple circuit. Similarly, if you connect a resistor directly across the terminals of a dry cell, you have a simple circuit since only one device is being used.

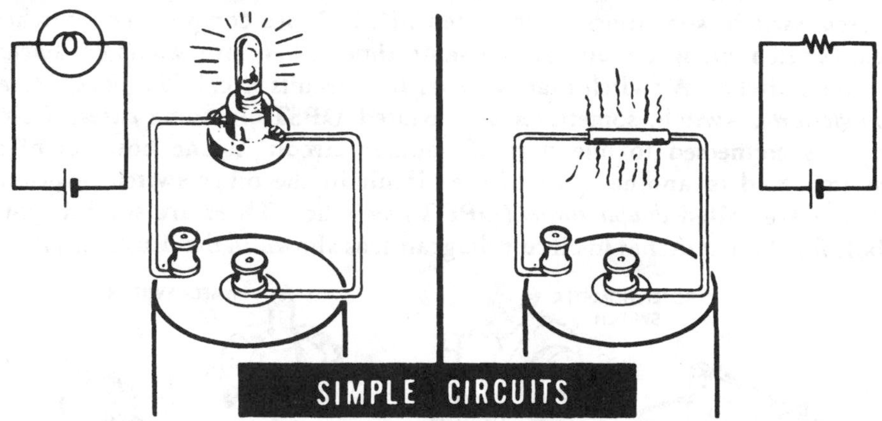

Simple circuits may have other devices connected in series with a lamp, but the nature of the circuit does not change unless more than one load is used. A switch and an ammeter inserted in series with the lamp do not change the type of circuit since they have *negligible* (practically no) resistance and, hence, are not additional loads.

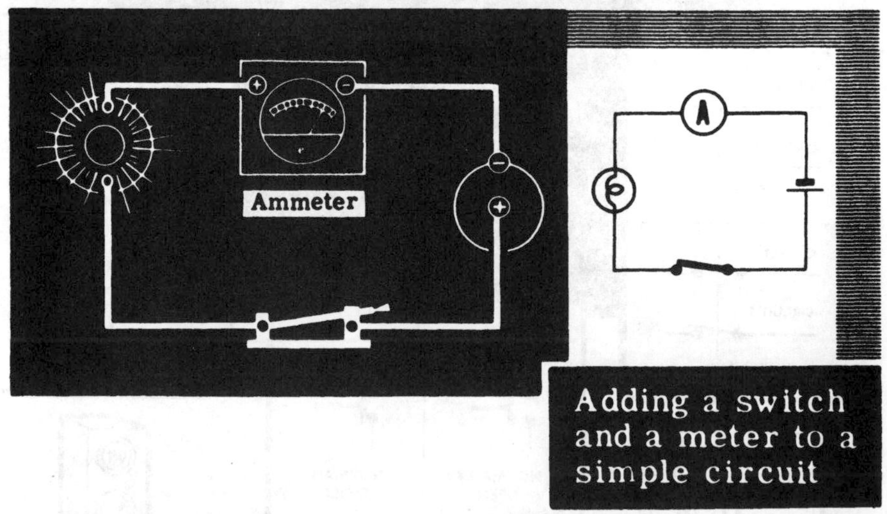

Whenever you use more than one load in the same circuit, they will be connected to form either a *series* or *parallel* circuit, or a combination *series-parallel* circuit.

Review of Electric Circuits

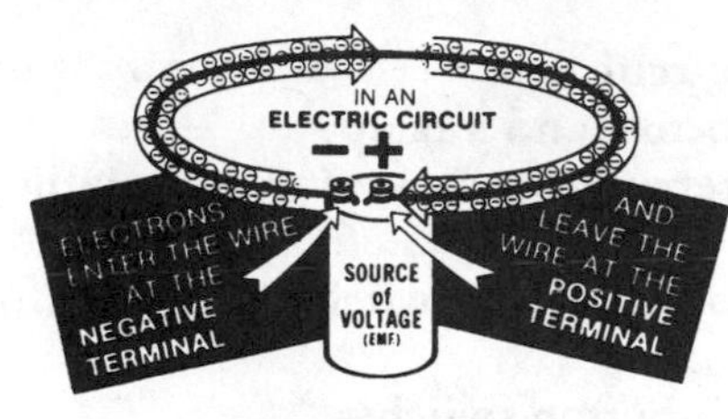

1. ELECTRIC CIRCUIT—A combination of a source of electricity and a conductor that allows electrons to travel in a continuous stream.

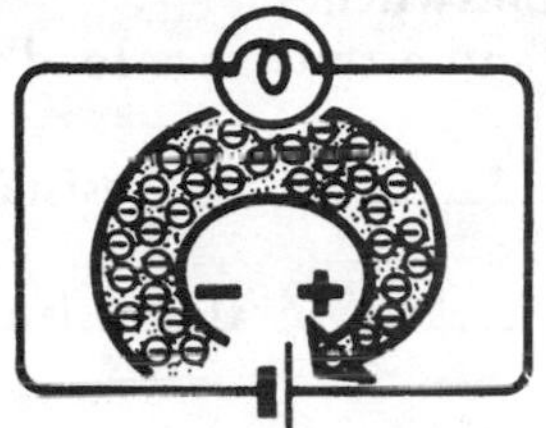

2. CLOSED CIRCUIT—A circuit whose path (loop) is unbroken and current can flow.

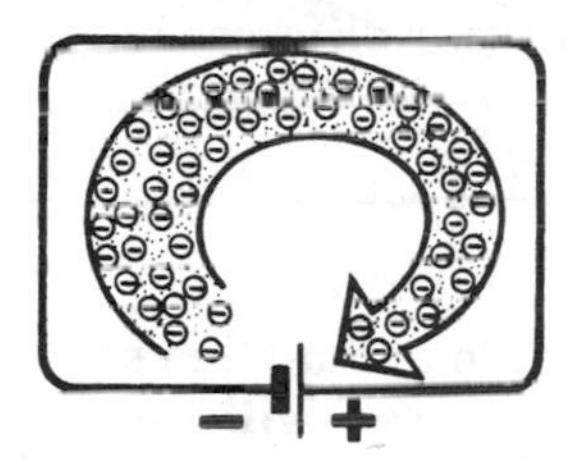

3. RESISTANCE, SMALL—When the resistance is *small*, *large* currents flow.

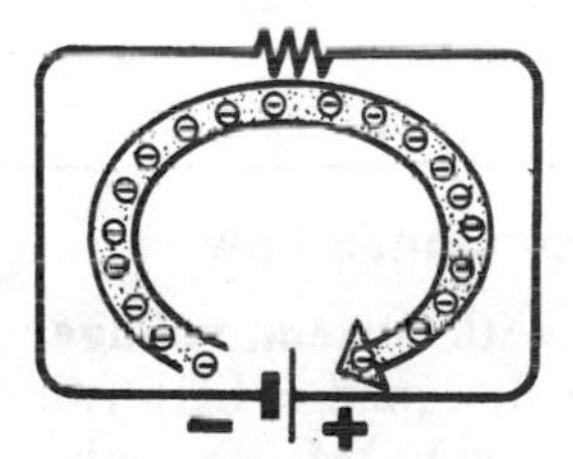

4. RESISTANCE, LARGE—When the resistance is *large*, *small* currents flow.

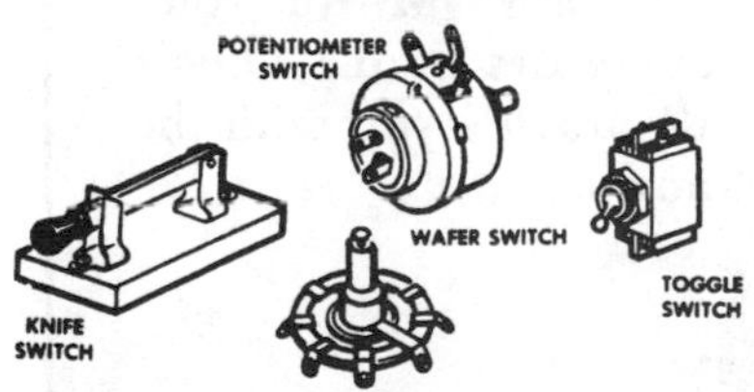

5. SWITCHES—Devices that open and close circuits and, thus, control the flow of electricity.

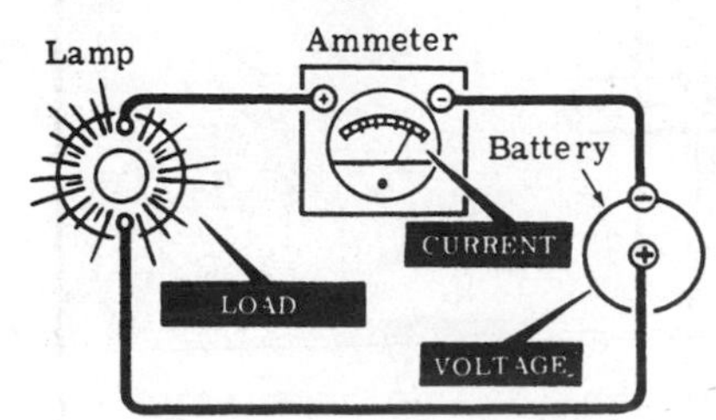

6. LOAD—The device that uses electricity for some function.

Self-Test—Review Questions

1. What are the essential elements in a circuit?
2. Draw a circuit using a battery, conductor, and a lamp.
3. Show a sketch of how you would interrupt the flow of current without a switch.
4. Draw the schematic representation of an SPST switch. Show both open and closed positions.
5. Repeat question 4 for DPST and push-button switches.
6. Draw a circuit using a battery, conductor, and a resistance load with an SPST switch to control current flow.
7. For a constant voltage, the current ____________ as the resistance decreases.
8. For a constant voltage, the current ____________ as the resistance increases.
9. In the circuit in question 2, which element is the load?
10. Draw a circuit with the battery switchable to two different loads using a DPDT switch.

Learning Objectives—Next Section

Overview—In the next section you will learn about Ohm's Law, one of the most important things that you will use throughout your career in electricity and electronics.

The VOM—The Test Instrument for Ohm's Law

As you will learn, Ohm's Law is concerned with current, voltage, and resistance—$I = E/R$. A test instrument designed to help you obtain the knowns for the above equation is the VOM—the volt/ohm/milliameter. Measuring any two quantities of the Ohm's Law equation by means of the VOM will provide you with the means for calculating the third—the unknown.

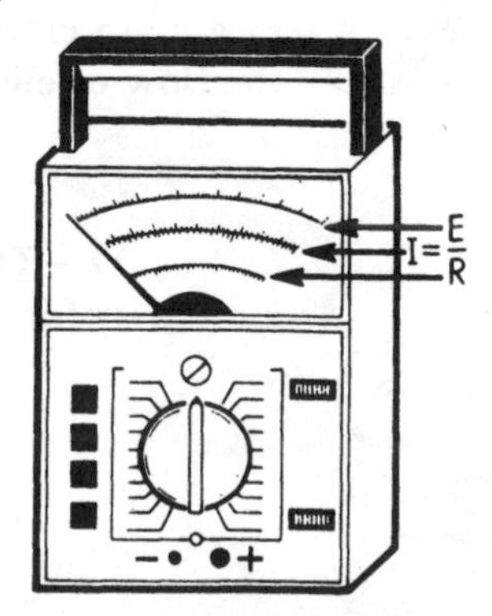

The Relationship of Voltage, Current, and Resistance

You learned in Volume 1 that there exists a fixed relationship of the voltage driving electrons through a circuit to the resistance of that circuit and the rate of current flow through the circuit. It would be wise for you to review these important concepts again.

Given a *constant resistance* in a circuit, you learned that the current flow increases as the voltage applied to the circuit increases. Given a *constant voltage* (emf) applied to the circuit, current flow decreases as the resistance of the circuit increases. You can combine these concepts as follows: *Current flow in a circuit increases as the voltage is increased, and decreases as the resistance is increased.*

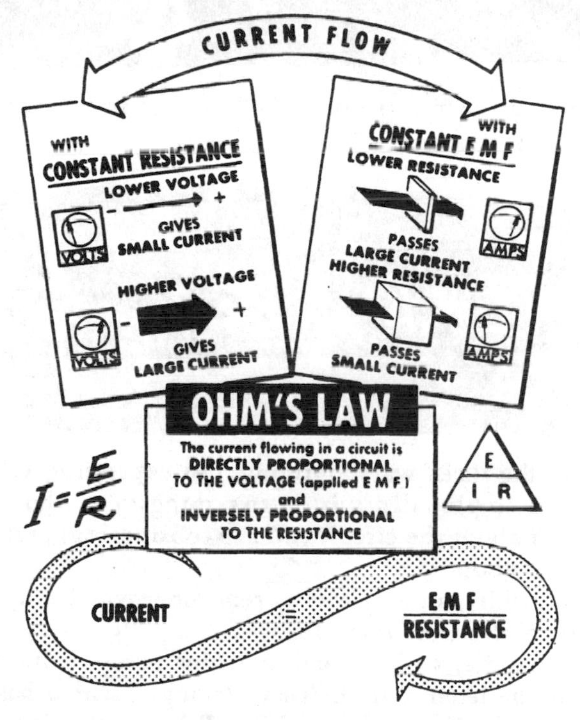

You also learned in Volume 1 that the relationship of voltage, current, and resistance was studied by a German physicist, George Simon Ohm. His statement of this relationship, called *Ohm's Law,* is one of the fundamental laws of physics. You will constantly be using Ohm's Law throughout your work in electricity, as well as later on, whether you intend to study power and light, telephone, electrical machinery, electronics, radar, computers, microwaves, etc.—or, indeed, anything else in which the flow of an electric current is involved.

What, then, does this vital law state? One of the simplest ways of expressing it is given in the illustration above.

It is also possible to express Ohm's Law as a *mathematical equation* (relationship) as further indicated in the illustration above.

The Relationship of Voltage, Current, and Resistance (continued)

In electrical terms (notation), *current* is always represented by the letter "I," *resistance* by the letter "R," and *voltage* by the letter "E." You can, therefore, rewrite the statement of Ohm's Law, at the bottom of the illustration on the last page, as follows:

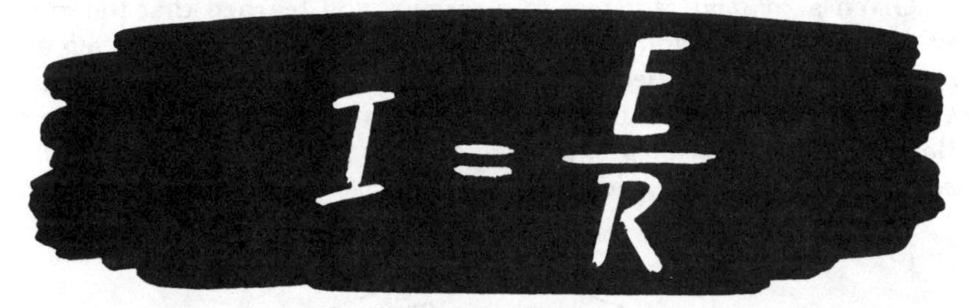

With the help of very simple algebra, this important equation can also be written as:

$$E = I \times R \text{ or as } R = \frac{E}{I}$$

Which of the three ways (formulas) of expressing Ohm's Law you might choose to employ depends on two things: (a) what *facts* you *know* to start off with about the circuit you are considering, and (b) what *facts* you *need to know* about it.

There is, luckily, an easy way to remember which way or formula to use. Call it, if you like, the *magic triangle*!

Draw a triangle, with a horizontal line across it half-way up from its base. Write the letter E in the small triangle, which has been formed above the line, and write the letters I and R below the line, like so.

The Magic Triangle

Now consider a circuit in which you know the values of any two of the three factors—voltage, current, and resistance—and want to find out the third. The rule for working the *magic triangle* to give you the correct formula is as follows:

Put your thumb over the letter in the triangle whose value you want to know—and the formula for calculating that value is given by the two remaining letters.

Here is how this useful memory-aid works in practice:

1. You know the values of current and resistance in a circuit, but you lack the means (a voltmeter) to measure the *voltage*. So you draw the magic triangle, put your thumb on the value you want to calculate, which in this case is E—and you are left with the formula you need—I × R.

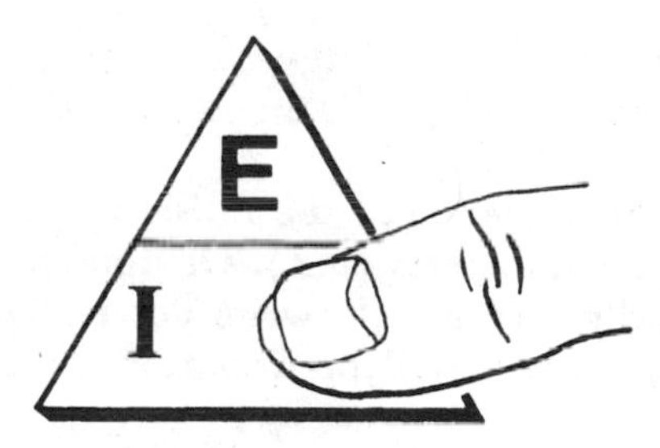

2. You know the values of current and voltage, but in this case you have no ohmmeter to measure the *resistance*. Put your thumb over the letter R and you are left with the formula $\frac{E}{I}$. Substitute the known values for E and I, and your answer is R.

3. The voltage and resistance of a circuit are known to you; but in this case the ammeter you need to measure the *current* is lost or broken. Put your thumb over the symbol I; and read off the formula you need: $\frac{E}{R}$.

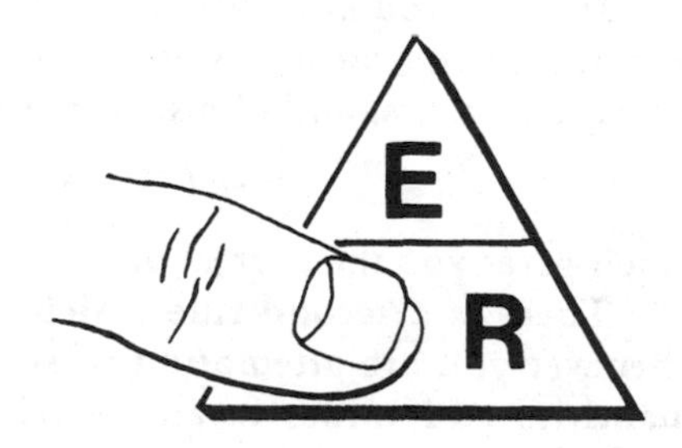

A little thought will show you that the Ohm's Law formula cannot work properly unless all values are expressed *in the correct units of measurement*. The simple rule for this is given on the next page.

Ohm's Law Rules

Ohm's Law will work for you and give you the correct answers to any problem situations which you may try to solve with its help, if you remember that in the Ohm's Law equation, the first rule is that:

CURRENT **is *ALWAYS* expressed in *AMPERES***
VOLTAGE **is *ALWAYS* expressed in *VOLTS***
RESISTANCE **is *ALWAYS* expressed in *OHMS***

Take a circuit in which you have measured the resistance to be 10 ohms, and the current to be 300 milliamperes (mA). Obviously, if you use the Ohm's Law equation blindly and merely write down that $E = I \times R = 10 \times 300 = 3{,}000$, your answer will be *wrong* by a factor of 1,000.

Instead, you must use the *conversion tables* you learned about in Volume 1, and you must rewrite *all* the factors in the simple equation above in amperes, volts, and ohms. When you do this, you will get:

$$E = I \times R = 10 \times 0.3 = 3 \text{ volts}$$

which gives you the correct answer.

There is a second rule which you should apply right from the start whenever you are attempting to solve an Ohm's Law problem involving quantities and values in an electric circuit. The rule is: *Always sketch a rough diagram of the circuit you are considering, before you start making calculations based on the values in the circuit which are already known to you.* This rule, you will find, becomes absolutely essential later on when circuits become more complex.

The sooner you get into the habit of *always* sketching out an Ohm's Law circuit *before* you begin trying to solve it, the better.

Ohm's Law Examples

Ohm's Law, and the correct way to use it, are so central to your electrical training that you should now work through the solution of three simple problems; and then embark on some practice work of your own on Ohm's Law drill.

Example 1

Problem. You have an unknown resistor connected across a battery, and you find by measurement that the voltage across it is 12 volts. You measure the current flowing as 3 amperes. You want to know the resistance of the resistor; but you have no ohmmeter.

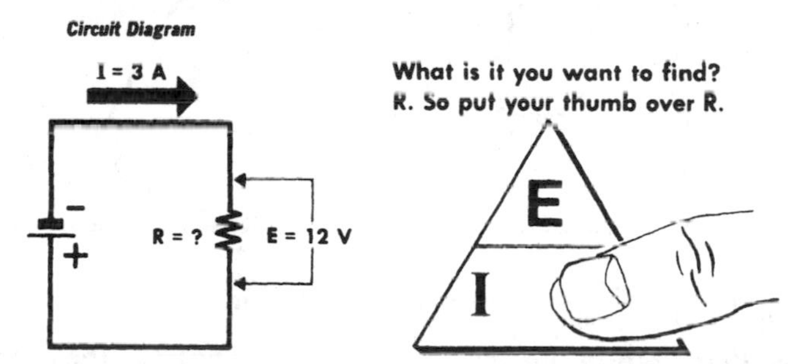

Solution. First draw the circuit diagram, and fill it in with the information you already have. Sketch out the *magic triangle*. The *magic triangle* tells you that $R = \frac{E}{I}$. Into this equation you substitute known values and get

$$R = \frac{12}{3} = 4, \text{ which is the value of the resistance in ohms.}$$

Example 2

Problem. What is the voltage across a resistor of 25 ohms when a current of 200 milliamperes is flowing through it?

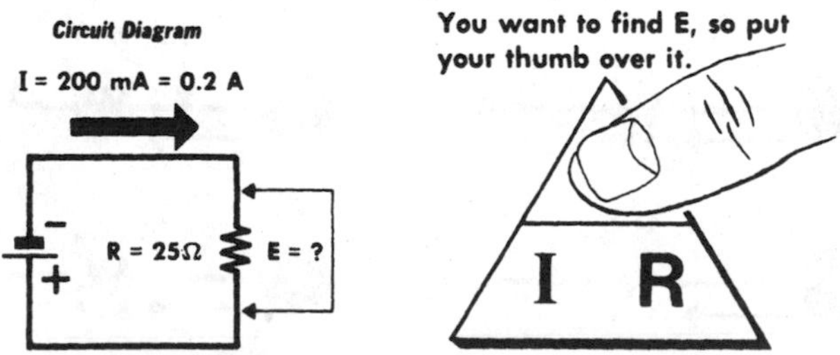

Solution. Sketch the circuit diagram. Draw the *magic triangle*. Convert milliamperes to amperes: $\frac{200}{1,000} = 0.2$ ampere. Then

$$E = IR = 0.2 \times 25 = 5 \text{ volts}$$

Ohm's Law Examples (continued)

Example 3

Problem. A voltage of 60 kilovolts is measured across a resistance of 12 megohms. What current is flowing?

Solution. Recall that a kilovolt is 1,000 volts, so 60 × 1,000 = 60,000 volts. Also, a megohm is a million ohms, so 12 × 1,000,000 = 12 million ohms.

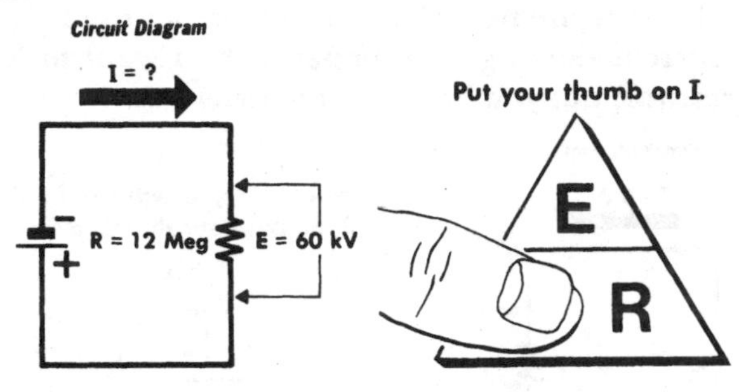

So $I = \frac{E}{R} = \frac{60{,}000}{12{,}000{,}000} = 0.005$ ampere $= 5$ mA.

(You see that despite the very large voltage applied, the current flowing is a very small one, thanks to the enormous resistance. A circuit with values such as this would seldom be set up in practice; but it is a good one for showing you that Ohm's Law works with *any* values of current, voltage, and resistance—*provided* that you use them correctly.)

Ohm's Law Drill

1. Solve the following:

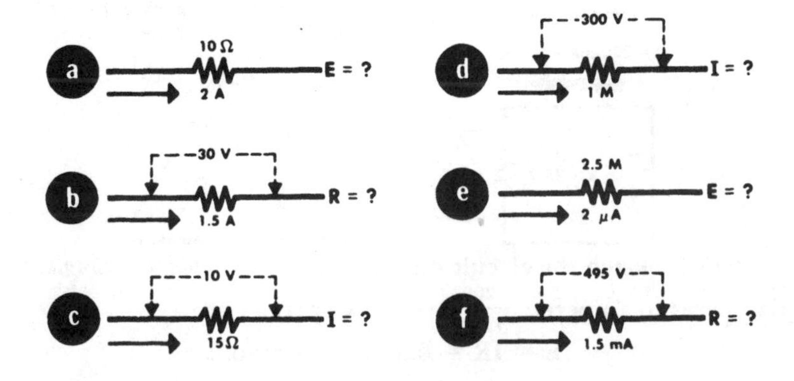

Ohm's Law Drill (continued)

2. You have determined from what is printed on its base, that a bulb from one of the headlights on your car is rated at 12 volts and 4 amperes. What is its resistance?

3. An electromagnet requires a current of 1.5 amperes to make it work properly, and you measured the resistance of its coil to be 24 ohms. What voltage must you apply to make it operate?

4. An electric soldering iron takes 2.5 amperes from a 240-volt supply when it is working. What is the resistance of its element?

5. What is the current through a 68-kilohm resistor when the voltage drop across the resistor is measured at 1.36 volts?

6. What resistance is needed to restrict a current driven by an emf of 10 volts to a flow of only 5 milliamperes?

(Answers to these questions are on page 2-149.)

A Valuable, Practical Tip

It so happens that when you work with practical electronic circuits later on, you will find *two patterns* of values for current, voltage, and resistance tending to recur rather frequently. The values are:

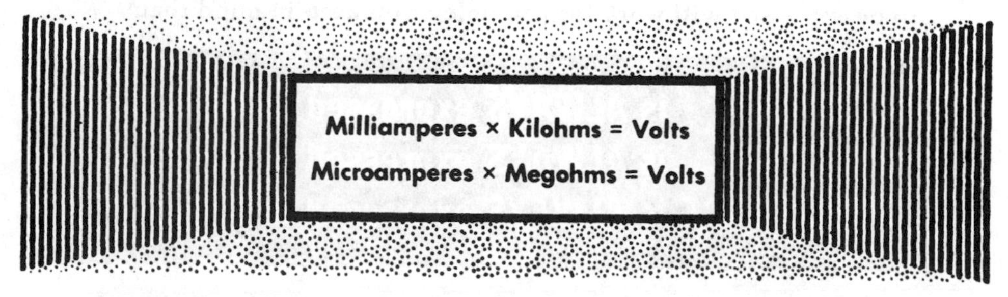

If you are going on to study the *Basic Electronics* series after you have finished your work in *Basic Electricity*, you will find it useful to memorize these two *relationships—milliamperes* with *kilohms* and *microamperes* with *megohms*—and the fact that they both multiply out into *volts*.

Review of Ohm's Law

Ohm's Law can be stated as a mathematical tool which is of the greatest use in determining an *unknown* factor of current, voltage, or resistance in an electric circuit in which the other two factors are *known*. It can, therefore, be used to take the place of an ammeter, voltmeter, or ohmmeter, respectively, when you are trying to resolve a circuit value in which you already know the two other values.

1. Ohm's Law can be stated in several ways. One of the most useful is this: *The current flowing in a circuit is directly proportional to the voltage applied to the circuit, and inversely proportional to the resistance of the circuit.*

2. This can be stated as an equation:

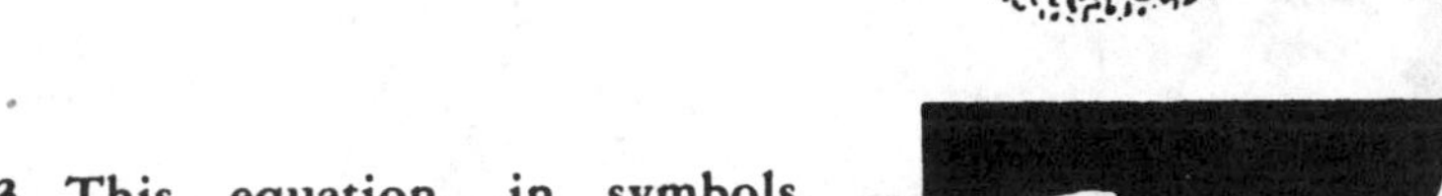

3. This equation, in symbols, reads:

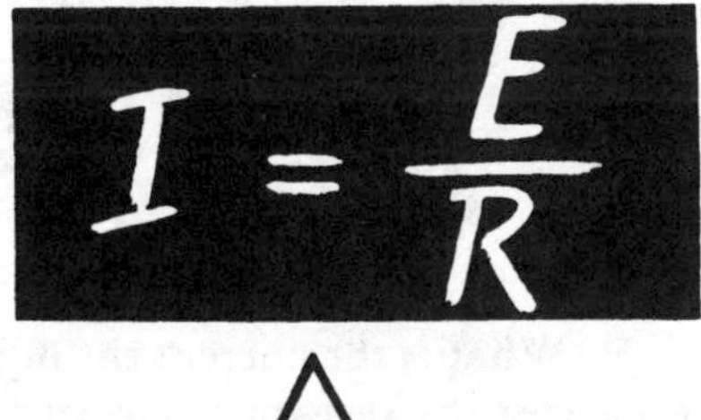

4. Remember the use of the *magic triangle* in helping you to decide the formula to use. (E on top, I and R below the line. Put your thumb on the quantity you do *not* know, and read off the formula for finding it.)

5. Remember that none of the forms in which the Ohm's Law formula can be expressed will work for you unless you keep in mind that:

CURRENT is ALWAYS expressed in AMPERES
VOLTAGE is ALWAYS expressed in VOLTS
RESISTANCE is ALWAYS expressed in OHMS

The Ohm's Law formula is a basic tool for all who work with electric circuits in any shape or form. And, like any tool, it becomes easier to use with practice; and the more often you use it, the more skilled you will find yourself becoming in its application.

Experiment/Application—Ohm's Law

To show how you can apply your understanding of Ohm's Law to find the resistance needed, suppose you connect four dry cells to form a 6-volt battery. Then, if you choose desired values of current such as 0.3, 0.6, and 1 ampere, you can determine the resistance, using Ohm's Law, which will give these currents when connected across the 6-volt battery. The voltage—6 volts—is divided by the desired currents—0.3, 0.6, and 1 ampere—giving required resistances of 20, 10, and 6 ohms, respectively. To check these values, connect two 3-ohm resistors in series to form a 6-ohm resistance, and connect it in series with an ammeter across the 6-volt battery. You will see that the resulting current is approximately 1 ampere. By adding more resistors in series to form 10- and 20-ohm resistances, you can show that these resistance values also result in the desired currents.

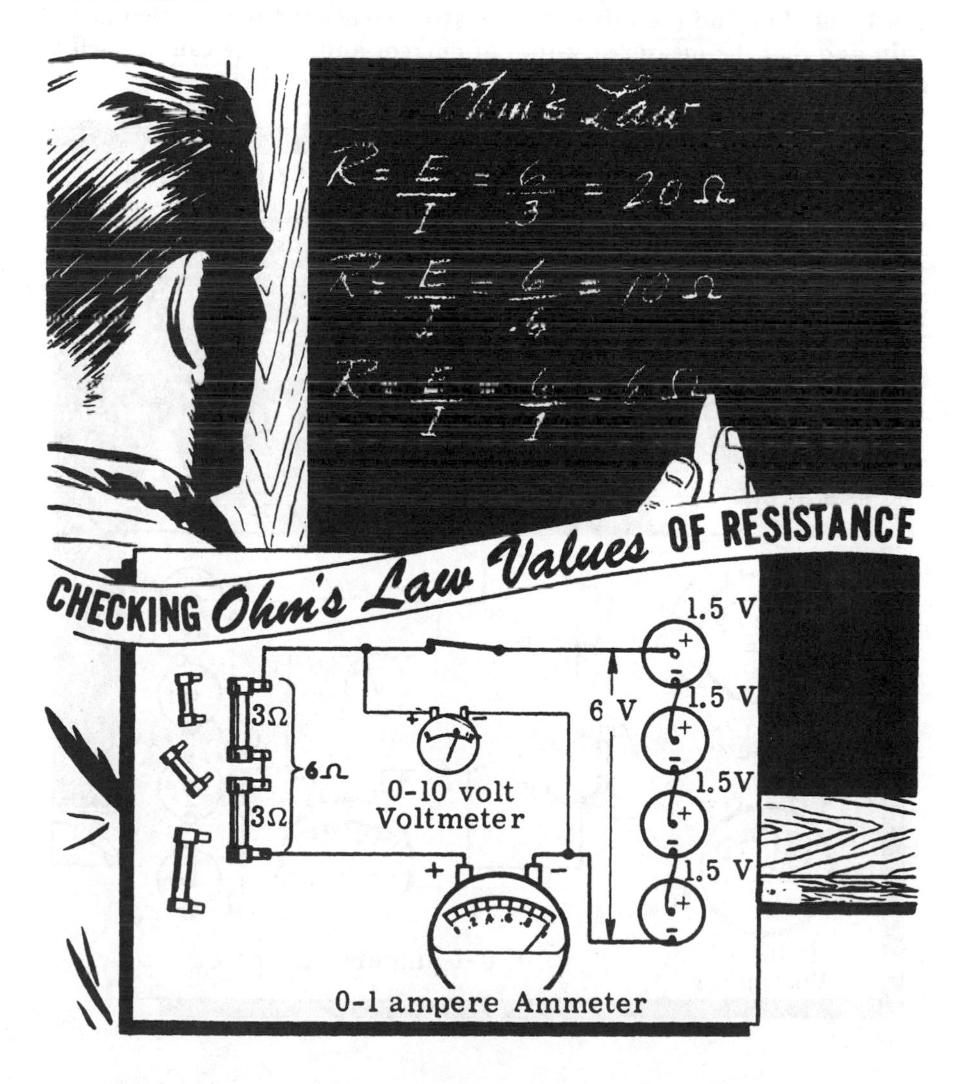

Experiment/Application—Ohm's Law (continued)

Current and voltage may also be used to find the value of a resistance in a circuit when the resistance is unknown. To see for yourself this use of Ohm's Law, suppose you connect two resistors, having no resistance marking, to form a series circuit across a 6-volt battery with an ammeter connected to measure the current flow. When the voltage across each resistor is read, you will see that these two voltages, when added, equal the battery voltage. By dividing the voltages across the resistors by the circuit current, you can obtain the resistance value of the resistors.

To show that your answers are correct, the resistances can be measured with an ohmmeter; it will be found that the values obtained by Ohm's Law equal those indicated on the ohmmeter. As several such problems are worked out, you will see that the rated current and voltage can be used to find the value of the resistance needed in a particular circuit, and that the measured values of current and voltage can be used to find the value of an unknown resistance in a particular circuit.

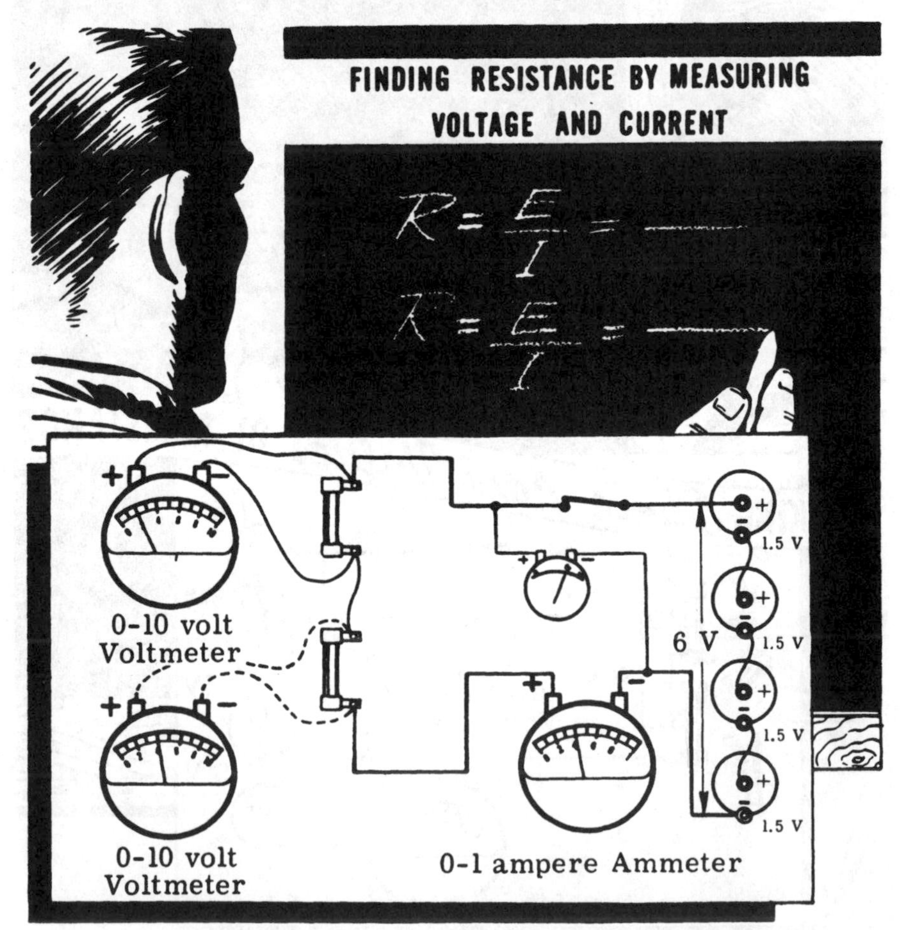

Experiment/Application—Ohm's Law (continued)

You will now see how you can use Ohm's Law to find the voltage required to give the correct current flow through a known resistance. Using a 10-ohm resistance consisting of two 2-ohm resistors and two 3-ohm resistors in series, you can determine the voltages needed to obtain 0.3, 0.6, and 0.9 ampere of current flow by multiplying 10 ohms by each current in turn. The voltage values obtained are 3, 6, and 9 volts, respectively.

To check these values, the 10-ohm resistance can be connected in series with an ammeter across cells connected to give these voltages. With the 3-volt battery of cells, you will see that the current is 0.3 ampere; with the 6-volt battery, it is 0.6 ampere; and with the 9-volt battery, it is 0.9 ampere—showing that the Ohm's Law values are correct.

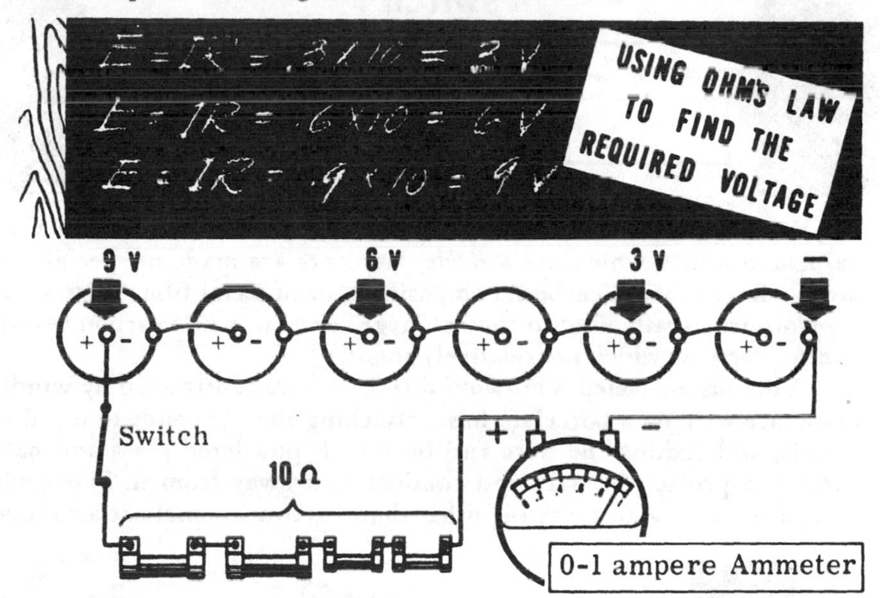

To confirm that Ohm's Law can be used to find the current in a circuit, you can connect six 3-ohm resistors in series across the terminals of a 9-volt battery of dry cells. Using Ohm's Law, determine the circuit current: $I = E/R = 9/18 = 0.5$ ampere. Now break the circuit and insert a 1.0-ampere ammeter in series with the resistors. You will see that the meter reading is 0.5 ampere—exactly the value of current determined by your Ohm's Law calculation.

Learning Objectives—Next Section

Overview—Now that you have learned Ohm's Law, and some of its applications, you can learn about the construction of resistors, some of their other properties, and how resistance is measured.

Resistors—Use, Construction, and Properties

There is a certain amount of resistance in all of the electrical equipment which you use. However, sometimes this resistance is not enough to control the flow of current to the extent desired. When *additional* control is required—for example, when starting a motor—resistance is purposely added to that of the equipment. In the circuit shown, a switch and a current-limiting resistor are used to control the flow of current through the motor. When starting the motor, the switch is kept open and the *resistance* is thereby *added* into the circuit to control the flow of current. After the motor has started, the switch is then closed in order to *bypass* the current-limiting resistor. Before you continue your study of circuits, you need to know more about resistance and resistors.

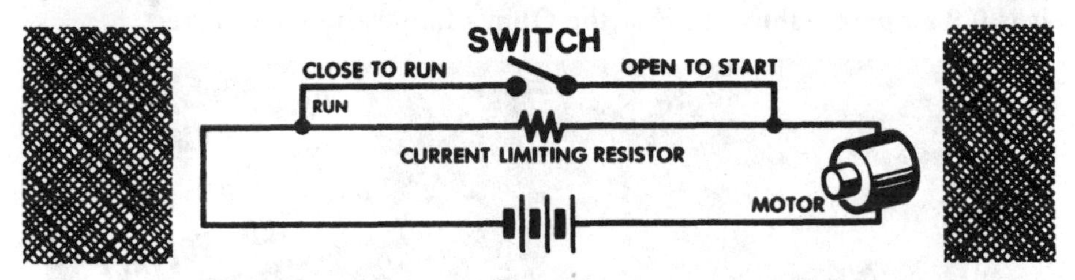

You will use a wide variety of resistors, some of which have a *fixed* value and others which are *variable.* Resistors are made of special resistance wire, graphite (carbon) composition, or of metal film. Wire-wound resistors are usually used to control large currents, while carbon resistors control currents which are relatively small.

Vitreous enameled wire-wound resistors are constructed by winding resistance wire on a porcelain base, attaching the wire ends to metal terminals, and coating the wire and base with powdered glass and baked enamel to protect the wire and conduct heat away from it. Fixed wire-wound resistors with a coating other than vitreous enamel are also used.

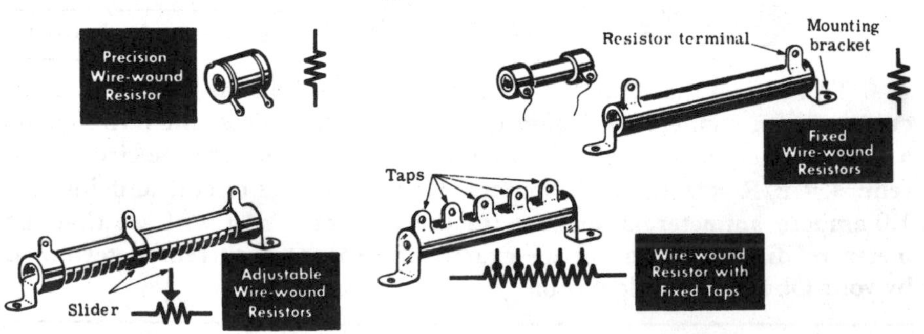

Wire-wound resistors may have fixed taps, which can be used to change the resistance value in steps, or sliders, which can be adjusted to change the resistance of any fraction of the total resistance.

Precision wound resistors of Manganin wire (a special wire that does not change resistance very much with temperature) are used where the resistance value must be very accurate, such as in test instruments.

Resistors—Use, Construction, and Properties (continued)

Generally, carbon resistors are used for low current applications. They are made from a rod of compressed graphite (carbon) that is mixed with clay and binders. By varying the amount of each component, it is possible to vary the resistance values obtained over a very wide range. Two lead wires called *pigtails* are attached to the end of the resistance rod, and the rod is embedded in a ceramic or plastic covering, leaving the pigtails protruding from the ends.

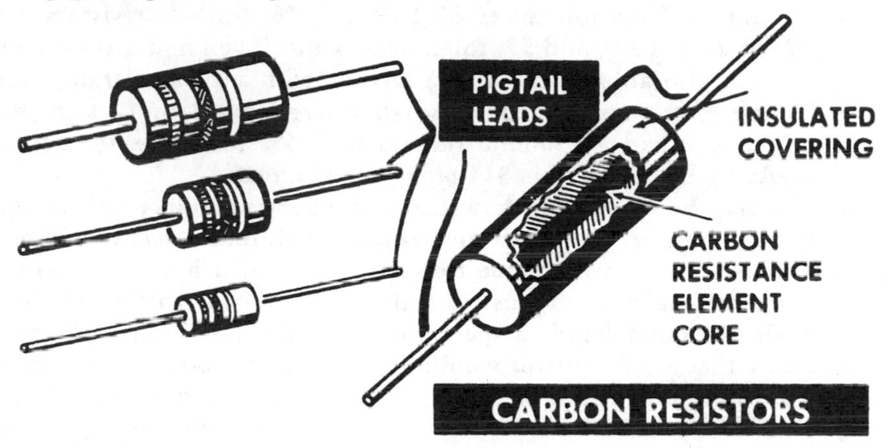

CARBON RESISTORS

Occasionally, you will find a type of resistor called a *deposited film resistor* used for special applications. These resistors are made by depositing a thin film of resistance metal or carbon on a ceramic core and then coating the resistor with either a ceramic or enamel protective coating. In many cases, you will find that these resistors have radial leads; that is, the leads come off at right angles to the body of the resistor. Also, in some cases the deposited film is laid down on the core as a spiral, similar to winding a wire around the tube, in order to increase the length of the resistance element without making the resistor too long.

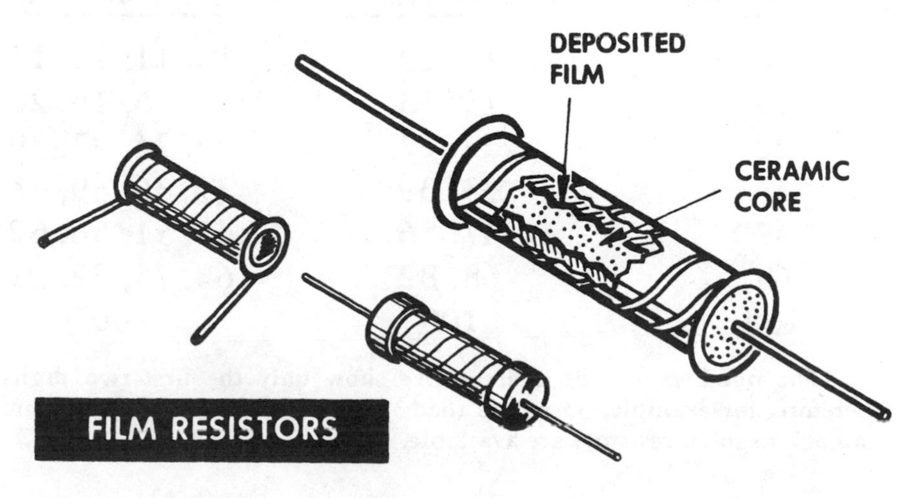

FILM RESISTORS

Resistor Tolerance and Values

Before you go on to the Color Code for resistors, you will need to find out something about resistor tolerances and something about the preferred values of resistance that you will find in circuits. It is very difficult to make a resistor to an *exact* value. Fortunately, in most cases, an approximate value of resistance will do very well. While special resistors may have tolerances of as little as 1%, 0.1%, or even 0.01%, most resistors that you will see have much greater tolerances. Large wire-wound resistors usually have tolerances of 10% or 5%. Carbon resistors are available in 20%, 10%, and 5% tolerances. Thus, if you had a 10-kilohm (10,000-ohm, also abbreviated 10-K) resistor with a 20% tolerance, the actual value of the resistor could be anywhere from 8 to 12 kilohms. Similarly, if you had a 330-ohm resistor with 5% tolerance, the actual value could be anywhere from 314 ohms to 347 ohms.

You may be wondering how many different resistance values one can get for a resistor. As it turns out, this depends on the *tolerance*. Since a 10-K resistor can have a value from 8 to 12 K if it has a 20% tolerance, it doesn't make much sense to make a 9-K resistor with a 20% tolerance. On the other hand, if the tolerance of the 10-K resistor is 5%, you can see that a 9-K resistor would not overlap the tolerance values of a 10-K resistor, and would be useful if such tight tolerances were necessary. Considerations such as this have led to the establishment of a set of preferred values of resistance in each tolerance where the highest tolerance of one value is about equal to the lowest tolerance of the next highest value. These preferred resistance values are shown in the table below. Later, when you learn about power ratings, you will find that resistors are available in different power ratings as well.

PREFERRED CARBON RESISTOR VALUES

20% Tolerance	*10% Tolerance*	*5% Tolerance*
10	10, 12	10, 11, 12, 13
15	15, 18	15, 16, 18, 20
22	22, 27	22, 24, 27, 30
33	33, 39	33, 36, 39, 43
47	47, 56	47, 51, 56, 62
68	68, 82	68, 75, 82, 91
100	100	100

The numbers on the chart above show only the first two digits; therefore, for example, 33 means that 3.3, 330, 3.3-kilohm, 330-kilohm, and 3.3-megohm resistors are available.

Resistor Color Code

You can find the resistance value of any resistor by using an ohmmeter; but in some cases, it is easier to find the value of a resistor by its marking. Most wire-wound resistors have the resistance value printed in ohms on the body of the resistor. If they are not marked in this manner, you must use an ohmmeter. Precision wire-wound resistors usually have all of the data printed directly on the resistor body, often including such information as tolerance, temperature characteristics, and exact resistance value. Carbon resistors usually do not have the data on their characteristics marked directly on them; instead they have a *color code* by which they can be identified. The reason for this is that some carbon resistors are so small that the written data would be impossible to read. In addition, carbon resistors are often mounted so that it would be very difficult to read printed values.

Carbon resistors are of two types, radial and axial, which differ only in the way in which the wire leads are connected to the body of the resistor. Both employ the same color code, but the colors are painted in a different manner on each type. Radial-lead resistors are not found in modern equipment, although they were widely used in the past.

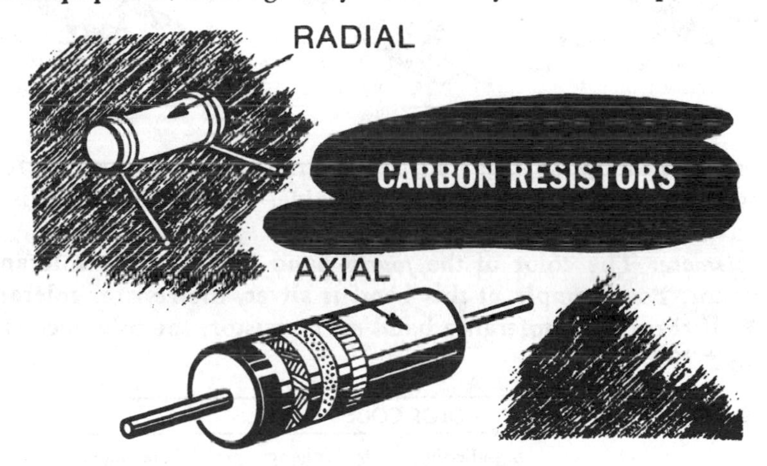

Axial-lead resistors are made with the leads molded into the ends of the carbon rod of the resistor body. The leads extend straight out from the ends and in line with the body of the resistor. The carbon rod is completely coated with a material which is a good insulator.

In the color code system of marking, three colors are used to indicate the resistance value in ohms, and a fourth color is sometimes used to indicate the tolerance of the resistor. By reading the colors in the correct order and by substituting numbers from the color code, you can immediately tell all you need to know about a resistor. As you practice using the color code shown on the next page, you will soon get to know the numerical value of each color, and you will be able to tell the value of a resistor at a glance.

Resistor Color Code (continued)

First Significant Figure: On the resistor, the color of the *first* band indicates the *first* digit of the resistance value. For example, as shown in the Color Code Table below, if this band is brown, the first digit is 1.

Multiplying Value: The color of the *third* band indicates the value by which the first two digits are to be multiplied to obtain the resistance value. For example, again using the Color Code Table, if this band is yellow, the first two digits are multiplied by 10,000. (Therefore, with first and second significant digits of 15, the value is 150,000.) This band can also be thought of as indicating the number of zeros to be added after the second digit. When used this way, the number of zeros shown in the Significant Figures column of the Color Code Table is the number of zeros to add. For example, if this band is blue, add six (6) zeros after the *second* digit; but if the band is black, no zeros are added. If the third band is gold or silver, the multiplying value must still be used.

Examples of the use of this table are as follows:

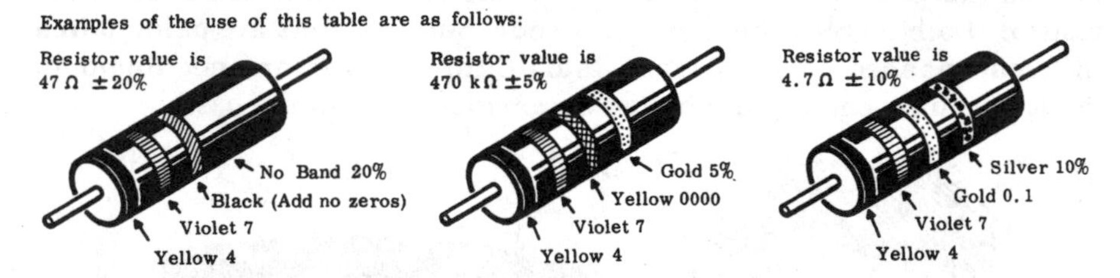

Second Significant Figure: The color of the *second* band on the resistor indicates the *second* digit of resistance value. For example, if this band is green, the second digit is 5.

Tolerance: The color of the *fourth* band indicates the tolerance of the resistor. For example, if this band is silver, the resistor tolerance is ± 10%. If there is *no* tolerance band on a resistor, the tolerance is *automatically* ± 20%.

COLOR CODE TABLE

COLOR	Significant Figures	Multiplying Value	Tolerance —
Black	0	1	—
Brown	1	10	—
Red	2	100	—
Orange	3	1000	—
Yellow	4	10000	—
Green	5	100000	—
Blue	6	1000000	—
Violet	7	10000000	—
Gray	8	100000000	—
White	9	1000000000	—
Gold	—	0.1	± 5%
Silver	—	0.01	±10%
No Band	—	—	±20%

How Resistance Is Measured

Voltmeters and ammeters are meters you are familiar with and may have used to measure voltage and current. Meters used to measure resistance are called *ohmmeters.* These meters differ from ammeters and voltmeters particularly in that the scale divisions are not *equally* spaced, and the meter requires a built-in battery for proper operation. When using the ohmmeter, no voltage should be present across the resistance being measured except that of the ohmmeter battery; otherwise, the ohmmeter will be damaged.

Ohmmeter ranges usually vary from 0-1,000 ohms to 0-10 megohms. The accuracy of the meter readings decreases at the maximum end of each scale, particularly for the megohm ranges, because the scale divisions become so closely spaced that an accurate reading cannot be obtained. Unlike other meters, the zero end of the ohmmeter scale is at full-scale deflection of the meter pointer.

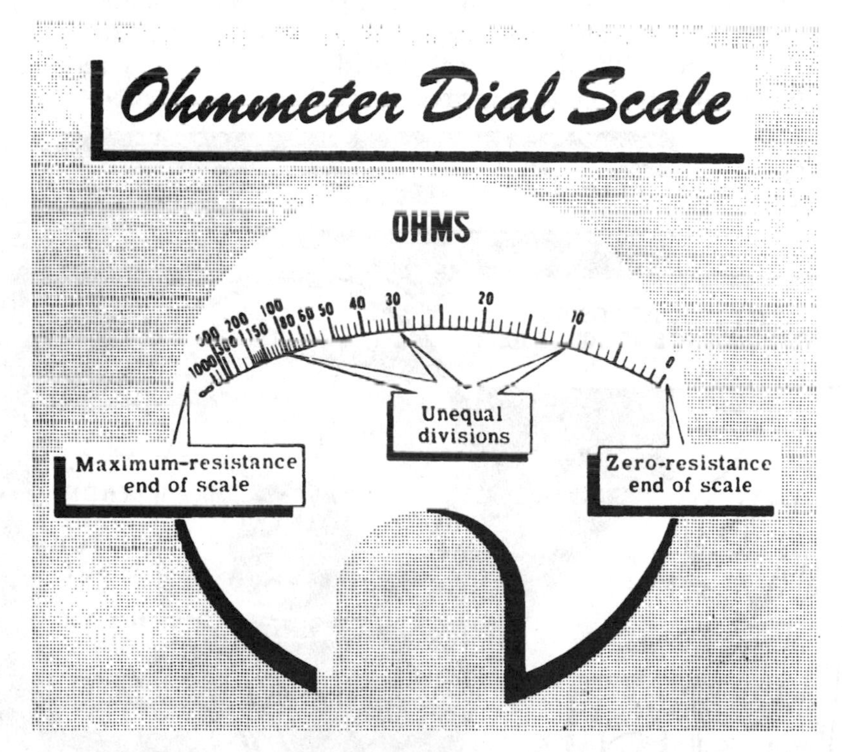

Special ohmmeters called MEGGERS® are required to measure values of resistance over 10 megohms, since the built-in voltage required is very high for ranges above 10 megohms. Some MEGGERS® use high-voltage batteries and others use a special type of hand generator to obtain the necessary voltage. While ohmmeters are used to measure the resistance values of resistors, the most important use of these megohmmeters is to measure and test the resistance of insulation.

How Resistance Is Measured (continued)

You will learn more about the way the ohmmeter works in detail later, when you have learned more about voltage, current flow, and resistance—and their relationship to each other. For now, we will concentrate on how the ohmmeter is used to measure resistance. The principle of the ohmmeter is quite simple. The current through the unknown resistor is measured under conditions where a *known* voltage is applied across the *unknown* resistor. If you have had an opportunity to see an ohmmeter, you may have noticed that in addition to the meter calibrated in ohms, there is a zero adjustment control and a range selector switch. The range selector switch is marked R, R × 10, R × 100, R × 1,000 (or R×1 K), etc. The function of the range selector switch is very much like that of the selector switch on the multirange voltmeters and ammeters that we studied earlier; that is, it allows the selection of the range that gives a reading on the useful part of the scale. It differs, however, in that the position of the range selector switch gives a *multiplying* factor to the values read on the ohmmeter scale. For example, if the range selector switch is on R × 100, then the value read on the meter is multiplied by 100 to get the actual value of the resistance that is being tested.

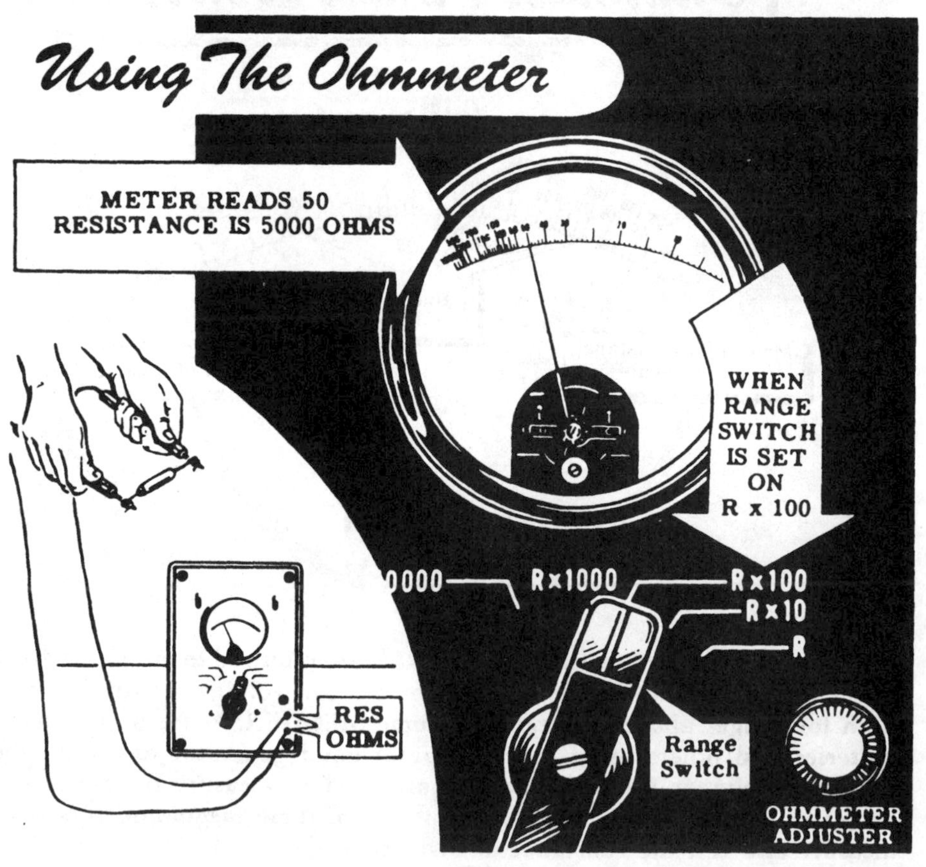

How Resistance Is Measured (continued)

Using the ohmmeter is very simple and proceeds in two steps. First, the voltage must be set to the proper value. This is done with the zero adjustment by shorting out (connecting together) the leads from the ohmmeter and setting to zero ohms on the meter with the zero adjustment. This must be done whenever the meter range selector switch is changed to a different scale. The meter is now calibrated for the given range scale since, with the leads shorted out, the meter reads zero ohms (no resistance between the test leads); and with the test leads open, the meter will indicate infinity (or open circuit). When the unknown resistance is connected between the test leads, the resistance can be read directly from the meter and multiplied by the multiplying factor from the range selector switch.

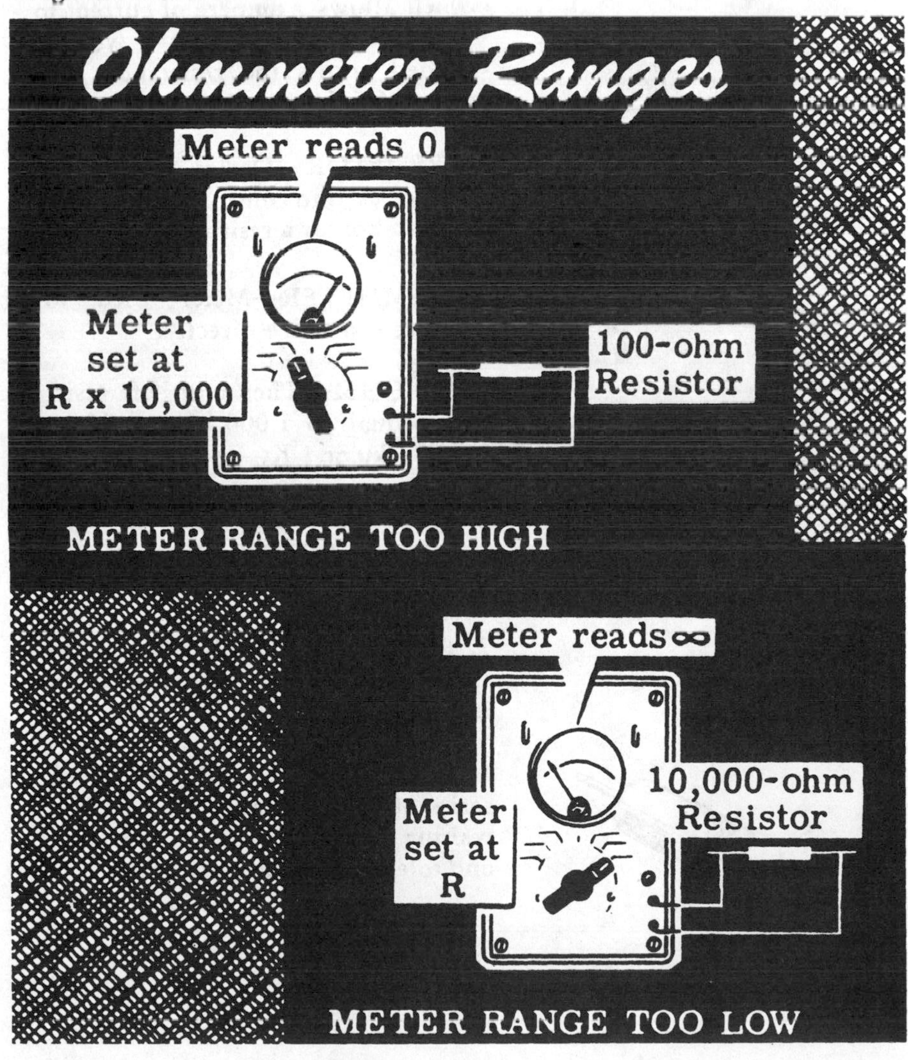

Review of Resistance (Including Material from Volume 1)

You have now learned about the fundamentals of voltage, current, and resistance; and you are ready to go on and see how electric circuits work. Before we go on, let's briefly review what you have learned about resistance and how it is measured.

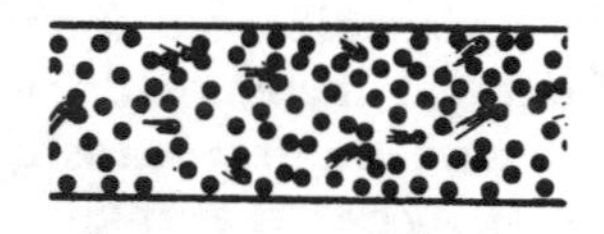

1. RESISTANCE—Opposition offered by a material to the flow of current.

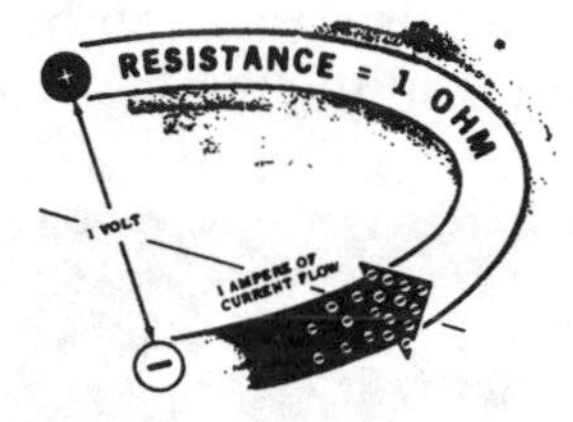

2. OHM—Basic unit of resistance measure equal to that resistance which allows 1 ampere of current to flow when an emf of 1 volt is applied across the resistance. The symbol for the ohm is Ω.

3. RESISTOR—Device having resistance used to control current flow. The symbol for a resistor is "R."

4. OHMMETER—Meter used to measure resistance directly.

1 K = 1,000 ohms

5. KILOHM—The unit of resistance equal to 1,000 ohms, abbreviated 1 kΩ or 1 K.

1 M = 1,000,000 ohms

6. MEGOHM—The unit of resistance equal to 1,000,000 ohms, abbreviated 1 MΩ or 1 M or 1 meg.

7. RESISTOR TOLERANCE—The spread in value of a resistor about its designated value.

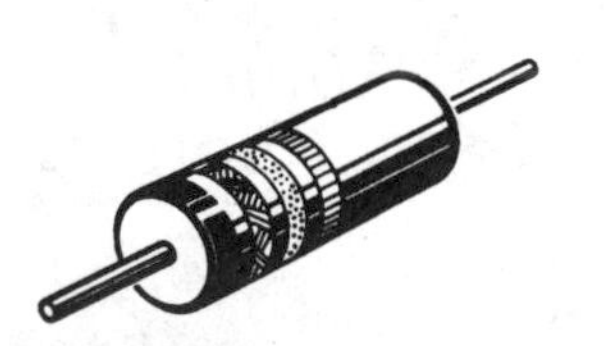

8. RESISTOR COLOR CODE—A series of colored bands on a carbon resistor that tells what the resistance and tolerance of the resistor are.

Self-Test—Review Questions (Including Material from Volume 1)

1. Define what resistance is. What is a resistor? What is the symbol used to designate a *fixed* resistor? A *variable* resistor?
2. In a circuit with *constant voltage*, what happens to the current when the resistance is doubled? Halved? Tripled?
3. In a circuit with *constant resistance*, what happens to the current when the voltage is doubled? Halved? Quadrupled? Tripled?
4. Define the unit of resistance. What symbol is used to designate it?
5. What factors determine the resistance of a resistor? Give examples of their effect.
6. Calculate the following conversions using appropriate symbols where applicable.

Convert to Ohms	*Convert to Kilohms*	*Convert to Megohms*
6.2 kilohms	4,700 ohms	1,000 kilohms
6.2 megohms	8.2 megohms	120,000 ohms
270 milliohms	100,000 ohms	92,000 ohms
3.3 kilohms	0.1 megohm	68 kilohms
9.1 kilohms	0.39 megohm	470,000 ohms
4.7 megohms	24,000 ohms	330 kilohms

7. What are the preferred values of resistance in 20% tolerance?
8. What is the device used to measure resistance? Describe very briefly how it is used.
9. What are the values of resistance indicated by the following color codes?

Band 1	*Band 2*	*Band 3*	*Band 4*	*Resistance Value*
red	red	red	gold	
white	brown	yellow	gold	
brown	black	orange	none	
brown	black	black	silver	
violet	green	blue	gold	
grey	red	brown	silver	

10. Describe the steps involved in calibrating and using an ohmmeter.

Learning Objectives—Next Section

Overview—You are now prepared to start using Ohm's Law seriously to solve for current, voltage and/or resistance. In the next section you will start by learning how to solve the simple series circuit.

The Series Circuit

A *series* circuit is formed when two or more resistors are connected *end-to-end* in a circuit in such a way that there is *only one path* for current to flow.

You already know how to connect cells in series so as to form a battery. Connecting resistors in series so as to form a series circuit is even easier. Resistors (unlike cells) have *no* polarity, so you do not have to worry about not connecting two positive or two negative terminals to one another.

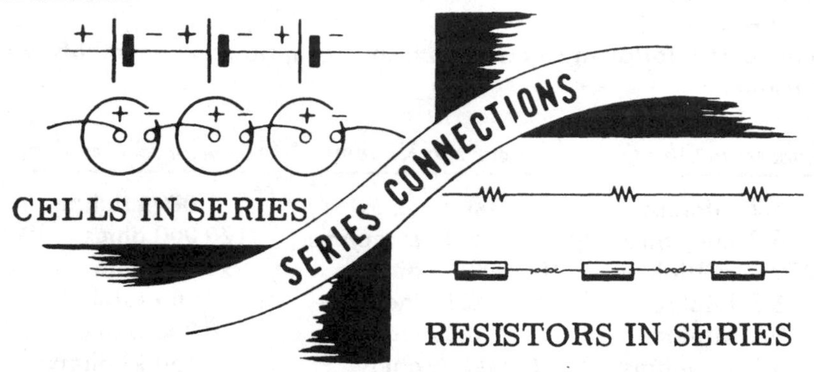

Note that if you connect a terminal of one lamp socket to a terminal on another socket, leaving one terminal on each socket unconnected, lamps placed in these sockets would be *series-connected*—but you would not have a series circuit. To complete the series circuit, you would have to connect the lamps *across a voltage source*, such as a battery, using the unconnected terminals to complete the circuit.

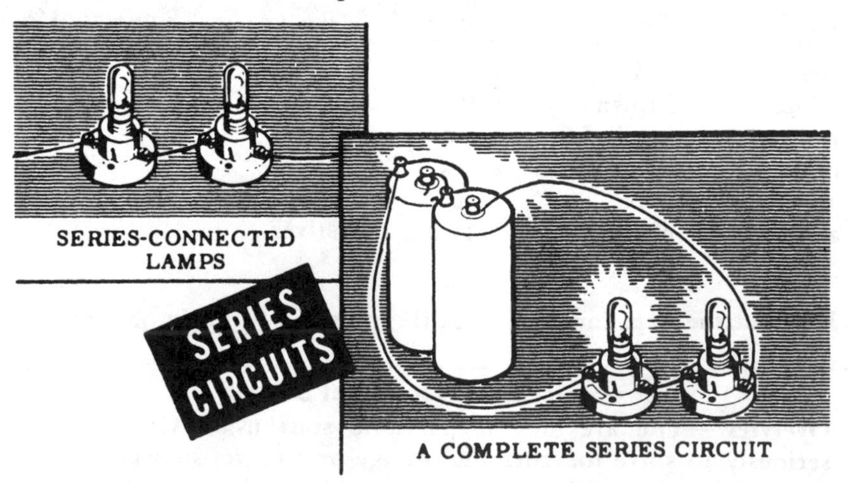

Any number of lamps, resistors, or other devices having resistance can be used to form a series circuit, provided they are connected end-to-end across the terminals of a voltage source and offer only one path for current flow between these terminals.

Resistance in Series Circuits

The important thing to remember about resistances connected in series is that *their values add.*

You already know that the resistance of a conductor increases as the length of the conductor increases. It is easy to see, then, that if you connect one length of wire to another, the resistance of the full length of wire will be equal to the sum of the resistances of the original lengths.

For example, if two lengths of wire—one having a resistance of 4 ohms and the other a resistance of 5 ohms—are connected together, the total resistance between the unconnected ends is 9 ohms. Similarly, when other types of resistances are connected in series, the total resistance always equals the sum of the individual resistances.

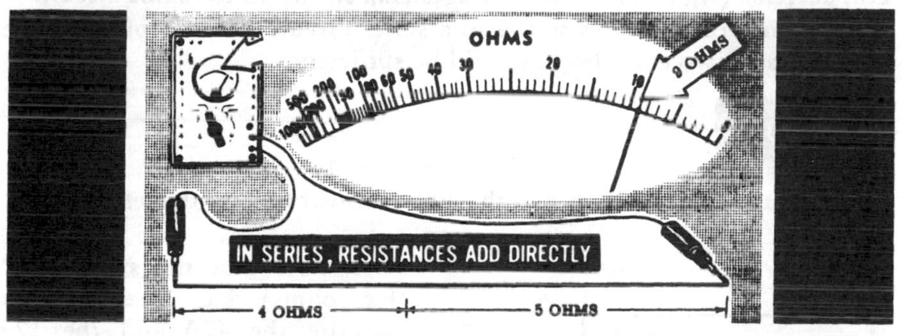

Whenever you use more than one of the same device or quantity in an electric circuit, some method of identifying each individual device or quantity is necessary. For example, if three resistors of different values are used in a series circuit, something other than just R is needed to distinguish them from each other.

To meet this need, a system of numerical identification is used. It consists of following the symbol of the device or quantity by an identification number (or reference). These numbers are sometimes called *subscripts*, because you will occasionally find them written somewhat smaller and slightly offset (below the line), particularly in older drawings. In most modern circuit diagrams, these numbers are written on the line; thus R1 is the same as R_1. R1, R2, and R3 are all symbols for resistors, but each identifies only one *particular* resistor. Similarly, E1, E2, and E3 are all different reference designations for values of voltage used in the same circuit, with the number identifying the particular voltage referred to.

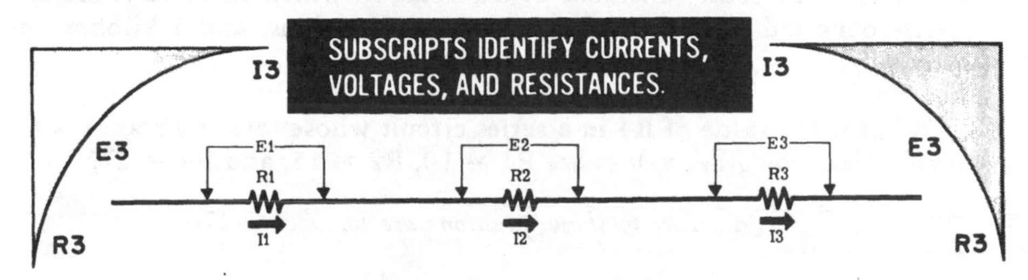

Resistance in Series Circuits (continued)

A small subscript letter *t* is often used to indicate the *total* resistance (R_t) of a series circuit in which two or more individual resistors are included. You know that the total resistance of a series circuit is equal to the sum of the individual resistances in the circuit. In other words,

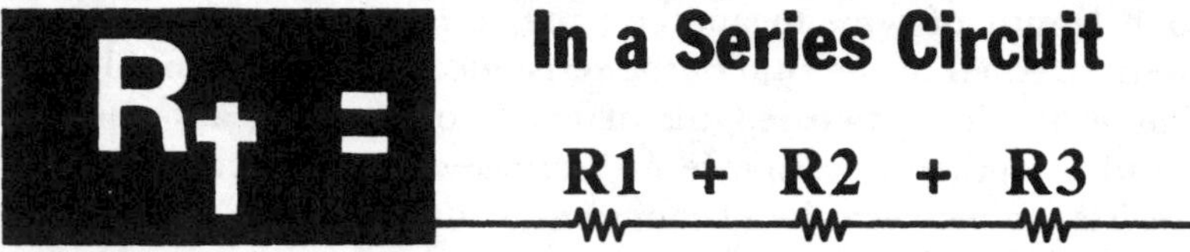

You will find that the subscript t is also used to express the total of several currents (I_t), or of several voltages (E_t), shown in the same circuit.

Finding the total resistance in a series circuit is so simple that one example, and two trial drills, should be sufficient.

Example

Find the total resistance of the circuit illustrated on the left, below.

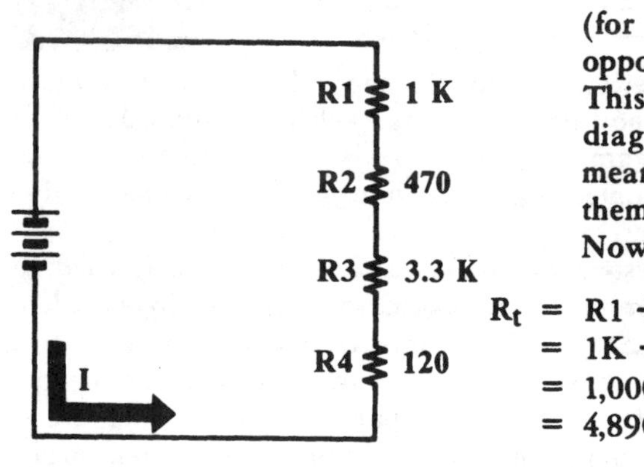

First, note that the symbol "Ω" (for ohms) has been omitted opposite the 470 and the 120. This is often done in circuit diagrams to save space, when the meaning of the figures by themselves is obvious.

Now write your formula:

$$\begin{aligned} R_t &= R1 + R2 + R3 + R4 \\ &= 1K + 470 + 3.3K + 120 \\ &= 1{,}000 + 470 + 3{,}300 + 120 \text{ ohms} \\ &= 4{,}890 \text{ ohms} \end{aligned}$$

Drill

1. What is the total resistance of a circuit in which three resistors are series-connected with values of 220 ohms, 680 ohms, and 1 kilohm, respectively?

2. What is the value of R4 in a series circuit whose total resistance is 67 ohms, where the other values are R1 = 10, R2 = 15, and R3 = 27?

(Answers to these questions are on page 2-150.)

Current Flow in Series Circuits

You already know that in a series circuit there is only *one* path for current flow, and this means that *all* the current must flow through *every* component in the circuit. If you connect an ammeter onto either end of every resistor in a series circuit, it will show that *an identical amount of current is flowing through every component in the circuit.*

This will not surprise you if you remember how an electric cirrent resembles the electron streaming orbit revolving through the completed circuit, but it is the central fact to grasp before you start applying the principles of Ohm's Law to a series circuit.

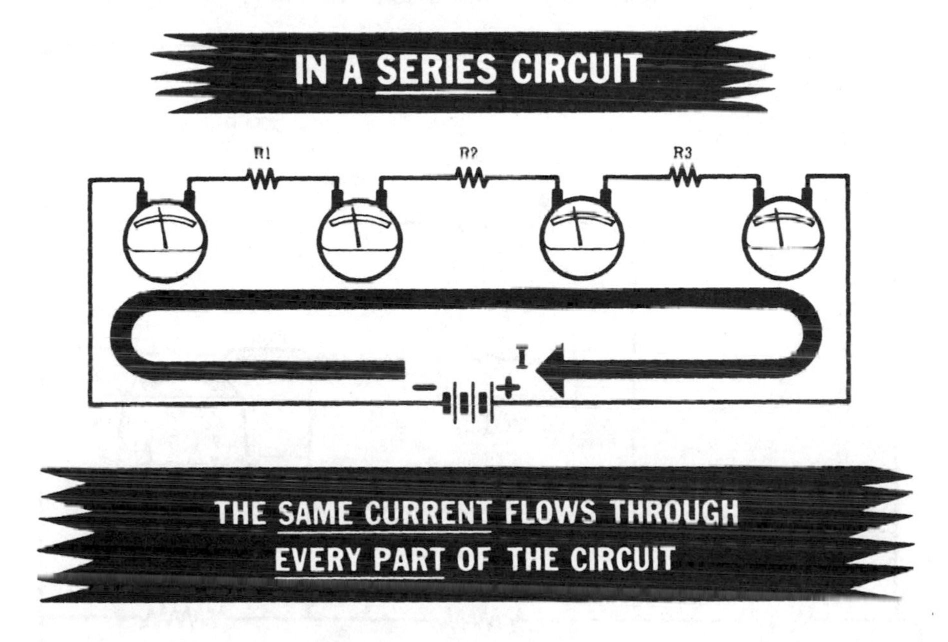

IN A SERIES CIRCUIT
R3
I
THE SAME CURRENT FLOWS THROUGH
EVERY PART OF THE CIRCUIT

Note, in passing, an important practical consequence of the principle illustrated above. The fact that *all* the circuit current must pass through *every* part of the series circuit means that *every component* connected into such a circuit must be capable of passing, without being damaged, the current which will flow through the circuit.

Lamps connected in series, for example, must all be rated to pass the full circuit current. When rated too low, they will light very brightly, and they will also be liable to burn out because of the excessive current flowing through them.

It is important to remember that *exactly the same thing can happen* if the circuit, instead of lamps, contains resistors, whose shorting out could be even more inconvenient or costly. A resistor required to pass more than its rated current will get extremely hot. Eventually, it will fail completely and become an open circuit. The equipment of which it forms a part will then probably cease to function, and trouble is at hand!

Voltages in Series Circuits—Kirchhoff's Second Law

You know that when a voltage moves electrons through a resistance, some of the available emf is used up. Such a loss of emf is called a *potential drop* or a *voltage drop* across that resistance. You will now find out how this voltage drop is distributed between several resistors of equal value connected in series.

Connect three resistors of equal value in series across a 6-volt battery, and touch voltmeters across the points as shown in the diagram below. Since the current passing through each of the equal resistors is the same, the energy expended in pushing this equal amount of current through each individual resistor must also be the same.

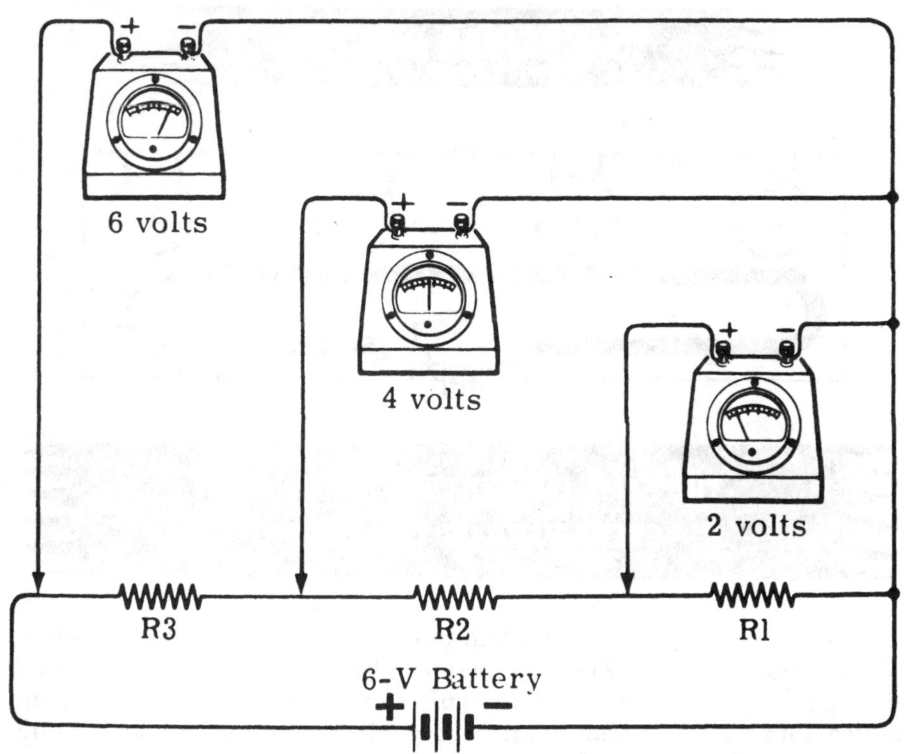

In other words, the voltage across each of the three resistors pictured above is 2 volts. The voltage drop across R1 will therefore read on your voltmeter as 2 volts; that across R1 and R2 in combination will read 4 volts; and that across R1, R2, and R3 (the complete circuit) will read 6 volts. If you add together the voltage drops across all three resistors, you will get exactly the original supply voltage (6 volts).

This important fact was expressed by the German physicist Kirchhoff (1824-1887) in what is known as Kirchhoff's Second Law. (We will be discussing his First Law shortly.)

Kirchhoff's Second Law states: *The sum of the voltage drops across the resistances of a closed circuit equals the total voltage applied to the circuit.*

Ohm's Law in Series Circuits

You now know three important facts about a series circuit:

1. The current flowing through it is the same everywhere. This can be expressed by the equation $I_t = I1 = I2 = I3$, and so on.
2. The total resistance of the circuit equals the sum of the individual resistances in it. This can be expressed by the equation $R_t = R1 + R2 + R3$, and so on.
3. When the voltage drops in a series circuit are added together, their total value is equal to the total applied voltage (Kirchhoff.) This can be expressed by the equation $E_t = E1 + E2 + E3$, and so on.

These three facts, used in conjunction with Ohm's Law, will be of constant help to you in determining the values of complete circuits, or parts of circuits, on the frequent occasions when you will either lack the correct meter to tell you the answer directly, or find it impossible to use a meter to obtain a value directly when constructing a circuit.

This will happen to you time and time again when you get on to the really interesting applications of electrical and electronic principles.

One useful simplification arises from the equation, $R_t = R1 + R2 + R3$, etc. Look at the two circuits below, and you will quickly see that the right-hand equation is a more convenient *equivalent* version of the one on the left.

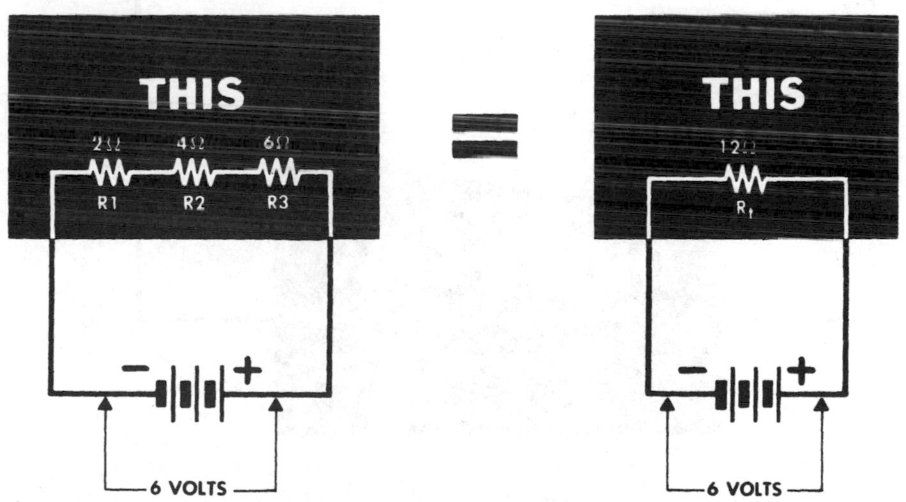

You can soon find the missing factor—the current through either circuit—by using Ohm's Law and the known facts that E=6 and R=12. The equation you want is (thumb on the I of the magic triangle!)

$$I = \frac{E}{R} = \frac{6}{12} = 0.5 \text{ ampere}$$

Always be on the lookout for a chance to simplify the resistances in a series circuit into the single resistance of an *equivalent circuit.*

Ohm's Law in Series Circuits (continued)

Ohm's Law can be usefully applied in series circuits either to the complete circuit itself or to individual parts of the circuit. Together with Kirchhoff's Second Law, for instance, it will enable you, by calculation, to insert a great many missing values in a series circuit of the following kind:

A circuit contains three resistors connected in series across 100 volts; the circuit current flow is 2 amperes. Two of the resistors (call them R1 and R2) have known values of 5 and 10 ohms, respectively. You wish to know the resistance of the entire circuit, the value of the third resistor, R3, and the voltage drops across each of the three resistors.

First, sketch the circuit on paper. Fill in the values you already know, leaving blanks opposite the ones you don't know.

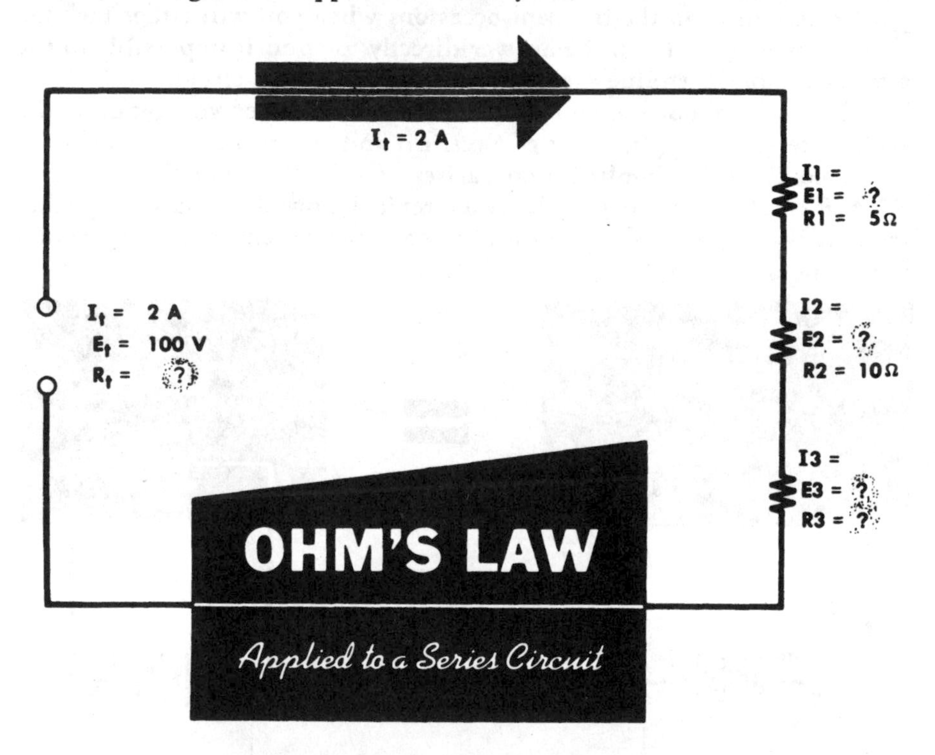

Now, for the missing values. The first and most obvious one is the current. You know that $I_t = I1 = I2 = I3. \ldots$ You are quite safe in filling in the value of 2 amperes for I wherever it appears on your sketch.

Next, take the group of figures on the left of your sketch. Finding the value of R_t is now a simple exercise with Ohm's Law. With your thumb *mentally* on R in the magic triangle, you get

$$R_t = \frac{E_t}{I_t} = \frac{100}{2} = 50 \text{ ohms}$$

Ohm's Law in Series Circuits (continued)

Now that you know the total circuit resistance is 50 ohms, you can easily fill in the missing link in the equation:

$$R_t\ (50\Omega) = R1\ (5\Omega) + R2\ (10\Omega) + R3\ (?)$$

Rewrite the equation as $R3 = R_t - R1 - R2$, and you get

$$R3 = 50 - 5 - 10 = 35 \text{ ohms}$$

Fill in this value on your sketch (see previous page) and you are clearly making progress.

Now, consider what you know about the resistor, R1. You know its value to be 5 ohms, and that the current through it is 2 amperes. Ohm's Law will give you the voltage drop across it. So *Thumb on E!* and

$$E1 = I1 \times R1 = 2 \times 5 = 10 \text{ volts}$$

Do the same thing for E2, and you get

$$E2 = I2 \times R2 = 2 \times 10 = 20 \text{ volts}$$

Fill both these values in on your sketch, and the only gap left in the entire list of circuit values is E3.

This last value can be calculated in two different ways. Use both methods, for they not only provide a useful check on one another, but will also prove that Ohm's Law and Kirchhoff's Second Law work accurately!

First, Ohm's Law tells you that

$$E3 = I3 \times R3 = 2 \times 35 = 70 \text{ volts}$$

Then, Kirchhoff's Second Law tells you that $E_t = E1 + E2 + E3$. Transposing, you get

$$E3 = E_t - E1 - E2 = 100 - 10 - 20 = 70 \text{ volts}$$

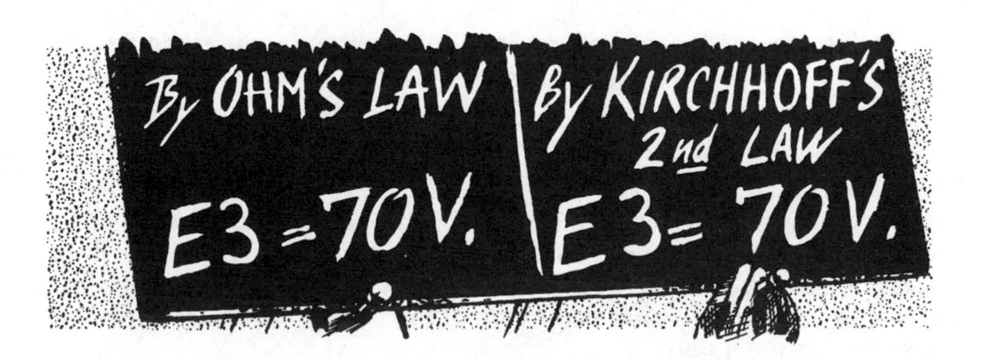

Ohm's Law in Series Circuits (continued)

Practice makes perfect—and it is so important that you grasp the use of Ohm's Law in solving series circuits that you should work through the exercise which follows (it is very similar to the one you did on the last two pages) and then tackle a page of problems like it on your own.

Example

In a circuit of which the relevant part is shown below, you measure a voltage drop of 5 volts across R1; but you cannot get your voltmeter leads across R2 and R3.

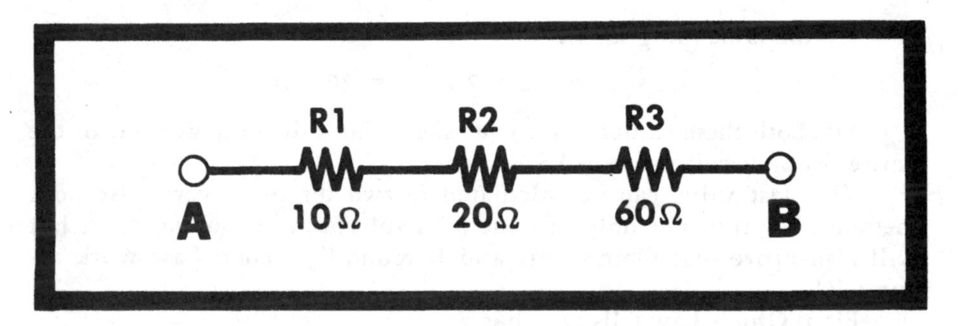

Calculate the voltage drops across R2 and R3. What is the total voltage applied across Points A-B?

You know two facts about R1—its value, and the voltage drop across it. So you can use Ohm's Law to find the current through it.

$$I = \frac{E}{R} = \frac{5}{10} = \frac{1}{2} \text{ ampere}$$

Given the current anywhere in a series circuit, you automatically know it everywhere throughout the circuit; so you can readily calculate E2 as

$$E2 = I_2 \times R2 = 1/2 \times 20 = 10 \text{ volts}$$
$$E3 = I_3 \times R3 = 1/2 \times 60 = 30 \text{ volts}$$

You have now calculated the voltage drops across E2 and E3 to be 10 and 30 volts, respectively, and you already know that the voltage drop across E1 is 5 volts. Kirchhoff's Law tells you that the voltage applied across the circuit resistance is the sum of the voltage drops across the individual resistors; thus,

$$E_t = E1 + E2 + E3 = 5 + 10 + 30 = 45 \text{ volts}$$

Further practice on your own will be provided on page 2-46.

Voltage Division in the Series Circuit

You often need to be able to take current from a given point in a series circuit at a stepped-down voltage—that is to say, at a voltage which is lower than the applied voltage.

A circuit widely used for this is shown in the diagram below. It is called a *voltage divider.*

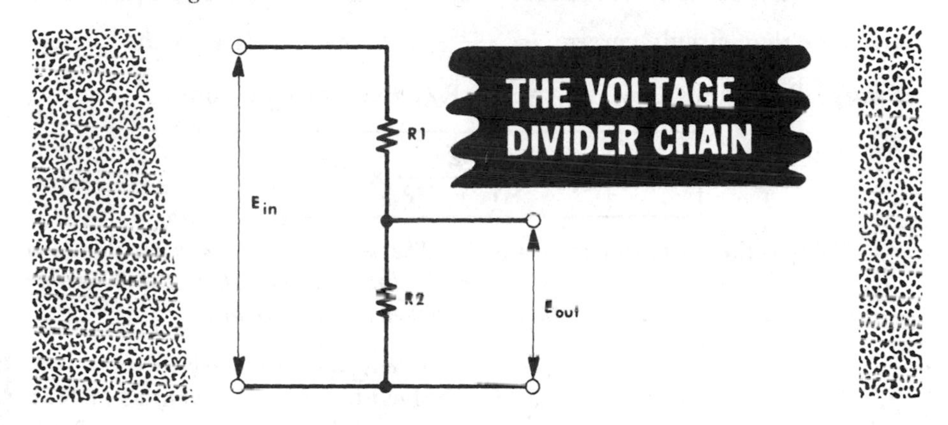

Assume that in this circuit the applied voltage (E_{in}, as it is written in electrical notation) is 100 volts, and that the values of R1 and R2 are 10 and 15 ohms, respectively. What sort of a voltage drop can you expect across R2? (Note that the notation for this voltage is E_{out}. Probably, it will be needed for the *input* voltage into another circuit, in which it will, of course, become E_{in} once more.)

You know that total resistance in this circuit is 10 + 15 = 25 ohms. You also know that the circuit voltage is 100 volts; so you can use Ohm's Law to find the circuit current.

$$I = \frac{E_{in}}{R_t} = \frac{100}{25} = 4 \text{ amperes}$$

Now look at R2. You know that its value is 15 ohms, and you have just calculated that the current through it is 4 amperes (remember that *the current anywhere is the current everywhere* throughout a series circuit). So you get

$$E_{out} = I \times R = 4 \times 15 = 60 \text{ volts}$$

Observe that by the appropriate choice of resistor values in a voltage divider chain, an input voltage of 100 volts has been *stepped down* to an output voltage of 60 volts. And Ohm's Law has enabled you to calculate in advance that this will be so.

Voltage Division in the Series Circuit (continued)

It is obviously possible, from what you have learned on the previous page, to work out a formula for the output voltage of a voltage divider chain which can be applied to a circuit containing a pair of resistors of *any* value.

You can see that total circuit resistance is R1 + R2 and Ohm's Law tells you that circuit current is $\frac{E_{in}}{R1+R2}$. Since this is also the current through R2, Ohm's Law ($E_{out} = I \times R2$) gives the equation:

$$E_{out} = \frac{R2}{R1 + R2} \times E_{in}$$

To put the equation into words: *The voltage across any resistor in a voltage divider chain can be calculated by multiplying the value of that resistor by the input voltage divided by the total resistance of the circuit.*

Note on Thevenin's and Norton's Theorems: It is easy to solve voltage divider problems when a load is involved by use of Thevenin's or Norton's theorem. These theorems, which are discussed on pages 2-133 through 2-136, are useful when you need to obtain a *particular* voltage under *load* conditions.

Variable Resistors

It is frequently convenient to be able to vary the value of a resistor in a circuit at will. (You have often done it yourself when you adjusted the volume control on your radio!)

A common means by which a resistor can be made *variable* in this way is for a sliding arm made of good conducting material to be arranged so it can be moved along the length of the resistor. The resistor is then connected into the circuit with one of its ends fastened to the sliding arm. By moving this sliding arm along the resistor, the value of the resistor can be varied at will between maximum and minimum (zero).

When a variable resistor is used in this way, it is called a *rheostat*. It is generally used to control *current flow* in a circuit.

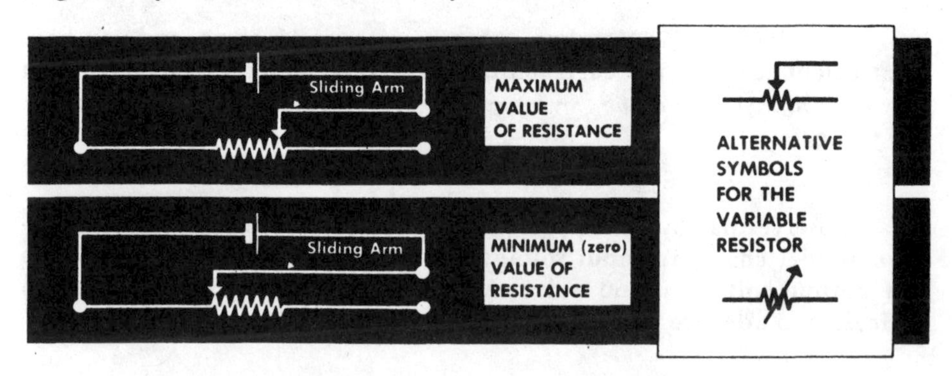

Variable Resistors (continued)

A variable resistor may have either two or three circuit connections. In the diagram below, A and C would always be connected; but B could either be connected or not, at will.

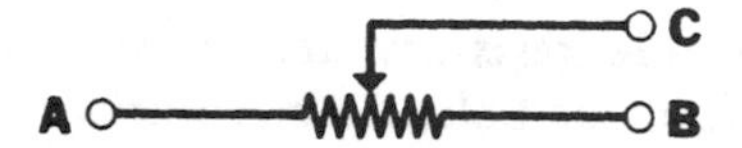

This is what two-terminal and three-terminal resistors, connected as rheostats, look like. The circuit diagram of each is shown below it.

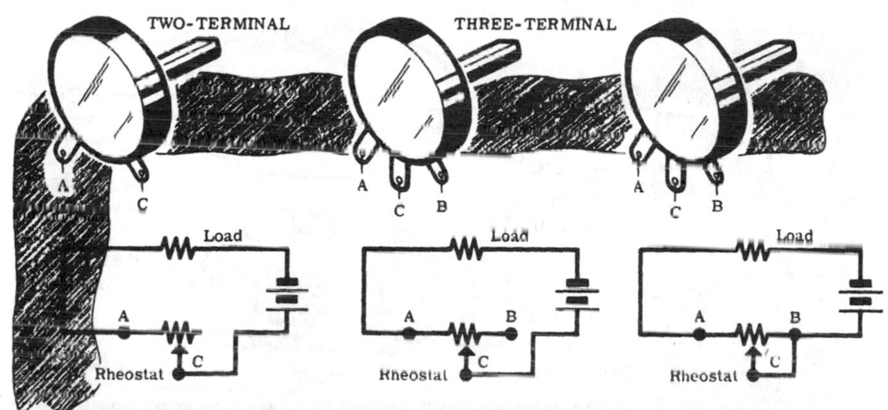

A three-terminal resistor in which all three terminals are connected into the circuit is a *potentiometer*. It is used to control circuit *voltage*.

Potentiometer Connections

The circuit diagram of a potentiometer is really no more than that of a voltage divider chain. R1-R2 is a single resistor, effectively divided by the sliding arm C, whose movement alters the relative values of R1 and R2.

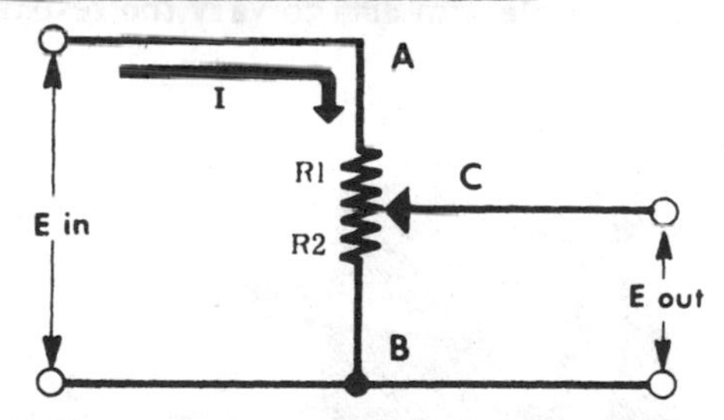

The output voltage can vary from zero (when C is *lowered* so that R2 = 0) to the full circuit voltage (when C is *moved up* so far that R1 = 0).

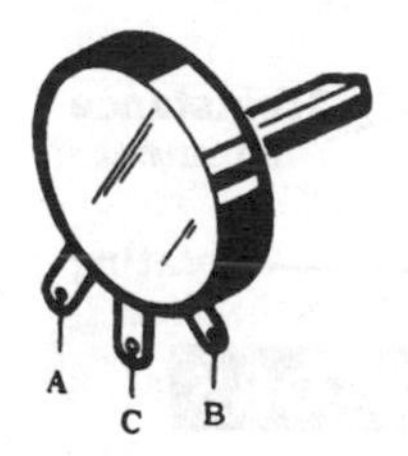

A typical potentiometer looks like this. Note how the connection of all three terminals into the circuit (at points corresponding to A, B, and C in the diagram at the top of the page) enables circuit voltage to be controlled.

Variable Resistors (continued)

Variable resistors, like fixed resistors, can be made with resistance material of carbon or can be wire-wound, depending on the amount of current to be controlled—wire-wound for *large* currents and carbon for *small* currents.

Wire-wound variable resistors are constructed by winding resistance wire on a porcelain or bakelite circular form, with a contact arm which can be adjusted to any position on the circular form by means of a rotating shaft. A lead connected to this movable contact can then be used, with one or both of the end leads, to vary the resistance used.

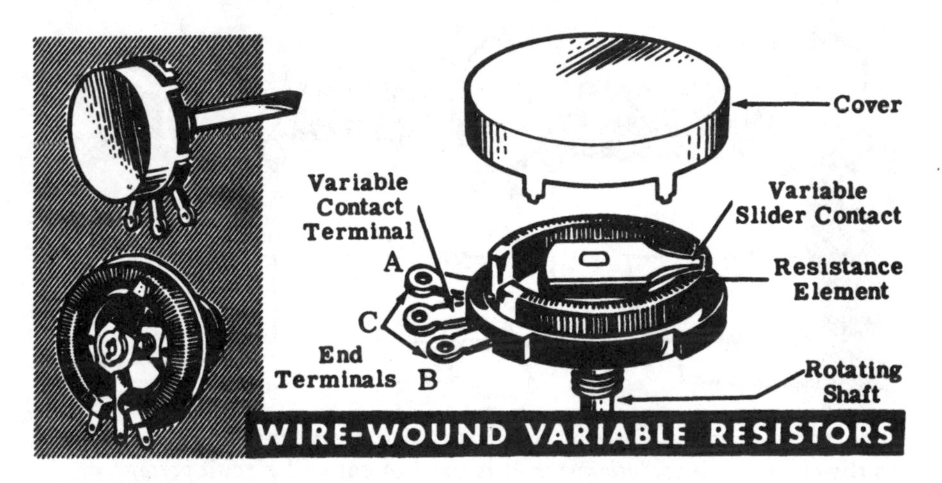

WIRE-WOUND VARIABLE RESISTORS

For controlling small currents, carbon variable resistors are constructed by depositing a carbon compound on a fiber disk. A contact on a movable arm acts to vary the resistance as the arm shaft is rotated.

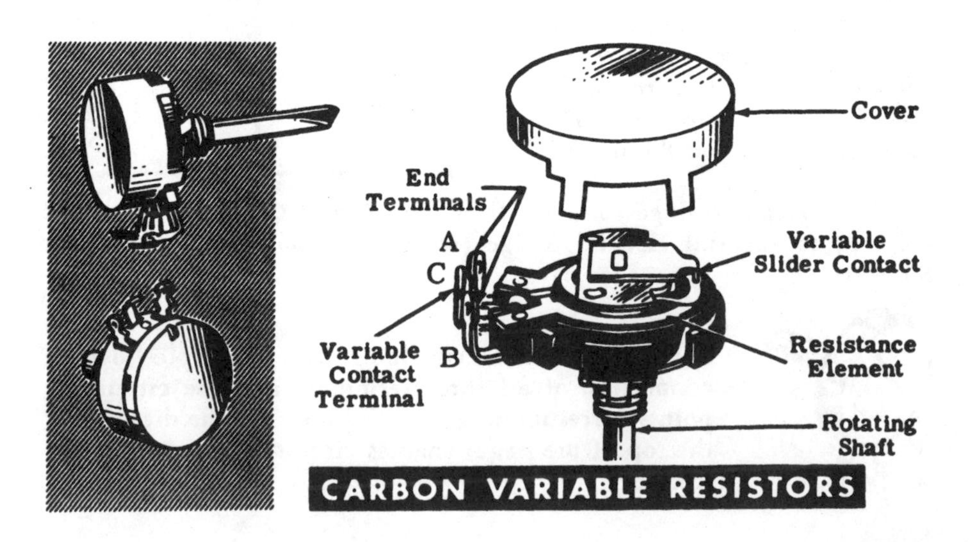

CARBON VARIABLE RESISTORS

Review of Ohm's Law in Series Circuits

1. CURRENT—The value of *current* flowing through a series circuit is always the *same* at every point in the circuit.

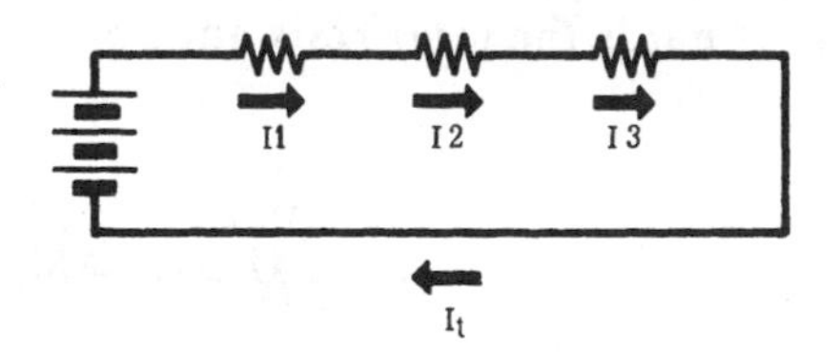

$$I_t = I1 = I2 = I3 = I4 = \ldots$$

2. RESISTANCE—The total *resistance* in a series circuit is always the sum of the individual values of resistance in the circuit.

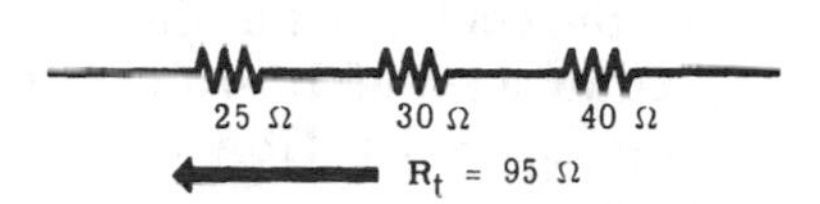

$$R_t = R1 + R2 + R3 + R4 + \ldots$$

3. VOLTAGE—The *voltage* applied across the circuit resistance of a series circuit is always equal to the *sum* of the voltage drops across the individual resistances.

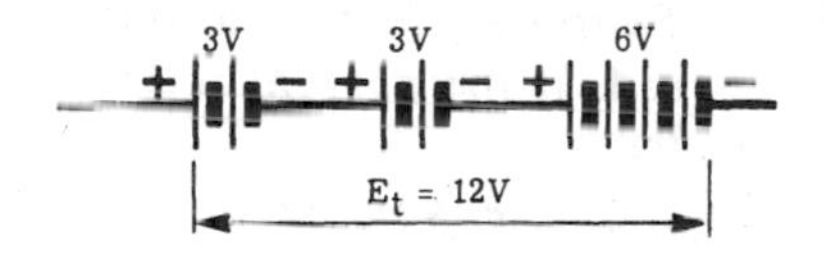

$$E_t = E1 + E2 + E3 + E4 + \ldots$$

4. UNKNOWNS—The way to find unknown quantities in a series circuit is as follows:

(a) Draw the circuit diagram.
(b) Insert on this diagram all the known facts.
(c) Look for resistances in the circuit about which you know any two values.
(d) Use Ohm's Law to find the third value.
(e) Carry on from your increasing store of known facts to calculate further *unknowns*—never forgetting that once you know the value of any I in the circuit you know the value of all of them.

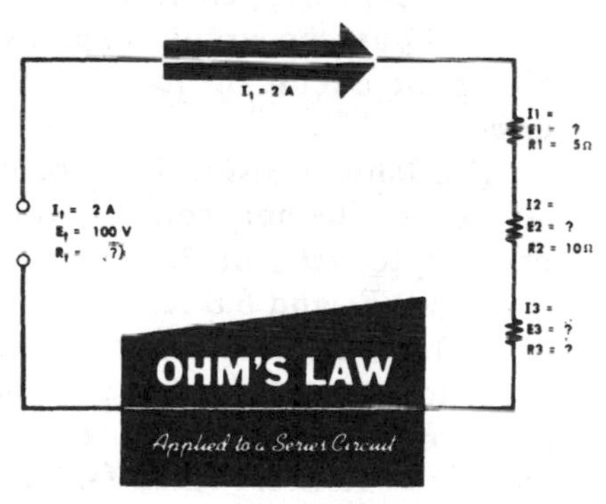

5. EQUIVALENT CIRCUIT—A series circuit containing two or more resistances can often be usefully simplified into an *equivalent circuit* containing a single theoretical resistance having a value equal to the sum of all the actual resistances in the circuit.

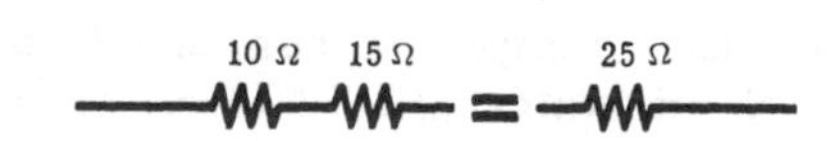

Self-Test— Review Questions

1. What is the total resistance of these four resistors connected in series?

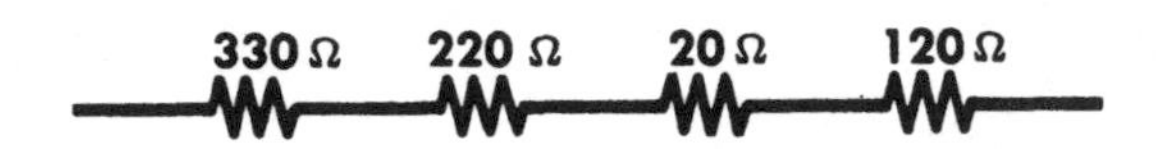

2. Draw the equivalent of the series resistance found in problem 1.
3. (a) Draw a set of 1.5-volt cells in a series circuit so that the total battery voltage is 9 volts. How many cells does it take? Why? Show polarity of connection.
   (b) Suppose you connected one of the cells in part (a) so that it was backwards. What would the total voltage across the battery be? Why?
4. (a) Suppose you had a 100-volt source, but you wanted to get 30 volts instead. If the current to be drawn is 1 A, calculate the divider resistors needed.
   (b) Suppose you wanted a 0-30 volt source. Show how you would change the circuit in part (a) to do this. What would be the resistance of the variable resistor?
5. Draw a series circuit showing two lamps and a battery of 12 volts. Calculate the currents in each lamp and the voltage across each lamp if the resistances are 40 and 60 ohms, respectively.
6. What would happen in question 5 if two 12-volt batteries were put in series? Draw the circuit. Explain.
7. Using the circuit in question 5, show how Kirchhoff's Second Law applies.
8. A 47-kilohm resistor in a piece of equipment fails and you have to replace it. The only resistors you can find are an assorted lot, having values, respectively, of 22 K, 120 K, 15 K, 220 ohms, 4.7 K, 4.7 K, 330 ohms, 1.2 K, and 6.8 K.
   (a) Which of these resistors would you choose to connect in series in order to come as close as possible to the value of 47 K, and what would be their combined resistance?
   (b) If the resistor which failed had a 10% tolerance, would the combination you have chosen be adequate to do its job?
9. Although your car has a 12-volt battery, an aunt gives you a handsome spotlight which takes 2 amperes of current, but which is, unfortunately, designed to work off a 6-volt battery. What values of resistance would you need to connect in series with the spotlight before you could safely switch it on? Draw the circuit diagram and show voltages and current.
10. Three lamps are connected in series. The resistances of the first and third lamps are 52 ohms each; the middle lamp is rated at 76 ohms. What current flows through the lamps when they are connected across a 120-volt supply?

Experiment/Application—Open Circuits

You already know that in order for a current to pass through a circuit, a *closed* path (complete loop) is required. Any break in the closed path causes an *open* circuit and stops current flow. Each time you open a switch, you are causing an open circuit.

Anything which causes an open circuit, other than actually opening a switch, interferes with the proper operation of the circuit, and must be corrected. An open circuit may be caused by a loose connection, a burned-out resistor or lamp filament, poor solder joints or loose contacts, or a broken wire.

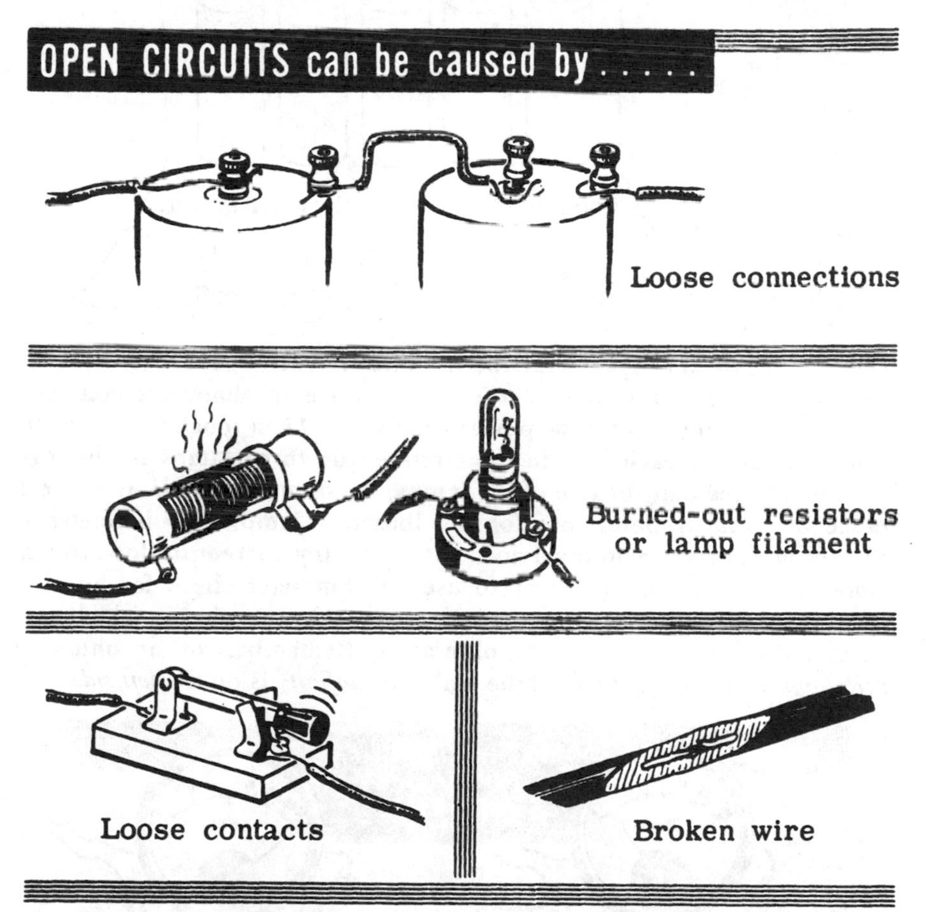
OPEN CIRCUITS can be caused by . . . . .
Loose connections
Burned-out resistors or lamp filament
Loose contacts
Broken wire

These *faults* or *troubles* can often be detected visually, and you may find that in your work you might encounter one or more of these *opens*.

In some cases it is not possible to detect visually the cause of an open circuit. An ohmmeter or a test lamp can then be used to find the cause of the trouble.

Experiment/Application—Open Circuits (continued)

Now, suppose you connect five dry cells, a knife switch, and three lamp sockets in series. Insert three 2.5-volt, 0.75-ampere lamps in the sockets. When you close the switch, the lamps light with normal brilliancy. If you then loosen one of the lamps, they all go out, indicating an *open* circuit. (A loosened lamp simulates a burned-out filament or other open.)

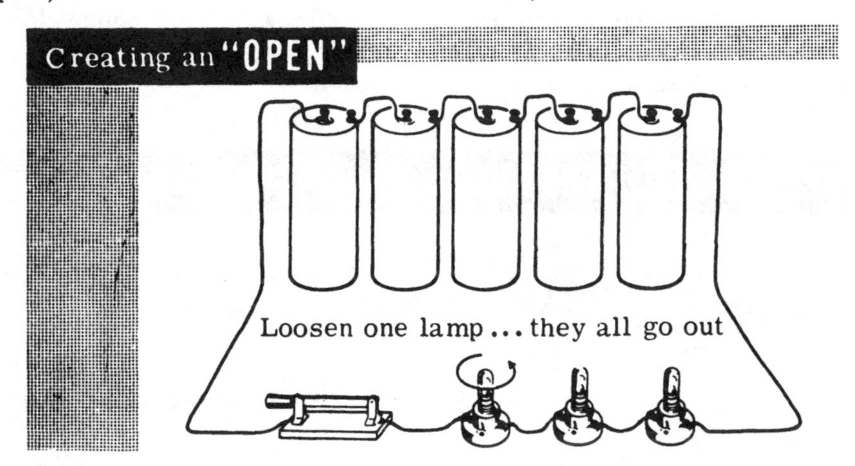
Creating an "OPEN"
Loosen one lamp... they all go out

To locate the open with the ohmmeter, you would first open the knife switch to remove the voltage source, since an ohmmeter must *never* be used on a circuit with the power connected. Then touch the ohmmeter test leads across each unit in the circuit—the three lamps in this case. You would see that for two of the lamps, the ohmmeter indicates a resistance of about 4 ohms; but for the loosened lamp, the ohmmeter indicates *infinity*. Since an open does not allow any current to flow, *its resistance must be infinite*. The way to use an ohmmeter check for an open, then, is to find the series-connected element in the circuit which measures infinite resistance on the ohmmeter. Remember, on an ohmmeter *zero ohms* is at the *right side* of the scale and *infinity* is on the *left side*.

Using the ohmmeter to test for an "OPEN"
OHMS
OHMS

Good bulb—resistance about 4 Ω "Open"— infinite resistance

Experiment/Application—Open Circuits (continued)

The second method used to locate an open is to test the circuit by means of a test lamp. A test lamp can be set up by attaching leads to the terminals of a lamp socket and inserting a 2.5-volt lamp. If you then close the circuit switch and touch the test lamp leads across each lamp in the circuit, the test lamp will not light until it is across the terminals of the loosened lamp. The test lamp then lights, indicating to you that you have found the open.

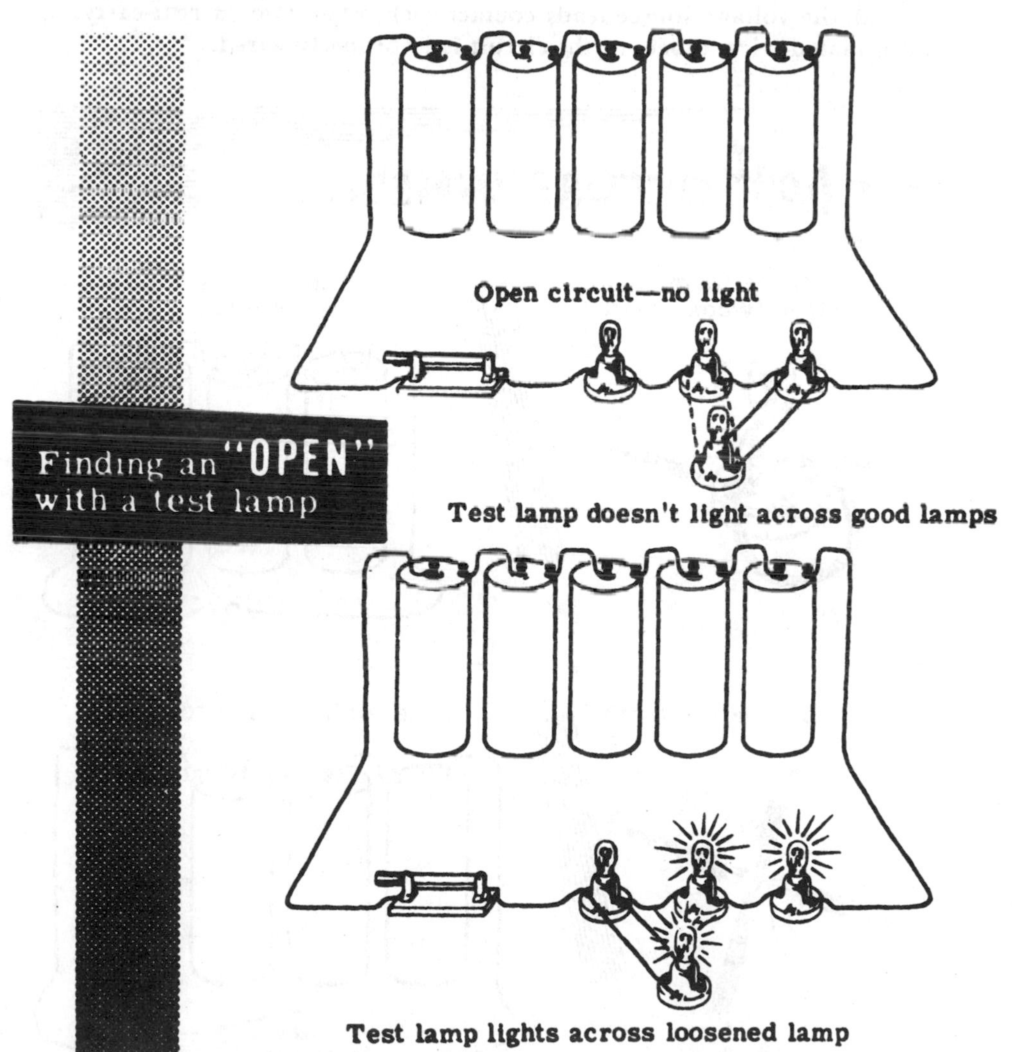
Finding an "OPEN" with a test lamp
Open circuit—no light
Test lamp doesn't light across good lamps
Test lamp lights across loosened lamp

The test lamp completes the circuit and allows current to flow, bypassing the open. (This also causes the other lamps to light since the open is being bypassed.) You will often use this method to detect opens which cannot be seen.

Experiment/Application—Short Circuits

You have seen how an open prevents current flow by breaking the closed path between terminals of the voltage source. Now you will see how a *short* produces just the *opposite* effect—creating a *short circuit* path of low resistance through which a larger than normal current flows.

A short occurs whenever the resistance of a circuit or part of a circuit drops from its normal value to a much lower or zero resistance. This happens if the two terminals of a resistance in a circuit are directly connected, the voltage source leads contact each other, two current-carrying uninsulated wires touch, or the circuit is improperly wired.

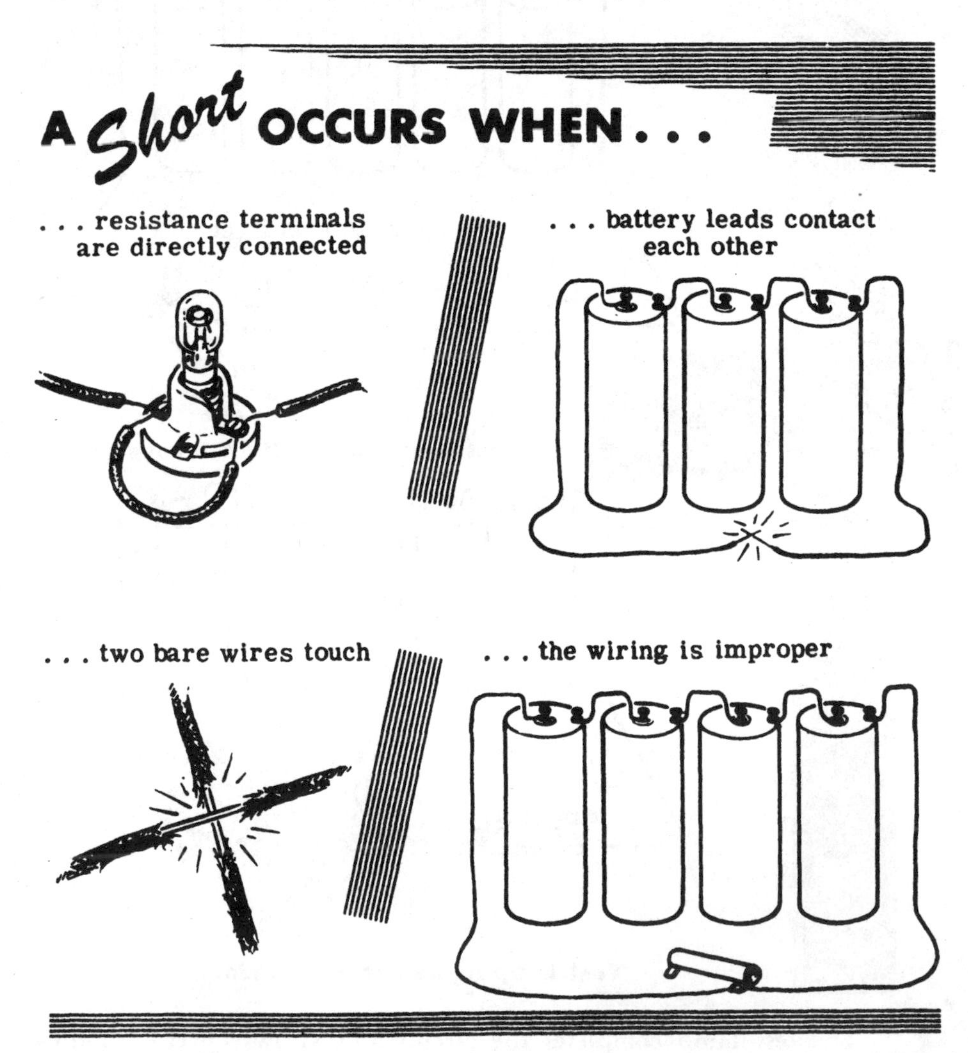

A Short OCCURS WHEN . . .
. . . resistance terminals are directly connected
. . . battery leads contact each other
. . . two bare wires touch
. . . the wiring is improper

These shorts are called *external shorts* and can usually be detected by visual inspection.

Experiment/Application—Short Circuits (continued)

When a short occurs in a simple circuit, the resistance of the circuit to current flow becomes very low, so that a very large current flows.

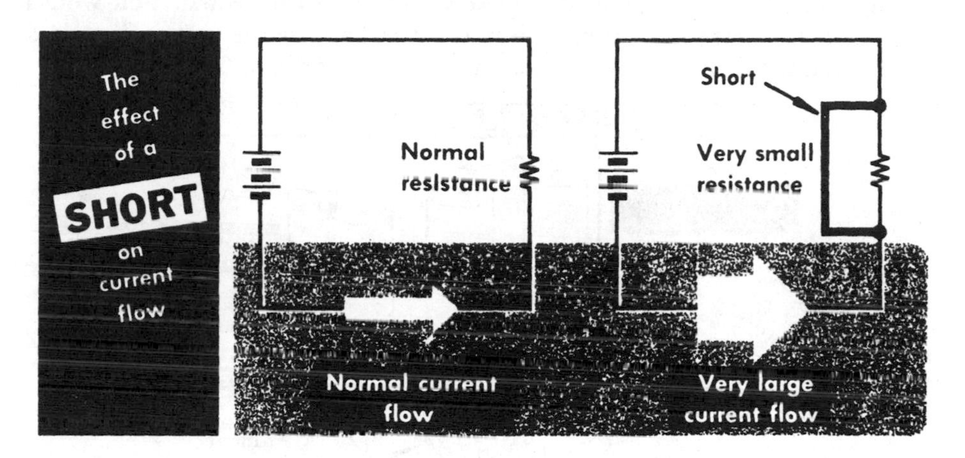

In a series circuit, a short across one or more parts of the circuit results in reduction of the total resistance of the circuit and correspondingly increased current flows, which may damage the other components in the circuit.

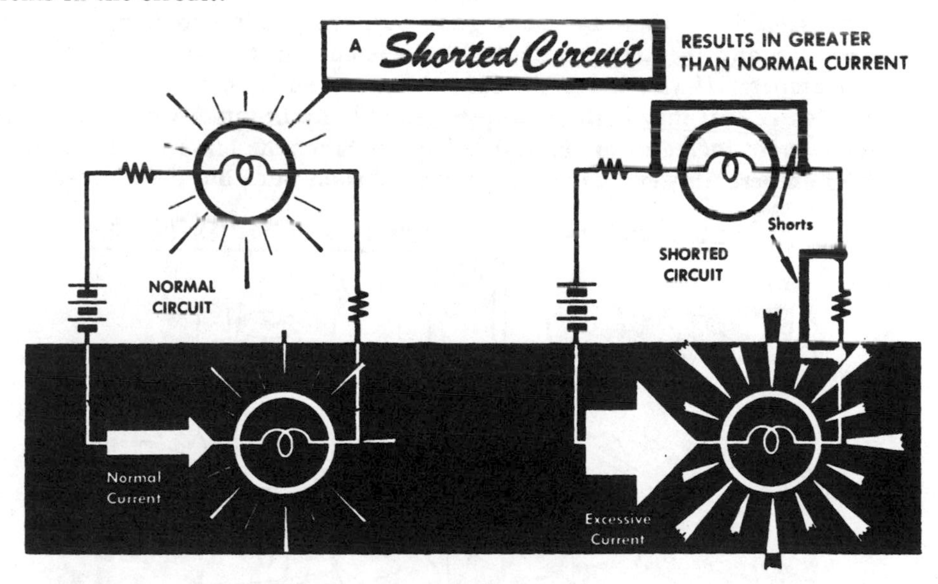

Circuits are usually protected against excessive current flow by the use of fuses, which you will learn about later. But it is important that you understand the reasons for and results of shorts, so that you can *avoid* accidentally shorting your circuits and causing damage to meters or other equipment or components.

Experiment/Application—Short Circuits (continued)

Again, suppose you connect three dry cells in series with a 0-1 ampere range ammeter and three lamp sockets. Then you insert three 2.5-volt, 0.75-ampere lamps in the sockets and close the switch. You would see that the lamps light equally but are dim because the voltage is only 6 volts and the ammeter indicates a current flow of about 0.5 ampere.

A NORMAL SERIES CIRCUIT

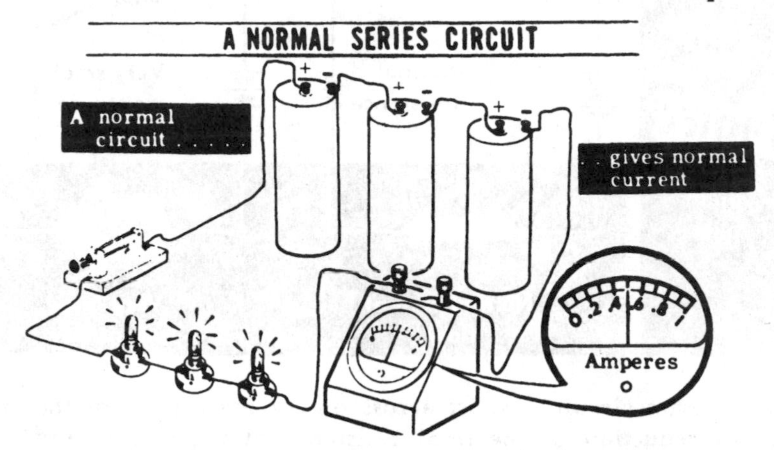

Now suppose you touch the ends of an insulated lead to the terminals of one of the lamps, *short-circuiting* the current around that lamp. You would see that the lamp goes out, the other two lamps become brighter, and the ammeter shows that the current has increased to about 0.6 ampere. If you move the lead to short out two of the lamps, you would see that they both go out, the third lamp becomes very bright, and the current increases to about 0.9 ampere. Since the lamp is rated at only 0.75 ampere, this excessive current would soon burn out the filament.

SEEING THE EFFECT OF A SHORT IN A SERIES CIRCUIT

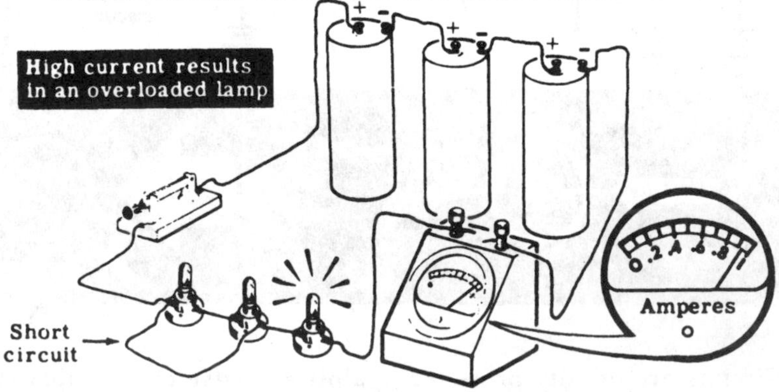

If you were to short out all three lamps, the lack of resistance of the circuit would cause a great amount of current to flow, which would *damage* the ammeter.

Experiment/Application—Series Circuit Resistance

You can see the effect of connecting resistances in series by measuring the resistance of three lamps individually, and then measuring the total resistance when they are connected in series.

Suppose you connected three lamp sockets in series and inserted a 6-volt, 0.5-ampere lamp in each socket. By using an ohmmeter to measure the resistance of each lamp, you would see that each lamp resistance measures about 12 ohms.

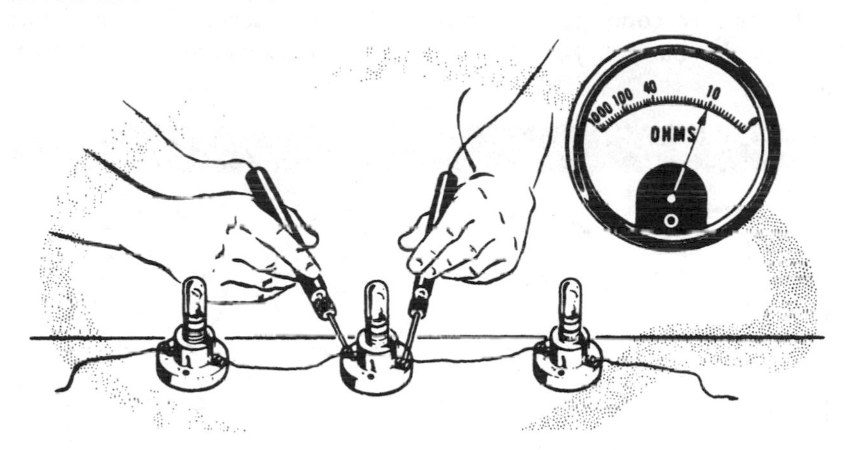

Next, if you measured the resistance of the three lamps in series, you would see that the total resistance is about 36 ohms. Thus, the total resistance of series-connected resistance is equal to the sum of all the individual resistances.

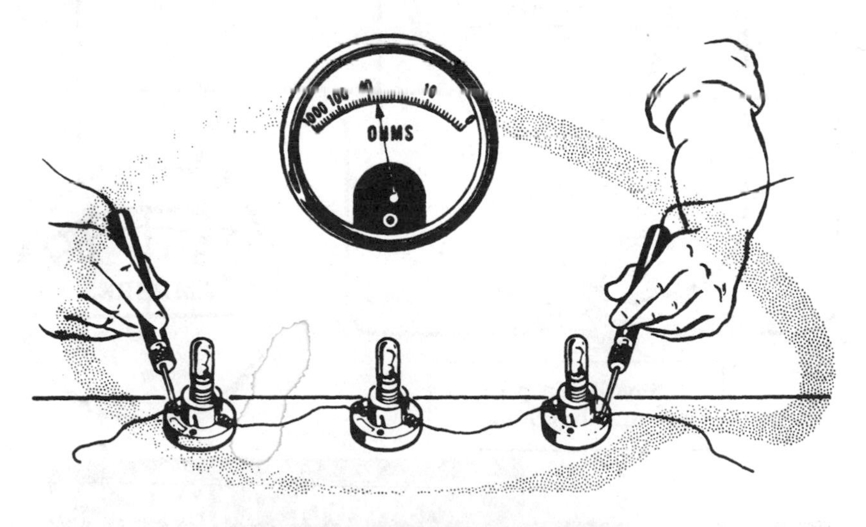

$$R_t = R1 + R2 + R3$$

Experiment/Application—Series Circuit Resistance (continued)

Now, suppose you connect four dry cells in series to form a 6-volt battery as a voltage source. Connect the battery, one lamp socket, a 0-1 ampere ammeter, and a switch in series; then connect a 0-10 volt voltmeter across the battery (see illustration below). You would notice that the voltmeter reads 6 volts. If you inserted a 6-volt, 0.5-ampere lamp in the socket and closed the switch, the ammeter would record a current flow of about 0.5 ampere and the lamp would light to normal brilliance.

If you now connected the voltmeter directly across the lamp, instead of across the battery, you would see the voltage across the lamp is 6 volts.

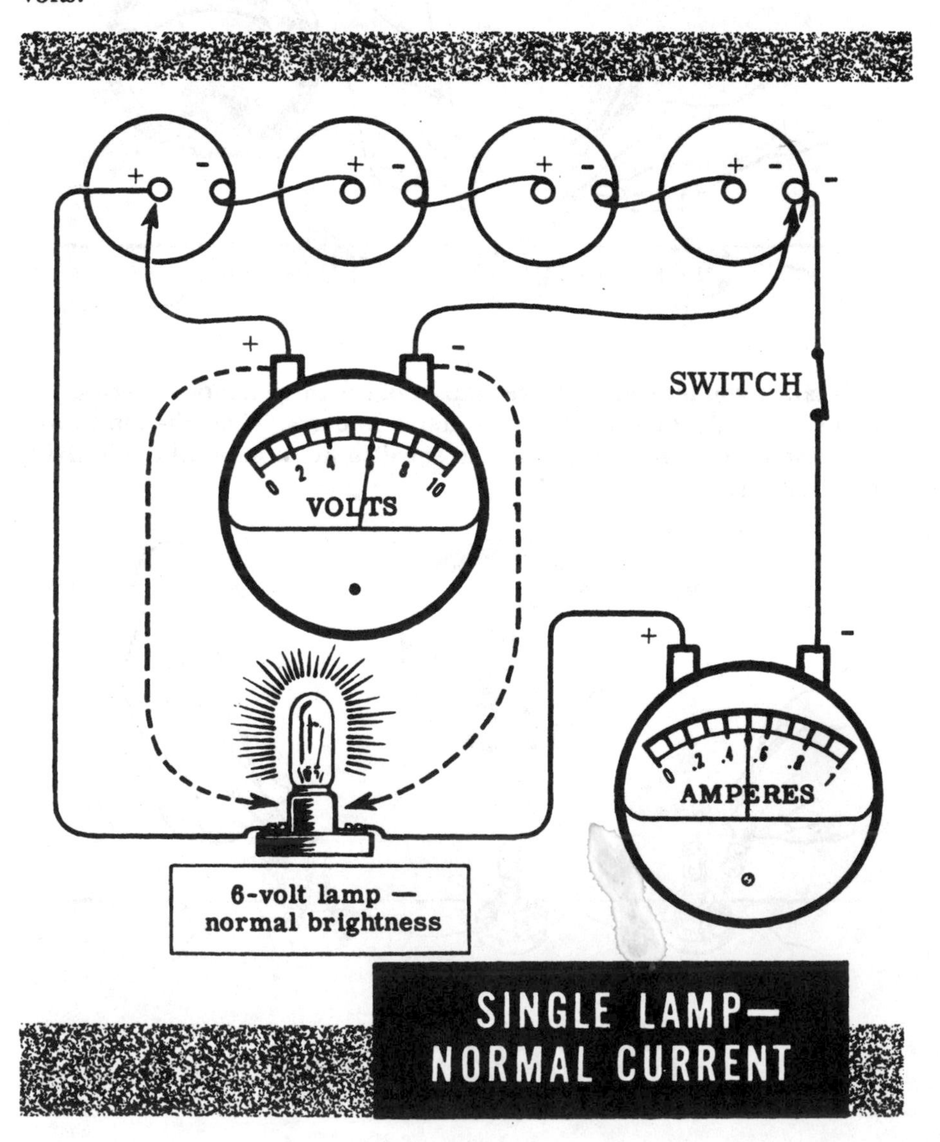

SINGLE LAMP—
NORMAL CURRENT

Experiment/Application—Series Circuit Resistance (continued)

Next, suppose you replace the single lamp socket by three sockets in series, and again insert 6-volt, 0.5-ampere lamps in the socket. The lamps now light well below normal brilliancy, and the ammeter reading would be about one-third of its previous value. A voltmeter reading taken across the total circuit reads 6 volts, and across each lamp the voltage is 2 volts.

Since the voltage from the battery is not changed, but the current is lower, it follows that the resistance must be greater.

If you measured and added the voltages across each lamp, you would find that the sum of the voltages across the individual resistances (lamps) still equals the total voltage.

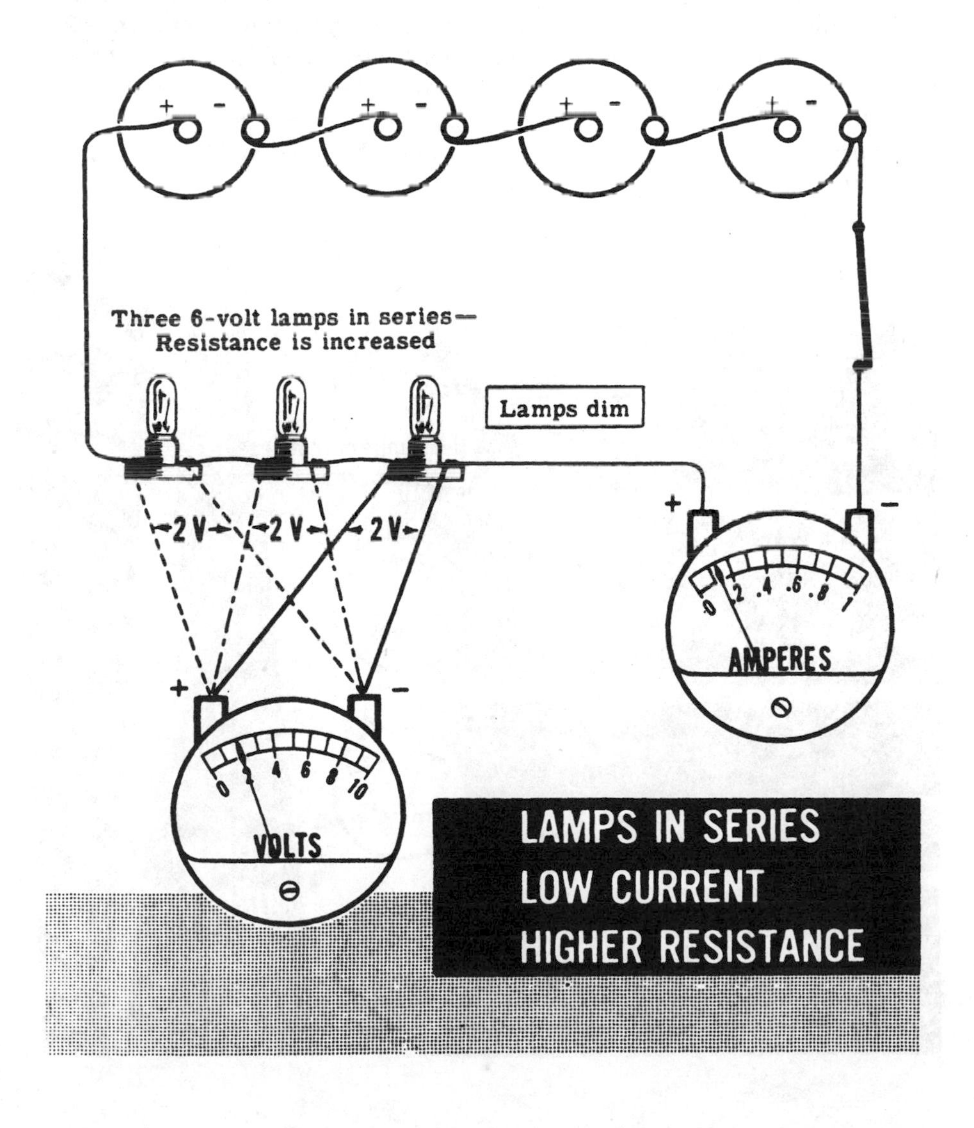

Experiment/Application—Series Circuit Current

To see the effect of changing resistances on the amount of current flow, and how different pieces of equipment require different amounts of current for proper operation, you could replace one of the 6-volt lamps by a 2.5-volt lamp drawing 0.75 ampere. The two 6-volt lamps will increase in brilliance by almost 50%, while the 2.5-volt lamp will light only dimly. The ammeter reading would show that the current has increased—which proves that decreasing the resistance of one part of the circuit decreases the total opposition to current flow and, hence, *increases* the total circuit current.

Replacement of another 6-volt lamp with a 2.5-volt lamp would further decrease the total resistance and increase the total circuit current.

The brilliance of the lamps increases as the current flow increases; and if the last 6-volt lamp were to be replaced by a lower-resistance 2.5-volt lamp, you would see that the circuit current for the three 2.5-volt lamps is approximately the same as that for a single 6-volt lamp. You would also observe that the three lamps light at about normal brilliance, because the current is only slightly less than the rated value of the lamps, as is the voltage measured across each lamp.

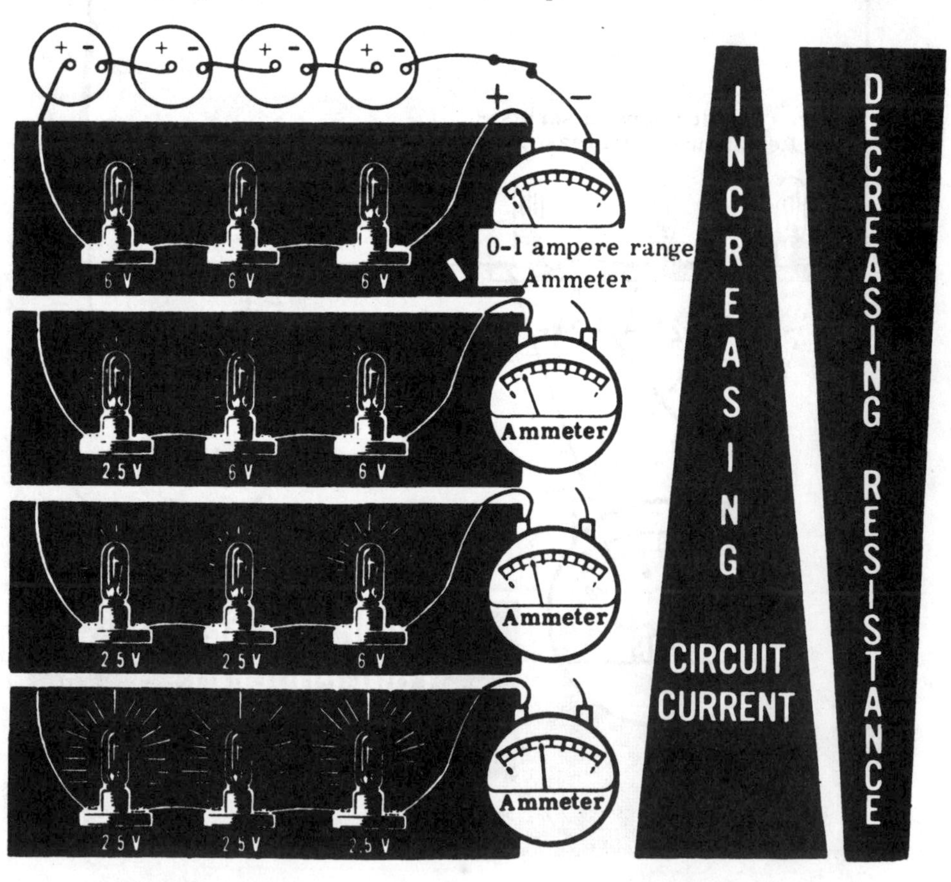

Experiment/Application—Series Circuit Voltage/Kirchhoff's Second Law

The rated voltage of three 2.5-volt lamps in series is 7.5 volts, so that a 6-volt battery does not cause the rated current to flow. Adding one more dry cell would increase the circuit voltage without changing the resistance, thereby causing a greater current flow, as indicated by the increased brilliance of the lamps and the increased current reading. Voltage readings taken across the lamps show that the voltage across each lamp is the rated voltage of 2.5 volts. If you measured the total battery voltage it would also equal 7.5 volts.

Removal of one cell of the battery at a time, followed by taking voltage readings across the lamps, and across the battery, will once again show that the voltages across the lamps are about equal, and always add up to the total battery voltage.

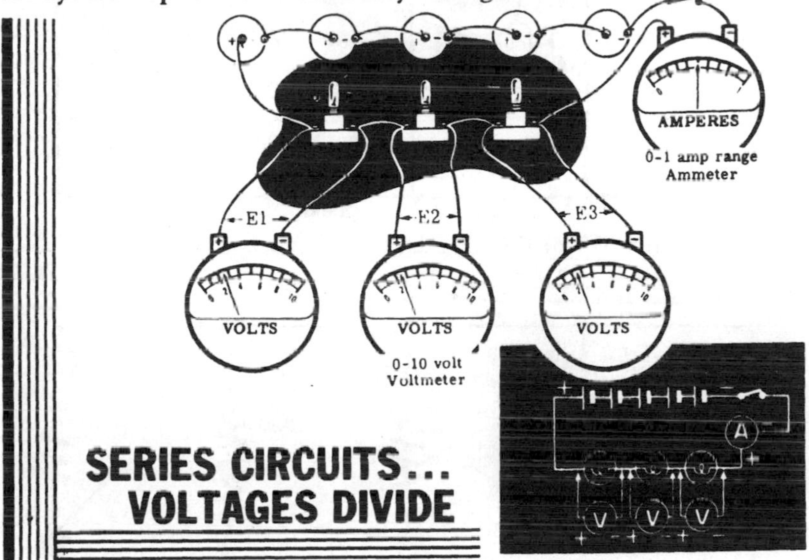

With five cells connected to form a 7.5-volt battery, replace one of the 2.5-volt lamps with a 6-volt lamp having greater resistance. Voltmeter readings across the lamps would still total 7.5 volts when added together, but would not all be equal. The voltages across the lower resistance 2.5-volt lamps would be equal but less than 2.5 volts, while the voltage across the higher resistance 6-volt lamp would be greater than 2.5 volts. You can see that for resistors in series the voltage divides in proportion across the various resistances connected in series, with *more* voltage drop across the *larger* resistance and *less* voltage drop across the *smaller* resistance; the total voltage (E_t) is exactly equal to the voltage across each resistance. To put it another way,

$$E_t = E1 + E2 + E3$$

which is *Kirchhoff's Second Law!*

Learning Objectives—Next Section

Overview—Aside from the series circuit that you already can solve, you will learn about *parallel* circuits in the next section. When you can solve both types of circuits, you will find that you can solve any circuit because all circuits are made of combinations of series and/or parallel circuits.

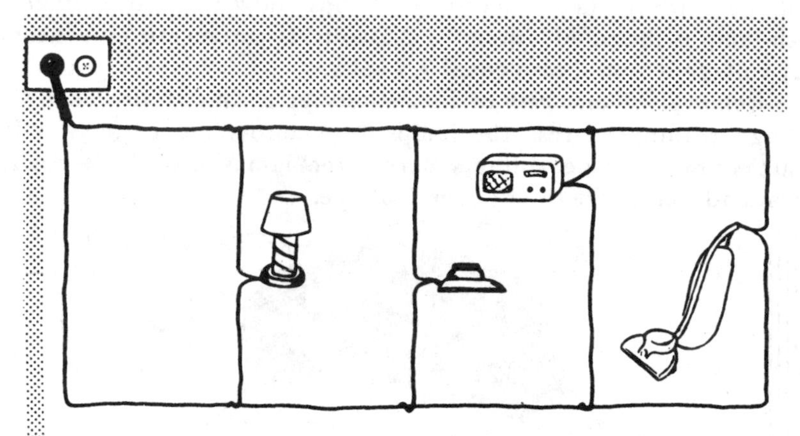

DIFFERENT TYPES OF ELECTRICAL EQUIPMENT IN PARALLEL DIVIDE . . . .

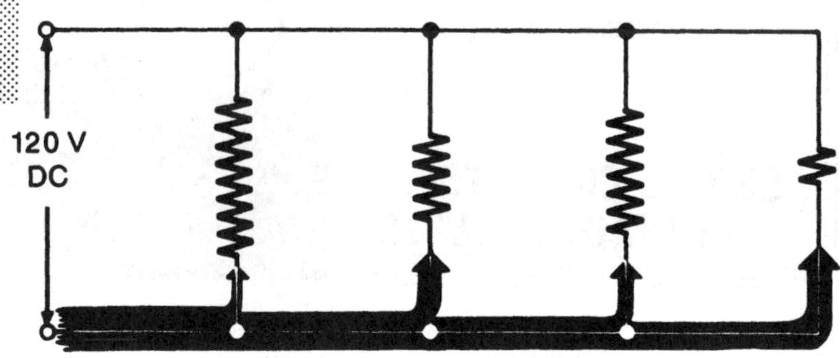

. . . . THE TOTAL CURRENT UNEQUALLY

The Parallel Circuit

When resistances, instead of being connected *end-to-end* as in a series circuit, are connected *side-by-side* so that there exists more than one path through which current can flow, the resistances are said to be *parallel-connected* or *connected in parallel*; and the circuit of which they form a part is called a *parallel circuit.*

Two lamp sockets, for instance, connected terminal-to-terminal by two pieces of wire, are *parallel-connected.* When any two terminals are connected across a voltage source, the whole arrangement—both lamps, the voltage source, and the wires connecting them together—forms a complete parallel circuit.

In the same way, cells in a battery connected so that there is *more than one path* for current flow through the battery are said to be parallel-connected. Each individual cell in a battery whose cells are so connected furnishes only a *part* of the total current drawn from the battery. As you might suspect, when you connect batteries or cells in parallel, you must connect the *positive* terminals *together* and the *negative* terminals *together.*

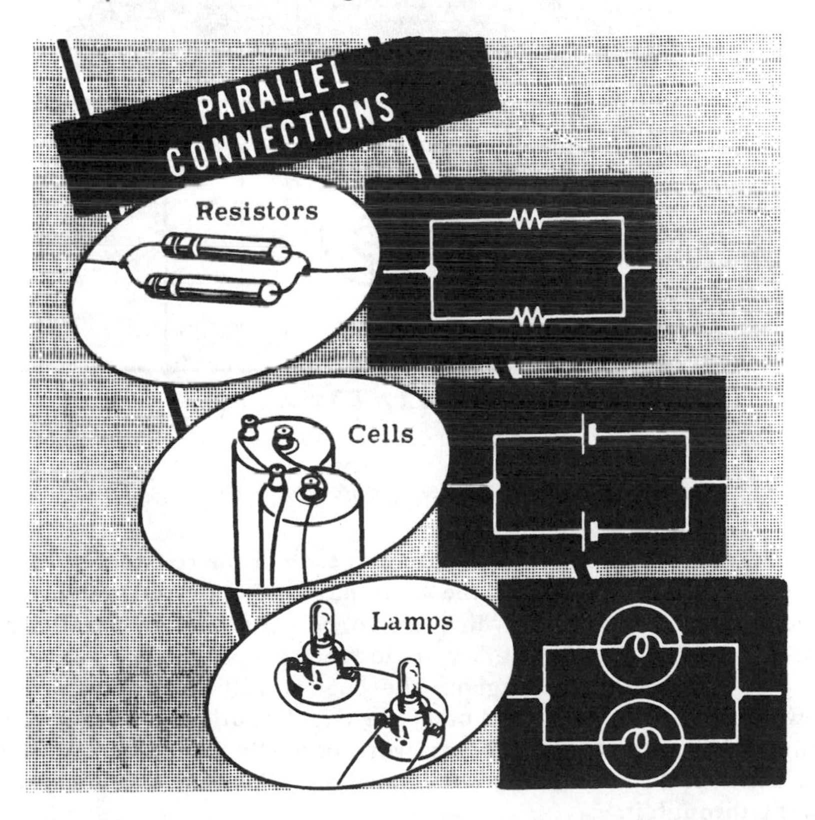

An obvious example of parallel connection is found in the electrical wiring of an ordinary house, in which every one of the various electrical appliances used in the house is connected in parallel across the line.

Voltage in Parallel Circuits

When resistances in parallel are placed across a voltage source, the *voltage* across each of the resistors is always the same. The *current* through each resistor, however, will vary according to the value of the individual resistance.

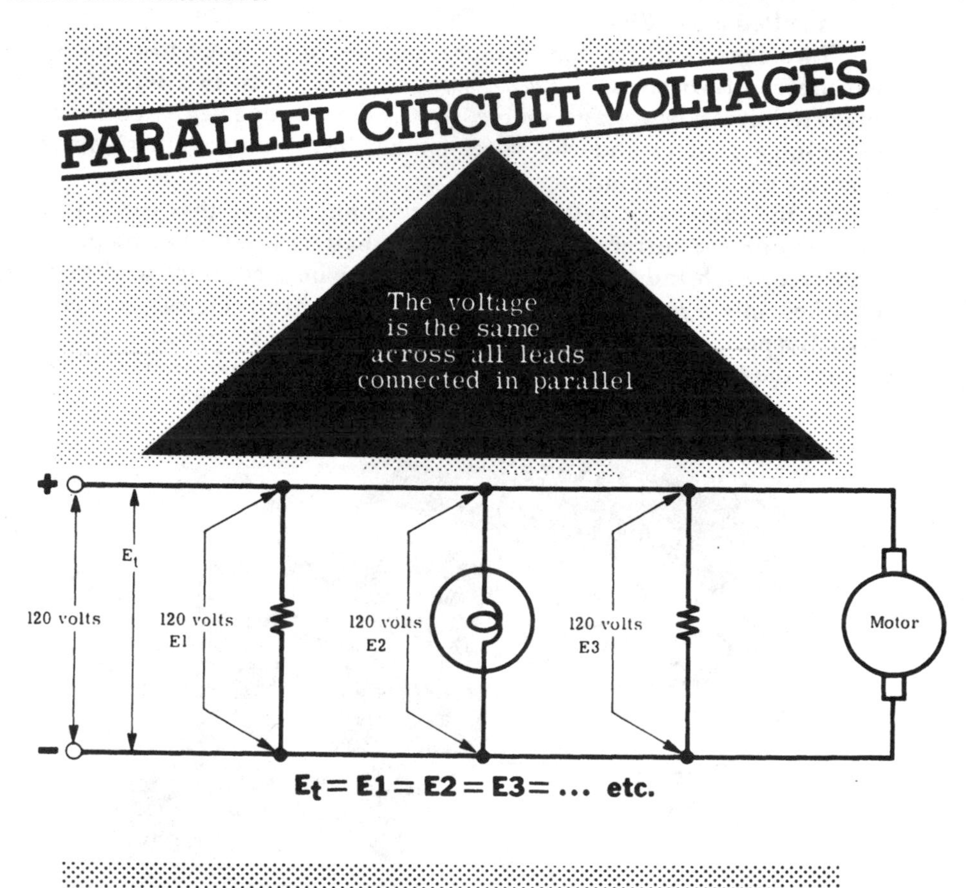

The fact that the voltages applied to each of the resistors or loads in a parallel circuit are always the same has an important practical consequence. All components which are to be connected in parallel must have the *same voltage rating* if they are to work properly.

The line voltage throughout the U.S. is 120 volts. You probably know by now that lamps and other electrical appliances with a voltage rating of 120 volts, or thereabouts, work perfectly well, whereas a lamp-bulb rated at 12 volts burns out immediately because excess current is flowing through it.

The reason is that, since all the appliances are connected across the same voltage source, the same voltage is applied across each. All must, therefore, be of the proper rating to handle this voltage.

Current Flow in Parallel Circuits

Current flowing through a parallel circuit divides to flow through each of the parallel paths. Take a circuit containing three branches—call them AB, CD, and EF, respectively—connected in parallel. In such a circuit:

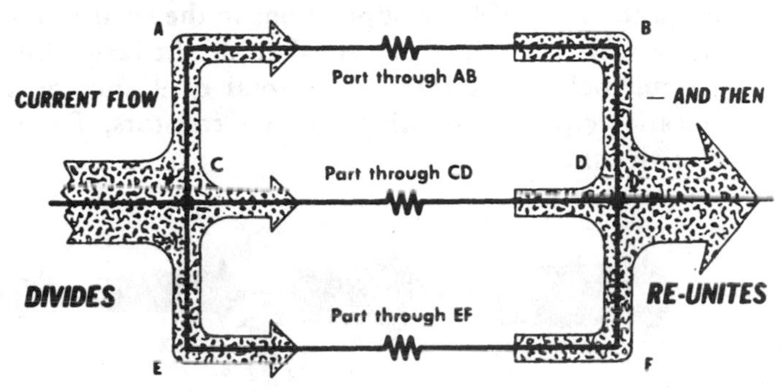

The current flowing through the several branches of a parallel circuit divides in inverse proportions, governed by the comparative resistance of the individual branches. Thus, the lower the resistance of any branch in proportion to the resistance of other branches in the same parallel circuit, the higher will be the proportion of total current flow which that branch will take.

You can put the same thing more simply, and almost as accurately, by saying: *In a parallel circuit, branches having low resistance draw more current than do branches having high resistance.*

In the illustration below, the differing values of the four resistors in the parallel circuit are indicated by the length of the resistor symbol used in each case. Note that the higher the resistance, the smaller the proportion of current flowing through it, and vice versa.

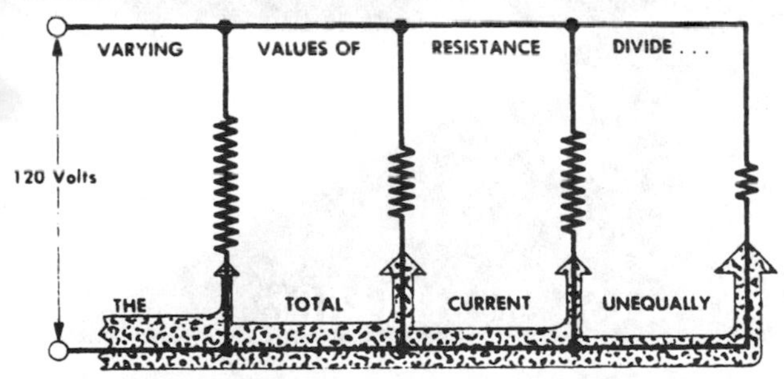

The way in which current divides in a parallel circuit is of great practical importance. For instance, since every electrical appliance used in a house, is (as you know) connected in parallel across the line, current will divide unequally through the differing values of resistance which these appliances present, *the highest current flowing through the lowest resistance.* You will learn how protection against excessive flow is given when you come to read about *fuses* later on.

Current Flow in Parallel Circuits (continued)

It is worth repeating the important facts you have just learned.

When unequal resistances are connected in parallel, opposition to current flow is not the same in every branch of the circuit. A *small* value of resistance offers *less* opposition to current flow. Current flow is always greatest through the path of least opposition; so the smaller resistors in a parallel circuit always pass more current than do the larger ones.

In the circuit below, for instance, a total of 9 amperes is flowing through a parallel circuit consisting of two resistors, R1 and R2, of which R1 has twice the value of R2.

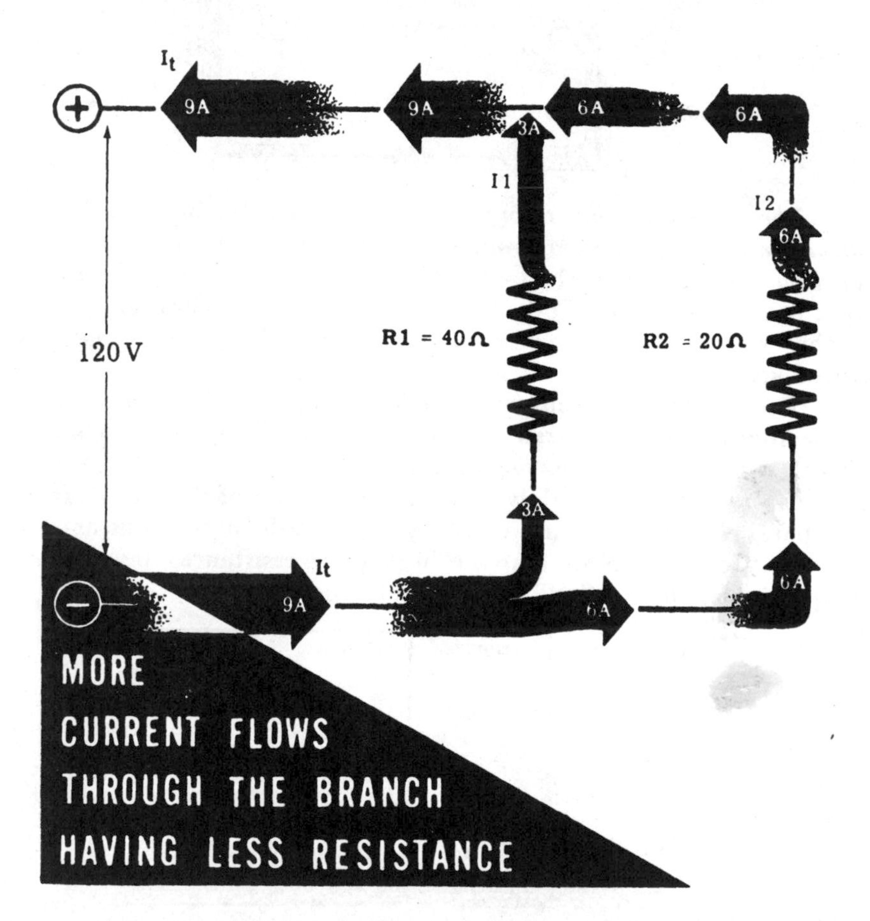

Note that *the current divides in inverse proportion to the values of the two resistors*—only 3 amperes flow through the 40-ohm resistance of R1, while 6 amperes flow through the 20-ohm resistance of R2. If the value of R1 were to be quadrupled to 160 ohms, the current through R1 would be reduced from 3 amperes to 3/4 ampere while the current through R2 would remain *unchanged*. Thus, the total current would be 6 amperes + 3/4 ampere = 6.75 amperes.

Current Flow in Parallel Circuits (continued)

You may have noticed on the previous page that the total current in a parallel circuit seems to be equal to the sum of the current in each component. This is always true, and can be expressed as

$$I_t = I1 + I2 + I3 + \ldots$$

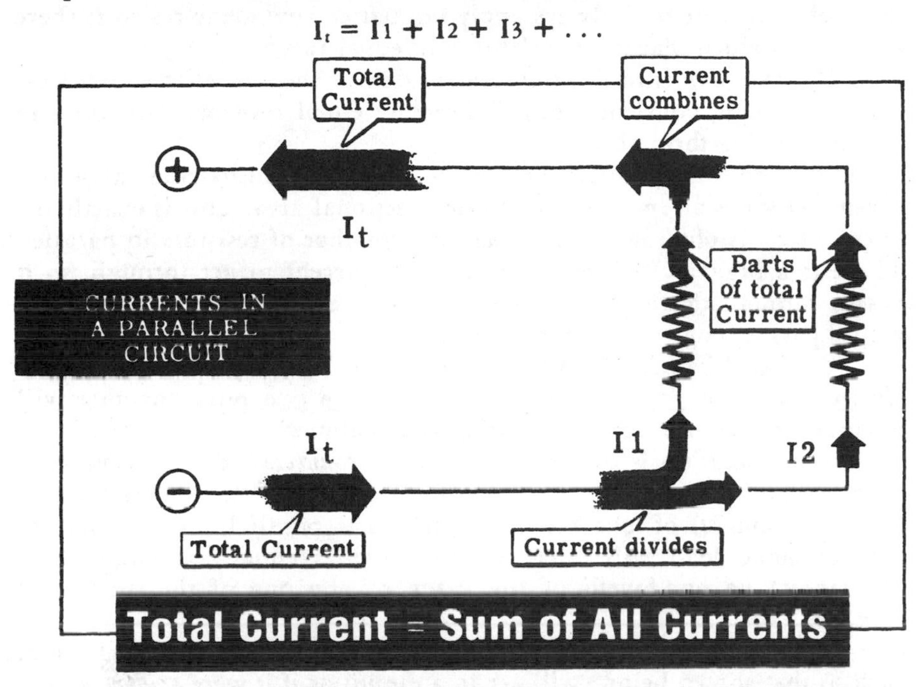

When a circuit consists of *equal* resistances in parallel, the current flowing through each and all of the resistances will be equal; and the current, seeking always for the line of least resistance, will find equal opposition in every path it can take.

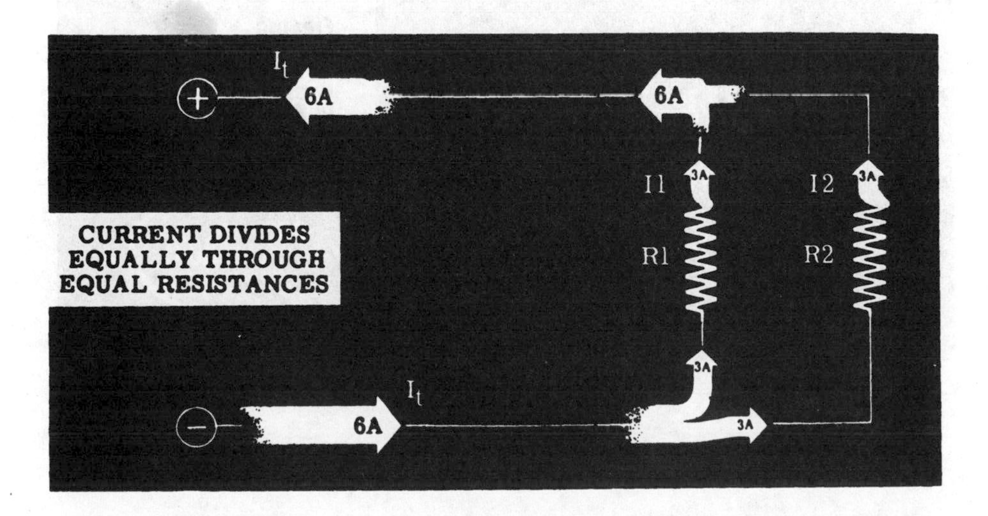

Equal Resistors in Parallel Circuits

In a circuit such as the one illustrated at the bottom of the preceding page, the current flow through each of the two resistors is plainly equal to the total current divided by the number of resistors connected in parallel. This rule will always apply no matter how many resistors there may be, provided that every resistor is of equal value.

The next step is to determine how great is the *total effective resistance* which a circuit, consisting of a number of equal resistors, will offer to current flowing through it.

You know that one of the factors which diminishes the value of a given resistor is an increase in its cross-sectional area. This is exactly the effect which is obtained by connecting a number of resistors in parallel. There is a bigger cross-sectional area for current to get through, so it passes through the parallel combination of resistors *more easily* than it would pass through a single one of them.

It is just like having two water pipes laid side-by-side in a system. Given a constant pressure (head) of water, the two pipes together will obviously pass more water than either could alone.

The conclusion is that *resistors or loads connected in parallel present a lower combined resistance or load than does any one of them individually.* If your load consists of four 200-ohm resistors in parallel, the resistance of the combined load will (since current divides equally through equal resistances) be one-fourth of the value of any one of the individual resistors.

Thus, the total load will be 50 ohms; and a parallel connection, such as that shown below, will act in a circuit as if it were *equivalent* to a single resistance of 50 ohms. It will often be convenient to combine parallel loads into a single one for purposes of calculation.

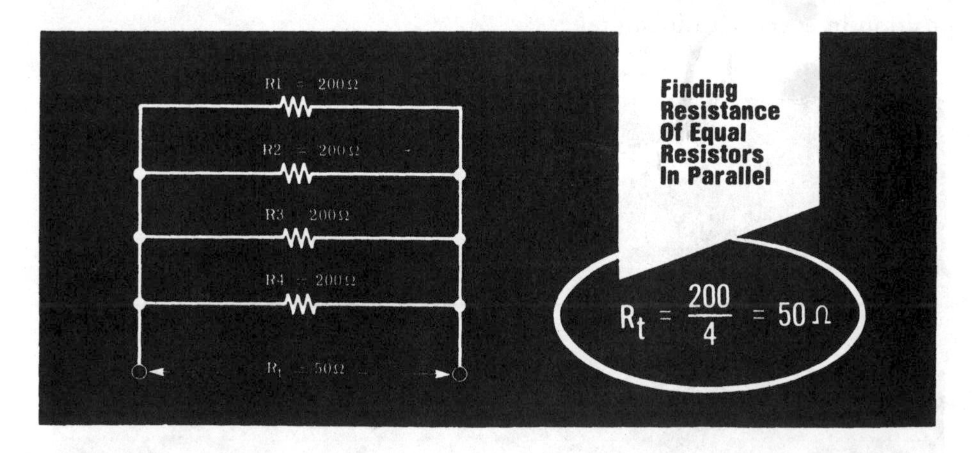

Unequal Resistors in Parallel Circuits

When a circuit contains resistors in parallel whose values are *unequal*, the problem of assessing total resistance becomes more difficult.

In favorable circumstances, you can find out the equivalent value of resistance offered by two unequal resistors in parallel by using an ohmmeter. In a parallel circuit consisting of two resistors, R1 and R2, whose values are 60 and 40 ohms, respectively, you would, in fact, get an ohmmeter reading of 24 ohms for total resistance.

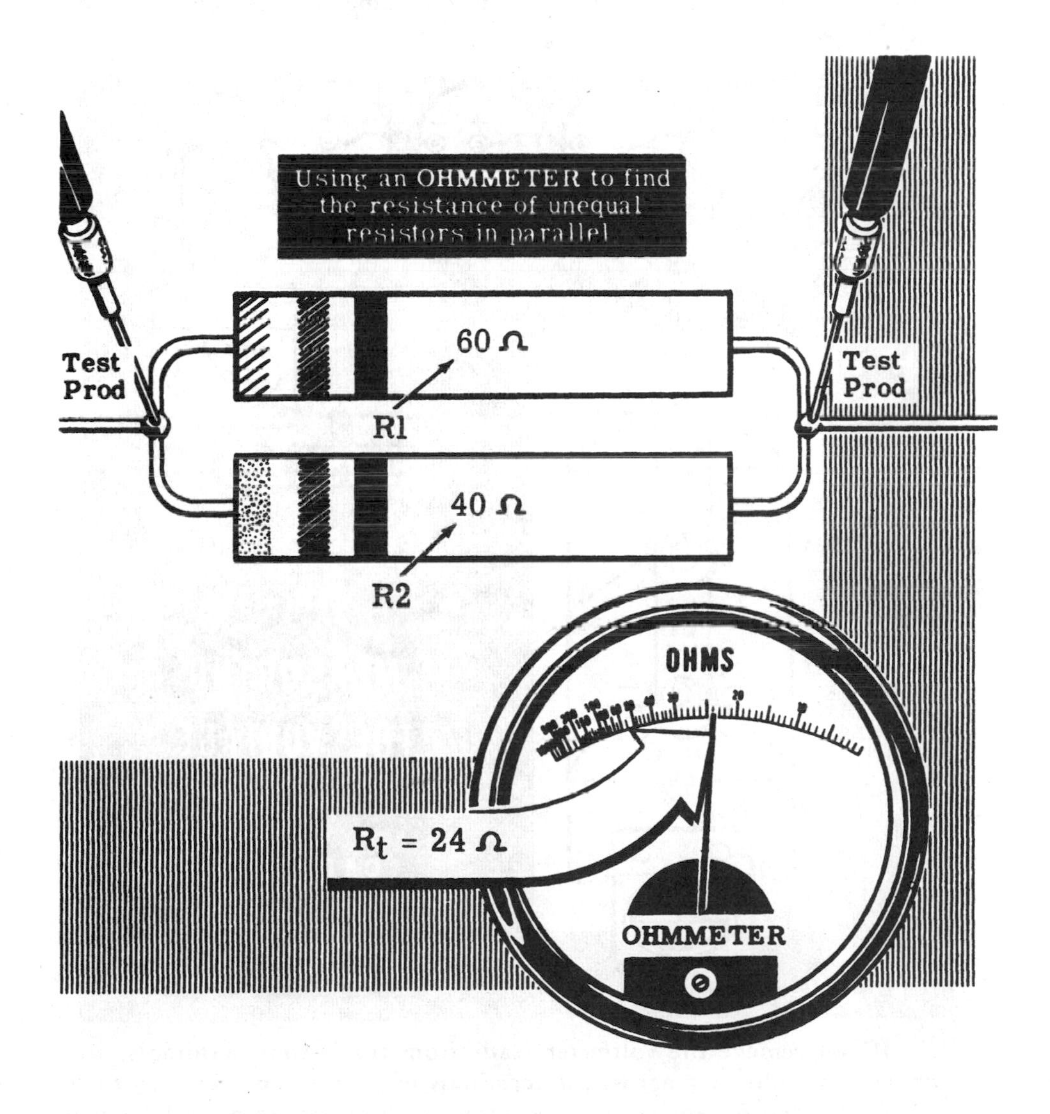

And now for some experiments to demonstrate further and apply your understanding of dc parallel circuits.

Experiment/Application—Parallel Circuit Voltage

While the current through the various branches of a parallel circuit is not always the same, the voltage across each branch resistance is equal to that across the others. If you connect three lamp sockets in parallel and insert 6-volt, 250-mA (0.25 ampere) lamps in each of the three parallel-connected sockets, you will see that each lamp lights with the same brilliance as when only a single lamp is used, and that for three lamps the circuit current is 750 mA. Also, you will see that the voltmeter reading across the battery terminals is the same whether one, two, or three lamps are used.

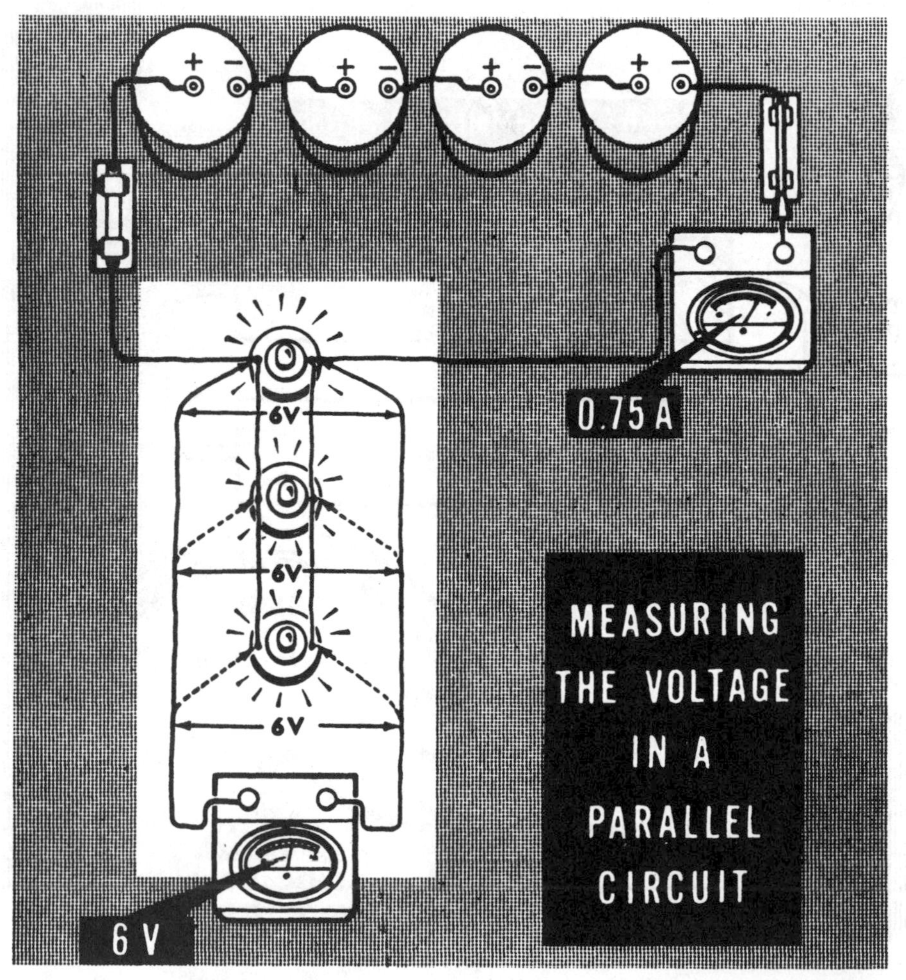

If you remove the voltmeter leads from the battery terminals, and connect the voltmeter across the terminals of each lamp socket in turn, you see that the voltage is the same across each of the lamps and is equal to that of the voltage source—the battery. This demonstrates that the voltage across each circuit element is the same in a parallel circuit.

Experiment/Application—Parallel Circuit Current

To demonstrate the division of current, you could replace a 250-mA lamp with a 150-mA lamp, as shown. The ammeter now shows that the total circuit current is 400 mA. By connecting the ammeter first in series with one lamp, then in series with the other, you will see that the 400-mA total current divides, with 250 mA flowing through one lamp and 150 mA through the other. Then connect the ammeter to read the total circuit current at that end of the parallel combination opposite the end at which it was originally connected. You will see that the total circuit current is the same at each end of the parallel circuit, the current dividing to flow through the parallel branches of the circuit and combining again after passing through these branches.

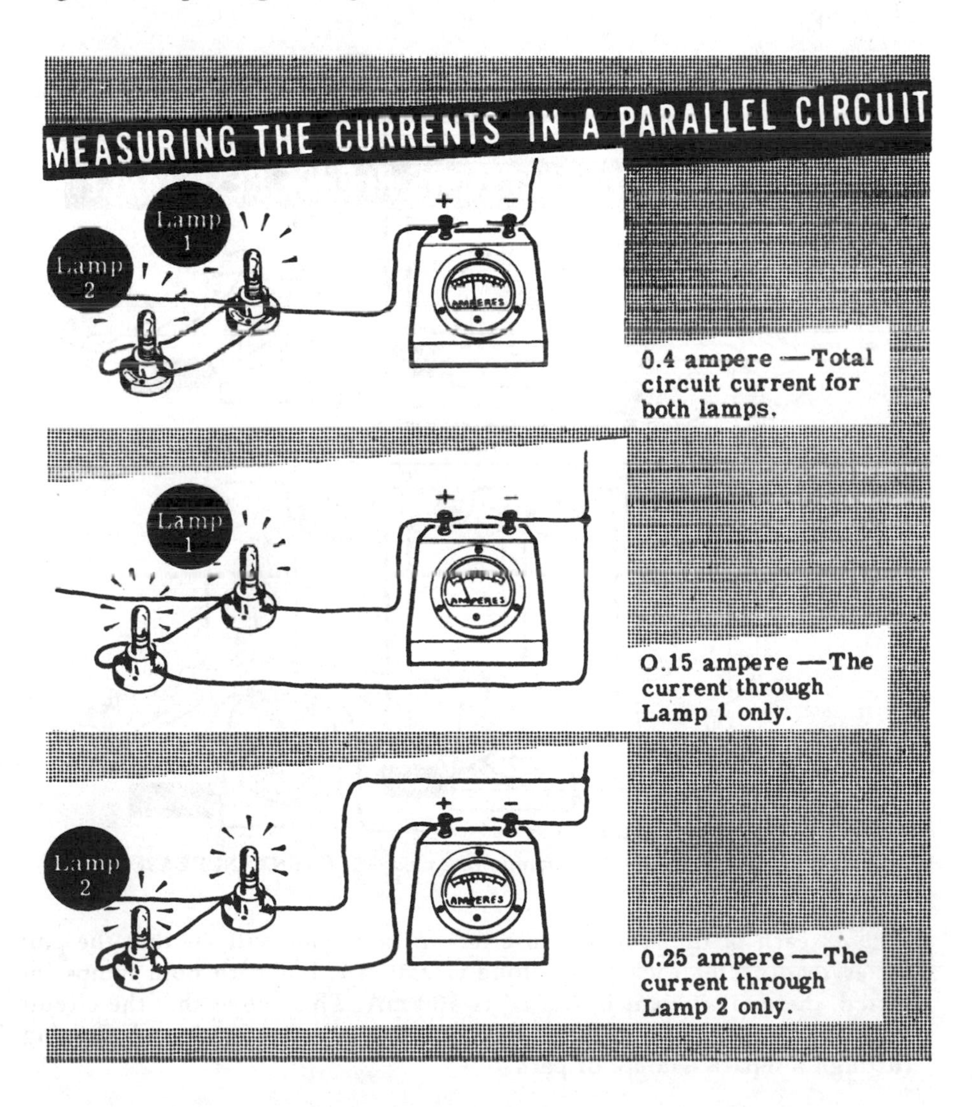
MEASURING THE CURRENTS IN A PARALLEL CIRCUIT
Lamp 1
Lamp 2
+
−
0.4 ampere —Total circuit current for both lamps.
Lamp 1
+
−
0.15 ampere —The current through Lamp 1 only.
Lamp 2
+
−
0.25 ampere —The current through Lamp 2 only.

Experiment/Application—Parallel Circuit Resistance

When the total current flow in a circuit increases with no change in the voltage, a decrease in total resistance is indicated. To show this effect, you could connect two lamp sockets in parallel. This parallel combination is connected across the terminals of a 6-volt battery of dry cells with an ammeter inserted in one battery lead to measure the total circuit current. As a voltmeter is connected across the battery terminals you will see that the voltage is 6 volts. When only one 250-mA, 6-volt lamp is inserted, the ammeter indicates a current flow of approximately 250 mA, and the voltmeter reads 6 volts. With both lamps inserted, the current reading increases to 500 mA (0.5 ampere), but the voltage remains at 6 volts—indicating that the parallel circuit offers less resistance than a single lamp.

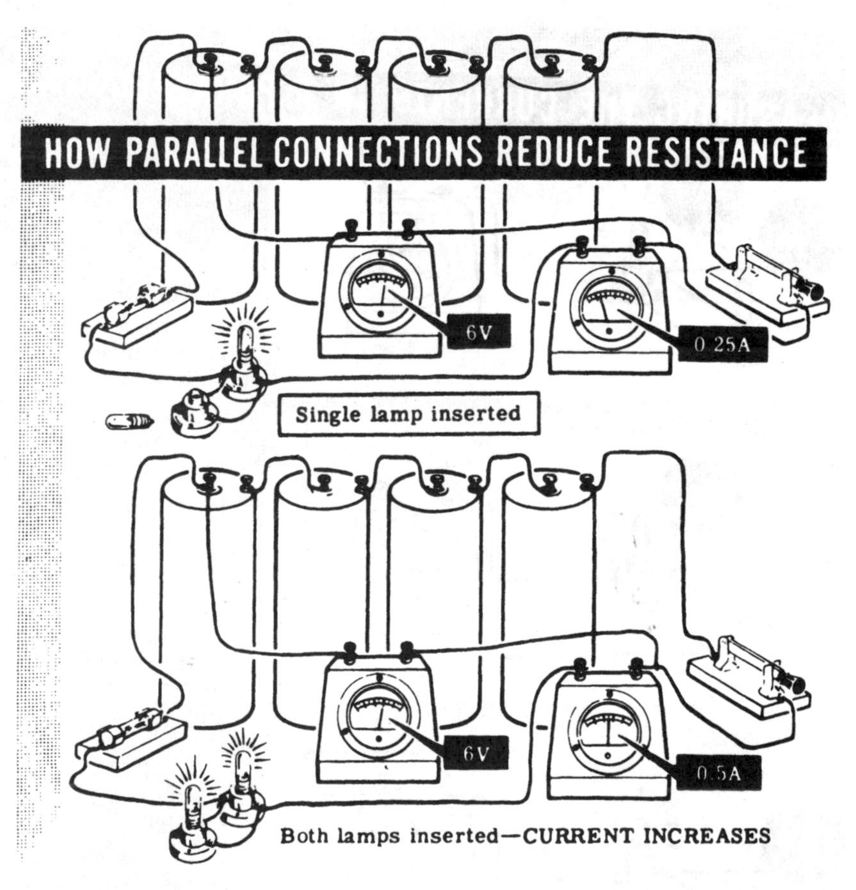

As each of the lamps is inserted in turn, you will see that the ammeter reading for each lamp alone is 250 mA; but with both lamps inserted, the total current indicated is 500 mA. This shows that the circuit current of 500 mA divides into two 250-mA currents, with each flowing through a separate lamp, or path.

Experiment/Application—Parallel Resistances

To show how parallel connection of resistances decreases the total resistance, you could measure the resistance of three 330-ohm resistors individually with the ohmmeter. When two of the 330-ohm resistors are paralleled, the total resistance should be 165 ohms; this is shown by connecting them and measuring the parallel resistance with the ohmmeter. As another 330-ohm resistor is connected in parallel, you will see that the resistance is lowered to a value of 110 ohms. This not only shows that connecting *equal* resistances in parallel reduces the total resistance, but also that the total resistance can be found by dividing the value of a single resistance by the number of resistances used.

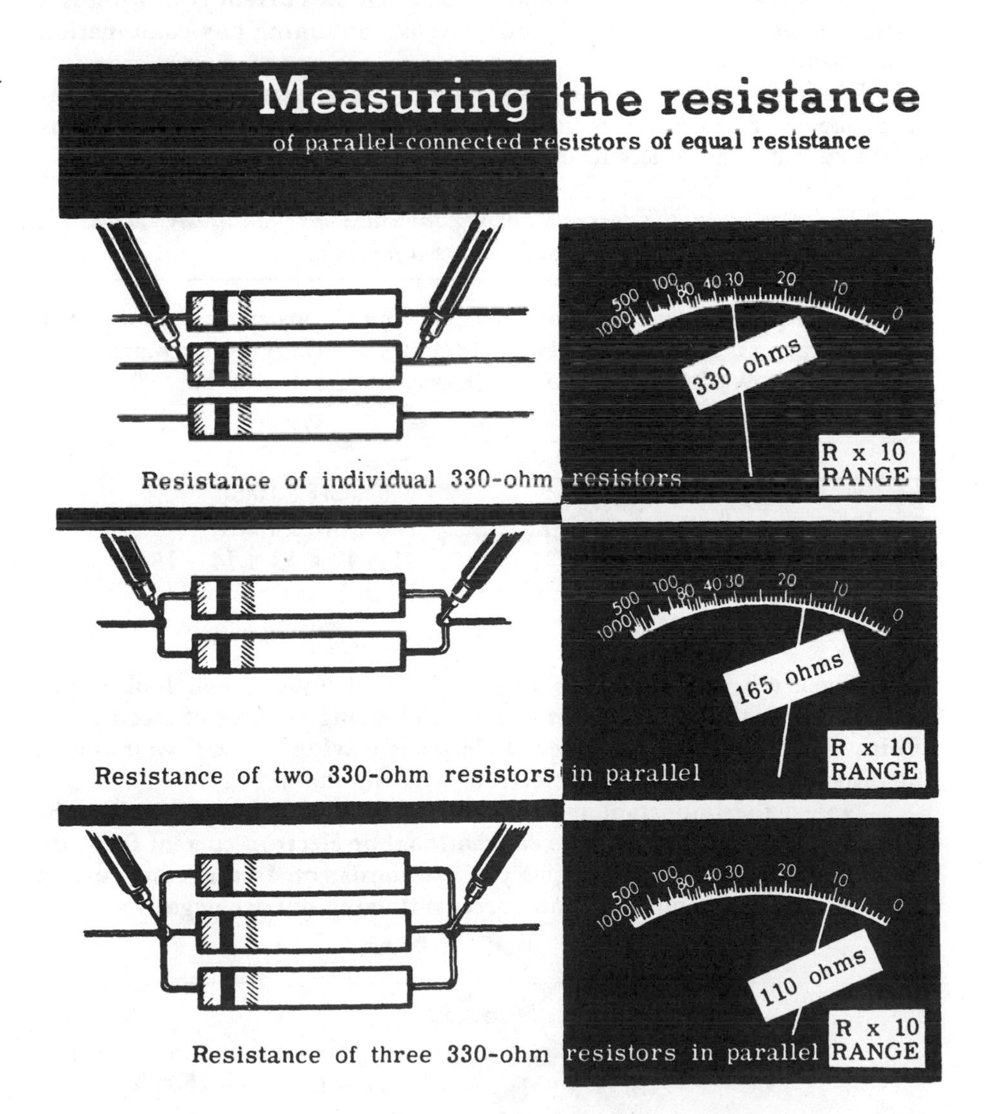

Kirchhoff's First Law

It is often not possible to use an ohmmeter to get resistance readings —and you ought, in any case, to know how such answers can be found by calculation. You will not be surprised to hear that it is quite easily done by intelligent use of Ohm's Law—but you also will need help from another equation which you have been waiting for some time to hear about. It is known as *Kirchhoff's First Law.*

You already know that in a series circuit the current entering the circuit is exactly equal to the current leaving the circuit. If you have grasped correctly the idea of current flow as a stream of electrons traveling around their circuit, you will see at once that the statement above must be true whether the circuit around which the current is flowing is a series circuit, a parallel circuit, or a circuit containing any combination of the two.

Kirchhoff's First Law is, thus, true of every type of circuit. It concerns, however, not the circuit as a whole but only individual junctions where currents combine within the circuit itself. It states:

> *The sum of all the currents flowing toward a junction always equals the sum of all the currents flowing away from that junction.*

Suppose you have a circuit, part of which consists of a junction of five conductors, and that all five conductors are carrying currents in the directions shown in the illustration below.

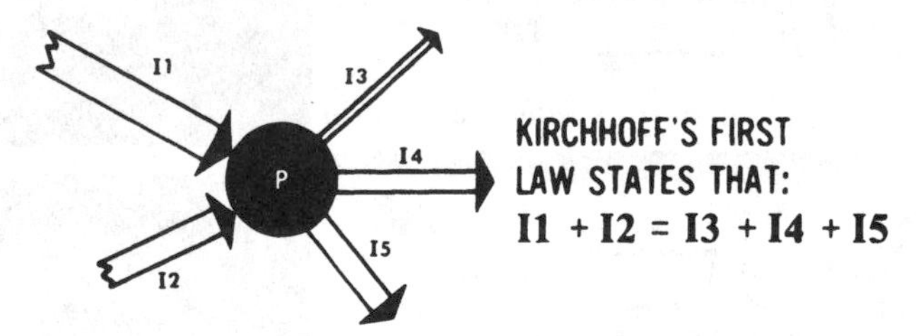

The truth of Kirchhoff's First Law is obvious if you look at the drawing above. Currents I_1 and I_2 are delivering streams of electrons to Point P. Therefore, the number of electrons leaving Point P *must* always be the same as the number of electrons arriving there.

Notice the important point that *direction* has been assigned to the current flow. Whether you use conventional or electron-current flow, direction is unimportant as long as you are consistent. In this case, current into the junction is positive, and currents flowing out are negative.

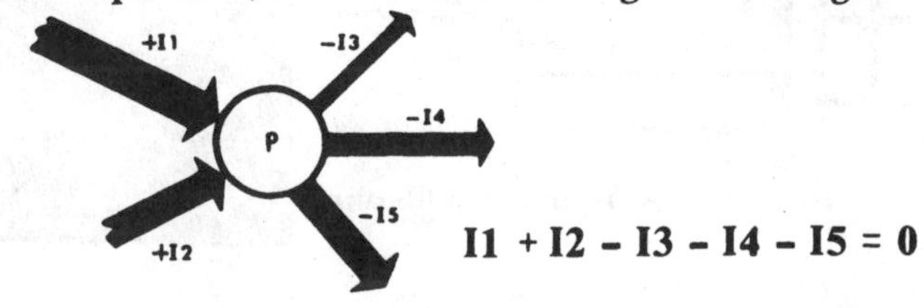

Kirchhoff's First Law (continued)

To use Kirchhoff's First Law in a complete circuit, the rule is (as always): *First draw the circuit.* Then indicate on the circuit diagram the direction of current flow through every resistance in the circuit. Then determine which of these currents flows toward, and which away from, every junction in the circuit. Mark this information in on the circuit diagram. The value and direction of flow of *unknown* currents can then often be determined by applying the Law.

At the circuit junction pictured below, neither the direction nor the value of I1 is known.

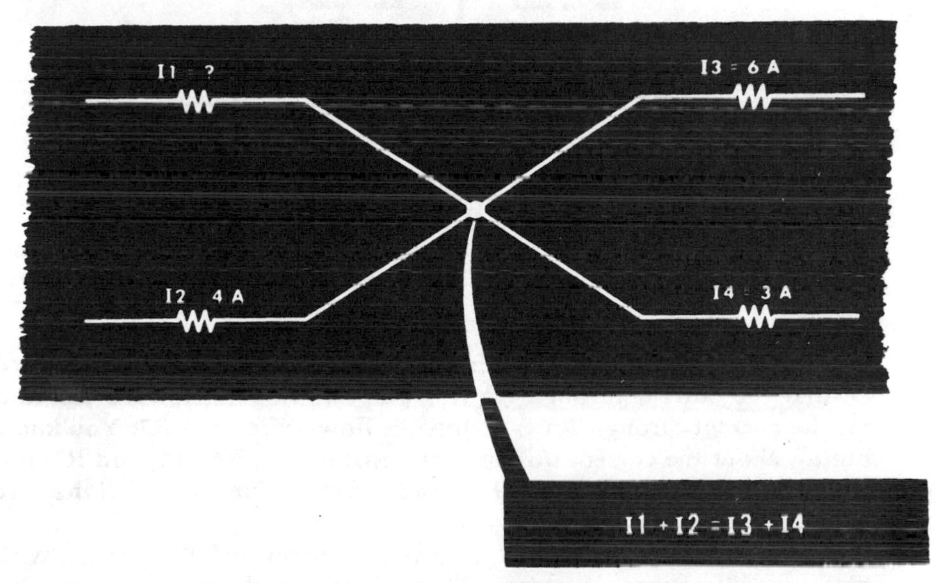

The direction of the unknown current is first determined by adding together all the known currents flowing toward the junction and all those flowing away from it, and then comparing the two. Here, the sum of the currents flowing away from the junction (I_3 = 6 amperes plus I_4 = 3 amperes, making a total of 9 amperes) is greater than is the value of the one known current (I_2 = 4 amperes) flowing into the junction.

It follows that the unknown current (I_1) must also be flowing *into* the junction, or else the balance of electron flow at the junction could not be maintained. Its value can be determined by substituting known values in the Kirchhoff equation, $I_1 + I_2 = I_3 + I_4$.

Therefore $$I_1 + 4\text{ A} = 6\text{ A} + 3\text{ A} = 9\text{ A}$$
$$I_1 = 9\text{ A} - 4\text{ A} = 5\text{ A}$$

Also using $$I_1 + I_2 - I_3 - I_4 = 0$$
$$I_1 + 4\text{ A} - 6\text{ A} - 3\text{ A} = 0$$
$$I_1 - 5\text{ A} = 0$$
$$I_1 = 5\text{ A}$$

Kirchhoff's First Law (continued)

Now take a slightly more complicated example of using Kirchhoff's First Law to find out the values and directions of flow of unknown currents in a circuit.

Suppose you have a circuit consisting of seven resistors, connected as shown in the diagram below. This is a series-parallel circuit, a type about which you will learn much more later.

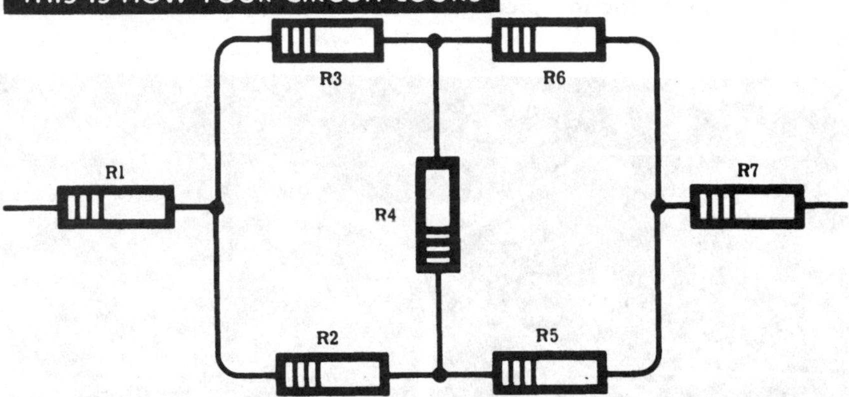

You know that the current through R2 is 7 amperes flowing toward R5; that the current through R3 is 3 amperes flowing toward R6; and that the current through R5 is 5 amperes flowing toward R7. You know nothing about the current through the resistors R1, R4, R6, and R7; but you need to know both their values and the directions in which they are flowing. Here is how it is done.

Draw the circuit in symbolic form, designating all currents, with values and directions, if known. Then identify each junction of two or more resistors with a letter.

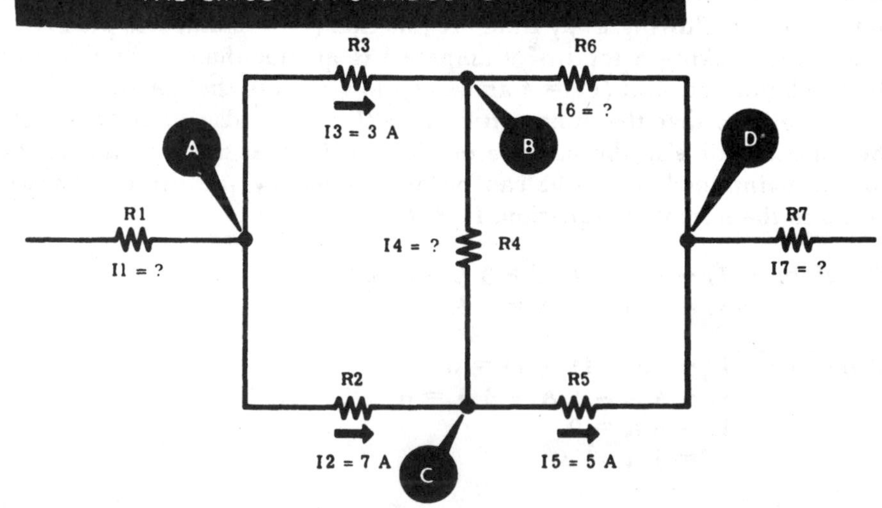

Kirchhoff's First Law (continued)

Find the unknown currents at all junctions where only one current is unknown; then you can use these new values to find unknown values at other junctions.

From the circuit you can see that junctions A and C have only one unknown. So start by finding the unknown current at junction A:

Of the three currents at junction A—I1, I2, and I3—both I2 and I3 are known, and flow away from the junction. I1 must, therefore, flow toward the junction, and its value must be equal to the sum of I2 and I3.

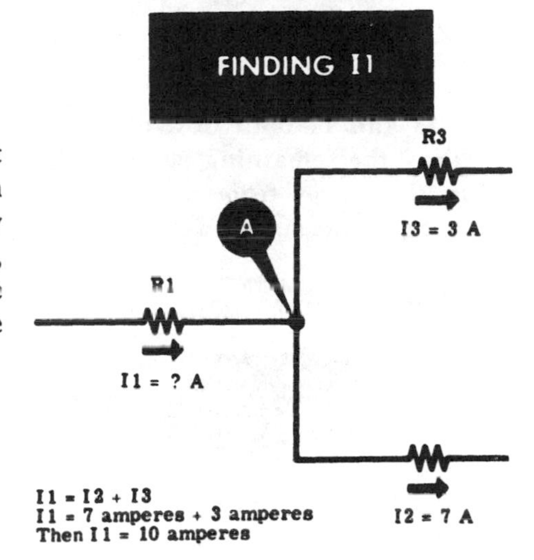

Next find the unknown current at junction C:

At C two currents—I2 and I5—are known, and only I4 is unknown. Since I2, flowing toward C, is greater than I5, flowing away from C, then the third current I4 also must flow away from C. Also, since the current flowing toward C equals that flowing away from it, it follows that I2 equals I4 plus I5.

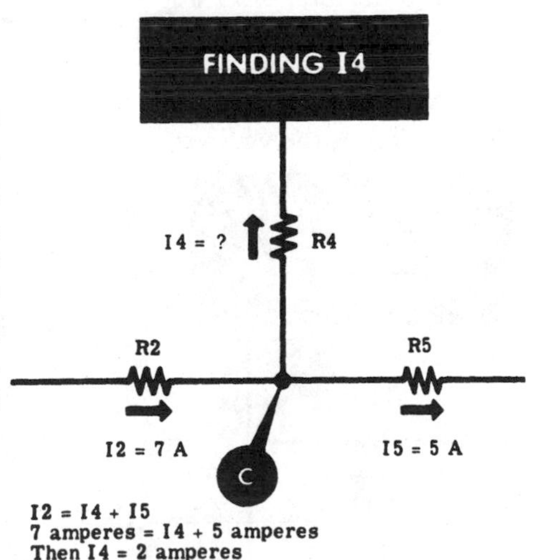

Kirchhoff's First Law (continued)

Now that the value and direction of I4 are known, only I6 is unknown for junction B. You can find the amount and direction of I6 by applying the law for current at B.

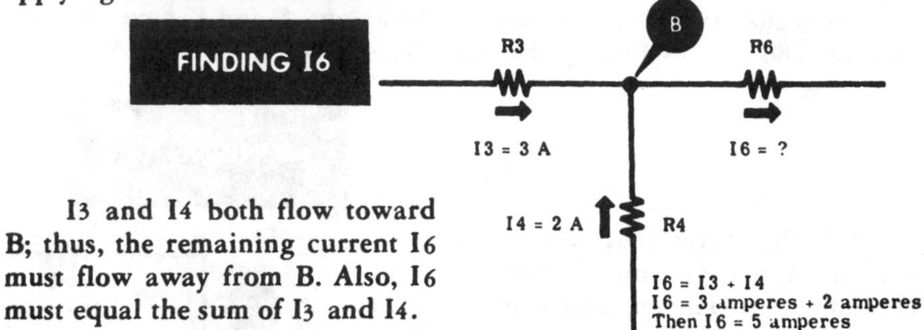

I3 and I4 both flow toward B; thus, the remaining current I6 must flow away from B. Also, I6 must equal the sum of I3 and I4.

With I6 known, only I7 remains unknown, at junction D.

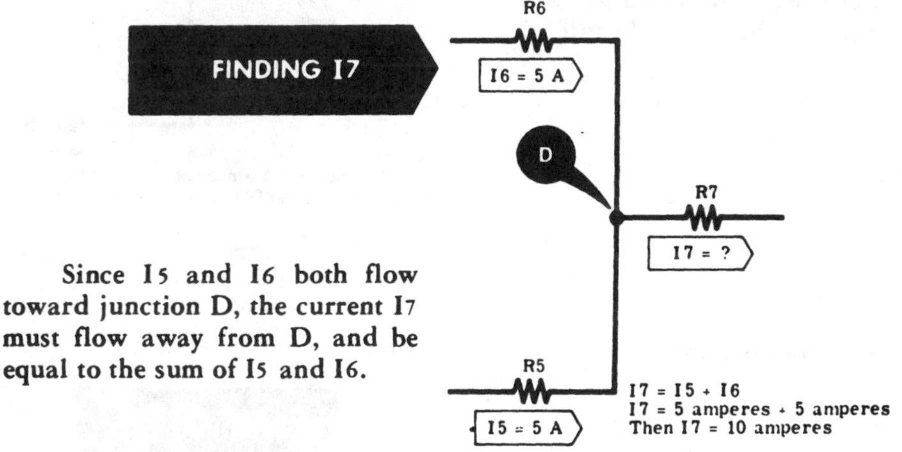

Since I5 and I6 both flow toward junction D, the current I7 must flow away from D, and be equal to the sum of I5 and I6.

You now know all of the circuit currents and the directions of their flow through the various resistors.

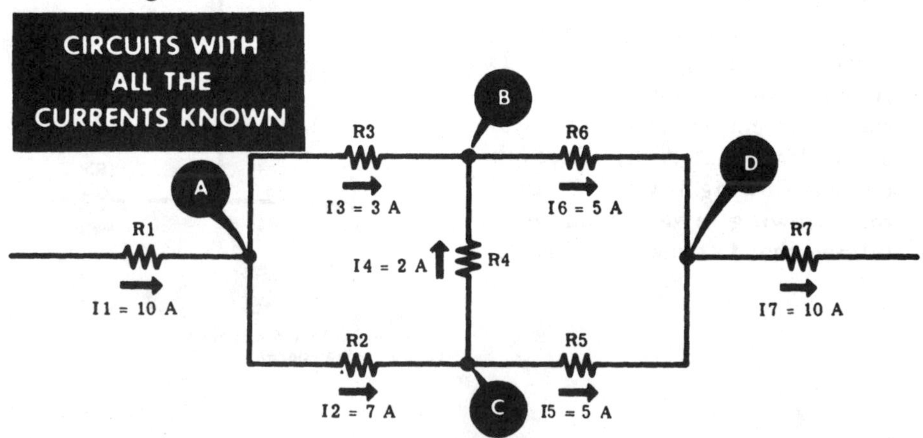

Experiment/Application— Kirchhoff's First Law

Check for yourself Kirchhoff's First Law—the law of circuit currents. Suppose you connect a 15-ohm resistor in series with a parallel combination of three 15-ohm resistors, and then connect the entire circuit across a 9-volt dry cell battery with a switch and fuse in series.

The total resistance of the circuit is 20 ohms, resulting (by Ohm's Law) in a total circuit current of 0.45 ampere. This total current must flow through the circuit from the negative (−) to positive (+) battery terminals (see the circuit diagram below). At junction (a), the circuit current—0.45 ampere—divides to flow through the three parallel resistors toward junction (b). Since the parallel resistors are all equal, the current divides equally, with 0.15 ampere flowing through each resistor. At junction (b), the three parallel currents combine, to flow away from the junction through the series resistor.

If you connect an ammeter to read the current in each lead at junction (b), you will see that the sum of the three currents flowing toward the junction is equal to the current flowing away from the junction.

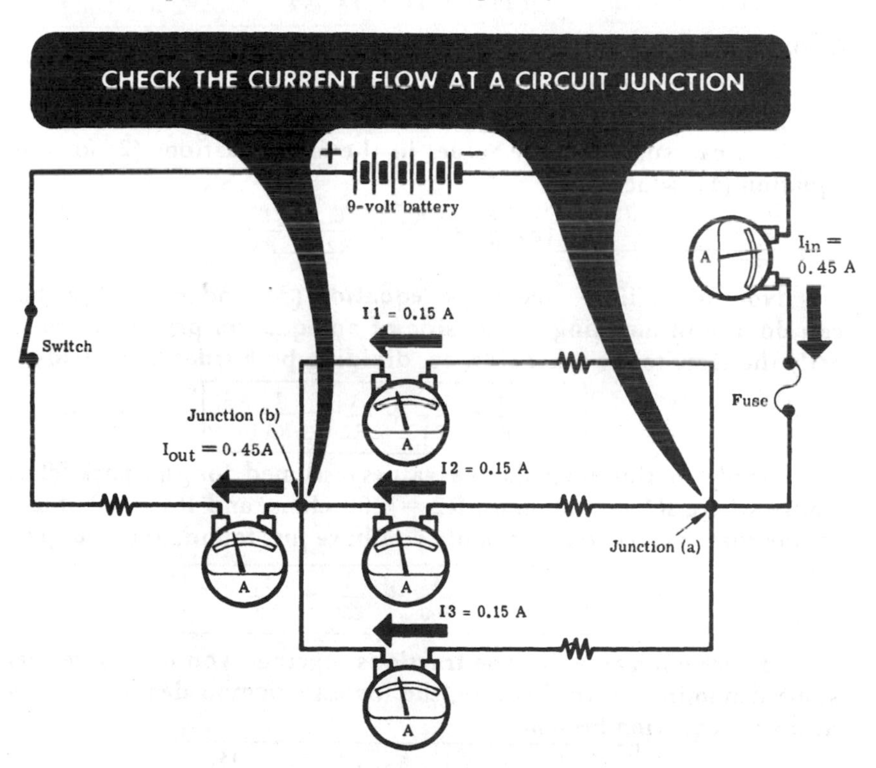

Current flowing toward the junction	=	Current flowing away from the junction

$$I_{in} = I_1 + I_2 + I_3$$

$$I_1 + I_2 + I_3 = I_{out}$$

$$0.15 + 0.15 + 0.15 = 0.45 \text{ ampere}$$

Unequal Resistors in Parallel Circuits (continued)

With this understanding of Kirchhoff's First Law you are now ready to return to the subject of unequal resistors in parallel circuits and to find out how to calculate the effective resistance of any number of unequal resistors so connected in parallel.

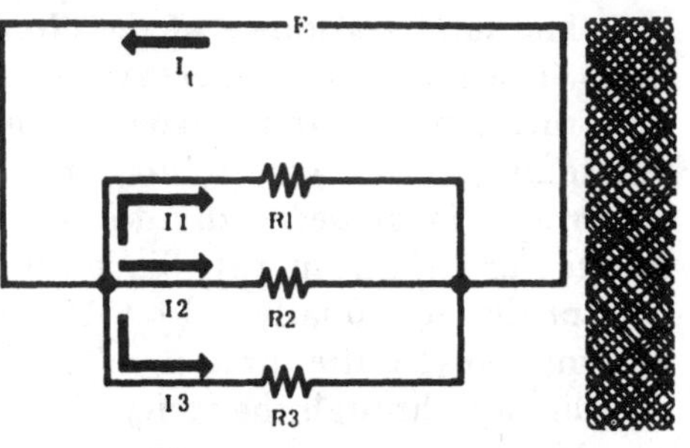

In the circuit opposite, the current divides to pass through three resistors. Apply Kirchhoff's First Law and Ohm's Law to the circuit exactly as you did before.

You know from Kirchhoff's First Law that, in a circuit like the one above,

$$(1)\ I_t = I1 + I2 + I3$$

Also, from Ohm's Law you know that

$$(2)\ I_t = \frac{E}{R_t},\ I1 = \frac{E}{R1},\ I2 = \frac{E}{R2},\ I3 = \frac{E}{R3}$$

You can substitute the values in the four equations (2) for those in equation (1), which gives

$$(3)\ \frac{E}{R_t} = \frac{E}{R1} + \frac{E}{R2} + \frac{E}{R3}$$

Now use a little algebra on equation (3), and remember that you can do almost anything to one side of an equation provided you do exactly the same to the other side. So, dividing both sides by E yields

$$\frac{1}{R_t} = \frac{1}{R1} + \frac{1}{R2} + \frac{1}{R3}$$

Apply to this formula the values assumed for the parallel circuit above where R1 = 300 ohms, R2 = 200 ohms, and R3 = 60 ohms. Substitute these values in the formula you have just found, and you get

$$\frac{1}{R_t} = \frac{1}{300} + \frac{1}{200} + \frac{1}{60}$$

Before you can add these fractions together, you must give them the same denominator. In this case, the lowest common denominator is 600; and your equation becomes

$$\frac{1}{R_t} = \frac{2}{600} + \frac{3}{600} + \frac{10}{600} = \frac{15}{600}$$

Now turn (invert) both sides of the equation ($\frac{1}{R_t} = \frac{15}{600}$) upside down, and you get

$$R_t = \frac{600}{15} = 40 \text{ ohms}$$

Unequal Resistors in Parallel Circuits (continued)

The formula you have just worked out can easily be extended to take account of any number of resistors in a parallel combination. If you had five resistors in your parallel circuit, for instance, the formula for their composite resistance would be

$$\frac{1}{R_t} = \frac{1}{R1} + \frac{1}{R2} + \frac{1}{R3} + \frac{1}{R4} + \frac{1}{R5}$$

A simplified formula for two resistors of unequal value can be derived by extending the formula from the previous page for any number of resistors in parallel.

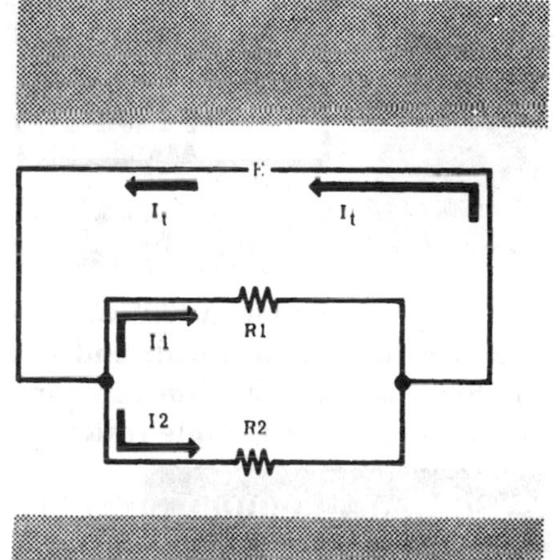

Take a simple circuit like the one opposite, with a current flowing through a pair of resistors connected in parallel, and then back to the voltage source. You know that

$$\frac{1}{R_t} = \frac{1}{R1} + \frac{1}{R2}$$

A simple mathematical formula for finding the least common denominator is

$$\frac{1}{A} + \frac{1}{B} = \frac{B+A}{AB}$$

Use of this rule in the equation for two parallel resistors yields

$$\frac{1}{R_t} = \frac{R2+R1}{R1 \times R2}$$

Since we have found a least common denominator, we can turn both sides of the third equation upside down (invert), and get

$$R_t = \frac{R1 \times R2}{R1+R2}$$

You can express the above formula in words as follows: *Total resistance in a parallel circuit composed of two resistors of unequal value is found by MULTIPLYING the value of one resistor by the value of the second; by ADDING the value of one resistor to the value of the second; and then by DIVIDING the first result by the second result.* You will find this a very useful equation to know.

Unequal Resistors in Parallel Circuits (continued)

You can now apply this formula to the parallel circuit containing two resistors with values of 60 and 40 ohms which you were considering on page 2-65. Recall that an ohmmeter reading of the total resistance of such a circuit would be 24 ohms. Let us see if the formula gives the same result.

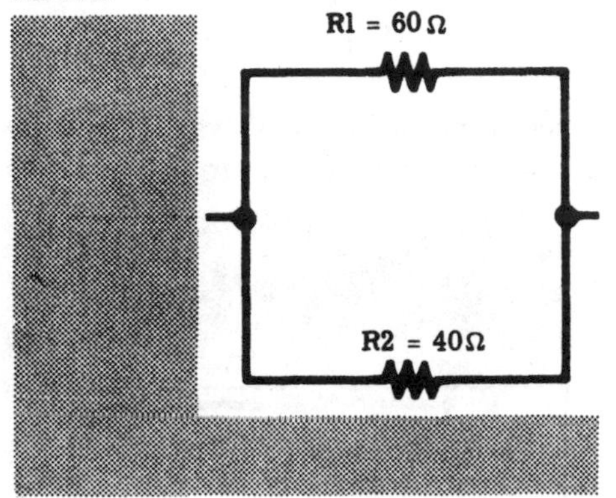

1. *Multiply* the values of the two resistors:

$$60 \times 40 = 2{,}400$$

2. *Add* the values of the two resistors:

$$60 + 40 = 100$$

3. *Divide* the product by the sum:

$$\frac{2{,}400}{100} = 24 \text{ ohms}$$

So you see that the *calculated* value of the total resistance and its *indicated* value are identical; and you can confidently say that the parallel combination of a 60-ohm resistor and a 40-ohm resistor will always act as though it were a single resistor with a value of 24 ohms.

Another Example

This formula is so useful that it is worthwhile working through one more example to fix it firmly in your mind.

When two resistors, R1 = 120 ohms and R2 = 60 ohms, are connected in parallel, what is their total resistance?

Draw the circuit diagram, and write down the formula:

$$R_t = \frac{R1 \times R2}{R1 + R2}$$

Substitute known values in the formula:

$$R_t = \frac{120 \times 60}{120 + 60}$$

$$= \frac{7{,}200}{180}$$

$$= 40 \text{ ohms}$$

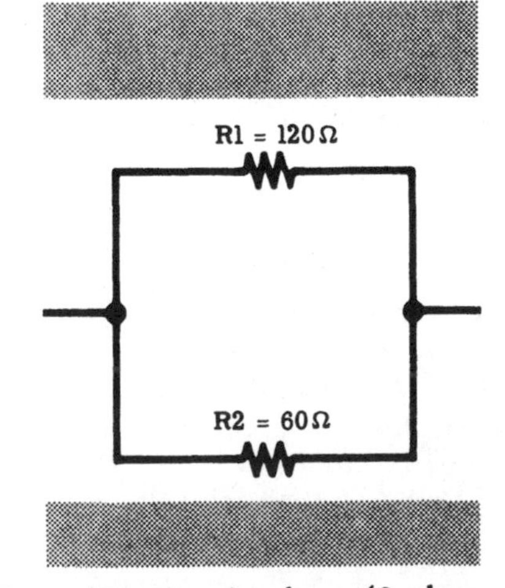

The total resistance of the parallel combination is, thus, 40 ohms, and the combination will act as if it were a single resistor of that value.

Unequal Resistors in Parallel Circuits (continued)

You can use the formula for finding the value of two resistors in parallel to solve more complicated problems by successive application of the formula to reduce the circuit to simpler terms.

Suppose you have three resistors connected in parallel—R1 with a value of 300 ohms, R2 with a value of 200 ohms, and R3 with a value of 60 ohms. What is the effective resistance of the combination?

Take, first, R1 and R2 alone. Substitute their known values in the formula $R_t = \frac{R1 \times R2}{R1 + R2}$ and you will get a value for the *equivalent resistance* of the R1 and R2 combination, which we will call R_a. Then combine R_a with R3 in the same way and your answer will be the effective resistance of the complete circuit.

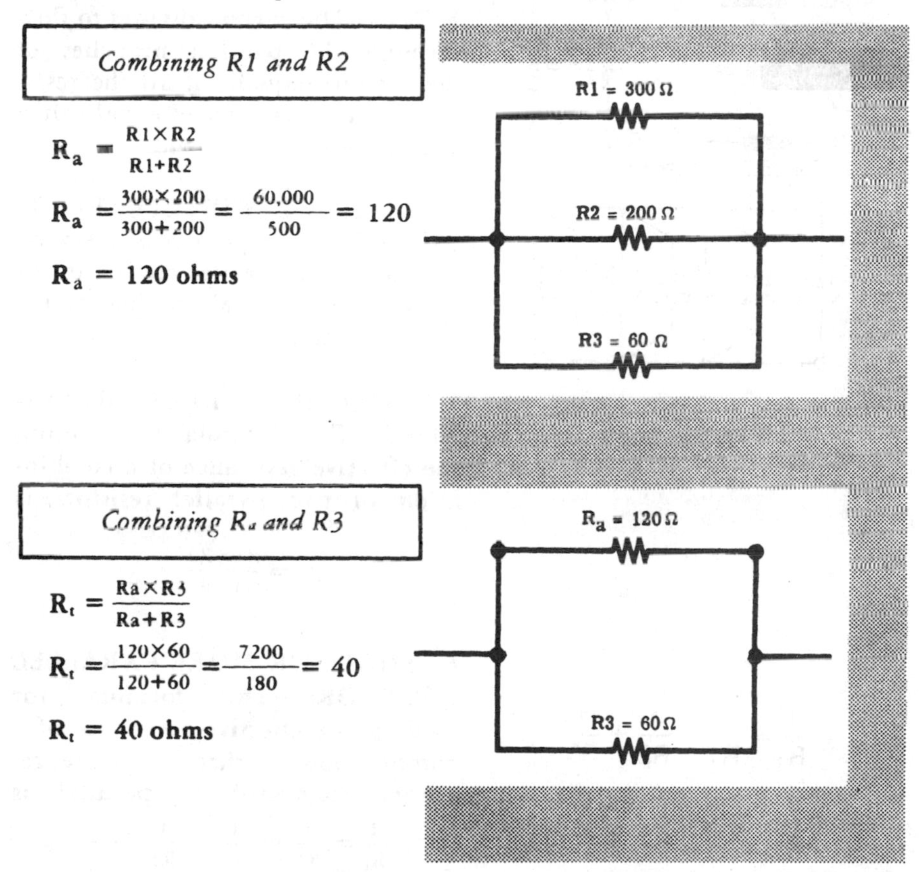

Combining R1 and R2
Ra = R1×R2 / R1+R2
Ra = 300×200 / 300+200 = 60,000 / 500 = 120
Ra = 120 ohms
R1 = 300 Ω
R2 = 200 Ω
R3 = 60 Ω
Combining Ra and R3
Rt = Ra×R3 / Ra+R3
Rt = 120×60 / 120+60 = 7200 / 180 = 40
Rt = 40 ohms
Ra = 120 Ω
R3 = 60 Ω

The total resistance of the three resistors connected in parallel is 40 ohms, and the combination will act as a single 40-ohm resistor in the circuit.

You will note that this is the same answer that you got on page 2-76 by applying the basic formula for resistance in parallel.

Review of Parallel Circuits

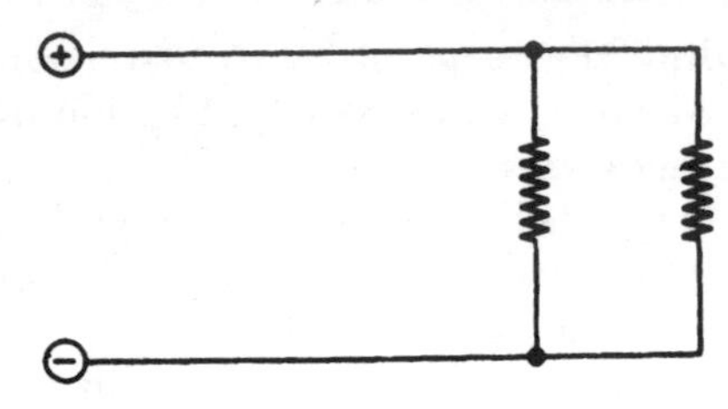

1. PARALLEL CIRCUIT — The circuit formed when resistances are connected side-by-side across a voltage source.

2. PARALLEL CIRCUIT RESISTANCE—The total resistance in a parallel circuit is lower than is that of the smallest individual resistance in the circuit.

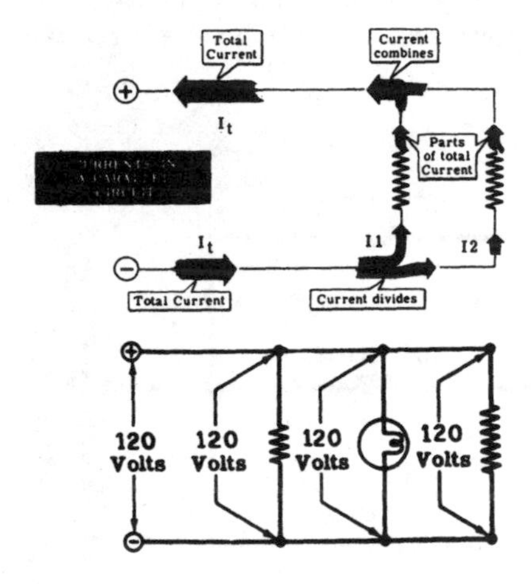

3. PARALLEL CIRCUIT CURRENT—The current divides to flow through the parallel branches of the circuit—equally if all the resistors in the circuit are of equal value; unequally, if they are not.

4. PARALLEL CIRCUIT VOLTAGE—The voltage across every resistance in a parallel circuit is the same, and is equal to that of the voltage source.

$$R_t = \frac{R1 \times R2}{R1 + R2}$$

5. TWO PARALLEL RESISTORS—The formula for finding the effective resistance of a combination of two parallel resistors is

$$R_t = \frac{R1 \times R2}{R1 + R2}$$

$$\frac{1}{R_t} = \frac{1}{R1} + \frac{1}{R2} + \frac{1}{R3}$$

6. THREE OR MORE PARALLEL RESISTORS—The formula for finding the effective resistance of a combination of three or more resistors connected in parallel is

$$\frac{1}{R_t} = \frac{1}{R1} + \frac{1}{R2} + \frac{1}{R3} \cdots$$

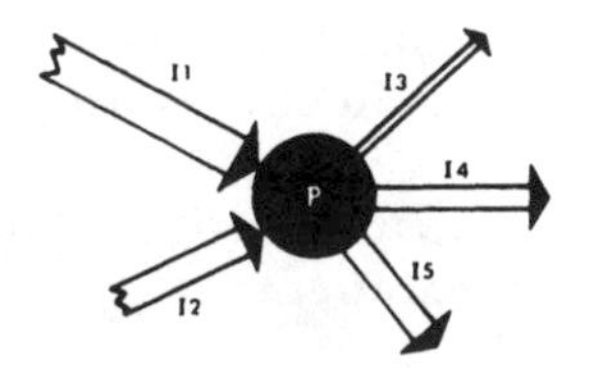

7. KIRCHHOFF'S FIRST LAW—The sum of all the currents flowing toward a junction always equals the sum of all the currents flowing away from that junction.

Self-Test— Review Questions

1. Calculate the equivalent value of resistance in each of the following parallel connections:
   (a) Two resistors of 12 and 8 ohms, respectively.
   (b) Six resistors, each valued at 4.8 ohms.
   (c) Two resistors of 20 and 4 K, respectively.
   (d) Two resistors of 1 M and 1.5 M, respectively.
2. Calculate the equivalent value of resistance for each of the following parallel connections:
   (a) Three resistors of 20, 30, and 40 ohms, respectively.
   (b) Four resistors of 20, 30, 40, and 50 ohms, respectively.
   (c) Three resistors of 20, 30, and 40 K, respectively.
3. Sketch the *equivalent circuit* of the parallel connection shown below:

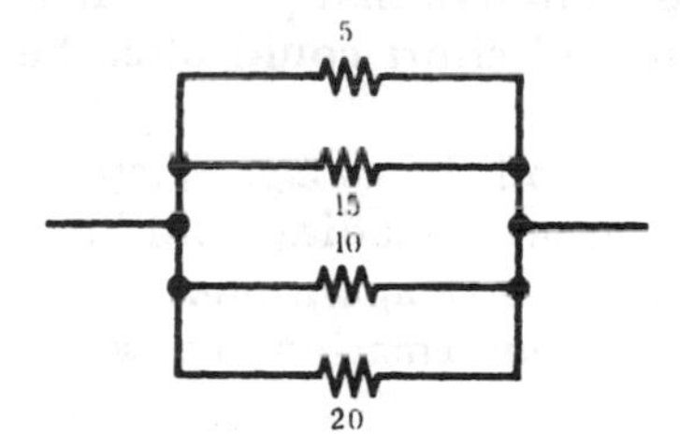

4. What value of resistance would you need to connect in parallel with a 20-ohm resistor to produce an equivalent resistance of 12 ohms?
5. A 22-K resistor having a tolerance of 10% has failed. You have on hand five resistors, valued at 120, 220, 47, 33, and 120 K, respectively. What combination of these would you choose to connect in parallel to replace the failed component? Give your answer by sketching the circuit diagram of the resultant.
6. In the circuit shown, what is value and direction of I3?

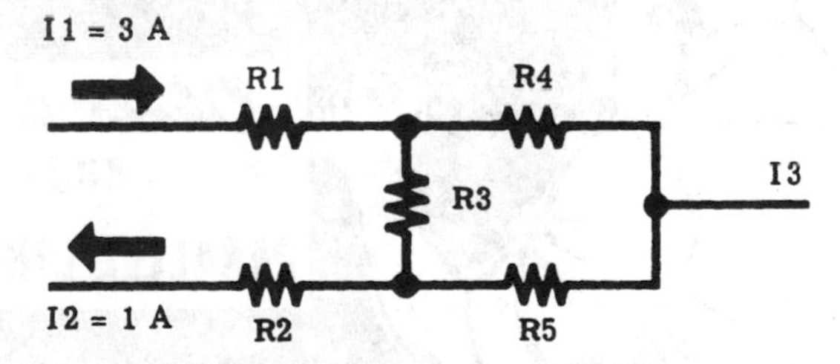

7. Sketch the equivalent circuit of

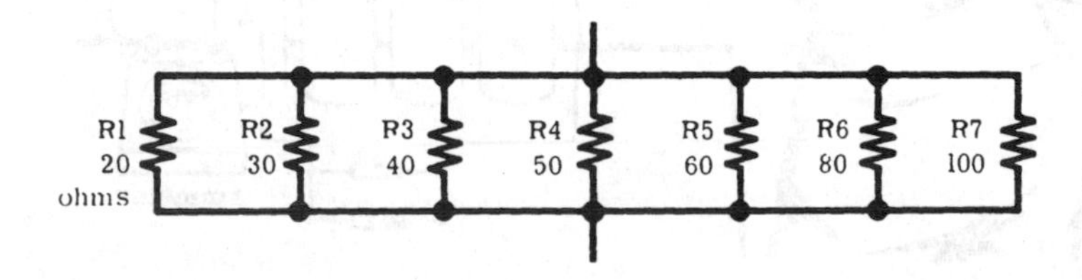

Applying Ohm's Law in Parallel Circuits

You have already seen something of the use of Ohm's Law in a parallel circuit. In practice, a good many more unknown quantities of current, voltage, and resistance in such circuits can generally be calculated by using this law.

Suppose you wanted to find out the *resistance* of a resistor connected in parallel with one or more other resistors. If you used an ohmmeter, you would first have to disconnect the resistor to be measured from the circuit; otherwise the ohmmeter would read the total resistance of the parallel combination of resistors.

Again, if you were to set about measuring with an ammeter the *current flow* through one particular resistor of a combination of parallel resistors, you would first have to disconnect it, and insert the ammeter to read only the current flow through that particular resistor.

In either case, time and effort could often be saved by the use of Ohm's Law.

If you were trying to find the *voltage* existing across a parallel circuit, of course, a direct voltmeter reading could be obtained without any need for a disconnection; but here again intelligent application of Ohm's Law will often give you the information you want without the need for actual measurement.

Solving Unknowns in Parallel Circuits

Six facts, all of them known to you by now, are all the equipment you need for finding out the unknown values in a dc parallel circuit.

In the circuit opposite, a voltage E is applied across three resistors —R1, R2, and R3—connected in parallel. What are the facts you know about this circuit?

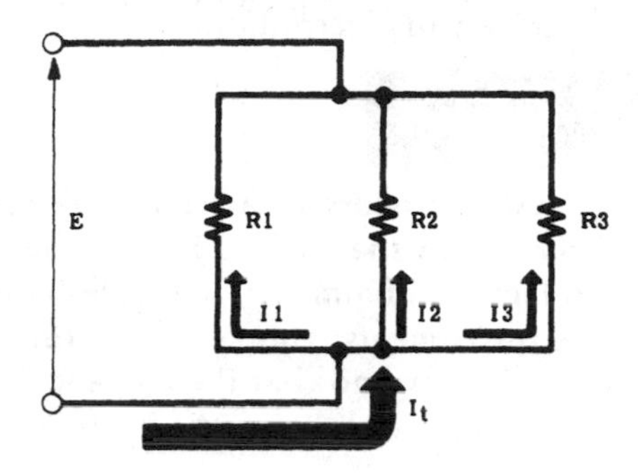

FACT 1: Kirchhoff's First Law tells you that

$$I_t = I1 + I2 + I3$$

FACT 2: You know that the full circuit voltage (E) appears across each one of the three parallel resistors.

FACT 3: Fact 2, plus Ohm's Law, tells you that

$$I1 = \frac{E}{R1} \quad I2 = \frac{E}{R2} \quad I3 = \frac{E}{R3}$$

FACT 4: Any parallel circuit can be reduced to an equivalent circuit.

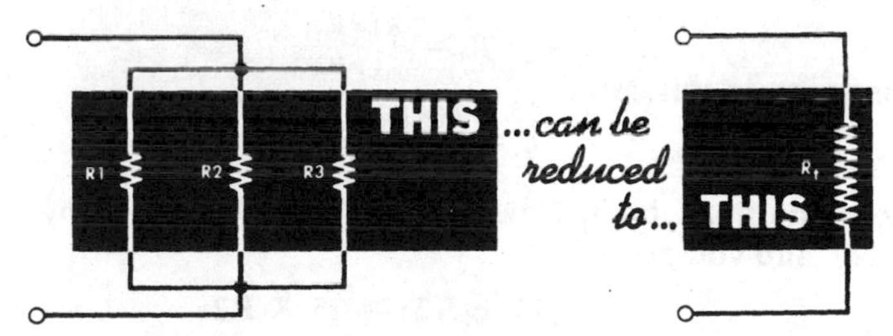

FACT 5: Ohm's Law can then be applied to the equivalent circuit.

$$I_t = \frac{E}{R_t} \text{ or } R_t = \frac{E}{I_t} \text{ or } E = I_t R_t$$

FACT 6: The total resistance of any parallel circuit can be found by applying the formula

$$\frac{1}{R_t} = \frac{1}{R1} + \frac{1}{R2} + \frac{1}{R3} + \ldots$$

and for 2 resistors in parallel

$$R_t = \frac{R1 \times R2}{R1 + R2}$$

Solving Unknowns in Parallel Circuits (continued)

Now watch how these six facts can be used to solve the sort of problem which you will be constantly meeting in your practical work in electricity and electronics.

Problem 1

Two resistors in a circuit are connected in parallel. One of them is marked with the value *15 ohms*, but the other is unmarked in any way. You lack an ohmmeter, but have both a voltmeter and an ammeter. With these instruments, you read that the voltage across the parallel combination is 6 volts and that the current flowing into the combination is 1 ampere. You need to know the value of the unmarked resistor.

Solution. First, sketch the circuit, and fill in the values you know.

You see at once that you know two of the three Ohm's Law quantities, and that $R_t = \frac{E}{I_t} = \frac{6}{1}$. The total resistance of the parallel circuit is, therefore, 6 ohms.

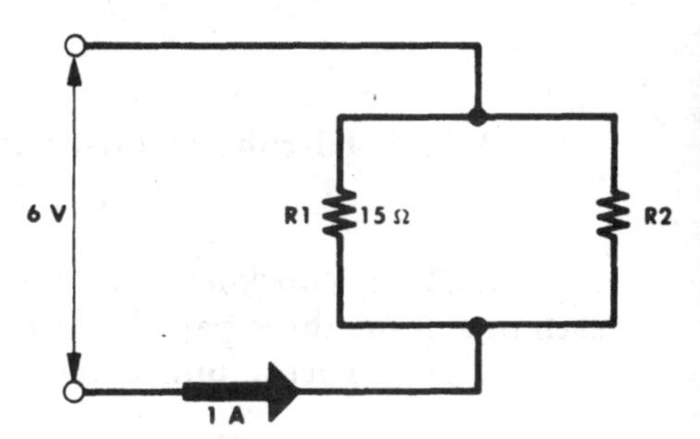

You already know the value of R1 (it is marked *15 ohms*). So take Fact 6 (the Parallel Resistance Formula), and substitute known values wherever you can.

$$R_t = \frac{R1 \times R2}{R1 + R2}$$

Therefore, in this case,

$$6 = \frac{15 \times R2}{15 + R2}$$

Now multiply both sides of this last equation by the factor (15 + R2), and you get

$$6 \times (15 + R2) = 15 \times R2$$

Multiply out the left-hand side of the equation:

$$90 + 6R2 = 15R2$$

Work this equation out according to the simple laws of algebra, and you see that

$$15R2 - 6R2 = 90$$

Therefore, $9R2 = 90$

and $R2 = 10$ ohms

You know now that the value of the unmarked resistor must be 10 ohms—and you have solved a typical problem of the kind you will often meet in practice by selecting one or more of the six facts which you needed, and by applying to them a little simple mathematics.

Solving Unknowns in Parallel Circuits (continued)

Try another practical problem of the same kind.

Problem 2

You are faced with a circuit having three resistors with values of 3, 9, and 12 ohms, respectively, connected in parallel. You find it possible to measure with an ammeter that the current flowing through the 9-ohm resistor is 8 amperes, but the other two resistors are inaccessible. You need to know the circuit current (possibly because you want to connect into the circuit a load which will burn out if too much current flows through it, and you need to find out the value of the resistor you will have to insert into the circuit to protect this load).

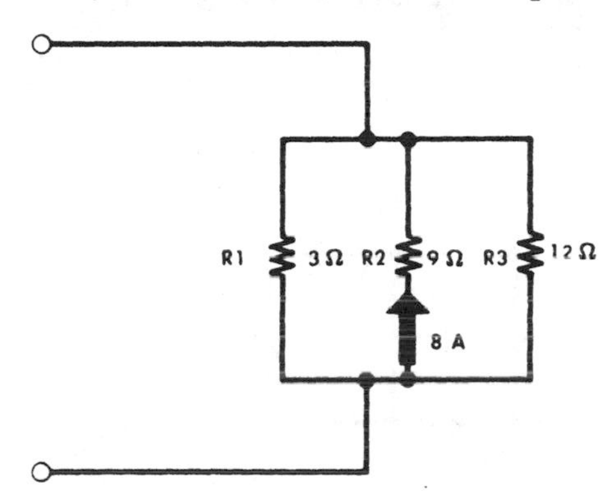

Sketch the circuit, and fill in on it the values you know. You have two of the essential facts about R2, so use Ohm's Law to calculate the third:

$$E = I_2 \times R2$$
$$= 9 \text{ ohms} \times 8 \text{ amperes}$$
$$= \textit{72 volts}$$

The voltage across one branch of a parallel circuit is the voltage across all of them; so you now have the facts you need to know about R1 and R3 to use Ohm's Law to find the current flowing through each of them.

Take R1 first. You know that its value is 3 ohms, and that the voltage across it is 72 volts. So,

$$I_1 = \frac{E}{R1} = \frac{72}{3} = 24 \text{ amperes}$$

Then deal with R3. Its value is 12 ohms, and the voltage through it is still 72 volts. So,

$$I_3 = \frac{E}{R3} = \frac{72}{12} = 6 \text{ amperes}$$

You now know the current flowing through all three branches of the parallel circuit. Kirchhoff's Second Law tells you that the circuit current itself is the sum of these three currents ($I_t = I_1 + I_2 + I_3$). So, in this case,

$$I_t = 8 + 24 + 6 = 38 \text{ amperes}$$

which is the value of the circuit current you are looking for.

Review of Ohm's Law and Parallel Circuits

If a circuit consists of two or more resistors—R1, R2, R3, etc.—in parallel, the following rules for using Ohm's Law apply:

R_t, I_t, and E are used together.
R1, I1, and E are used together.
R2, I2, and E are used together; and so on.

Only quantities having the same, or no, subscript can be used together to find an unknown by means of Ohm's Law.

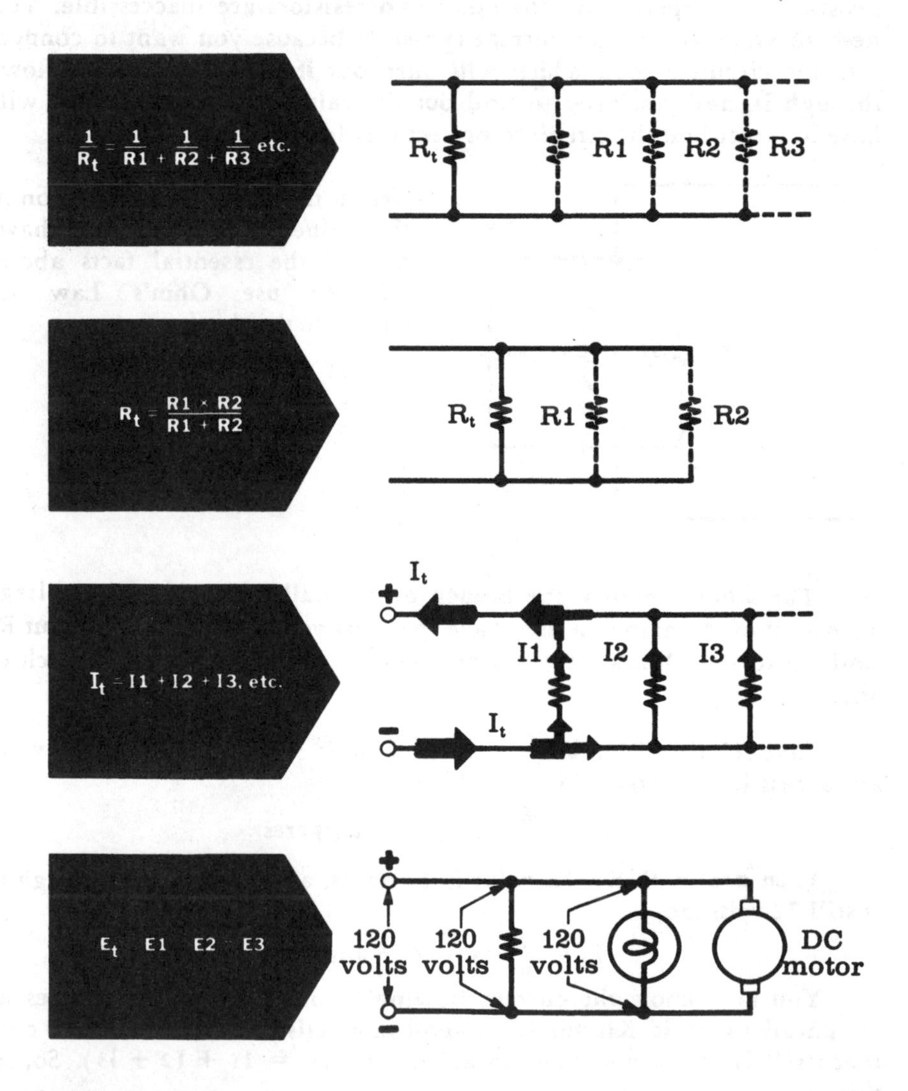

and since E = IR
$I1 \times R1 = I2 \times R2 = I3 \times R3$ etc.

Self-Test—Review Questions

1. Three resistors—R1 = 16 ohms, R2 = 24 ohms, and R3 = 32 ohms—are connected in parallel; the total current passing through them is 5.2 amperes.
   (a) What is the current passing through each resistor?
   (b) What is the voltage across the parallel combination?
2. Four resistor loads are connected in parallel across a 120-volt line. Their values are 500, 200, 100, and 50 ohms, respectively. What is the total current drawn? What are the individual currents?
3. A divider chain has been built to work into a known load as follows:

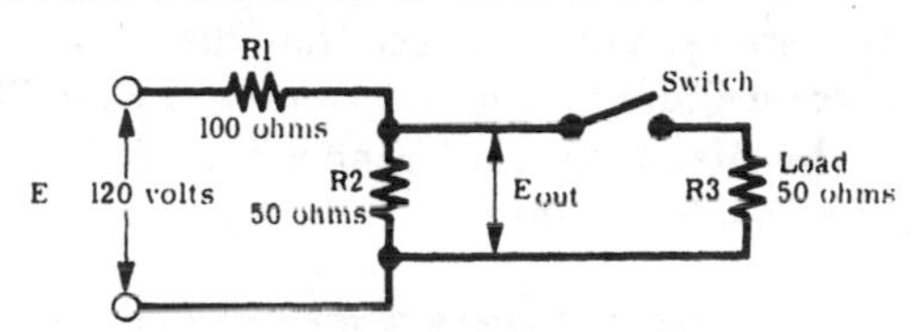

   (a) What is E_{out} with the switch open?
   (b) What is E_{out} with the switch closed?
4. Solve the circuits shown below for the unknowns:
   (a)

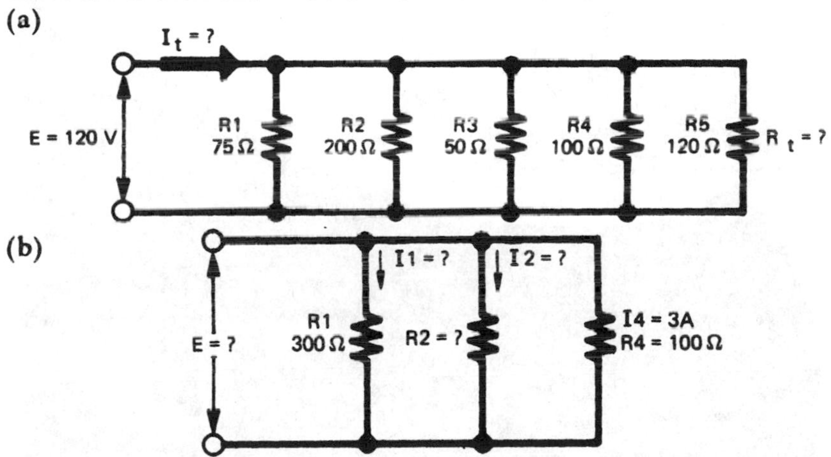

   (c) Draw the equivalent circuits for (a) and (b) above.
5. A moving-coil meter movement has a resistance of 10 ohms and needs a current of 40 mA to give it full-scale deflection. How would you adapt such an instrument in order to use it as an ammeter reading up to 0.2 ampere?
6. A moving-coil meter movement requires a current of 25 mA to give it full-scale deflection. The voltage required to produce this deflection is 25 mV. How would you adapt this meter for use as
   (a) a millivoltmeter capable of reading 0-100 mV?
   (b) a voltmeter capable of reading 0-200 V?
   (c) a milliammeter capable of reading 0-50 mA?
   (d) an ammeter capable of reading 0-50 A?

Experiment/Application—Ohm's Law and Parallel Resistances

To see how an ammeter and voltmeter may be used as a substitute for an ohmmeter to find the values of the individual and total resistances in a parallel combination, four dry cells can be connected in series to be used as a voltage source. Then connect a voltmeter across the dry cell battery to make certain the voltage remains constant at 6 volts, and connect an ammeter in series with the negative (−) terminal of the battery to read the current.

Now if you connect a fuse, a resistor with one orange band and two black ones, and a switch in series between the positive (+) terminals of the ammeter and the battery, you will see that the voltage remains at 6 volts, and that the current indicated is 0.2 ampere. From Ohm's Law, the resistance value must be 30 ohms, and a check of the color code shows that this is, indeed, the correct value.

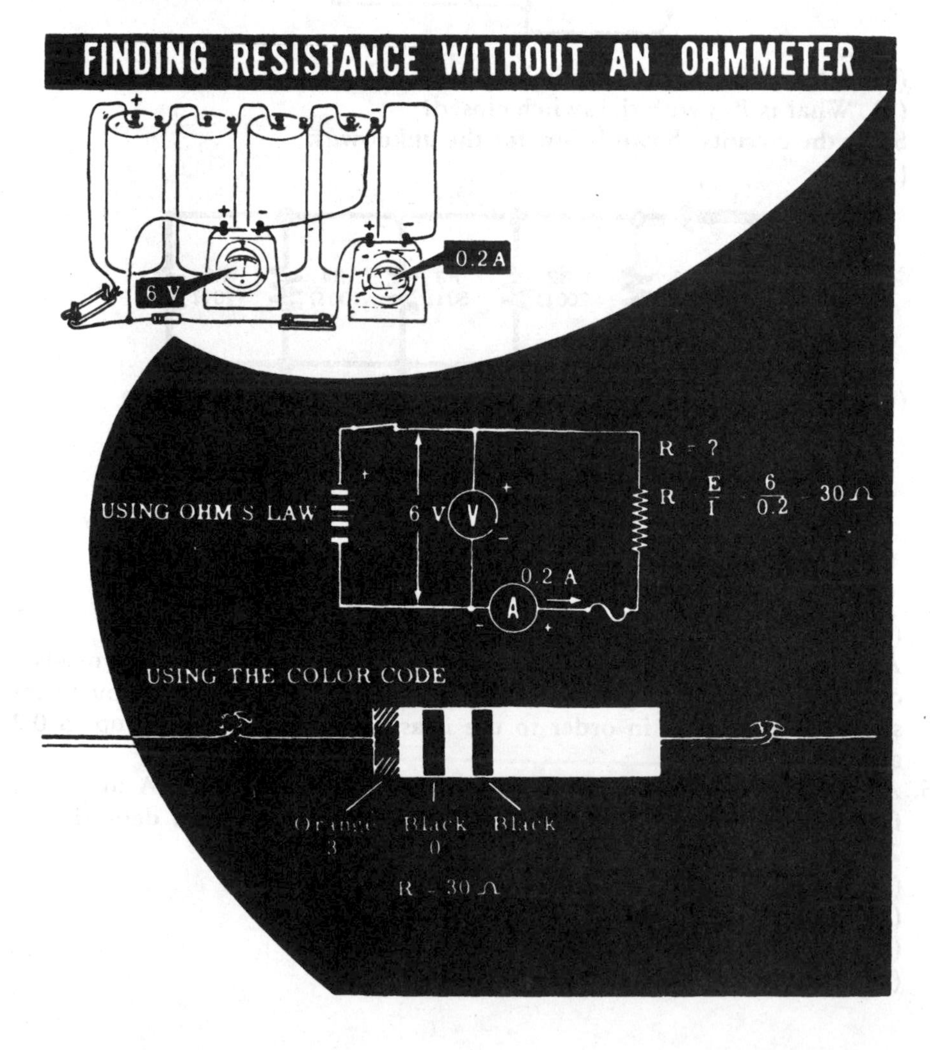

Experiment/Application—Ohm's Law and Parallel Resistances (continued)

If another resistor carrying one brown, one green, and one black band is added in parallel, you see that the current reading is 0.6 ampere, with no change in voltage. Since the first resistor passes 0.2 ampere, the current through the second resistor must be 0.4 ampere. So the Ohm's Law value of the second resistor is 6 volts divided by 0.4 ampere, or 15 ohms.

The total resistance of the parallel combination is equal to 6 volts divided by the total current, 0.6 ampere—or 10 ohms.

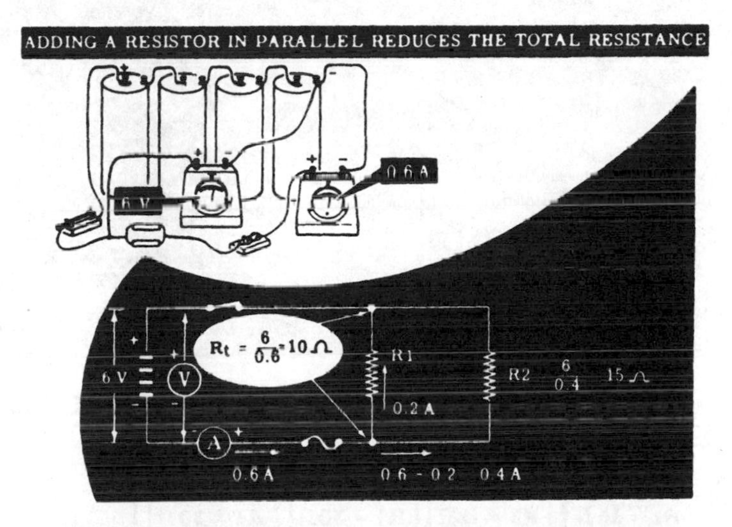

With still another resistor (one orange and two black bands) added in parallel, you see a further 0.2-ampere increase in current—showing that the Ohm's Law value of the added resistor is 30 ohms. The total current is now 0.8 ampere, resulting in a total resistance value of 7.5 ohms for the parallel combination.

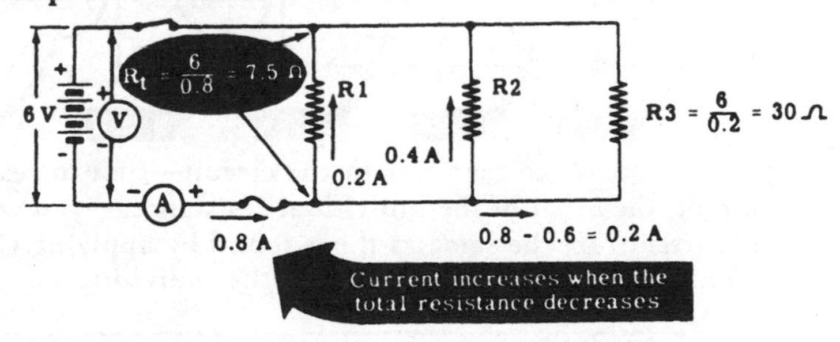

When you have disconnected the battery and the various resistors, you can check the total and the individual resistances with an ohmmeter; you will see that the Ohm's Law and color code values are identical with the values indicated by the ohmmeter.

Experiment/Application—Ohm's Law and Parallel Circuit Current

Using only three series-connected cells as a voltage source, you could connect the voltmeter across the battery. Then you could connect four resistors—two 15-ohm and two 30-ohm resistors—across the battery in parallel.

By Ohm's Law, the current through each 15-ohm resistor will be 0.3 ampere, and through each 30-ohm resistor, 0.15 ampere. The total current will be the sum of the currents through the individual resistances, or 0.9 ampere.

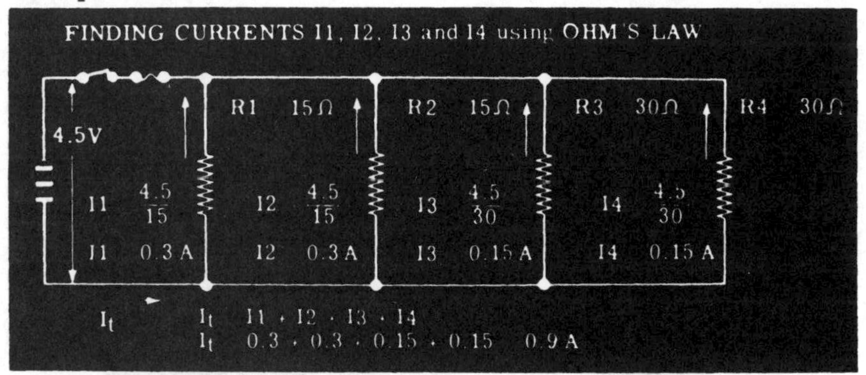

MEASURING CURRENTS IN PARALLEL CIRCUITS

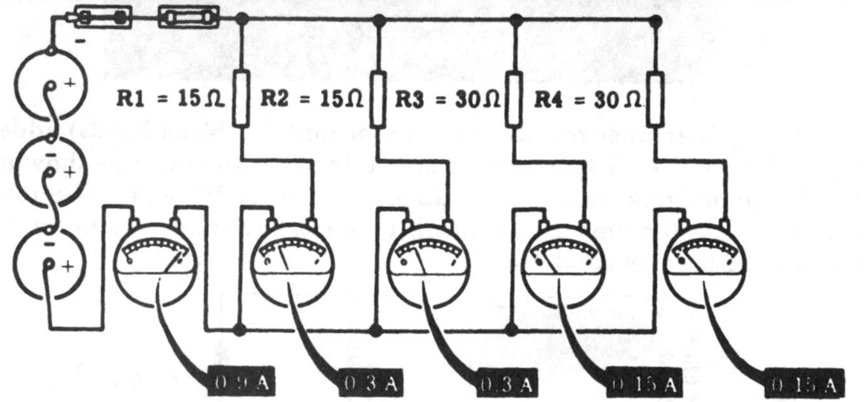

If you now insert an ammeter in the circuit—first to read the total circuit current, then that of the individual resistances—you will see that the actual currents are the same as those found by applying Ohm's Law, and that total current equals the sum of all the individual currents.

Learning Objectives—Next Section

Overview—You are now ready to proceed with solving complex series-parallel circuits. You will see that any complex circuit can be reduced to solvable combinations of series and parallel circuits.

Series-Parallel Circuits

Circuits consisting of three or more resistors may be connected in a complex circuit, partially series and partially parallel.

There are two basic types of series-parallel circuits: one in which a resistance is connected in series with a parallel combination; and the other in which one or more branches of a parallel circuit consist of resistances in series.

If you were to connect two lamps in parallel (side-by-side connection) and connect one terminal of a third lamp to one terminal of the parallel combination, the three lamps would be connected in series-parallel. Resistances other than lamps may also be connected in the same manner to form series-parallel circuits.

You can connect the three lamps to form another type of series-parallel circuit by first connecting two lamps in series, then connecting the two terminals of the third lamp across the series lamps. This forms a parallel combination with one branch of the parallel circuit consisting of two lamps in series.

Such combinations of resistance are frequently used in electric circuits, particularly in electric motor circuits and in control circuits for electrical equipment.

TWO WAYS OF CONNECTING LAMPS IN SERIES-PARALLEL

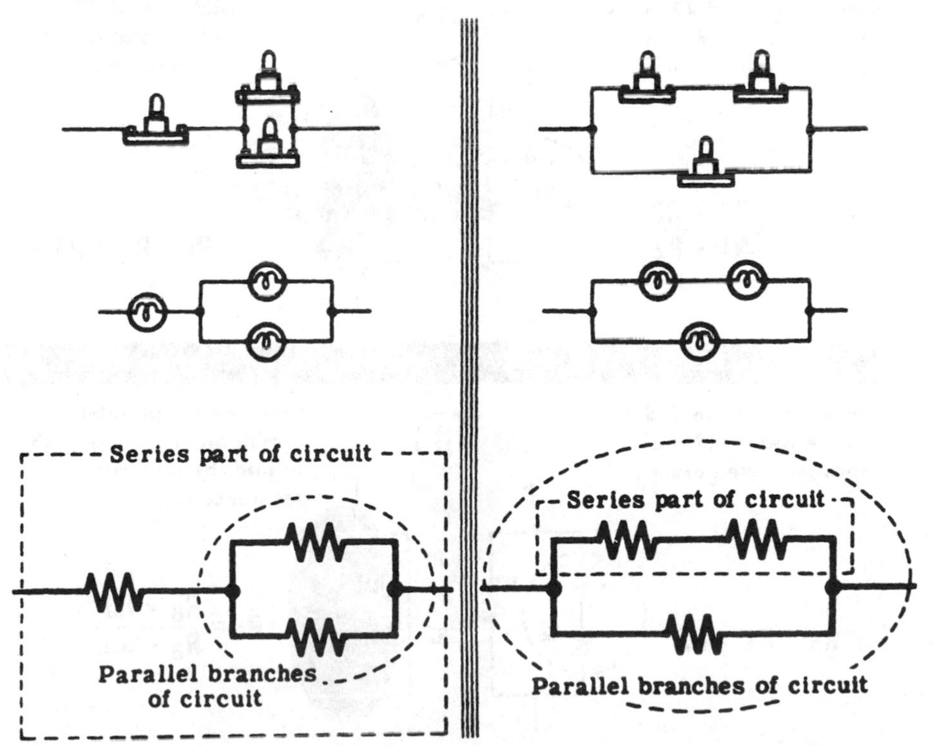

Resistors in Series-Parallel

No new formulas are needed to find the total resistance of resistors connected in series-parallel. What you need to do is to break the complete circuit into parts, each consisting of simple series and simple parallel circuits. Then solve each part separately and combine the answers. But before using the rules for series and parallel resistances, you must first decide how best to simplify the circuit.

Suppose your problem is to find the total resistance of three resistors—R1, R2, R3—connected in series-parallel, with R1 and R2 connected in parallel, and R3 connected in series with the parallel combination. To simplify the circuit, you would break it down into two parts—the parallel circuit of R1 and R2, and the series resistance R3. First you find the equivalent resistance of R1 and R2, using the formula for parallel resistances. This value is then added to the series resistance R3 to find the total resistance of the series-parallel circuit.

If the series-parallel circuit consists of R1 and R2 in series, with R3 connected across them, the steps are reversed. The circuit is broken down into two parts—the series circuit of R1 and R2, and the parallel resistance R3. First you find the total resistance of R1 and R2 by adding; then combine this value with R3, using the formula for parallel resistance.

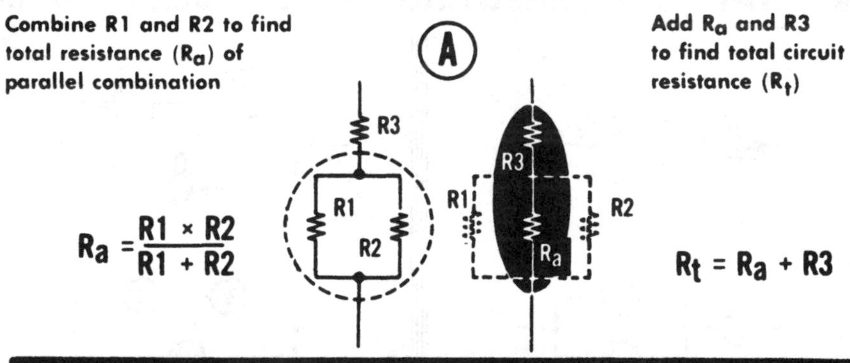

FINDING THE TOTAL RESISTANCE OF SERIES-PARALLEL CIRCUIT

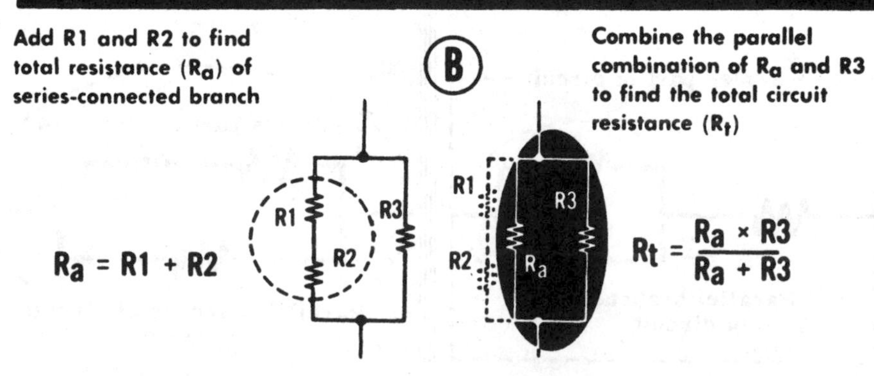

Add R1 and R2 to find
total resistance (Ra) of
series-connected branch
B
R1
R2
R3
Ra
$R_a = R1 + R2$
Combine the parallel
combination of Ra and R3
to find the total circuit
resistance (Rt)
$R_t = \frac{R_a \times R3}{R_a + R3}$

Resistors in Series-Parallel (continued)

Complex circuits may be simplified and their breakdown made easier by redrawing the circuits before applying the steps to combine resistances.

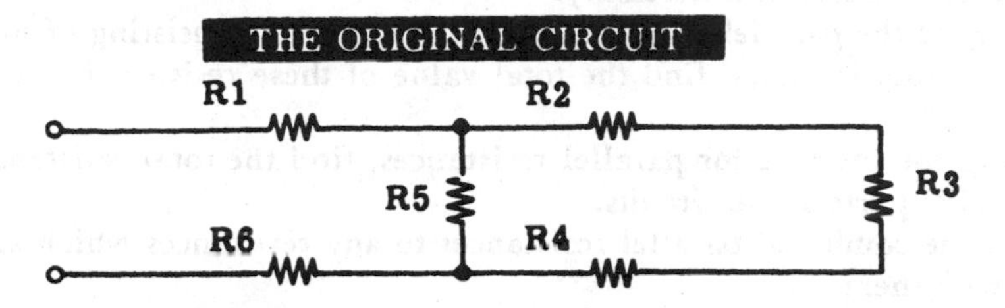

1. Start at one end of the circuit and draw all series resistances in a vertical straight line until you reach a point where the circuit has more than one path to follow. At that point draw a horizontal line across the end of the series resistance.

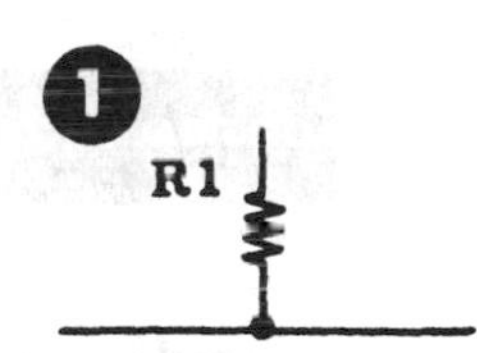

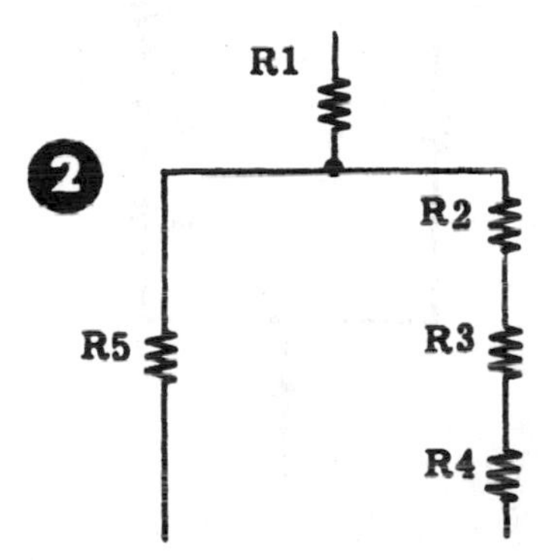

2. Draw the parallel paths from this line in the same direction as the series resistances—that is, vertically.

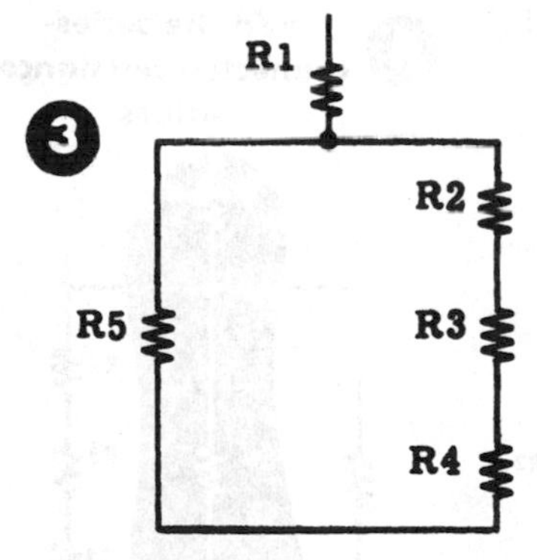

3. Where the parallel paths combine, draw another horizontal line across the ends to join the paths.

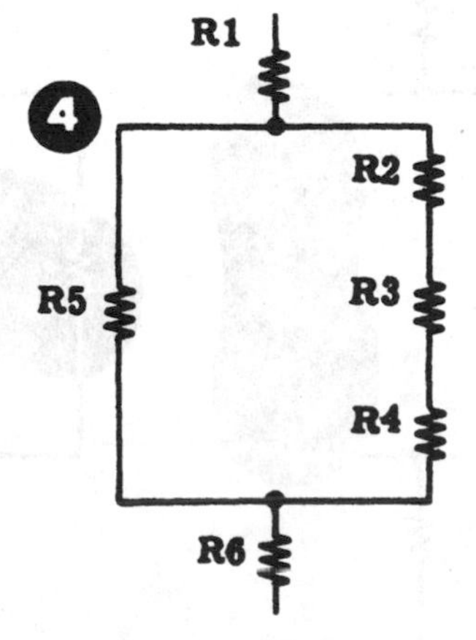

4. Continue the circuit from the center of the parallel connecting line, adding the series resistance to complete the redrawn circuit.

Resistors in Series-Parallel (continued)

The basic steps in finding the total resistance of a complex series-parallel circuit are therefore as follows:

1. Redraw the circuit if necessary.
2. If any of the parallel combinations have branches consisting of two or more resistors in series, find the total value of these resistors by adding them.
3. Using the formula for parallel resistances, find the total resistance of the parallel parts of the circuits.
4. Add the combined parallel resistances to any resistances which are in series with them.

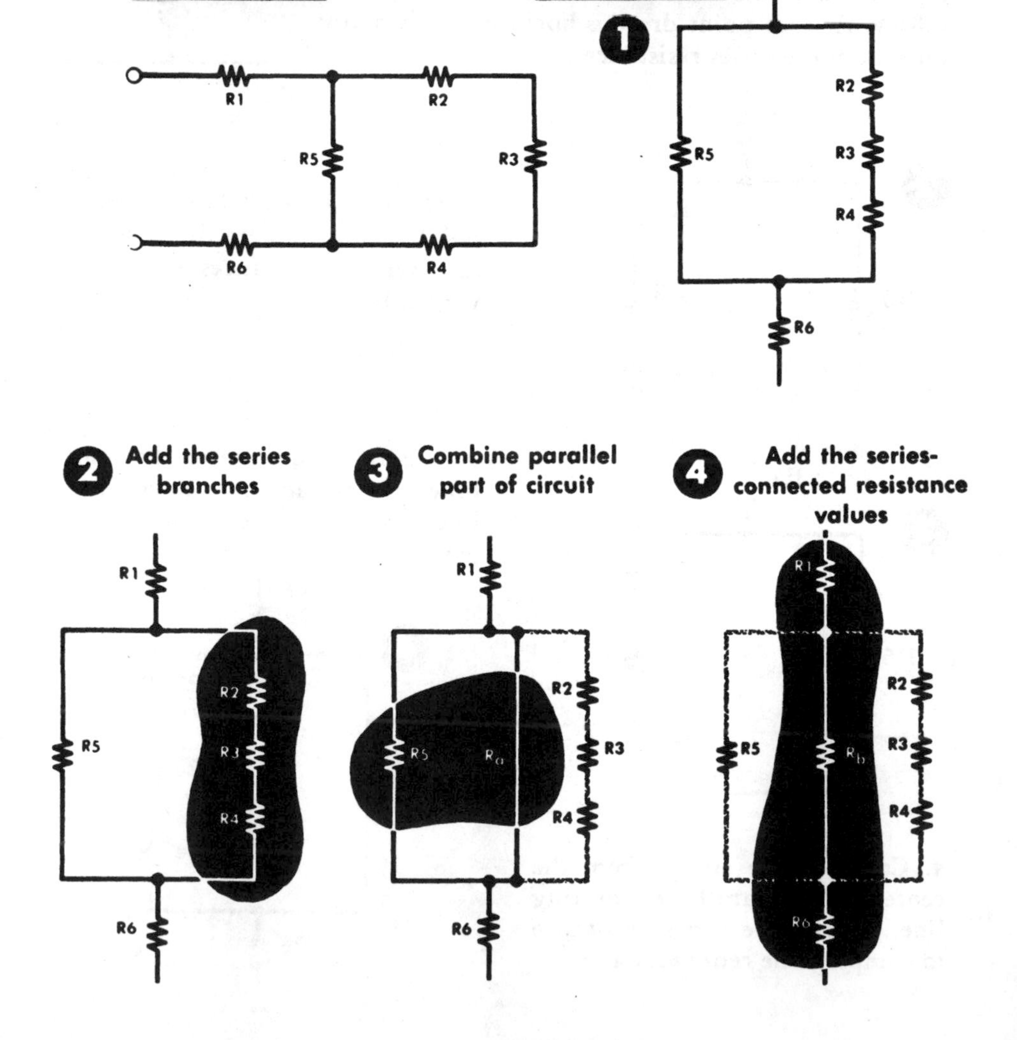

Resistors in Series-Parallel (continued)

Here is a practical example of how to break down complex circuits to find the total resistance:

Suppose your circuit consists of four resistors—R1, R2, R3, and R4—connected as shown. You want to find the total resistance of the circuit.

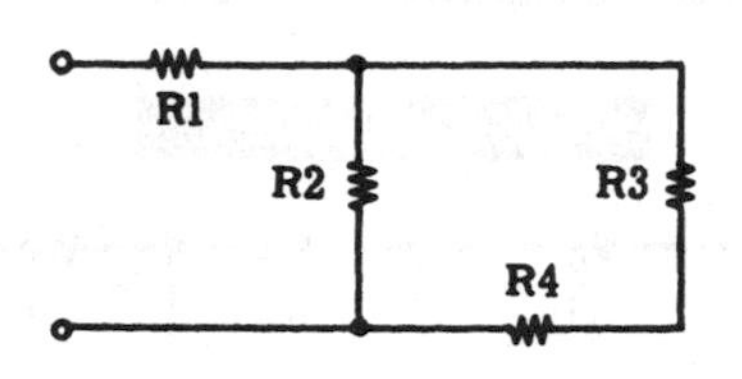

Suppose also that R1 = 7 ohms, R2 = 10 ohms, R3 = 6 ohms, and R4 = 4 ohms.

First, the circuit is redrawn and the series branch resistors R3 and R4 are combined by addition to form an equivalent resistance R_a.

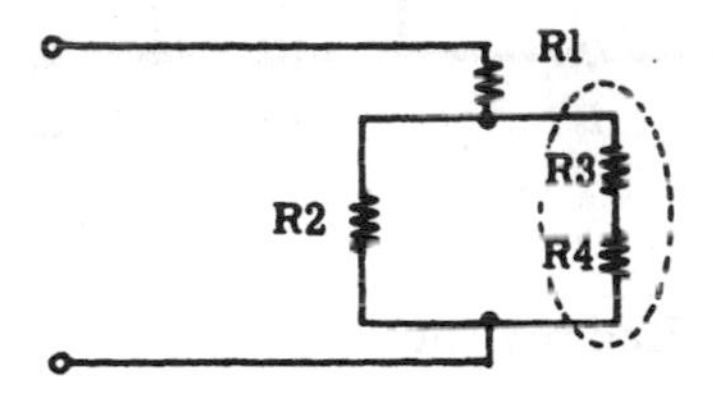

$$R_a = R3 + R4$$
$$= 6 + 4$$
$$= 10 \text{ ohms}$$

Next, the parallel combination of R2 and R_a is combined (using the parallel resistance formula) as an equivalent resistance, R_b.

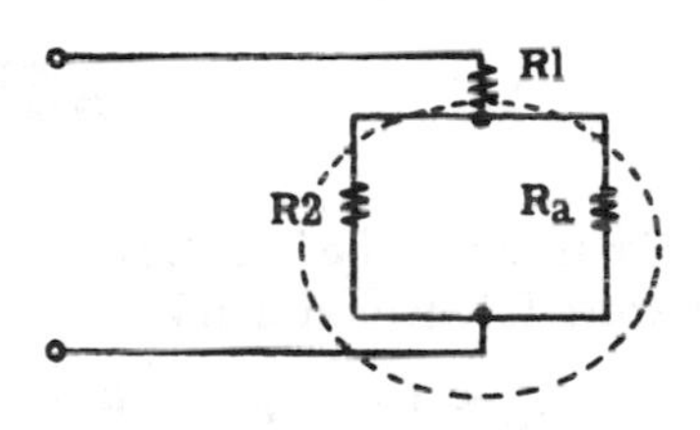

$$R_b = \frac{R2 \times R_a = 10 \times 10}{R2 + R_a = 10 + 10}$$
$$= 5 \text{ ohms}$$

Last, the series resistor R1 is added to the equivalent resistance —R_b—of the parallel combination to find the total circuit resistance, R_t.

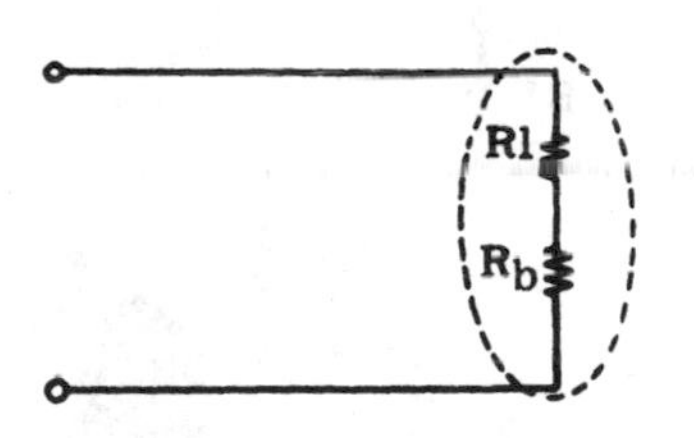

$$R_t = R1 + R_b$$
$$= 7 + 5$$
$$= 12 \text{ ohms}$$

In other words, the whole complex circuit can be broken down and simplified until R_t = total resistance of series-parallel circuit = 12 ohms.

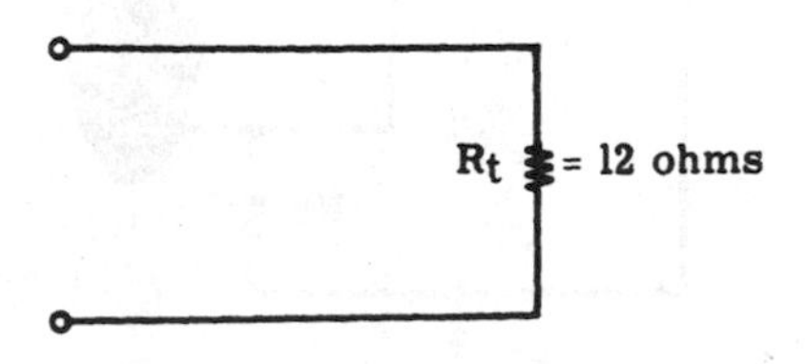

Resistors in Series-Parallel (continued)

More complicated circuits only require more steps; they do not require any additional formulas. For example, the total resistance of a circuit consisting of nine resistors may be found as shown:

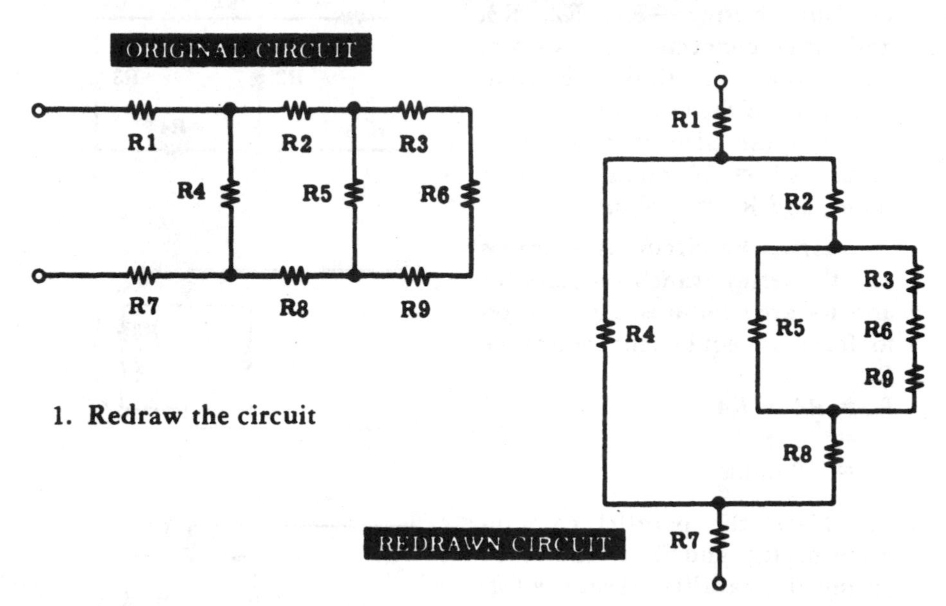

1. Redraw the circuit

2. Combine the series branch resistors R3, R6, and R9.

$$R_a = R3 + R6 + R9$$

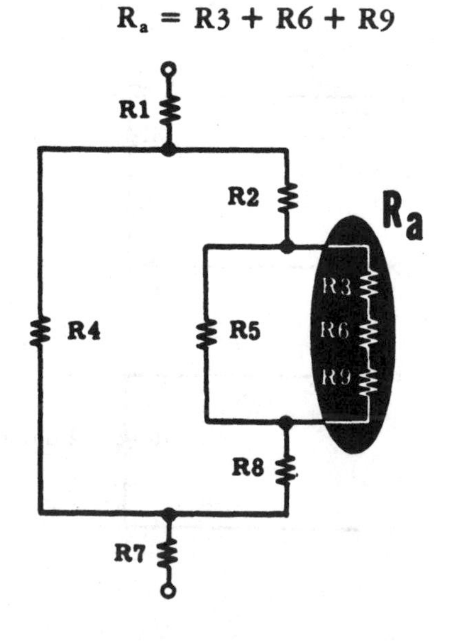

3. Combine the parallel resistances R5 and R_a.

$$R_b = \frac{R5 \times R_a}{R5 + R_a}$$

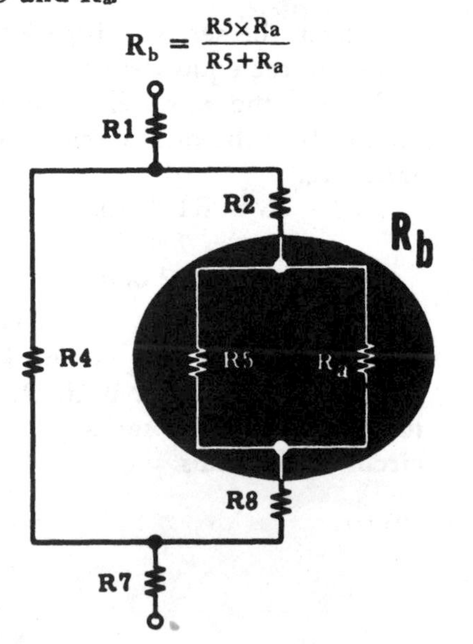

Resistors in Series-Parallel (continued)

4. Combine the series resistances R2, R_b, and R8.

$$R_c = R2 + R_b + R8$$

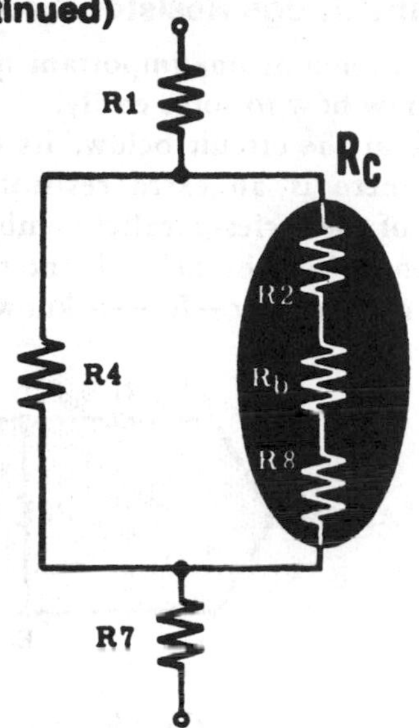

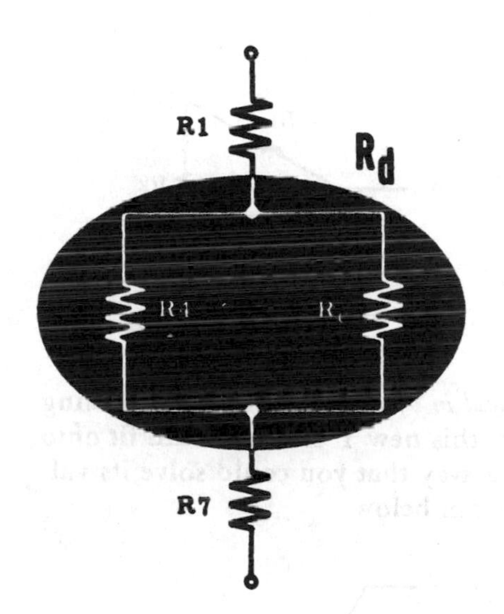

5. Combine the parallel resistances R4 and R_c.

$$R_d = \frac{R4 \times R_c}{R4 + R_c}$$

6. Combine the series resistances R1, R_d, and R7.

$$R_t = R1 + R_d + R7$$

7. R_t is the total resistance of the circuit, and the circuit will act as a single resistor of this value when connected across a voltage source.

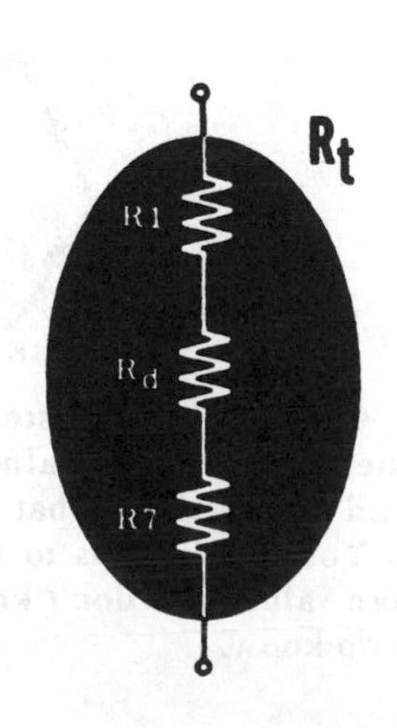

Solving the Bridge Resistor Circuit

There remains one important type of complex circuit which you do not yet know how to solve easily.

Look at the circuit below. Its outline is familiar enough—but you see that there is an extra resistor (R2) connecting the two parallel branches of the series-parallel combination in such a way that the series connection in both branches is interrupted by the leads to the new resistor. This new resistor—R2—is known as a *bridge*.

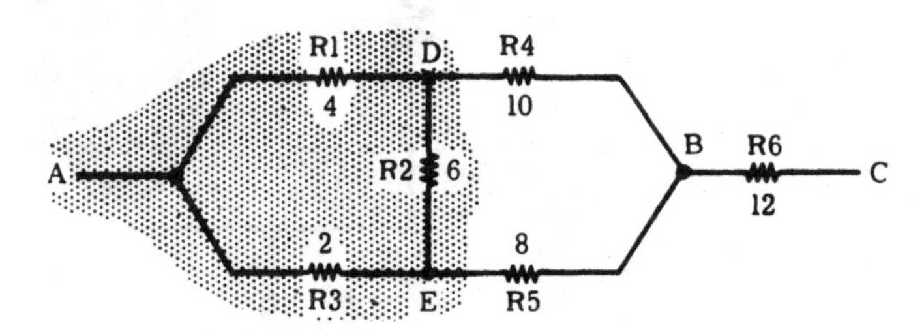

If you look at the shaded part of the circuit above you see that it is essentially the shape you see opposite. This arrangement, from its similarity to the shape of the Greek letter D (delta), is said to be *delta-connected.*

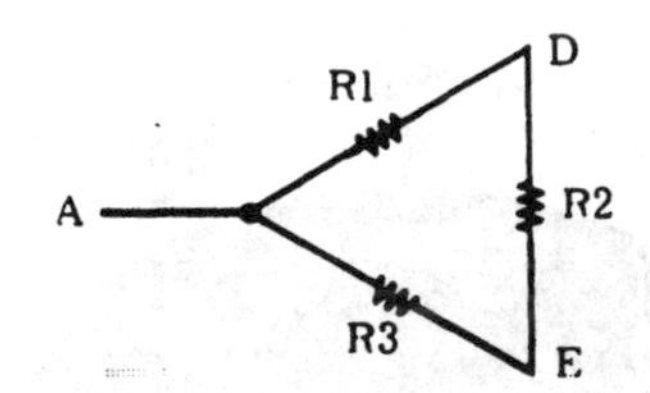

You also observe, however, that if you could devise a circuit shaped like a Y (wye) *such that its terminal resistances at D and E were identical in value* with the corresponding terminal resistances in the delta circuit, this new Y circuit would fit onto the rest of the original circuit in such a way that you could solve its values without difficulty. Look at the diagram below.

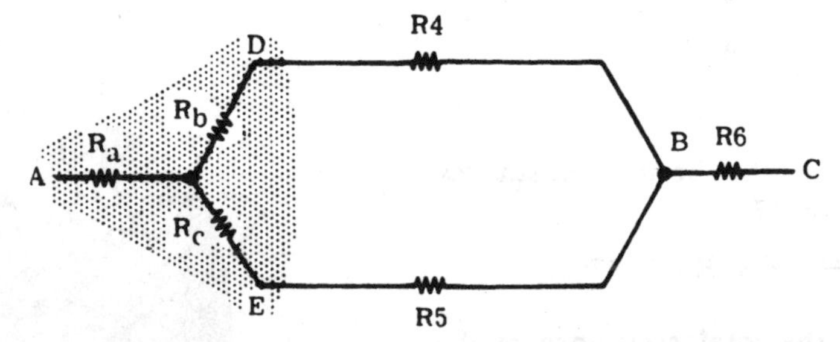

Call the three resistors in your proposed Y circuit R_a, R_b, and R_c. Remember that their values must be such that the terminal resistances at D and E are exactly what they were in the original circuit.

Your problem is to find a formula for expressing R_a, R_b, and R_c, whose values you don't know, in terms of R1, R2, and R3, whose values you do know.

Solving the Bridge Resistor Circuit (continued)

Redraw the delta circuit and the Y circuit from the last page together so that you can look at them conveniently side-by-side. Then mark in firmly the *equals* sign between them to remind you that both circuits must give exactly the same values of resistance across every corresponding pair of terminals. You are now all set for an operation called the *delta-Y conversion.*

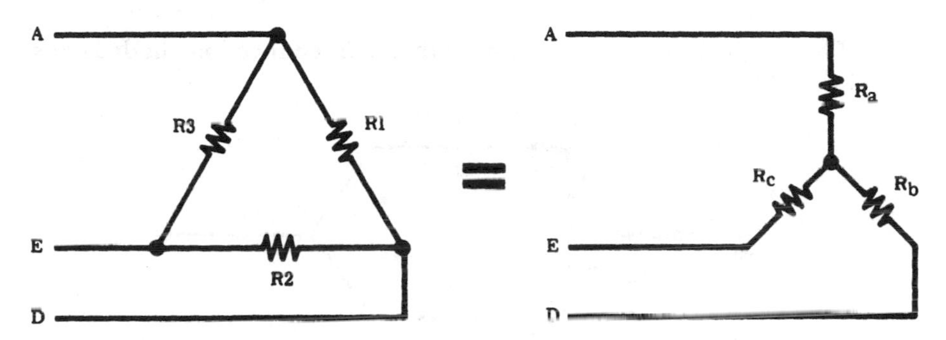

Consider, first, the sum of the resistances between A and E, assuming D to be disconnected. In the delta combination you will see that between these two points there is effectively a series combination of R1 and R2 in parallel across R3. You can, therefore, from the knowledge you already have, express the resistance A-E as

$$\frac{R3\,(R1 + R2)}{R1 + R2 + R3}$$

In the Y circuit, the total resistance between A and E is obviously $R_a + R_c$. Since you know that these two resistances must be equal, you can write down as your first equation:

$$R_a + R_c = \frac{R3\,(R1+R2)}{R1+R2+R3}$$

In exactly the same way, you can express the total resistances between A-D and between D-E in terms of R1−R2−R3 and of $R_a-R_b-R_c$. Work it out for yourself, and you will get two more equations as follows:

$$R_a + R_b = \frac{R1\,(R2+R3)}{R1+R2+R3}$$

$$R_c + R_b = \frac{R3\,(R1+R2)}{R1+R2+R3}$$

Now do a little simple algebra (beginning by subtracting equation (2) from equation (1), to get equation (4); then adding equation (4) to equation (3), to get a value for R_c in terms of R1 - R2 - R3; and, lastly, substituting for R_c in equations (1) and (3) to get similar values for R_a and R_b). You find that:

$$R_a = \frac{R1 \times R3}{R1+R2+R3}; \quad R_b = \frac{R1 \times R2}{R1+R2+R3}; \quad R_c = \frac{R2 \times R3}{R1+R2+R3}$$

Solving the Bridge Resistor Circuit (continued)

Now go back to the original circuit, and fill in the known values of R1, R2, and R3. You get:

$$R_a = \frac{4\times 2}{4+6+2} = 0.67 \text{ ohm}; \; R_b = \frac{4\times 6}{4+6+2} = 2 \text{ ohms};$$

$$R_c = \frac{2\times 6}{4+6+2} = 1 \text{ ohm}$$

The original circuit (redrawn, with the Y connection shaded) now looks like this:

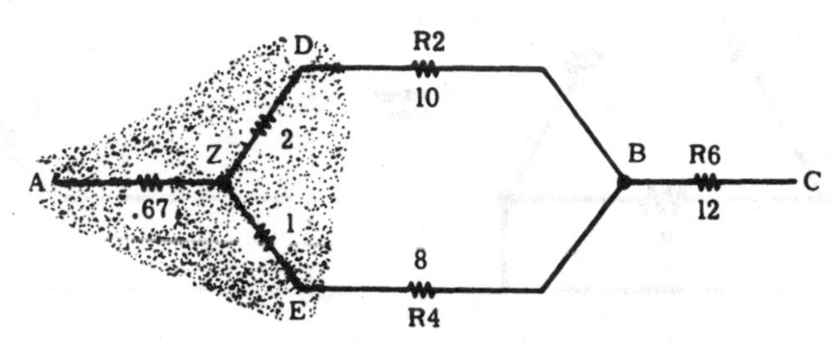

Here you have, between Z and B, a simple parallel circuit whose equivalent resistance you already know how to calculate.

$$R_{zb} = \frac{(2+10)\times(1+8)}{2+10+1+8} = \frac{108}{21} = 5.14 \text{ ohms}$$

The equvalent resistance of the entire bridge circuit is therefore:

$$AZ + ZB + BC = 0.67 + 5.14 + 12 = 17.81 \text{ ohms}$$

correct to two decimal places. Your problem is solved.

It is quite easy to convert any delta connection into its Y equivalent if you get the numbering of the various resistors right. Look at the drawing below, in which the equivalent Y is dotted in on top of an original delta:

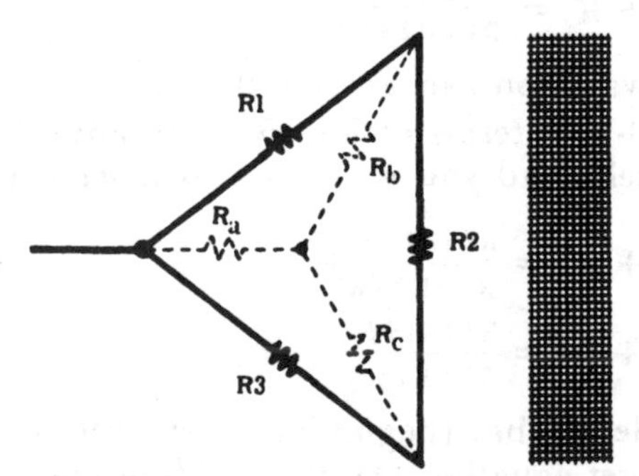

$$R_a = \frac{R1\times R3}{R1+R2+R3}$$

$$R_b = \frac{R1\times R2}{R1+R2+R3}$$

$$R_c = \frac{R2\times R3}{R1+R2+R3}$$

Label as R_a the Y connection which bisects the angle R1-R3; label as R_b the connection which bisects R1-R2; and label as R_c the connection which bisects R2-R3. If you do this, the formula almost remembers itself. The denominator is always R1 + R2 + R3. The numerator is the product of the two delta resistors which your Y resistor connection bisects.

Ohm's Law in Series-Parallel Circuits—Current

The total circuit current for a series-parallel circuit depends on the total resistance offered by the circuit when connected across a voltage source. Current flow in the circuit will divide to flow through all parallel paths, and come together again to flow through series parts of the circuit. It will divide to flow through a branch circuit, and then repeat this division if the branch circuit itself subdivides into secondary branches.

As in parallel circuits, the current through any branch resistance is inversely proportional to the resistance of the branch—the greater current flows through the least resistance. However, all of the branch currents always add up to equal the total circuit current.

The total circuit current is the same at each end of a series-parallel circuit, and is equal to the current flow through the voltage source.

HOW CURRENT FLOWS IN A SERIES-PARALLEL CIRCUIT

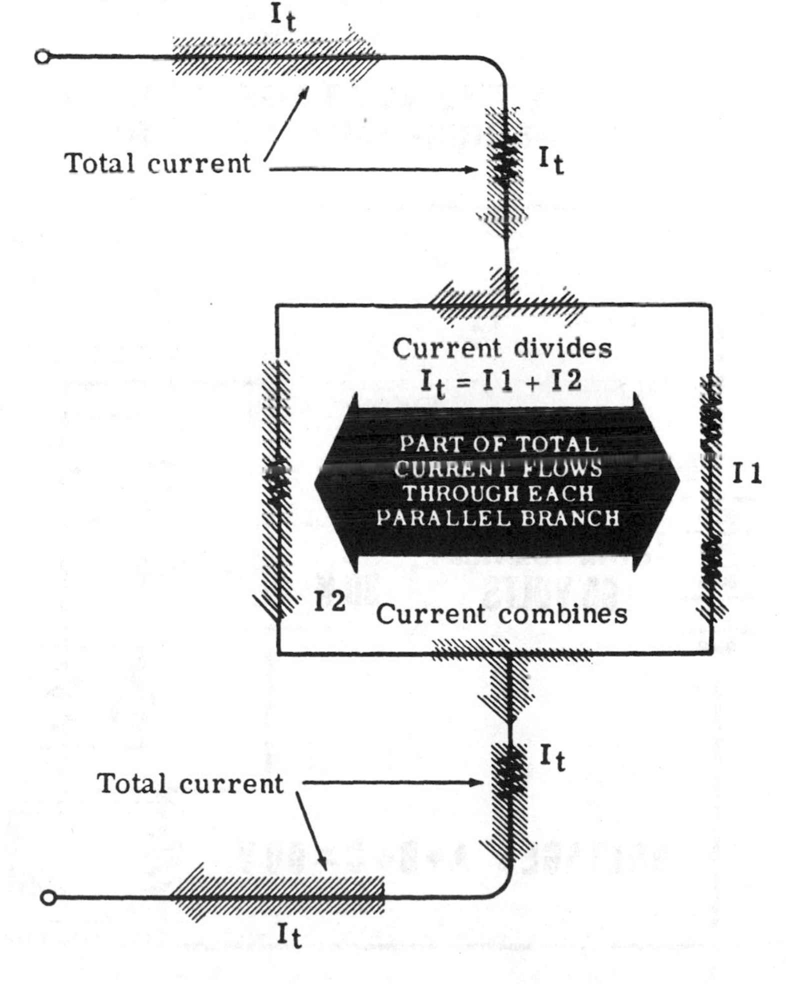

Ohm's Law in Series-Parallel Circuits—Voltage

Voltage drops across a series-parallel circuit occur in the same way as they do in series and parallel circuits. In the series parts of the circuit, the voltage drops across the resistors depend on the individual values of the resistors. In the parallel parts of the circuit, every branch has the same voltage across it, and carries a current which is dependent on the resistance in that particular branch.

Series resistances forming a branch of a parallel circuit will divide the voltage across the parallel circuit. In a parallel circuit consisting of a branch with a single resistance and a branch with two series resistances, the voltage across the single resistance is equal to the sum of the voltages across the two series resistances. The voltage across the entire parallel circuit is exactly the same as that across either of the branches.

The voltage drops across the various paths between the two ends of the series-parallel circuit always add up to the total voltage applied to the circuit.

HOW THE VOLTAGE DIVIDES IN A SERIES-PARALLEL CIRCUIT

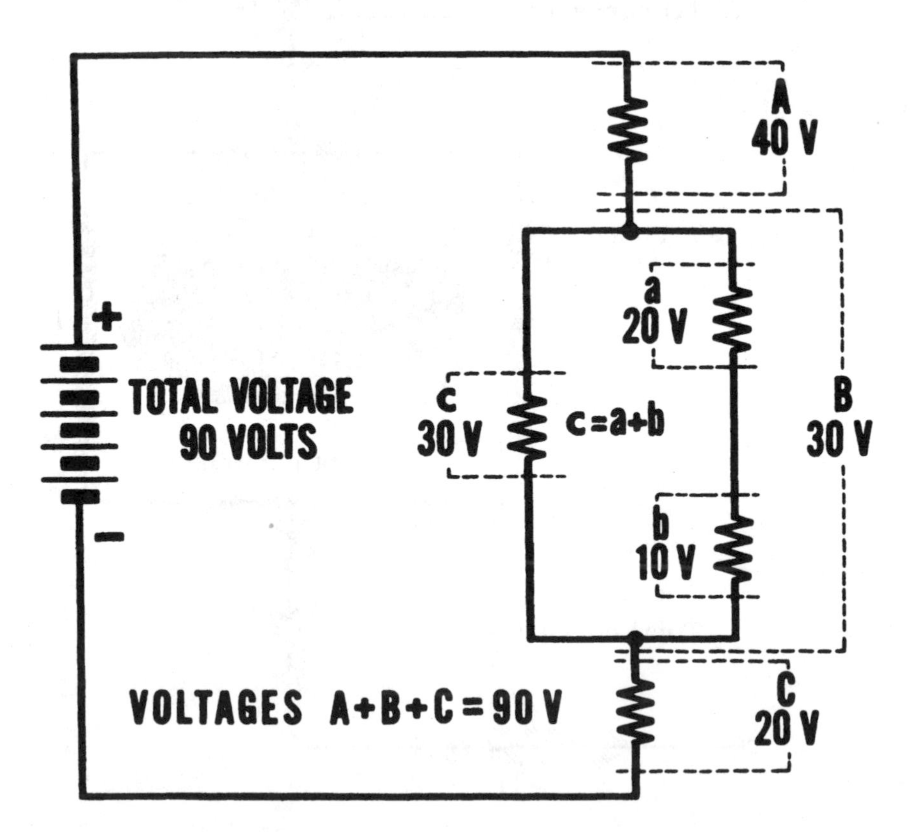

Ohm's Law in Series-Parallel Circuits

Example 1

Consider the circuit of page 2-92:

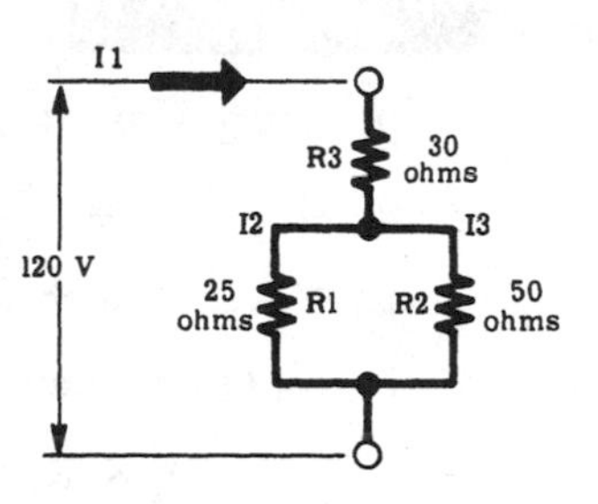

To find the current I_1 through this circuit

$$R_a = \frac{R1 \times R2}{R1 + R2} = \frac{25 \times 50}{25 + 50} = \frac{1{,}250}{75} = 16.67 \text{ ohms}$$

$$R_t = 46.67 \text{ ohms}$$

$$I_1 = \frac{120}{46.67} = 2.57 \text{ amperes}$$

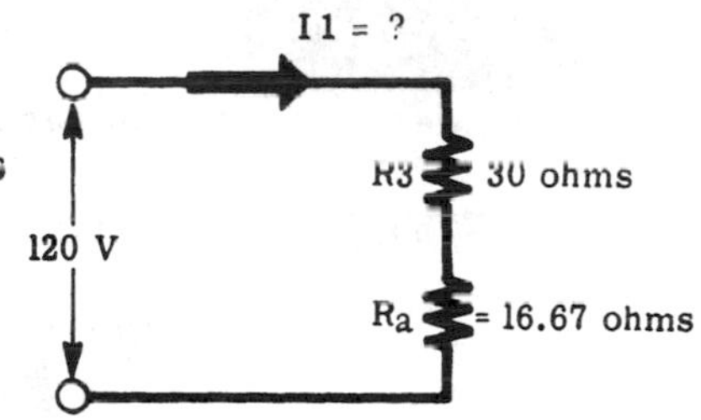

Example 2

Consider the circuit of pages 2-93 and 2-94:

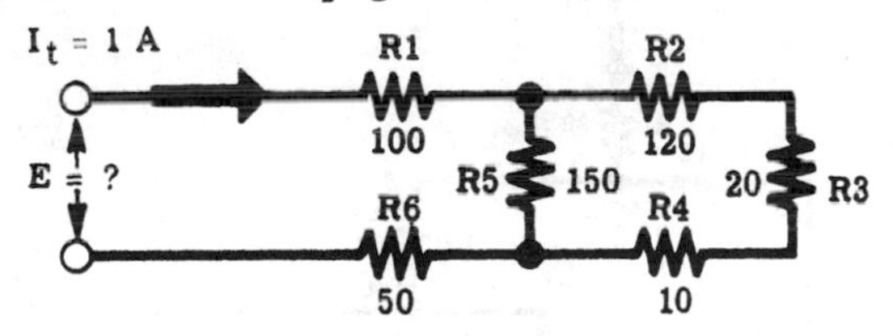

Redrawn, the circuit looks like

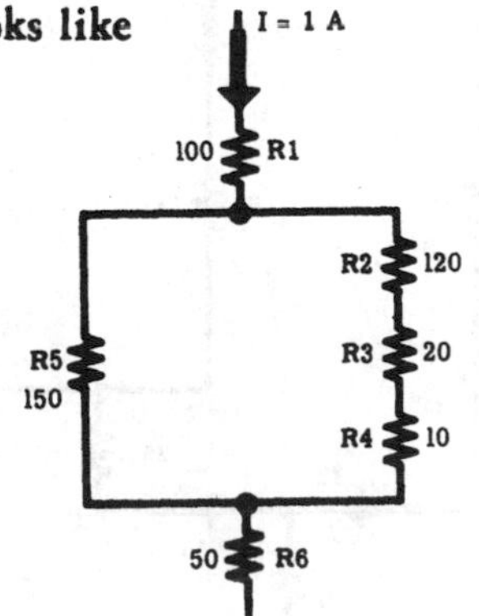

Ohm's Law in Series-Parallel Circuit (continued)

Example 2 (cont.)

$$R_a = R2 + R3 + R4 = 120 + 20 + 10 = 150$$

$$R_b = \frac{R_a \times R5}{R_a + R5} = \frac{150 \times 150}{150 + 150} = 75$$

$$R_t = R1 + R_b + R6 = 100 + 75 + 50 = 225$$

Using Ohm's Law, the voltage can be calculated as

$$E = I_t \times R_t = 1A \times 225 = 225 \text{ volts}$$

Example 3

Consider the circuits of pages 2-96 and 2-97:

(A) What is E_{out} if E_{in} is 100 volts?

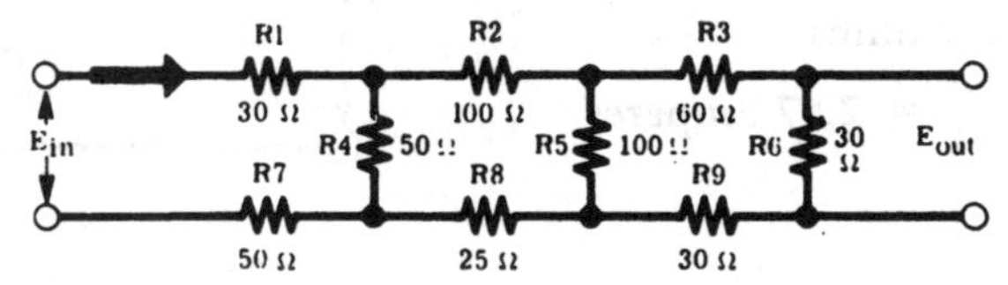

(B) Redrawn circuit (see page 2-97).

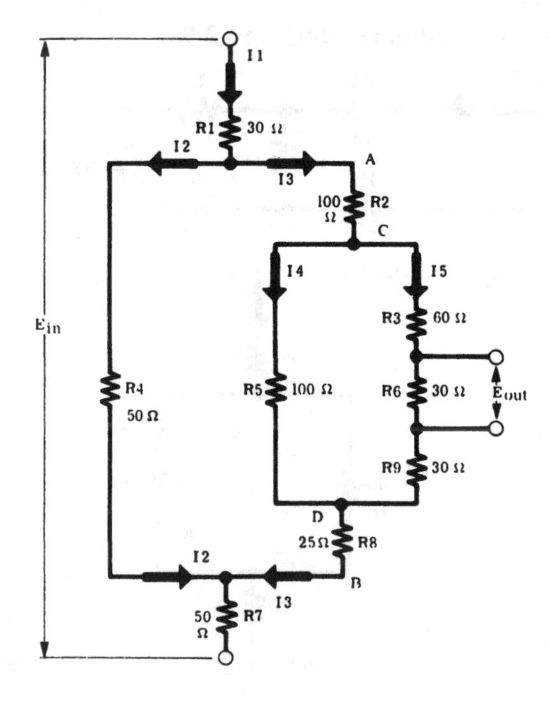

Ohm's Law in Series-Parallel Circuits (continued)

Example 3 (cont.)

To solve this, we must find the current through R6 to find E_{out}. Initially, we must find the currents which require us to find the resistance of the various paths. For the series combination of R3, R6, and R9,

$$R_a = R3 + R6 + R9 = 120 \text{ ohms}$$

Then R5 in parallel with R_a can be calculated as

$$R_b = \frac{R5 \times R_a}{R5 + R_a} = \frac{100 \times 120}{100 + 120} = \frac{12{,}000}{220} = 54.5 \text{ ohms}$$

The circuit can now be drawn as

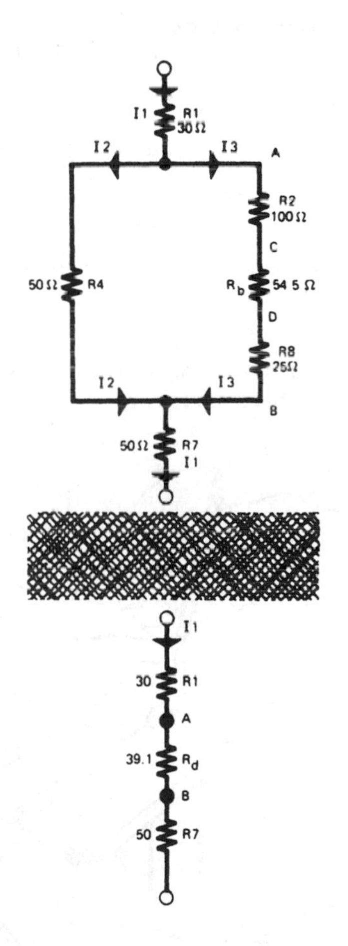

(C)

Combining the series resistors R2, R_b, and R8 yields

$$R_{ab} = R_c = R2 + R_b + R8 = 100 + 54.5 + 25 = 179.5$$

The parallel combination of R4 and R_c is

$$R_d = \frac{R_c \times R4}{R_c + R4} = \frac{179.5 \times 50}{179.5 + 50} = \frac{897.5}{229.5} = 39.1$$

The circuit can now be drawn as

(D)

And the current is calculated as

$$I1 = \frac{E_{in}}{R_t}, \text{ where}$$

$$R_t = R1 + R_d + R7 = 30 + 39.1 + 50 = 119.1$$

$$I1 = \frac{100}{119.1} = 0.84 \text{ ampere}$$

Ohm's Law in Series-Parallel Circuits (continued)

Example 3 (cont.)

We can now back track to drawing (C) of our circuit and calculate the currents I_2 and I_3.

The voltage drop across R_d—from point A to point B in drawing (C)—can be calculated as

$$E_{ab} = I_1 \times R_d = 0.84 \times 39.1 = 32.8 \text{ volts}$$

As shown in drawing (C), we can calculate the current I_3 as

$$I_3 = \frac{E_{ab}}{R_c} = \frac{32.8}{179.5} = 0.183 \text{ ampere}$$

But from drawing (B),

$$I_3 = I_4 + I_5$$

We can calculate E_{cd}, which we need to find the current I_5.

$$E_{cd} = I_3 \times R_b = 0.183 \times 54.5 = 9.97 \text{ volts}$$

Thus, $I_5 = \frac{E_{cd}}{R_a} = \frac{9.97}{120} = 0.83$ ampere

We can now finally solve for E_{out} since

$$E_{out} = I_5 \times R6 = 0.083 \text{ ampere} \times 30 = 2.49 \text{ volts}$$

While this may seem complicated, it is just the application of what we have been learning.

Measuring the Total Resistance of the Circuit...

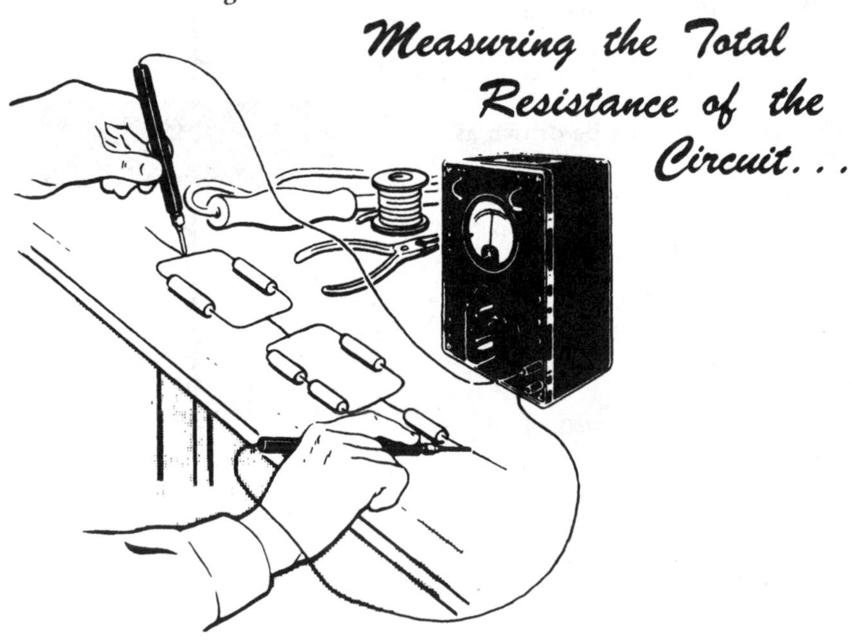

Review of Series-Parallel Circuits

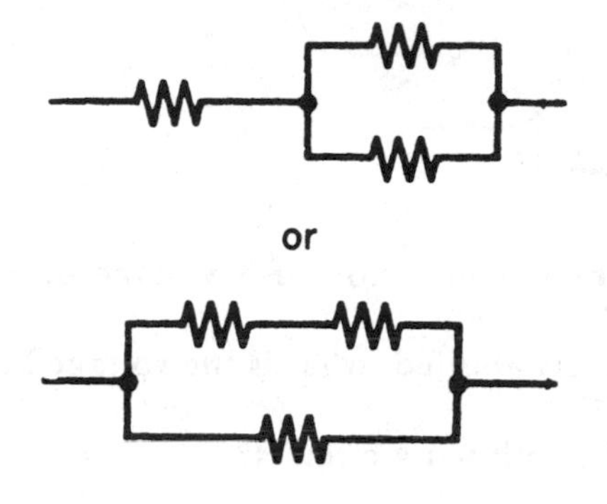

1. A SERIES-PARALLEL CIRCUIT—A series-parallel circuit has both parallel and series elements combined.

2. CIRCUIT REDUCTION—The formulas for determining series and parallel resistor combinations are used to reduce complex circuits.

3. CIRCUIT SIMPLIFICATION—Redrawing the circuit often results in simplification.

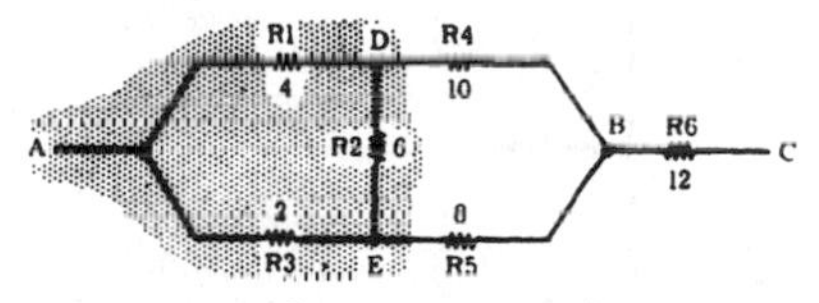

can be converted to

4. BRIDGE CONFIGURATION—

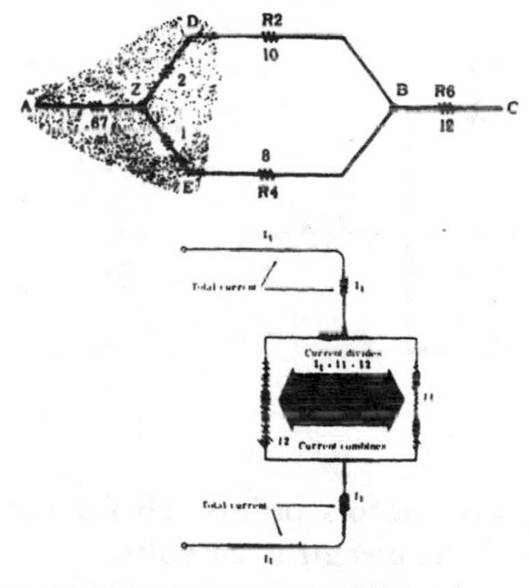

5. THE CURRENT IN A SERIES-PARALLEL CIRCUIT—The current in series-parallel circuit divides in the parallel paths and comes together in the series portion.

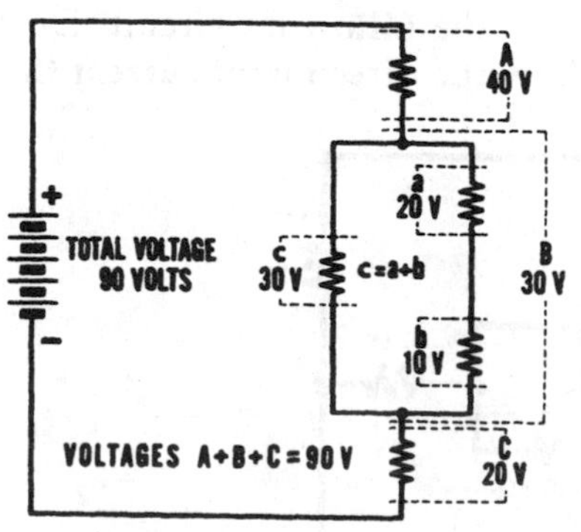

6. THE VOLTAGE IN A SERIES-PARALLEL CIRCUIT—The voltages in series-parallel circuits divide up so that the sum of the voltages in the series portion and the parallel portion is equal to the total voltage.

Self-Test— Review Questions

1. What is the equivalent resistance of the following circuit?

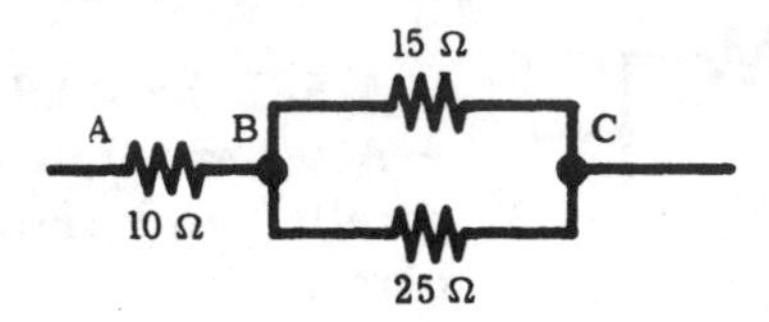

2. In the circuit above, how much current would flow if a voltage of 12 volts were connected across points AC?
3. In the circuit of question 1, with 12 volts applied, what is the voltage between AB? Between BC? Between AC?
4. What is the equivalent resistance of the following circuits?

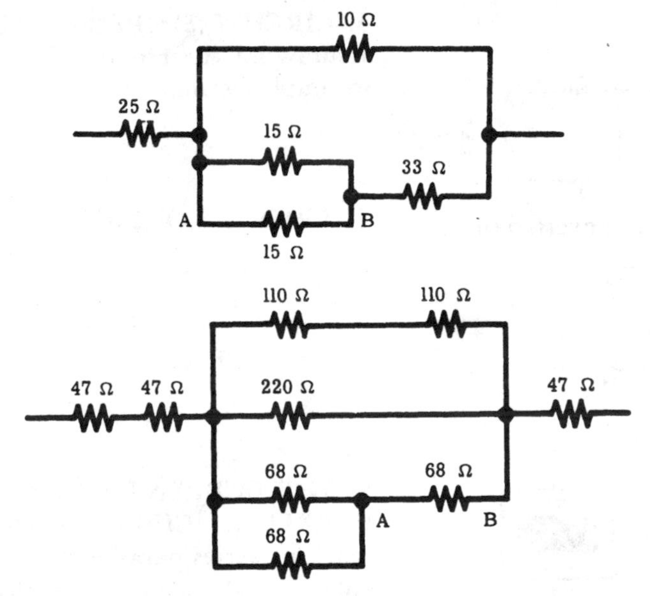

5. In the circuits of question 4, find the voltage across points AB for each circuit. Assume that the total voltage across the circuit is 24 volts.
6. Solve for all the voltages and currents in the following circuit. Do the results conform to Kirchhoff's Laws? Show the direction of current flow.

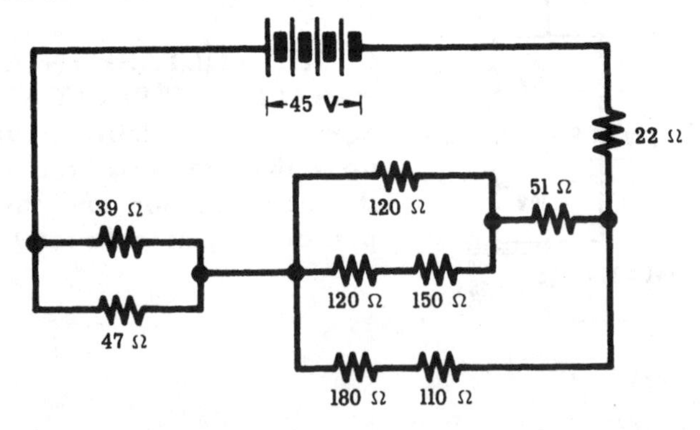

Experiment/Application—Series-Parallel Connections

Suppose that three 30-ohm resistors are connected together, with one resistor in series with a parallel combination of the other two, thus forming a series-parallel circuit. The total resistance is found by combining the parallel 30-ohm resistors to obtain their equivalent resistance, which is 15 ohms, and by adding this value to the 30-ohm resistor in series—making a total resistance of 45 ohms. If the resistors are checked with an ohmmeter, it will read 45 ohms across the entire circuit.

Next, suppose two 30-ohm resistors are connected in series, and a third resistor of the same value is connected in parallel across the series combination. The total resistance can be found by adding the two resistors in the series branch, to obtain an equivalent value of 60 ohms. This value is in parallel with the third 30-ohm resistor, and combining them results in a value of 20 ohms for the total resistance. If you check this value with an ohmmeter, you will see that the meter reading is, indeed, 20 ohms.

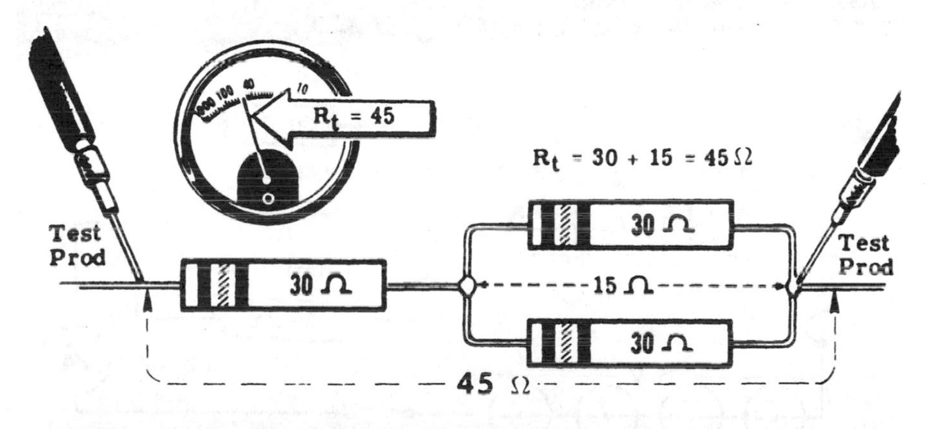

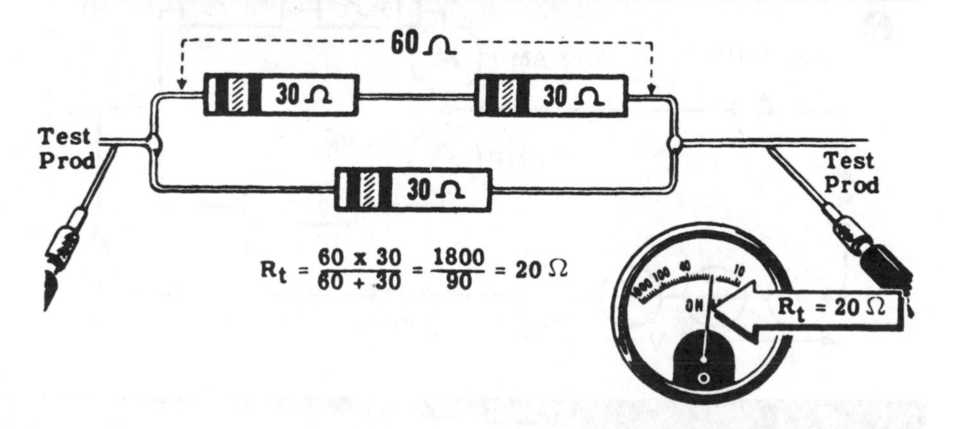

Experiment/Application—Current in Series-Parallel Circuits

Next, consider a series-parallel circuit consisting of two 30-ohm resistors connected in parallel and a 15-ohm resistor in series with one end of the parallel resistors across a 6-volt dry cell battery.

If you connected an ammeter in series with each resistor, in turn, to find the current flow through each, you would see that the current for the 15-ohm series resistor is 0.2 ampere, as is the current at each battery terminal; the current through the 30-ohm resistors, however, is 0.1 ampere each.

Now, suppose the circuit connections are changed so that the 15-ohm and one 30-ohm resistor form a series-connected branch in parallel with the other 30-ohm resistor. An ammeter would now show that the battery current is 0.33 ampere, the 30-ohm resistor current is 0.2 ampere, and the current through the series branch is 0.13 ampere.

SEEING HOW THE CURRENT FLOWS THROUGH SERIES-PARALLEL CIRCUITS

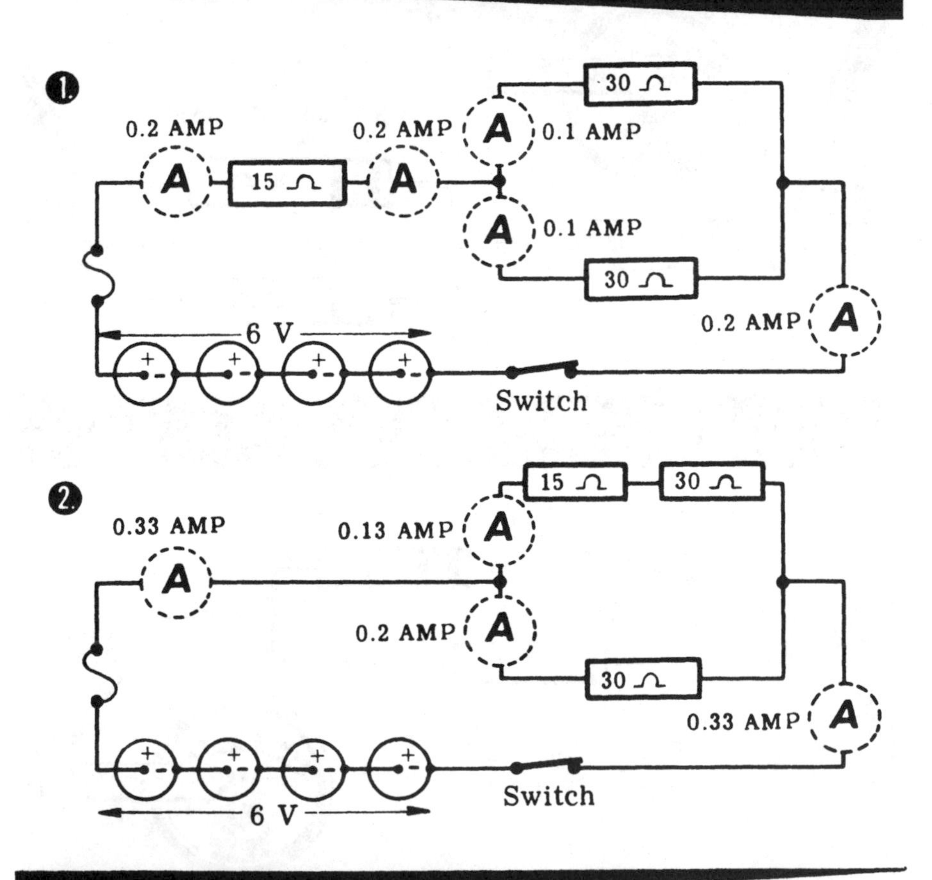

Experiment/Application—Voltage in Series-Parallel Circuits

The division of voltage across series-parallel circuits can be shown by connecting several resistors to form a complex circuit having more than one complete path between the battery terminals, as diagrammed below. As you trace several possible paths across the circuit and measure the voltage across each resistance, you can see that—regardless of the path chosen—the sum of the voltages for any one path always equals the battery voltage. Also, you see that the voltage drop across resistors of equal value differs, depending on whether they are in a series or parallel part of the circuit, and depending also on the total resistance of the path in which they are located.

SEEING HOW VOLTAGE DIVIDES IN A SERIES-PARALLEL CIRCUIT

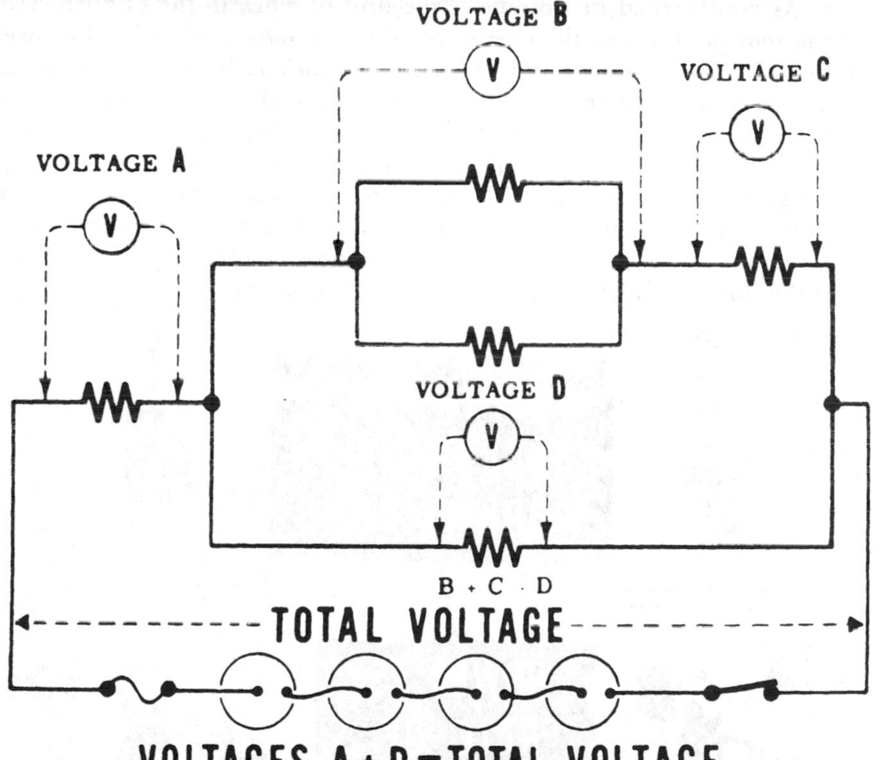

VOLTAGES A + D = TOTAL VOLTAGE
VOLTAGES A + B + C = TOTAL VOLTAGE

Learning Objectives—Next Section

Overview—Now that you know how to solve all kinds of dc circuits, you will learn about electric power. You will learn how it is calculated and how to find the power in dc circuits.

What Electric Power Is

As you know, whenever a force of any kind causes motion, *work* is said to be done. When a mechanical force, for instance, is used to lift a weight, work is done. A force which is exerted *without* causing motion—for example, such as the force of a spring held under tension between two objects which do *not move*—that force does *not* cause work to be done.

You also know that a difference in potential between any two points in an electric circuit gives rise to a voltage which, when the two points are connected, causes electrons to move and current to flow. Here is an obvious case of a force causing motion, and thus of causing work to be done. Whenever a voltage causes electrons to move, therefore, work is done in moving them.

As you learned in Volume 1, the unit of work in the English system is the foot-pound, i.e., the energy required to *raise* 1 pound a distance of 1 foot. In the metric system the unit of work is based on meters and grams and is called the *joule*, with 1 joule equal to about 3/4 of a foot-pound. We can get work back by letting the weight of 1 pound *fall* a distance of 1 foot after connecting the weight to something to get an output of work as the weight falls. Thus, we can draw an analogy between the lifting of the weight and the generation of the potential difference or voltage at a power station, and the lowering of the weight to do work and the flow of electrons to do work at the load end.

The *rate* at which the work of moving electrons from point to point is done is called *electric power*. It is represented by the symbol *P*, and the unit of power is the *watt*, usually represented by the symbol *W*. *The watt can be practically defined as the rate at which work is being done in a circuit in which a current of 1 ampere is flowing when the voltage applied is 1 volt.*

The Power Formula

As you learned in Volume 1, it is the ease with which electric power can be transmitted from place to place, and converted to other forms of energy, that makes it so valuable. For example, electrical energy can be converted to heat, light, or accoustical and mechanical energy. The rate of energy conversion is what the engineer really means by the word *power.*

The rate at which work is done in moving electrons through a resistor obviously depends on how many electrons there are to be moved. In other words, the *power consumed in a resistor is determined by the voltage measured across it, multiplied by the current flowing through it.* Expressed in units of measure, this becomes

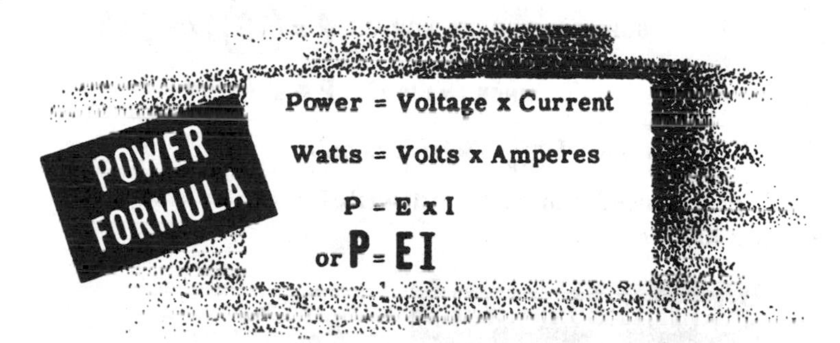

In the circuit below, a 15-ohm resistor is connected across a supply of 45 volts. How much power is used up when a current of 3 amperes flows through the resistor?

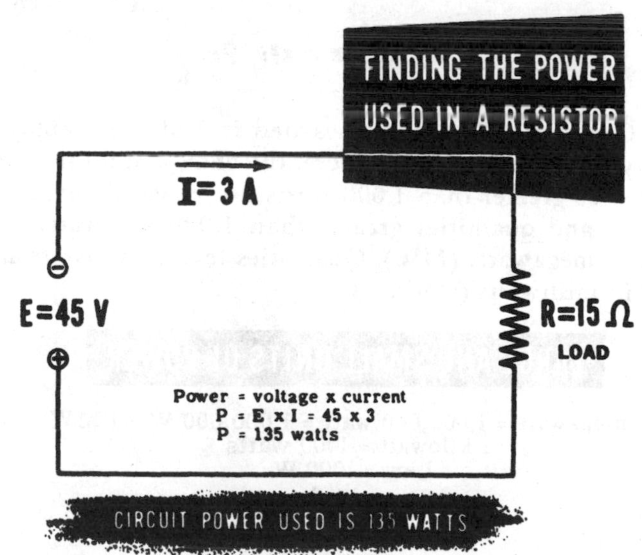

For direct-current (dc) circuits, you can always find the power in a circuit by using the *power formula.*

The Power Formula (continued)

The power formula you learned on the last page, P = EI, can obviously be expressed alternatively in terms of current and resistance, or of voltage and resistance, by the use of our old friend, Ohm's Law. (As E or V can be used interchangeably, this formula can also be stated as P = VI. We will use E.) Since E = IR, the E in the power formula can be replaced by its equal value of IR, and the power used can be calculated *without* the voltage being known.

SUBSTITUTING (IR) FOR E: $P = (IR)I$ OR $I \times R \times I$

SINCE I x I IS I^2: $P = I^2R$

Equally, of course, $I = \frac{E}{R}$. So if E is substituted in the power formula for I, the power used can be found with only the voltage and the resistance being known.

ANOTHER VARIATION

P=EI

SUBSTITUTING $\frac{E}{R}$ FOR I: $P = E\left(\frac{E}{R}\right)$ OR $\frac{E \times E}{R}$

SINCE E x E IS E^2: $P = \frac{E^2}{R}$

The Conversion Tables you learned in Volume 1 apply to the watt just as they do to the volt, the ampere, the ohm, and all the others. Quantities of power greater than 1,000 watts are generally expressed in kilowatts (kW), and quantities greater than 1,000,000 watts are generally expressed as megawatts (MW). Quantities less than 1 watt are generally expressed in milliwatts (mW).

LARGE AND SMALL UNITS OF POWER

1 megawatt = 1,000,000 watt = 1,000,000 W = 1 MW

1 kilowatt = 1000 watts

1 kw = 1000 W

1 milliwatt = $\frac{1}{1000}$ watt

1 mw = $\frac{1}{1000}$ w

1 microwatt = $\frac{1}{1,000,000}$ watt = $\frac{1}{1,000,000}$ W = 1 μW

Power Rating of Equipment

You have probably found from your own experience that most electrical equipment is rated for both voltage and power—volts and watts. Electric lamps rated at 120 volts for use on 120-volt lines are also rated in watts, and are usually identified by wattage rather than by voltage.

Perhaps you have wondered what this rating in watts means and indicates. *The wattage rating of an electric lamp or other electrical equipment indicates the rate at which electrical energy is changed into another form of energy, such as heat or light.* The faster a lamp changes electrical energy to light, the brighter the lamp will be; thus, a 100-watt lamp furnishes more light than a 75-watt lamp.

Electric soldering irons are made in various wattage ratings, with the higher wattage irons changing electrical energy to heat faster than those of a lower wattage rating. Similarly, the wattage rating of motors, resistors, and other electrical devices indicates the rate at which they are designed to change electrical energy into some other form of energy. Motors are often classified in terms of horsepower—which you will learn more about later. Horsepower is another unit for measuring the rate at which work is done, and 1 horsepower is equal to 746 watts. You will learn more about horsepower when we study motors.

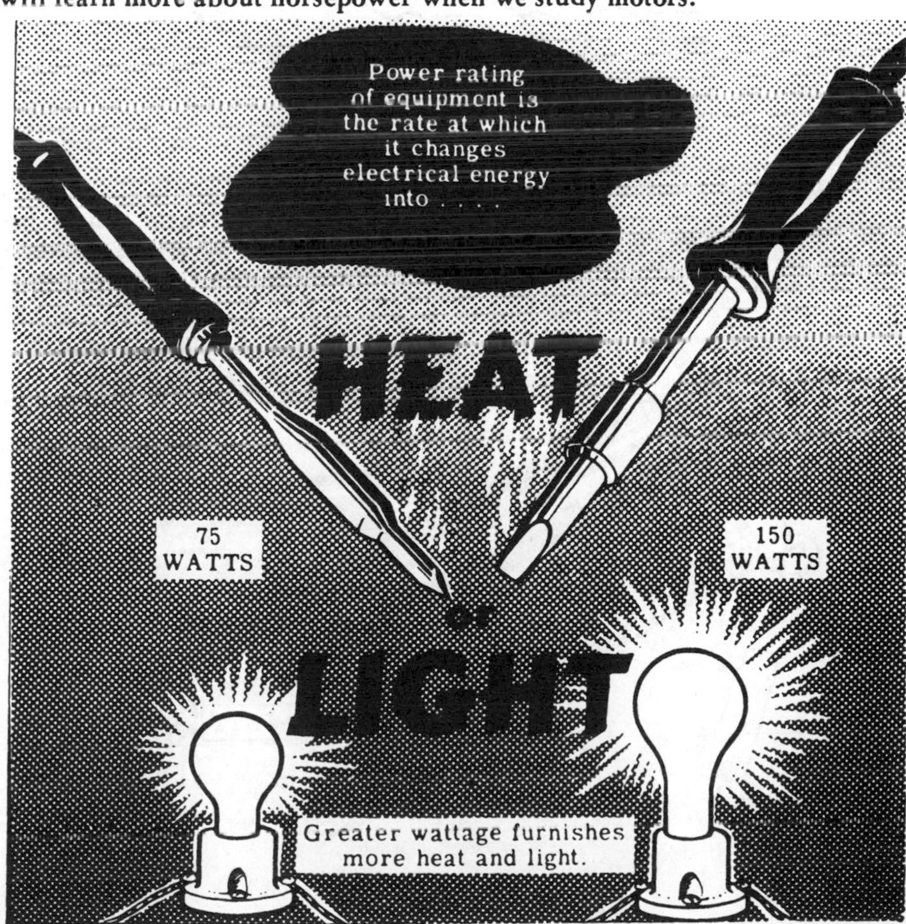
Power rating of equipment is the rate at which it changes electrical energy into . . . .
HEAT
LIGHT
75 WATTS
150 WATTS

Greater wattage furnishes more heat and light.

Power Rating of Equipment (continued)

When power is used in a material having resistance, electrical energy is changed into heat. When more power is used in the material, the rate at which electrical energy is changed to heat increases, and the temperature of the material rises. If the temperature rises too high, the material may change its composition, expand, contract, or burn. For that reason, all types of electrical equipment are rated for a maximum wattage. This rating may be in terms of watts, or in terms of maximum voltage and current—which effectively gives the rating in watts.

Resistors are rated in watts as well as in ohms of resistance. Resistors of the same resistance value are available in different wattage values. Carbon resistors, for example, are commonly made in wattage ratings of 1/4, 1/2, 1, and 2 watts. The larger the size of carbon resistor, the higher its wattage rating, since a larger amount of material will absorb and give up heat more easily.

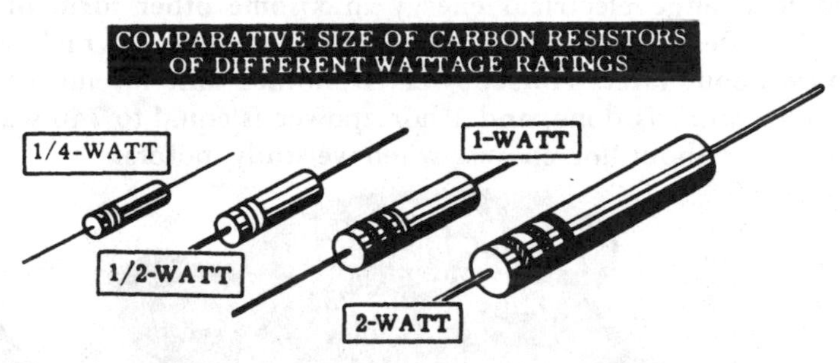

When resistors of wattage ratings greater than 2 watts are needed, wire-wound resistors are used. Such resistors are made in ranges between 5 and 200 watts, with special types being used for power in excess of 200 watts.

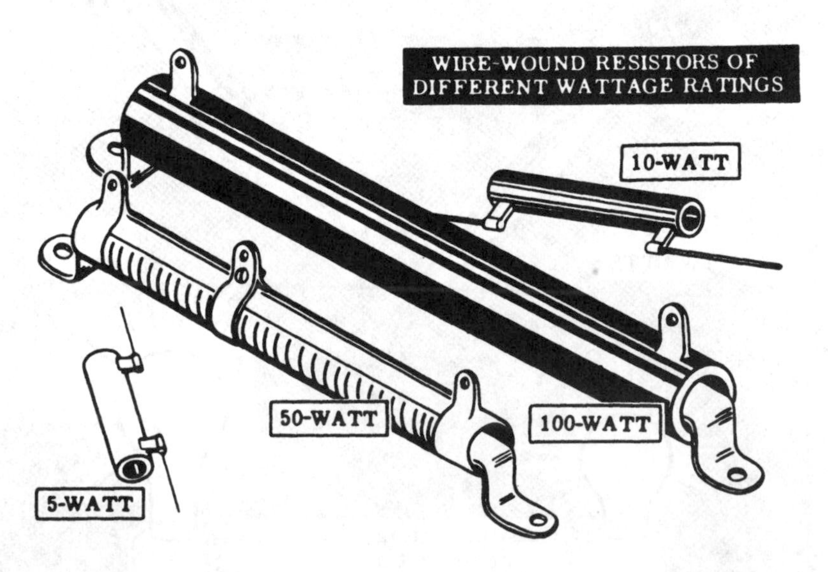

Fuses

You know that when current passes through a resistor, electrical energy is transformed into heat, which raises the temperature of the resistor. If the temperature rises too high, the resistor may be damaged. The metal wire in a wound resistor may melt, thereby opening the circuit and interrupting current flow. This effect is used to advantage in fuses.

Fuses are resistors using special metals with very low resistance values and a low melting point, which are designed to *blow out* and thus open the circuit when the current exceeds the fuse's rated value. When the power consumed by the fuse raises the temperature of the metal too high, the metal melts and the fuse *blows.* Blown fuses can usually be identified by a broken filament and darkened glass. If you are uncertain, you can remove the fuse and check it with an ohmmeter.

You have already learned that excessive current may seriously damage electrical equipment—motors, test instruments, radio receivers, etc. Fuses are cheap, yet the other equipment is much more expensive.

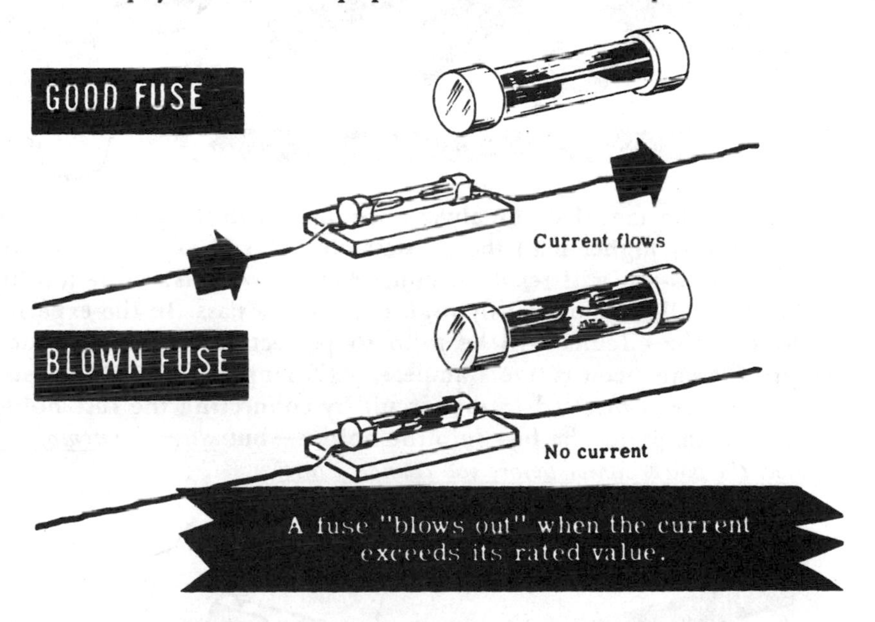

A fuse "blows out" when the current exceeds its rated value.

There are two types of fuses in use today—*conventional* fuses, which blow immediately when the circuit is overloaded, and *slow-blowing* fuses. Slow-blowing (slo-blo) fuses can accept momentary overloads without blowing, but if the overload continues, they will open the circuit. These slo-blo fuses are used in circuits that have a sudden rush of high current when turned on, such as motors and some appliances. If such circuits used a conventional fuse with a high enough value to handle the high starting currents, there would be little protection under normal running conditions. It is important that you replace fuses with the proper type, whether conventional or slow blowing.

Fuses (continued)

Although it is the power used by a fuse which causes it to blow, fuses are rated by the *current* which they will conduct without burning out, since it is high current which damages equipment. Since various types of equipment use different currents, fuses are made in many sizes, shapes, and current ratings.

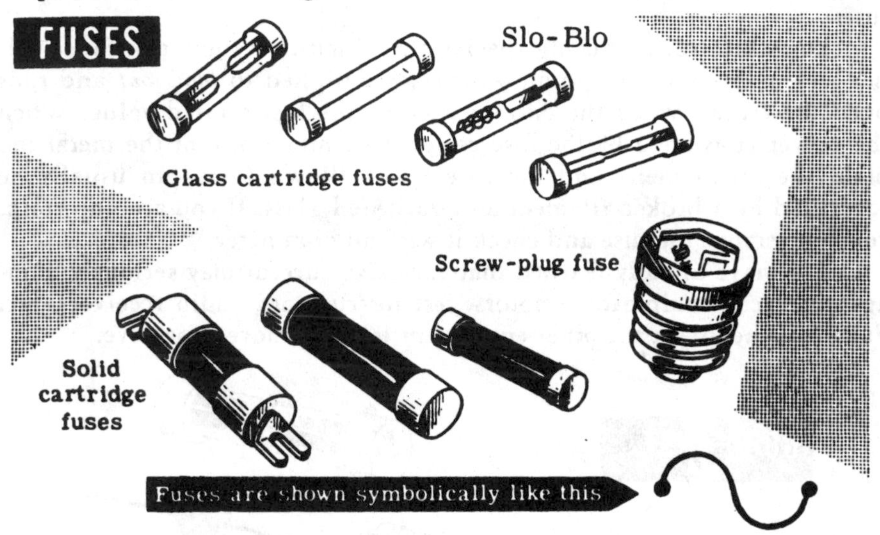

It is important that you always use fuses with the proper current rating—slightly higher than the greatest current you expect in the circuit. Too low a rating will result in unnecessary blowouts, while too high a rating may allow dangerously high currents to pass. In the experiments to follow, the circuits will be *fused* to protect the ammeter. Since the range of the ammeter is 0 to 1 ampere, a 1.5-ampere fuse will be used.

The fuse is inserted in the circuit by connecting the fuse holder in series and snapping the fuse into the holder—but *always remember to disconnect the power source before you change a fuse!*

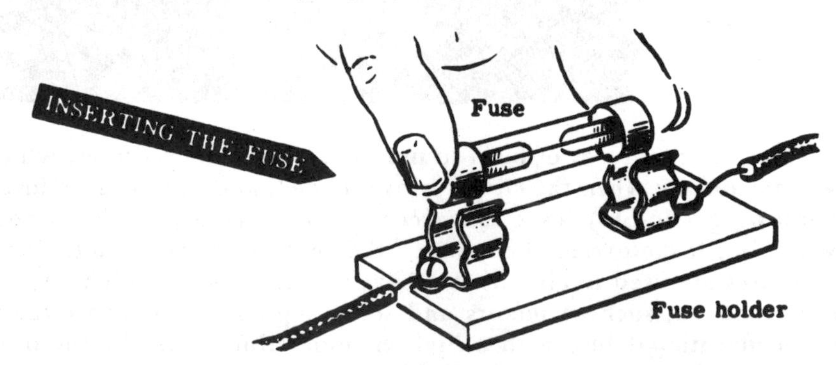

Later, you will learn about another protective device called the *circuit breaker* that provides protection without the inconvenience of changing fuses.

Power in Series Circuits

The total power consumed in a series circuit is the sum of the power used in all the individual circuit elements and is easily found.

Consider a circuit in which three resistors—R1 = 20 ohms, R2 = 16 ohms, and R3 = 12 ohms—are connected in series across a power supply of 72 volts. You now need to know how much power will be consumed by the circuit.

First, sketch the circuit diagram, and fill in the known values.

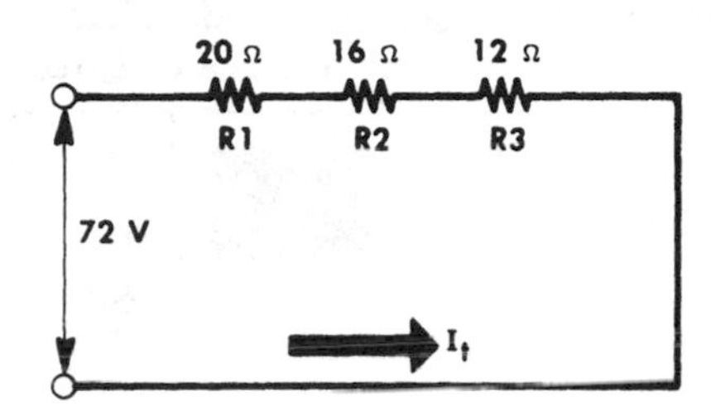

Then calculate circuit current, which you can do as soon as you have the circuit resistance, R_t. Here you see that R_t = 20 + 16 + 12 = 48 ohms.

With the voltage and circuit resistance both known, Ohm's Law tells you that the circuit current is:

$$I_t = \frac{E}{R} = \frac{72}{48} = 1.5 \text{ amperes}$$

You can now use the variant of the power formula which gives you P when you know only I and R. It is, you will remember, $P = I^2R$.

$$P1 = I^2R1 = 1.5 \times 1.5 \times 20 = 45 \text{ watts}$$
$$P2 = I^2R2 = 1.5 \times 1.5 \times 16 = 36 \text{ watts}$$
$$P3 = I^2R3 = 1.5 \times 1.5 \times 12 = 27 \text{ watts}$$

Since the power taken by a series circuit is the sum of the power taken by the individual resistors in the circuit, you find that

$$P_t = 45 + 36 + 27 = 108 \text{ watts}$$

And, since Ohm's Law tells us that P = EI—and P = 72 volts and I = 1.5 amperes, then

$$P = 72 \times 1.5 = 108 \text{ watts}$$

Another way of attacking this problem would be to simplify the circuit before you start to calculate I, and to draw the equivalent circuit like this.

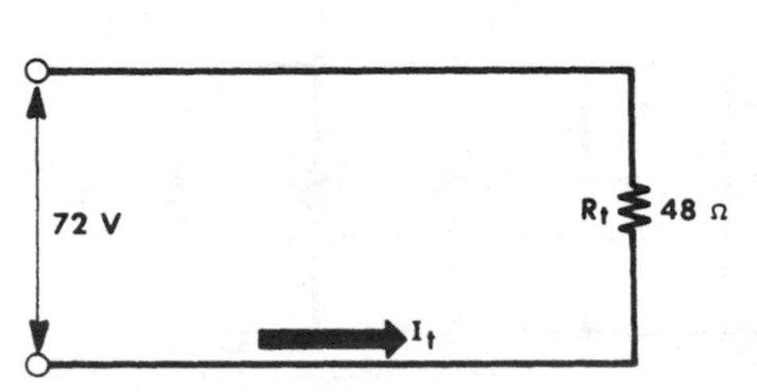

Now calculate I_t by using Ohm's Law exactly as you did before:

$$I_t = \frac{E}{R_t} = \frac{72}{48} = 1.5 \text{ amperes}$$

Then the same variant of the power formula gives you the power consumed by the circuit:

$$P_t = I^2R_t = 1.5 \times 1.5 \times 48$$
$$= 108 \text{ watts}$$

which is, as you would expect, the same answer as you found before.

Power in Parallel Circuits

You have seen that the total power taken by a series circuit is equal to the sum of the power taken by all the individual resistors in the circuit.

The same thing is true of all parallel circuits. *The total power taken by a parallel circuit is equal to the sum of the power taken by all the individual resistors in the circuit.* It can be found by multiplying the total circuit current by the voltage across the circuit.

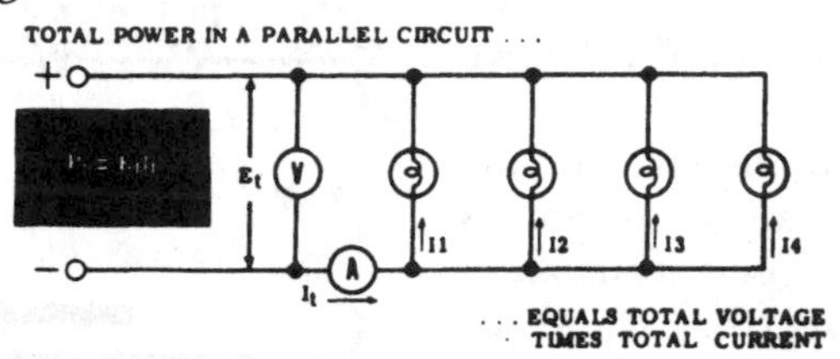

$$\text{Total power} = 100 + 100 + 100 + 100 = 400 \text{ watts}$$

By the power formula

$$P_t = E_t \times I_t = 125 \times 3.2 = 400 \text{ watts}$$

If either circuit current or circuit voltage is unknown, circuit power can still be found by applying the rules for parallel circuits to calculate total circuit resistance—but only, of course, if the value of every individual resistor in the circuit is known.

It is then only a matter of selecting the correct variation of the power formula to take advantage of the particular set of facts you already know. The two possibilities are shown in the double diagram below.

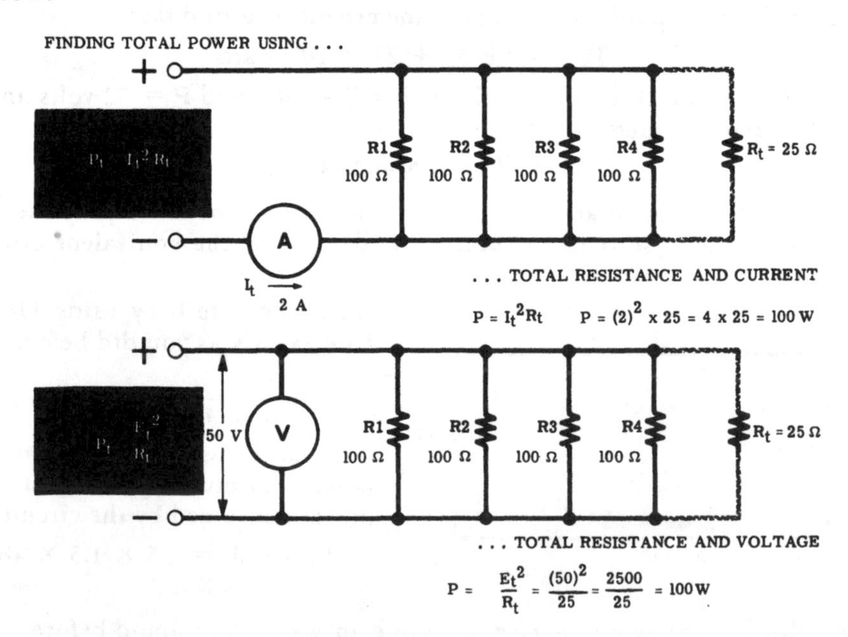

Power in Complex Circuits

You have seen that the power taken in both series and parallel circuits is equal to the sum of the power taken by all the individual resistors or loads.

The same thing is true of complex circuits involving series and parallel parts. The *total power taken by a series-parallel circuit is equal to the sum of the power taken by all the individual loads in the circuit.* It can be found by multiplying the total current by the voltage across the circuit.

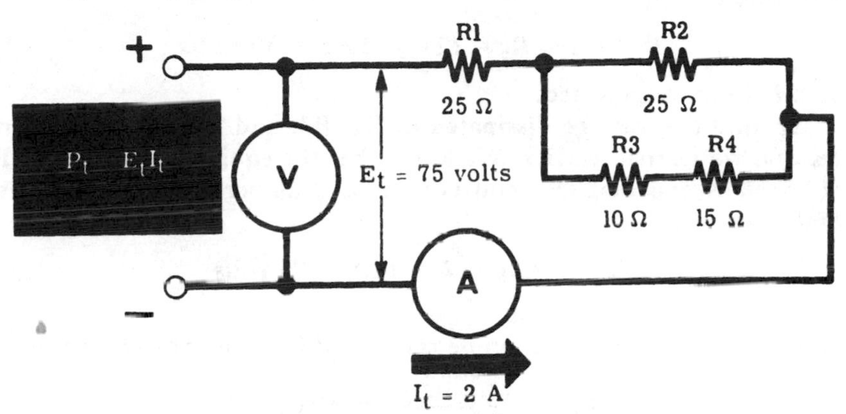

TOTAL POWER = $P_t = E_tI_t = 75 \times 2 = 150$ watts

Since we know that the total circuit current is 2 amperes, and this current must flow through R1, then using the right version of the power formula for current and resistance will give us

$$P_{R1} = I_t^2 \times R1 = (2)^2 \times 25 = 100 \text{ watts}$$

Since the total wattage is 150 watts, then the total wattage in R2, R3, and R4 must be 50 watts (150 − 100 = 50). Let us check this by combining R2, R3, and R4 into their equivalent resistance.

$$R_a = R3 + R4 = 10 + 15 = 25 \text{ ohms}$$

The circuit is now

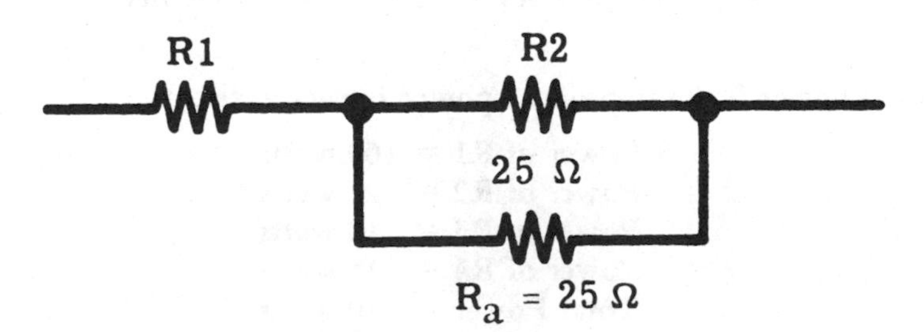

The parallel combination of R_a and R2 is equal to

$$R_b = \frac{R2 \times R_a}{R2 + R_a} = \frac{25 \times 25}{25 + 25} = 12.5 \text{ ohms}$$

Power in Complex Circuits (continued)

The total current is now

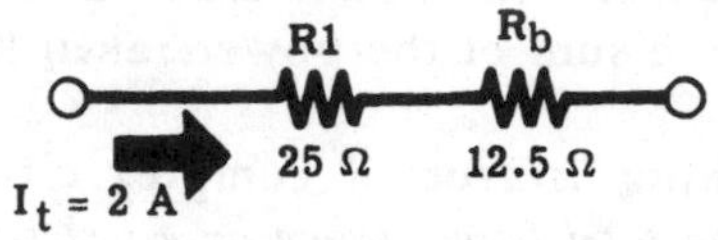

R1 Rb

25 Ω 12.5 Ω

$I_t = 2$ A

The power in R_b (parallel combination) can be calculated in a number of ways. The simplest way is to use the power formula for current and resistance,

$$P_{Rb} = I_t^2 \times R_b = (2)^2 \times 12.5 = 50 \text{ watts}$$

which is what we expected.

To find the wattage dissipated in R2, R3, and R4 we can use some facts that we learned earlier. We know that the equivalent resistance R_b is 12.5 ohms and that the total current is 2 amperes. Therefore, using Ohm's Law,

$$E_b = I_t \times R_b = 2 \times 12.5 = 25 \text{ volts}$$

Therefore, the power in R2 can be calculated from the power formula.

$$P_{R2} = \frac{E2}{R} = \frac{(25)2}{25} = 25 \text{ watts}$$

Since R_a is 25 watts, then R3 and R4 together dissipate 25 watts also. The current in the branch of the parallel circuit containing R3 and R4 can be calculated by Ohm's Law as

$$I_{R3+R4} = \frac{E_{R3+R4}}{R_{R3 \times R4}} = \frac{25}{25} = 1 \text{ ampere}$$

The power in R3 is

$$P_{R3} = (I_{R3})^2 \times R3 = (1)^2 \times 10 = 10 \text{ watts}$$

The power in R4 is

$$P_{R4} = (I_{R4})^2 \times R4 = (1)^2 \times 15 = 15 \text{ watts}$$

The total power is the sum of the power in each resistor.

Power of R1 = 100 watts
Power of R2 = 25 watts
Power of R3 = 10 watts
Power of R4 = 15 watts
Total Power = 150 watts

Thus, again, it is clear that no matter how complex the circuit, the total power is the sum of the individual wattage and is equal to the input voltage times the input current.

Review of Electric Power

Whenever an electric current flows, work is done in moving electrons through the conductor. The electrons to be moved may be moved in either a short or a long period of time, and the rate at which the work is done is called electric power.

1. ELECTRIC POWER—The rate of doing work in moving electrons through a material is electric power, "P." The basic unit of power is the watt, "W." 1 watt is used to make a current of 1 ampere flow through a resistance of 1 ohm.

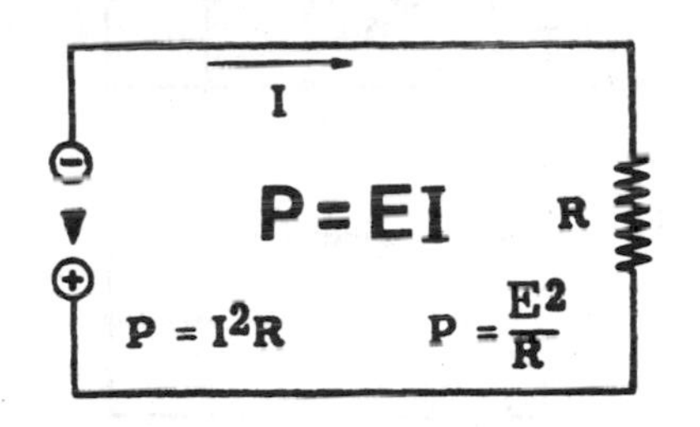

2. POWER FORMULA—Electric power used in a resistance equals the voltage across the resistance terminals times the current flow through the resistance. It is also equal to the current squared, times the resistance, or to the voltage squared, divided by the resistance.

3. POWER RATING—Electrical equipment is rated according to the rate at which it uses electric power. The power used is converted from electrical energy into some other form of energy, such as heat or light.

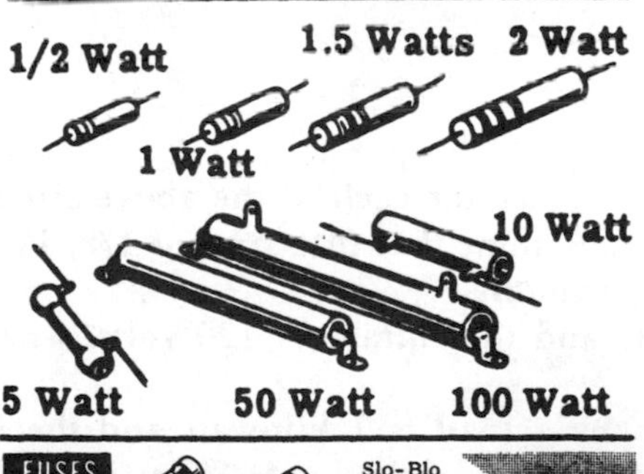

4. RESISTOR POWER RATINGS—Resistors are rated both in ohms of resistance and by reference to the maximum power which can safely be used in the resistor. High-wattage resistors are constructed larger than low-wattage resistors, so as to provide a greater surface for dissipating heat. They are also made of materials that can withstand greater quantities of heat.

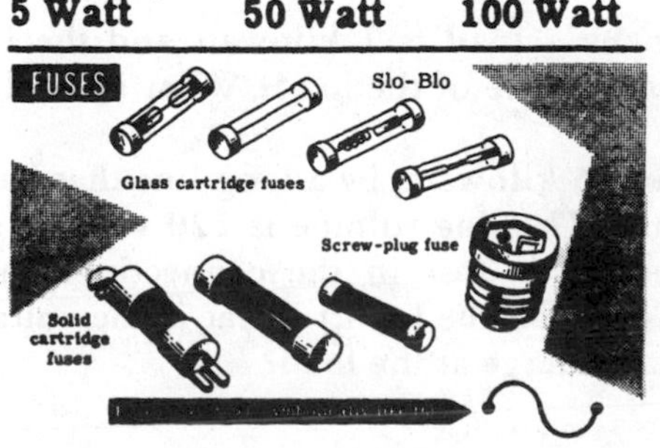

5. FUSES—Fuses are metal resistors designed to open an electric circuit if the current through them exceeds their rated value.

Self-Test—Review Questions

1. Calculate the power in the circuits shown below:

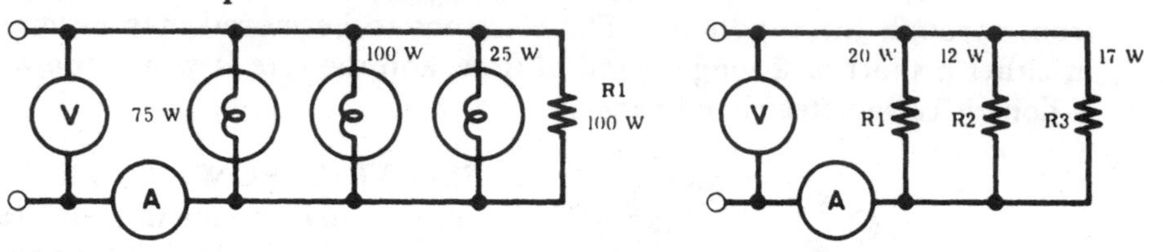

2. Calculate the power—total and in each circuit element—for the circuits shown below:

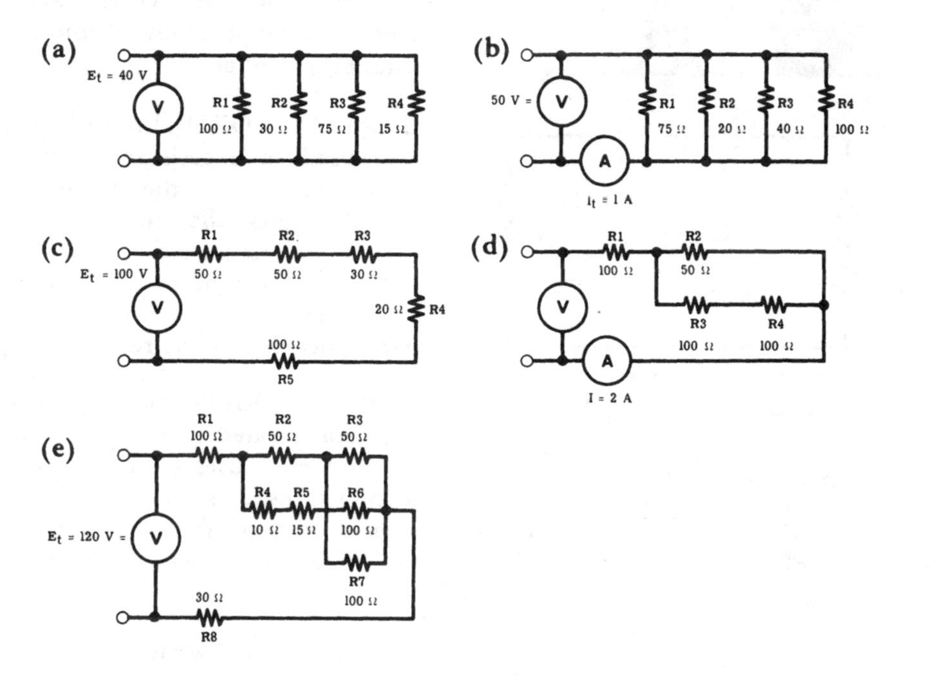

3. Calculate and specify the proper fuse size for each of the above circuits, assuring about a 50% safety factor. Assume that fuse sizes of 1/8, ¼, ½, 1, 2, 3, 5, 7.5, and 10 amperes are available.
4. If the power in a line is 200 watts and the voltage is 120 volts, what is the line current?
5. You know that the power drawn by a load is 1 kilowatt and the line voltage is 120 volts. What is the resistance of the load? What is the line current?
6. You must connect a load that draws 5 kilowatts by a long line that has a resistance (in each leg) of 0.5 ohm. The line voltage is 120 volts at the input to the line. How much power is lost in the connecting lines? (Neglect the effect of the voltage drop on the load.) What is the voltage drop in the line? What is the actual voltage at the load?

Experiment/Application—The Use of Fuses

You have seen how a resistor overheats when it uses more power than its power rating. Now you will see how this effect is put to use to protect electrical equipment from damage due to excessive currents.

Suppose you connect four dry cells in series to form a 6-volt battery and then connect a 15-ohm, 10-watt resistor, a knife switch, a fuse holder, and an ammeter in series across the battery, and insert a 1/8-ampere fuse in the fuse holder. When you close the switch, the fuse will *blow*, thus opening the circuit so that no current can flow, as shown by a zero reading of the ammeter.

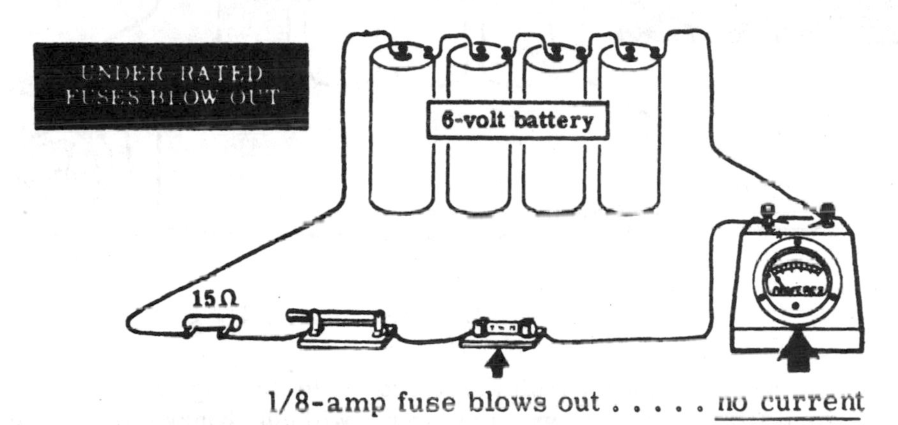

However, if you insert a 1/2-ampere fuse in the fuse holder, it will not blow, and the ammeter will show a current flow.

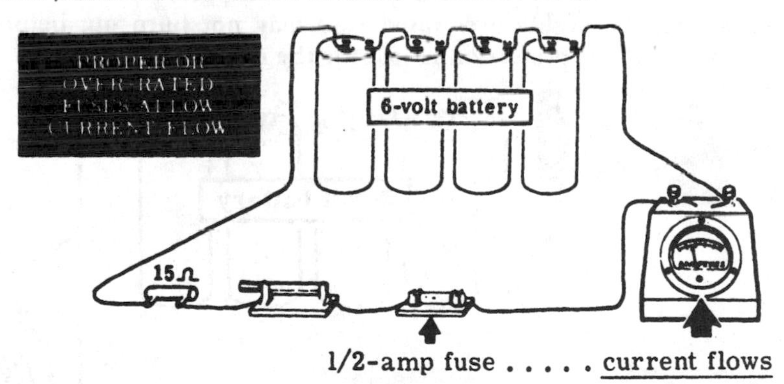

Since the resistance of the circuit is 15 ohms and the voltage is 6 volts, the current flow by Ohm's Law is about 0.4 ampere ($\frac{6 \text{ volts}}{15 \text{ ohms}}$). The 1/8-ampere (0.125-ampere) fuse *blows out* because the current exceeds its rating, and it will not carry 0.4 ampere. However, the 1/2 ampere (0.5 ampere) fuse carries the current without blowing, since its rating exceeds the actual current flow.

Experiment/Application—How Fuses Protect Equipment

Using the circuit shown previously, note that the 15-ohm resistor limits the current through the circuit sufficiently to keep a 1/2-ampere fuse from burning out. The circuit operates without damage to the ammeter.

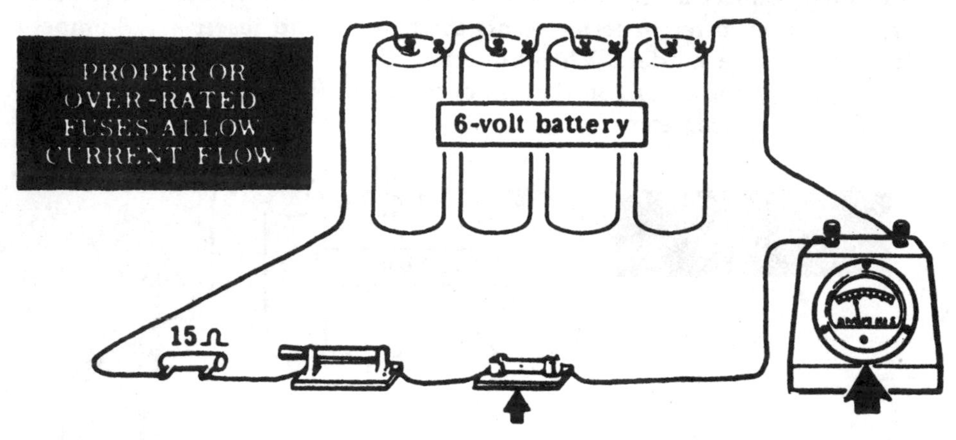

1/2-amp fuse . . . . current flows

If you were to short-circuit the resistor as in the diagram below, the fuse would burn out and open the circuit without damage to the ammeter. Because the fuse serves as the *predetermined* weakest link in this circuit, it is the electrical safety device. In choosing fuses, be sure not to choose one whose rating is too high for the expected current flow. If trouble occurs, the highly over-rated fuse may not burn out before the meter does so that all protection is lost for the meter.

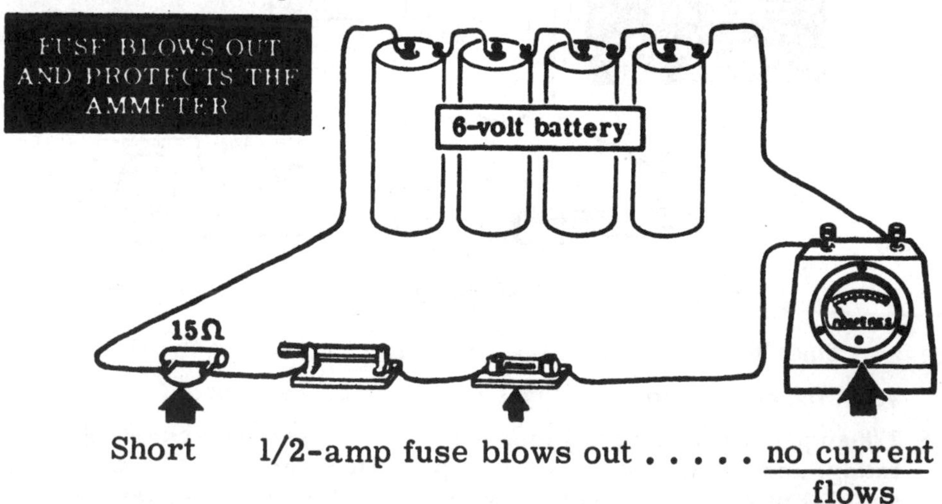

1/2-amp fuse blows out . . . . . no current flows

Experiment/Application—Power in Series Circuits

To show that power can be determined when any two of the circuit variables—current, voltage, and resistance—are known, connect three 15-ohm, 10-watt resistors in series across a 9-volt dry cell battery.

After measuring the voltage across each resistor, you can apply the power formula $P = \frac{E^2}{R}$ to find the power for each resistor. You see that the power used by each resistor is about 0.6 watt and that the total power is about 1.8 watts.

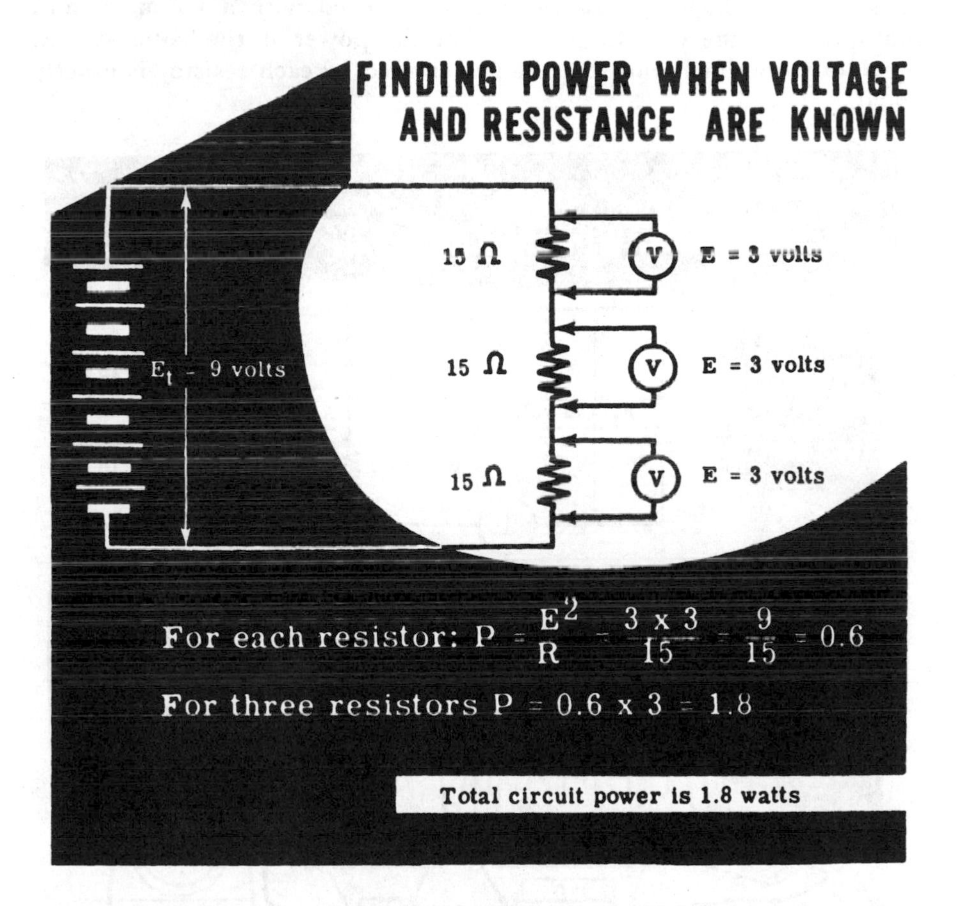

To show that the same results are obtained using current and resistance or current and voltage, connect an ammeter in the circuit to measure current. The power used by each resistor is then found by using the power formula in two ways: $P = I^2R$ and $P = EI$. Notice that the power in watts is nearly the same for each variation of the power formula used, with the negligible difference due to meter inaccuracies and slight errors in meter readings.

Experiment/Application—Power in Series Circuits (continued)

To show the effect of the power rating of a resistor on its operation in a circuit, suppose that two 15-ohm resistors—one rated at 10 watts and the other rated at 1 watt—are connected in a series circuit as shown below. The ammeter reads the circuit current and, using the power formula, $P = I^2R$, you find that the power used in each resistor is approximately 1.35 watts. This is slightly more than the power rating of the 1-watt resistor, and you will see that it heats rapidly, while the 10-watt resistor remains relatively cool. To check the power used in each resistor, the voltages across them are measured with a voltmeter and multiplied by the current. Notice that the power is the same as that previously obtained, and that the power used by each resistor is exactly equal.

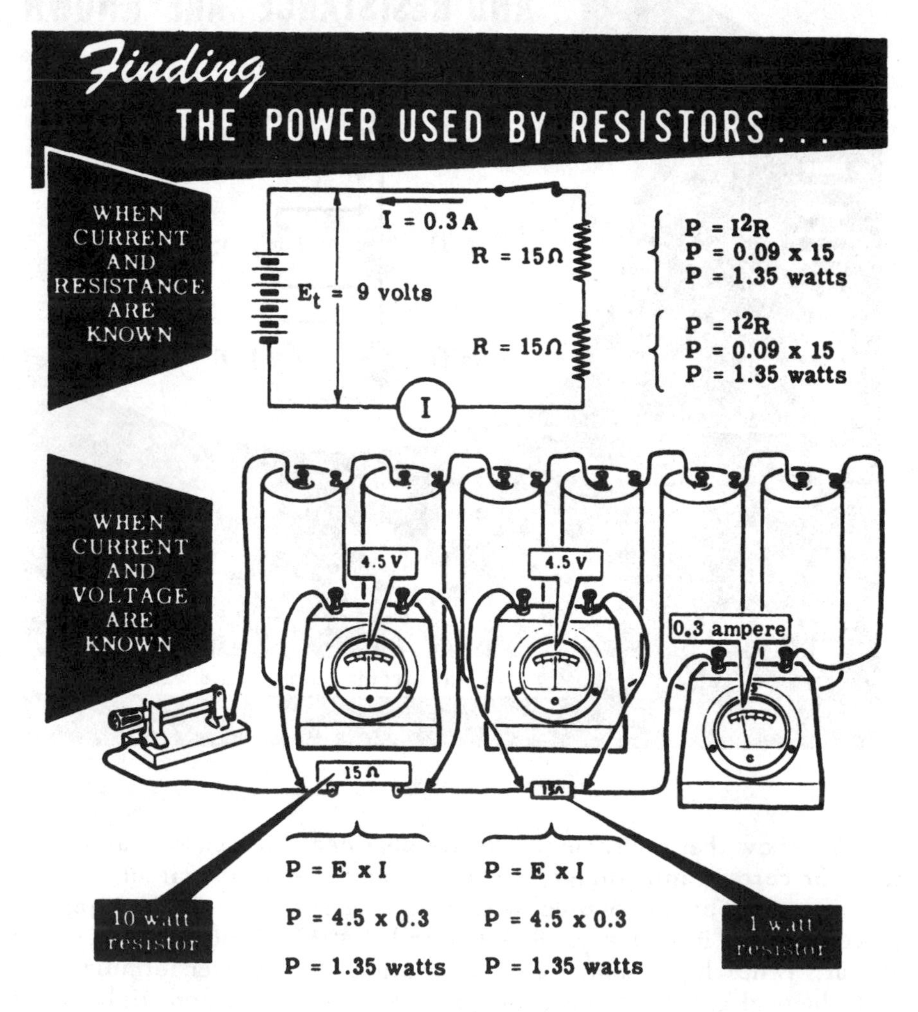

Experiment/Application—Power in Series Circuits (continued)

Next, the 1-watt resistor is replaced by one rated at ½ watt. Observe that it heats more rapidly than the 1-watt resistor and becomes very hot, indicating that the power rating has been greatly exceeded. As the power for each resistor is found (using current and resistance, then voltage and current as a check), you will see that each resistor is using the same amount of power. This shows that the power rating of a resistor does not determine the amount of power used in a resistor. Instead, the power rating only indicates the maximum amount of power that may be used *without damaging* the resistor.

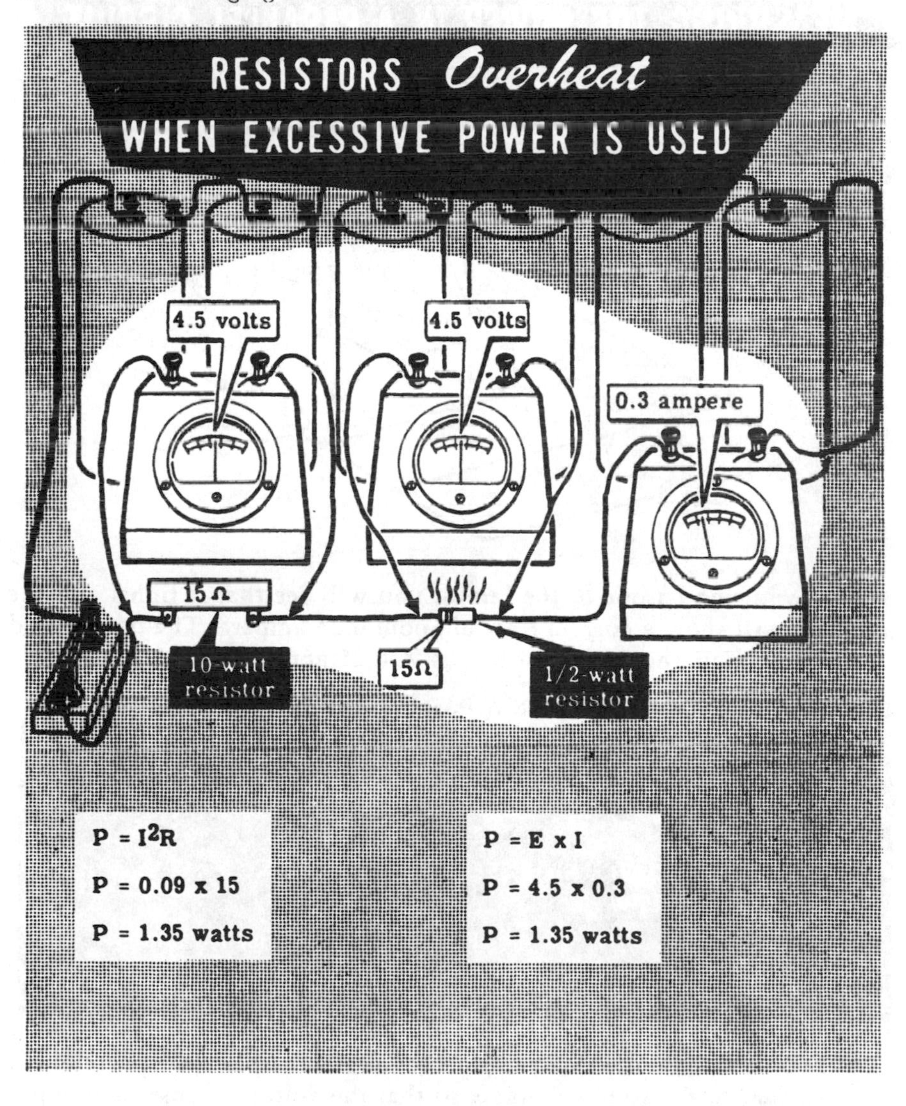

Experiment/Application—Power in Parallel Circuits

To show that the power used by a parallel circuit is equal to the power used by all of the parts of the circuit, connect three lamp sockets in parallel across a 6-volt battery, with a 0-1 ampere range ammeter in series with the battery lead, and a 0-10-volt range voltmeter across the battery terminals. Next, insert 6-volt, 250-milliampere lamps in the lamp sockets, but do not tighten them. When you close the switch, you will see that the voltmeter indicates battery voltage, but the ammeter shows no current flow, since no power is being used by the circuit.

MEASURING THE POWER IN A PARALLEL CIRCUIT

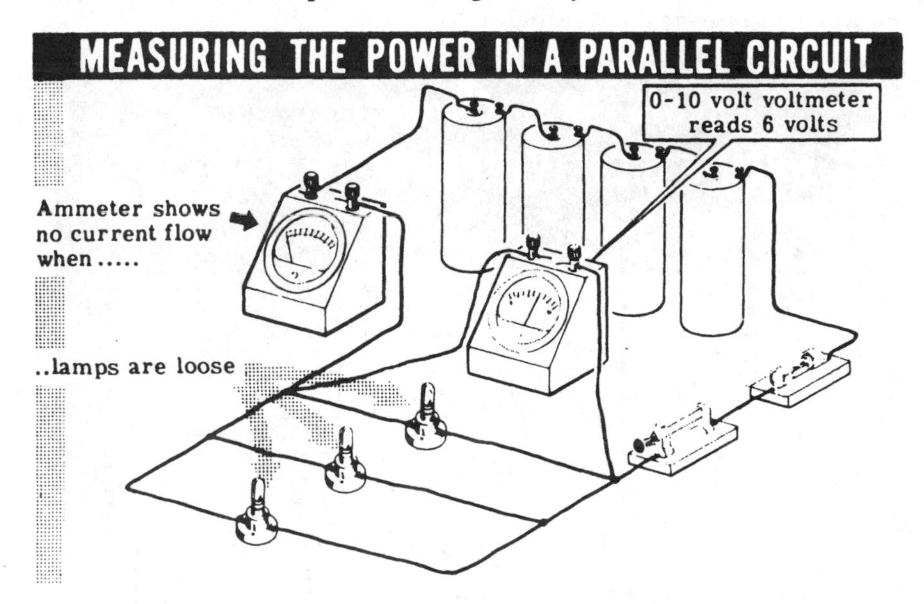

As you tighten one of the lamps, you will see that it lights and the ammeter will show a current flow of about 0.25 ampere. The power used by this one lamp, then, is about 6 volts × 0.25 ampere, or 1.5 watts.

POWER USED BY ONE LAMP IN A PARALLEL CIRCUIT

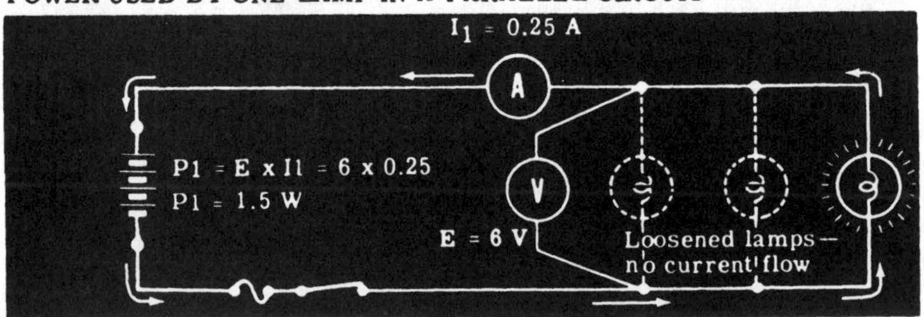

You already know that the voltage across any part of a parallel circuit is equal to the source voltage, so that the voltage across the lamp is equal to the battery voltage.

Experiment/Application—Power in Parallel Circuits (continued)

As you loosen the first lamp and tighten each of the other lamps in turn, you will see that the current—and hence the power used by each lamp—is about the same. The current measured each time is the current through only the one tightened lamp.

MEASURING POWER IN INDIVIDUAL LAMPS

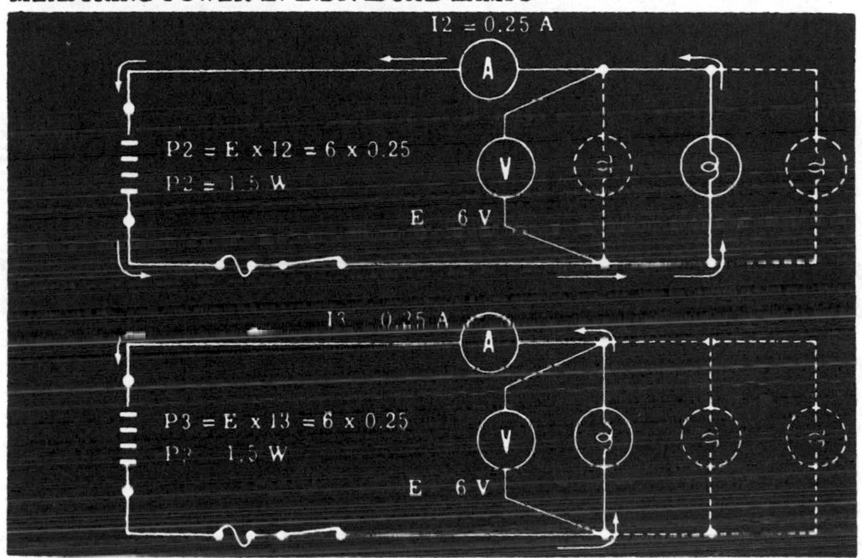

Next, suppose you tighten all three lamps in their sockets. You will see that they all light and that the ammeter shows a circuit current of about 0.75 ampere. The voltage is still about 6 volts, so that the circuit power (P_t) equals 0.75 × 6, or about 4.5 watts.

MEASURING TOTAL CIRCUIT POWER

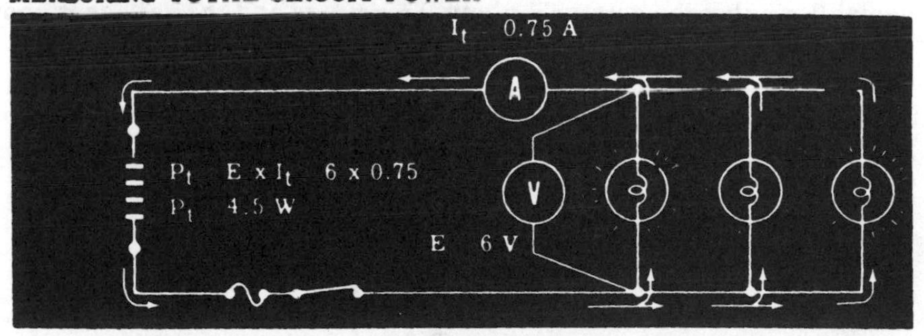

The total circuit power is found to be about 4.5 watts. If you add the power used individually by each of the lamps, the sum is equal to 4.5 watts (1.5 + 1.5 + 1.5 = 4.5). Therefore, you can see that the total power used by a parallel circuit is equal to the sum of the power used by each part of the circuit.

Experiment/Application—Power in Parallel Circuits (continued)

Now replace the three lamp sockets with 30-ohm resistors. If you then remove the voltmeter leads from the battery and close the switch, you will notice that the ammeter shows a current flow of about 0.6 ampere.

The total resistance found by applying the rules for parallel circuits is 10 ohms, so that the circuit power is equal to $P = I^2R$ or $(0.6)^2 \times 10 = 3.6$ watts.

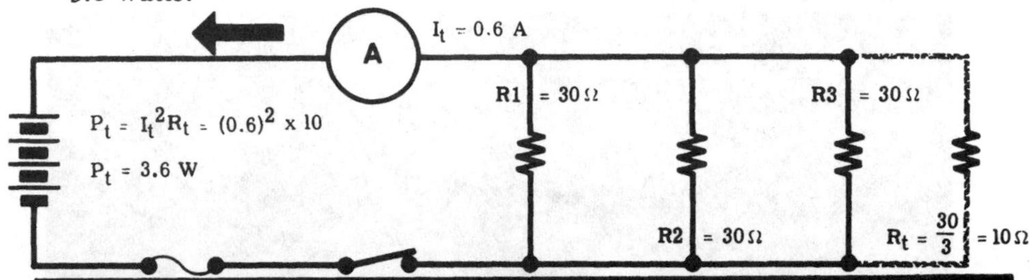

USING TOTAL CURRENT AND RESISTANCE TO MEASURE CIRCUIT POWER

Now remove the ammeter and connect the voltmeter to the battery leads. When the switch is closed, the voltage registers about 6 volts, so that the circuit power is equal to $P = \frac{E2}{R}$ or $\frac{(6)^2}{10} = 3.6$ watts.

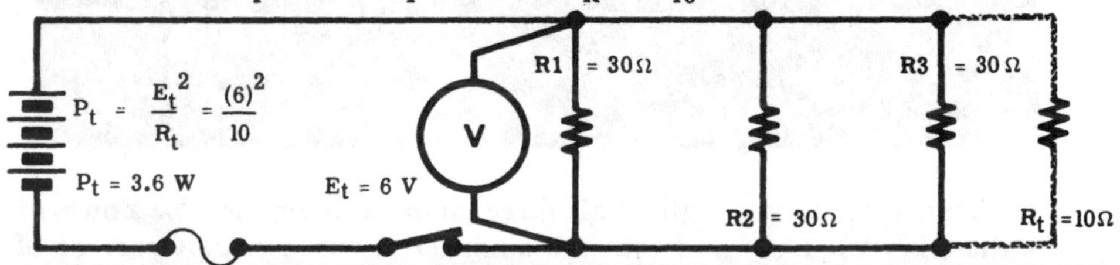

USING TOTAL VOLTAGE AND RESISTANCE TO MEASURE CIRCUIT POWER

Finally, replace the ammeter in the circuit; when power is applied, you will see that the current is about 0.6 ampere and the voltage is about 6 volts, so that the total circuit power is equal to $P = EI$ or $6 \times 0.6 = 3.6$ watts.

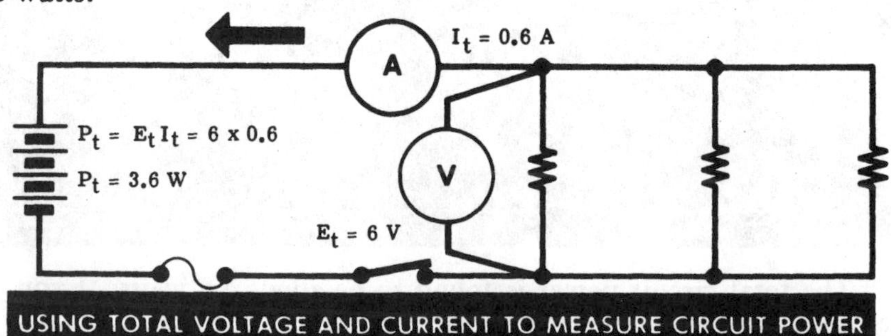

USING TOTAL VOLTAGE AND CURRENT TO MEASURE CIRCUIT POWER

Thus, you see that the total power in a parallel circuit may be determined, as in a series circuit, whenever any *two* of the factors—current, voltage, or resistance—are known.

Thevenin's Theorem—Voltage Division in the Series Circuit (continued from page 2-42)

If you load a voltage divider with an external resistance, the voltage will decrease. It is necessary to take this into consideration sometimes. You could do this by using Ohm's Law and calculating the parallel resistance of R2 and R_{load} and then the voltage divider itself. A simpler method is to use Thevenin's theorem, which enables you to calculate quickly the effect of any load. With Thevenin's theorem you can replace the circuit shown in 1 below (as shown on page 2-41) with that shown in 2A. That is, the voltage source is the old E_{out} without load and the series resistance is the parallel combination of R1 and R2. Illustration 2B shows the Thevenin equivalent circuit.

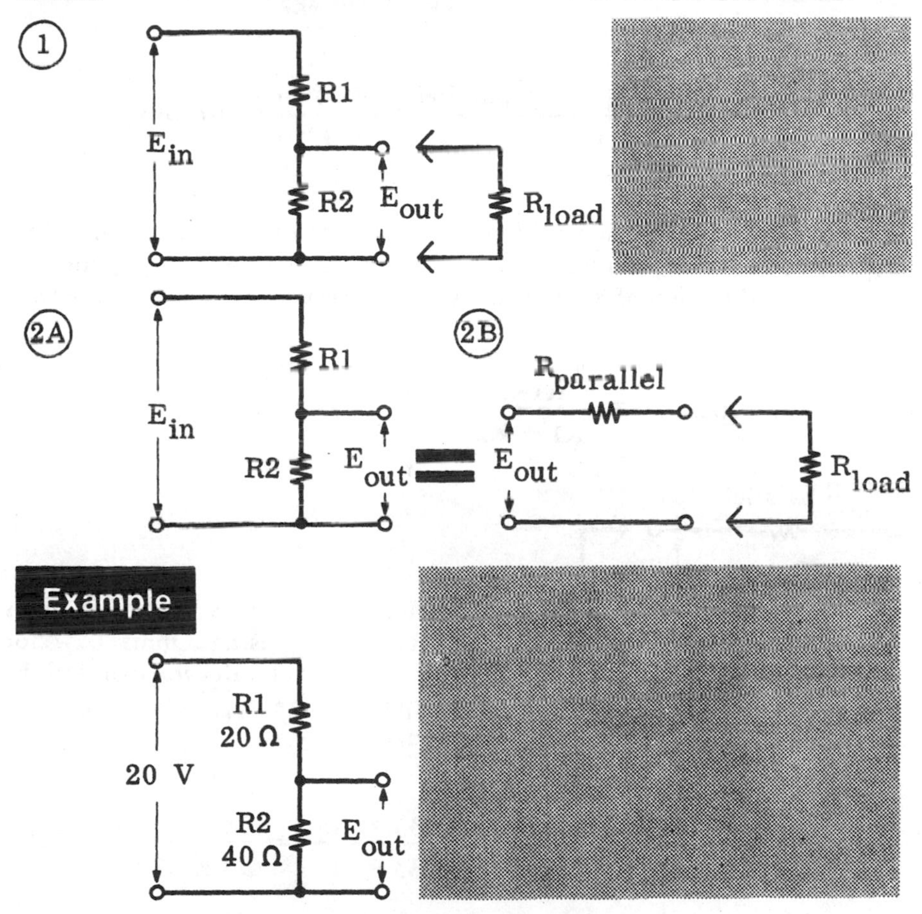

As you know from the voltage divider calculations on pages 2-41 and 2-42, the voltage across R2 with no load is

$$E_{out} = \frac{E_{in}\,R2}{R1+R2} = \frac{(20)\times(40)}{20+40} = \frac{800}{60} = 13.33 \text{ volts}$$

Thevenin's Theorem—Voltage Division in the Series Circuit (continued)

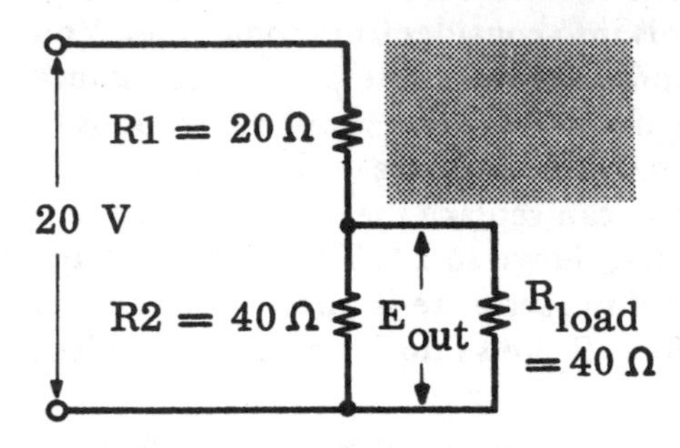

Now, when the load is added, the circuit is changed to that shown at the left. The output voltage is now developed across the parallel combination of R2 and R_{load}, and is equal to 20 ohms. Therefore, the output voltage E_{out} is

$$E_{out} = \frac{(E_{in}) \times R_{parallel}}{R1 + R_{parallel}} = \frac{20 \times 20}{20 + 20} = 10 \text{ volts}$$

You can calculate the *same* thing easier by use of Thevenin's theorem, replacing the voltage source by E_{out} (new E_{in}) without R_{load} and the parallel combination of R1 and R2 as a series resistance as shown below:

$$R_{parallel} = \frac{R1R2}{R1 + R2} = \frac{20 \times 40}{60} = 13.33 \text{ ohms}$$

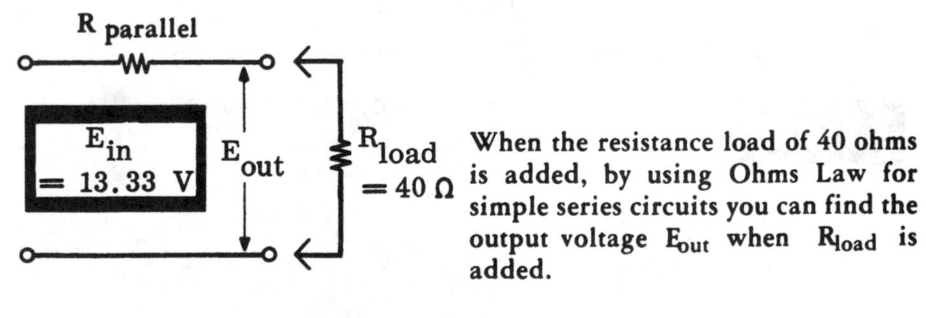

When the resistance load of 40 ohms is added, by using Ohms Law for simple series circuits you can find the output voltage E_{out} when R_{load} is added.

$$E_{out} = \frac{(E_{in}) \times R_{load}}{R_{parallel} + R_{load}} = \frac{13.33 \times 40}{13.33 + 40} = \frac{533.3}{53.33} = \textit{10 volts!}$$

The answer is the *same!* But you can see that if you wanted to know the effects for many different loads, using Thevenin's theorem will make it much easier to do the calculations. And more importantly, when you are calculating complex circuits, the circuit can often be reduced quickly by applying Thevenin's theorem.

Norton's Theorem—Voltage Division in the Series Circuit (continued)

A similar theorem—Norton's theorem—can be used in situations where a *current* source, rather than a *voltage* source, is more convenient. To use Norton's theorem, the load terminals (across R2) are shorted and the current is calculated.

$$I\ (R2_{\text{shorted}}) = \frac{E_{\text{in}}}{R1} = \frac{20}{20} = 1 \text{ ampere}$$

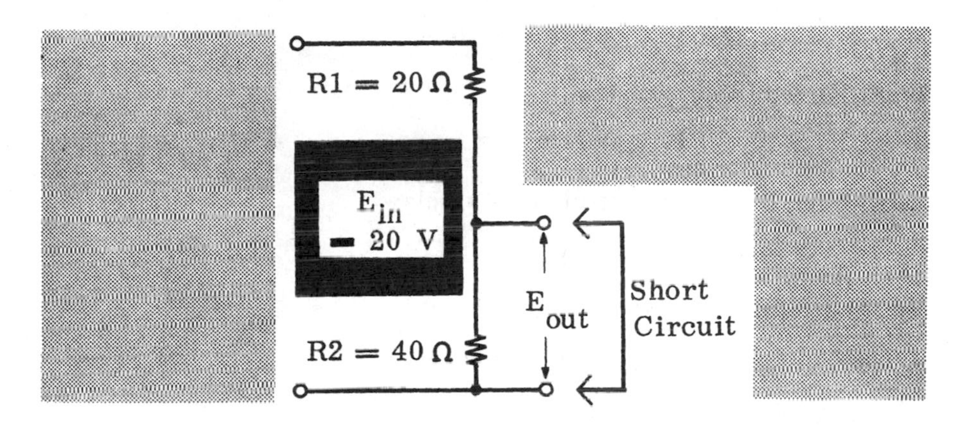

which is the *equivalent current source*. As before, the parallel combination of R1 and R2 is calculated as 13.33 ohms. The Norton equivalent circuit is:

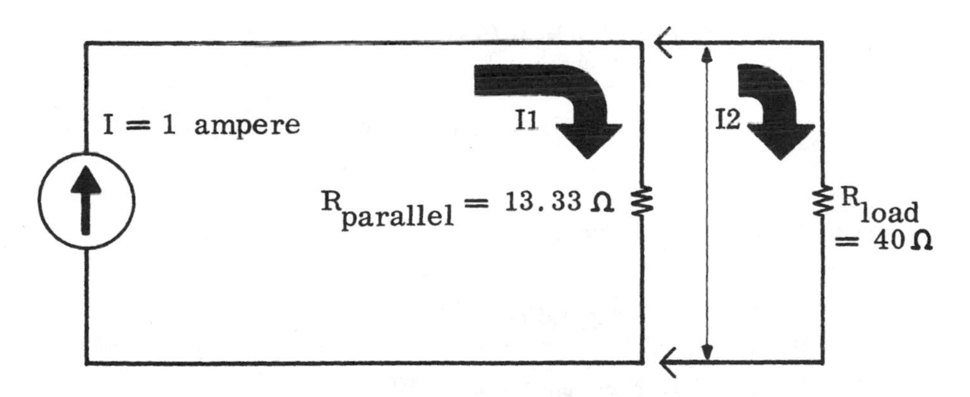

Now without the load, the voltage across E_{out} is 13.33 volts (1 amp and 13.33 ohms), as was calculated using Thevenin's theorem. When the load is added, the current divides into R_{parallel} and R_{load} but the total current is still 1 ampere. In addition, you know that the voltage across R_{parallel} and R_{load} must be the same.

Norton's Theorem—Voltage Division in the Series Circuit (continued)

Therefore, the voltage E_{out} across $R_{parallel}$ is

$$\text{(1)}\quad E_{out} = I_1 \times R_{parallel} = I_1 \times 13.33$$

and also E_{out} is

$$\text{(2)}\quad E_{out} = I_2 \times R_{load} = I_2 \times 40$$

and therefore we can equate 1 and 2

$$\text{(3)}\quad I_1 \times 13.33 = I_2 \times 40$$

From Kirchoff's current law

$$I_{total} = I_1 + I_2 = 1 \text{ amp}$$

Solving for I_2 gives the result shown below

$$I_2 = I_{total} - I_1 = 1 - I_1$$

By substituting the above equation 3

$$I_1 \times 13.33 = (1 - I_1)\,40$$

Rearranging terms and solving for I_1 yields

$$13.33\, I_1 = 40 - 40\, I_1$$

$$53.33\, I_1 = 40$$

$$I_1 = \frac{40}{53.3} = 0.75 \text{ amp}$$

Since the total current is 1 ampere, then I_2 must be .25 ampere.

You can calculate the output voltage as either:

$$I_1 \times R_{parallel} \text{ or } I_2 \times R_{load}$$

$$I_2 \times R_{load} = 0.25 \times 40$$

$$= \mathit{10\ volts!}$$

This is the *same* as was calculated earlier using Thevenin's theorem! Again, Norton's theorem can be useful in solving complex circuits.

Troubleshooting DC Circuits—Basic Concepts

One of the most important things that you must learn to do is to troubleshoot electric circuits. As you proceed in your study of electricity, there will be troubleshooting sections in each volume of *Basic Electricity* that will prepare you for work with various electric circuits. In preparation, you should review the beginning of this volume on what electric circuits are and the meaning of open and short circuits. You then will be able to learn how to troubleshoot dc series, parallel, and series-parallel circuits. If you can troubleshoot these, you then will have the foundation to fix any type of malfunctioning dc circuit.

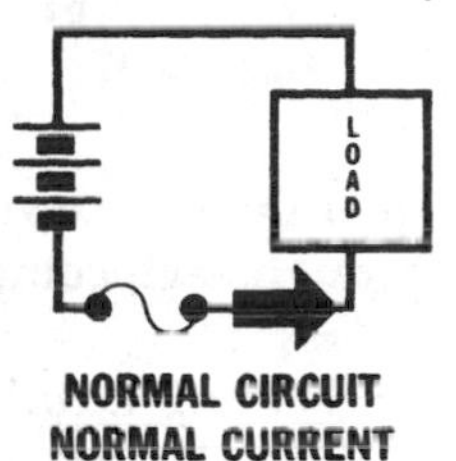

NORMAL CIRCUIT NORMAL CURRENT

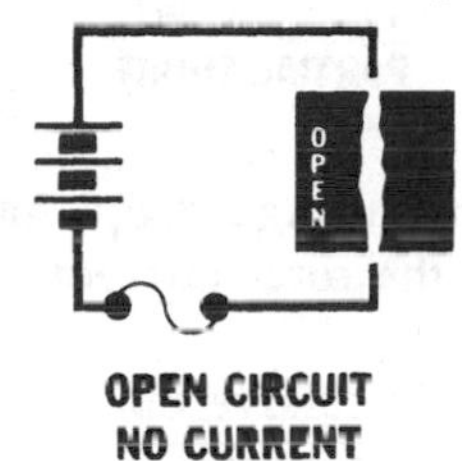

OPEN CIRCUIT NO CURRENT

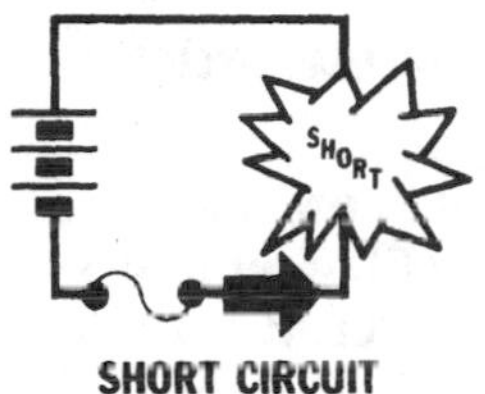

SHORT CIRCUIT HIGH CURRENT FLOW

As you already know, the two most common problems (faults) in circuits are open circuits and shorts. A short in all or part of the circuit causes excessive current flow. This may blow fuses or burn out components, so a short initially can result in an open circuit.

One of the most important things in acquiring troubleshooting skill is to learn to *use your head* and proceed *logically* through a circuit. Also, to learn to use your senses—to look for loose connections, frayed wires, evidence of overheating, blown fuses, open switches, or plugs not installed. In most cases, you will *see* the problem; you may also *smell* the problem; or even determine the fault by *touch*. If not, then by using your head, that is, using a logical procedure and voltmeters, ohmmeters, and ammeters, you can find the trouble. Remember, a blown fuse means that excessive current was drawn and that indicates a full or partial short circuit. If you replace the fuse and it blows again, you know for sure that you are looking for a short. If you replace a blown fuse and nothing happens, chances are that a short blew out a component of the circuit.

THE BASIS OF TROUBLESHOOTING LIES IN USING A LOGICAL PROCEDURE

By using your senses, your head, and test instruments, you can troubleshoot any electric circuit.

Troubleshooting DC Series Circuits

Suppose you were asked to troubleshoot the series circuit shown:

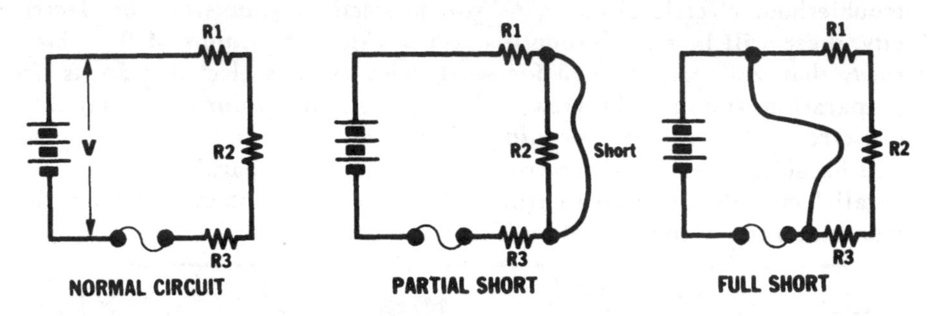

You know that the sum of the voltage drops has to equal the source voltage (E). You also know that the total current flows through each component (R1, R2, and R3).

Suppose the symptom is excessive current flow, indicating a full or partial short. If inspection of the circuit shows no frayed or shorted wiring (and you are sure the wiring is correct), the next thing to do would be to determine what potential drop there should be across each resistance. If you calculate these voltages and measure them, the one that is *too low* indicates that this resistance is *low* and should be *replaced.* If the resistances appear to be normal, it is possible the short lies across the entire circuit; measure the resistance across the source (disconnect the power first!), if it is lower than it should be, some connection or misconnection is giving a full short.

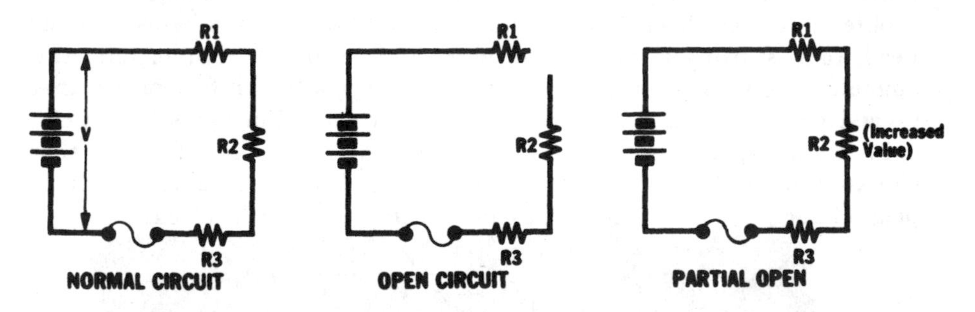

Suppose the symptom is that no current or too little current flows. If no current flows, you should check the voltage source with a voltmeter, and the fuse with an ohmmeter. If these are good, then you must look at the circuit. Inspect the circuit for loose connections or broken wires. After this, proceed to test the voltage drops across each resistance as before; if there is no voltage drop across any resistor, the problem is in the wiring connections. If one voltage is higher than calculated or all the voltage appears across it, then this is the defective component!

Troubleshooting DC Series Circuits (continued)

Example

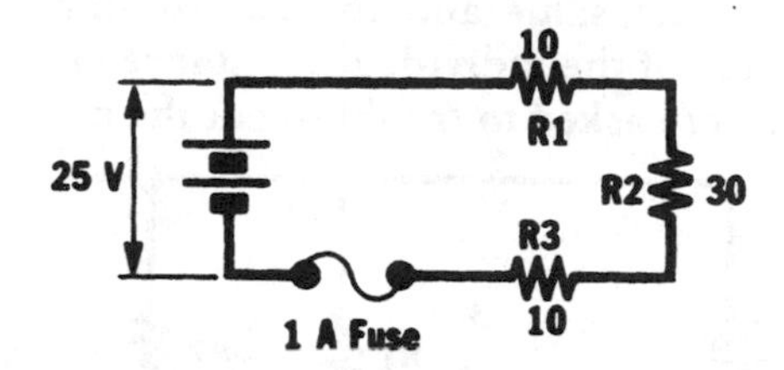

SYMPTOM—*The 1 ampere fuse keeps blowing.*

Inspection shows that the circuit is properly connected and the voltage source is checked with the voltmeter to be 25 volts. By doing the following calculation, you can show that:

The total current should be:

$I_t = 25/R_t; R_t = 10 + 30 + 10 = 50$ ohms
$I_t = \frac{25}{50} = 1/2$ A.

The voltages across resistances should be:

ER1 = (1/2) × (10) = 5 volts
ER2 = (1/2) × (30) = 15 volts
ER3 = (1/2) × (10) = 5 volts

Measurement of voltages (or resistances) shows that the voltage R2 is zero. The resistances also measure zero. The current under these conditions is 1.25 amperes and blows the fuse. The corrective or remedial action is to replace resistor R2.

Example

SYMPTOM—*No current flow.*

Inspection shows the circuit to be properly connected. The voltage source is checked with the voltmeter to be 25 volts. Measurement of the voltage across each resistor shows the following:

ER1 = 0 volts
ER2 = 0 volts
ER3 = 25 volts

Thus, it is apparent that R3 is defective.

If the voltages were all measured as zero, then you would have to inspect the wiring since an open lead is indicated.

Troubleshooting DC Parallel Circuits

Troubleshooting parallel circuits is a bit different than for the series circuits shown earlier since for parallel circuits, the voltage across each component is the same and the current through the component depends on the value of the individual resistance of the component.

Suppose you were asked to troubleshoot the parallel circuit shown.

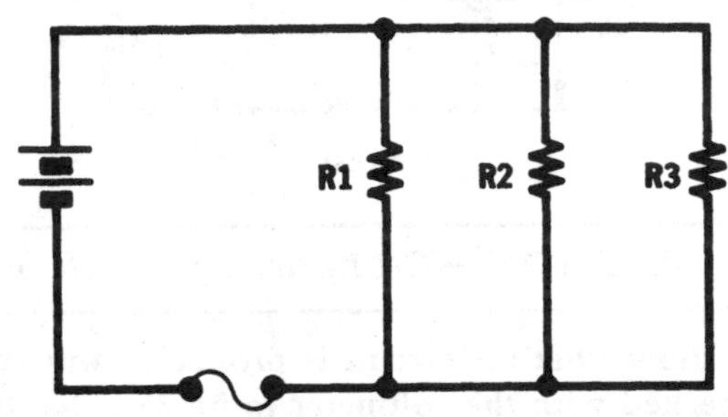

Assume the symptom was excessive current flow, indicating a partial or complete short. Inspection of the circuit shows the wiring to be correct and there are no frayed wires. In the case of parallel circuits, a full short in any component will lead to a full short on the line. It is now necessary to determine *which* component draws excessive current. The quickest thing to do is to check for excessive heat. If this doesn't work, then it is necessary to break the circuit, one component at a time, until the bad (faulty) component is found. You could also put an ammeter in the circuit for each component and find the faulty component that way.

Suppose the symptom was no current flow or less than rated current flow. As with the series circuit, you should check the voltage source and the connections first, to make sure they are correct and that there are no loose or open connections.

Since this is a parallel circuit, an open resistance results only in reduced current flow if there is more than one resistance in the circuit. The presence of the other resistances in parallel makes it difficult to use an ohmeter without disconnecting the individual resistances. Often a parallel circuit will have switches connected to each resistance (or load).

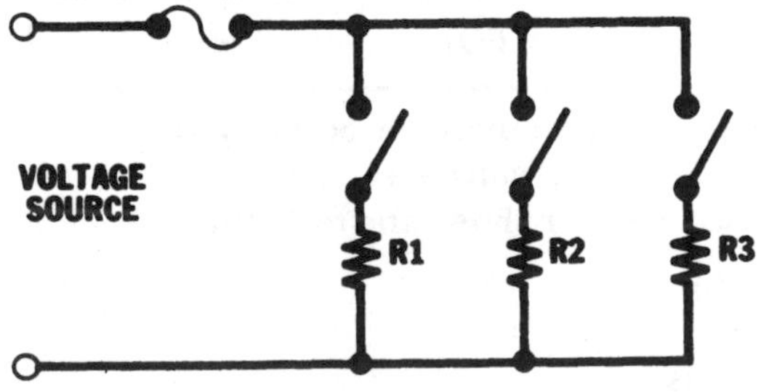

In this case you can isolate the resistances or loads for measurement to determine which one is incorrect (faulty).

Troubleshooting DC Parallel Circuits (continued)

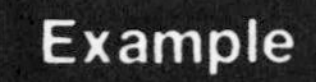

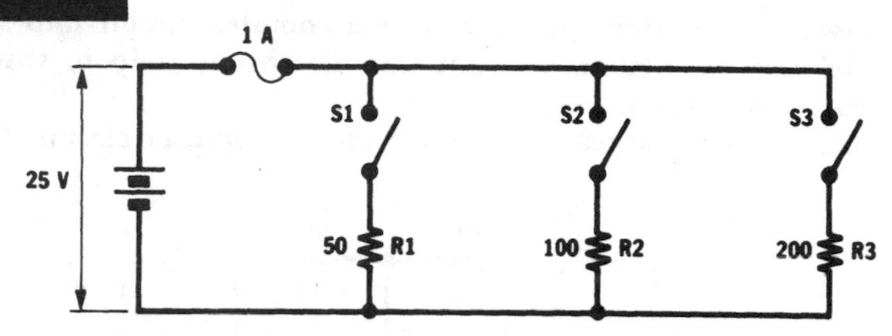

SYMPTOM	*The 1 ampere fuse blows when S2 is closed but not when S1 and/or S3 are closed.*

Inspection shows the circuit is properly connected and does not have miswiring or frayed wires. It should be apparent that R2 should only draw 1/4 amp when connected and the total load should not blow the fuse. The problem, as you have probably figured out, is that R2 has dropped very much in value or is shorted. You could verify this by opening switch S2 and measuring the resistance of R2.

Example

SYMPTOM—*No increase in current flow when S1 is closed.*

Examination of the circuit shows no broken leads or defective or loose connections associated with S1 and R1. You know by Ohm's Law that the current should increase by 1/2 amp when S1 is closed. Also, the parallel resistance with all switches closed should be:

$$\frac{1}{R_t} = \frac{1}{R1} + \frac{1}{R2} + \frac{1}{R3} = \frac{1}{50} + \frac{1}{100} + \frac{1}{200} = 0.035$$

$R_t = 28.57$ ohms

Measurement of the parallel resistance (switches closed and power removed!) shows:

$$R_t = 66.6 \text{ ohms}$$

This doesn't change when S2 is either closed or open, telling you that either the switch is bad or R1 is open. The corrective or remedial action is to replace R1.

Troubleshooting DC Series-Parallel Circuits

Troubleshooting series-parallel circuits is simply an extension of what you already know about series and parallel circuits. You know that you can always break down or rearrange a complex circuit into series and parallel sets of components. You can do this to help in troubleshooting complex circuits.

Suppose you were asked to troubleshoot the complex circuit shown below:

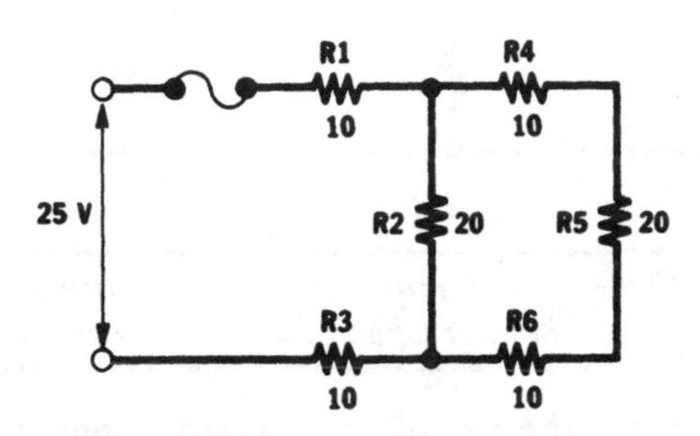

The symptom is that excessive current is drawn. Inspection of the circuit shows frayed insulation on the wires connecting R2 into the circuit. Removal of the short caused by this frayed insulation causes the circuit to act normally.

As you know, the circuit shown above can be connected to an equivalent series circuit by combining R2 in parallel with the series combination R4 + R5 + R6.

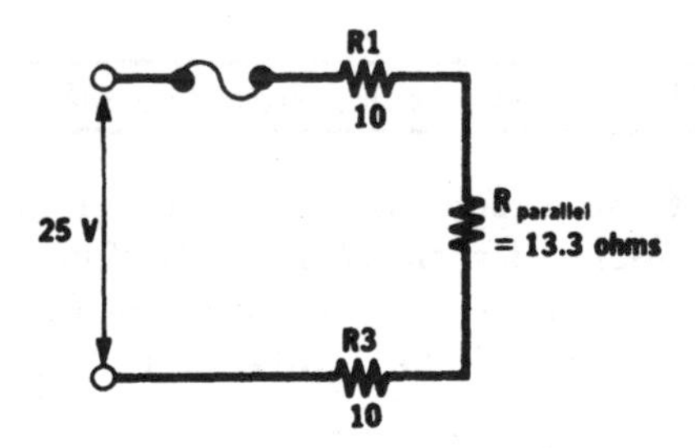

As you can see, this is a series circuit, so you can troubleshoot it as you did for simple series circuits.

Drill in Troubleshooting DC Circuits

1. You have been asked to troubleshoot the series circuit shown below. The symptom is that the fuse blows when the switch is closed. The wiring is correct and there are no obvious short circuits.

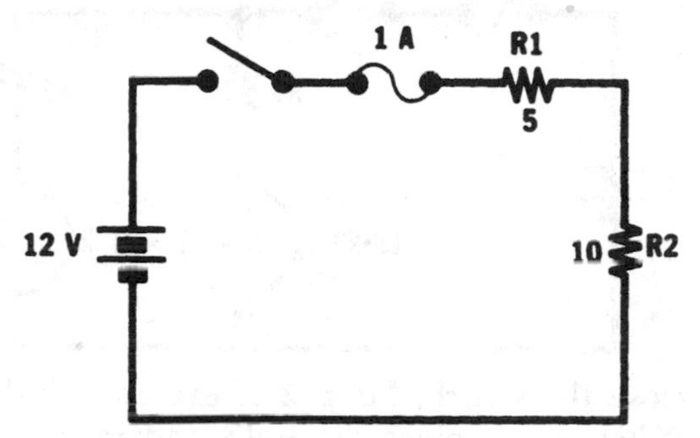

   (a) What current should flow?
   (b) What would you do with an ohmmeter to determine what the problem is?
   (c) What would you do to correct it?

2. In the circuit of question 1, the symptom is that no current flows, the wiring is correct, and there does not appear to be any loose or broken connection. Voltage measurements across the resistance are as follows:

   ER1 = 0
   ER2 = 12 V

   (a) What is the probable difficulty?
   (b) What measurement could you make to verify your diagnosis?

3. The parallel circuit shown below is part of an automobile electrical system. The symptom is that there are no lights or heat (heater fan) when the switch S2 is closed; however, there is ignition.

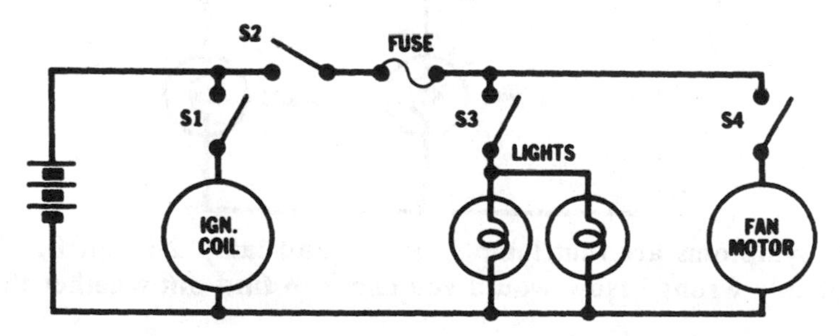

   You find that the fuse is blown. If you replace the fuse and close S2 with S3 and S4 open, the fuse does not blow. Closing S3 makes the lights go on normally. Closing S4 causes the lights to go out and the fan does not run. Inspection shows that the fuse is blown. What do you think the trouble is?

Drill in Troubleshooting DC Circuits (continued)

4. You have the following circuit to operate two 6 V lamps from a 12 volt source.

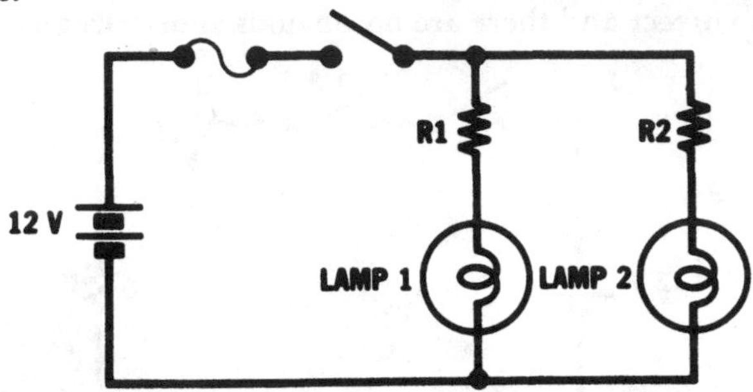

When you close the switch, lamp 2 is exceedingly bright and soon burns out. When you replace the bulb and make voltage measurements, you have the following:

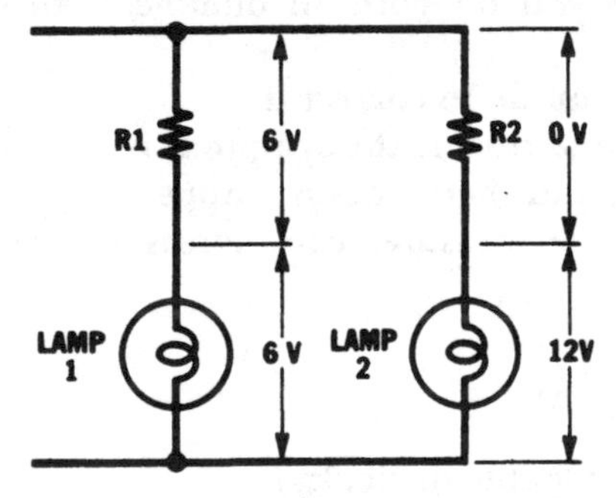

What is wrong? Why?

5. You have been asked to troubleshoot the following complex circuit.

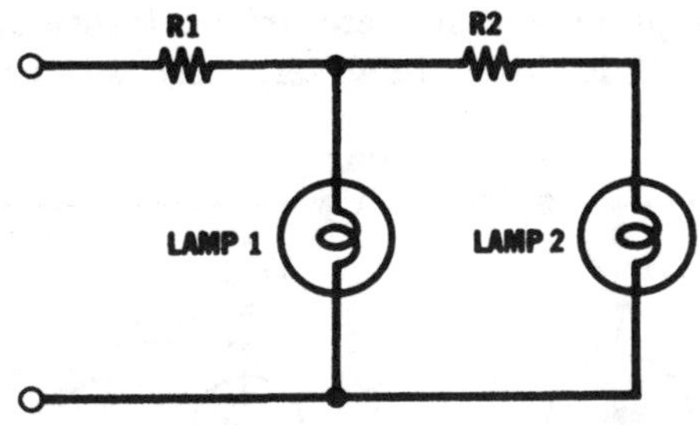

The symptoms are that lamp 1 is dim and lamp 2 is not lit. What is probably wrong? How would you check to find out whether this were so?

General Review of DC Fundamentals

Any combination of a conductor and of a source of electromotive force (emf) which permits free electrons to travel around in a continuous stream from the negative terminal of the voltage source to its positive terminal, and back through the source to the negative terminal again, constitutes an *electric circuit.* As long as this electrical pathway remains *unbroken*, it forms a *closed circuit* and current will flow through it. But if the pathway is *broken* at any point, an *open circuit* results, and no current will flow.

The number of electrons in the electron stream in the closed circuit is dictated by the strength of the emf—voltage—forcing the electrons to move. The magnitude of the electron stream can be restricted and controlled by inserting into the external circuit at any point any kind of *resistor.* The emf is dissipated by the effort of forcing the electron stream through the resistor, and a *voltage drop* takes place across any resistor connected into a circuit.

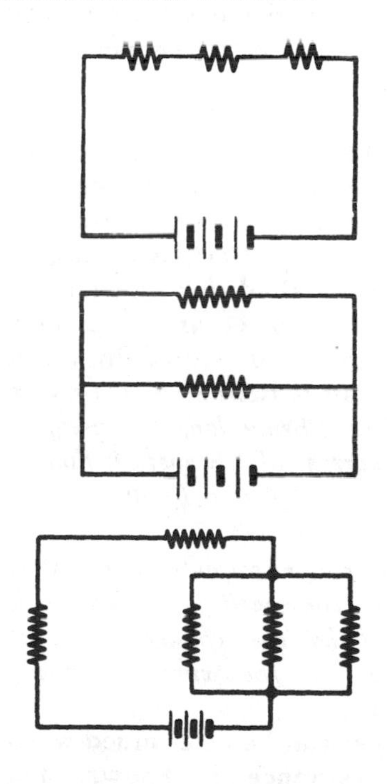

1. SERIES CIRCUIT—When two or more resistors are connected *end-to-end* across a voltage source so that the same current flows through all resistors, the circuit is called a *series circuit.*

2. PARALLEL CIRCUIT—When two or more resistors are connected *side-by-side* across a voltage source so that the current divides between them, the circuit is called a *parallel circuit.*

3. COMPLEX CIRCUIT—When a number of resistors are connected into a circuit, some in series and some in parallel, the circuit is called a *series-parallel* or *complex circuit.*

4. OHM'S LAW—There exists a fixed relationship between the voltage, the total resistance of that circuit (or the individual resistance of any resistor connected into it), and the value of current flow through the circuit (or through the individual resistor, as the case may be). This relationship is stated in *Ohm's Law,* which says that *the*

$$\text{CURRENT} = \frac{\text{VOLTAGE}}{\text{RESISTANCE}}$$

General Review of DC Fundamentals (continued)

$$\text{CURRENT} = \frac{\text{VOLTAGE}}{\text{RESISTANCE}}$$

OR $$I = \frac{E}{R}$$

current flowing in a circuit is directly proportional to the applied voltage, and inversely proportional to the circuit resistance.

In other words, *current flow increases as the voltage is increased, and decreases as the resistance is increased.*

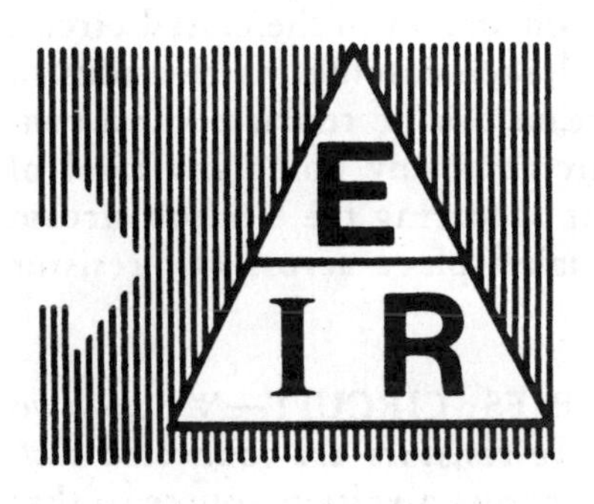

5. THE MAGIC TRIANGLE—You can always find the formula for determining the value of I, E, or R when you know two of the three values, by using the *magic triangle.* Remember the rule: *Put your thumb over the value you don't know—and the formula you want is what's left.*

WATTS = VOLTS × AMPERES

or (in symbols)

$$P = EI$$

6. POWER—Whenever a voltage causes electrons to move, work is done. The rate at which work is done in moving electrons through a conductor is called *electric power.*

Electric power is represented by "P" and is measured in watts, "W"; 1 watt is defined as *the rate at which work is being done in a circuit in which a current of 1 ampere is flowing when the emf applied is 1 volt.*

The *power formula* states that *the power consumed in a resistor is determined by the voltage across it multiplied by the current flowing through it.*

$$P = I^2R \quad \text{and} \quad P = \frac{E^2}{R}$$

Power can be determined when circuit resistance is known, but either current or voltage is not known, by substituting Ohm's Law for the unknown factor in the equation, P = EI. The rewritten equations are:

$$P = I^2R \quad \text{and} \quad P = \frac{E^2}{R}$$

Learning Objectives—Next Volume

Overview—You have learned about dc circuits. In the next volume—Volume 3—you will learn about alternating current or ac circuits. You will see that what you know about dc circuits can be applied to ac circuits.

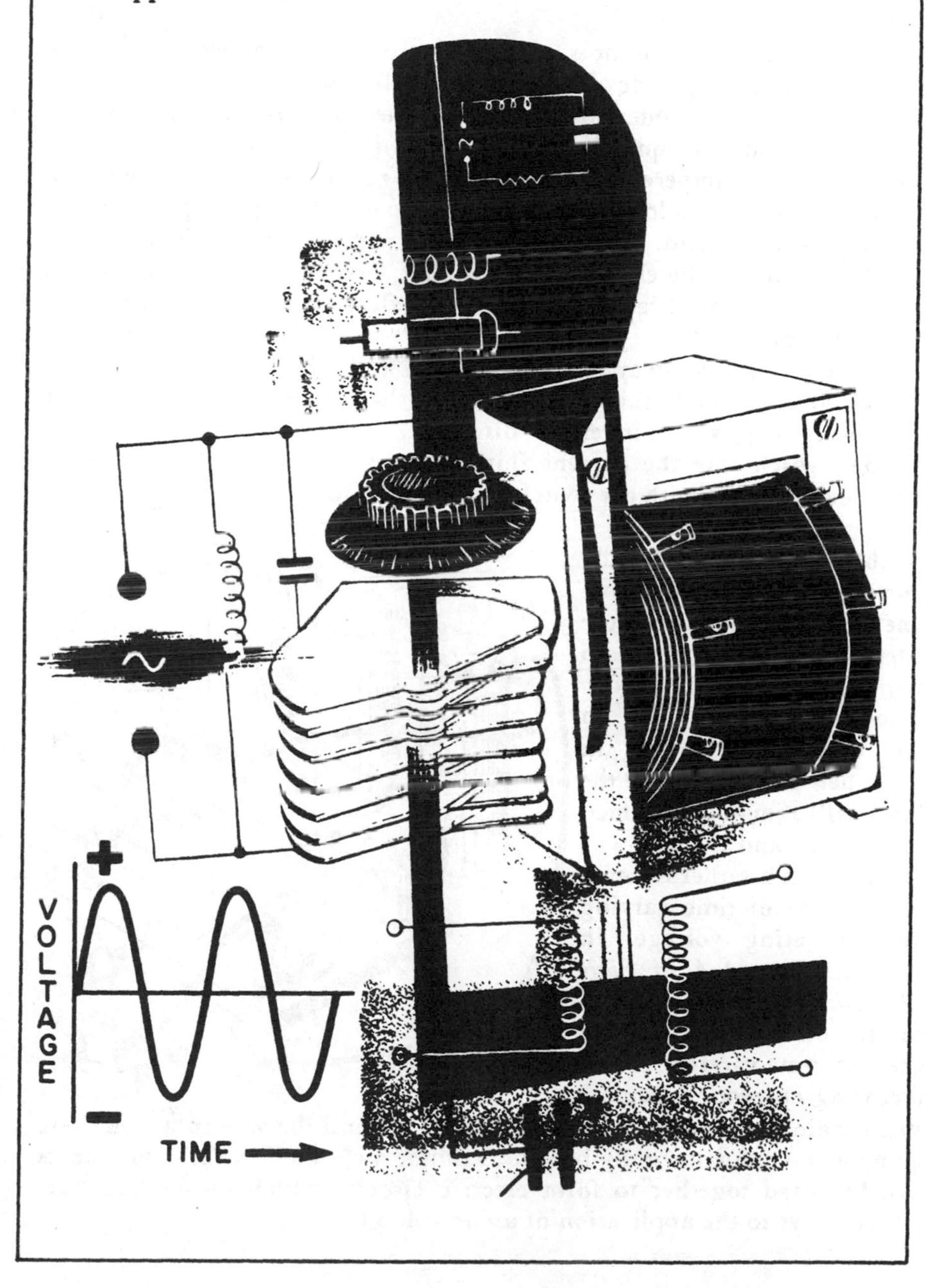

Alternating Current

The circuits you have been studying so far have all been direct current (dc) ones. In practice, however, very little direct current is used to supply electric light or power; and, furthermore, nearly all electronic circuits make at least as much use of what is called *alternating current* (ac) as they do of dc. So you now must start to learn what this ac is and how it behaves.

Alternating current does not flow through a conductor always in the same direction, as dc does. Instead, it flows back and forth in the conductor at regular intervals, continually reversing its direction of flow and can do so very quickly. It is measured in amperes, just as dc is measured. One ampere of current is said to be flowing, you will remember, when 1 coulomb of electrons is passing a given point in the conductor in 1 second. This definition also applies when ac is flowing—only now some of the electrons during that 1-second flow past the given point going in one direction, and the rest flow past it going in the opposite direction.

If a lamp, a two-way (DPDT) switch, and a battery of dry cells are connected as shown in the diagram below, the lamp can be made to light by closing the switch in either direction. The position of the switch decides which way the current flows through the lamp, but the lamp lights equally well with the switch in either position.

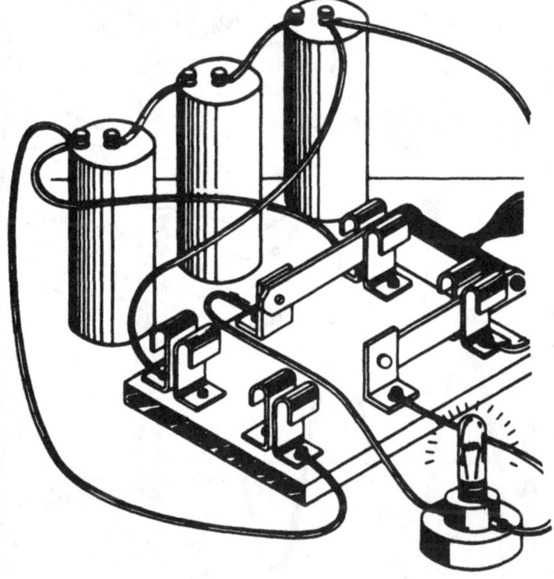

If the switch is repeatedly moved very quickly from one position to the other, the current through the lamp will *alternate* or *vary with time*—that is, first flow in one direction and then in the other direction—and the lamp will remain lit. In fact, the battery and the two-way switch form a rather elementary source of time varying or alternating voltage. In Volumes 3 and 4 of *Basic Electricity*, you will be studying the nature, behavior and uses of time-varying or alternating current. Volume 3 deals mainly with ac terms and components, and the way in which these components fundamentally behave. Volume 4 shows how the components can be fitted together to form electric circuits which respond in particular ways to the application of an ac voltage.

Alternating Current (continued)

Also in Volume 3, you will meet for the first time two components —the *inductor* and the *capacitor*—which are frequently used to control direct as well as alternating current and voltage. The resistors with which you have been working so far have all acted in such a way as to restrict the flow of current *directly*. In other words, the bigger the resistor you put in, the more you restrict current flow. The inductor and the capacitor, on the other hand, act to control current and voltage in rather different ways, and you will see that what they do depends on how often the current is reversed. These three components—the resistor, the inductor, and the capacitor—are basic elements of all electric and electronic circuits.

Answers to Drill Questions

Page 2-16

1. (a) $E = IR = 2 \times 10 = 20$ volts

   (b) $R = \frac{E}{I} = \frac{30}{1.5} = 20$ ohms (Ω)

   (c) $I = \frac{E}{R} = \frac{10}{15} = 0.667$ ampere

   (d) $I = \frac{E}{R} = \frac{300}{1{,}000{,}000} = 0.0003$ ampere $= 300\ \mu A$

   (e) $E = IR = 2{,}500{,}000 \times 0.000002 = 5$ volts

   (f) $R = \frac{E}{I} = \frac{495}{0.0015} = 330{,}000 = 330$ K

2. $R = \frac{E}{I} = \frac{12}{4} = 3\Omega$ is the lamp resistance.

3. To make rated current flow, the voltage required is

   $E = IR = 1.5 \times 24 = 36$ volts

4. The element resistance must be

   $R = \frac{E}{I} = \frac{240}{2.5} = 96$ ohms

5. $I = \frac{E}{R} = \frac{1.36\ V}{68\ K} = \frac{1.36}{68{,}000} = 0.00002$ ampere $= 0.02$ mA $= 20\ \mu A$

6. $R = \frac{E}{I} = \frac{10\ V}{5\ mA} = \frac{10}{0.005} = 2{,}000$ ohms (Ω) $= 2$ K

Answers to Drill Questions

Page 2-34

1. In a series circuit, the total resistance is the sum of the resistances.

$$R_t = R1 + R2 + R3 \ldots$$
$$= 220 \text{ ohms} + 680 \text{ ohms} + 1{,}000 \text{ ohms}$$
$$= 1{,}900 \text{ ohms} = 1.9\text{ K}$$

2. $R_t = R1 + R2 + R3 + R4$

   Therefore, $67 = 10 + 15 + 27 + R4$

   and $R4 = 67 - 10 - 15 - 27$

   $= 15 \text{ ohms}$

Basic Electricity

REVISED EDITION

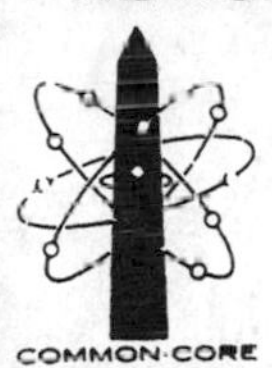

VAN VALKENBURGH,
NOOGER & NEVILLE, INC.

VOL. 3

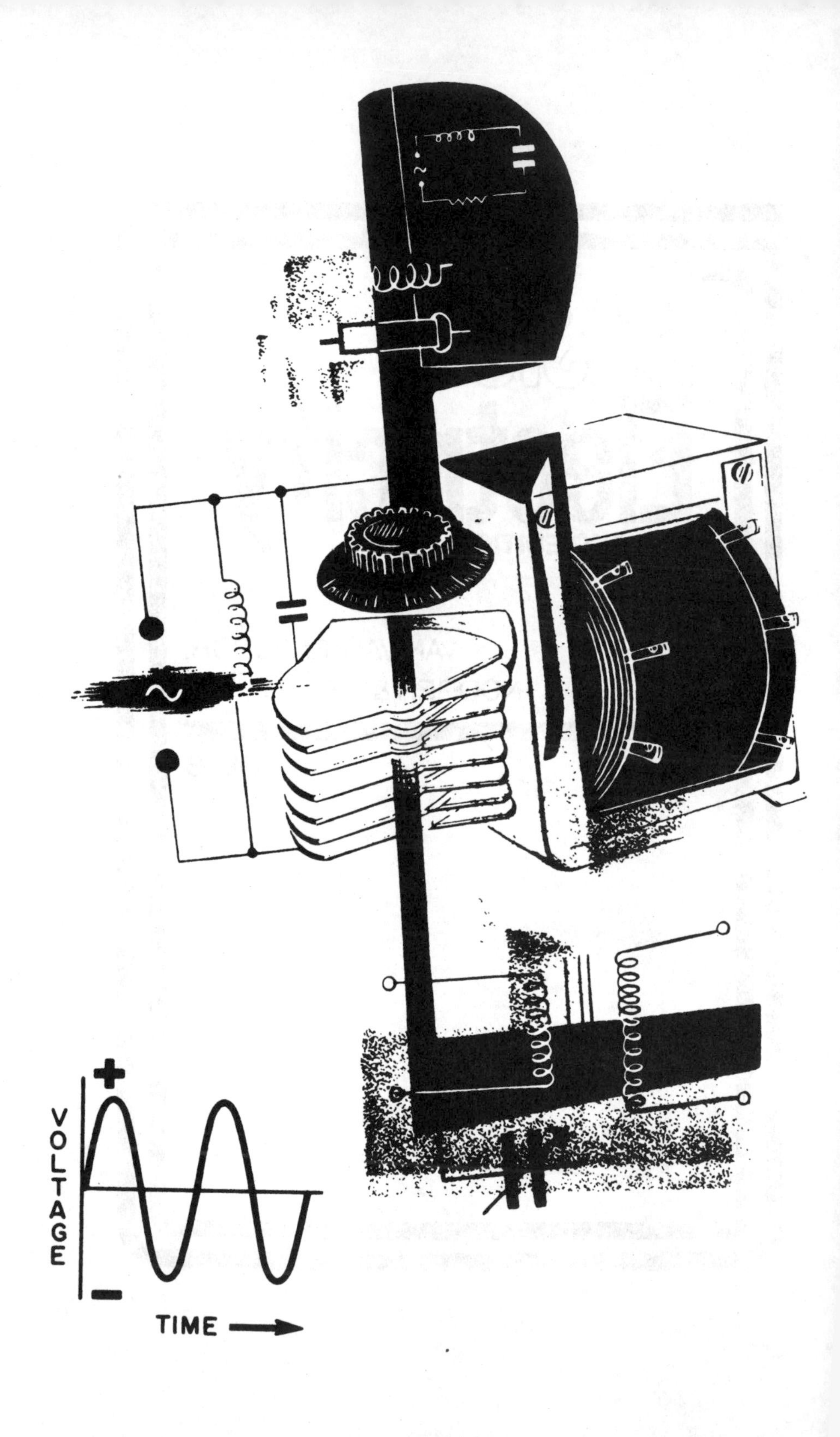
VOLTAGE
TIME
+
−

Direct Current and Alternating Current

In your study of electricity, you will work with both *direct current* (*dc*) and *alternating current* (*ac*). In Volume 2, you learned that in dc circuits, the current always flows in one direction—a *constant* direction. In ac circuits, the direction of current flow *reverses periodically*—at one instant it will flow in *one* direction; at the next instant, it will flow in the *opposite* direction.

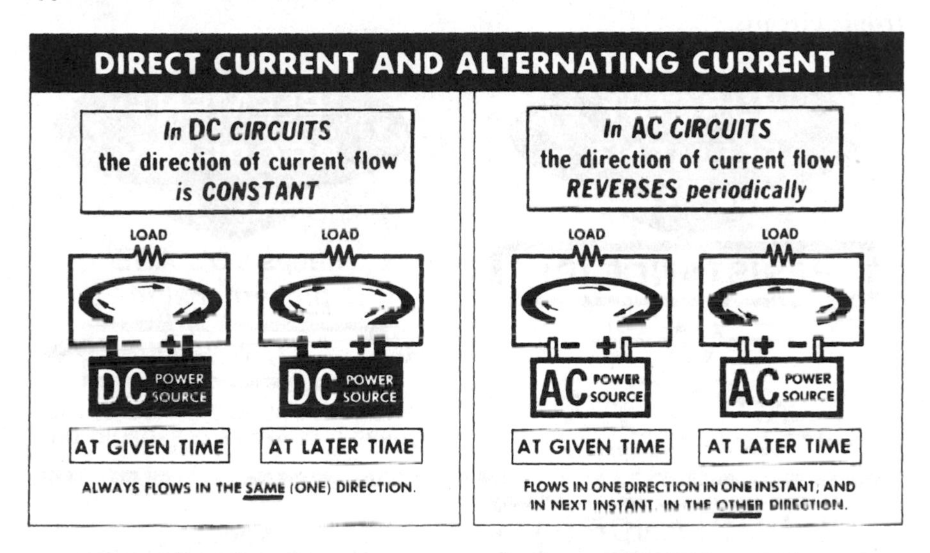
DIRECT CURRENT AND ALTERNATING CURRENT
In DC CIRCUITS the direction of current flow is CONSTANT
LOAD
LOAD
DC POWER SOURCE
DC POWER SOURCE
AT GIVEN TIME
AT LATER TIME
ALWAYS FLOWS IN THE SAME (ONE) DIRECTION.
In AC CIRCUITS the direction of current flow REVERSES periodically
LOAD
LOAD
AC POWER SOURCE
AC POWER SOURCE
AT GIVEN TIME
AT LATER TIME
FLOWS IN ONE DIRECTION IN ONE INSTANT, AND IN NEXT INSTANT IN THE OTHER DIRECTION.

In Volume 2, you also learned about the nature, behavior, and function of dc resistive circuits (circuits containing only resistance) using Ohm's Law and Kirchhoff's Laws as tools for analysis and for understanding the relationships of current, voltage, and resistance.

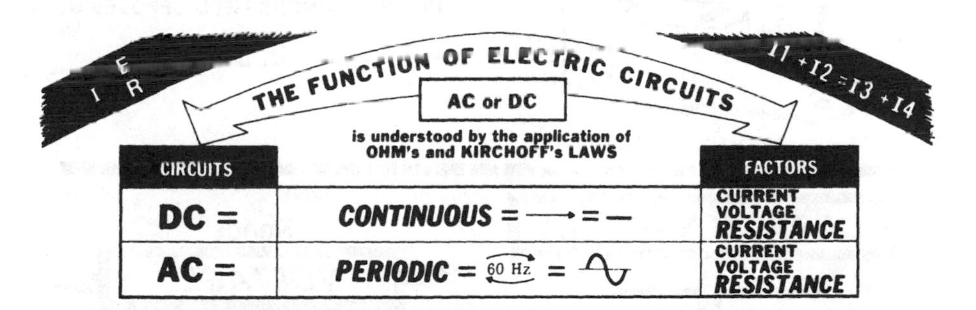
E
I R
THE FUNCTION OF ELECTRIC CIRCUITS
AC or DC
is understood by the application of OHM's and KIRCHOFF's LAWS
I1 + I2 = I3 + I4
CIRCUITS
FACTORS
DC =
CONTINUOUS =
CURRENT VOLTAGE RESISTANCE
AC =
PERIODIC =
60 Hz
CURRENT VOLTAGE RESISTANCE

In this volume, you will learn about the nature, behavior, and function of ac circuits. Your present understanding of the voltage, current, and resistance in dc circuits can now be applied to help you understand the operation and control of ac circuits. In addition to learning about resistance in ac circuits, you will learn about *inductance* and *capacitance*, the additional circuit elements that control the flow of electricity in both dc and ac circuits.

The Three Circuit Elements—Resistance/Inductance/Capacitance

All electric and electronic circuits are made up of circuit elements consisting of *resistance, inductance,* and *capacitance.* When you know how resistance (R), inductance (L), and capacitance (C) behave in dc and ac circuits, you will be ready to learn in Volume 4 how these three elements can be used to control and influence current flow in electric and electronic circuits.

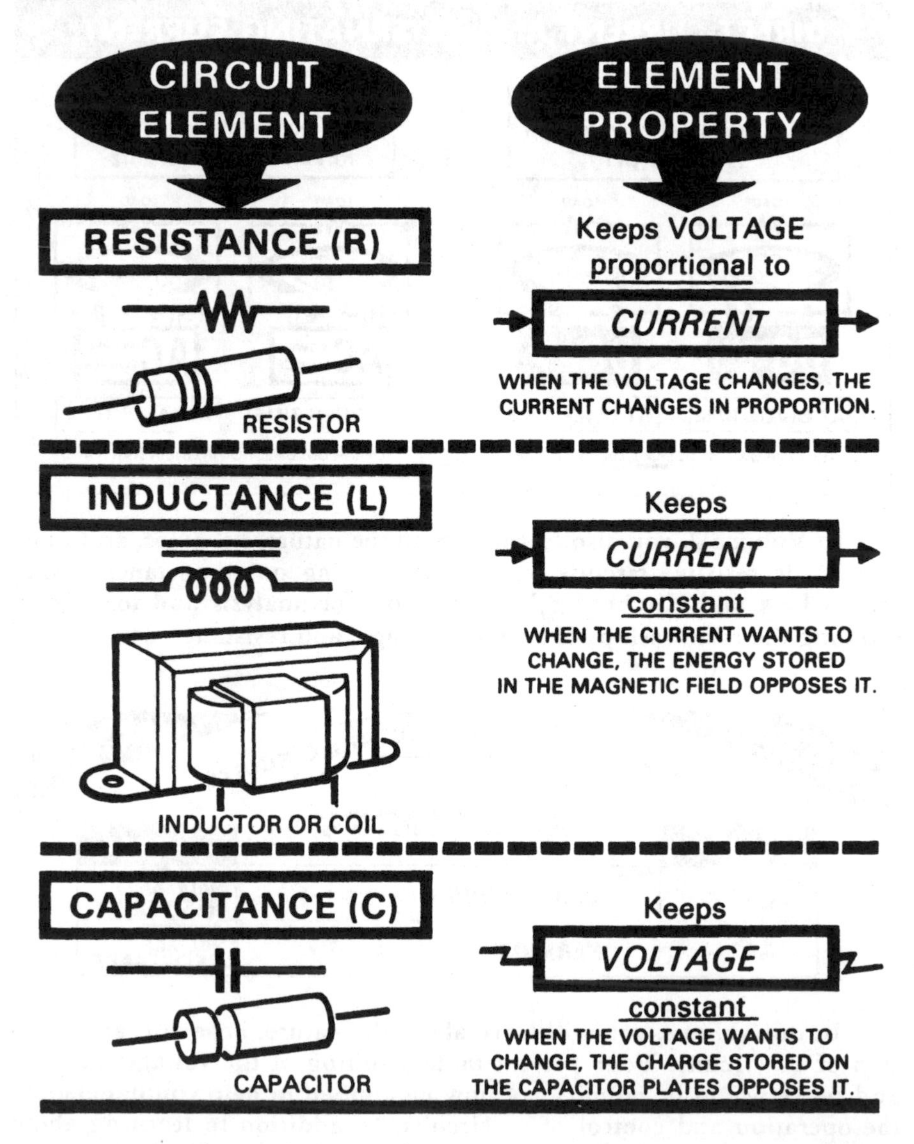

R, L, and C are called *passive* circuit elements because their behavior is independent of the direction of current flow. Devices such as rectifiers and transistors, which you will learn about later, are called *active* devices because their behavior differs for *different directions* of current flow.

Basic Electric Circuit Configurations

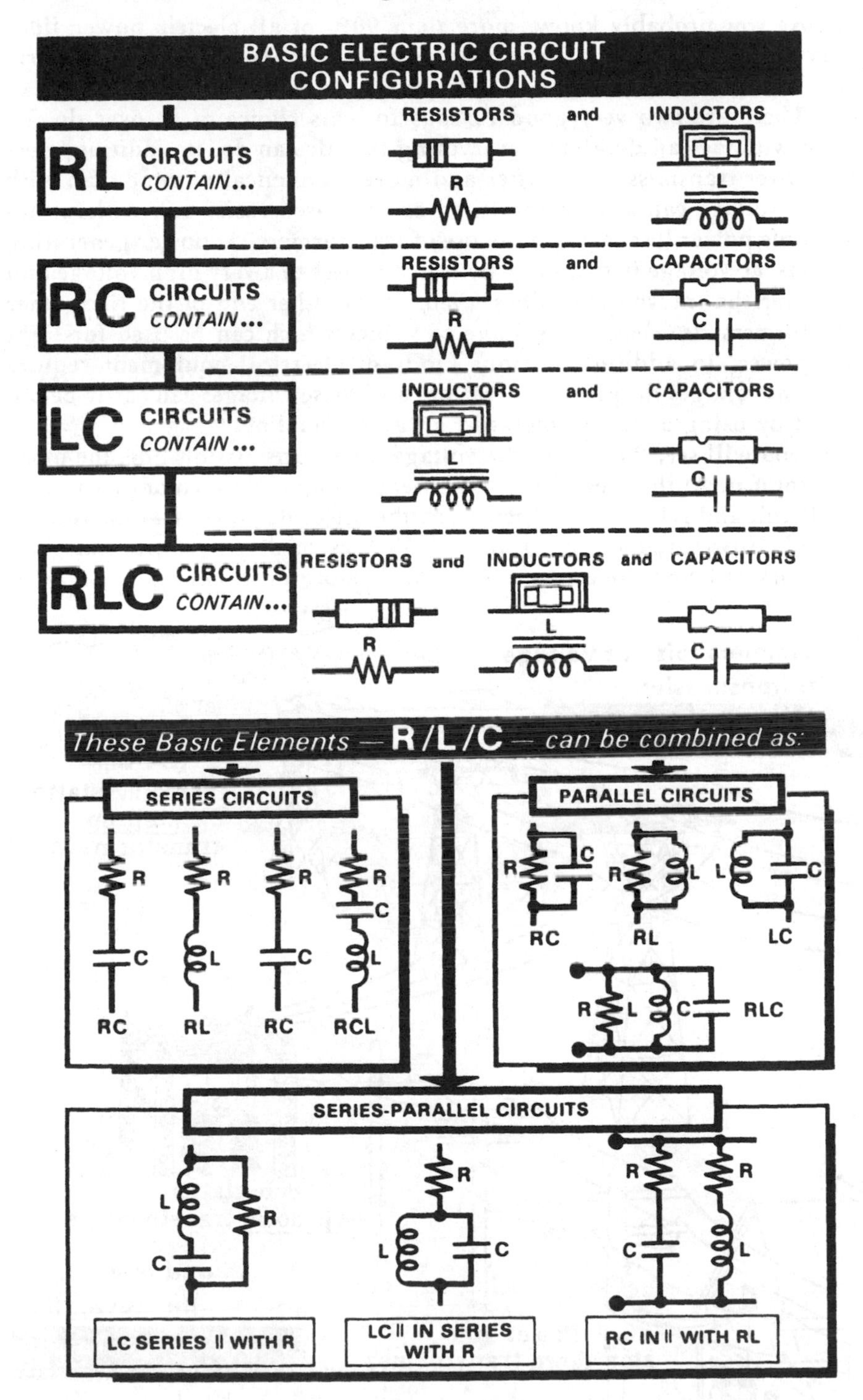

Why Alternating Current Is Used

As you probably know, more than 90% of all electric power lines carry alternating current. Very little direct current is now used for electric light and power. However, dc is important in electronic circuits.

There are two very good reasons for this choice of ac over dc. To begin with, ac can do almost everything that dc can do. In addition, electric power transmission is easier and more economical with ac than with dc. Ac voltage can also be increased or decreased easily and without appreciable power loss through the use of transformers. At power generating stations, ac voltage is *stepped-up* by transformers to a very high voltage and sent over the transmission lines; then, at the other end of the line, other transformers *step-down* the voltage to values which can be used for light and power. In addition, various kinds of electrical equipment require different voltages for proper operation, and these voltages can easily be obtained by using a transformer and an ac power line.

As you will see, the higher the voltage on a transmission line, the more efficient it is. At the present time, the stepping up and down of dc voltages is difficult and relatively inefficient, so the use of dc power transmission is limited. However, there are some advantages to dc power transmission, and work is being done to try to make it more practical.

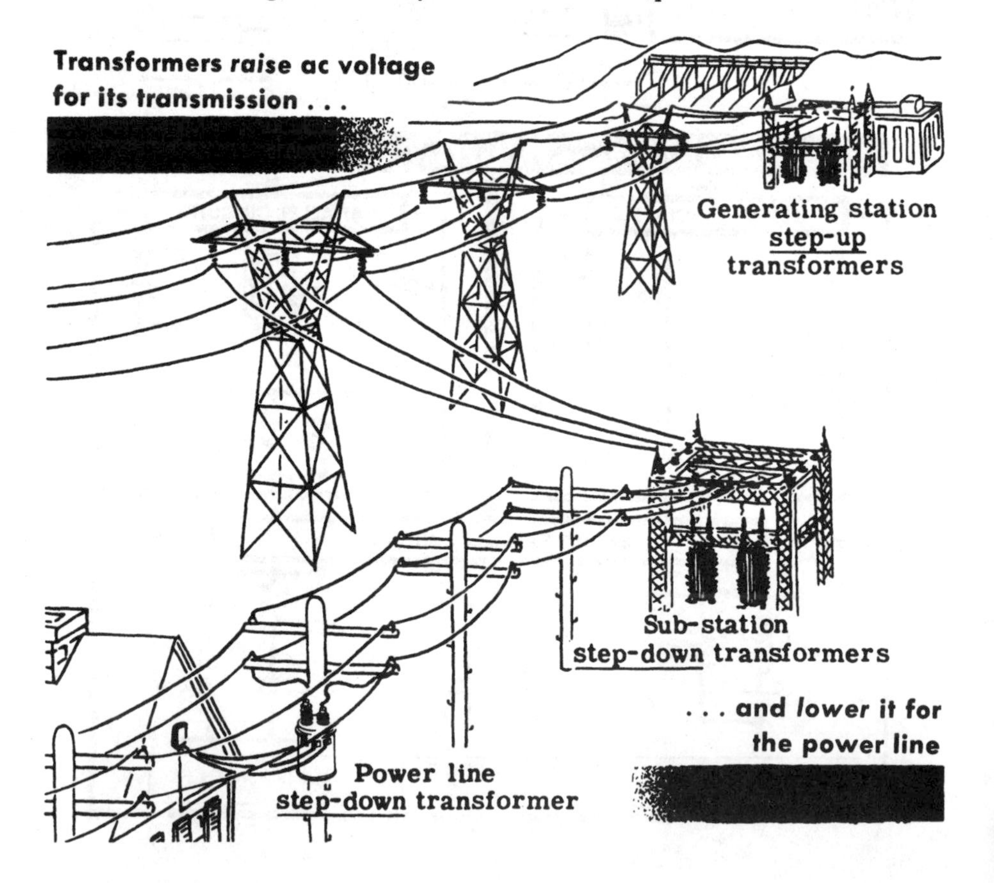
Transformers raise ac voltage for its transmission . . .
Generating station step-up transformers
Sub-station step-down transformers
. . . and lower it for the power line
Power line step-down transformer

AC Power Transmission

As you know, the power (P) transmitted by a power line is the product of the voltage (E) at the termination of the line and the current (I) in the line (P = EI). For maximum power transmission, E and I must be as large as possible. The current (I) is limited by the size of the wire used to make the transmission line; the voltage (E) is limited by the insulation of the line. It is easier and more economical to make a line with very good insulation, permitting the use of a very high voltage, than it is to make a line with large wire capable of carrying very large currents.

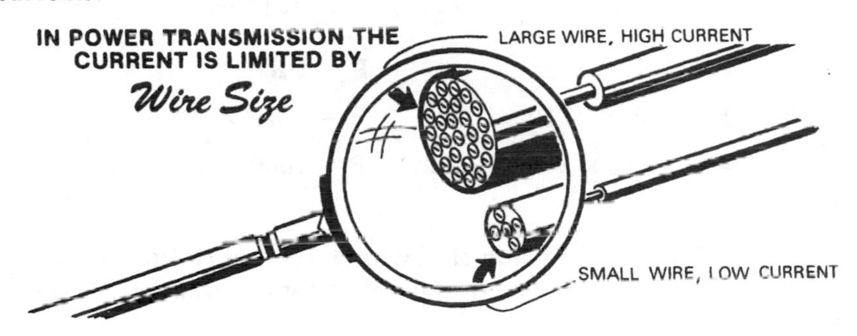

As you will recall, when current flows through a wire to reach the electrical devices (loads) using power, there is a power loss in the wire proportional to the square of the current ($P = I^2R$, where P is the power loss, I is the current in the wire, and R is the resistance of the wire). All wire has resistance. Any reduction in the amount of current flow required to transmit power results in a reduction in the amount of power lost in the transmission line. By using high voltage, lower current is required to transmit a given amount of power.

Efficient transmission of electric power, then, demands the use of very high voltages; and the output from a distant generating station has to be *stepped-up* for transmission and *stepped-down* again for use. This is most economically achieved if the current *supply* is ac, which permits the use of transformers. You will learn, later, how transformers *step-up* and *step-down* voltages.

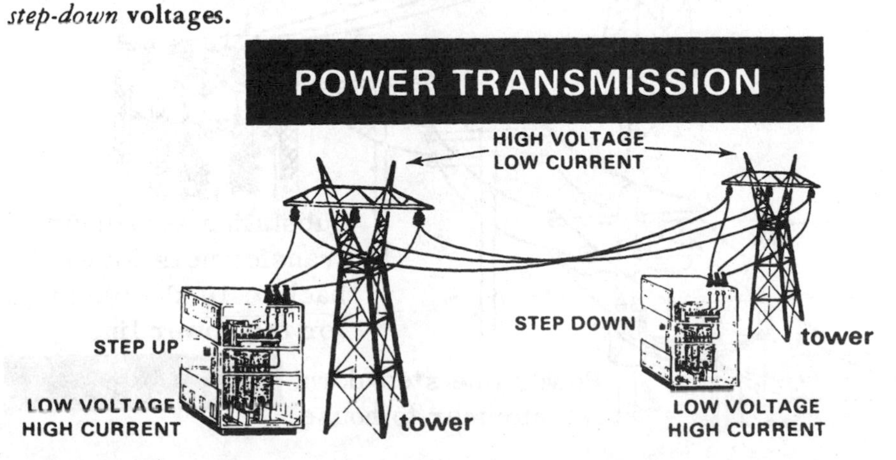

AC Power Transmission (continued)

To illustrate the importance of using high-voltage, low-current power transmission, consider the following example. You can see that 1 megawatt (1,000,000 watts or 1 MW) of power can be obtained from a 10-kilovolt (10,000 volts or 10 kV) source when 100 amperes of current flow. You can also obtain the same power with 100 kV when 10 amperes of current flow. If you had a transmission line with a 10-ohms resistance, the power loss in each case would be:

10 kV line	*100 kV line*
$P = I^2R$	$P = I^2R$
$P = (100)^2 \times 10$	$P = (10)^2 \times 10$
$P = 100{,}000$ watts!	$P = 1000$ watts

Thus, a 10-kV transmission line with a 10-ohm resistance has *100 times* the loss of the same line at 100 kV. Another practical scheme is shown below.

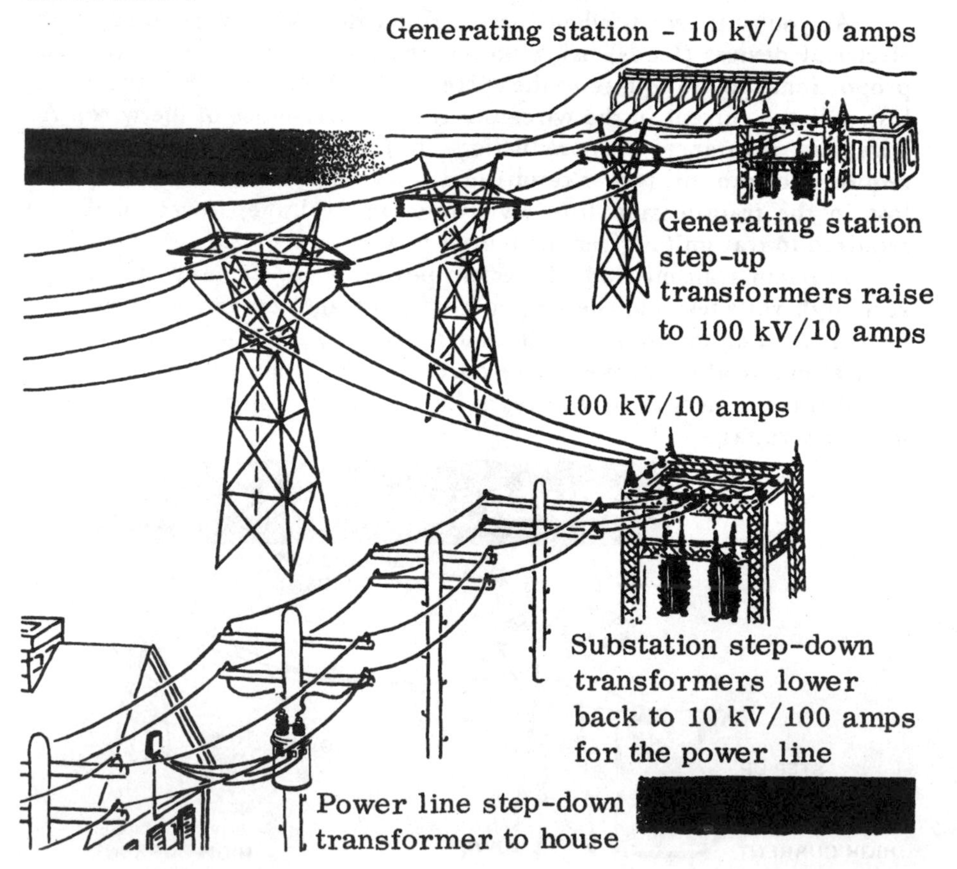
Generating station - 10 kV/100 amps
Generating station step-up transformers raise to 100 kV/10 amps
100 kV/10 amps
Substation step-down transformers lower back to 10 kV/100 amps for the power line
Power line step-down transformer to house

AC Current Flow

Alternating current flows back and forth in a wire, usually at regular intervals, going first in one direction and then in the other. You know that current is measured by counting the number of electrons flowing past a point in a circuit in a given time. Suppose 1 coulomb of electrons moves past a point in a wire in 1 second with all of the electrons moving in the same direction; the current flow is then 1 ampere dc. If 1/2 coulomb of electrons moves in one direction past a point in 1/2 second, then *reverses* direction and moves past the same point in the *opposite* direction during the next 1/2 second, a total of 1 coulomb of electrons passes the point in 1 second, and the current flow is 1 ampere ac. Thus, it is the *total flow* of electrons that is electric current, regardless of the direction of the electron flow.

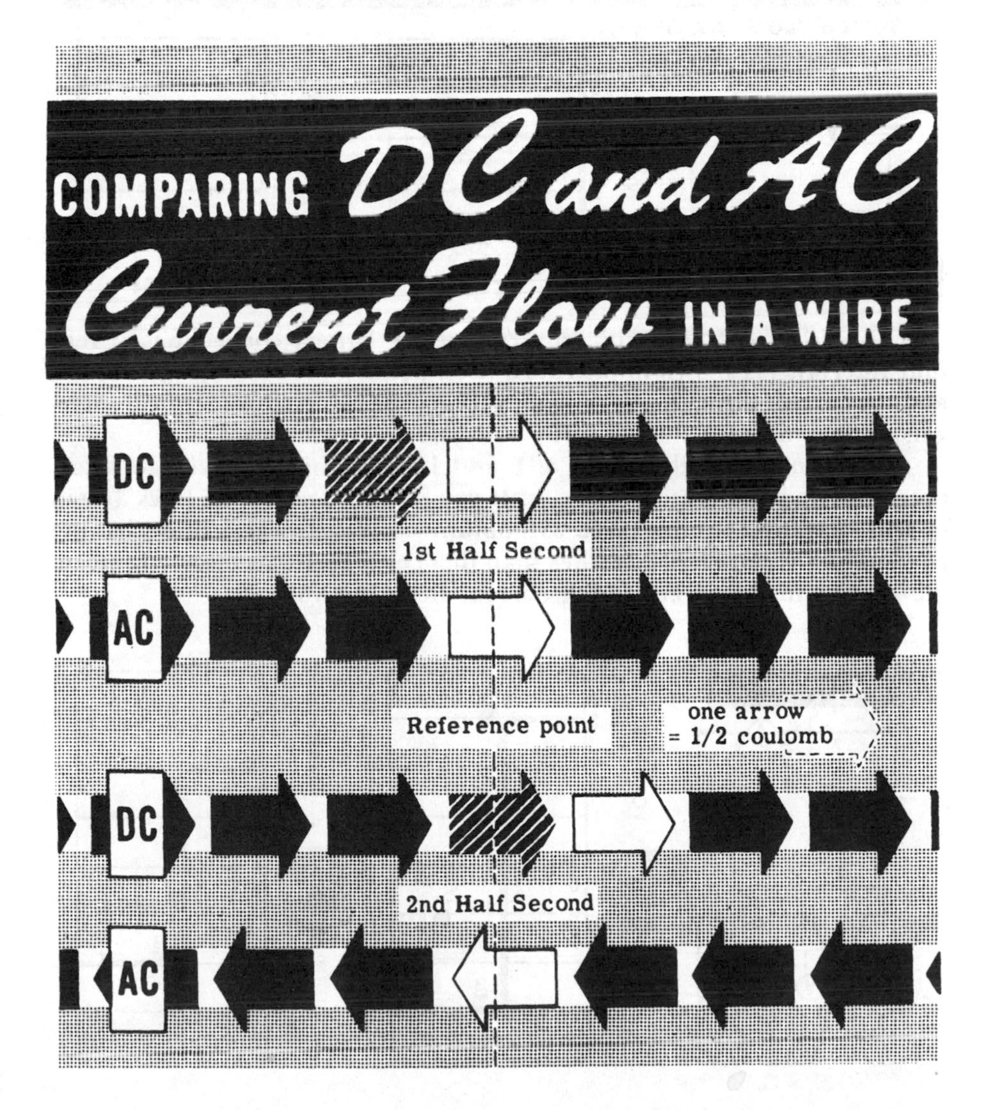

Waveforms

Waveforms are pictures that show how voltages and currents vary over a period of time.

The waveforms for direct current are, for the most part, *straight lines*, and change very slowly, as neither the voltage nor the current varies rapidly in a given circuit. If you connect a resistor across a battery and take measurements of the voltage across, and the current through, the resistor at regular intervals of time, you will find that there is no change in their values. If you plot the values of E and I, each against time, you will obtain straight lines—the waveforms of the circuit voltage and current.

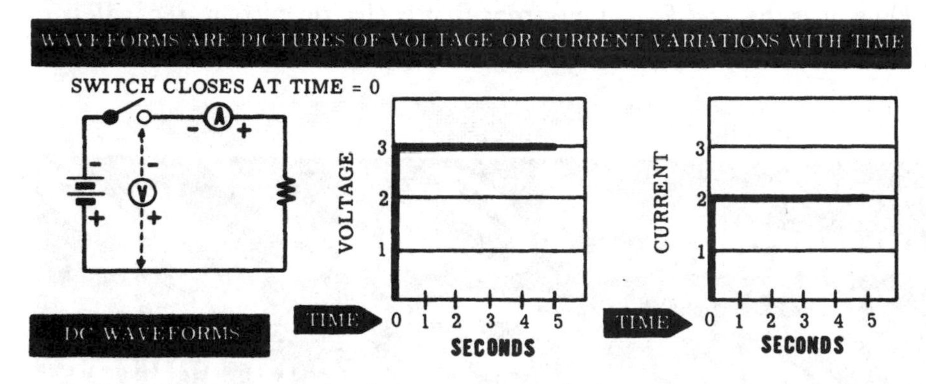

If, with the same circuit, you reverse the battery connections at regular intervals (as shown in the illustration below), then the voltage across, and the current through, the resistor will also be reversed at identical intervals. You can prove this by using *zero-center* instruments.

If you now plot the values of E and I once again, you will get alternate positive and negative readings. If you join the ends of the lines drawn to represent these readings, you will obtain waveforms which show that the current and voltage are ac rather than dc. The waveforms clearly indicate the change in direction of both current flow and voltage.

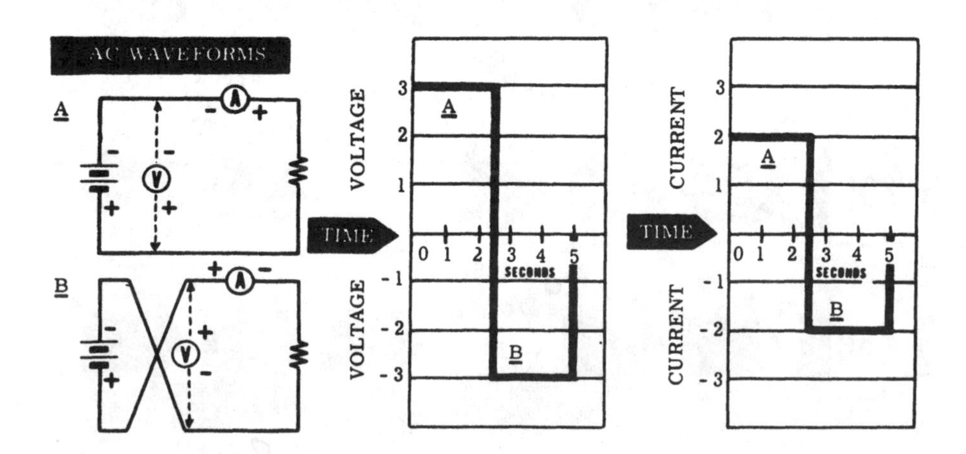

Waveforms (continued)

If you could reverse the polarity *very* quickly, at uniform time intervals like 1 second, then you would have a steady state ac waveform.

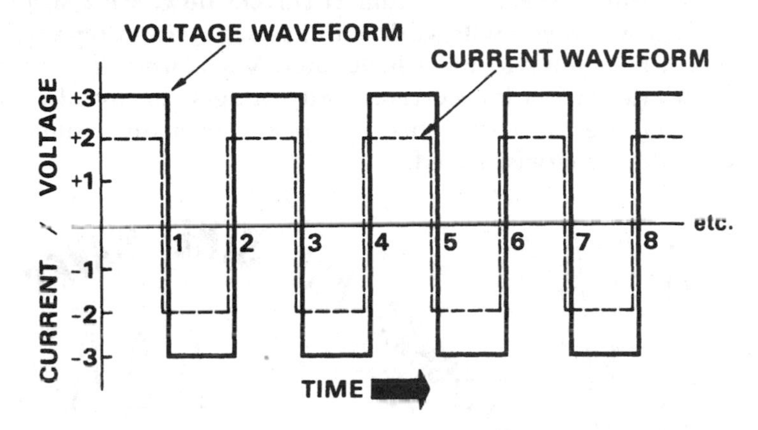

Waveforms of voltage and current are not always seen as straight lines with connecting points. In most cases, waveforms are smooth curves representing gradual changes in voltage and current. This is particularly true of pulsating dc waveforms, which vary with time but always have the same polarity.

Also, pulsating direct current does not always vary between zero and a maximum value but may vary over any range between these values. The waveform of a dc generator is pulsating dc and does not fall to zero but, instead, varies only slightly below the maximum value.

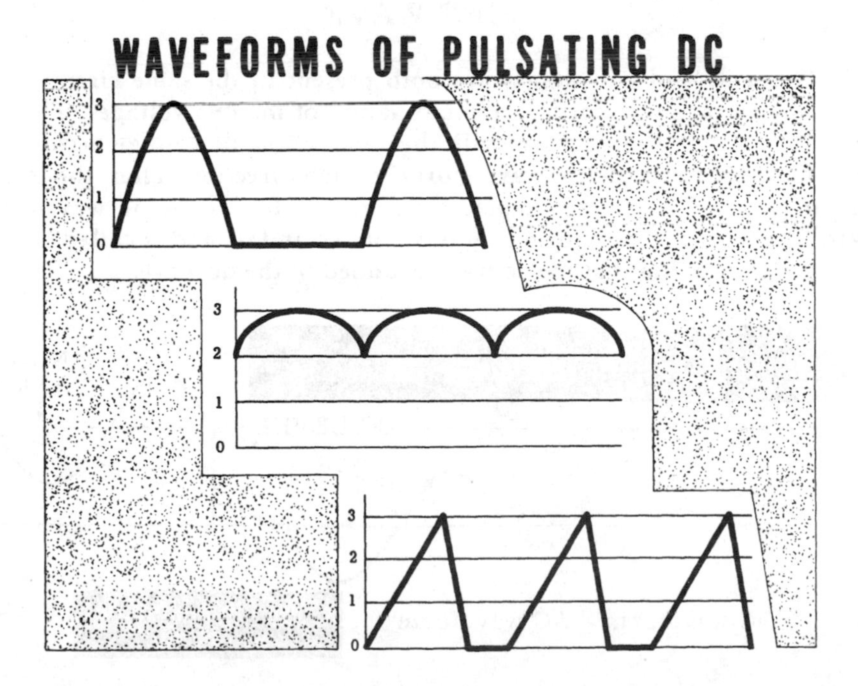

Waveforms (continued)

The waveforms of most alternating currents are smooth curves that represent gradual changes in voltage and current, first increasing, then decreasing in value for each direction of current flow. Most of the alternating current which you will work with has a waveform represented by a *sine curve*, which you will use a little later. Waveforms of this type are *sinusoidal*. While alternating currents and voltages do not always have waveforms that are exact sine curves, they are normally assumed to be sinusoidal unless otherwise stated.

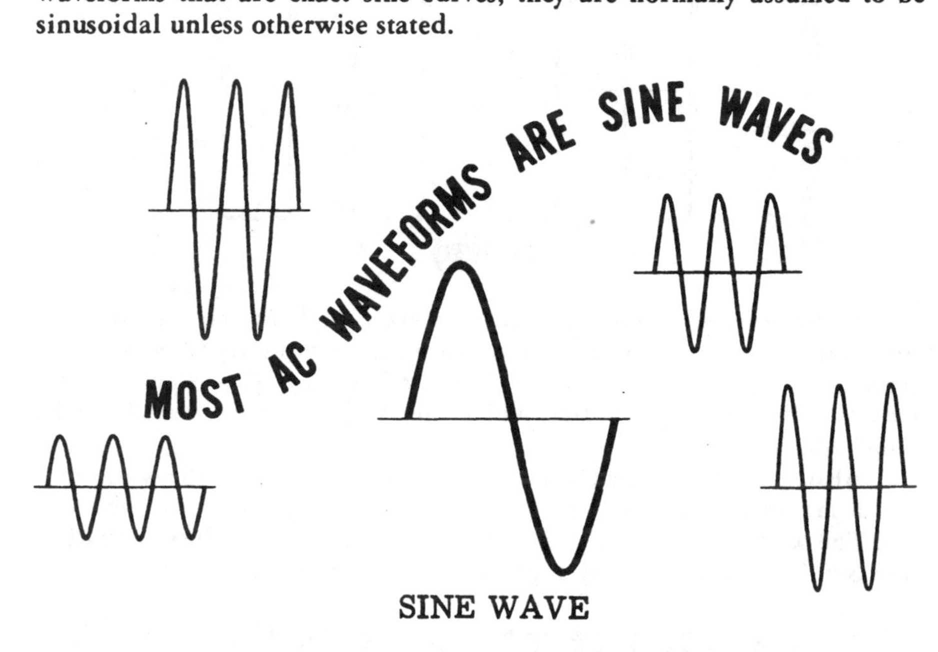

When dc and ac voltages are both present in the same circuit, the resulting voltage waveform is a combination of the two voltages. The ac wave is added to the dc wave, with the value of the dc voltage becoming the axis from which the ac wave moves in each direction. Thus, the maximum point of dc voltage replaces the zero value as the ac waveform axis. The resulting waveform contains both ac and dc and is called *superimposed ac*, meaning that the ac wave is added to the dc wave.

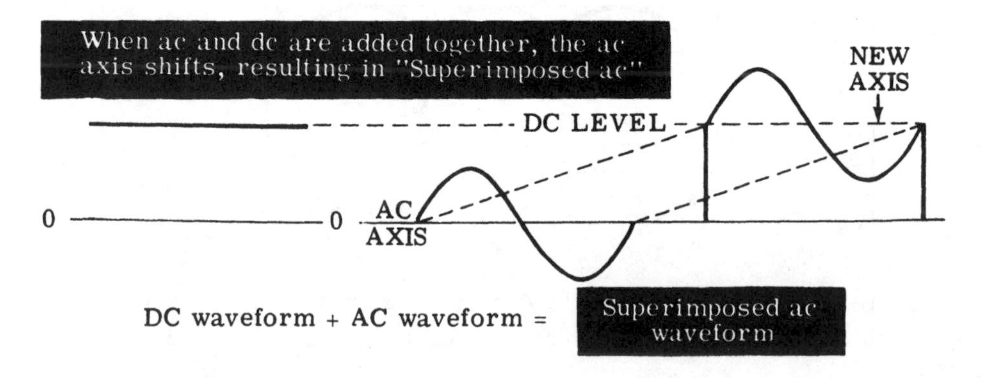

Alternating Current Cycles

When an ac voltage or current passes through a complete set of positive and negative values, we say it has *completed a cycle*. The ac current first rises to a maximum and falls to zero in one direction, then rises to a maximum and falls to zero in the opposite direction. This completes a cycle of ac current, and the cycle repeats as long as the current flows. Similarly, ac voltage first rises to a maximum and falls to zero in *one polarity*, then rises to a maximum and falls to zero in the *opposite polarity* to complete the cycle. Each complete set of both positive and negative values of either voltage or current is a cycle.

On the next page you will see that an ac generator consists of a coil of wire rotating in a magnetic field between two opposite magnetic poles. Each time a side of the coil passes from one pole to the other, the current flow generated in the coil reverses its direction. When passing two opposite poles, the current flows first in one direction and then in the other, completing a cycle of current flow with one revolution of the coil.

A CYCLE IS A COMPLETE SET OF POSITIVE AND NEGATIVE VALUES

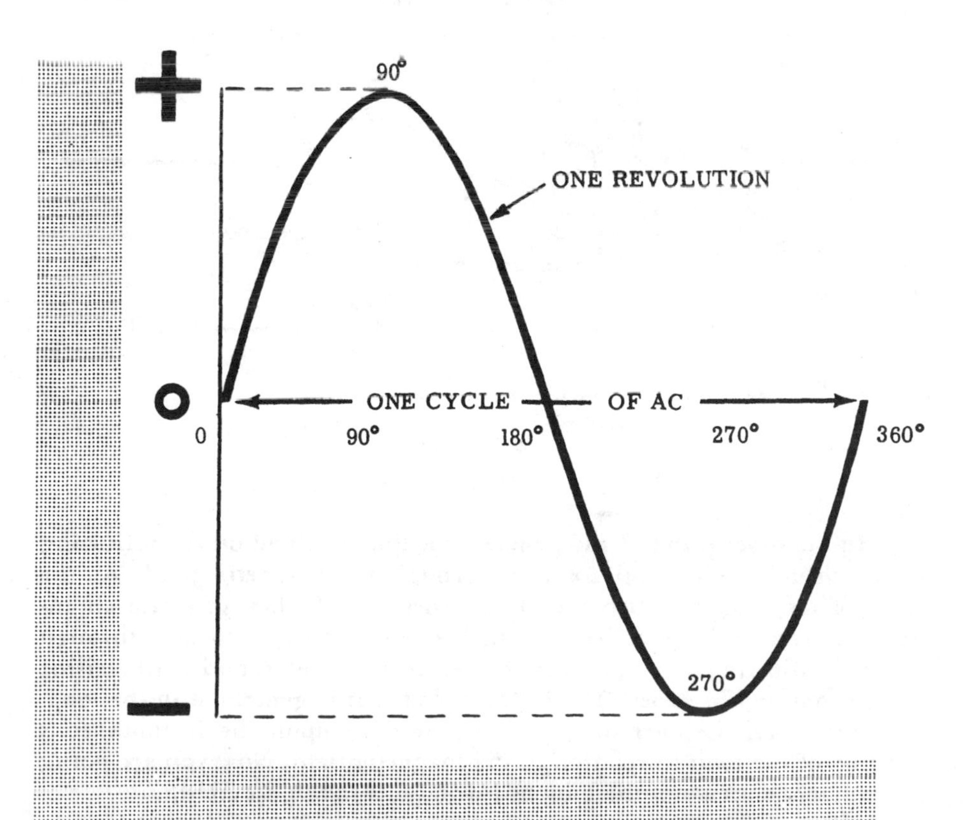

Elementary Generator Construction

As you know, a current is induced in a wire moving through a magnetic field. An elementary generator or alternator consists of a loop of wire placed so that it can be rotated in a stationary magnetic field to cause an induced current in the loop. Sliding contacts called *slip rings* are used to connect the loop to an external circuit in order to use the generator output.

The pole pieces are the North and South poles of the magnet that supplies the magnetic field. The loop of wire which rotates through the field is called the *armature*. The ends of the armature loop are connected to the slip rings, which rotate with the armature. Brushes ride against the slip rings to pick up the electricity generated in the armature and carry it to an external circuit.

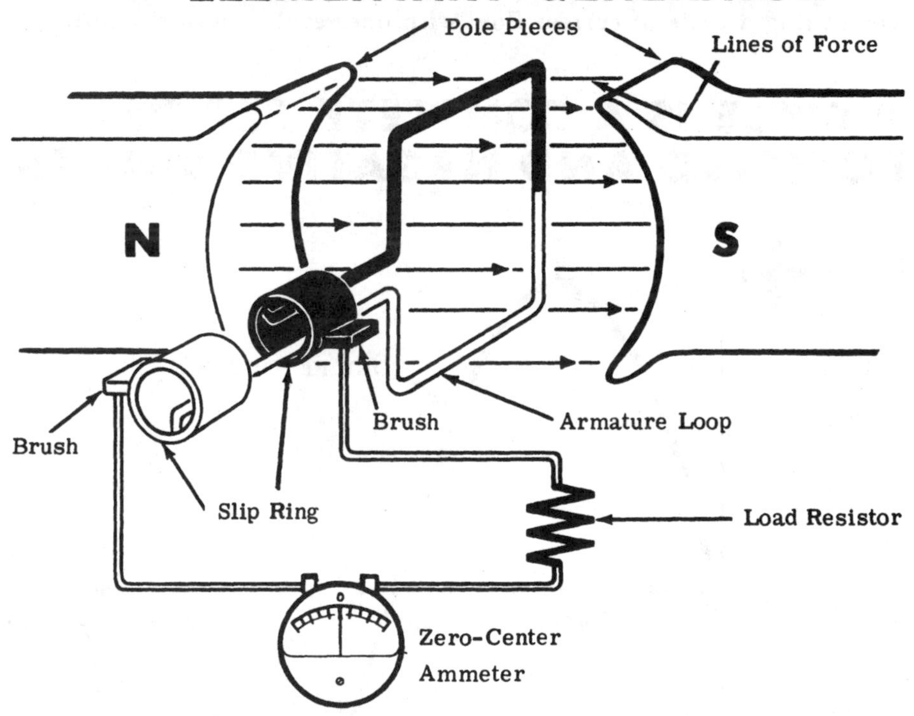

In the description of the generator action outlined on the following pages, visualize the loop rotating through the magnetic field. As the sides of the loop cut through the magnetic field, they generate an induced electromotive force (emf) which causes a current to flow through the loop, slip rings, brushes, zero-center current meter, and load resistor —all connected in series. The induced emf that is generated in the loop, and, therefore, the current that flows, depend upon the instantaneous position of the loop in relation to the magnetic field. Now you are going to analyze the action of the loop as it rotates through the field.

Elementary Generator Operation

Here is the way the elementary generator works. Assume that the armature loop is rotating in a clockwise direction, and that its initial position is at A (0°). In position A, the loop is perpendicular to the magnetic field and the black and white conductors of the loop are moving parallel to the magnetic field. If a conductor is moving parallel to a magnetic field, it does *not* cut through any lines of force and *no* emf can be generated in the conductor. This applies to the conductors of the loop at the instant they go through position A—*no* emf is induced in the conductors and, therefore, *no* current flows through the circuit. The current meter registers *zero*.

As the loop rotates from position A to position B, the conductors are cutting through more and more lines of force until, at 90° (position B), they are cutting through a maximum number of lines of force. In other words, between 0° and 90°, the induced emf in the conductors builds up from zero to a *maximum* value. Observe that from 0° to 90°, the black part of the conductor moves down through (cuts) the lines of force while at the same time, the white part moves up through (cuts) the lines of force. The induced emfs in both conductors are therefore in *series-adding*, and the resultant voltage across the brushes (the terminal voltage) is the *sum* of the two induced emfs. The current through the circuit will vary just as the induced emf varies—being zero at 0° and rising up to a maximum at 90°. The current meter deflects increasingly to the right between positions A and B, indicating that the current through the load is flowing in the direction shown. The direction of current flow and polarity of the induced emf depend on the direction of the magnetic field and the direction of rotation of the armature loop. The waveform shows how the terminal voltage of the elementary generator varies from position A to position B. The illustration of the simple generator on the right shows it has shifted in position to illustrate the relationship between the loop position and the generated waveform.

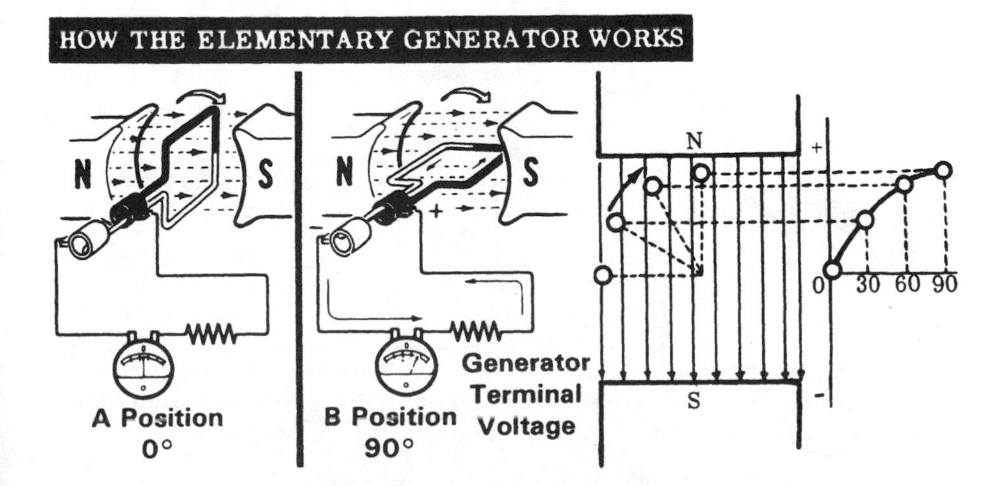

Elementary Generator Operation (continued)

As the loop continues to rotate from position B (90°) to position C (180°), the conductors, which are cutting through a maximum number of lines of force at position B, cut through fewer lines, until at position C they are moving parallel to the magnetic field and no longer cut through any lines of force. The induced emf, therefore, will decrease from 90° to 180° in the same manner as it increased from 0° to 90°. Similarly, the current flow will follow the voltage variations. The generator action at positions B and C is illustrated.

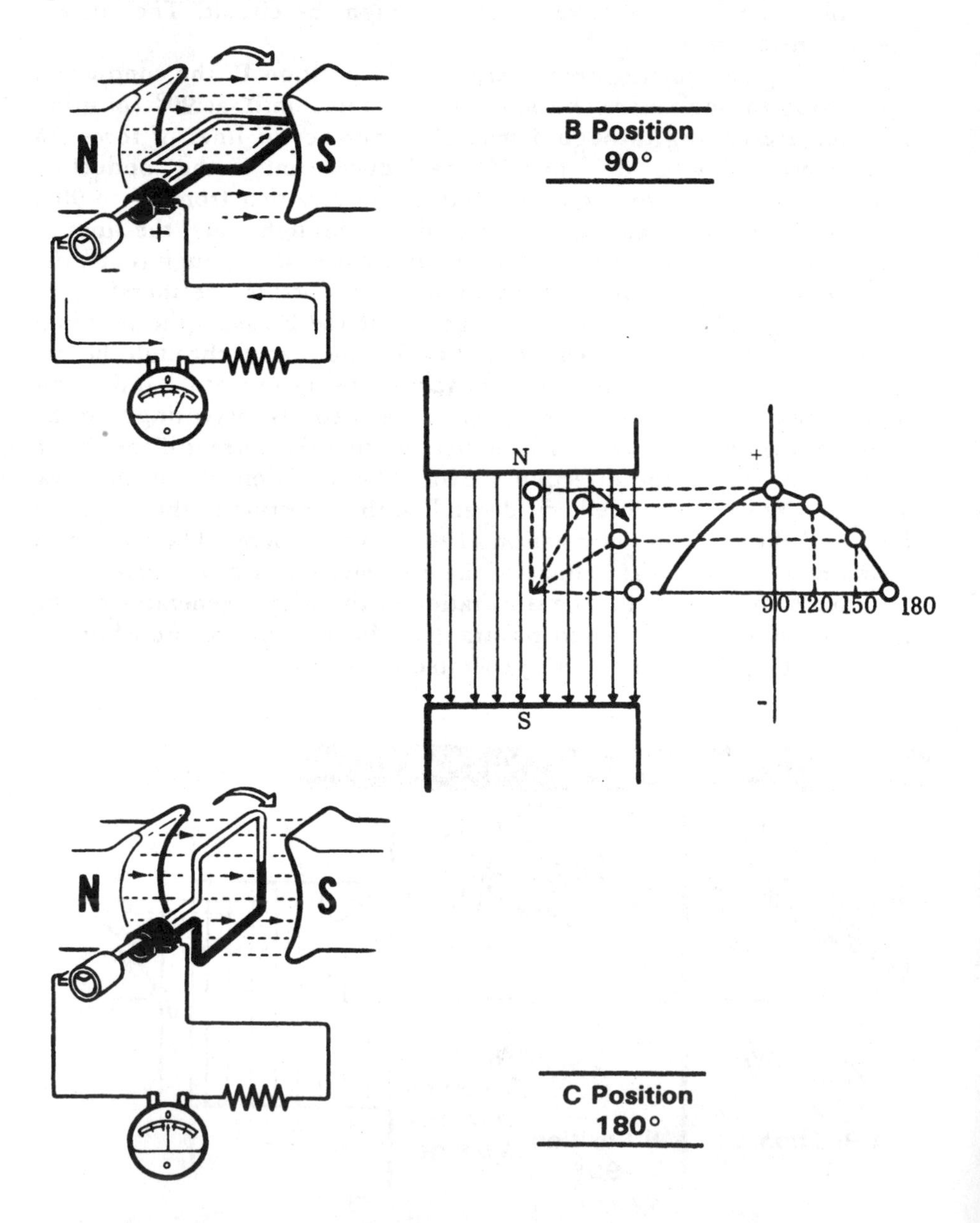

Elementary Generator Operation (continued)

From 0° to 180° the conductors of the loop have been moving in the same direction through the magnetic field and, therefore, the polarity of the induced emf has remained the *same.* As the loop starts rotating beyond 180°, back to position A, the direction of the cutting action of the conductors through the magnetic field reverses. Now the black conductor moves *up* through the field, and the white conductor moves *down* through the field. As a result, the polarity of the induced emf and the current flow are *reversed.* From position C through D and back to position A, the current flow will be in the *opposite* direction than from positions A to C. The generator terminal voltage will be the same as it was from positions A to C, except for its *reversed* polarity. The voltage output waveform for the complete revolution of the loop is as shown.

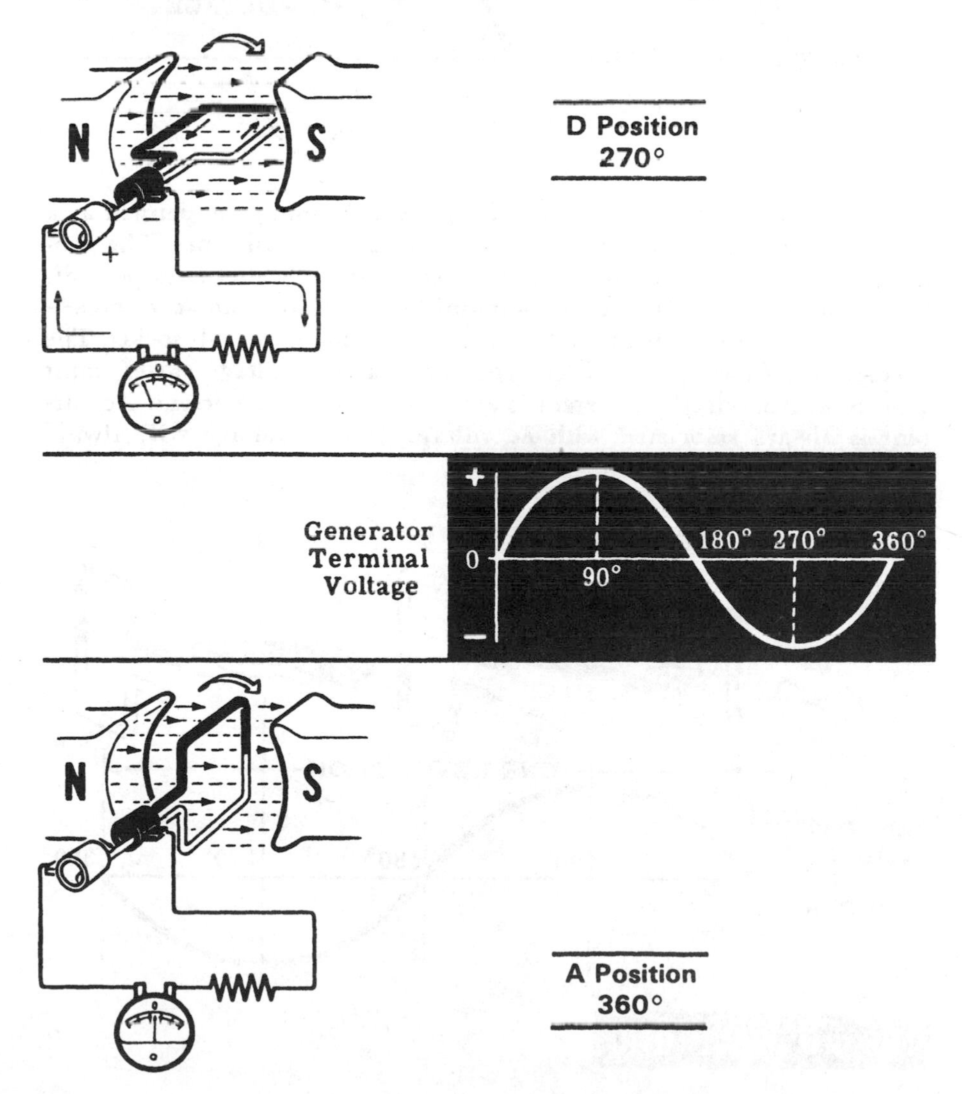

Elementary Generator Output

Take a closer look at the output waveform of the elementary generator and study it for a moment. How does it compare to the voltages you have been dealing with up to this time? The only voltages you have used so far are dc voltages, like those obtained from a battery. A dc voltage can be represented by a straight line whose distance above the zero reference line depends upon its value. The diagram shows the dc voltage waveform next to the voltage waveform put out by the elementary ac generator. The generated waveform does not remain constant in value and direction, as it does for dc. In fact, the voltage varies continuously in value and is as high in the negative direction as it is in the positive.

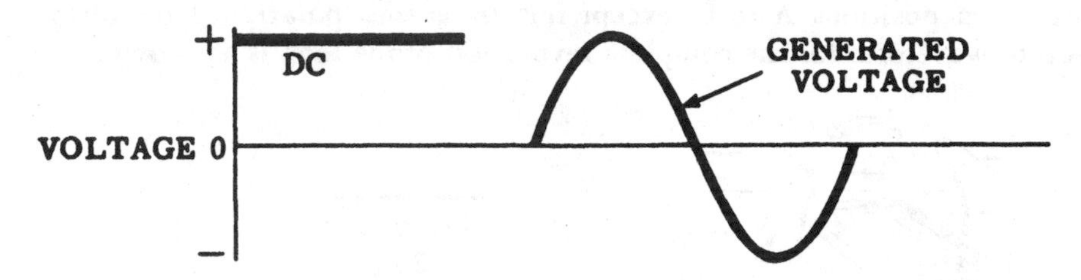

The generated voltage is not dc, since a dc voltage is defined as a voltage which maintains the same polarity output at all times. The generated voltage is called an *alternating voltage*, since it alternates periodically from plus to minus. It is commonly referred to as an ac voltage—the same type of voltage that you get from your home ac wall socket. The current that flows, since the current varies as the voltage varies, must also be alternating. The current is referred to as an ac current. Ac current is always associated with ac voltage—an ac voltage will always cause an ac current to flow.

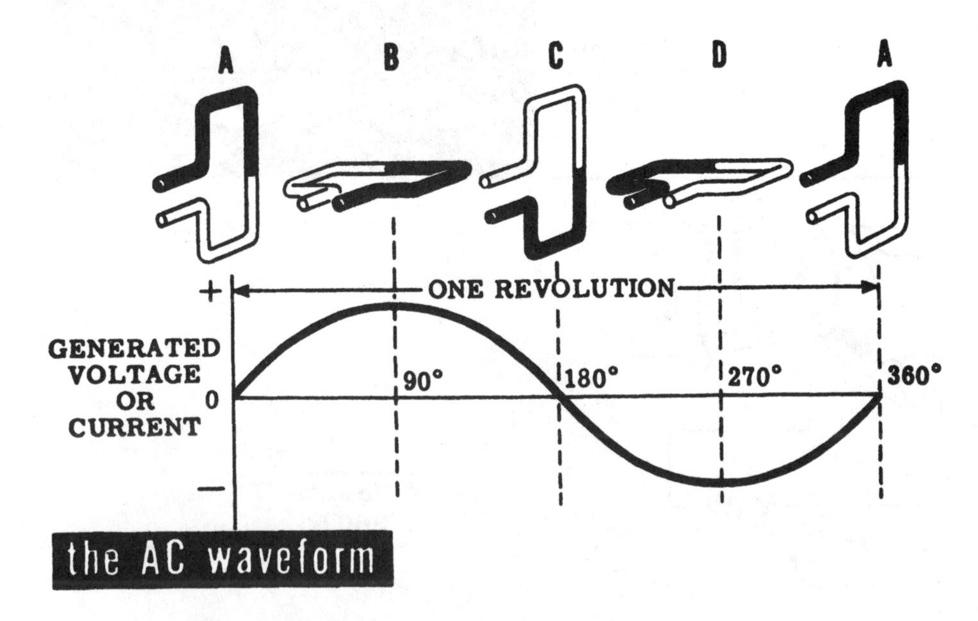

the AC waveform

Phase Relationships

As you know, the output from a simple generator varies as a sinusoidal wave. If you had two generators going at exactly the same speed, and they were started at the same time with their rotors in the same position, then the waves from generators 1 and 2 would correspond *exactly*—going through the peak values and zero values at the *same* time. If, however, you had the second rotor *advanced* 30° with respect to the first when the generators were started, the output of generator 1 would go through its peak values and zero values later than the rotor of generator 2. In the first case, you would say the waveforms are *in phase*; and in the second, that the generator 2 waveform was *leading* generator 1 by a *phase angle* of 30°. Since the phase is *relative*, we can say that generator 1 is *lagging* generator 2 by 30°. Since the phase of the generators, when started, can be any value, you can see that relative phase can have any value from 0° to 360°, with one waveform either leading or lagging the other.

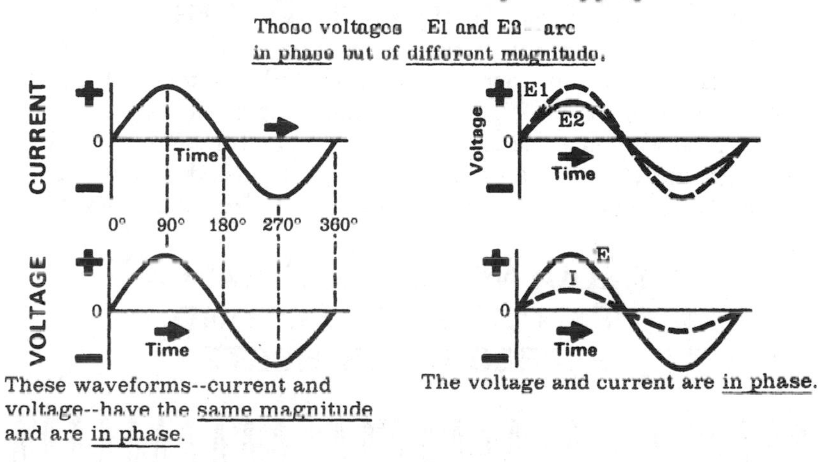

The waveforms are in phase when they both reach their maximum and minimum values and go through zero at the same time.

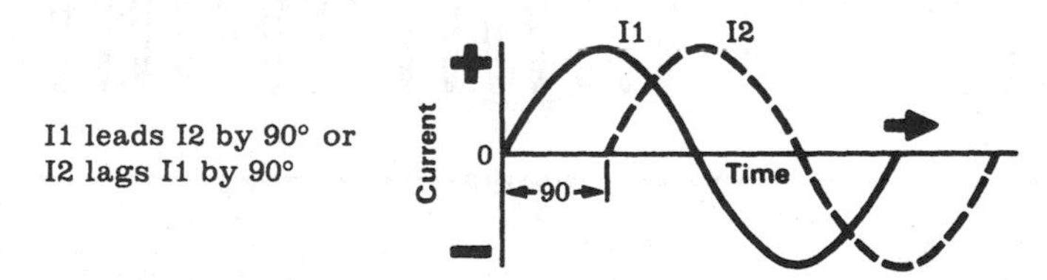

If maximum, minimum, and zero values of one waveform occur before the corresponding values of another waveform, the two are not in phase. When such a phase difference exists, one of the waveforms leads, or lags, with respect to the other.

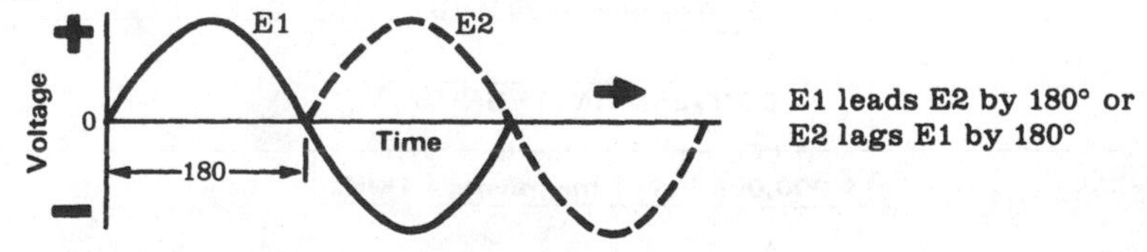

Alternating Current Frequency

You have seen that as the loop of the elementary generator rotated through 360°—one complete revolution—the generated voltage completed one cycle.

If the loop rotates at a speed of 60 revolutions per second, the generated voltage will complete 60 cycles in 1 second. It can then be said that the generated voltage has a *frequency* of 60 *hertz*. The word *hertz* is used to honor Heinrich Hertz, the discoverer of radio waves, and stands for *cycles per second*. It is abbreviated *Hz*. Frequency is always the number of cycles completed per second, expressed in Hz. Older equipment and textbooks also use cycles per second, abbreviated *cps*.

Frequency is important since most ac electrical equipment requires a specific frequency, as well as a specific voltage and current, for proper operation.

The standard commercial frequency used in North America is 60 Hz. Frequencies lower than 60 Hz would cause a flicker when used for lighting. The reason for this is that every time the current changes direction, it falls to zero and, therefore, momentarily switches off an electric lamp as it does so. At 60 Hz, however, the lamp switches on and off 120 times per second—once for each half cycle. The human eye cannot react fast enough to detect this and so receives the impression that the lamp is permanently lit. Some countries use 50 Hz, and aircraft use 400 Hz.

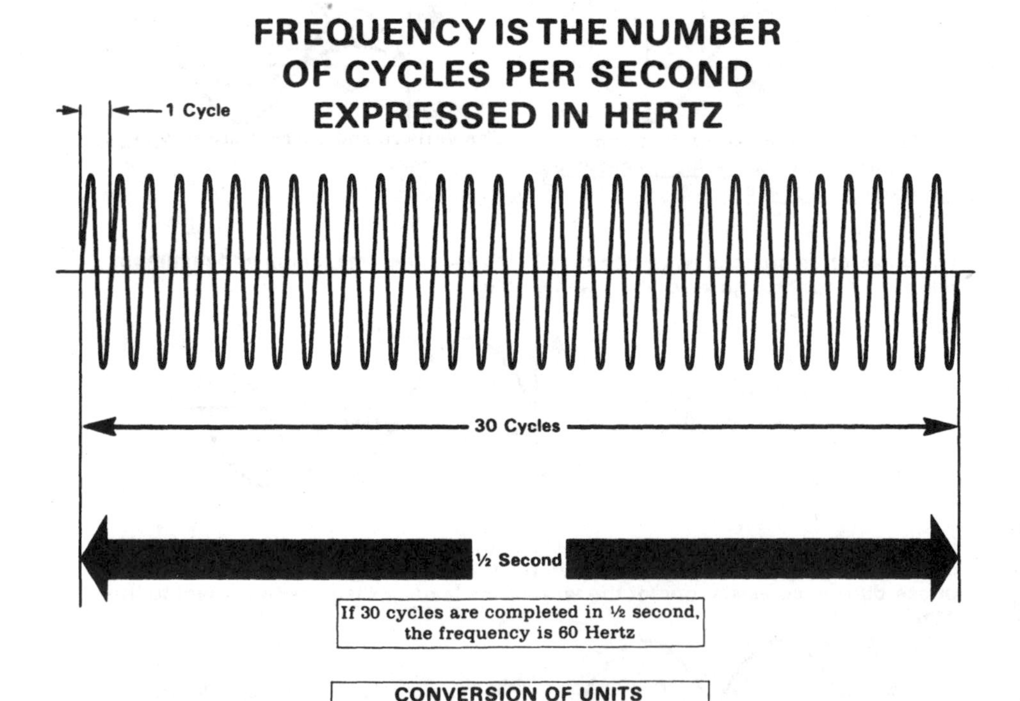
FREQUENCY IS THE NUMBER
OF CYCLES PER SECOND
EXPRESSED IN HERTZ
1 Cycle
30 Cycles
½ Second
If 30 cycles are completed in ½ second, the frequency is 60 Hertz

CONVERSION OF UNITS
1,000 Hz = 1 kilohertz = 1 kHz
1,000,000 Hz = 1 megahertz = 1 MHz

Maximum and Peak-to-Peak Values of a Sine Wave

Suppose you compare a half-cycle of an ac sine wave to a dc waveform for the same length of time. If the dc starts and stops at the same moment as the half-cycle sine wave and each rises to the same maximum value, the dc values are greater than the corresponding ac values at all points, except the point at which the ac sine wave passes through its maximum value. At this point, the dc and ac values are equal. This point on the sine wave is the *maximum* or *peak value.*

COMPARISON of DC and AC WAVEFORMS

Maximum Value of DC Waveform

Maximum (Peak) Value of AC Waveform

0° 90° 180°

There are two maximum or peak values for each complete cycle of ac; one for the positive half-cycle and the other for the negative half-cycle. The difference between the peak *positive* value and the peak *negative* value is called the *peak-to-peak* value of a sine wave. This value is *twice* the peak value of the sine wave and is sometimes used for measurement of ac voltages. Oscilloscopes and certain types of ac voltmeters measure peak-to-peak values of ac voltages in electronic circuits. Usually, though, ac voltages and currents are expressed in *root mean square* (rms) or *effective values* rather than peak-to-peak values. You will hear more about root mean square and effective values in the latter part of this section.

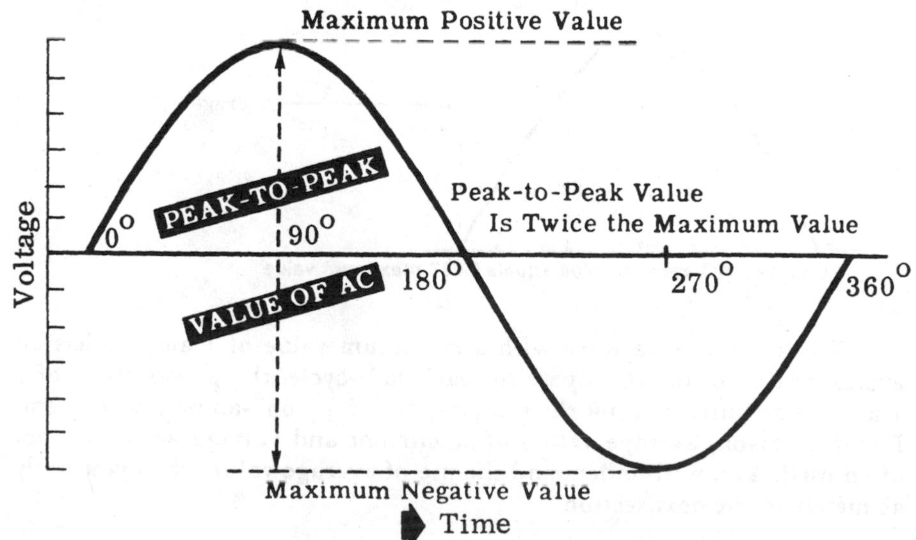

Average Value of a Sine Wave

When you compared a half-cycle ac sine wave to a dc waveform, you found that the ac instantaneous values were all less than the dc values, except at the peak value of the sine wave. Since all points of the dc waveform are equal to the maximum value, this value is also the average value of the dc wave. The average value of a half-cycle of the ac sine wave is less than the peak value, since all but one point on the waveform are lower in value. The average value of a half-cycle for all sine waves is 0.637 of the maximum or peak value. This value is obtained by averaging all the instantaneous values of the sine wave for a half-cycle. Since the shape of the sine wave does not change, even though its maximum value changes, the average value of any sine wave is always 0.637 or 63.7% of the peak value.

AVERAGE VALUES OF WAVEFORMS

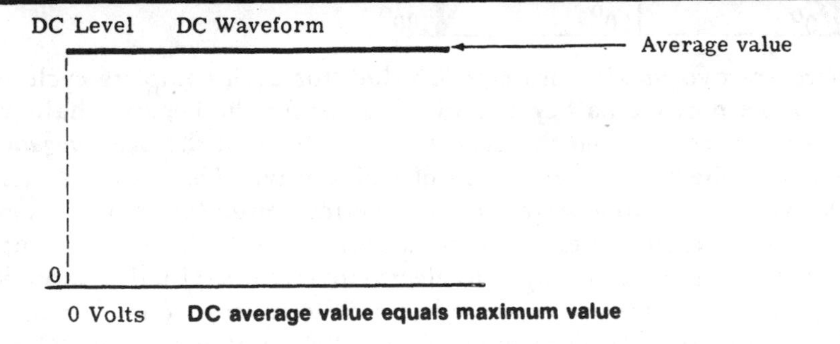

DC average value equals maximum value

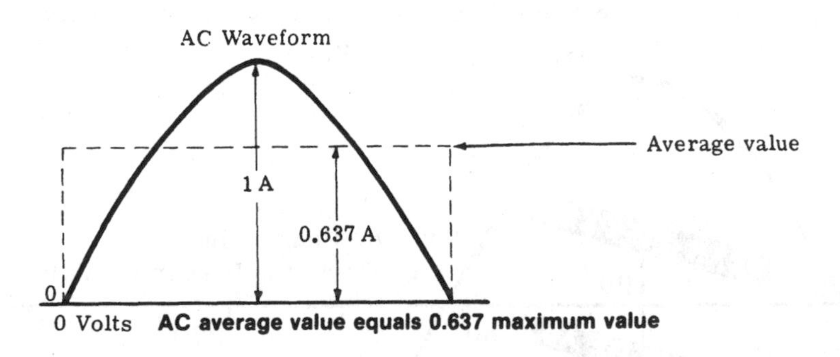

AC average value equals 0.637 maximum value

While an ac sine wave with a maximum value of 1 ampere has an average value of 0.637 ampere for each half-cycle, the power effect of a 1-ampere ac current is not the same as that of a 0.637-ampere dc current. For this reason, average values of ac current and voltage waves are not often used. You will understand the use of average value when you study ac meters in the next section.

RMS or Effective Value of a Sine Wave

You know that when any type of current—dc or ac—flows through a resistance, electric energy is converted into heat. The *rate* at which energy is converted and the power used, however, will be lower in the case of ac. This current varies continuously between maximum values and zero and is lower than the steady dc with a value equal to the peak value of the ac.

Some means must be found of relating dc and ac so that their relative efficiency in the conversion of energy can be determined. A convenient way of doing this is to compare the heating effect in a resistor of given value when dc is passed through it and when ac of maximum value equal to the value of the dc is passed through it for the same period of time. The *increase* in temperature produced by the ac in the resistor is then compared with the increase in temperature produced by the dc, and from their ratio the *effective value* or *power used* can be calculated.

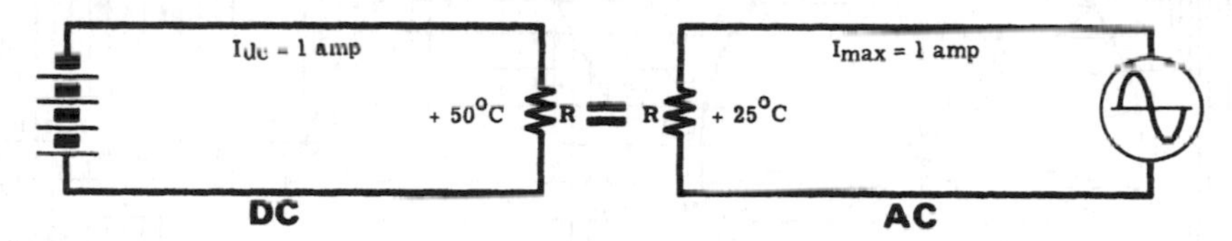

Consider the two circuits above, both of which have a resistor R of identical value. In the dc circuit on the left, a current of 1 ampere raises the temperature of the resistor by +50° C. In the ac circuit on the right, where the maximum value of the current (I_{max}) is also 1 ampere, the temperature of the resistor is raised only by + 25° C.

Therefore, our question is, "*What is the effective value of the ac current, expressed as a fraction of I_{max} (or of I_{dc})?*" The effective value of an alternating current is also known as its *rms value*. This stands for *root mean square* value and is explained on the next page.

You know that the power used to heat a resistor is calculated by using the formula, $P = I^2R$. In the circuits above, the power loss ($I^2_{dc} \times R$) caused by the flow of 1 ampere of dc raised the temperature of the resistor to 50°C, while in the ac circuit, the heating ($I^2_{ac} \times R$) caused by an I_{max} of 1 ampere was only *half* the dc heating, since it raised the temperature to only 25°C. It follows that:

$$I^2_{ac} \times R = \tfrac{1}{2}I^2_{dc} \times R = \tfrac{1}{2}I^2_{max} \times R$$

simplifying,

$$I^2_{ac} = \tfrac{1}{2}I^2_{max}$$

and,

$$I_{ac} = \frac{1}{\sqrt{2}} I_{max} \text{ or } 0.707\ I_{max}$$

In other words, the effective value of the alternating current, I_{ac}, is only 0.707 times I_{max}; and I_{max}, *in the ac circuit, will have to be increased to $I_{ac} \times \sqrt{2}$ (1.414 amperes) before it will produce the same heating effect as will 1 ampere of direct current*. Similarly, the peak voltage is 1.414 times the rms voltage.

RMS or Effective Value of a Sine Wave (continued)

You may now want to confirm that the rms value (I) of an ac current of sinusoidal waveform is 0.707 (I_{max}). Therefore, you must first draw a sine wave with a maximum value of 25 mm. You can do this by drawing a circle with a 25-mm radius and, with a protractor, dividing it into 10° arcs. Then, number the lines from 0 to 35. Observe the illustration below.

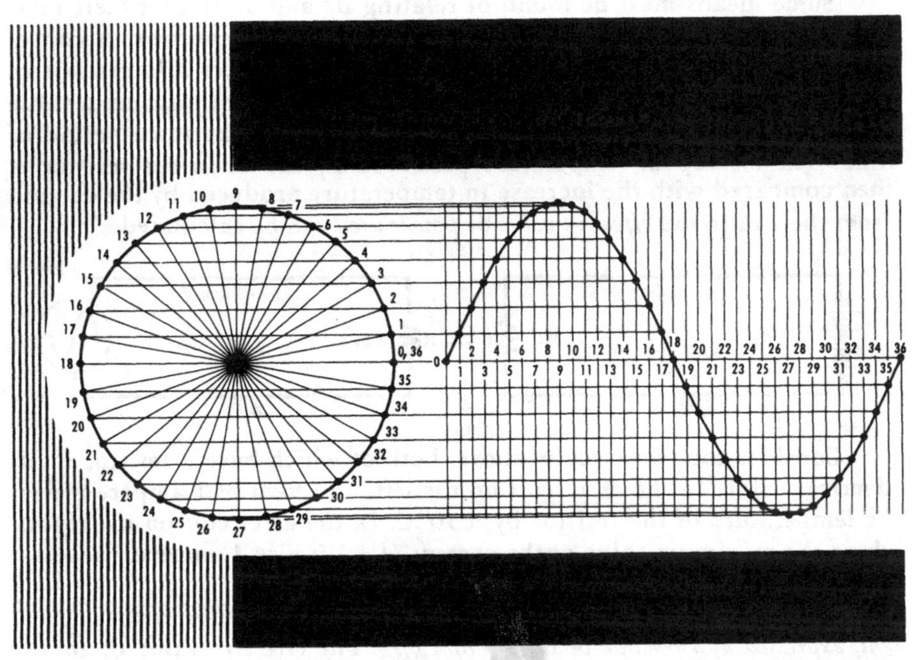

Draw the axis of your sine wave on the right of the circle so that if it were extended, it would pass through the center of the circle. Mark this axis off in 36 equal divisions and number them from 0 to 36. Draw a vertical line through each division mark.

Now draw lines parallel to the axis from each of the numbered points on the circle until they cut the vertical line with the same number. The line from *point one* on the circle should meet the line from *division one* on the axis and so on.

If a curve were drawn joining the resulting intersections, you would recognize it as a sine wave, as shown in the diagram.

Suppose the sine wave you have drawn represents an ac current with maximum value of 1 ampere. You can find the rms value of the current by (a) measuring the value of the current, which is the height of the sine wave above the axis at each of the divisions on the axis, (b) squaring these instantaneous values and finding the average of all the squared values (remember that the heating effect of a current is proportional to the average value of the square of the current), and (c) finding the square root of the average of the squared values. This is the *root mean square* or *rms value* of the current.

RMS or Effective Value of a Sine Wave (continued)

EFFECTIVE VALUE OF A SINE WAVE

N

S

1 2 3 4 5 6 7 8 9

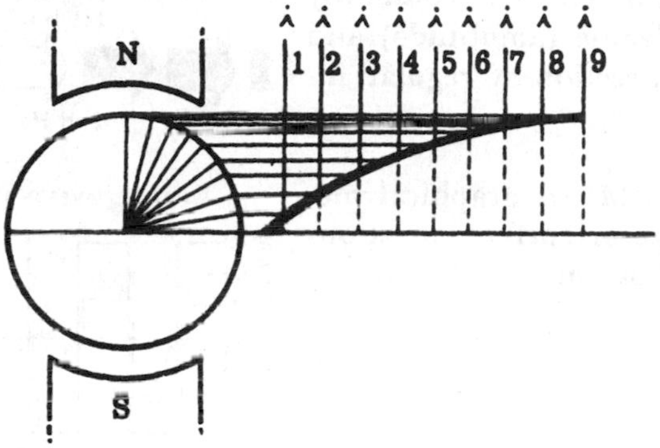

$I_{rms} = \sqrt{\text{Average of the Sum of the Squares of } I_{ins}}$

$I_{rms} = 0.707 \times I_{max}$ $I_{max} = 1.414 \times I_{rms}$

RMS VALUE

$I_{rms} \sqrt{\text{Sum of Instantaneous } I^2}$

$E_{rms} \sqrt{\text{Sum of Instantaneous } E^2}$

When we specify ac voltages and current, we mean the rms value, unless otherwise specified.

You will find that the rms value of an ac current with a maximum value of 1 ampere is 0.707 ampere.

Since alternating voltages cause alternating currents to flow, the ratio between effective and maximum values of emfs is the same as that for currents. The effective, or rms, value (E) of a sine wave emf is 0.707 times the maximum value (E_{max}).

When an alternating current or voltage is specified, it is always the rms value that is meant, unless there is a definite statement to the contrary. It should be noted that all meters, unless marked to the contrary, read *rms* values of current and voltage.

Review of Alternating Current

Alternating current differs from direct current not only in its waveform and electron movement, but also in the way it reacts in electric circuits. Before finding out how it reacts in circuits, you should review what you have already discovered about ac and the sine wave.

1. ALTERNATING CURRENT—A current flow which is constantly changing in value (amplitude) and reverses its direction at regular intervals.

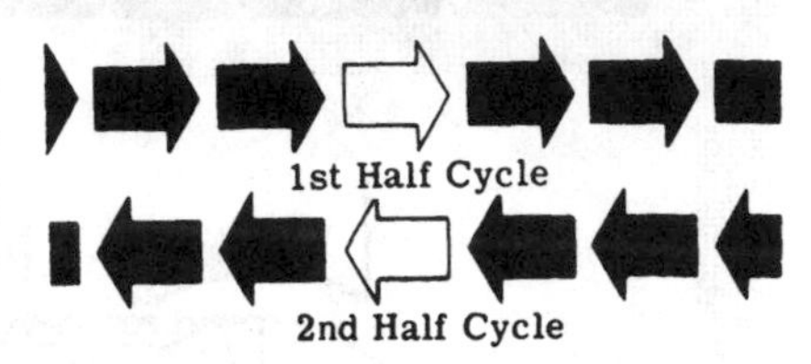

2. WAVEFORM—A graphical picture of voltage or current variations over a period of time.

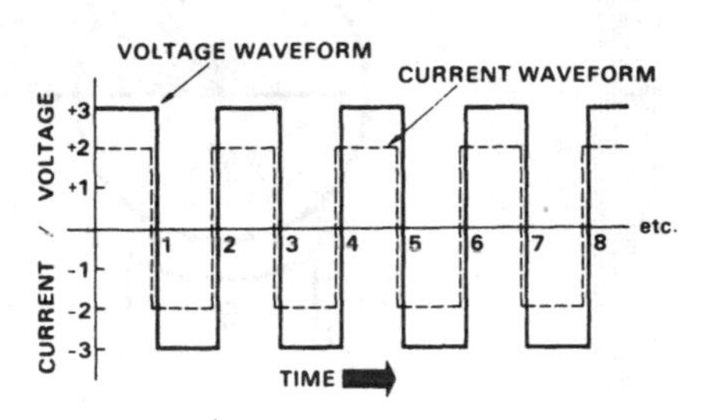

3. SINE WAVE—A continuous curve of all the instantaneous values of an ac current or voltage.

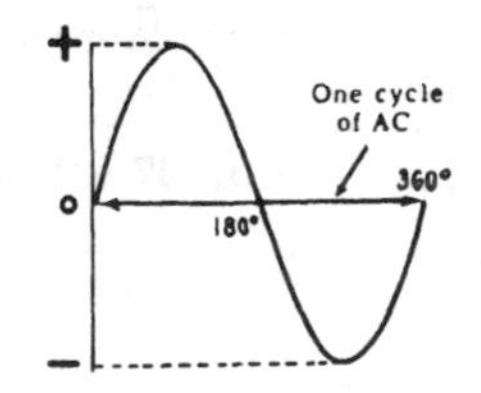

4. CYCLE—A complete set of positive and negative values of an ac current or voltage wave.

5. FREQUENCY—The number of cycles per second. It is expressed in hertz, abbreviated Hz. 1 Hz = 1 cycle/second.

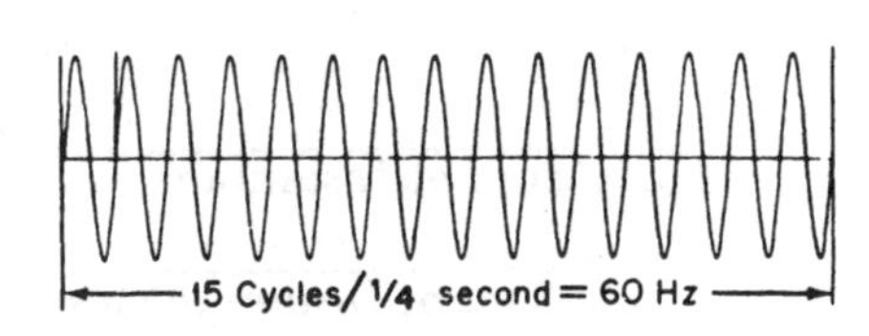

6. PHASE—The relative time difference between the same points on two waveforms.

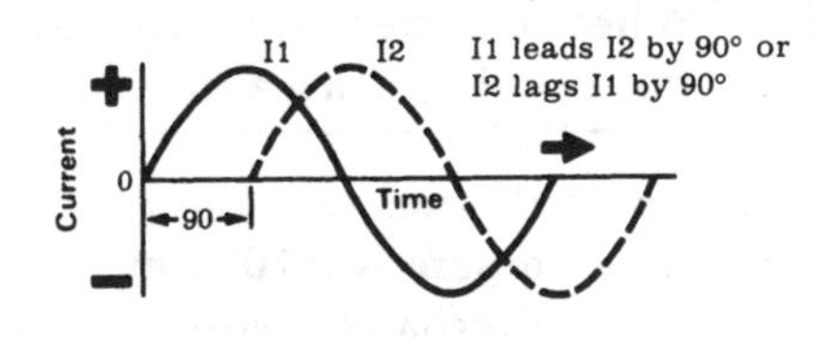

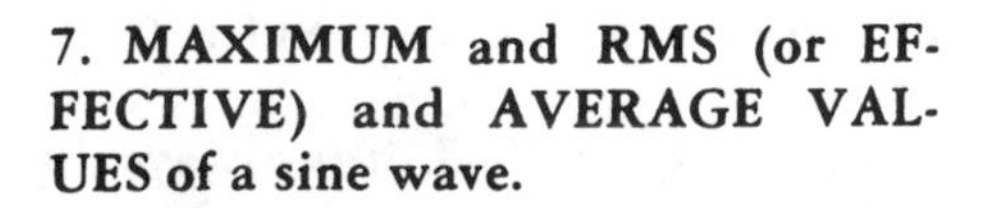
7. MAXIMUM and RMS (or EFFECTIVE) and AVERAGE VALUES of a sine wave.

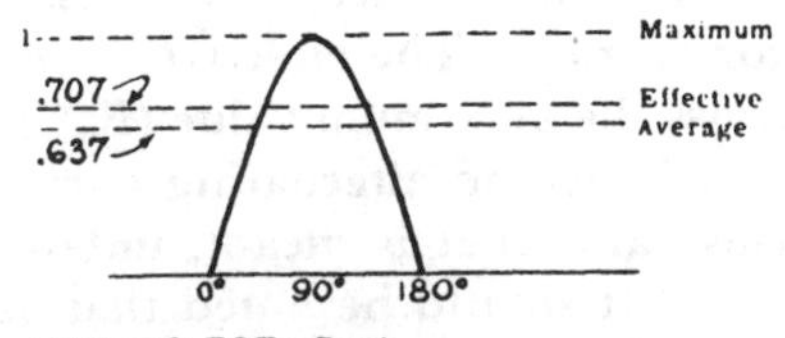

Self-Test—Review Questions

1. What are the major advantages of using ac?
2. Draw several graphs showing the difference between ac and dc waveforms.
3. Draw a sine wave and label the illustration to show where the peak values occur and where the value is zero.
4. Draw an elementary generator and show how it produces an ac current.
5. Define phase. Draw waveforms showing phase relationships of 0°, 30° 90°, 180°, and 300°.
6. If you had a resistor and put 1 ampere ac (rms) through it, would the resistor get as hot if you put 1 ampere dc through it?
7. Given the following currents, find the corresponding peak or rms (effective) values.
   (a) 1.5 A (c) 3.6 A (e) 4,200 A peak
   (b) 10.5 A peak (d) 9.8 A (f) 1,000 A
8. Given the following voltages, find the corresponding peak or rms (effective) values.
   (a) 130 kV peak (c) 1,180 V (e) 120 V
   (b) 440 V (d) 1,600 V peak (f) 240 V
9. Define the frequency of an ac waveform. How is it usually expressed?
10. Given the following values of hertz (numbers of cycles) in the times specified, calculate the frequency, and express it in proper units.
    (a) 60 Hz per second (c) 20,000 Hz per minute
    (b) 800 Hz per 2 seconds (d) 10 Hz per ¼ second

Learning Objectives—Next Section

Overview—In the next section, you will learn how ac meters operate. You will find that they can be completely different in construction, or they can be the same as dc meters with an added part called a *rectifier*.

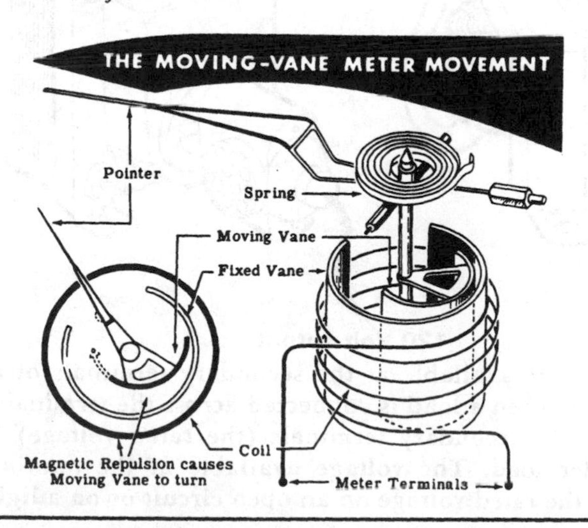

Using Transformers

You have already learned that transformers are used to *step-up* and *step-down* voltages at the ends of an electricity supply line. In the experiment application that follows, it is necessary to step-down the line voltage to a low value suitable for experimental work. You will use a transformer to do this.

A transformer consists of two coils of wire on an iron core, with each coil insulated from the other. If an alternating voltage is applied to one of the coils, an alternating voltage is induced in the other. By choosing coils of suitable sizes, it can be arranged for the induced voltage to be either equal to, greater than, or smaller than the applied voltage.

The coil to which the voltage is applied is called the *primary*, and the coil in which the voltage is induced is called the *secondary*. You will learn much more about the principles of the transformer in Volume 4.

The symbols used to indicate a transformer in circuit diagrams are shown below. The transformers you will use for your experiments are also illustrated.

TRANSFORMER SYMBOLS

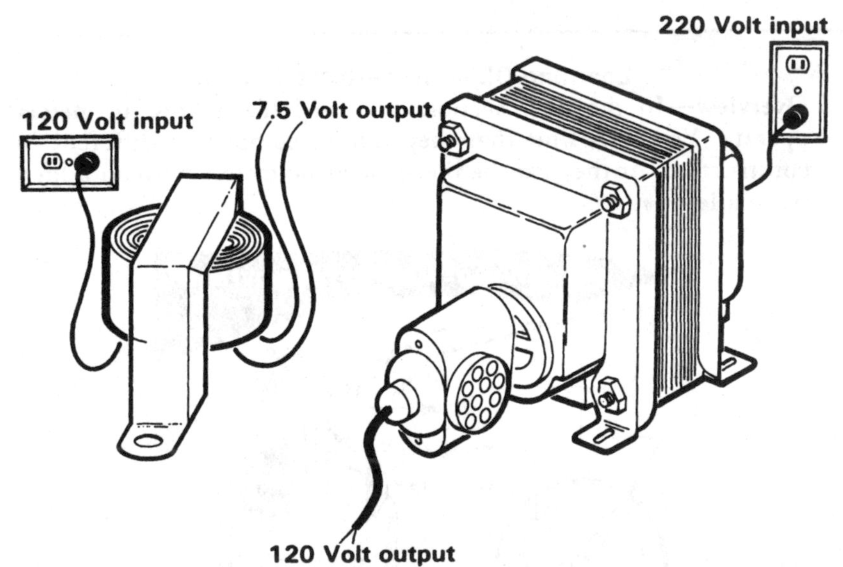

The voltage available at the secondary terminals of a transformer drops slightly when a load is connected across the terminals. The voltage indicated at the secondary terminals (the rated voltage) is usually the voltage under load. The voltage available at the terminals is a little higher than the rated voltage on an open circuit or on a light load.

Experiment/Application—RMS Effective Value of AC Voltage

To show that the 7.5-volt (rms) ac has the same effect as 7.5-volt dc, a 7.5-volt dc source and a 7.5-volt ac (rms) source are each used to light the same type of lamp. A transformer will be used to obtain 7.5 volts ac.

One lamp socket is connected across the battery and another is connected across the secondary leads of the transformer. If you then insert identical lamps in each socket, you will see that the brightness of the two lamps is the same. This shows that the effect of the two voltages is the same and the power in both circuits is the same.

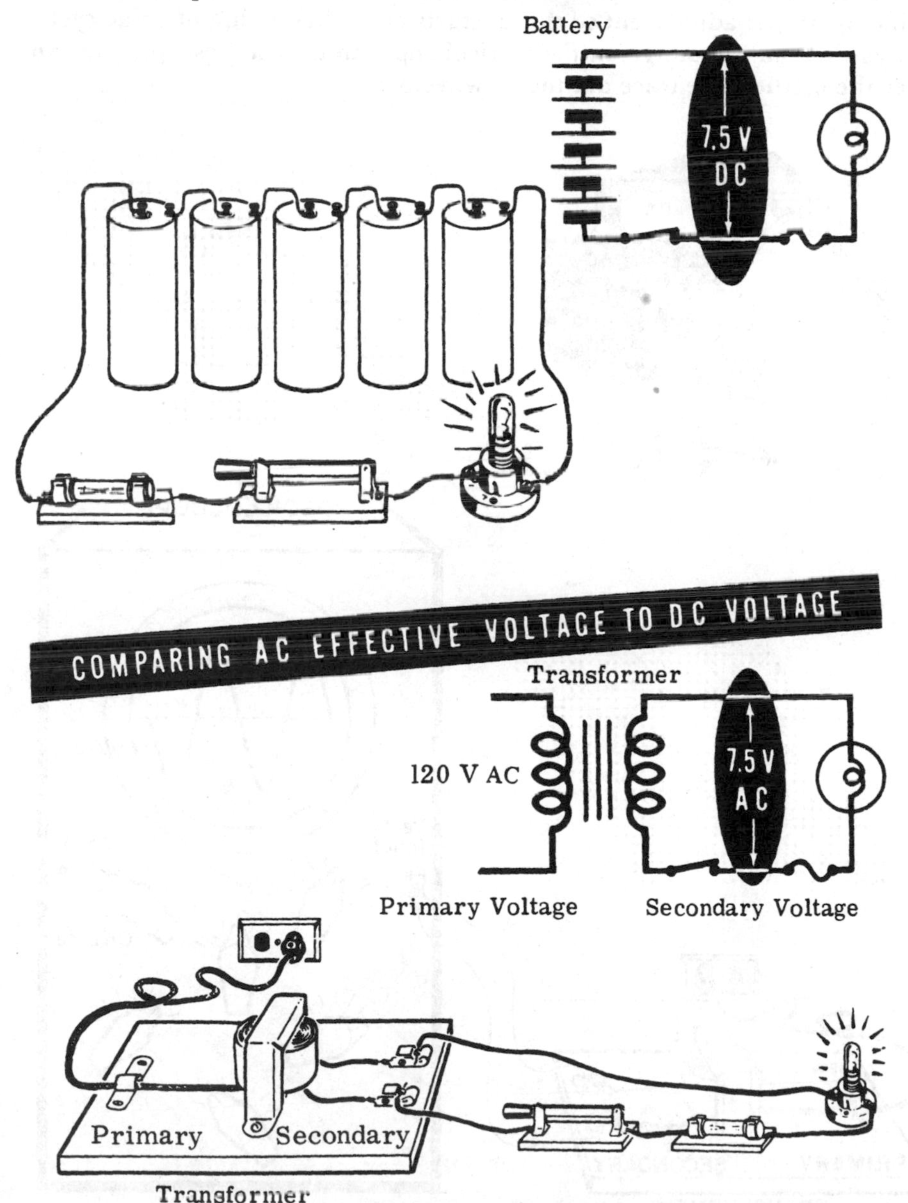

Experiment/Application—Use of the Oscilloscope to Observe AC Waveforms

An *oscilloscope* is an instrument used to observe and measure waveforms. This instrument uses a *cathode ray tube*—like the one in a TV set—to observe fast-moving waveforms. An electron beam, driven by an ac voltage *sawtooth*, makes the beam travel horizontally. Since the sawtooth is repetitive, the beam is retraced and the result is that the beam traces a horizontal line on the oscilloscope, with each position of the beam on the line representing a change in time. This trace can be triggered by the ac line by proper adjustment so that it starts at the beginning of an ac cycle. If you put an ac voltage on the vertical input to the oscilloscope, you can see the oscilloscope trace out the ac waveform.

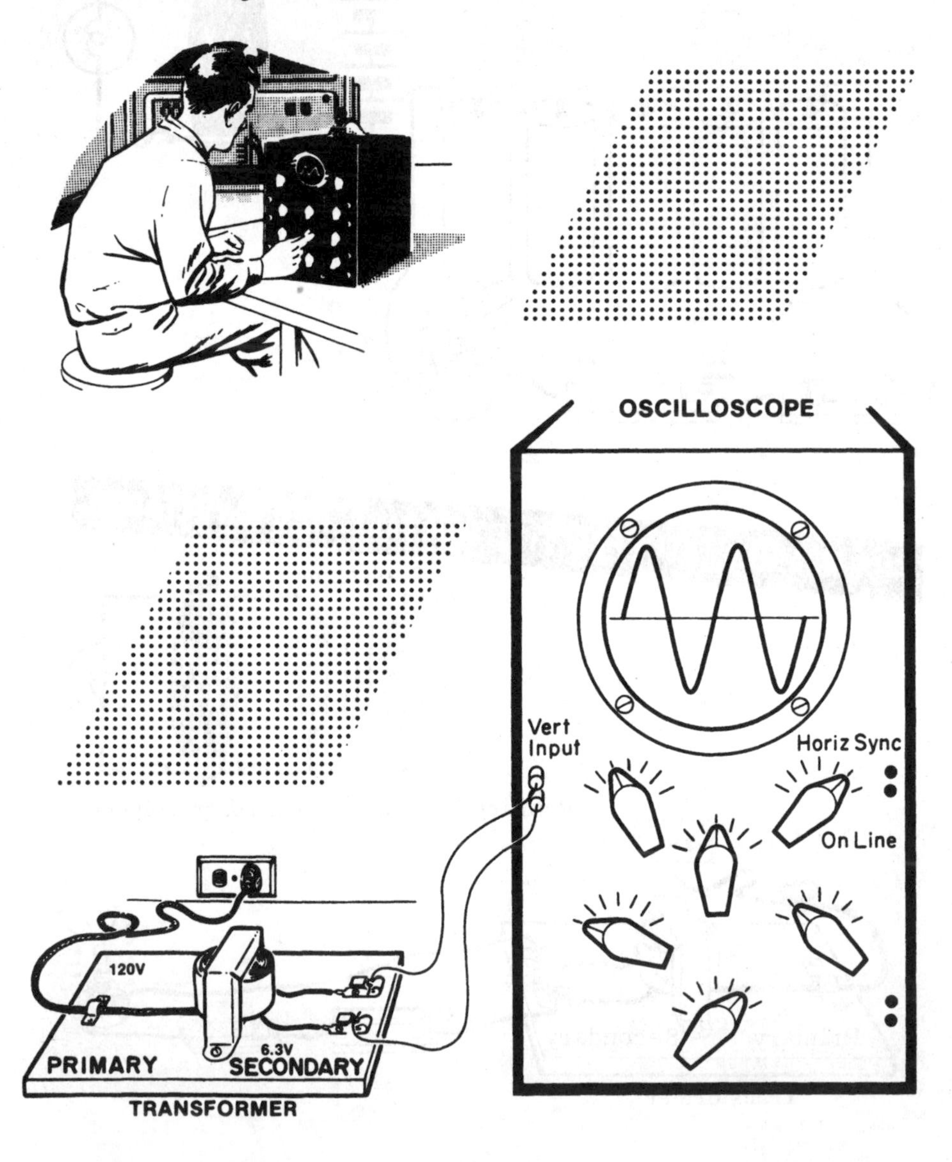

Why DC Meters Cannot Measure AC

There are noticeable differences, particularly in the scales, between dc and ac voltmeters. There may also be a basic difference in the meter movements.

As you know, dc meters use a moving-coil meter movement in which the moving coil is suspended in the magnetic field between the poles of a permanent magnet. Current flow through the coil in the correct direction (polarity) causes the coil to turn, moving the meter pointer up-scale. However, you will recall that a reversal of polarity causes the moving coil to turn in the opposite direction, moving the meter pointer down below zero.

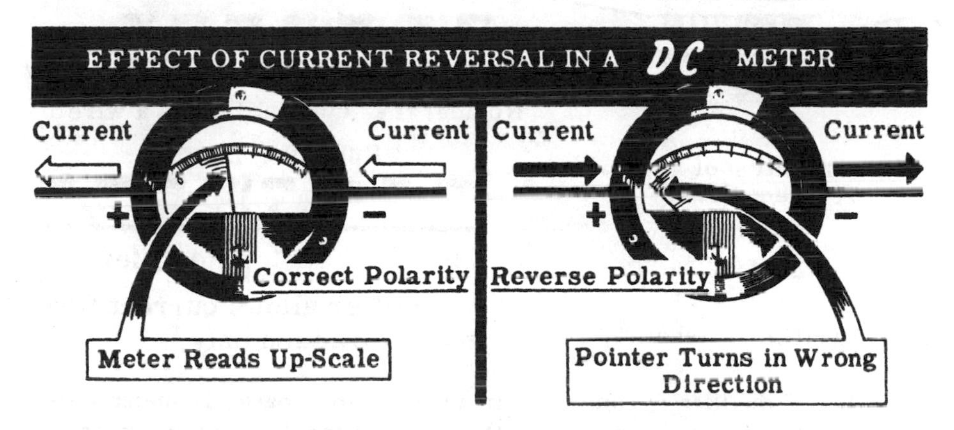

If an ac current were passed through a basic dc meter movement, the moving coil would turn in one direction for a half cycle; then, as the current reversed direction, the moving coil would turn in the opposite direction. For ordinary 60 Hz, the pointer would be unable to follow the reversal in current fast enough and would vibrate back and forth at zero. The greater the current flow, the farther the pointer would attempt to swing back and forth, and, in a short time, the excess vibration would break it. Even if the pointer could move back and forth fast enough, the speed of movement would prevent your obtaining a meter reading.

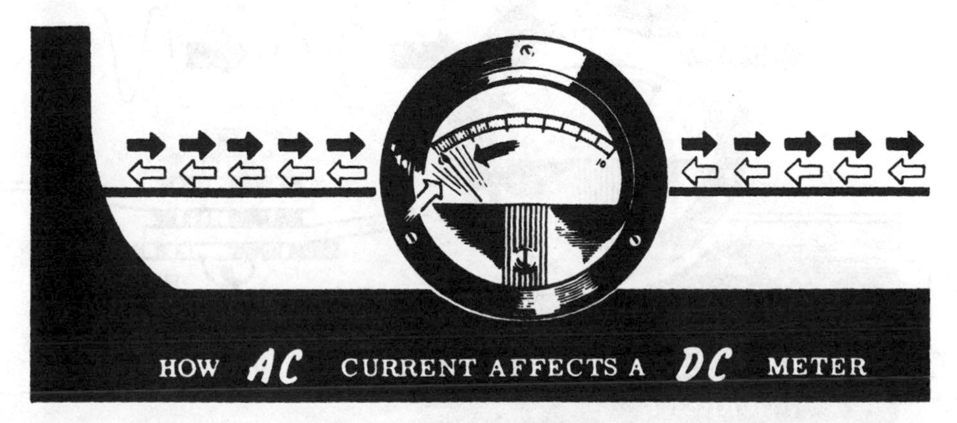

Rectifier Type AC Voltmeters

A basic dc D'Arsonval meter movement may be used to measure ac through the use of a *rectifier*—a device which converts ac to pulsating dc. The rectifier permits current to flow in only one direction so that, when ac tries to flow through it, current will flow for only half of each complete cycle. The rectifier is an *active device,* as mentioned on page 2, and its effect on ac current flow is illustrated below.

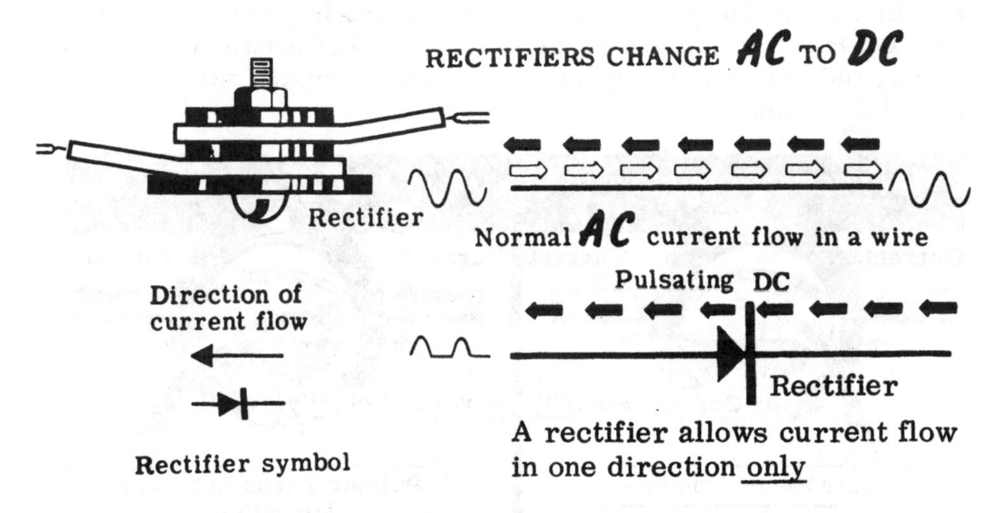
RECTIFIERS CHANGE AC TO DC
Rectifier
Normal AC current flow in a wire
Pulsating DC
Direction of current flow
Rectifier
A rectifier allows current flow in one direction only
Rectifier symbol

If the rectifier is connected in series with a basic dc meter movement so that it permits current to flow only in the direction necessary for correct meter polarity, the meter current will flow in pulses. Since these pulses of current are all in the *same* direction, each causes an up-scale deflection of the meter pointer. The pointer cannot move rapidly enough to return to zero between pulses, so that it continuously indicates the average value of the current pulses.

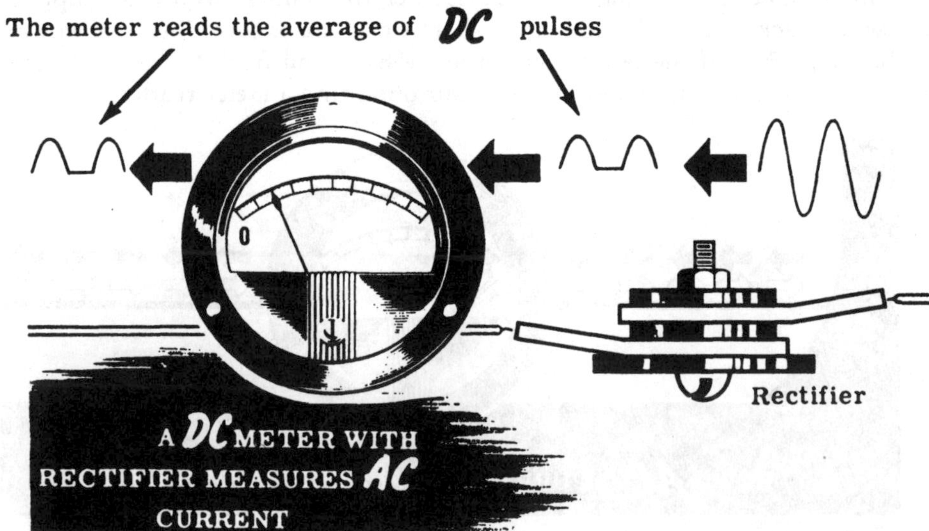
The meter reads the average of DC pulses
0
Rectifier
A DC METER WITH RECTIFIER MEASURES AC CURRENT

Rectifier Type AC Voltmeters (continued)

When certain metallic materials are pressed or alloyed together to form a junction, the combination acts as a *rectifier* having a *low* resistance to current flow in *one* direction, and a very *high* resistance to current flow in the *opposite* direction. This action is due to the physical properties of the combined materials. The combinations usually used in meter rectifiers are copper and copper oxide, or iron and selenium. Copper-oxide rectifiers consist of disks of copper coated on one side with a layer of copper oxide; selenium rectifiers are constructed of iron disks coated on one side with selenium.

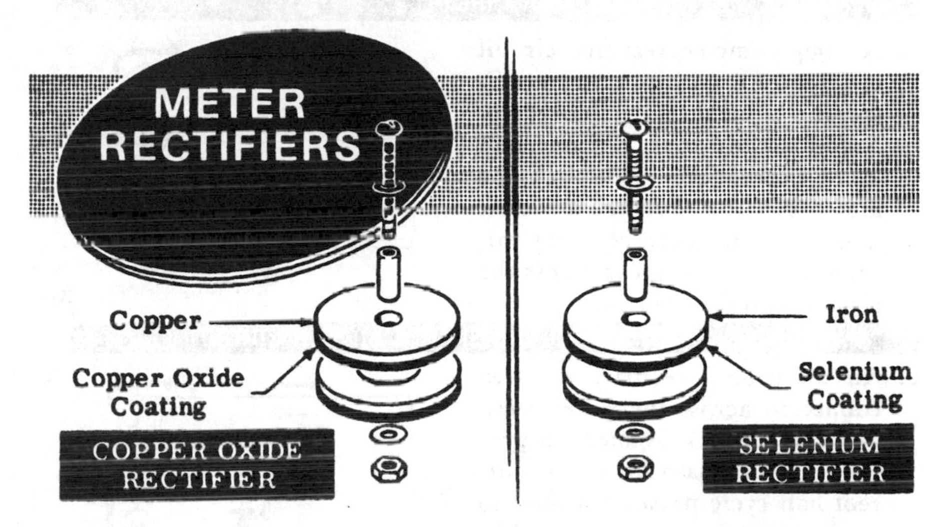

Metal rectifier elements (an *element* is a single disk) are generally made in the form of washers assembled on a mounting bolt in any desired series or parallel combination to form a rectifier unit. The symbol shown below is used to represent a rectifier of any type. Since these rectifiers were used before the electron theory was developed to determine the direction of current flow, the arrow points in the direction of conventional current flow, but in the direction opposite to the electron flow. Thus, as used in electronics, the arrow points in the *opposite* direction to that of the current flow.

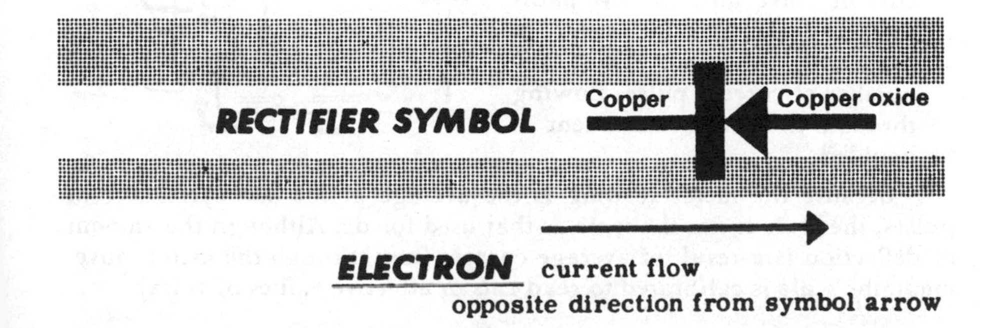

Rectifier Type AC Voltmeters (continued)

Rectifier type ac meters are used only as voltmeters, and the meter range is determined and changed in the same manner as that of a dc voltmeter. These meters cannot be used to measure current, since ammeters are connected in series with the line current, and a rectifier type meter connected in that manner would change the ac circuit current to dc, which is not desirable. These types of voltmeters can, however, be used with an ammeter shunt, as you learned to do with dc meters. Various ac rectifier type meter circuits are illustrated below.

SIMPLE METER RECTIFIER CIRCUIT ....

1. A simple meter rectifier circuit consists of a multiplier, a rectifier, and a basic meter movement connected in series. For one half cycle, current flows through the meter circuit. During the next half cycle, no current flows, although a voltage exists across the circuit including the rectifier.

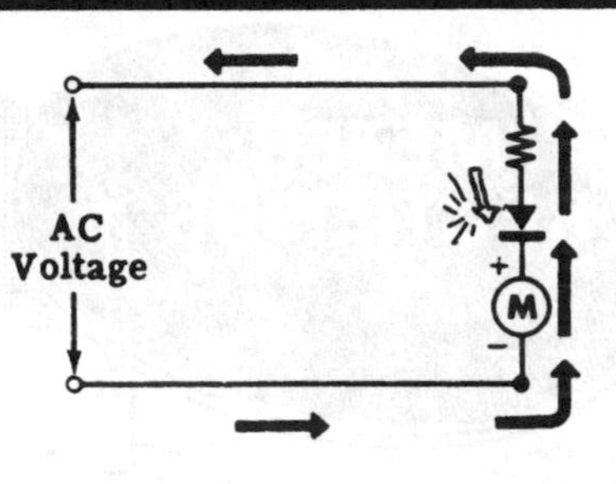

ADDING A RECTIFIER TO THE SIMPLE METER CIRCUIT

2. An additional rectifier is often connected across the meter rectifier and meter movement to provide a return path for the ac current half-cycle pulses not used to operate the meter movement. The unused ac pulses flow through this branch, not through the meter.

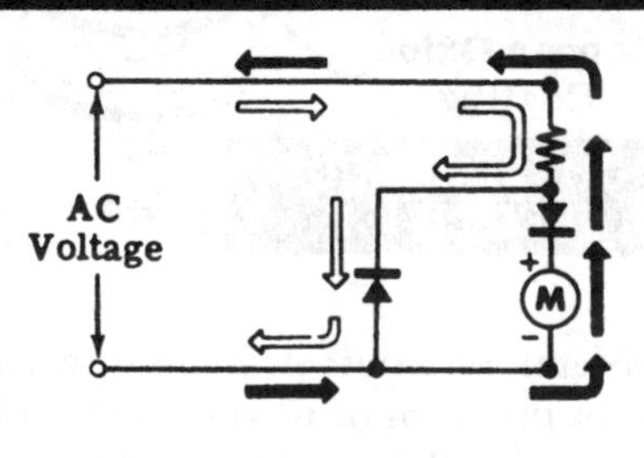

BRIDGE RECTIFIER CIRCUIT ....

3. A bridge circuit with four rectifiers is most often used because it makes use of both halves of the ac cycle. This circuit is connected so that both halves of the ac current wave must follow paths that lead through the meter in the same direction. Thus, the number of current pulses flowing through the meter movement is doubled.

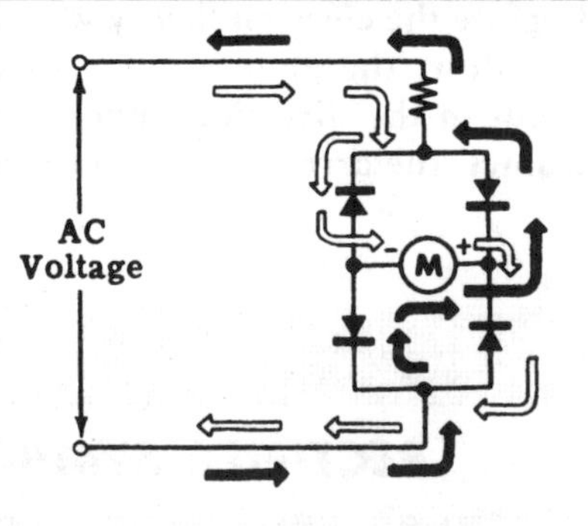

Because the meter reading is the average of the half-cycle current pulses, the scale is not the same as that used for dc. Although the amount of deflection is a result of average current flow through the meter movement, the scale is calibrated to read rms or effective values of voltage.

Moving-Vane Meter Movements

A meter that can measure both ac current and voltage is the *moving-vane meter.* The moving-vane meter movement operates on the principle of magnetic repulsion between like poles. The current to be measured flows through a field coil, producing a magnetic field proportional to the strength of the current. Suspended in this field are two iron vanes—one fixed in position, the other movable and attached to the meter pointer. The magnetic field magnetizes these iron vanes with the same polarity regardless of the direction of current flow in the coil. Since like poles *repel,* the movable vane pulls away from the fixed vane, moving the meter pointer. This motion exerts a turning force against a spring. The distance the vane moves against the force of the spring depends on the strength of the magnetic field, which depends on the coil current.

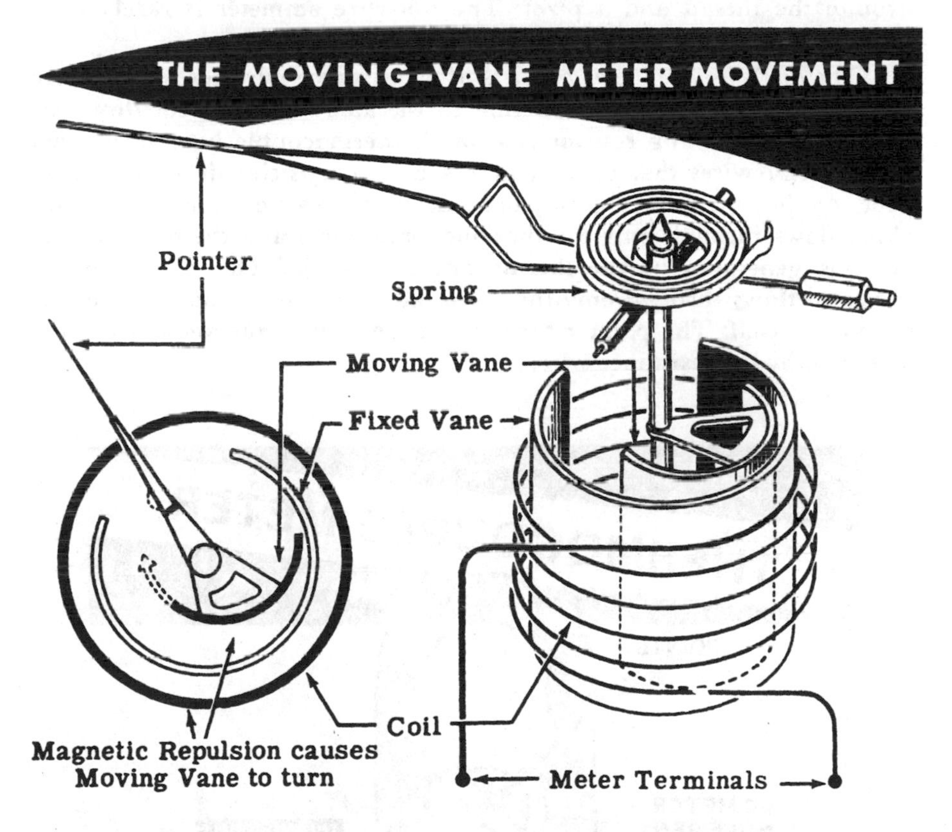

Moving-vane meters may be used for voltmeters, in which case the field coil consists of many turns of fine wire which generate a strong field with only a small current flow. Ammeters of this type use fewer turns of a heavier wire and depend on the larger current flow to obtain a strong field. These meters are generally calibrated at 60 Hz ac but may be used at other ac frequencies. By changing the meter scale calibration, moving-vane meters will also measure dc current and voltage.

Thermocouple Meters and Hot-Wire Meters

Thermocouple meters and hot-wire meters both utilize the heating effect of current flowing through a resistance to cause meter deflection, but each uses this effect in a different manner. Since their operation depends only on the heating effect of current flow, they may be used to measure direct current and alternating current of any frequency.

The *hot-wire* ammeter deflection depends on the expansion of a high resistance wire. This is caused by the heating effect of the wire itself as current flows through it. A resistance wire is stretched taut between the two meter terminals with a thread attached at a right angle to the center of the wire. A spring connected to the opposite end of the thread exerts a constant tension on the resistance wire. Current flow heats the wire, causing it to expand. This motion is transferred to the meter pointer through the thread and a pivot. The hot-wire ammeter is rarely used today.

The *thermocouple* meter consists of a resistance wire across the meter terminals which heats in proportion to the amount of current flow. Attached to this heating resistor is a small thermocouple junction of two unlike metal wires that connect across a very sensitive dc meter movement. As the current being measured heats the resistor, a small current, which flows through only the thermocouple wires and the meter movement, is generated by the thermocouple junction. The current being measured flows through only the resistance wire, not through the meter movement itself. The pointer turns in proportion to the amount of heat generated by the resistance wire.

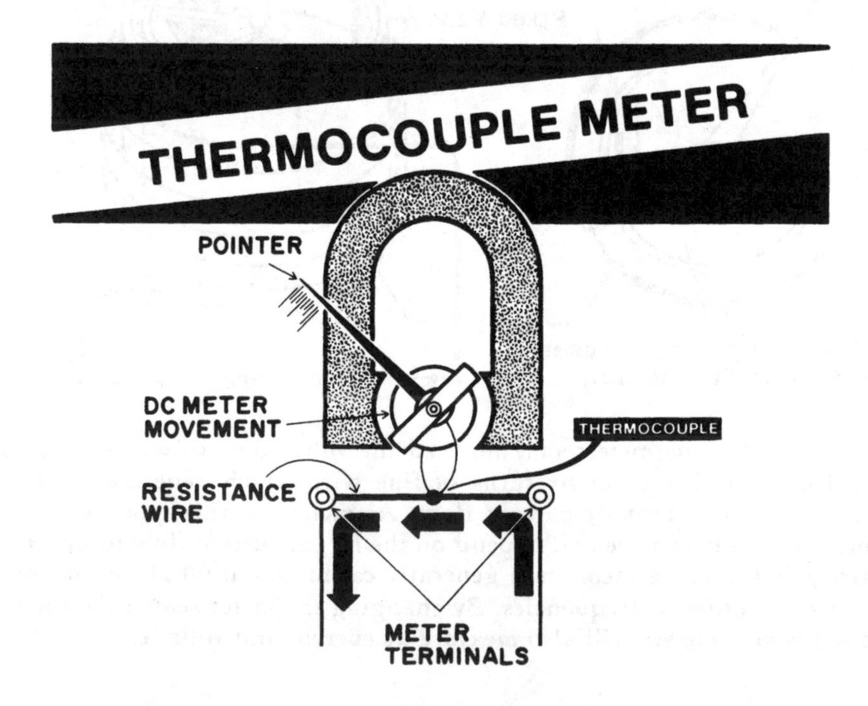

AC Ammeters—Current Transformers

As mentioned earlier, an ac voltmeter (usually the rectifier type) can be connected across a shunt as can dc meters.

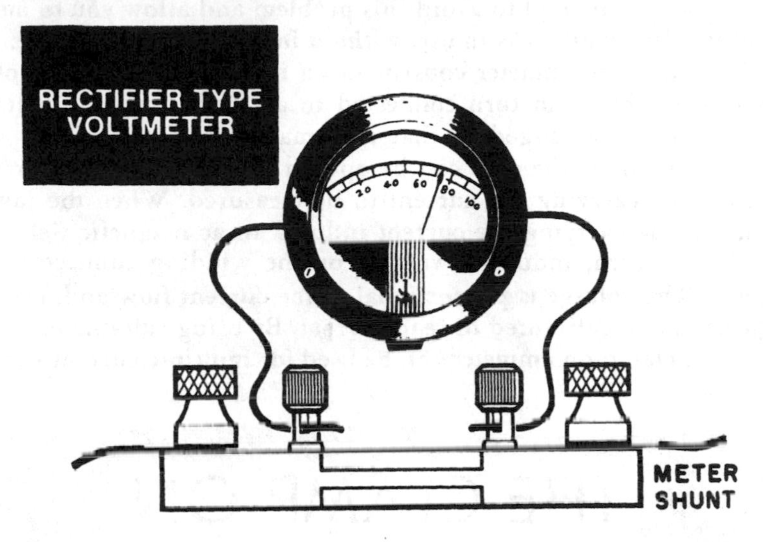

Usually, however, ac ammeters are constructed using the transformer principle, where the line carrying the current is the primary coil, and a many-turn coil is the secondary coil, connected to a rectifier type voltmeter. These transformers are arranged so that the voltage in the winding connected to the voltmeter is proportional to the current flow. The meter face is then calibrated in amperes ac.

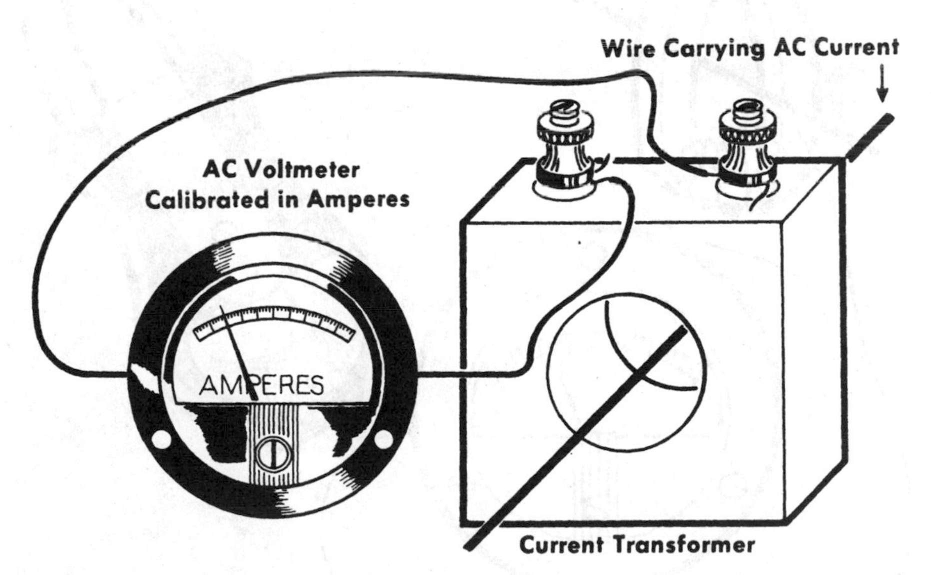

Current transformers are usually used in permanent installations and can be obtained for any desired current rating.

AC Ammeters—The Clamp-on Ammeter

As you learned from the previous discussion on ammeters, you must open the line to put an ammeter in the circuit. An ac ammeter called the *clamp-on ammeter* is used to avoid this problem and allow you to measure current in a line while it is in use, without interrupting the service.

The clamp-on ammeter consists of an iron core with a coil of wire wound on it; this is in turn connected to a rectifier type ac voltmeter. The iron core is arranged so that it is made up of two pieces with a hinge (jaws). Thus, the core can be opened up to allow for insertion of the conductor carrying the current to be measured. When the jaws are closed, the line carrying the current induces an ac magnetic field in the core; that, in turn, induces a voltage on the winding connected to the voltmeter. The voltage is proportional to the current flow and, therefore, the meter can be calibrated to read current. By using suitable meter multipliers, the clamp-on ammeter can be used for multiple current ranges.

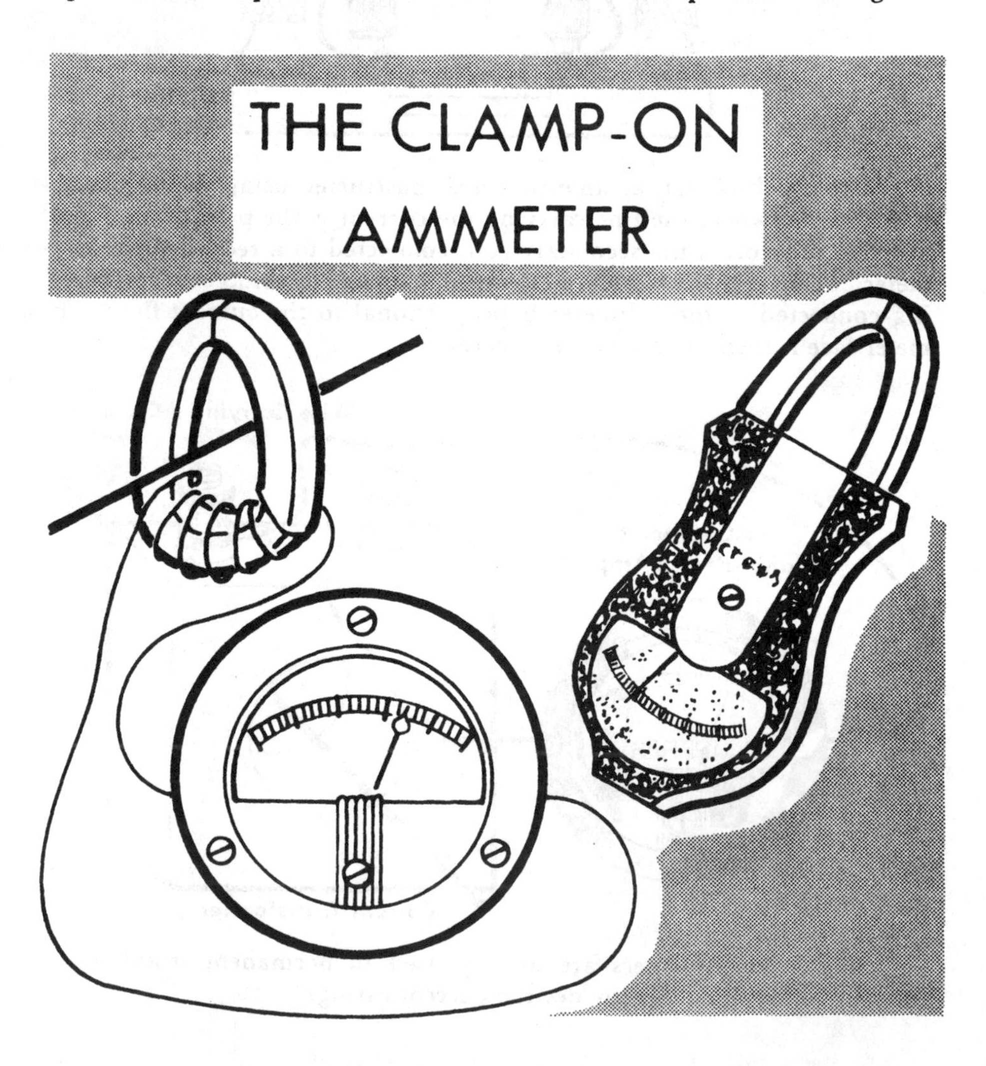

Electrodynamometer Movements

An *electrodynamometer movement* utilizes the same basic operating principle as the basic moving-coil dc meter movement, except that the permanent magnet is replaced by fixed coils to generate the fixed magnetic field. A moving coil to which the meter pointer is attached is suspended between two field coils and connected in series with these coils. The three coils (two field coils and the moving coil) are connected in series across the meter terminals so the same current flows through each.

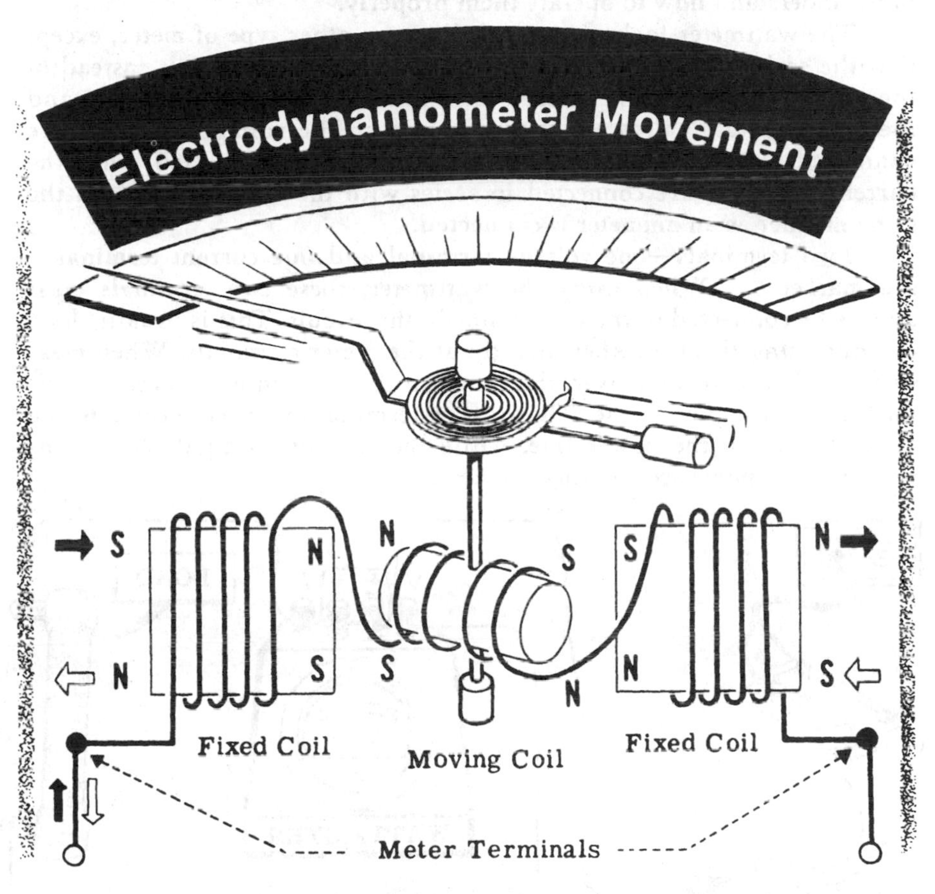

Current flow in either direction through the three coils causes a magnetic field to exist between the field coils. The current in the moving coil causes it to act as a magnet and exert a turning force against a spring. If the current is reversed, the field polarity and the polarity of the moving coil reverse simultaneously, and the turning force continues in the original direction. Since reversing the current direction does not reverse the turning force, this type of meter can be used to measure both ac and dc current. While some voltmeters and ammeters use the dynamometer principle of operation, its most important application is in the wattmeter, about which you will find out next.

Wattmeters

While power may be computed from the measured effective values of E and I in ac circuits containing only resistance, it can also be measured directly with a *wattmeter*. Wattmeters are not used as frequently as the meters with which you are familiar—voltmeters, ammeters, and ohmmeters—but in order to find out about ac circuits, you will need to use them. Wattmeters work differently than the meters you have already used and are easily damaged if connected incorrectly. Therefore, you must understand how to operate them properly.

The wattmeter looks very much like any other type of meter, except that the scale is calibrated in watts and it has four terminals instead of the usual two. Two of these terminals are called the *voltage terminals* and the other two are called the *current terminals*. The voltage terminals are connected across the circuit exactly as a voltmeter is connected, while the current terminals are connected in series with the circuit current in the same manner as an ammeter is connected.

Two terminals—one voltage terminal and one current terminal—are marked ±. When using the wattmeter, these two terminals must always be connected to the same point in the circuit. This is usually done by connecting them together directly at the meter terminals. When measuring either ac or dc power, the common ± junction is connected to either side of the power line. The voltage terminal (V) is connected to the opposite side of the power line, while the current terminal (A) is connected to the power-consuming load.

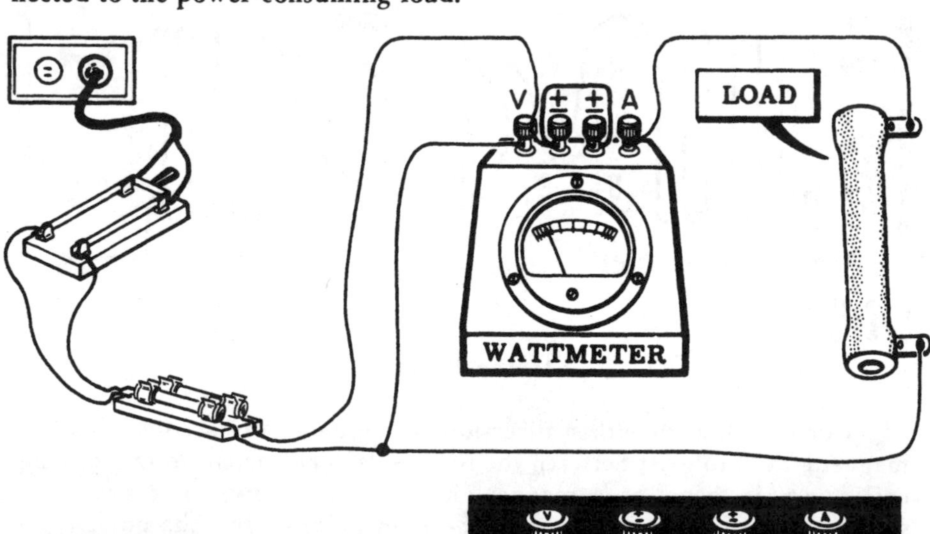

The Wattmeter...
CONNECTED TO MEASURE POWER USED IN LOAD RESISTANCE

Wattmeters (continued)

Wattmeters are not constructed with a D'Arsonval or Weston basic meter movement. Instead, they use a *dynamometer* type movement, which differs from the other types in that it has no permanent magnet to furnish the magnetic field, as you just learned.

These field coils are connected in series across the wattmeter current terminals, so that the circuit current flows through the coils when measurements are being made. A large circuit current makes the field coils act as strong magnets, while a small circuit current makes them act as weak magnets. Since the strength of the meter's magnetic field depends on the value of the circuit current, the wattmeter reading varies as the circuit current varies.

The current in the moving coil—the voltage coil—is dependent on the circuit voltage. This turning force is dependent on both the moving coil current and the field coil current for a fixed current in the moving coil, and the turning force and meter reading depend only on the circuit current.

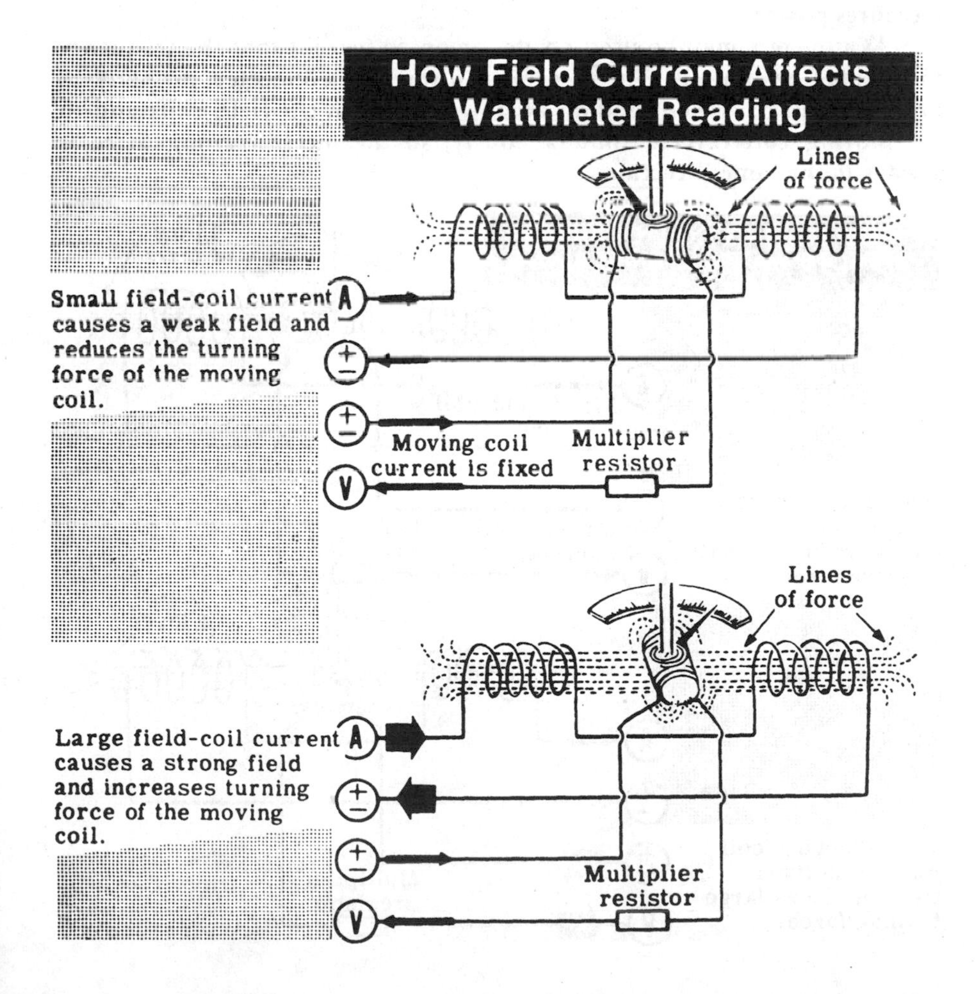
How Field Current Affects Wattmeter Reading
Lines of force
Small field-coil current causes a weak field and reduces the turning force of the moving coil.
A
V
Moving coil current is fixed
Multiplier resistor
Lines of force
Large field-coil current causes a strong field and increases turning force of the moving coil.
A
V
Multiplier resistor

Wattmeters (continued)

The moving coil of a wattmeter is like those coils used in the basic meter movement and is connected in series with an internal multiplier resistor to the voltage terminals of the wattmeter. The voltage terminals are connected across the circuit voltage in the same manner as a voltmeter, and the multiplier resistor limits the current flow through the moving coil. Since the resistance of the multiplier is fixed, the amount of current flow through it and the moving coil varies with the circuit voltage. A high voltage causes more current to flow through the multiplier and moving coil than a low voltage does.

For a given magnetic field, determined by the amount of circuit current (that which flows through the load), the turning force of the moving coil depends on the amount of current flowing through the moving coil. Since this current depends on the circuit voltage, the meter reading will vary as the circuit voltage varies. Thus, the meter reading depends on both the circuit current and the circuit voltage and will vary if either changes. Since power depends on both voltage and current, the meter measures power.

Wattmeters may be used on dc, or on ac, except that the frequency range may be limited. But they must always be connected properly to prevent damage. When used on ac, the currents in the field coils and in the moving coil reverse simultaneously, so the meter turning force is always in the same direction.

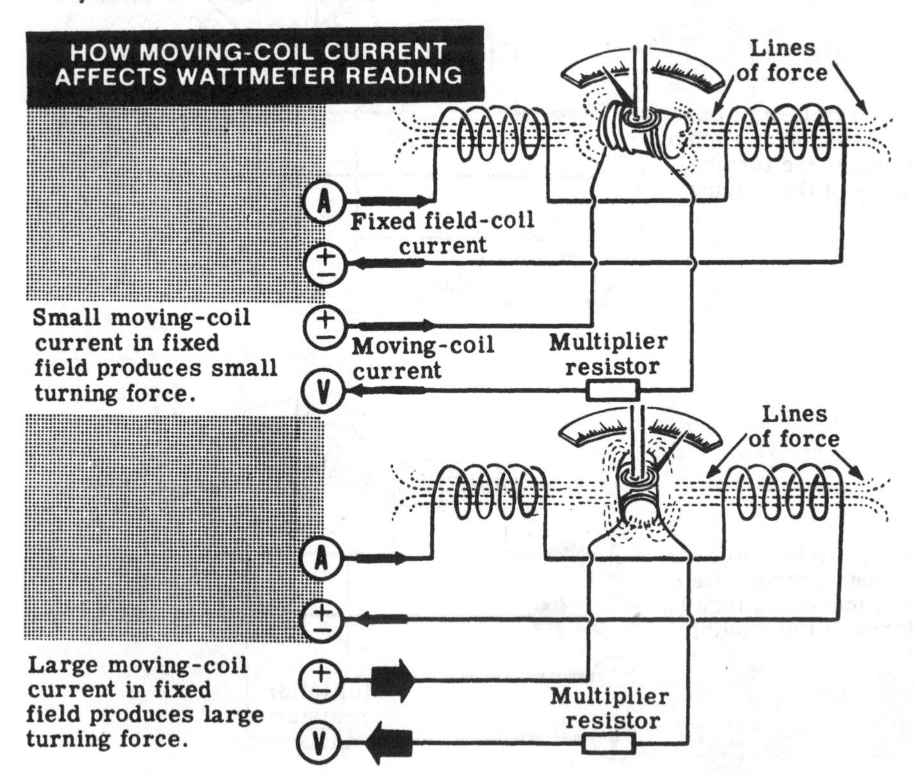

Review of AC Meters

To review the principles and construction of ac meters, suppose you compare the various meter movements and their uses. Although there are other types of meters used for ac, you have found out about those most commonly used.

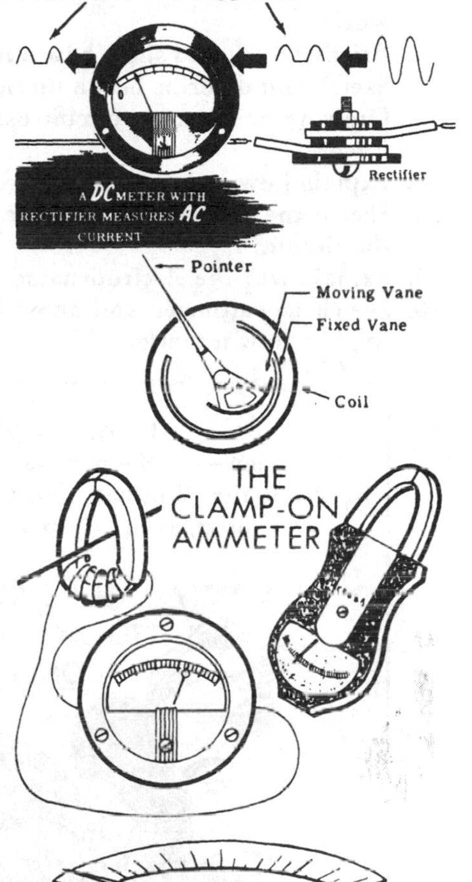

1. RECTIFIER TYPE AC METER—A basic dc meter movement with a rectifier connected to convert ac to dc. Most commonly used ac voltmeter.

2. MOVING-VANE METER—Meter which operates on the magnetic repulsion principle, using one movable and one fixed vane. Can be used on ac or dc to measure either voltage or current.

3. AC AMMETER—An ac voltmeter operating with a shunt, as with dc ammeters.
4. CLAMP-ON AMMETER—An ac ammeter that couples to a line without opening the line.

5. ELECTRODYNAMOMETERS—A device commonly used in wattmeters rather than voltmeters and ammeters. Basic principle is identical to that of a D'Arsonval movement except that field coils are used instead of a permanent magnet.

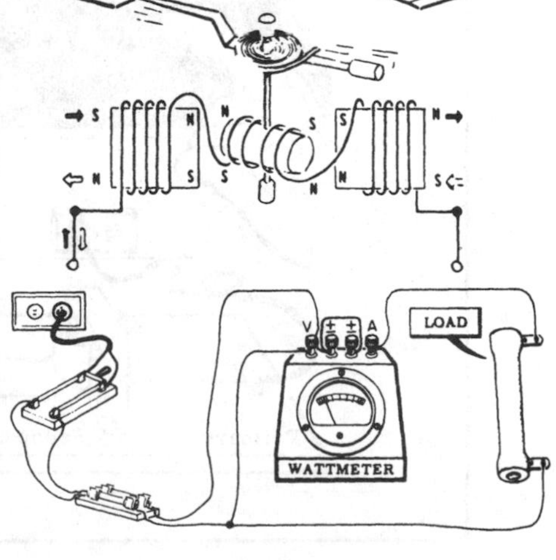

6. WATTMETER—A meter that uses the electrodynamometer movement with the windings separated into current and voltage coils. Can measure both dc and ac power.

Self-Test—Review Questions

1. Sketch the basic ac voltmeter circuit using a single rectifier.
2. Show how the circuit shown can use a second rectifier to equalize the loading on the second half cycle.
3. Show the construction of a copper oxide rectifier. Describe how it works.
4. Briefly sketch and show how a moving-vane meter works.
5. Sketch and describe how a thermocouple meter works.
6. Draw an ac ammeter circuit using a current transformer. Describe how it works.
7. Explain how a clamp-on ammeter works. Where would it be used?
8. Sketch the electrodynamometer movement. Show how it connects with the circuit.
9. Explain why the electrodynamometer works on ac.
10. Sketch a wattmeter and show how it is connected to measure power. Explain how it works.

Learning Objectives—Next Section

Overview—In the next section, you will learn about *resistance* in ac circuits. You will find that Ohm's Law and Kirchhoff's Laws can be used, as in dc circuits.

ALL DC RULES AND LAWS APPLY TO AC CIRCUITS CONTAINING ONLY RESISTANCE
E=IR
I=E/R
R=E/I
I1+I2=I3
E1+E2+E3=ET
Primary
Secondary

Experiment/Application—AC Voltmeter

Although calibrated to read the effective value of ac voltages, ac voltmeters can also be used to measure the approximate value of a dc voltage. To show how the effective value of an ac voltage compares to a dc voltage, you can use an ac voltmeter to measure both the dc voltage of a 7.5-volt battery and the effective ac voltage output of a 6.3-volt transformer.

Five dry cells are connected to form a 7.5-volt battery, and the 0-25 volt ac voltmeter is used to measure the voltage across the battery terminals. You see that the meter reading is approximately 7.5 volts, but the reading is not as accurate as it would be if a dc voltmeter were used.

MEASURING A BATTERY VOLTAGE WITH AN AC VOLTMETER

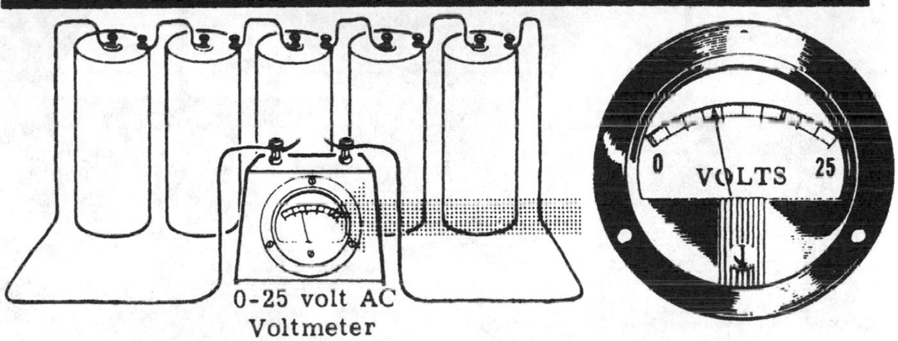

Next, you could connect the 120-volt primary lead of the transformer across the ac power line. The voltage across the secondary leads is then measured with the ac voltmeter, and you see that it is approxmately 7.5 volts. Although the transformer is rated at 6.3 volts ac, the secondary voltage will always be higher than its rated value when the transformer is not furnishing power. The size of the load determines the exact value of the secondary voltage. When comparing the measured voltages—7.5 volts dc and 7.5 volts ac—you find that the two meter readings are nearly the same. Some difference in the readings should be expected, as the ac voltage is approximately 7.5 volts rms, while the dc voltage is exactly 7.5 volts.

MEASURING THE TRANSFORMER'S SECONDARY VOLTAGE

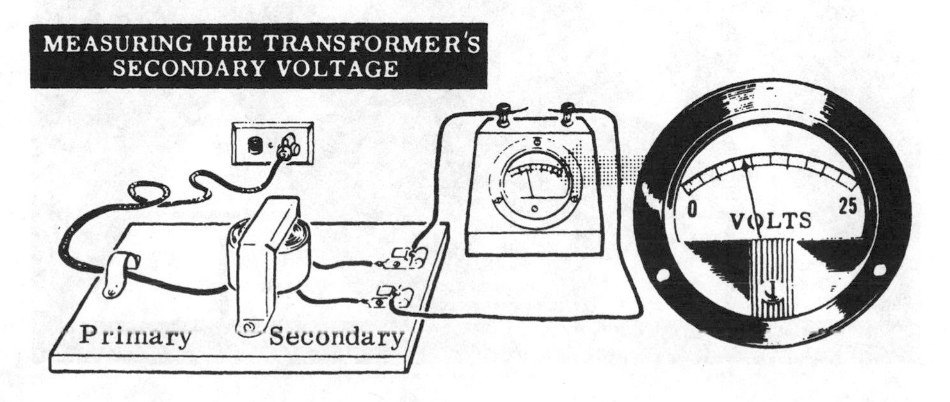

AC Circuits Containing Resistance Only

In ac circuits that contain *only resistance*, the *same* relationship of current, voltage, and resistance that we saw in Ohm's Law and Kirchhoff's Laws for dc circuits will continue to apply.

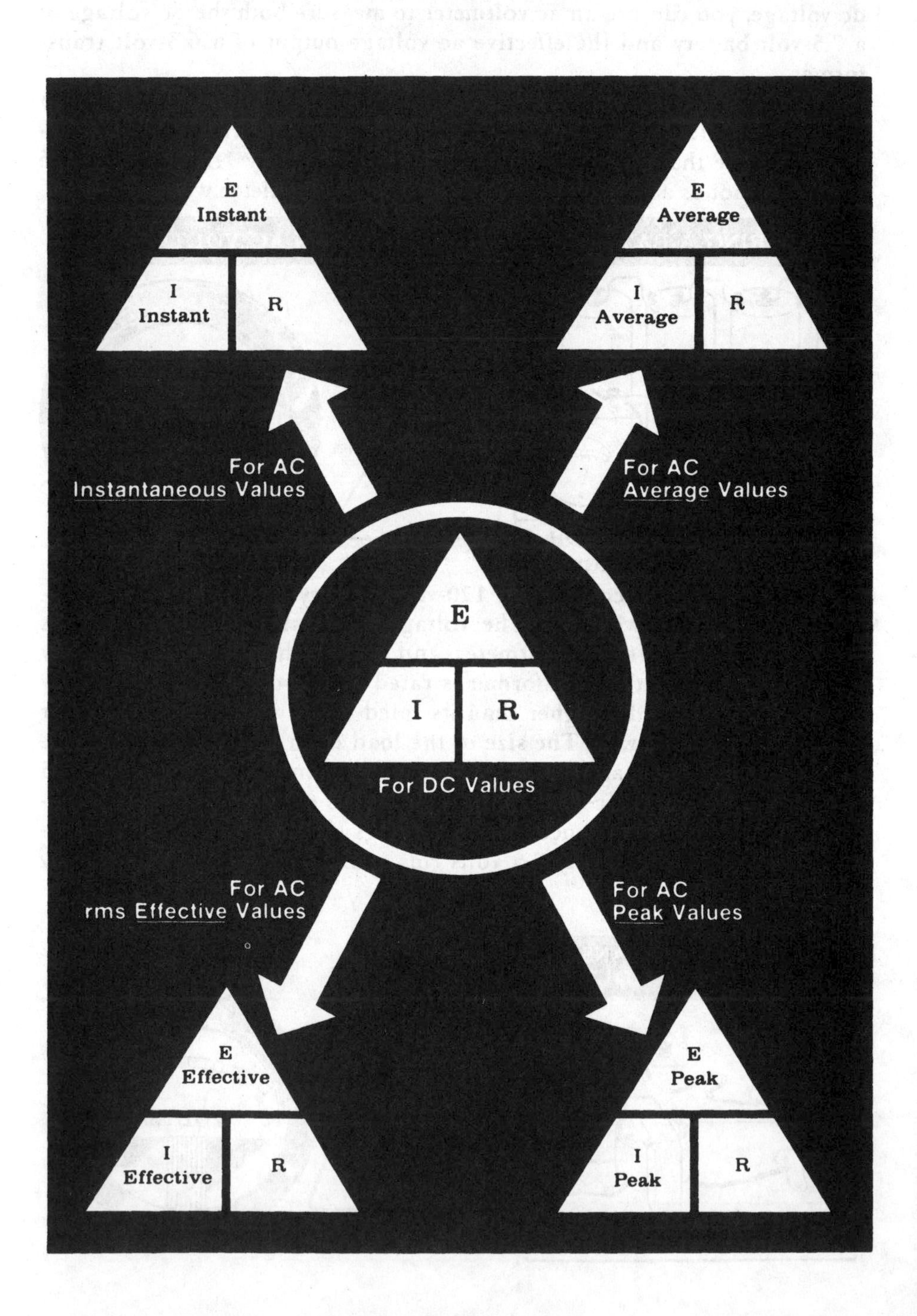

AC Circuits Containing Resistance Only (continued)

Many ac circuits consist of *pure resistance*, and for such circuits the same rules and laws apply as for dc circuits. Pure resistance circuits are made up of electrical devices which contain no inductance or capacitance. You will find out more about inductance and capacitance a little later. Devices such as resistors, lamps, and heating elements have negligible inductance and capacitance, and for practical purposes are considered to be made up of pure resistance. When only these devices are used in an ac circuit, Ohm's Law, Kirchhoff's Laws, and the circuit rules for voltage, current, and power can be used exactly as in dc circuits.

When using the circuit laws and rules, you must use rms or effective values of ac voltage and current. *Unless otherwise stated, all ac voltage and current values are given as rms or effective values.* Other values, such as peak-to-peak voltages measured on an oscilloscope, must be changed to rms or effective values before being used in circuit computations.

ALL DC RULES AND LAWS APPLY TO AC CIRCUITS CONTAINING ONLY RESISTANCE
E=IR
I₁+I₂=I₃
I=E/R
E₁+E₂+E₃=Eₜ
R=E/I
Primary
Secondary

Current and Voltage in Resistive Circuits

When an ac voltage is applied across a resistor, the voltage increases to a maximum with one polarity, decreases to zero, increases to a maximum with the opposite polarity, and again decreases to zero. This completes a cycle of voltage. The current flow follows the voltage exactly: as the voltage increases, the current increases proportionally; when the voltage decreases, the current decreases proportionally; and at the moment the voltage changes polarity, the current flow reverses its direction. Because of this, we say that the voltage and current waves are *in phase*.

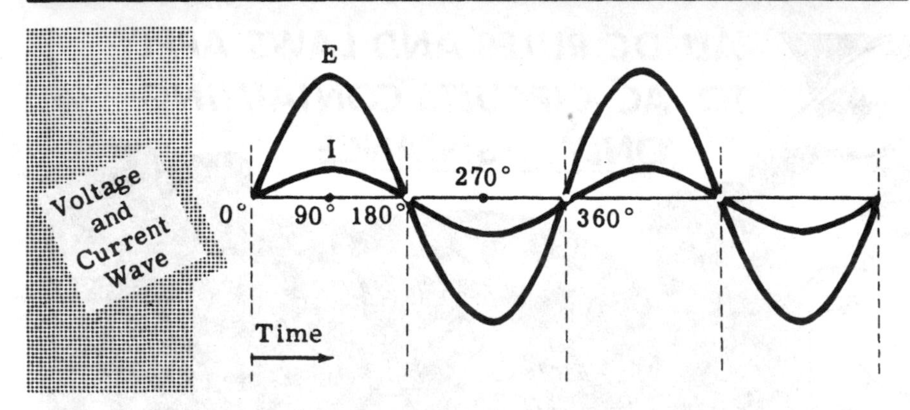

Sine waves of voltages or currents are in phase whenever they are of the same frequency and pass through zero simultaneously, both going in the same direction. The amplitude of two voltage waves or two current waves which are in phase are not necessarily equal, however. In the case of in phase current and voltage waves, they are not equal since they are measured in different units. In the circuit shown below, the voltage is 6.3 volts (8.6-V peak), resulting in a current of 2 amperes, and the voltage and current waves are in phase.

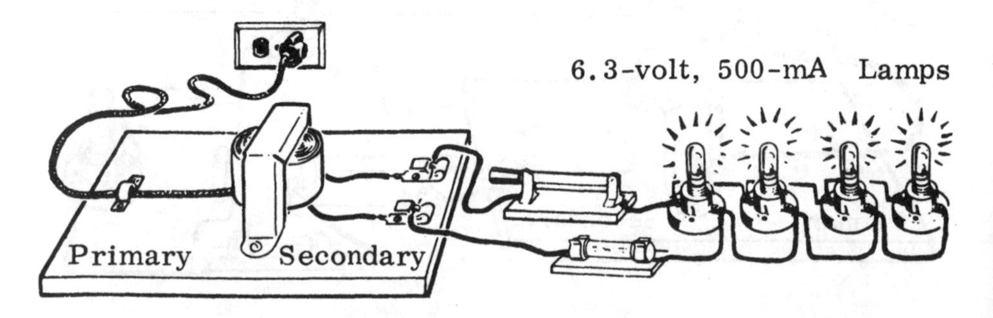

Power in AC Circuits

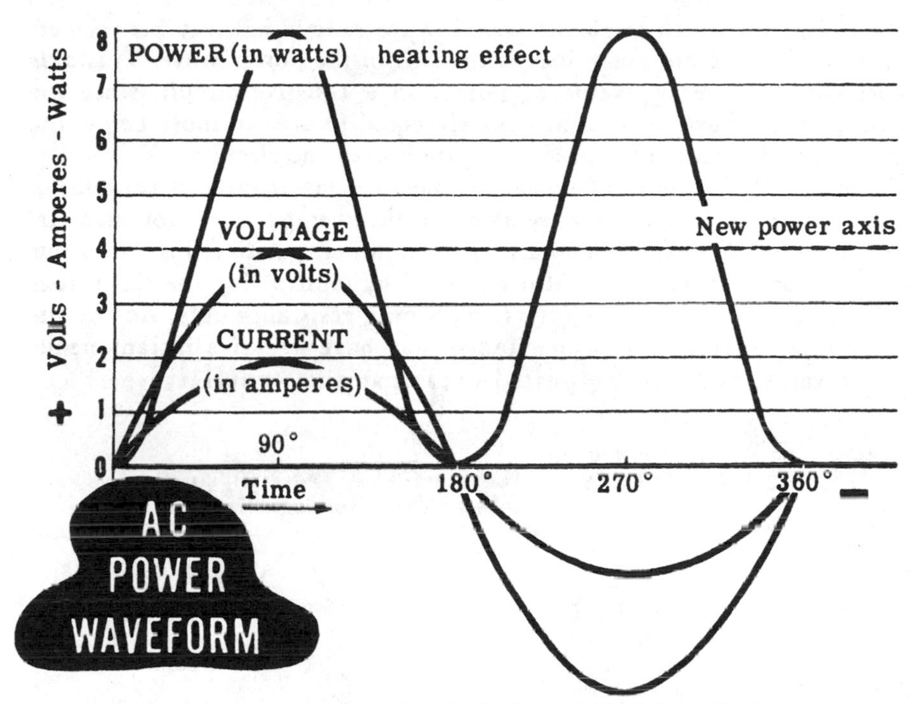

The power used in an ac circuit is the *average* of all the instantaneous values of power or heating effect for a complete cycle. To find the power, all of the corresponding instantaneous values of voltage and current are multiplied together to find the instantaneous values of power, which are then plotted for the corresponding times to form a *power curve.* The average of this power curve is the actual power used in the circuit.

For in-phase voltage and current waves, as in resistive circuits, all of the instantaneous powers are above the zero axis, and the entire power curve is above the zero axis. This is due to the fact that whenever two positive values are multiplied together the result is positive, and whenever two negative values are multiplied together the result is also positive. Thus, during the first half cycle of E and I, the power curve increases in a positive direction from zero to a maximum, and then decreases to zero just as the E and I waves do. During the second half cycle, the power curve again increases in a positive direction from zero to maximum, and then decreases to zero while E and I both increase and decrease in the negative direction. Notice that if a new axis is drawn through the power wave, halfway between its maximum and minimum values, the power wave frequency is twice that of the voltage and current waves.

Note that when two numbers—each being less than 1—are multiplied together, the result is a smaller number than either of the original numbers. For example. 0.5 volt × 0.5 ampere = 0.25 watt. For that reason, some or all of the instantaneous values of a power wave may be less than those for the current and voltage waves.

Power in Resistive Circuits

A line drawn through the power wave exactly halfway between its maximum and minimum values is the *axis* of the power wave. This axis represents the average value of power in a resistive circuit, since the shaded areas above the axis are exactly equal in area to those below the axis. *Average power* is the *actual power* used in any ac circuit.

Since all the values of power are positive for ac circuits containing only resistances, the power wave axis and the average power for such circuits is equal to exactly one-half the maximum, positive, instantaneous power value. This value can also be found by multiplying the rms values of E and I together for ac circuits containing resistance only. Ac circuits containing inductance or capacitance may have negative instantaneous power values and must be treated somewhat differently.

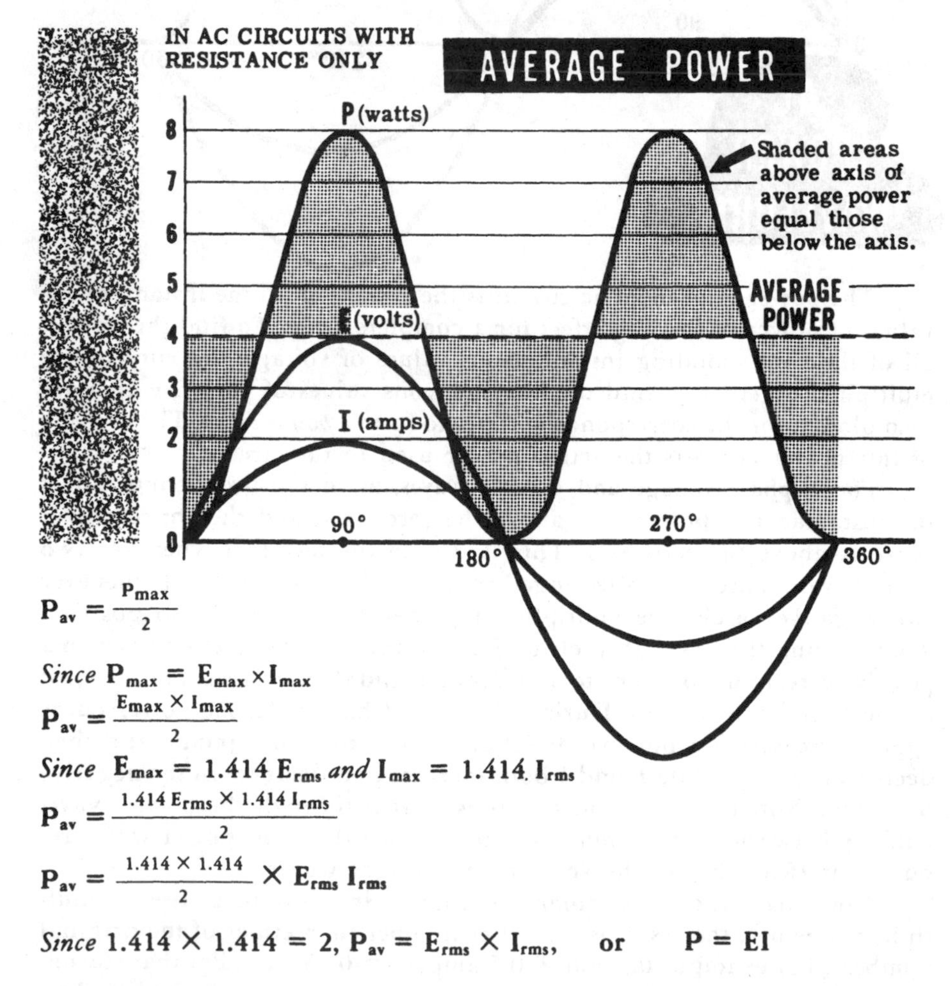

$$P_{av} = \frac{P_{max}}{2}$$

Since $P_{max} = E_{max} \times I_{max}$

$$P_{av} = \frac{E_{max} \times I_{max}}{2}$$

Since $E_{max} = 1.414\ E_{rms}$ *and* $I_{max} = 1.414\ I_{rms}$

$$P_{av} = \frac{1.414\ E_{rms} \times 1.414\ I_{rms}}{2}$$

$$P_{av} = \frac{1.414 \times 1.414}{2} \times E_{rms}\ I_{rms}$$

Since $1.414 \times 1.414 = 2$, $P_{av} = E_{rms} \times I_{rms}$, or $P = EI$

NOTE: The power can also be calculated as $(I_{rms})^2R$ for any circuit with resistance.

Power Factor—Volt-Amperes

When I_{rms} and E_{rms} are in phase, their product is power in *watts*, the same as in dc circuits. As you will find out later when you study inductance and capacitance, the product of I_{rms} and E_{rms} is not always power in watts, but is called *volt-amperes*, often abbreviated VA. This distinction is necessary in ac circuits because some of the current drawn by circuits with inductance and/or capacitance may not be used. The real power in watts, however, is always given by I^2R, E^2/R, or power used in the resistive part of the circuit.

As you will learn, while a source may use volts and amperes, the power in watts may be small or zero. This is because the phase between the current and voltage may not be the same when inductance and/or capacitance are added. The ratio between the power in watts in a circuit and the volt-amperes in a circuit is called the *power factor*. In a pure resistive circuit, power in watts is equal to $I_{rms} \times E_{rms}$, so power factor in a pure resistive circuit is equal to power in watts divided by volt-amperes which equals 1. Power factor is also expressed in percent or as a decimal.

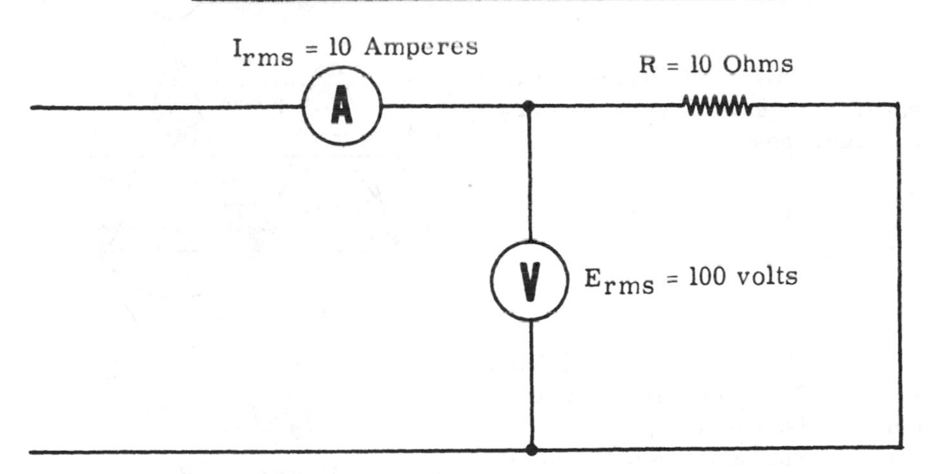

I^2R or $E^2/R = E_{rms} \times I_{rms}$
I^2R or E^2/R = Watts

$$\text{Power Factor} = \frac{I^2R}{I_{rms} \times E_{rms}} \text{ or } \frac{E^2/R}{I_{rms} \times E_{rms}} = \frac{1000}{1000} = 1.0 \text{ or } 100\%$$

$$\text{Power Factor} = \frac{\text{Watts}}{\text{Volts-Amperes}}$$

Power Factor = 1.0 or 100% in a pure resistive circuit.

Review of Resistance in AC Circuits

Suppose you review some of the facts concerning ac power and power in resistive circuits. You have already learned these facts and they will help you to understand other ac circuits.

1. AC CIRCUIT WITH RESISTANCE ONLY—A circuit that obeys all of Ohm's and Kirchhoff's Laws that you already know.

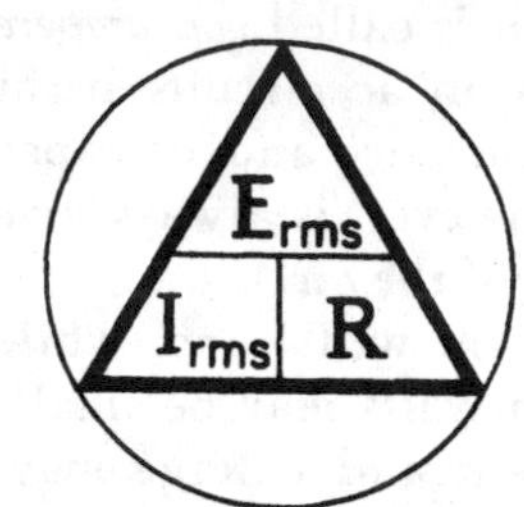

2. AC POWER WAVE—A pictorial graph of all values of instantaneous power.

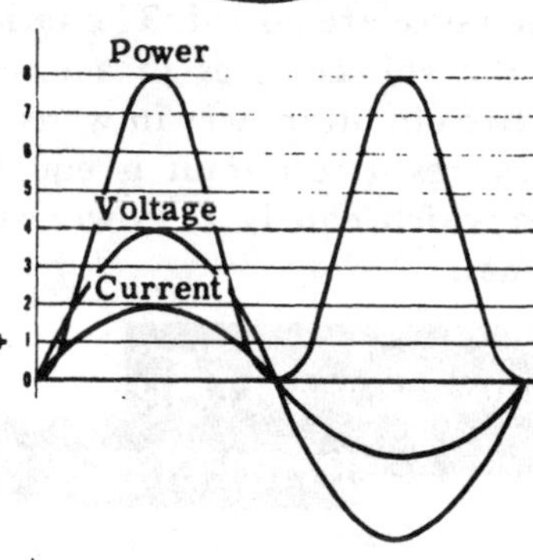

3. AVERAGE POWER—A value equal to the axis of symmetry drawn through a power wave.

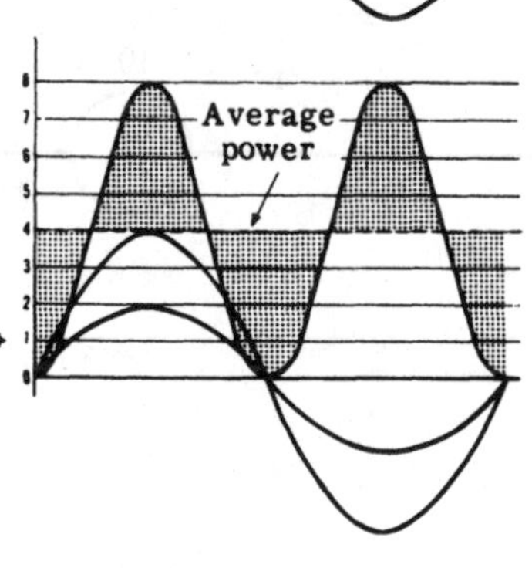

4. POWER IN RESISTIVE CIRCUITS—The power in resistive circuits is given by the product of the rms current and voltage, or the square of the rms current times the resistance.

$$P = E_{rms} I_{rms} \text{ or } I_{rms}^2 R$$

5. VOLT-AMPERES—The product of the circuit voltage and current.

$$VA = E_{rms} I_{rms}$$

6. POWER FACTOR—The ratio of the power to VA.

$$\text{Power Factor} = \frac{\text{Watts}}{\text{VA}}$$

$$= 1 \text{ or } 100\% \text{ in resistive circuits}$$

Self-Test—Review Questions

1. Calculate the current (rms) in the following circuits with 120 volts ac applied.

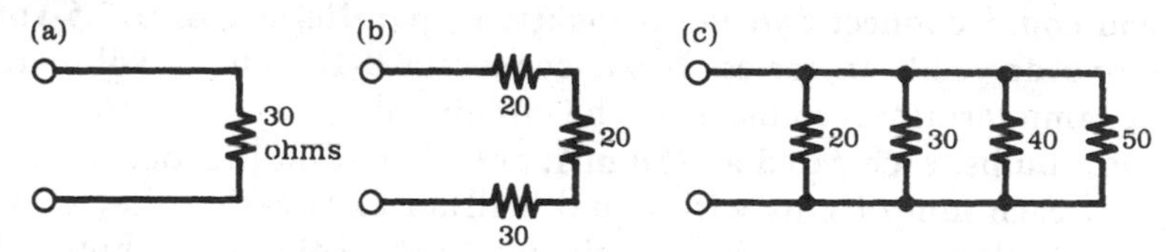

2. Given the values shown below, calculate the unknowns.

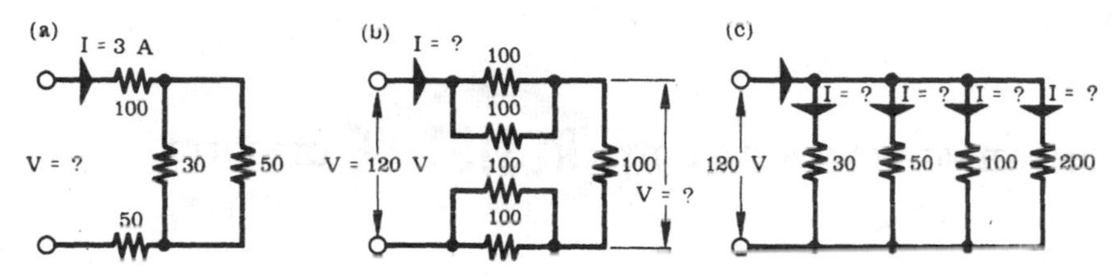

3. For the circuits in question 1, calculate the power in each resistance.
4. For the circuits in question 2, calculate the power in each resistance.
5. Define VA and power factor.
6. For the circuits in question 1, calculate the VA and total power. Show that the power factor is 1.
7. For the circuits in question 2, calculate the VA and total power.
8. Show the phase relationships between voltage and current in resistive circuits.
9. Given the circuits shown below, calculate the unknowns.

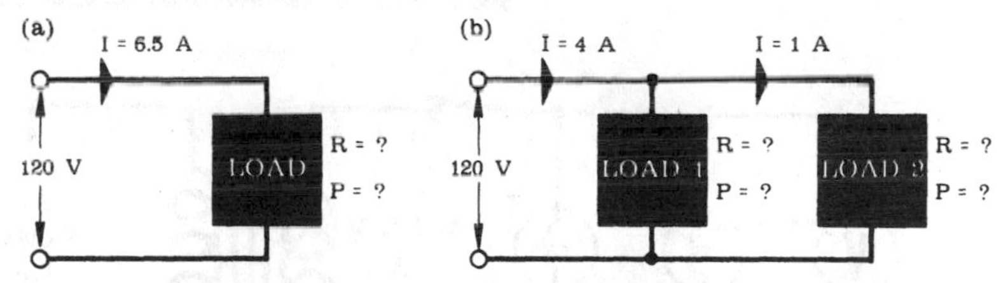

10. For the circuits in question 9, calculate VA, total power, and power factor.

Learning Objectives—Next Section

Overview—In the next section, you will learn about the property of wires and coils carrying current that is called *inductance*. You will learn that the generation or collapse of the magnetic field around these wires and coils strongly influences what happens when you try to change the current flow.

Experiment/Application—Power in Resistive AC Circuits

To show that rms values of ac voltage and current can be used to determine the power used in resistive circuits in the same manner as dc values, you could connect two lamp sockets in parallel across a 7.5-volt battery—five dry cells in series. Next, connect a 0-10-volt dc voltmeter across the lamp terminals to measure the circuit voltage.

Six-volt bulbs, each rated at 250 mA, are inserted in the sockets, and you see that each lamp lights with equal brilliance. Together they allow 0.5 ampere to flow through the circuit, while the voltage is about 7.5 volts. Using the power formula, P = EI, the power then is 7.5 × 0.5, or 3.75 watts.

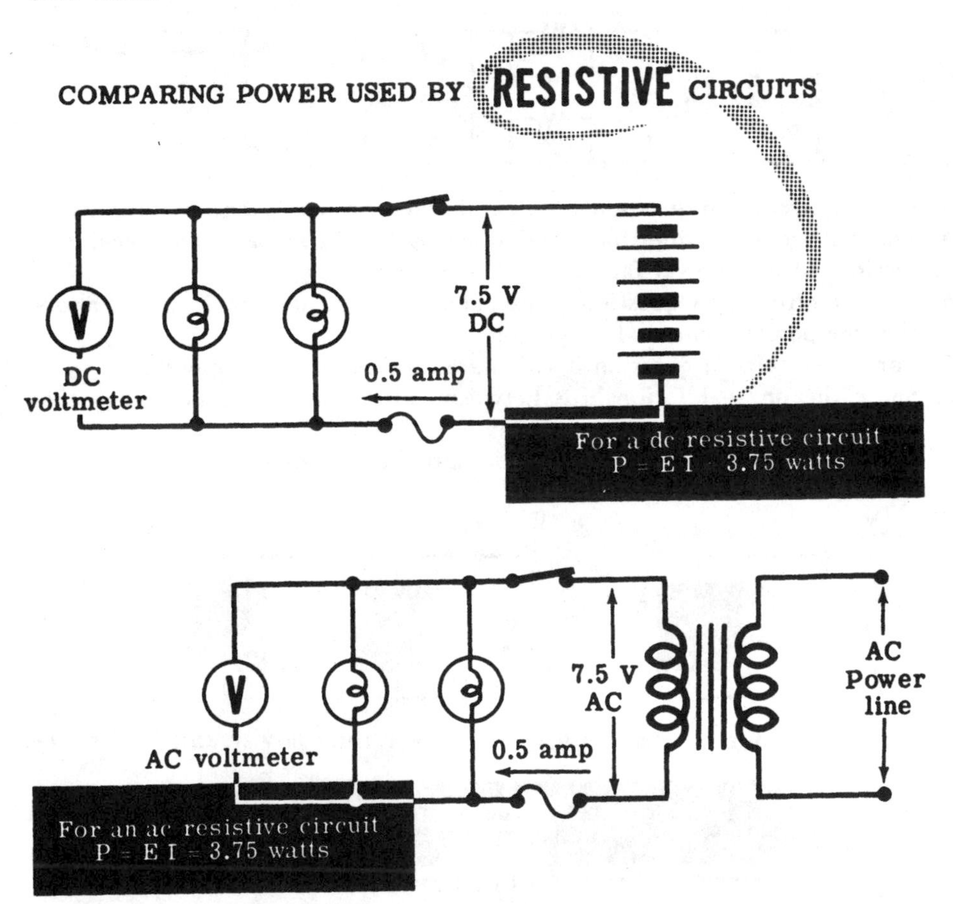

Next, the battery is disconnected and the dc voltmeter is replaced with an ac meter of the same range. The 7.5-volt (rms) transformer is used as an ac voltage source, and you see that the lamps light as brightly as they did in the dc circuit. Notice that the voltmeter reading is almost the same as that obtained using dc, about 7.5 volts.

Applying the power formula, the rms ac power is 7.5 × 0.5, or 3.75 watts, equal to the dc power and producing the same amount of light.

Experiment/Application—Power in Resistive AC Circuits (continued)

Wattmeters with a range of less than 75 watts are not generally available and, since it would be difficult to read 3 or 4 watts on a standard 0-75-wattmeter scale, a larger amount of power is used to demonstrate power measurement with a wattmeter. To obtain a larger amount of power, you could use the 120-volt ac power line as a power source through a step-down autotransformer, which provides a voltage of about 60 volts ac. You will measure the power used by a resistor, first using a voltmeter and milliammeter and then a wattmeter.

Suppose you connect the DPST knife switch and the DP fuse holder in the line cord, as shown below, and insert 1-ampere fuses in the fuse holder. With a 0-1-ampere ac ammeter connected in series with one of its leads, the line cord is connected across a 150-ohm, 100-watt resistor. Then a 0-150-volt range ac voltmeter is connected directly across the terminals of the resistor to measure resistor voltage. The line cord plug is inserted in the transformer outlet, and with the switch closed, the line voltage indicated on the voltmeter is about 60 volts; also, the 150-ohm resistor allows a current flow of about 0.40 ampere, as measured by the ammeter. The resistor becomes hot due to the power being used, so the switch is opened as soon as the readings have been taken. The current reading may vary slightly as the heated resistor changes in resistance value, so an average current reading is used.

Computing the power used by the resistor, you see that it is approximately 24 watts. Assuming that the voltage is 60 volts and the current is exactly 0.40 ampere, the power is then 60 × 0.40 or 24 watts. The actual results may be slightly different, depending on the exact voltage and current readings which are obtained. The power factor is also 1 since $I^2R/VA = 24/24 = 1$.

COMPUTING AC POWER USED BY A RESISTOR

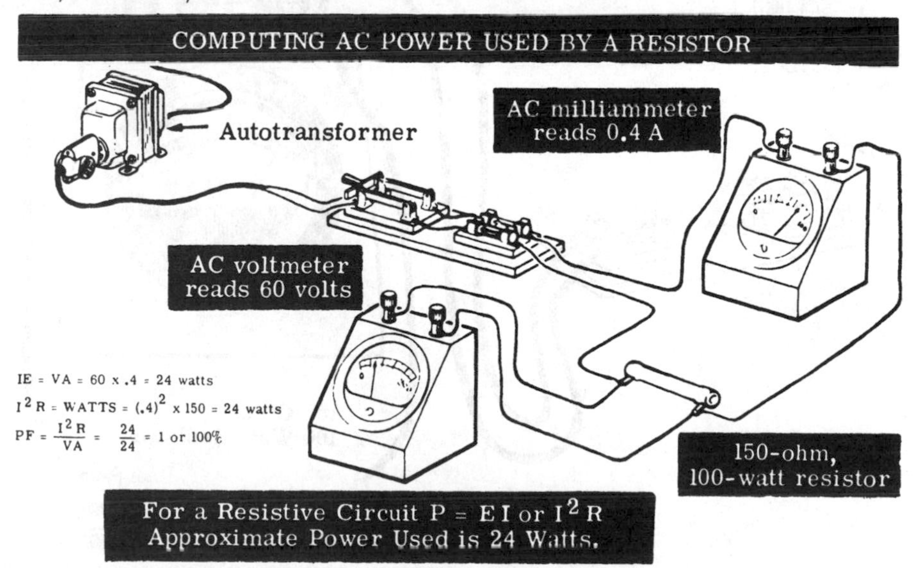

Experiment/Application—Power in Resistive AC Circuits (continued)

Now the ammeter and voltmeter are removed from the circuit and the wattmeter is connected to measure directly the power used by the resistor. The current and voltage ± terminals of the wattmeter are connected together with a short jumper wire to form a common ± terminal. One lead from the fuse block is then connected to this common ± terminal, and the other fuse block lead is connected to the remaining voltage terinal, marked "V." Wires are connected to each end of the resistor and these are, in turn, connected to the wattmeter—one to the voltage terminal V and the other to the current terminal A.

When the connections are completed, the autotransformer is connected to the ac power outlet and the switch is closed. You see that the wattmeter indicates that about 24 watts of power are being used. The wattage reading will vary slightly as the resistor heats and changes value, but will become steady when the resistor temperature reaches a maximum. Observe that the measured power is about the same as that obtained when using a voltmeter and ammeter. The two results can be considered equal for all practical purposes.

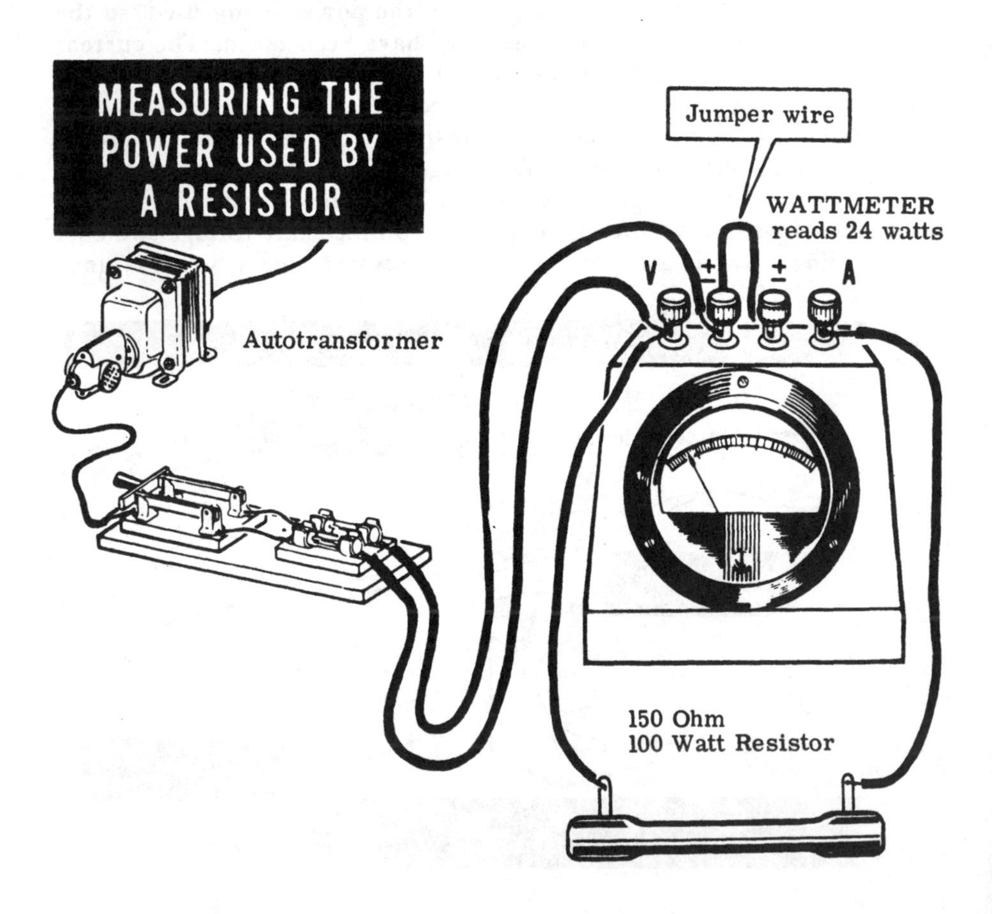

Magnetic Fields around a Conductor

As you know, an electric current is made up of electrons moving in a coordinated direction in a wire. Each of the free electrons has a magnetic field and since they are moving in the same direction at a given instant, these fields combine to produce the field around the wire. There is *no* magnetic field around a wire not carrying current because the electrons are moving at random and the magnetic fields from those electrons cancel each other out.

NO CURRENT FLOW

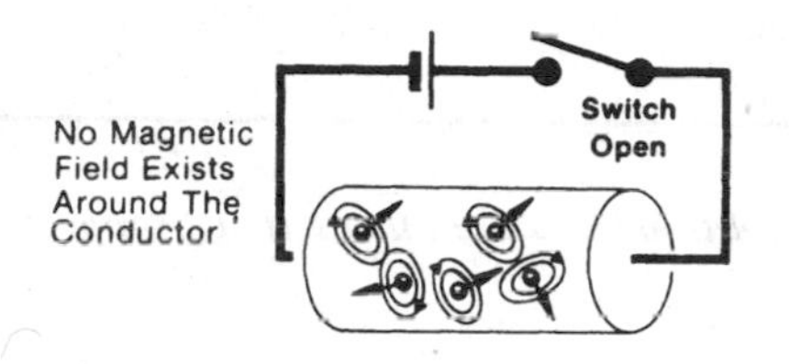

CURRENT FLOW

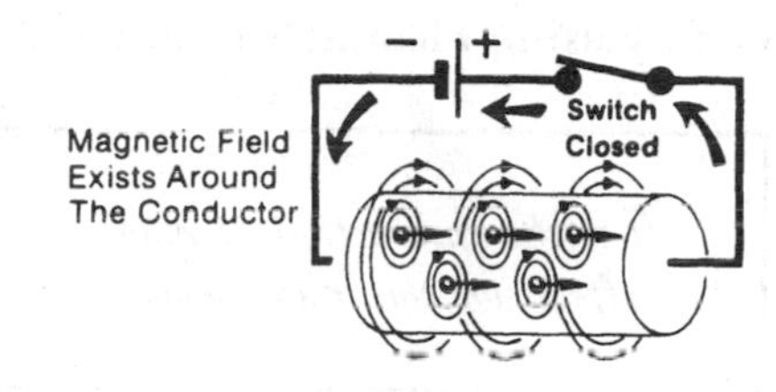

The direction of the magnetic field around a current carrying conductor can be found by the *left-hand rule* that you learned about in Volume 1, page 1-78. If your left hand is wrapped around the conductor with your thumb in the direction of the current flow, the magnetic field around the wire will be in the direction of your fingers.

You also know from Volume 1 that there is a current induced in a wire *moving* through a magnetic field. This occurs because the magnetic field exerts its influence on the free electrons to generate an orderly movement of electrons. This separation of charges induced by the relative motion between the conductor and the magnetic field is called *induced emf*. The magnitude of induced emf depends on the *strength* of the *field*, the *length* of the *conductor*, the *direction* of the *conductor* relative to the field, and the *rate* that the *conductor* moves through the field. If the conductor is moved *parallel* to the field so that *no* lines of force are cut, there is *no* induced emf. Fleming, an early scientist, also found that there is a definite relationship between the direction of the magnetic field, the direction of current in the conductor, and the direction in which the conductor tends to move. This relationship is called *Fleming's right-hand rule for motors*.

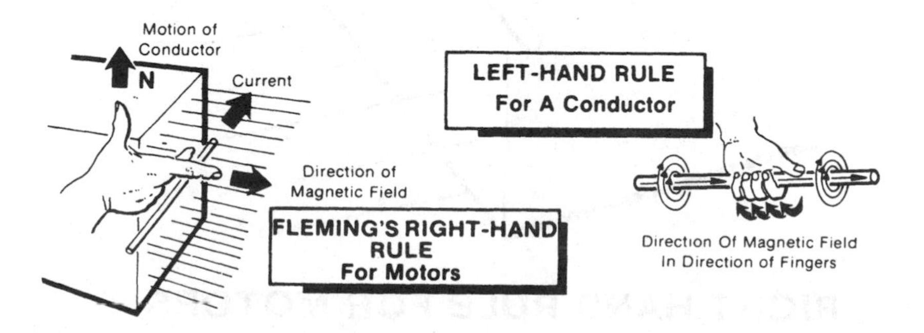

Magnetic Field around a Conductor—Fleming's Right-Hand Rule

As previously stated, Fleming found that there is a definite relation between the direction of the magnetic field, the directionof current in the conductor, and the direction in which the conductor tends to move.

If the thumb, index finger, and third finger of the right hand are extended at right angles to each other, and if the hand is placed so that the thumb points out the direction of motion of the conductor, and the third finger points out the direction taken by the current through the conductor, then the *index finger* will point out the *direction* taken by the *flux lines* of the *magnetic field.* You can also find the direction of induced emf by using Fleming's right-hand rule.

> *It is the interaction of magnetic fields with a conductor that produces the effects we call inductance.*

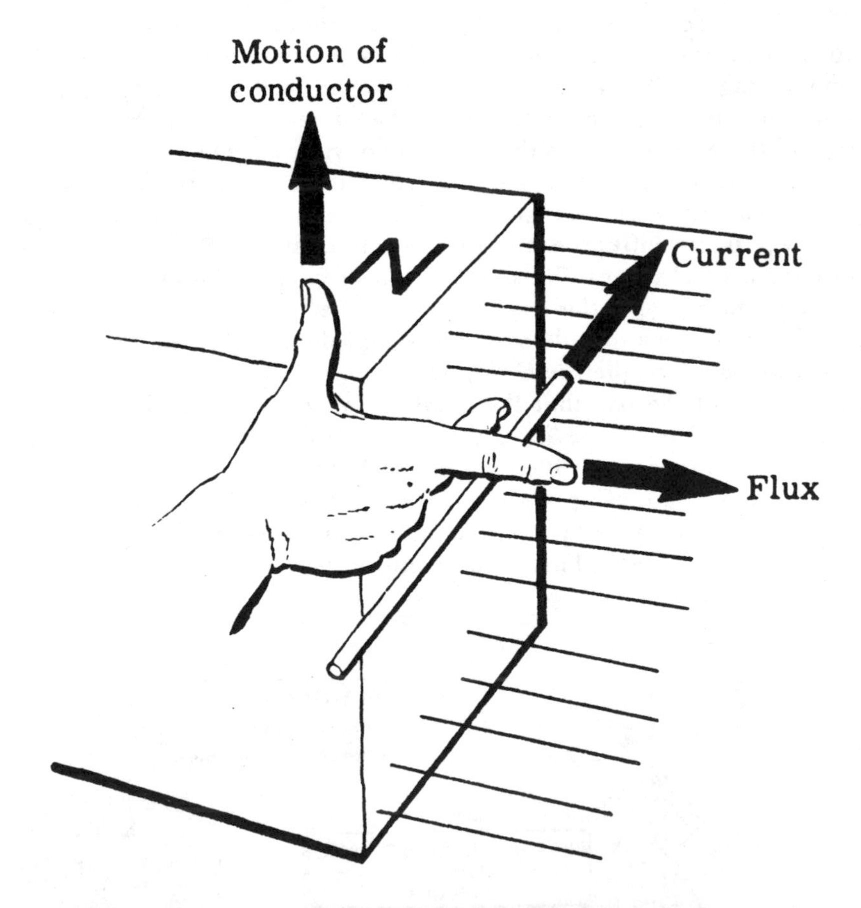

RIGHT HAND RULE FOR MOTORS

Self-Induction

When the current in an electric circuit changes, the circuit may *oppose* the change. The property of the circuit which opposes the change is called *inductance.*

An electric current always produces a magnetic field. The lines of force in this field encircle the conductor which carries the current, forming concentric circles around the conductor. The strength of the magnetic field depends on the amount of current flow, with a large current producing many lines of force, and a small current producing only a few lines of force.

CURRENT PRODUCES MAGNETIC FIELD AND THE AMOUNT OF CURRENT FLOW DETERMINES STRENGTH OF THE MAGNETIC FIELD

wire

wire

Small current flow | Small magnetic field | Increased current flow | Larger magnetic field

When the current increases or decreases, the magnetic field strength increases or decreases in the same way. As the field strength *increases*, the lines of force increase in number and *expand* outwards from the center of the conductor. Similarly, when the field strength *decreases*, the lines of force *contract* or decrease toward the center of the conductor.

This expansion and contraction of the magnetic field as the current varies causes an *emf of self-induction* which opposes any further change of current.

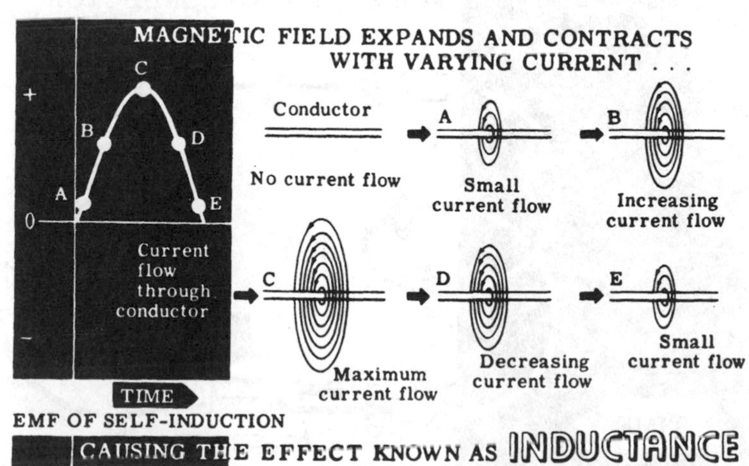

Inductance in a DC Circuit

To see how inductance behaves, suppose you had a circuit containing a coil like the one shown below. As long as the circuit switch is *open*, there is *no* current flow, and *no* field exists around the conductors.

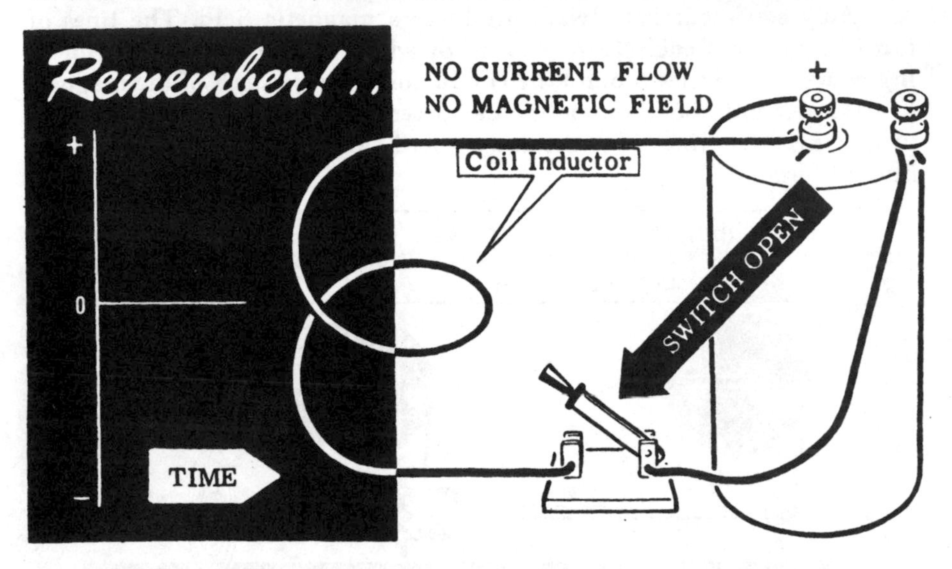

When the switch is *closed*, current flows through the circuit and lines of force *expand* outward around the circuit conductors, including the turns of the coil. At the instant the switch is closed, the current flow starts rising from zero toward its maximum value. Although this rise in current flow may be very rapid, it cannot be instantaneous. Imagine that you are actually able to see the lines of force in the circuit at the instant the current starts to flow. You see that they form a field around the circuit conductors.

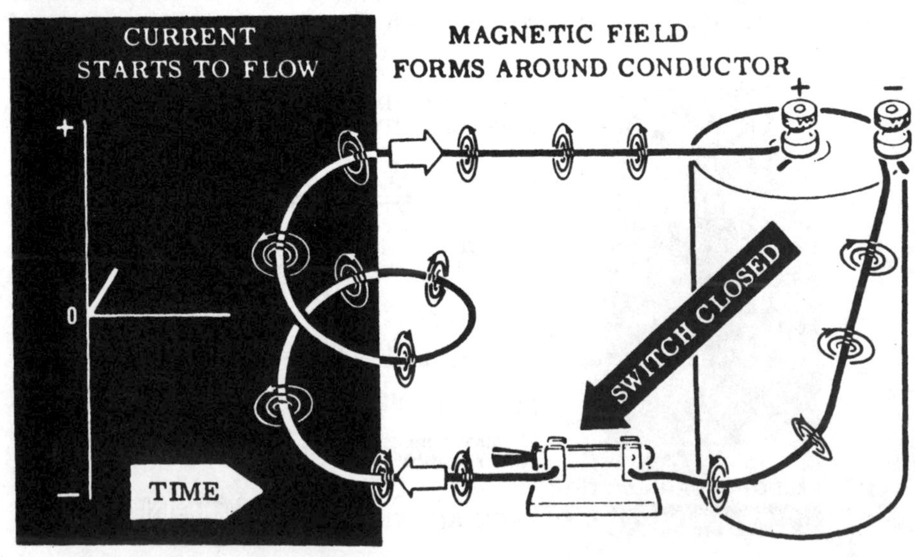

Inductance in a DC Circuit (continued)

As the current continues to increase, the lines of force continue to expand. The fields of adjacent turns of wire eventually interlace.

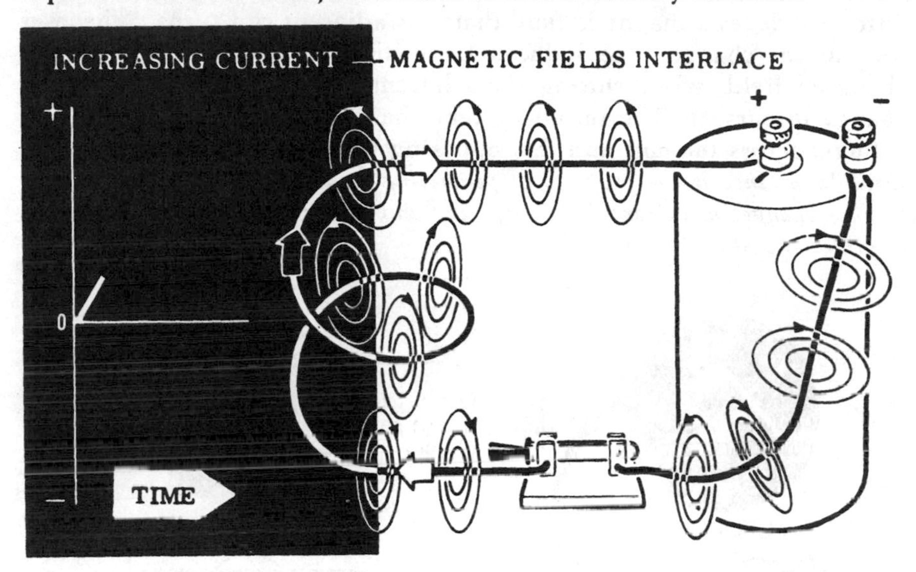

The lines of force around each turn of wire continue their expansion and, in so doing, cut across adjacent turns of the coil. This expansion continues as long as the circuit current increases, with more and more lines of force from the coil turns cutting across adjacent turns of the coil.

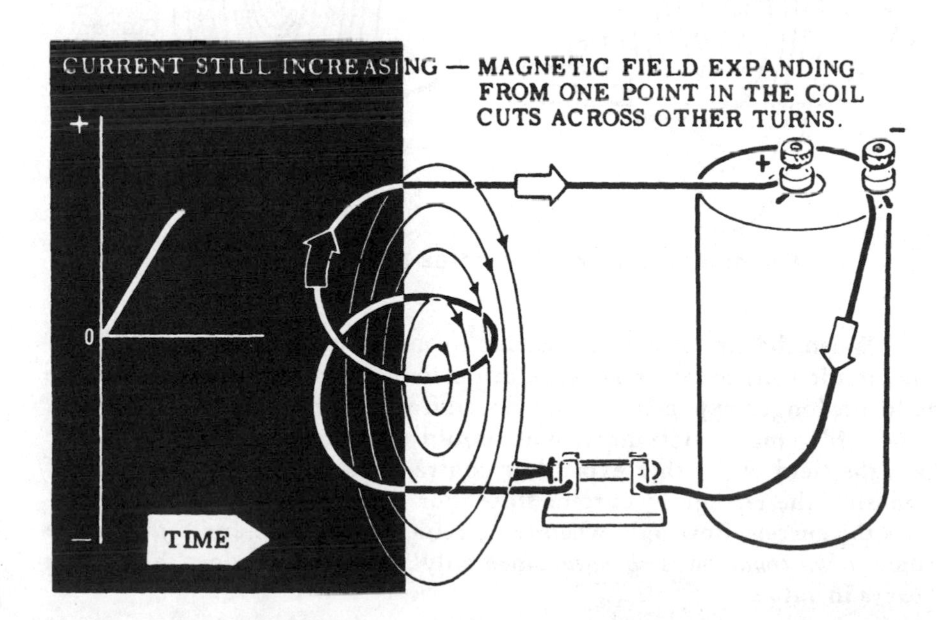

Inductance in a DC Circuit (continued)

As you know, whenever a magnetic field moves across a wire, it induces an emf in the wire. Whenever a current flows through a coil of wire, it induces a magnetic field that cuts adjacent coil turns. Whenever the current changes, the induced field changes and the effect of this changing field, when cutting the adjacent coil turns, is to *oppose* the change in current. The initial current change is caused by the emf, or voltage, across the coil and this opposing force is an emf of self-induction. *Inductance is the property of generating an emf of self-induction which opposes changes in current.* This *opposing emf of self-induction* is also called the *counter emf* or *back emf.*

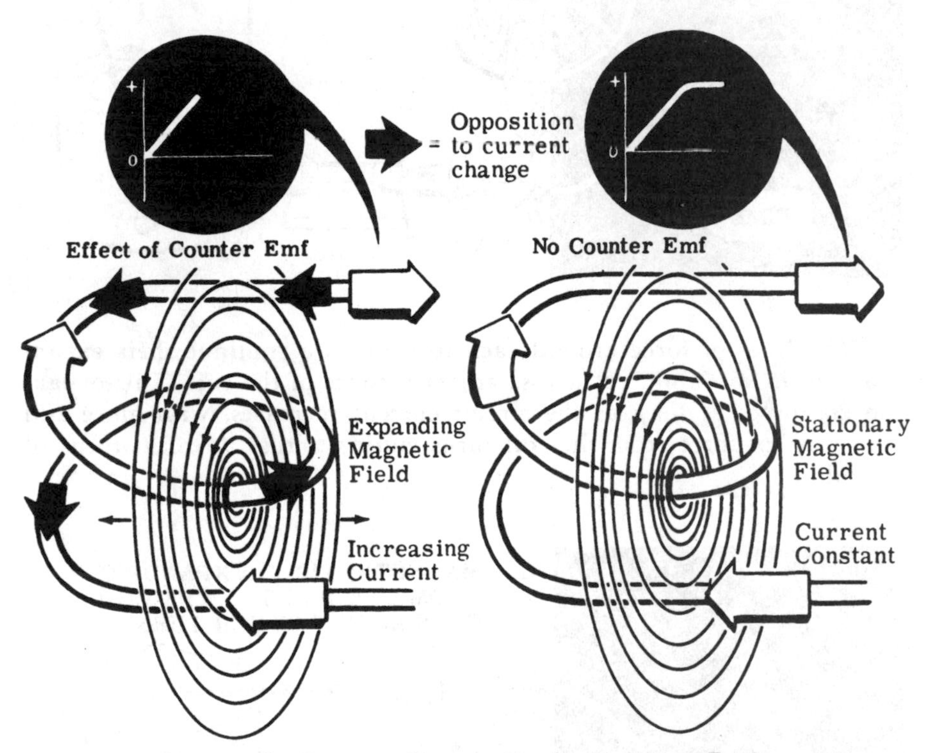

A Change in Current Generates An Induced Emf

When the circuit current reaches its maximum value, determined by the circuit voltage and resistance, it no longer changes in value and the field no longer expands, so that no emf of self-induction is generated. The field remains stationary, but, should the current attempt to rise or fall, the field will either expand or contract and generate a counter emf opposing the change in current flow. For direct current, inductance affects the current flow only when the circuit is turned *on* and *off*, or when some *circuit condition* is *changed*, since only at those times does the current *change* in value.

Inductance in a DC Circuit (continued)

With the current and magnetic field stationary at maximum (as determined by the circuit resistance and voltage by Ohm's Law), no counter emf is generated; but if you lower the source voltage or increase the circuit resistance, the current will decrease.

Suppose the source voltage decreases. The current drops toward its new Ohm's Law value. As the current decreases, the magnetic field also diminishes, with each line of force contracting inward toward the conductor. This contracting or collapsing field cuts across the coil turns in a direction opposite to that caused by increasing the circuit current.

Since the direction of change is reversed (decreasing, rather than increasing), the collapsing field generates a counter emf opposite to that caused by the expanding field; thus, it has the same polarity as the source voltage. This emf then increases the apparent source voltage, trying to prevent the drop in current. However, it cannot keep the current from falling indefinitely since the counter emf ceases to exist whenever the current stops changing. Thus, *inductance—the effect of counter emf—opposes any change in current flow, whether it be an increase or decrease, and slows down the rate at which the change occurs.*

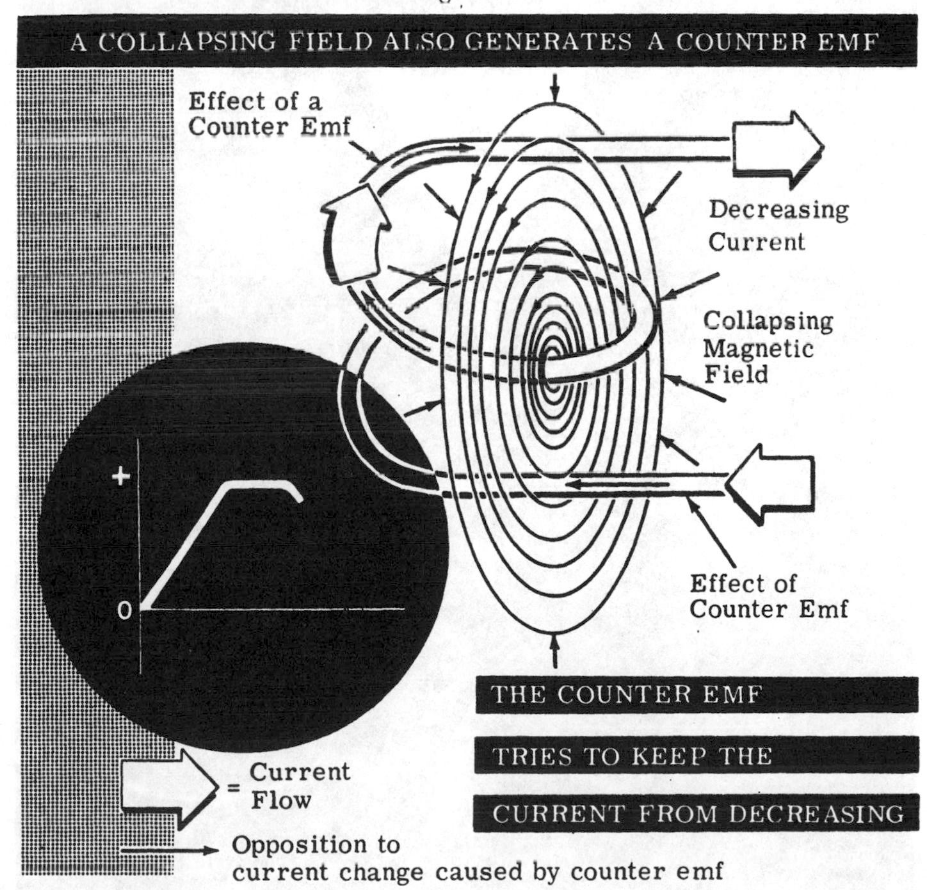
A COLLAPSING FIELD ALSO GENERATES A COUNTER EMF
Effect of a Counter Emf
Decreasing Current
Collapsing Magnetic Field
Effect of Counter Emf
+
0
THE COUNTER EMF
TRIES TO KEEP THE
CURRENT FROM DECREASING
= Current Flow
Opposition to current change caused by counter emf

Inductance in a DC Circuit (continued)

As long as the circuit is closed, the current remains at its Ohm's Law value and no induced emf is generated. Now suppose you open the switch to stop the current flow. The current should fall to zero and stop flowing immediately but, instead, there is a slight delay and a spark jumps across the switch contacts.

When the switch is opened, the current drops rapidly toward zero and the field also collapses at a very rapid rate. The rapidly collapsing field can generate a very high induced emf, which not only opposes the change in current, but can also cause an arc across the switch in an effort to maintain the current flow. Although only momentary, the induced emf caused by this rapid field collapse can be very high, sometimes many times that of the original source voltage. Shortly, you will learn how this emf can be calculated and you will see that it can be very high under some conditions. This action is often used to advantage in special types of equipment to obtain very high voltages.

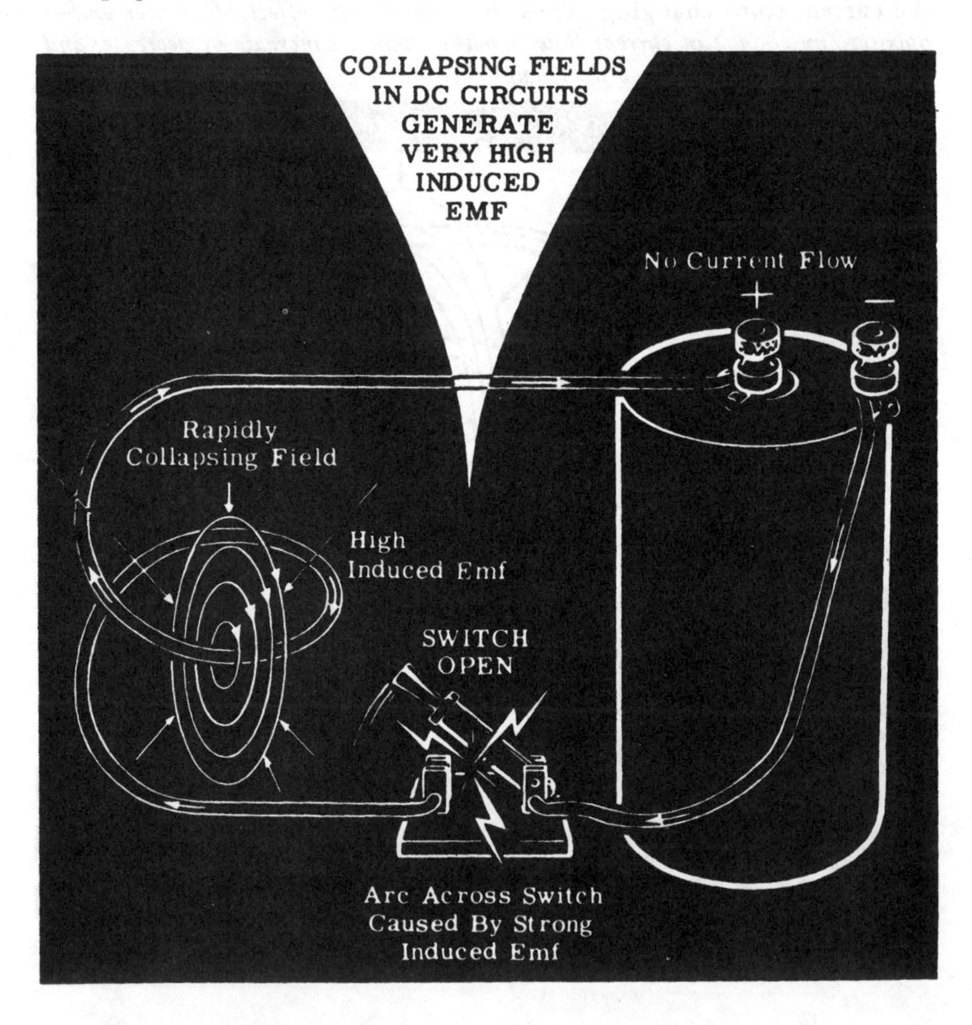

Inductance Symbols

While you cannot see inductance, it is present in every electric circuit and has an effect on the circuit whenever the circuit current changes. In electrical formulas, the letter "L" is used as a symbol to designate inductance. Because a coil of wire has more inductance than a straight length of the same wire, the coil is called an *inductor*. Both the letter and the symbol are illustrated below.

Since *direct current* is normally *constant* in value except when the circuit power is turned on and off to start and stop the current flow, inductance usually only affects dc current flow at these times and usually has little effect on the operation of the circuit. *Alternating current*, however, is *continuously changing* so that the *circuit inductance affects ac current flow at all times*. Although every circuit has some inductance, the value depends upon the physical construction of the circuit, and the electrical devices used in it. In some circuits, the inductance is so small that its effect is negligible, even for ac current flow.

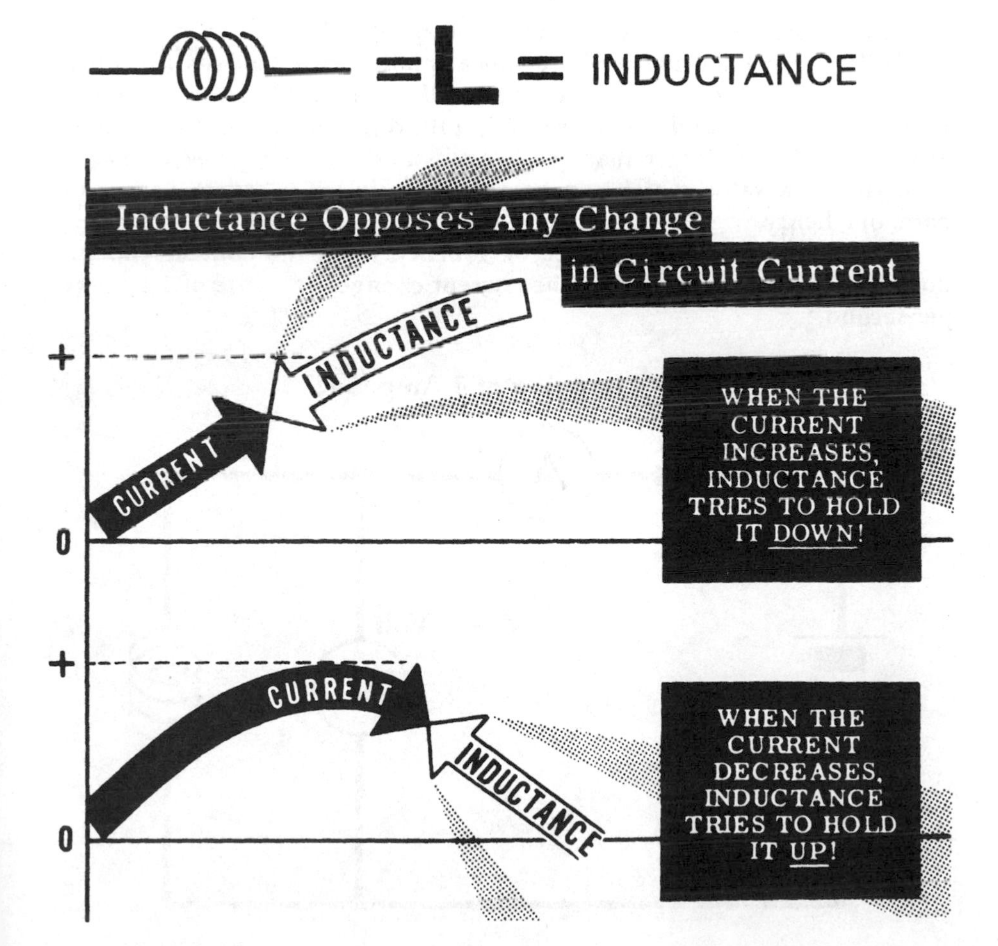
= L = INDUCTANCE
Inductance Opposes Any Change
in Circuit Current
INDUCTANCE
CURRENT
+
0
WHEN THE CURRENT INCREASES, INDUCTANCE TRIES TO HOLD IT DOWN!
CURRENT
INDUCTANCE
+
0
WHEN THE CURRENT DECREASES, INDUCTANCE TRIES TO HOLD IT UP!

Units of Inductance—Relationship to Counter EMF

Inductance opposes change of current in a circuit, whatever the direction of the change. While the current is increasing, energy is being *stored* by building up a magnetic field around the conductors in the circuit. This energy is given up by the magnetic field when the current decreases.

IN AN INDUCTIVE CIRCUIT
WHEN CURRENT INCREASES — THE CIRCUIT
STORES ENERGY IN THE MAGNETIC FIELD
WHEN CURRENT DECREASES — THE CIRCUIT
GIVES UP ENERGY FROM THE MAGNETIC FIELD

The basic unit of measure for inductance is the *henry* and the symbol is "H." For quantities of inductance smaller than 1 henry, the millihenry and microhenry are used. The millihenry (abbreviated mH) is 1/1000 of a henry and the microhenry (abbreviated μH) is 1/1,000,000 of a henry. A unit larger than the henry is not used, since inductance is normally of a value which can be conveniently expressed in henrys or parts of a henry.

A circuit has an inductance of 1 henry when the counter emf induced in it is 1 volt and when the current changes at a rate of 1 ampere per second.

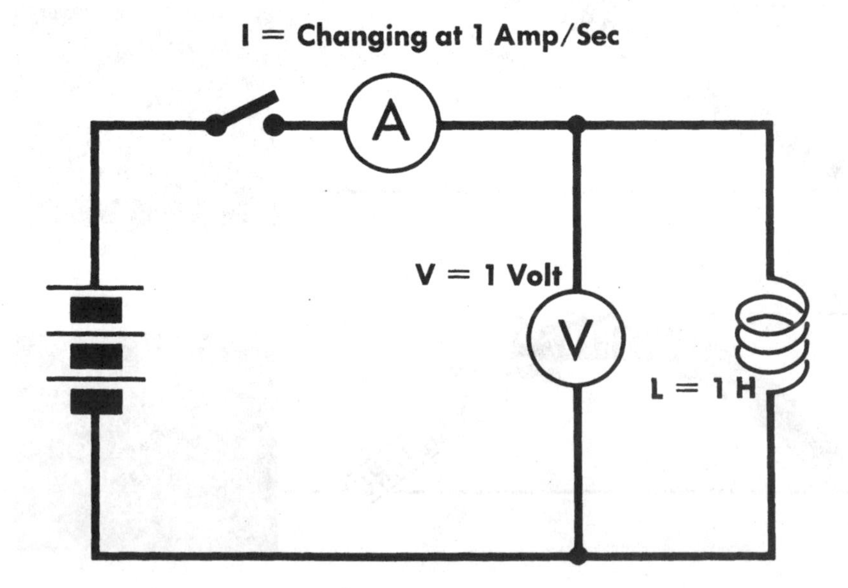

Units of Inductance—Relationship to Counter EMF (continued)

Since the definition of a henry is based on the counter emf produced, you can calculate counter emf knowing the inductance and the rate at which the current is changing. The counter emf can be calculated from the equation:

$$\text{Counter emf} = -L \frac{\Delta I}{\Delta T}$$

where L is the inductance in henrys, and $\frac{\Delta I}{\Delta T}$ (stated as "delta I over delta T") is the change in current per unit time. The minus sign indicates that the counter emf is opposite in direction to the applied voltage. Suppose you had a circuit with an inductance of 4 henrys and the current changed from 2 amperes to 6 amperes in 1 second.

$$\text{Then the counter emf} = -4 \times \frac{6-2}{1} = -16 \text{ volts}$$

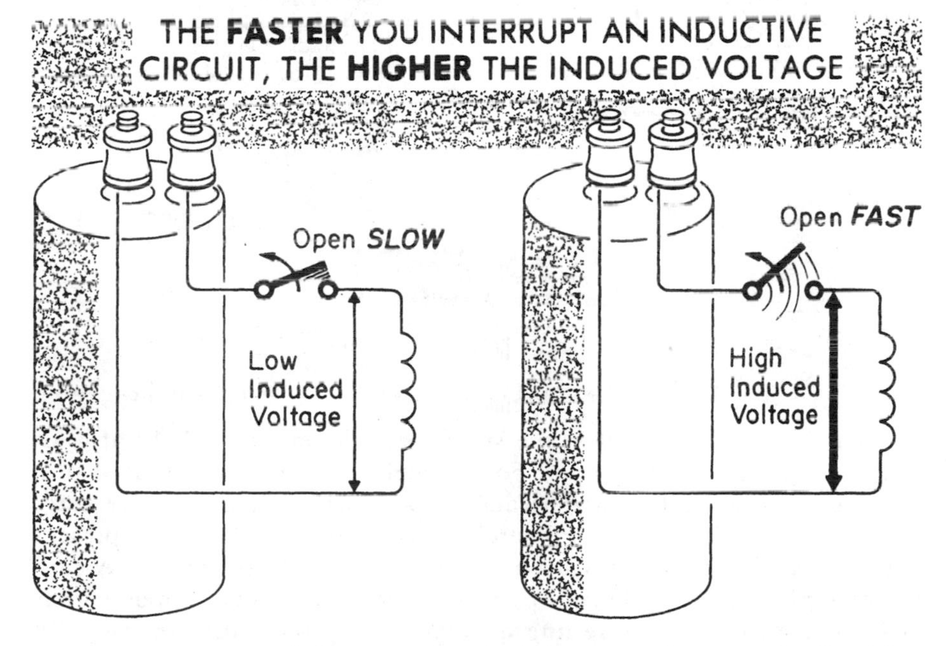

Suppose you had the same circuit and you tried to turn it off with a switch so that the current went from 2 amperes to zero and you could open the switch in 5 milliseconds (0.005 sec).

$$\text{Then the counter emf} = -4 \times \frac{(2-0)}{(0.005)} = -1600 \text{ volts}$$

As you can see, turning off inductive circuits can lead to some high voltages that can cause arcing and other problems that require special handling. You will learn more about this effect when you learn about electric control circuits.

Factors Affecting Inductance

Every complete electric circuit has some inductance since even the simplest circuit forms a complete loop or single-turn coil. An induced emf is generated even in a straight piece of wire by the action of the magnetic field expanding outward from the center of the wire or collapsing inward to the wire's center. The greater the number of adjacent turns of wire cut across by the expanding field, the greater the induced emf generated, so that a coil of wire having many turns has a higher inductance than a coil of wire having few turns.

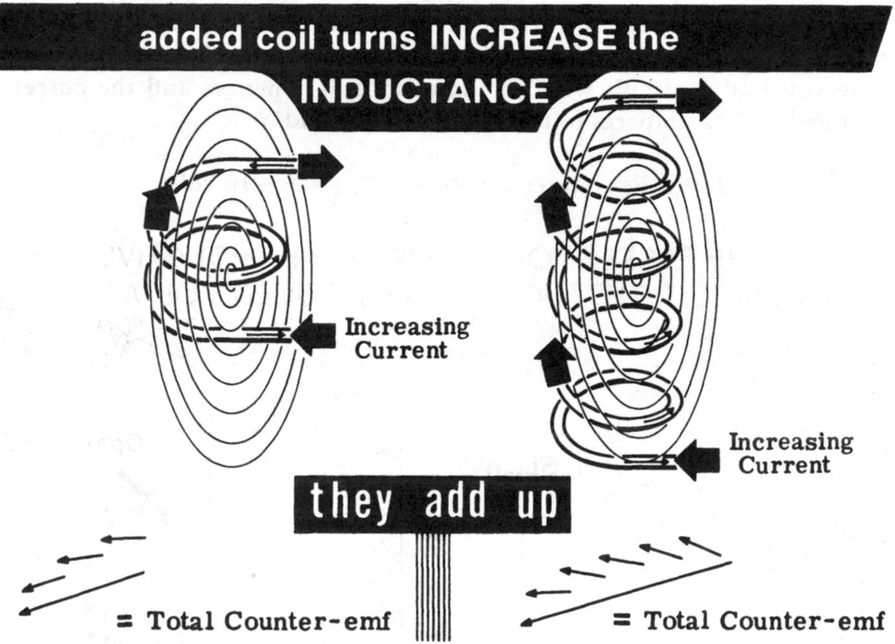

Any factors which affect the strength of the magnetic field also affect the inductance of a circuit. For example, an iron core inserted in a coil increases the inductance because it provides a better path for magnetic lines of force than air. Therefore, more lines of force are present that can expand and contract when there is a change in current. A copper core piece has exactly the opposite effect. Since copper opposes lines of force more than air, inserting a copper core piece results in less field change when the current changes, thereby reducing the inductance.

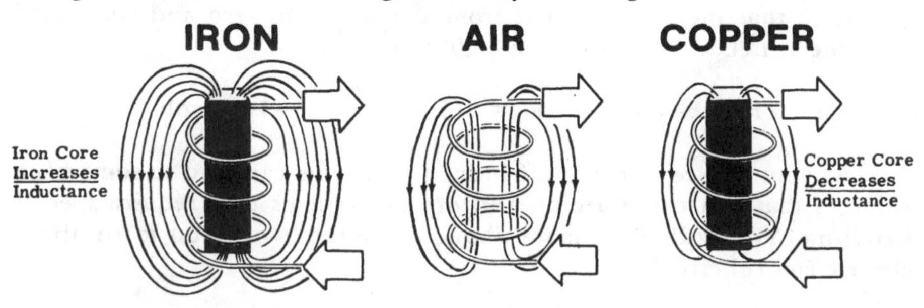

Factors Affecting Inductance (continued)

Inductance can be measured only with special laboratory instruments and depends entirely on the physical construction of the circuit. Some of the factors most important in determining the amount of inductance of a coil are: (a) the number of turns, (b) the spacing between turns, (c) the coil diameter, (d) the kinds of materials around and inside the coil, (e) the number of layers of wire, and (f) the type of coil winding and the overall shape of the coil. Wire size does not affect inductance directly, but it does determine the number of turns that can be wound in a given space. All of these factors are variable, and no single formula can be used to find inductance. Many differently constructed coils could have an inductance of one henry, and each would have the same effect in the circuit.

For a simple single layer coil closely wound on a core, the relationship between the physical construction and the inductance is given by the formula:

$$L \text{ (henrys)} = \frac{0.4\pi N^2 A \mu}{l}$$

where N is the number of turns, A is the cross sectional area of the core, l is its length, and μ is the magnetic permeability of the core—which is 1 for air, several hundred for iron, and less than 1 for nonmagnetic metals like copper.

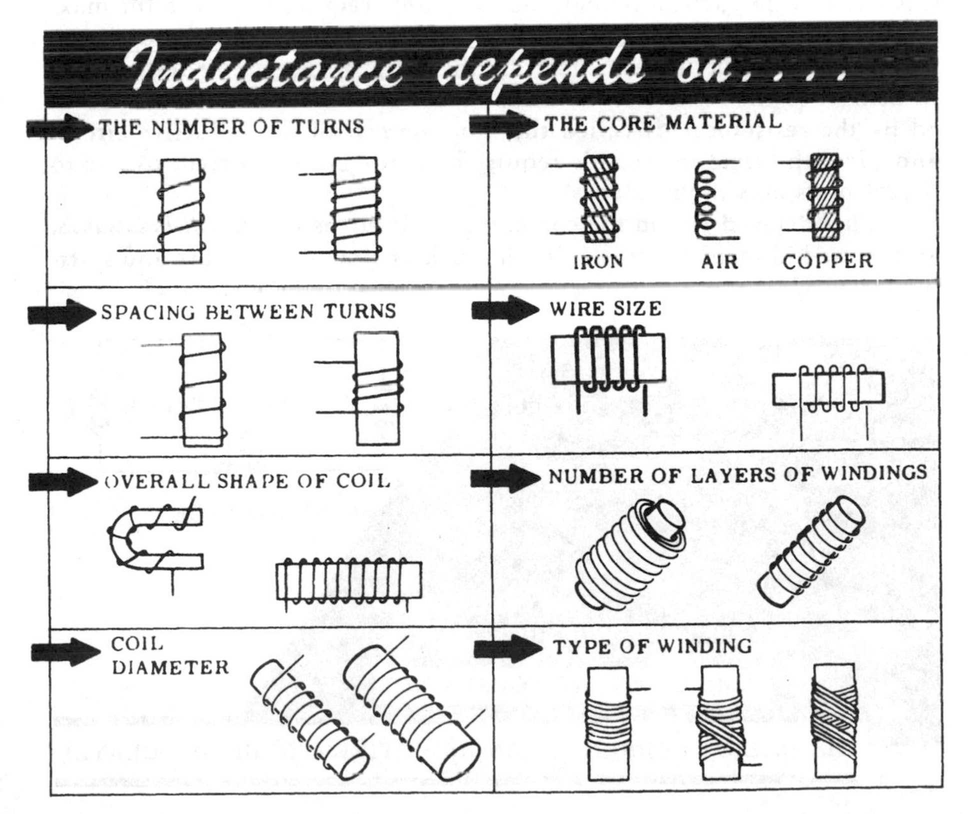

Inductive Time Constant in a DC Circuit

In a circuit consisting of a battery, switch, and a resistor in series, the current rises to its maximum value at once, whenever the switch is closed. Actually, it cannot change from zero to its maximum value instantaneously, but the time is so short it is considered instantaneous.

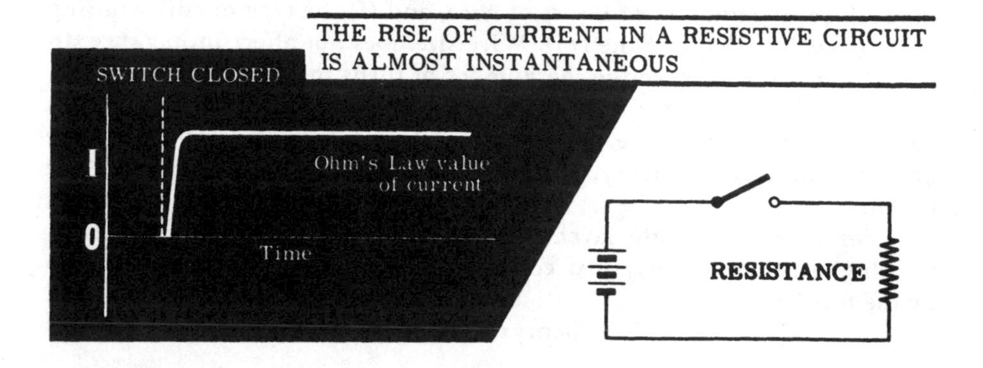

THE RISE OF CURRENT IN A RESISTIVE CIRCUIT IS ALMOST INSTANTANEOUS

If an inductor is used in series with the resistor, the current does not rise instantaneously; it rises rapidly at first, then more slowly as the maximum value is approached. For all inductive circuits, the shape of this curve is the same, although the total time required to reach the maximum current value varies. The time required for the current to rise to its maximum value is determined by the ratio of the circuit inductance, in henrys, to the resistance, in ohms. This ratio L/R—inductance divided by the resistance—is called the *time constant* of the inductive circuit and gives the time in seconds required for the circuit current to rise to 63.2% of its maximum value.

This delayed rise in the current of a circuit is called *self-inductance*, and is used in many practical circuits such as time-delay relays and starting circuits.

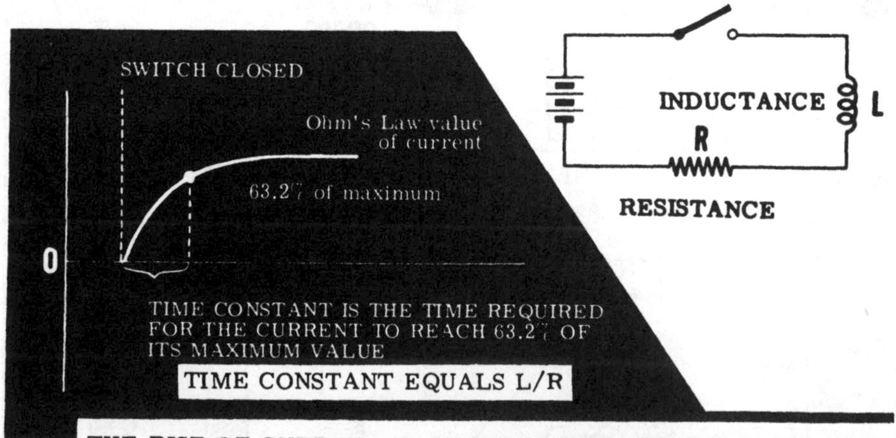

THE RISE OF CURRENT IN AN INDUCTIVE CIRCUIT IS DELAYED

Inductive Time Constant in a DC Circuit (continued)

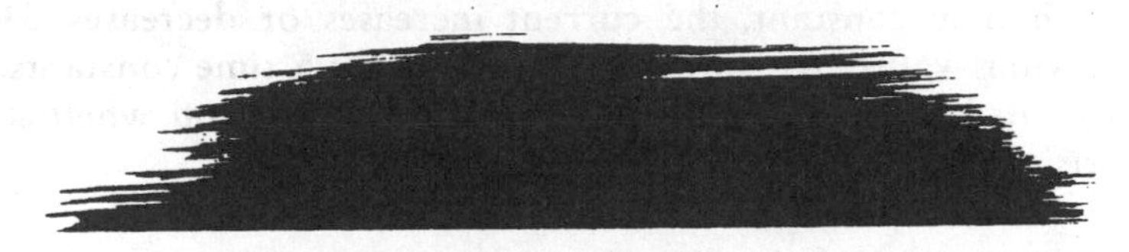

BUILD-UP AND DECAY OF CURRENT IN AN INDUCTIVE CIRCUIT

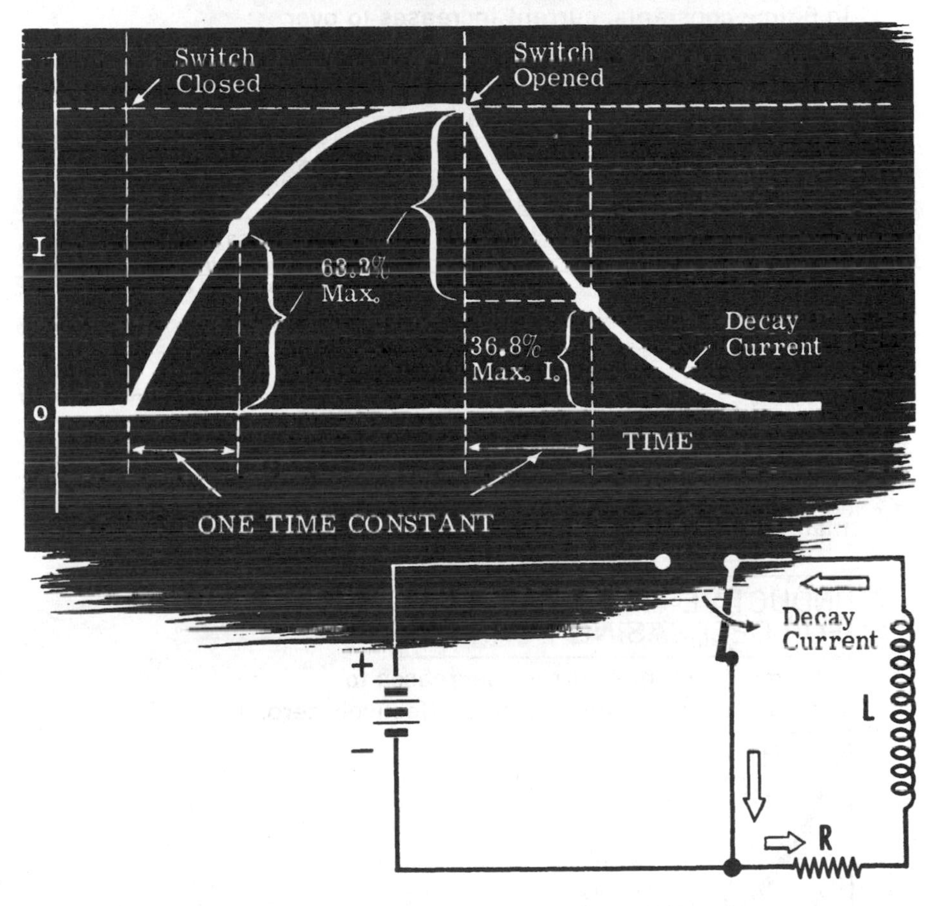

If the circuit terminals are shorted together at the same moment that the battery switch is opened, the current continues to flow due to the action of the collapsing field. The current falls in the same manner as the original rise in current, except the curve is in the *opposite* direction.

Again, the *time constant* (L/R) can be used to determine when the current has decreased by 63.2%, or has reached 36.8% of its original maximum value. For inductive circuits, the *lower* the circuit resistance, the *longer* the time constant for the same value of inductance.

Inductive Time Constant in a DC Circuit (continued)

For each time constant, the current increases or decreases 63.2% from its previous value. As shown below, in about 5 time constants, the current is at its maximum value or is zero, depending on whether the current is rising or falling.

INDUCTIVE TIME CONSTANT FOR INCREASING CURRENT

In 5 time constants, current increases to over 99% of maximum, or is effectively 100%.

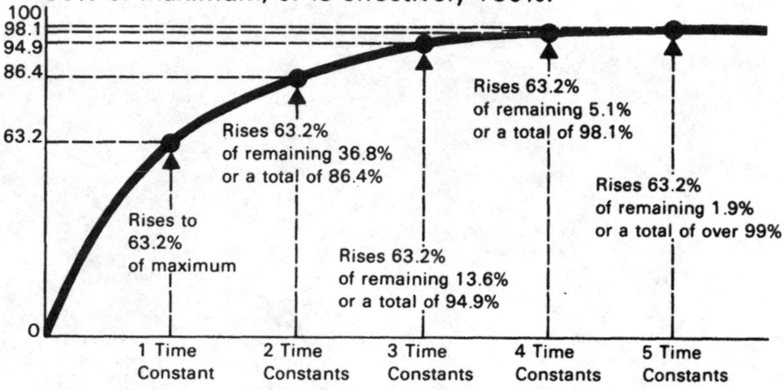

In each time constant, the current increases to a value 63.2% closer to its maximum value than the previous value.

INDUCTIVE TIME CONSTANT FOR DECREASING CURRENT

In 5 time constants, current decreases to less than 1% of maximum, or is effectively zero.

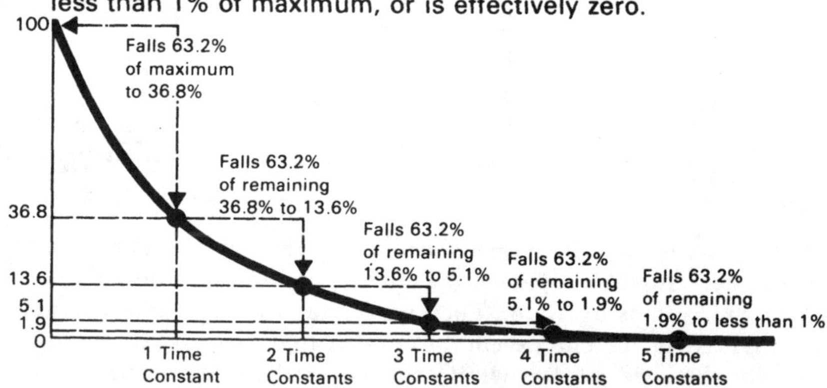

In each time constant, the current decreases to a value 63.2% closer to zero than the previous value.

Inductive Time Constant in a DC Circuit (continued)

The time constant of a given inductive circuit is always the same for both the build-up and decay of the current. If the maximum current value differs, the current may rise at a different rate but will reach its maximum in the same amount of time, and the shape of the curve is the same. Thus, if a greater voltage is used, the maximum current will increase but the time required to reach the maximum current is unchanged.

Every practical inductive circuit has resistance, since the wire used in a coil always has resistance except under the special condition of *superconductivity*, where the temperature is near absolute zero. Thus, a perfect inductance—an inductor with no resistance—can be found only in this special case.

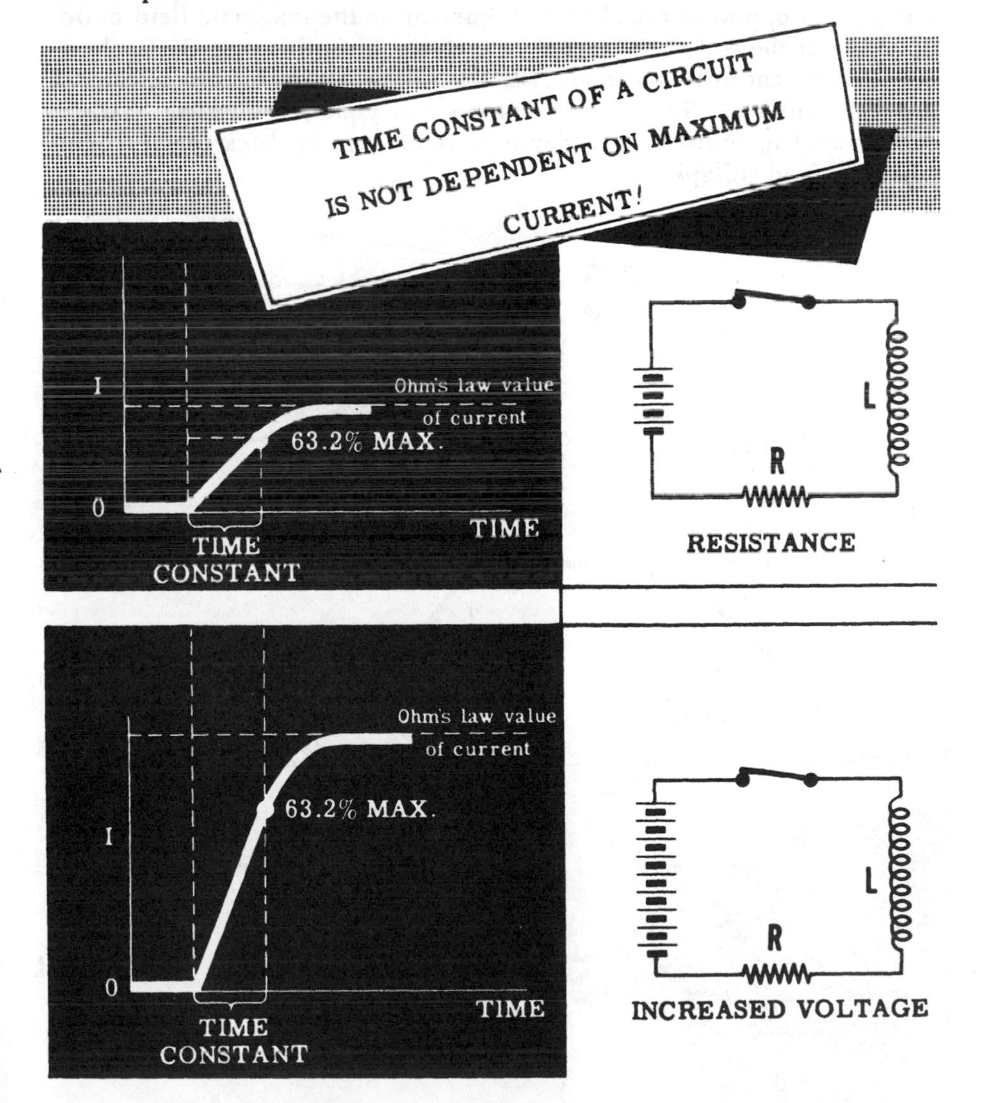

Inductive Reactance

Inductive reactance is the opposition to current flow offered by the inductance of a circuit. As you know, inductance only affects current flow while the current is changing, since the current change generates an induced emf. For *direct current* the effect of inductance is usually noticeable only when the current is turned on and off; but, since *alternating current* is continuously changing, a continuous induced emf is generated.

Suppose you consider the effect of a given inductive circuit on dc and ac waveforms. The time constant of the circuit is always the same, determined only by the resistance and inductance of the circuit.

For dc, the current waveforms would be as shown below. At the beginning of the current waveform, there is a shaded area between the maximum current value and the actual current flow which shows that inductance is opposing the change in current as the magnetic field builds up. Also, at the end of the current waveform, a similar area exists, showing that current flow continues after the voltage drops to zero because of the field collapse. These shaded areas are equal, indicating that the energy used to build up the magnetic field is given back to the circuit when the field collapses.

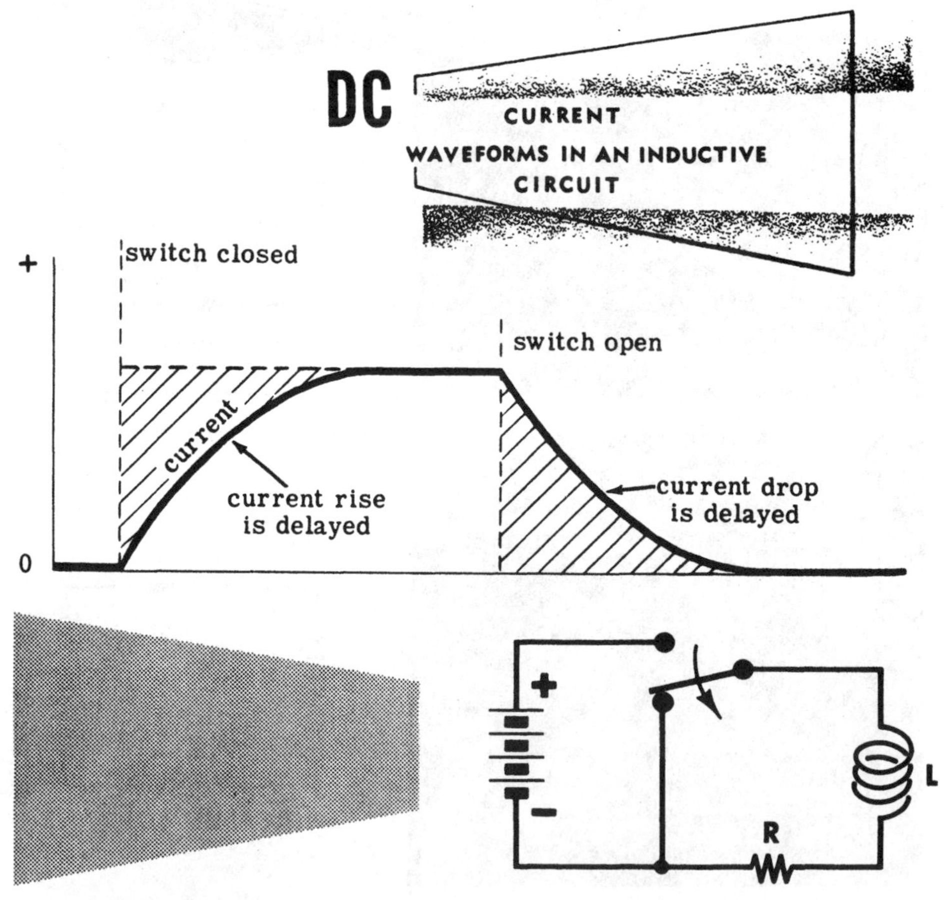

Inductive Reactance (continued)

The same inductive circuit would affect ac voltage and current waveforms as shown below. The current rises as the voltage rises, but the delay due to inductance prevents the current from ever reaching its maximum dc value before the voltage reverses polarity and changes the direction of current flow. Thus, in a circuit containing inductance, the maximum current will be much greater for dc than for ac.

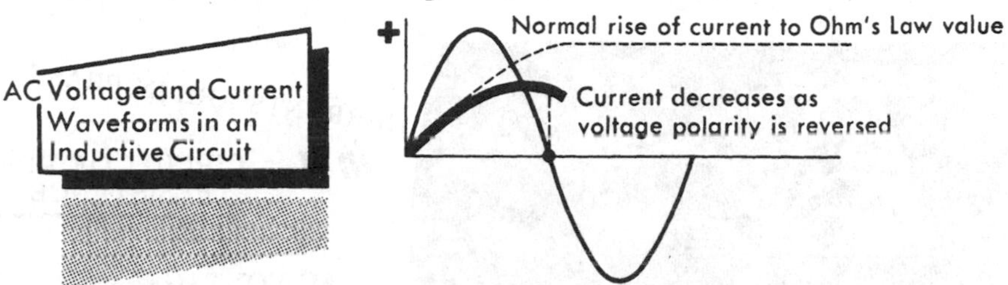
AC Voltage and Current Waveforms in an Inductive Circuit
+
Normal rise of current to Ohm's Law value
Current decreases as voltage polarity is reversed

If the frequency of the ac wave is low, the current will have time to reach a higher value before the polarity is reversed than if the frequency is high. Thus, the *higher* the *frequency*, the *lower* the *circuit current* through an inductive circuit. Frequency, then, affects the opposition to current flow as does circuit inductance. For that reason, inductive reactance—opposition to current flow offered by an inductance—depends on *frequency* and *inductance*.

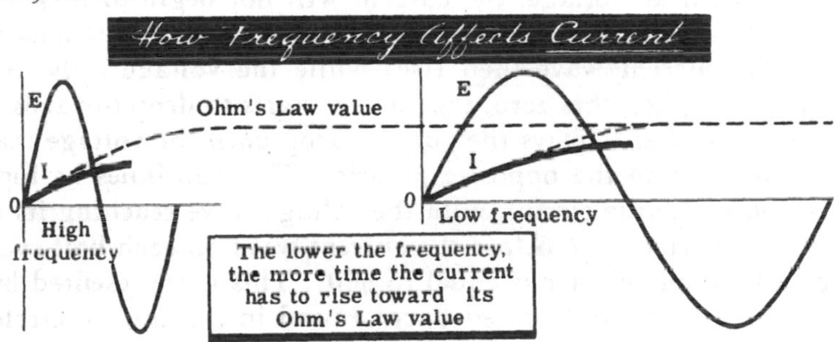
How Frequency Affects Current
E
Ohm's Law value
I
0
High frequency
E
I
0
Low frequency
The lower the frequency, the more time the current has to rise toward its Ohm's Law value

The formula used to calculate inductive reactance, designated by the symbol X_L, is

$$X_L = 2\pi fL$$

where X_L is the inductive reactance, f is the frequency in hertz, L is the inductance in henrys, and π is the constant 3.14. Since X_L represents opposition to current flow, it is expressed in ohms.

For example, suppose you had a 2-henry inductor in a circuit connected to a 60-Hz power line. The inductive reactance would be:

$$X_L = 2\ \pi\ fL = 2 \times 3.14 \times 60 \times 2 = 753.6 \text{ ohms}$$

We can use Ohm's Law to calculate the current. If the source were 120 volts, 60 hertz, the current would be:

$$I = \frac{E}{R} \text{ or } \frac{E}{X_L} = \frac{120}{753.6} = 0.158 \text{ ampere}$$

Phase Relationships in AC Inductive Circuits

Actually, the circuit current does not rise at the same time the voltage rises. The current is delayed to an extent depending on the amount of inductance in the circuit as compared to the resistance.

As you know, if an ac circuit has only pure resistance, the current rises and falls at exactly the same time as the voltage, and the two waves are in phase with each other, as shown below.

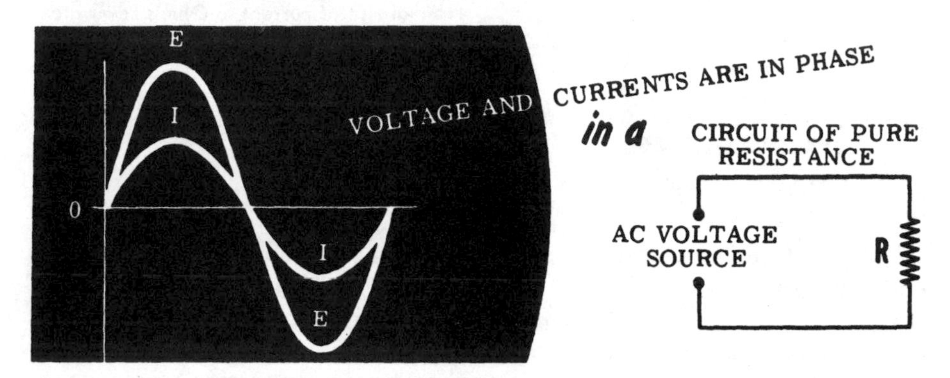

With a theoretical circuit of pure inductance and no resistance driven by a sinusoidal ac voltage, the current will not begin to flow in the same direction as the voltage until the voltage has reached its maximum value, and the current wave then rises while the voltage falls. At the moment the voltage reaches zero, the current starts to drop towards zero, but the collapsing field delays the current drop until the voltage reaches its maximum value in the opposite polarity. This continues as long as voltage is applied to the circuit, with the voltage wave reaching its maximum value a quarter cycle before the current wave on each half cycle. A complete cycle of an ac wave is equal to 360°. This is represented by the emf generated in the wire rotated once around in a complete circle between two opposite magnetic poles. A quarter cycle then is 90°, and in a purely inductive circuit, the voltage wave leads (or is ahead of) the current by 90° or, in opposite terms, the current wave lags the voltage wave by 90°.

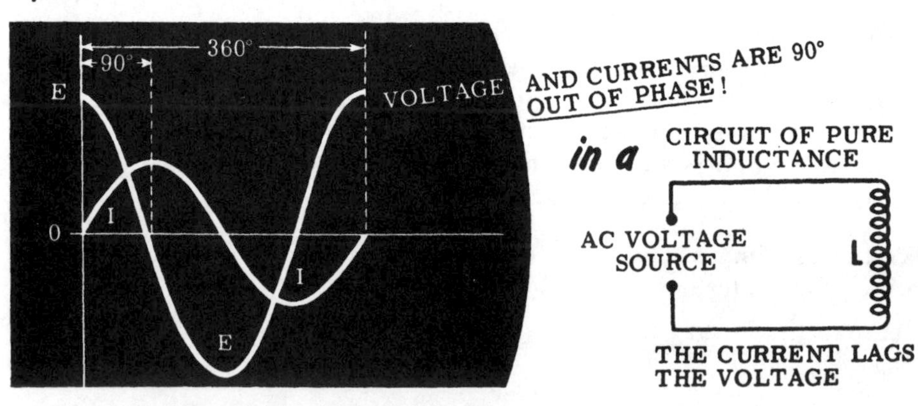

Phase Relationships in AC Inductive Circuits (continued)

We can see how the current lags the applied voltage if we suppose that the inductor (L) in the diagram on the previous page has no resistance, and then consider the effect of the alternating current flowing through it.

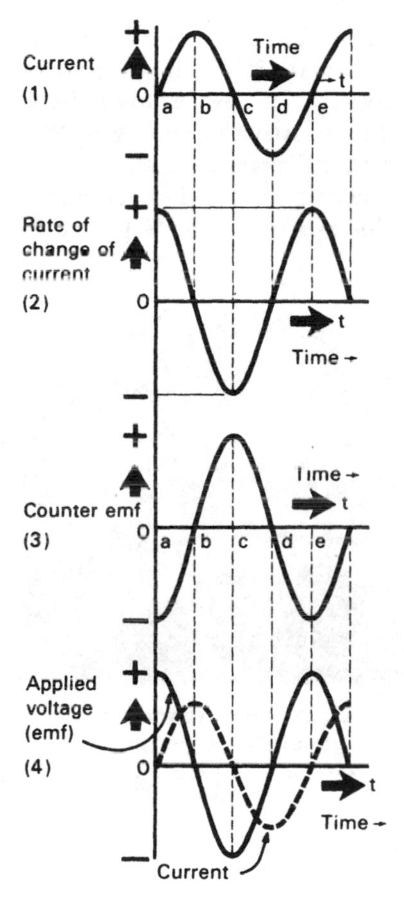

First draw the waveform of the current through the inductor—curve (1). We know that this alternating current induces a counter emf in the inductor which is greatest when the current is changing most rapidly, and least when the current is changing most slowly. If we plot a graph of the rate of change of current, we can see when the counter emf is greatest and when it is least.

If you examine the current waveform (1), you can see that at point "a" the current is changing very rapidly in the positive direction. At point "b" it is changing slowly and crossing over from the positive direction (increasing) to the negative direction (decreasing). At point "c" the current is again changing very rapidly but this time in the negative direction. By thus noting the change of rate of current as indicated by curve (1), you can draw a curve showing the rate of change of current, curve (2).

You know that the counter emf induced by the alternating current is a maximum when the rate of change of current is a maximum, and that it is of such polarity that it always opposes the change of current. From curve (2) we can therefore deduce the waveform of the counter emf; this is drawn at curve (3).

Now it is obvious that, for any current to flow at all, this counter emf must be more than overcome by the applied emf. It follows, therefore, that the applied emf must always be of *opposite phase* to the counter emf, and of greater magnitude.

The waveform for the applied emf is drawn at curve (4) and the current waveform is shown dotted for comparison. You see that the current reaches its maximum 90° later than the voltage—that is, it lags the voltage by 90°.

Phase Relationships in AC Inductive Circuits (continued)

In a circuit containing both inductive reactance and resistance, the ac current wave will lag the voltage wave by an amount between 0° and 90°; or, stated otherwise, it will lag somewhere between *almost in-phase* and *90° out-of-phase.* The exact amount of lag depends on the ratio of circuit resistance to inductance—the *greater* the resistance compared to the inductance, the nearer the two waves are to being *in phase*; the *lower* the resistance compared to the inductance, the nearer the two waves are to being a full quarter cycle (90°) *out-of-phase.*

When stated in degrees, the current lag is called the *phase angle.* If the phase angle between the voltage and the current is 45° lagging, then the current wave is lagging the voltage wave by 45°. Since this is halfway between 0° (the phase angle for a pure resistive circuit) and 90° (the phase angle for a pure inductive circuit), the resistance and the inductive reactance must be equal, with each having an equal effect on the current flow. You will learn more about this when you study power in ac inductive circuits and in Volume 4.

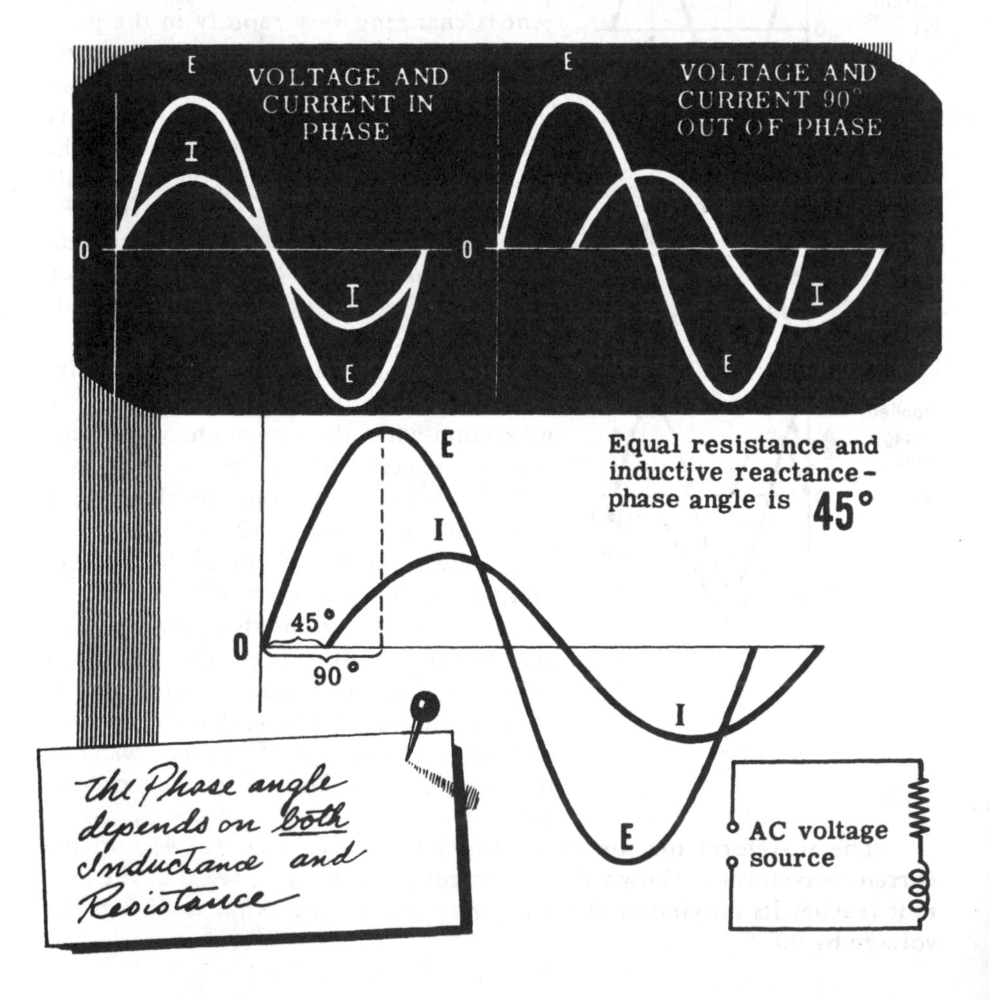

Inductances in Series and in Parallel

You learned in Volume 2 how to calculate the total resistance in series and parallel circuits. Total inductance in series and parallel circuits can be found by applying the *same* rules.

Suppose a series circuit consisting of three inductors (with negligible resistance) in series, is connected to an ac voltage source.

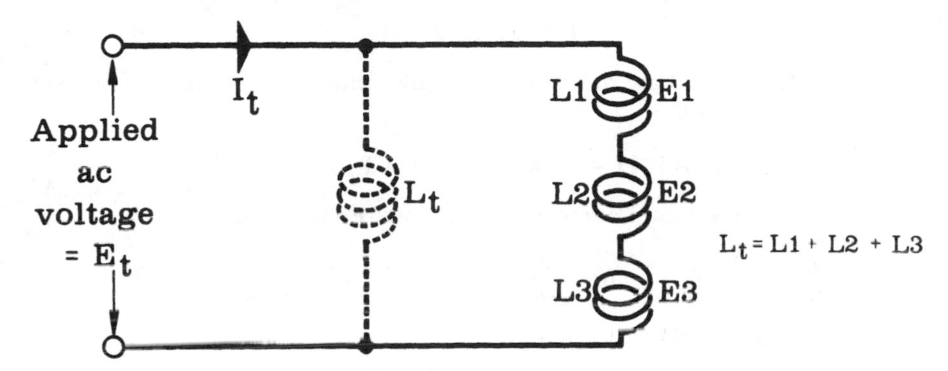

The current I flows through the circuit, through each of the inductors; the total voltage, $E_t = E1 + E2 + E3$.

We can apply Ohm's Law to an ac circuit, substituting reactance X for resistance R. So, $E_t = IX_t$, $E1 = IX1$, $E2 = IX2$ and $E3 = IX3$. Therefore, $IX_t = IX1 + IX2 + IX3$. Substituting for X_t, X1, X2, and X3, we can rewrite this formula as follows:

$$I(2\pi fL_t) = I(2\pi fL1) + I(2\pi fL2) + I(2\pi fL3)$$

or

$$(2\pi fI)L_t = 2\pi fI(L1 + L2 + L3)$$

Since 2π fI is on both sides of the equation, it can be deleted, leaving:

$$L_t = L1 + L2 + L3$$

So the total inductance in a series circuit is the sum of the separate inductances. Inductors can also be connected in parallel, as shown below.

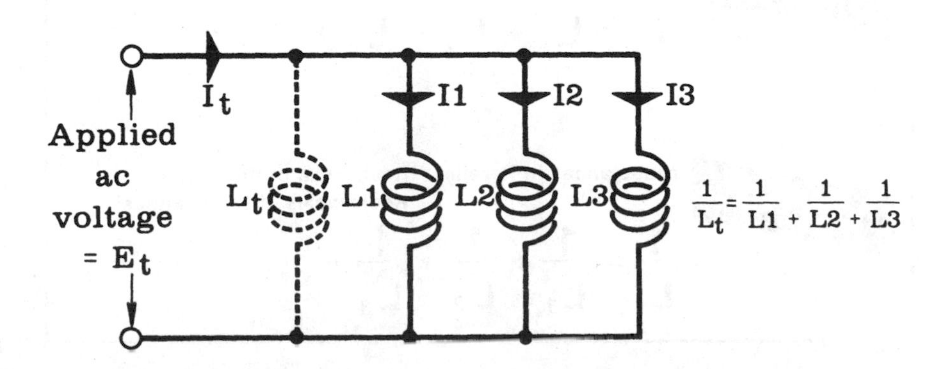

Inductances in Series and in Parallel (continued)

In the case of inductors with negligible resistance in a parallel circuit, the voltage across each inductor is the same, and the total current divides to flow through the separate branches. (The actual magnitude of the current through each branch, of course, depends on the opposition to ac current flow in each branch.) Therefore, the formula is:

$$I_t = I1 + I2 + I3$$

Now, apply Ohm's Law to this circuit, substituting reactance X for resistance R, and you have:

$I_t = E/X_t$, $I1 = E/X1$, $I2 = E/X2$, and $I3 = E/X3$

The expression for total current can now be rewritten as:

$$\frac{E}{X_t} = \frac{E}{X1} + \frac{E}{X2} + \frac{E}{X3} \quad \text{or} \quad \frac{E}{2\pi f L_t} = \frac{E}{2\pi f L1} + \frac{E}{2\pi f L2} + \frac{E}{2\pi f L3}$$

This can be put more conveniently as:

$$\frac{E}{2\pi f} \times \frac{1}{L_t} = \frac{E}{2\pi f}\left(\frac{1}{L1} + \frac{1}{L2} + \frac{1}{L3}\right)$$

and since $E/2\pi f$ is on both sides of the equation, it can be deleted, leaving:

$$\frac{1}{L_t} = \frac{1}{L1} + \frac{1}{L2} + \frac{1}{L3}$$

So, the reciprocal of the total inductance in a parallel circuit is equal to the sum of the reciprocals of the separate inductances.

remember

Inductances in Series = Sum of Inductances

$$L_t = L_1 + L_2 + L_3 + \cdots$$

Inductances in Parallel = Reciprocal of the Sum of Reciprocals of Inductances

$$\frac{1}{L_t} = \frac{1}{L_1} + \frac{1}{L_2} + \frac{1}{L_3} + \cdots$$

Power in Inductive Circuits—Review of Power Formulas

You will recall that the power in *any* circuit is *always equal* to the *square of the current* (dc or rms) *times* the circuit *resistance.*

$$P = I^2R$$

You also know that power in *dc circuits* and in *ac resistive circuits* is only *equal* to the *voltage* (dc or rms) *times* the *current* (dc or rms).

$$P = E \times I$$

You also know that power in ac and dc circuits is equal to the square of the voltage (dc or rms) across the resistances divided by the resistance.

$$P = \frac{E^2}{R}$$

You will notice that the above formulas are either restricted to pure resistive circuits ($P = E \times I$) or are involved with only the circuit resistance ($P = I^2R$ or $P = E^2/R$). What about inductance and capacitance? Where do they come in? You will learn in the next section that pure inductances consume *no* power—that is, the electrical energy that they take in is returned directly to the circuit as electrical energy, whereas the electrical energy taken into a resistance is converted to heat and cannot be returned directly as electrical energy.

As you know, an inductance uses energy to create a magnetic field and the magnetic field restores the energy to the line when it collapses. Since an ac ammeter reads current in either direction, there is a current indication in a purely inductive circuit. As you will learn, power calculated by voltage times current has little to do with actual power consumed, since no power is consumed in purely inductive (or capacitive) circuits. Such power is called *wattless* power. You will learn why this is so in the following pages of this section.

The Effect of Phase Difference on the AC Power Wave

As you know, in a theoretical circuit containing only pure inductance, the current lags the voltage by 90°. To determine the power wave for such a circuit, all of the corresponding instantaneous values of voltage and current are multiplied together to find the instantaneous values of power, which are then plotted to form the power curve just as you did for resistance in ac circuits.

As you already know, the power curve for *in-phase* voltages and currents (pure resistance) is entirely above the zero axis, since the result is positive when either two positive numbers are multiplied together or two negative numbers are multiplied together. This occurs when the voltage and current are in phase since they are positive or negative at the same time. When a negative number is multiplied by a positive number, however, the result is a negative number. Thus, in computing instantaneous values of power when the current and voltage are not in phase, some of the values are negative. If the phase difference is 90°, as in the case of a theoretical circuit containing only pure inductance, half of the instantaneous values of power are positive and half are negative, as shown below. For such a circuit, the voltage and current axis is also the power wave axis, and the frequency of the power wave is twice that of the current and voltage waves.

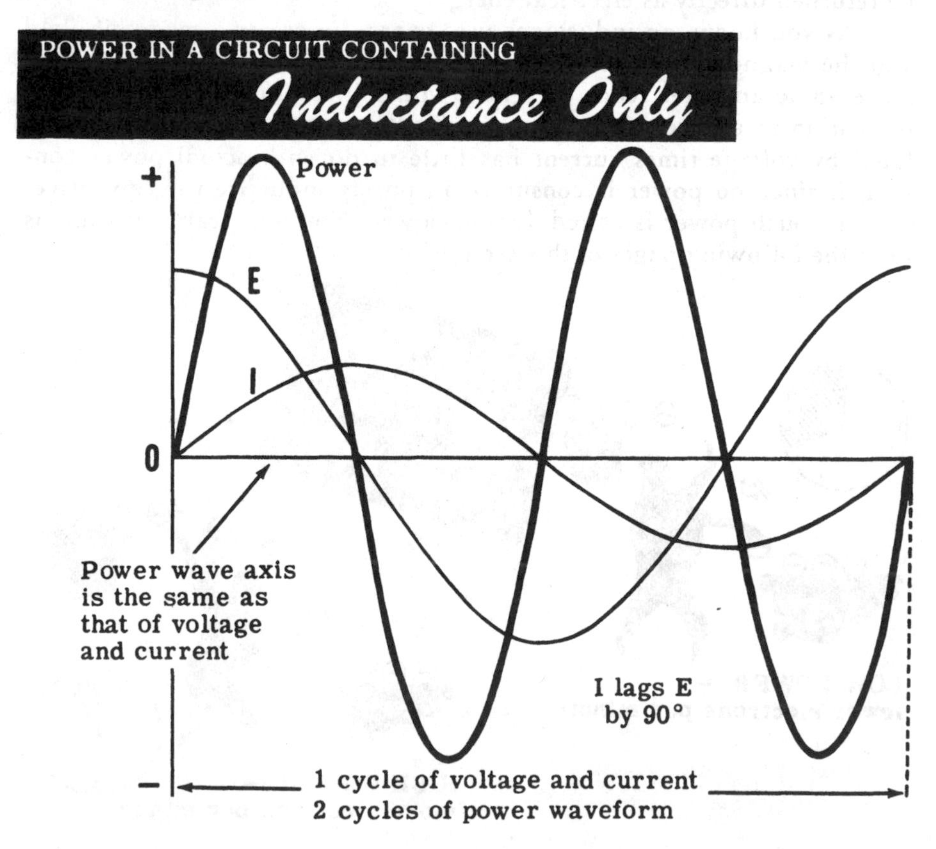

Positive and Negative Power in AC Inductive Circuits

That portion of a power wave which is above the zero axis is called *positive power*, and that which is below the axis is called *negative power*. Positive power represents power furnished to the circuit by the power source, while negative power represents power the circuit returns to the power source.

In the case of a pure inductive circuit, the positive power furnished to the circuit causes the magnetic field to build up. When this field collapses, it returns an equal amount of power to the power source. Since no power is used for heat or light in a circuit containing only pure inductance (if it were possible to have such a circuit), no actual power will be used, even though the current flow were large. The actual power used in any circuit is found by subtracting the negative power from the positive power.

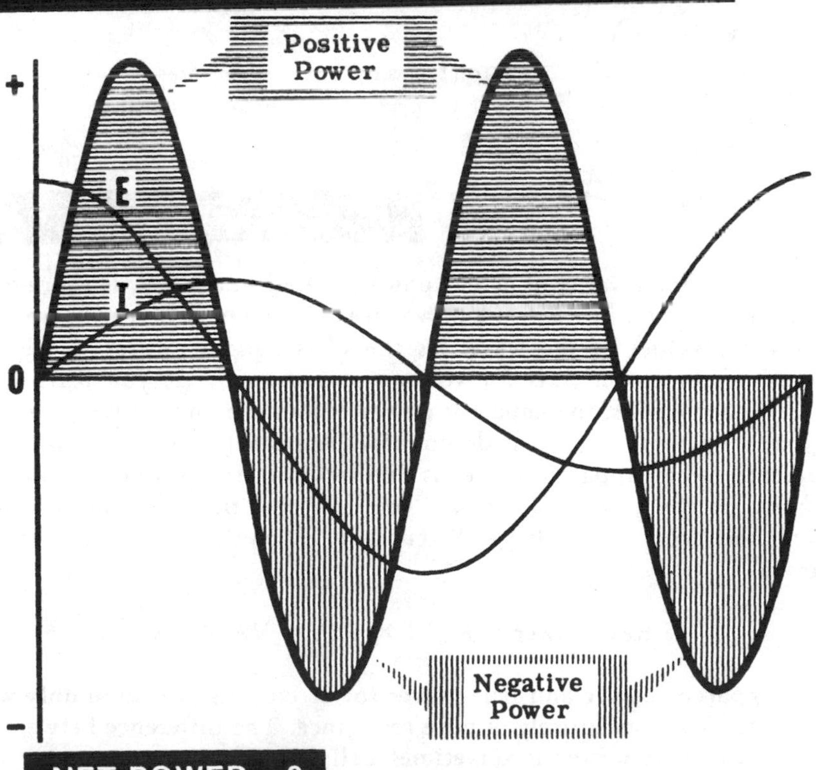

Apparent and True Power—Power Factor in AC Circuits

Any practical inductive circuit contains some resistance, and since the phase angle depends on the ratio between the inductive reactance and the resistance, it is always less than 90°. For phase angles different than 90°, the amount of positive power always exceeds the negative power, with the difference between the two representing the actual power used in overcoming the circuit resistance. For example, if your circuit contains equal amounts of inductive reactance and resistance, the phase angle is 45° and the positive power exceeds the negative power, as shown below.

90° PHASE ANGLE (NEGATIVE POWER EQUALS THE POSITIVE POWER)

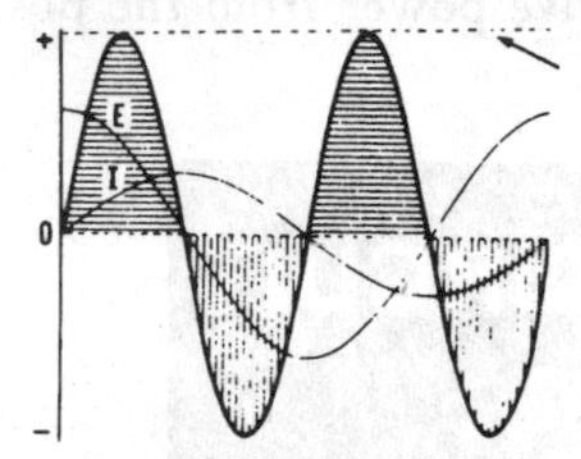

Apparent power E x I = Volt Amperes

Power wave axis--True power is zero

$$\text{Power Factor} = \frac{I^2 R \text{ or } \frac{E}{R}}{VA} = 0 \text{ at } 90°$$

45° PHASE ANGLE (POSITIVE POWER EXCEEDS NEGATIVE POWER)

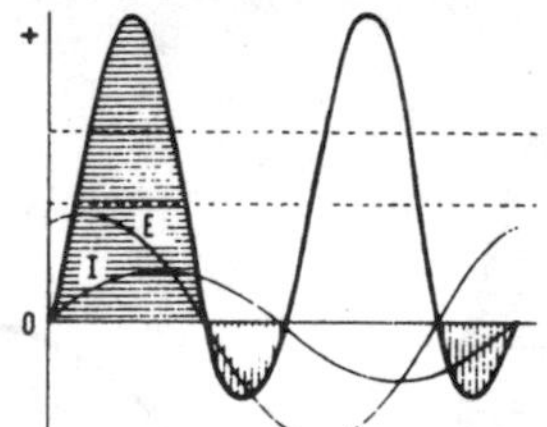

Apparent power ExI = Volt Amperes

Power wave axis--True power is ($I^2 R$)

$$\text{Power Factor} = \frac{I^2 R}{VA} = .70\% \text{ at } 45°$$

DECREASING THE PHASE ANGLE INCREASES THE TRUE POWER

The average value of actual power, called *true power*, is represented by an axis drawn through the power wave halfway between the opposite maximum values of the wave. As the phase angle increases, this axis moves nearer to the axis for voltage and current. As you learned for resistive ac circuits, the apparent power is found by multiplying the voltage and current, just as in dc circuits (Apparent Power = Voltage × Current). When apparent power is divided into true power, the resultant decimal is the power factor. As you can see, the real power can be calculated from the voltage × current × power factor if these are known.

$$\text{Real Power} = E \times I \times P.F. = VA \times P.F.$$

Apparent power and true power for ac circuits are equal only when the circuit consists entirely of pure resistance. The difference between apparent and true power is sometimes called *wattless power*, since it does not produce heat or light but does require current flow in a circuit.

Power Factor in AC Circuits

You have already learned that in a pure resistive circuit, I^2R or E^2/R (power in watts) is equal to $I_{eff} E_{eff}$ (apparent power), and that the power factor is equal to 1. Power factor is the ratio of true power to apparent power. This is true because the phase angle between the voltage and current waves in a pure resistive circuit is always zero.

In an inductive circuit, the phase angle is not zero and power in watts does not equal apparent power. As a result, the power factor will be between zero and 1.

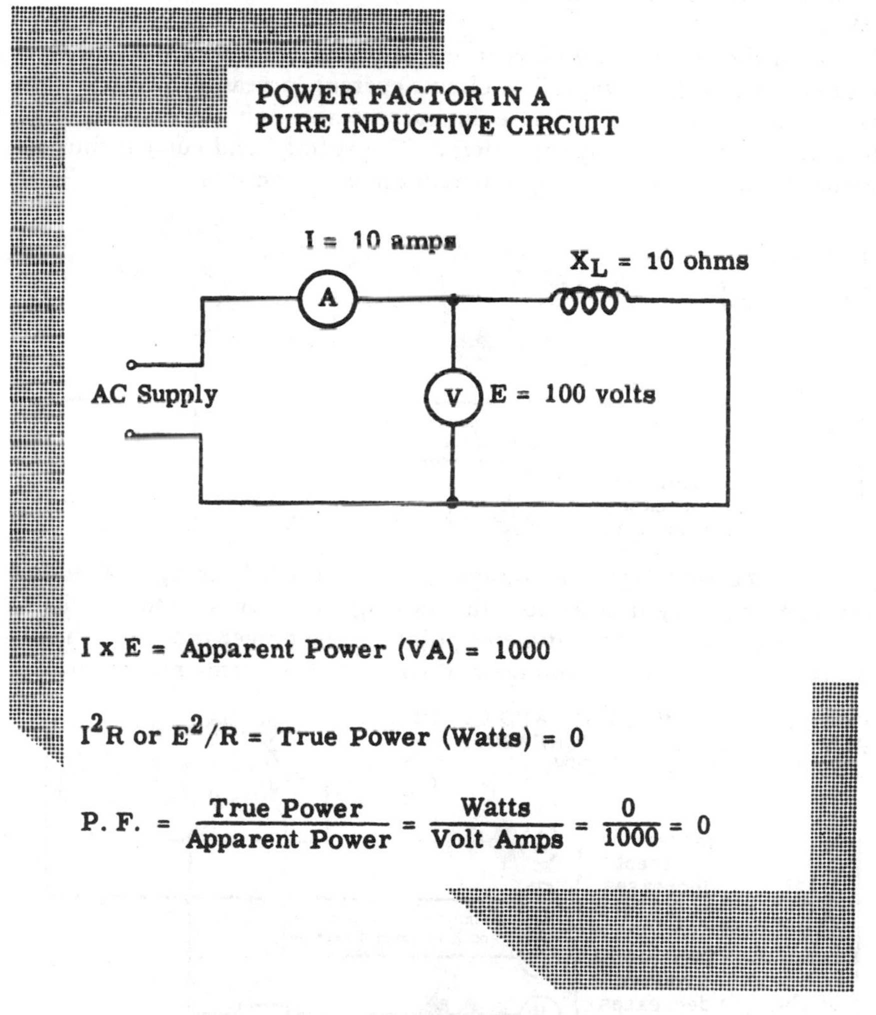

Power factor is used to determine what percentage of the supplied power is used in watts, and what percentage is returned to the source as wattless power.

Power factor in a pure inductive circuit is equal to zero since the phase angle is 90°.

Measurement of True Power in AC Circuits

Since the product of the circuit current and voltage is the apparent power and not the true power, you must either know the power factor to calculate true power or use some other device that is sensitive to the phase angle or power factor. A wattmeter is used to measure the true power used in an ac circuit. Voltmeter and ammeter readings are not affected by the phase difference between circuit current and voltage, since the voltmeter reading is affected only by voltage and the ammeter reading is affected only by current. The wattmeter reading is affected by both the circuit current and voltage, and the phase difference between them, as shown below.

When the voltage and current are in phase, the current increases at the same time as the voltage. The circuit current increases the meter field simultaneously with an increase in current through the moving coil, which is in turn caused by the voltage. The voltage and current thus act together to increase the turning force on the meter pointer.

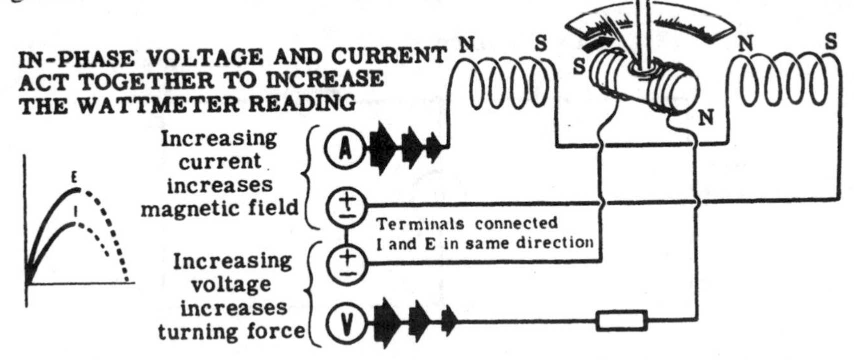

If the current lags the voltage, the meter field strength does not increase at the same time as does the moving coil current. This results in less turning force on the wattmeter pointer. The power indicated then is less than with an in-phase voltage and current of the same magnitude.

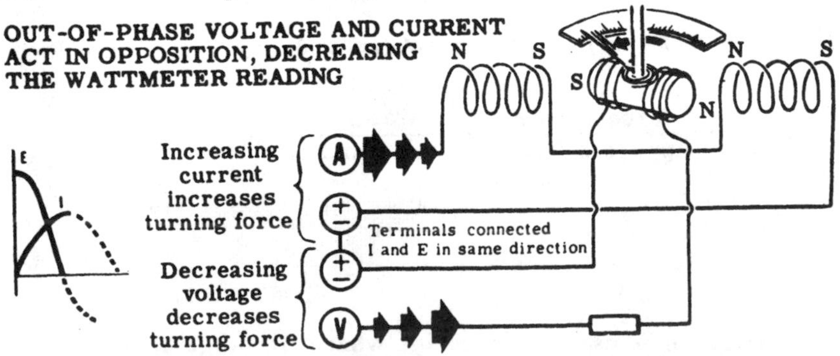

Similarly, if the current leads the voltage, the meter field strength and the moving coil current will not increase at the same time. This results in a lower wattmeter reading, the actual power used by the circuit again being less than the apparent power.

Review of Inductance in DC and AC Circuits

To review inductance, what it is, and how it affects current flow, consider these facts concerning both it and inductive reactance.

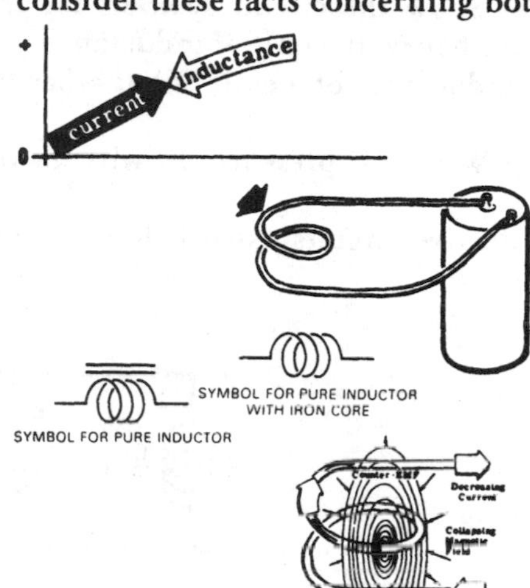

1. INDUCTANCE—The property of a circuit which opposes any change in the current flow; measured in henrys and symbolized L.

2. INDUCTOR—A coil of wire used to supply inductance in a circuit.

3. INDUCTANCE SYMBOL—The symbol for a pure inductor is a coil of wire.

4. INDUCED EMF—A voltage which is generated within a circuit by the movement of the magnetic field whenever the circuit current changes, and which opposes the current change.

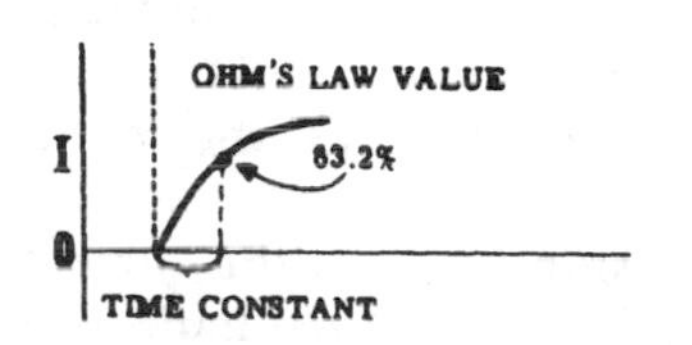

5. INDUCTIVE TIME CONSTANT—The ratio of L to R which gives the time in seconds required for the circuit current to rise to 63.2% of its maximum value, or to drop 63.2% from its maximum value.

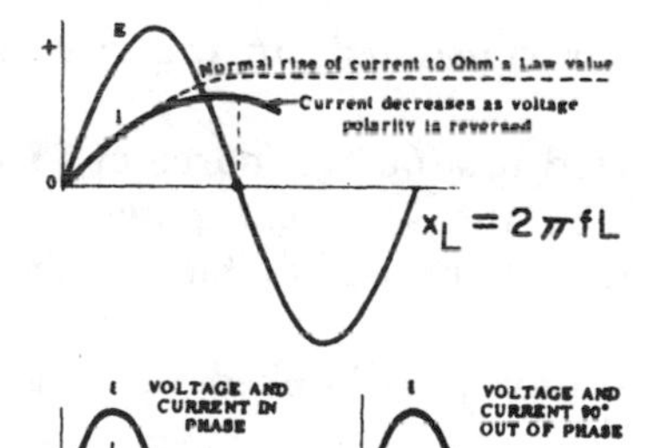

6. INDUCTIVE REACTANCE—The action of inductance in opposing the flow of ac current and in causing the current to lag the voltage; measured in ohms and symbolized by X_L.

7. PHASE ANGLE—The number of degrees by which the current wave lags or leads the voltage wave. In a pure inductive circuit, the phase angle is 90°.

$$PF = \frac{\text{True Power}}{\text{Apparent Power}}$$

8. POWER FACTOR—The ratio of the true power to the apparent power.

$$\text{APPARENT POWER} = E \times I = VA$$

9. TRUE AND APPARENT POWER IN PURELY INDUCTIVE CIRCUITS—The phase angle is 90°, the power factor is zero, and the true power is zero. The apparent power is often called volt-amperes (VA).

Self-Test—Review Questions

1. Why does a wire carrying current have a magnetic field?
2. What is the symbol for inductance? Show the circuit symbols for iron core and air core inductors. Define the henry, the unit of inductance.
3. Why does an iron core increase the inductance of a coil? What other factors affect inductance?
4. What is induced emf (counter emf)? Why is it present? On what factors does it depend?
5. Calculate the current flow in the inductive circuits listed below at 60 Hz and 400 Hz.

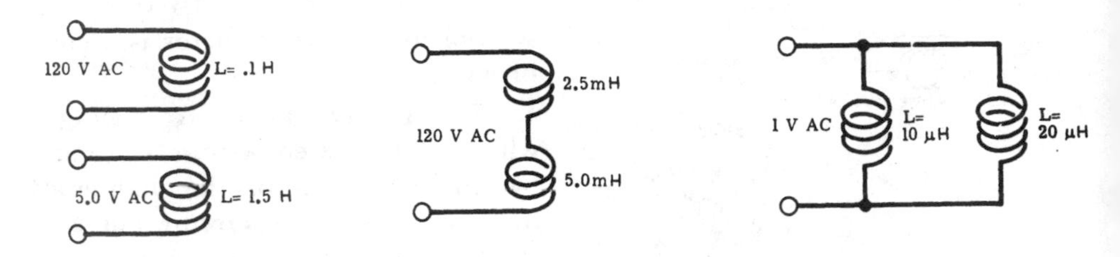

6. Suppose you had a circuit made up of an inductor and resistor in series. If you connected this to a battery through a switch and closed the switch at time = 0, what would be the current through the circuit at one, two, and four time constants if the resistance were 30 ohms, the inductor 1 henry, and the voltage source 15 volts?
7. Draw sketches showing the phase relationship between the current and voltage in an ac circuit containing pure inductance. What happens when some resistance is added to the circuit?
8. How much real power is dissipated (or consumed) in the circuits of question 6? Explain your answer.
9. If the circuit of question 7 were connected to a 60-Hz source of 15 volts rather than the battery, what would be the apparent power? What would be the true power? What would be the power factor? Explain your answers.
10. Explain how a wattmeter works. Show how it is hooked up in ac circuits. Why does it measure true power?

Learning Objectives—Next Section

Overview—Now that you know about inductance, you are ready to learn about the last major circuit characteristic—capacitance —and how it functions in dc and ac circuits.

Experiment/Application— Effect of Core Material on Inductance

Suppose you wired a series circuit consisting of an air-core coil and a 60-watt lamp. When the circuit is energized from the 115-volt ac line, the lamp brilliance is noted.

With the circuit energized, carefully insert a solid iron core into the coil. Note the *decrease* in lamp brilliance resulting from the *increased* inductance of the coil and, hence, increased inductive reactance. A larger percentage of the 115-volt source voltage is now dropped across the coil.

Next, remove the iron core and insert a copper core. Note the *increase* in lamp brilliance resulting from the *decreased* inductance of the coil. Large internal induced currents in the copper weaken the coil magnetic field, thus decreasing its inductance. You will learn about this *eddy current* effect later. A smaller percentage of the source voltage is now dropped across the inductor and, therefore, the lamp burns brighter.

Next, remove the copper core and insert a laminated core. Note that the lamp brilliance has dropped greatly. The laminated iron core has increased the coil inductance an even greater amount than the solid iron core because the laminations have greatly reduced the losses due to effects called *hysteresis* as well as *eddy currents*. Again, you will learn more about hysteresis later. Most of the source voltage is now dropped across the coil and, as a result, the lamp barely lights.

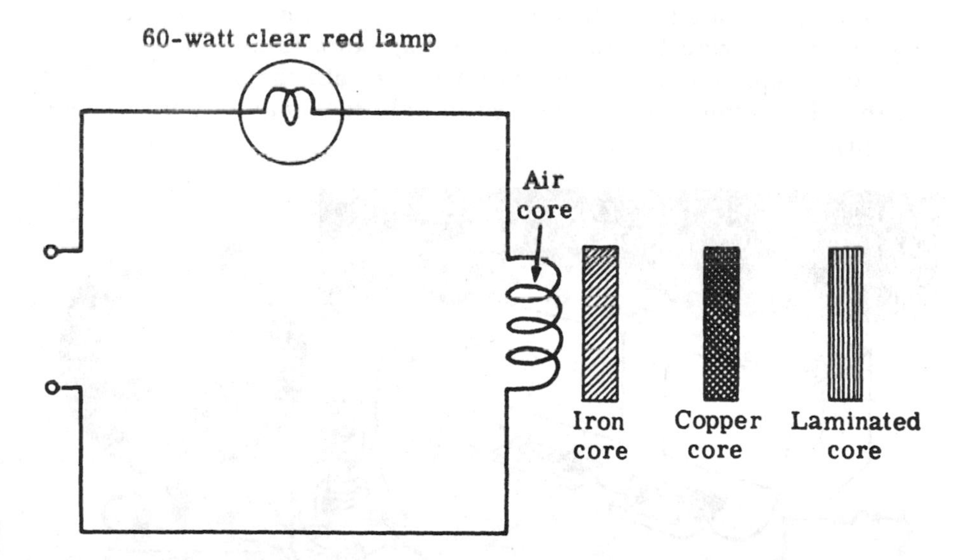

HOW INDUCTANCE VARIES WITH CORE MATERIAL

Experiment/Application—Generation of Induced EMF

When the current flow in a dc circuit containing inductance is stopped abruptly, by opening a switch, for example, the magnetic field of the inductance tries to collapse instantaneously. The rapid collapse of the field momentarily generates a very high voltage, and this induced emf may cause an arc at the switch. While the field collapse is too rapid to allow measurement of this voltage with a voltmeter, a neon lamp can be used to show that the voltage is much higher than the original battery voltage.

Neon lamps differ from ordinary lamps in that they require a certain voltage before they begin to light, but they will light abruptly at this voltage. This voltage, called the *starting voltage*, varies for different neon lamps. Its value can be determined by increasing the voltage applied across the lamp until it lights. The voltage applied at the time the lamp first lights is the starting voltage.

To find the starting voltage required for the neon lamp, you could connect two 45-volt batteries in series to form a 90-volt battery. Across the 90-volt battery, connect a variable resistor as a potentiometer, with the outside or end terminals of the variable resistor connected to the battery terminals. A lamp socket is connected between the center terminal of the variable resistor and one of the outside terminals, and a 0-100-volt range dc voltmeter is connected across the lamp socket terminals.

With the neon lamp inserted, you can vary the voltage applied to the lamp by varying the setting of the variable resistor. The correct starting voltage is found by lowering the voltage to a value which does not light the lamp, and then slowly increasing it until the lamp lights. You see that the starting voltage required to light the lamp is approximately 70 volts.

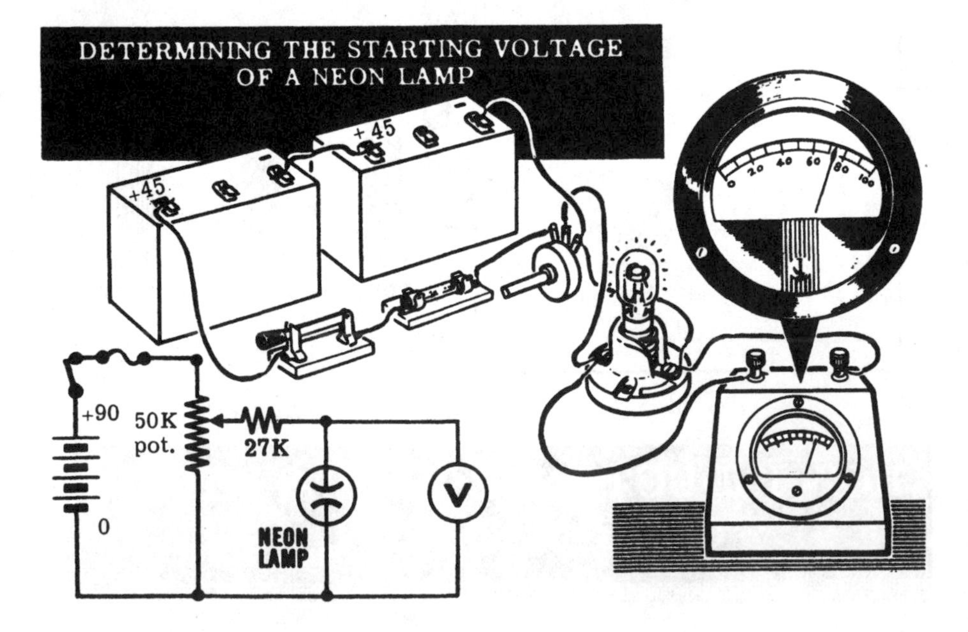

Experiment/Application— Generation of Induced EMF (continued)

Next, four dry cells are connected in series to form a 6-volt battery, with the lamp socket connected across its terminals through a fuse and a switch. A neon lamp is inserted in the socket, and a 5-henry inductor is connected across the lamp terminals.

When you close the switch, you see that the neon lamp does not light and the battery voltage measured with an 0-10-volt dc voltmeter is 6 volts. Since 6 volts is less than the starting voltage of the lamp, some means of obtaining a higher voltage is required to cause the lamp to light.

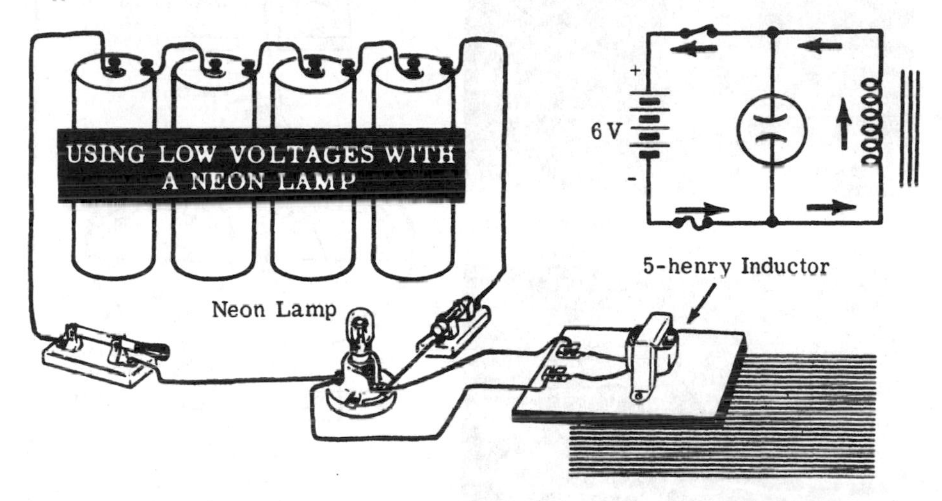

As the switch is opened rapidly, you see that the lamp flashes, indicating that the voltage across the lamp and coil (in parallel) is higher than the starting voltage required for the lamp. This voltage is the induced emf generated by the collapsing field of the inductor and is a visible effect of inductance.

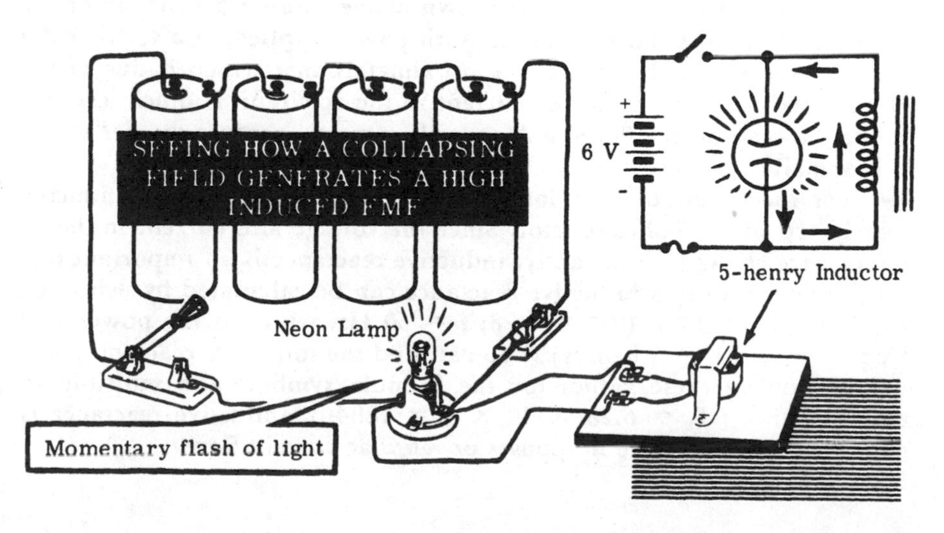

Experiment/Application—Current Flow in Inductive Circuits

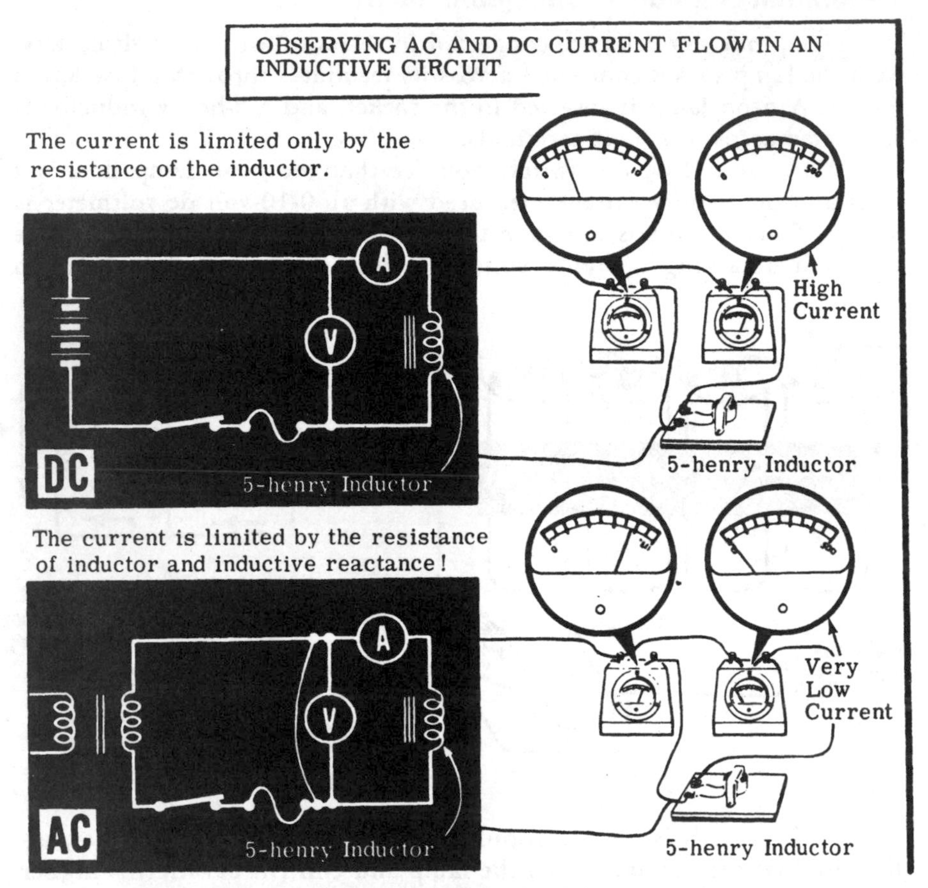
OBSERVING AC AND DC CURRENT FLOW IN AN INDUCTIVE CIRCUIT
The current is limited only by the resistance of the inductor.
A
V
DC
5-henry Inductor
High Current
5-henry Inductor
The current is limited by the resistance of inductor and inductive reactance!
A
V
AC
5-henry Inductor
Very Low Current
5-henry Inductor

Inductance holds back AC current

Suppose you had the circuit shown above using a 5-henry inductor with a 60-ohm internal resistance. With power applied, you see that the current flow in the dc circuit is approximately that which is due to the 60-ohm resistance *only*, but the current in the ac circuit is much less and cannot be read on the 0-500 mA ac milliammeter, because the deflection is too small to be observed.

For the dc circuit, the inductance has no effect, and the inductor merely acts as a 60-ohm resistor. Since the voltage and current in the ac circuits are changing constantly, inductive reactance is an important factor. The effect of this inductive reactance can be calculated by using the formula, $X_L = 2\pi fL$ ($2\pi = 6.28$; $f = 60$ Hz, which is the power line frequency; and $L = 5$ henrys). You can find the inductive reactance, X_L, by substituting these values for the formula symbols and multiplying them together ($X_L = 6.28 \times 60 \times 5 = 1884\Omega$). Inductive reactance is expressed in ohms, since it opposes or *resists* ac current flow.

Experiment/Application—Apparent and True Power

You can use a voltmeter, ammeter, and wattmeter to show the relationship between true and apparent power. Suppose you hooked up the circuit shown below. The current (I) will be 2.4 amperes and the apparent power will be E × I, or 288 watts.

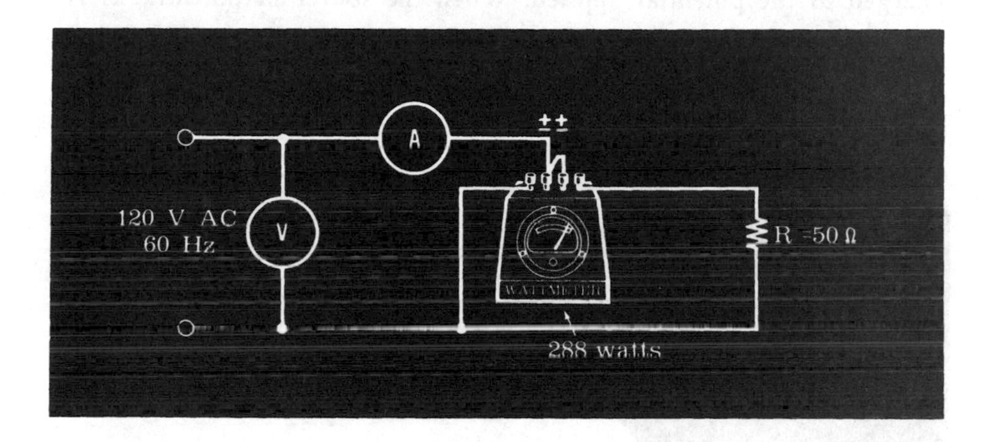

Examination of the wattmeter will also show a true power of 288 watts. Therefore, the true and apparent power will appear to be equal and the power factor is 1 since true power divided by apparent power = 288/288 = 1.

Suppose you replaced the resistor with a 0.13-henry inductor (X_L = 50 ohms at 60 Hz). You would have a circuit like the one shown below.

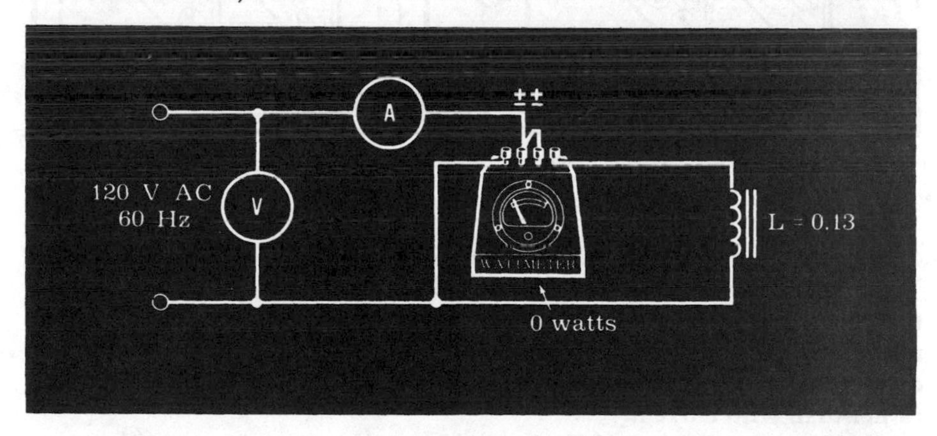

You would find that the apparent power would still be about 288 watts as measured by the product of E and I. The wattmeter, however, will read zero, showing that little or no true power is being consumed. (There may be a small output of power indicated by the wattmeter because of losses in the inductor due to resistance and the effect of the core.) Thus, except for these losses, an inductor consumes no power.

Capacitance—Storage of Charge on Conductors

As you know, *conductors* or *insulators* can be charged so that they may have an excess of electrons (negative charge) or a deficiency of electrons (positive charge). A *potential difference* (voltage) applied to a pair of conducting plates near each other, but not touching each other, will be charged to the potential applied. When the source of potential is removed, the charge will remain; but as you know, the charge can be removed by connecting the two plates together or by the use of an external circuit.

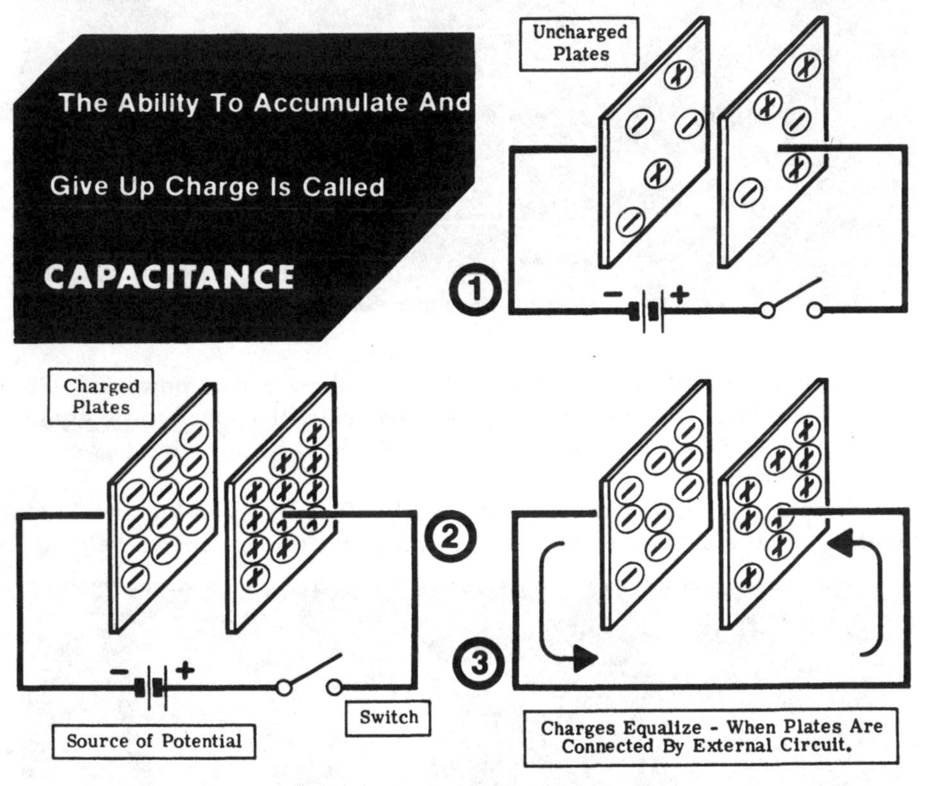
The Ability To Accumulate And
Give Up Charge Is Called
CAPACITANCE
Uncharged Plates
1
2
3
Charged Plates
Source of Potential
Switch
Charges Equalize - When Plates Are Connected By External Circuit.

Energy is taken from the circuit to supply the charge to the plates. This energy is returned to the circuit when the charge is removed from the plates. Thus, the charging and discharging of the plates is analogous to the generation of and the collapse of magnetic fields in an inductor or conductor, except that we are dealing with electric charge (potential) rather than with current.

A magnetic field can be enhanced by coiling the wire, as with an inductor, and the storage of charge can be enhanced by additional factors, such as increasing the area of the plates. As you know, there is an electrostatic field between the charged plates. This field increases when the plates are moved closer together, increasing the ability of the plates to accumulate charge. This ability to accumulate charge from the circuit and to give it back to the circuit is called *capacitance*.

Capacitance in DC Circuits

To see how capacitance affects the voltage in a circuit, let us say that your circuit contains a capacitor consisting of two plates, a knife switch, and a dry cell as shown below. Assuming both plates are uncharged and the switch is *open*, then *no* current will flow and the voltage between the two plates is *zero*.

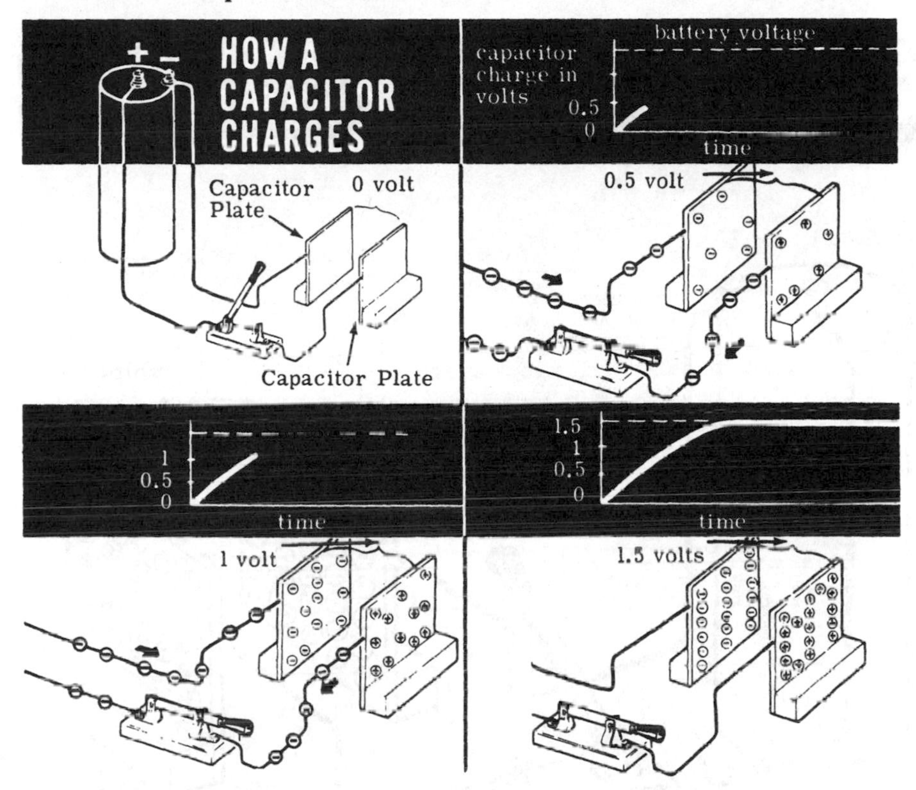

When the switch is closed, the battery furnishes electrons to the plate connected to the negative terminal and takes electrons away from the plate connected to the positive terminal. The voltage between the two plates should equal the voltage between the cell terminals, or 1.5 volts. However, this does not occur immediately. One plate must take the excess of electrons to become negatively charged, while the other must give up electrons to become positively charged, in order for a voltage of 1.5 volts to exist between the two plates. As electrons move onto the plate attached to the negative terminal of the cell, a negative charge is built up, opposing the movement of more electrons onto the plate; and, similarly, as electrons are taken away from the plate attached to the positive terminal, a positive charge is built up, opposing the removal of more electrons from that plate. This action on the two plates is called *capacitance* and it opposes the change in voltage from zero to 1.5 volts. Capacitance delays the change in voltage for a limited amount of time but it does not prevent that change.

Capacitance in DC Circuits (continued)

When the switch is opened the plates remain charged, since there is no path between the two plates through which they can discharge. As long as no discharge path is provided, the voltage between the plates will remain at 1.5 volts and, if the switch is closed again, there will be no effect on the circuit since the capacitor is already charged.

DC CURRENT FLOW STOPS WHEN THE CAPACITOR BECOMES CHARGED

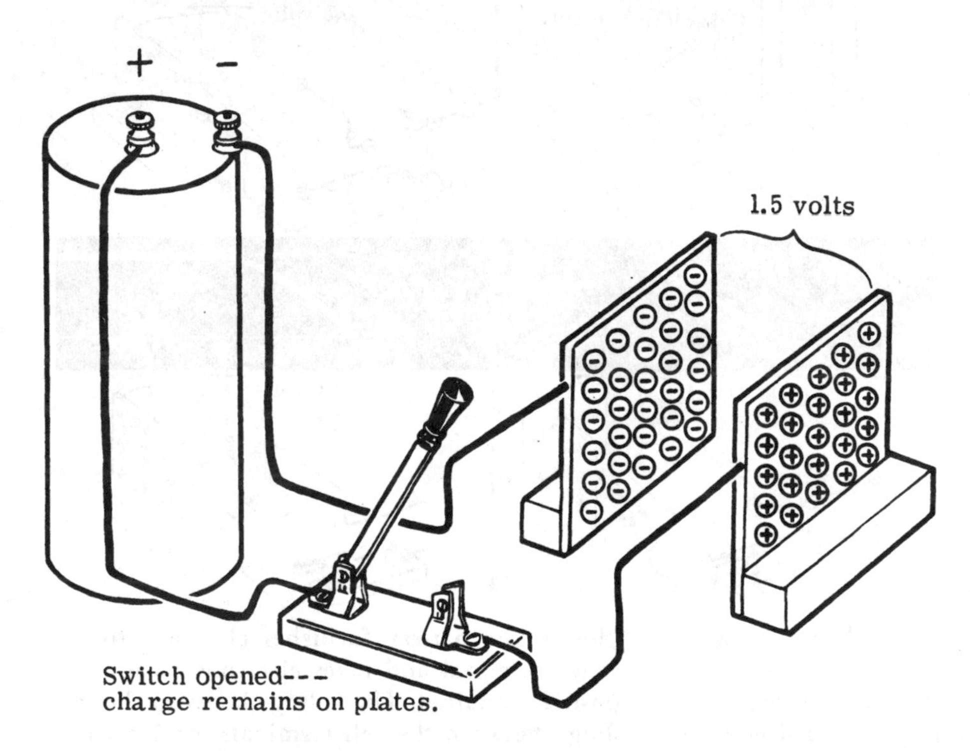

Current will flow in a capacitive circuit only long enough to charge the capacitor, as with a dc voltage source. When the dc circuit switch is closed, an ammeter connected in order to read circuit current will show that a very large current flows at first, since the capacitor plates are uncharged. Then as the plates gain additional charge, the charging current decreases until it reaches zero—at the moment the charge on the plates equals the voltage of the dc voltage source.

The current that charges a capacitor flows only for the first moment after the switch is closed. After this momentary flow the current stops, since the plates of the capacitor are separated by an insulator which does not allow electrons to pass through it. Thus, capacitors do not allow dc current to flow *continuously* through a circuit.

Capacitance in DC Circuits (continued)

When the voltage across an electric circuit changes, the circuit opposes this change. This opposition is called *capacitance*. Like inductance, capacitance cannot be seen, but its effect is present in every electric circuit whenever the circuit voltage changes.

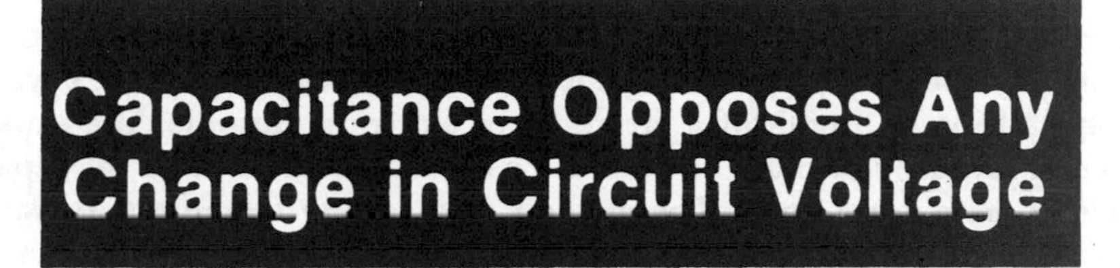

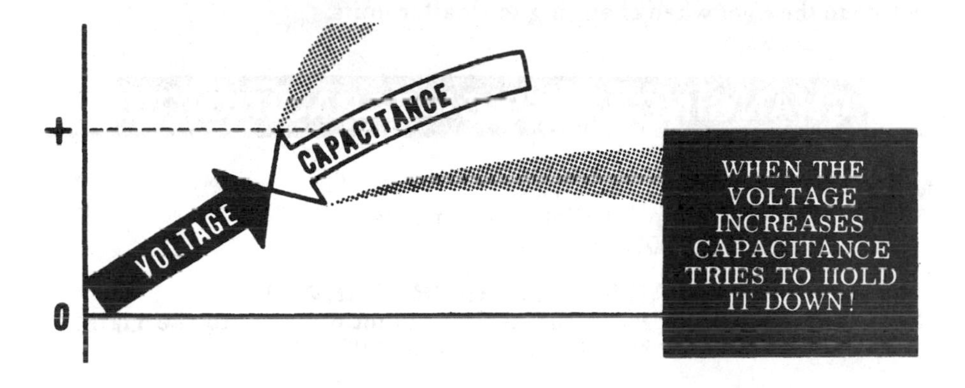

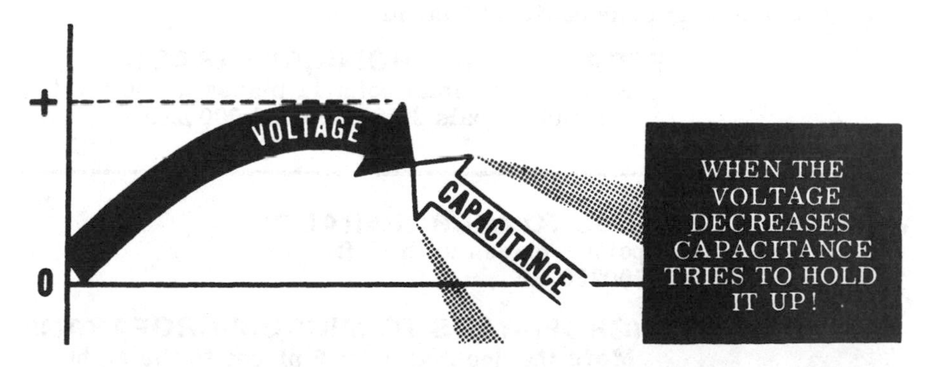

Capacitance affects dc circuits *only* when it is turned on and off. (In ac circuits, however, the voltage is continuously changing, so that the effect of capacitance is *continuous*.) The amount of capacitance present in a circuit depends on the physical construction of the circuit and the electrical devices used. The capacitance may be so small that its effect on circuit voltage is negligible.

Units of Capacitance

The basic unit of capacitance is the *farad.* A capacitor has a capacitance of 1 farad when a charging current of 1 ampere, flowing for 1 second, causes a change of 1 volt in the potential between its plates.

The farad is much too great to use as the unit of capacitance for practical electric circuits. Therefore, the units normally used are the *microfarad*, abbreviated μF, which is one-millionth of a farad, and the picofarad, abbreviated pF (sometimes called *micromicrofarad*, $\mu\mu$F), which is one-million-millionth of a farad. Since electrical formulas use capacitance in farads, you must be able to change the various units of capacitance to these other units. The method of changing these units is *exactly* like the method used for changing units of voltage, current, resistance, etc. The decimal point moves to the *left* when changing to larger units while to the *right* when changing to smaller units.

CHANGING UNITS OF CAPACITANCE

MICROFARADS TO FARADS

Move the decimal point 6 places to the left
120 μF equals 0.000120 farad

FARADS TO MICROFARADS

Move the decimal point 6 places to the right
8 farads equals 8,000,000 μF

MICROMICROFARADS TO FARADS

Move the decimal point 12 places to the left
1500 $\mu\mu$F equals 0.000000001500 farad

FARADS TO MICROMICROFARADS

Move the decimal point 12 places to the right
2 farads equals 2,000,000,000,000 $\mu\mu$F

MICROMICROFARADS TO MICROFARADS

Move the decimal point 6 places to the left
250 $\mu\mu$F equals 0.000250 μF

MICROFARADS TO MICROMICROFARADS

Move the decimal point 6 places to the right
2 μF equals 2,000,000 $\mu\mu$F

Symbols for Capacitance

In electrical formulas, the letter "C" is used to denote capacitance in farads. The circuit symbols for capacitance are shown below. You will find the terms *capacitor* and *condenser* used interchangeably, since they mean the same thing.

or

Fixed

Electrolytic (aluminum or tantalum)

or

Variable

Ganged Variable

Capacitor Construction

Three basic factors that influence the capacity of a capacitor are: (1) the *area* of the plates, (2) the *distance* between the plates (thickness of the dielectric), and (3) the *material* used for the dielectric.

Basically, capacitors consist of two conducting plates which can be charged; the plates are separated by an insulating material called the *dielectric*. While early capacitors were made with solid metal plates, newer types of capacitors use metal foil or plating, particularly aluminum, for the plates. Dielectric materials commonly used are air, mica, paper, ceramic, plastic, and metallic oxides.

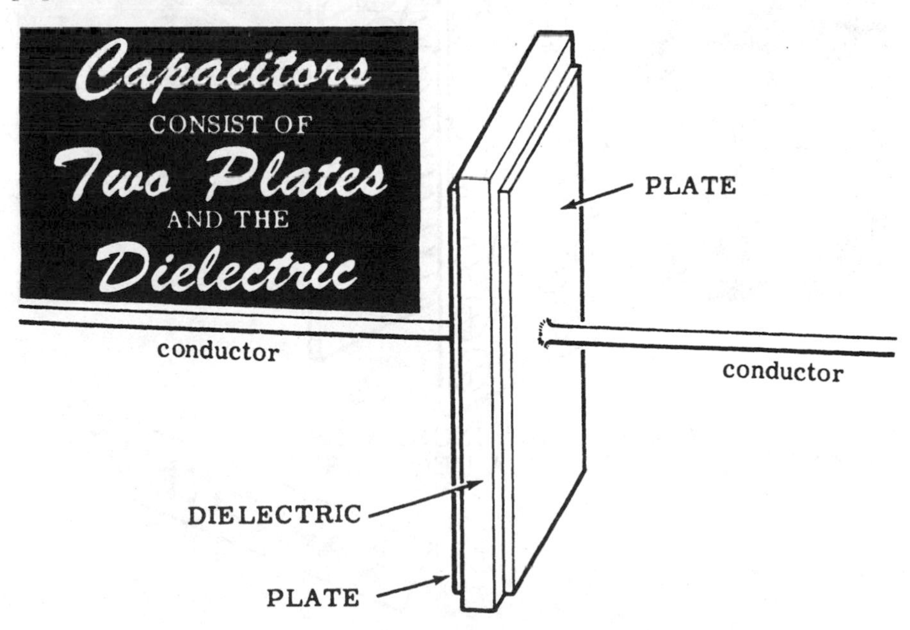

Factors which Affect Capacitance—Plate Area

Plate area is a basic factor in determining the amount of capacitance, since the capacitance varies *directly* with the area of the plates. A large plate area has room for more excess electrons than a small area, and thus it can hold a greater charge. Similarly, the large plate area has more electrons to give up and will hold a much larger positive charge than a small plate area. Thus, an *increase* in plate area *increases* capacitance, and a *decrease* in plate area *decreases* capacitance. So, we can say that capacitance is *directly proportional* to plate area.

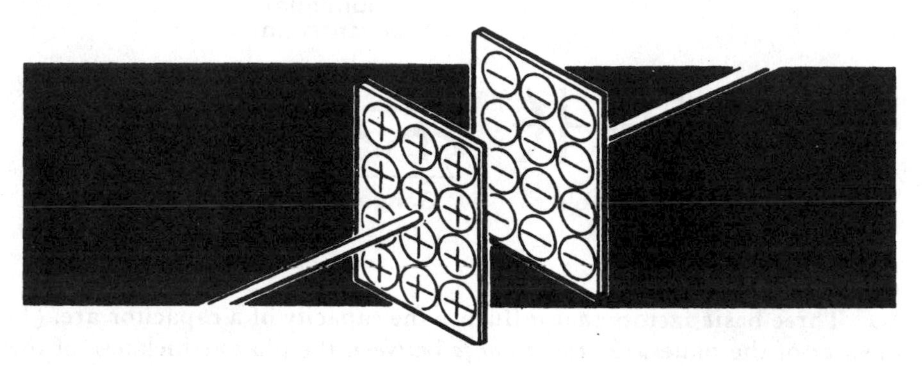

Larger plates hold more electrons

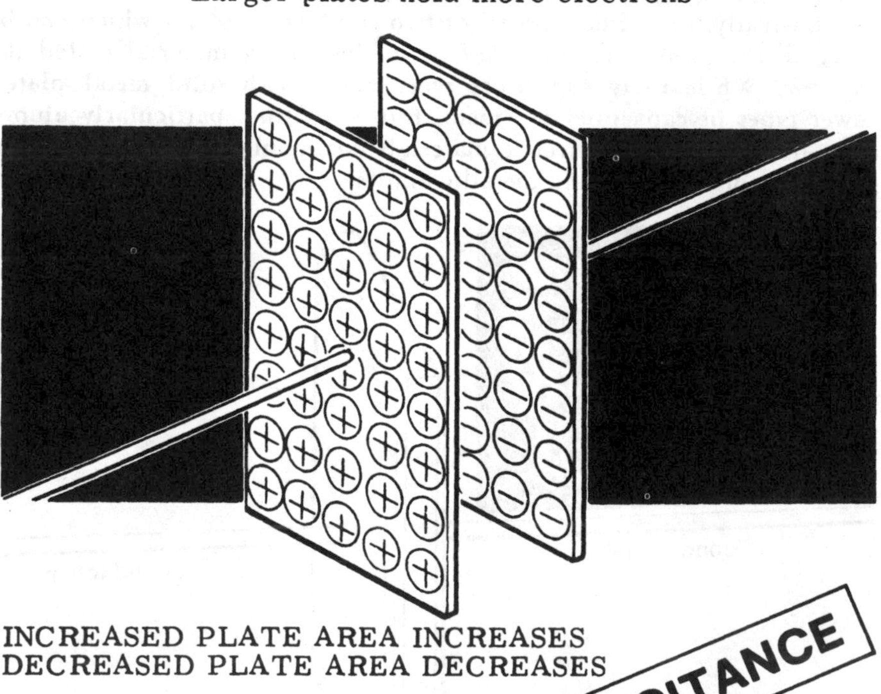

INCREASED PLATE AREA INCREASES
DECREASED PLATE AREA DECREASES
CAPACITANCE

Factors which Affect Capacitance—Plate Separation

The effect two charged bodies have on each other depends on the distance between them. Since capacitance depends on the ability to accumulate charge, the amount of capacitance changes when the distance between the plates changes. The capacitance between two plates *increases* as the plates are brought *closer* together and *decreases* as the plates are moved farther *apart*. Thus, the closer the plates are to each other, the greater the effect a charge on one plate will have on the charge of the other plate.

When an excess of electrons appears on one plate of a capacitor, electrons are forced off the opposite plate, inducing a positive charge on this plate. Similarly, a positively charged plate induces a negative charge on the opposite plate. The closer the plates are to each other, the stronger the force between them, and this force increases the capacitance of a circuit. Thus, we can say that the capacitance is *inversely proportional* to the spacing.

INCREASING THE DISTANCE BETWEEN THE PLATES
DECREASES CAPACITANCE

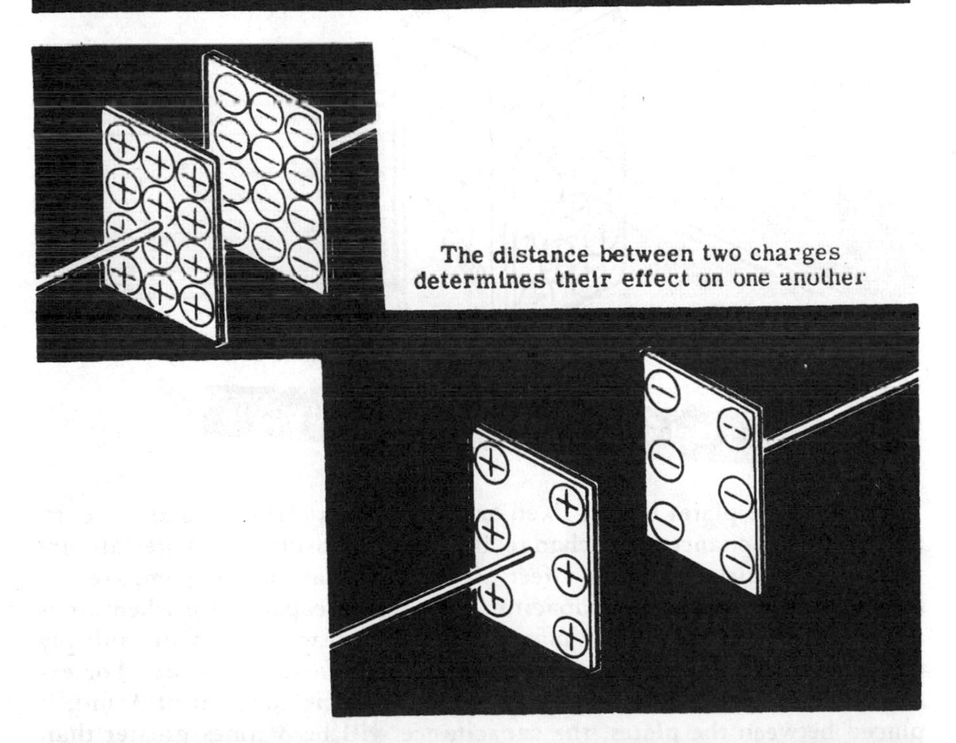

The distance between two charges determines their effect on one another

DECREASING THE DISTANCE BETWEEN THE PLATES
INCREASES CAPACITANCE

Factors which Affect Capacitance—Dielectrics

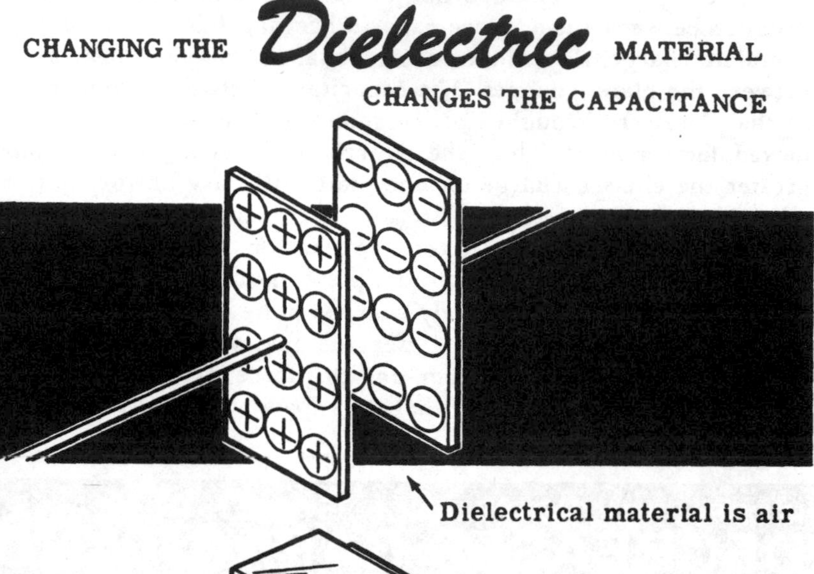
CHANGING THE Dielectric MATERIAL
CHANGES THE CAPACITANCE
Dielectrical material is air

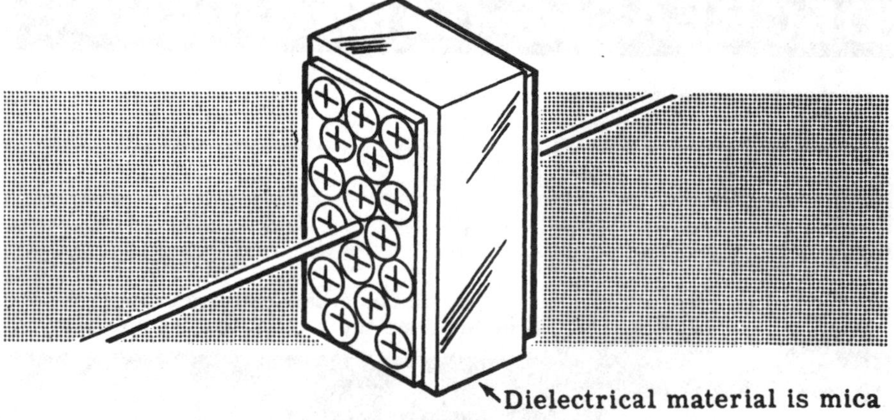
Dielectrical material is mica

Mica dielectric increases the capacitance.

When the plates have a fixed area, and are a certain distance apart, then the capacitance will change if different insulating materials are used for the dielectric. The effect of different materials is compared to that of air—that is, if the capacitor has a given capacitance when air is used as the dielectric, other materials used instead of air will multiply the capacitance by a certain amount called the *dielectric constant*. For example, some types of oiled paper have a dielectric constant of 3 and, if placed between the plates, the capacitance will be 3 times greater than when air is used as the dielectric. Different materials have different dielectric constants and thus will change the capacitance when they are placed between the plates to act as the dielectric.

Factors which Affect Capacitance—Dielectrics (continued)

In addition, the thickness of the dielectric is determined by the voltage rating of the capacitor. Obviously, the dielectric must not break down under the stress of the potential difference across the capacitor terminals. Thus, *voltage rating* of a capacitor is important and must not be exceeded, otherwise the capacitor will fail because of dielectric failure. The dielectric constants and dielectric strengths of some common materials are listed in the table below.

Material	*Dielectric Constant (K)*	*Dielectric Strength Volts per .001 inch*	*Volts per cm*
Air	1	80	32,000
Paper (oiled)	3-4	1500	600,000
Mica	4-8	1800	720,000
Glass	4-8	200	80,000
Porcelain	5	750	300,000
Titanates	100-200	100	40,000

Thus, the relationship between all of the factors affecting capacitance is:

$$C = \frac{KA}{D}$$

where K is the dielectric constant, A is the area of the plates, and D is the distance between the plates (dielectric thickness).

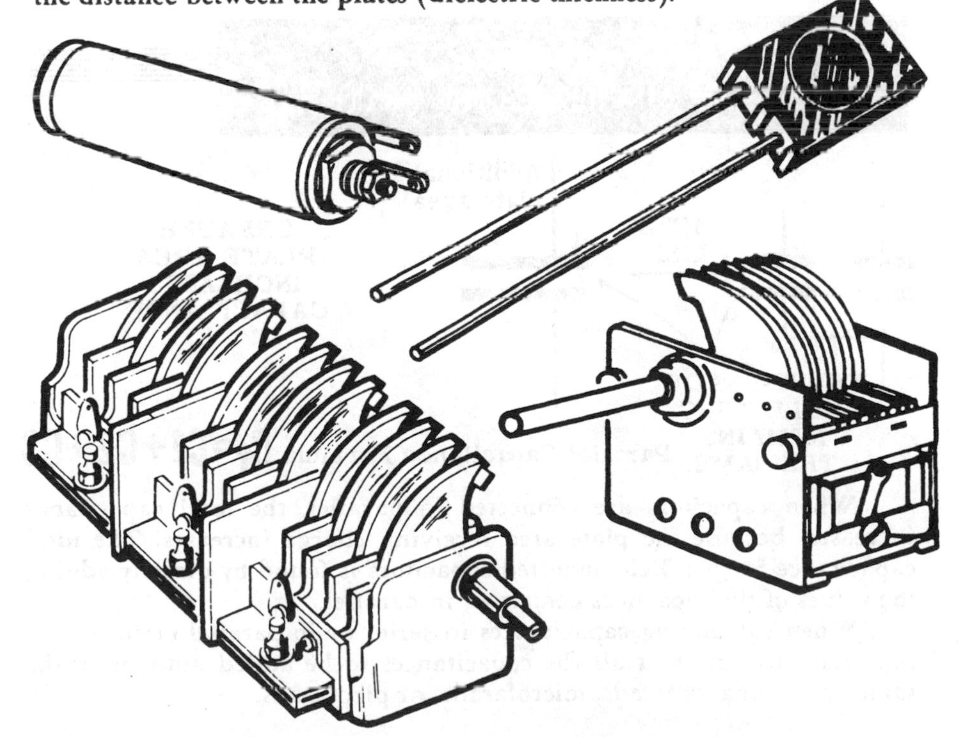

Capacitors in Series and Parallel

When you connect capacitors in series or in parallel, the effect on the total capacitance is as shown below.

Connecting capacitors in series decreases the total capacitance, because it effectively increases the spacing between the plates. To find the total capacitance of series-connected capacitors, a formula is used similar to the formula for parallel resistances.

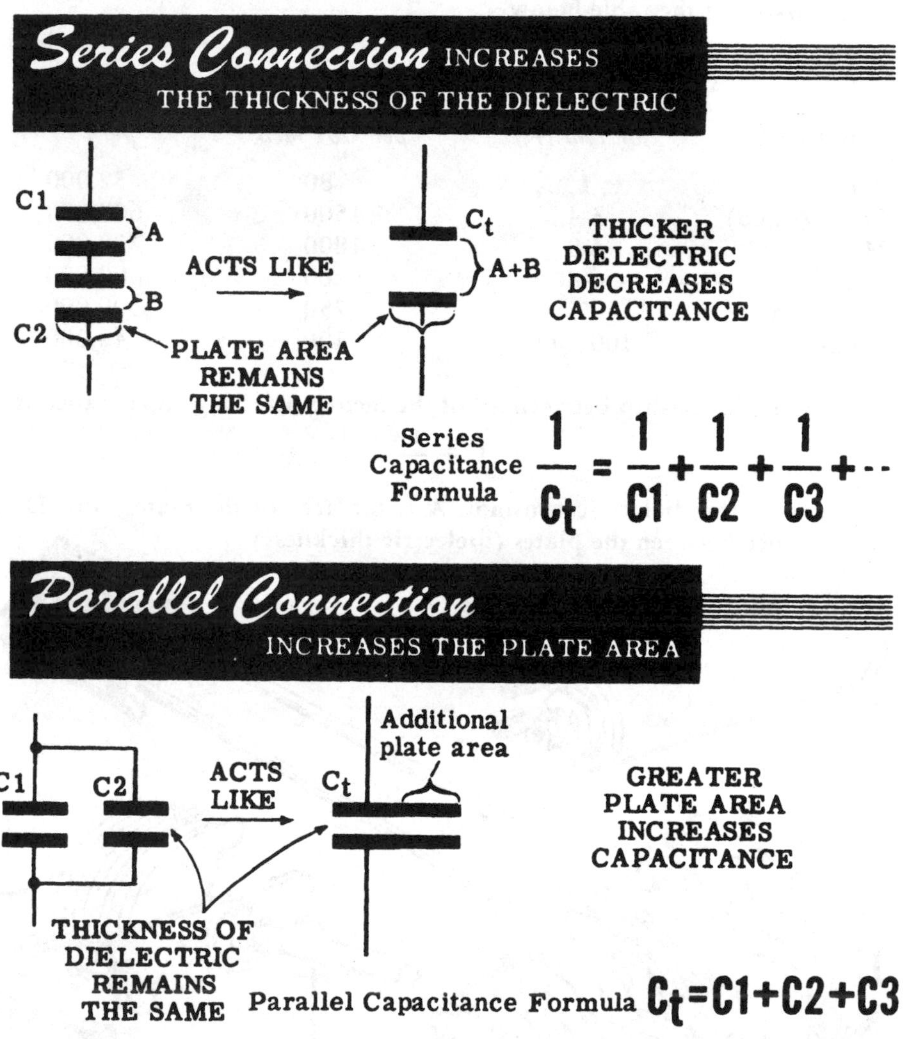

Series Connection INCREASES THE THICKNESS OF THE DIELECTRIC
C1
A
B
C2
ACTS LIKE
Ct
A+B
THICKER DIELECTRIC DECREASES CAPACITANCE
PLATE AREA REMAINS THE SAME
Series Capacitance Formula
1/Ct = 1/C1 + 1/C2 + 1/C3 + ...
Parallel Connection INCREASES THE PLATE AREA
C1
C2
ACTS LIKE
Ct
Additional plate area
GREATER PLATE AREA INCREASES CAPACITANCE
THICKNESS OF DIELECTRIC REMAINS THE SAME
Parallel Capacitance Formula Ct=C1+C2+C3

When capacitors are connected in parallel, the total capacitance increases, because the plate area receiving charge increases. The total capacitance for parallel-connected capacitors is found by directly adding the values of the capacitors connected in parallel.

When calculating capacitances in series or in parallel circuits, it is important to note that all the capacitances to be added must be in the same units—that is farads, microfarads, or picofarads.

Types of Capacitors—Variable

Many kinds of capacitors are used in electrical and electronic equipment. In order for you to choose the best type for a particular job, you should know how they are made and how they operate. You should also be familiar with the symbols used to indicate certain special types of capacitors. Capacitors are generally classified according to their dielectric material, and whether they are fixed or variable.

The most basic type of capacitor—which may be either *fixed* or *variable*—is constructed of metal plates with air spaces between them. A similar type is the vacuum capacitor, which consists of two plates separated by a vacuum—the vacuum being the dielectric. The capacitance of air and vacuum capacitors is low, usually between 1 pF and 500 pF, and the plates must be spaced far enough apart to prevent arcing.

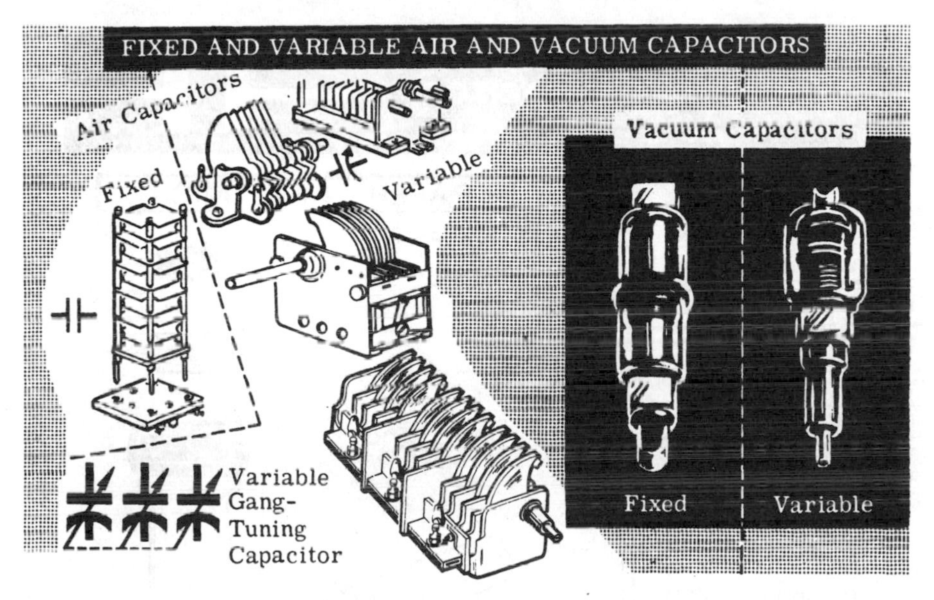

FIXED AND VARIABLE AIR AND VACUUM CAPACITORS

Another type of capacitor is the mica capacitor, which is variable, usually has a maximum value of less than 500 pF, and consists of two plates with a sheet of mica between them. A screw adjustment is used to force the plates together. The adjustment of this screw, therefore, varies the capacitance. Several layers of plates and mica are used in larger units of this type. These are often used in parallel with the larger variable capacitors and provide a finer adjustment of capacitance.

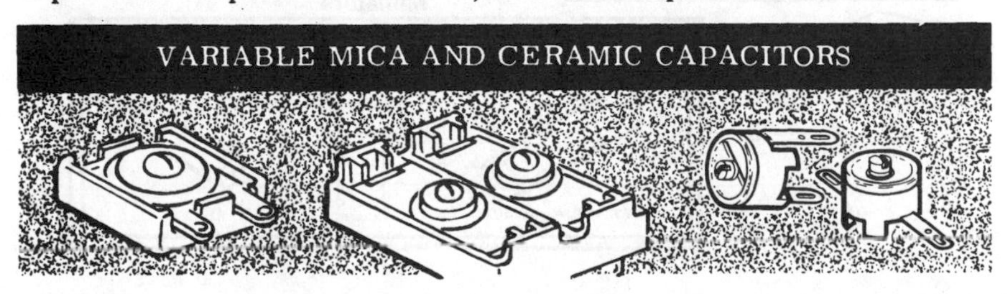

VARIABLE MICA AND CERAMIC CAPACITORS

Types of Capacitors—Fixed

Fixed capacitors can have their capacitance values predetermined by their composition and construction and these values cannot be changed. They differ from variable capacitors in that variable capacitors can have their capacitance values changed within certain limits.

Usually, fixed capacitors derive their names from the dielectric materials used in their construction. Examples of these are *ceramic* capacitors, *paper* capacitors, and *mica* capacitors. These capacitors have plates usually made of metal foil or a metallic plating directly on the dielectric. To keep their physical size to a minimum, the plates and the dielectric material are rolled into a tubular shape. At the present time, extensive use is made of high dielectric constant ceramic materials to provide high capacity in a small space.

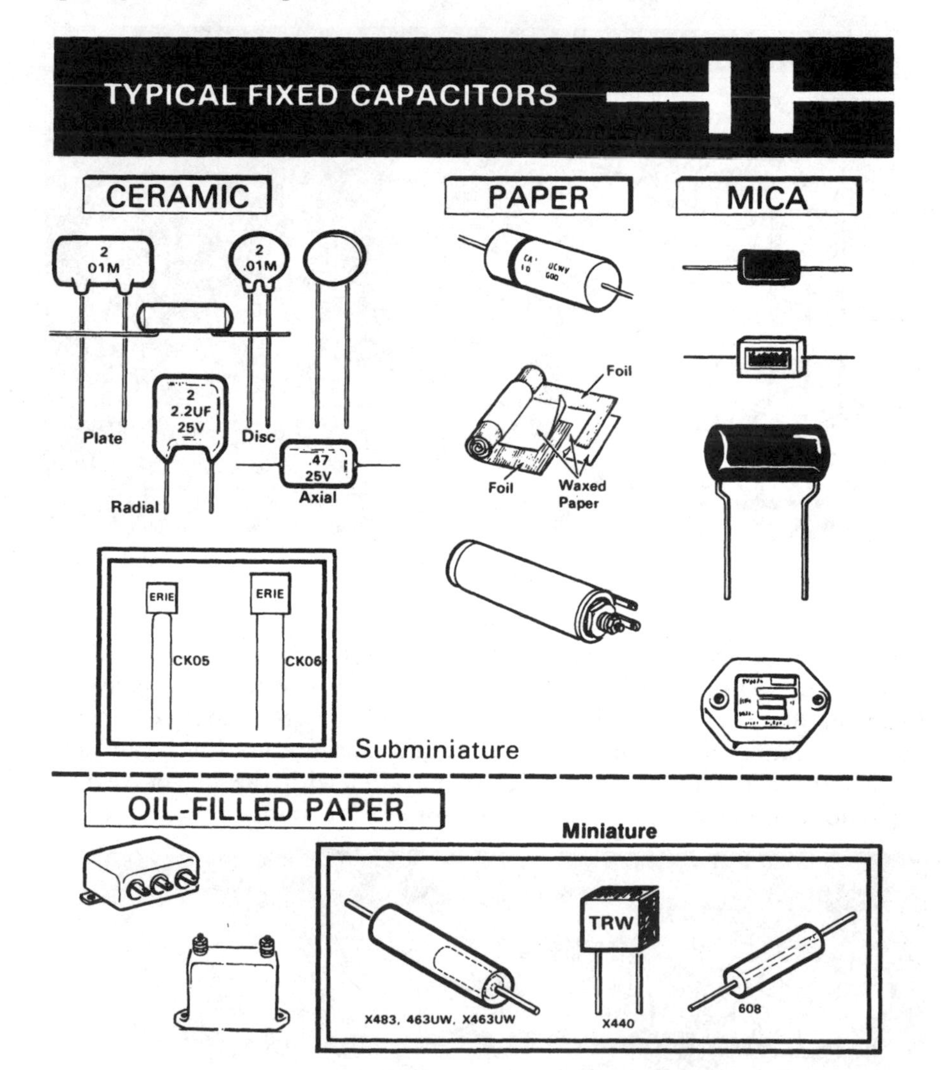

Types of Capacitors—Electrolytic/Tantalytic Capacitors

Due to size and cost, the small capacitors previously described are generally limited to values of less than 1 μF. *Electrolytic capacitors* are used when seeking larger values of capacitance. The dielectric material of an electrolytic capacitor is a very thin film of oxide that is formed by electrolytic action. The thin dielectric allows for high capacitance in a small space. Capacitances as large as many thousand microfarads can be obtained with electrolytic capacitors at reasonable size and cost. Electrolytic capacitors use aluminum foil and aluminum oxide as the dielectric.

The *tantalytic capacitors* use tantalum plates and tantalum oxide as the dielectric. Tantalytic capacitors are more rugged and stable but are much more expensive, and their use is confined mainly to military electronic equipment. Since their basic construction is the same, only electrolytic capacitors will be described.

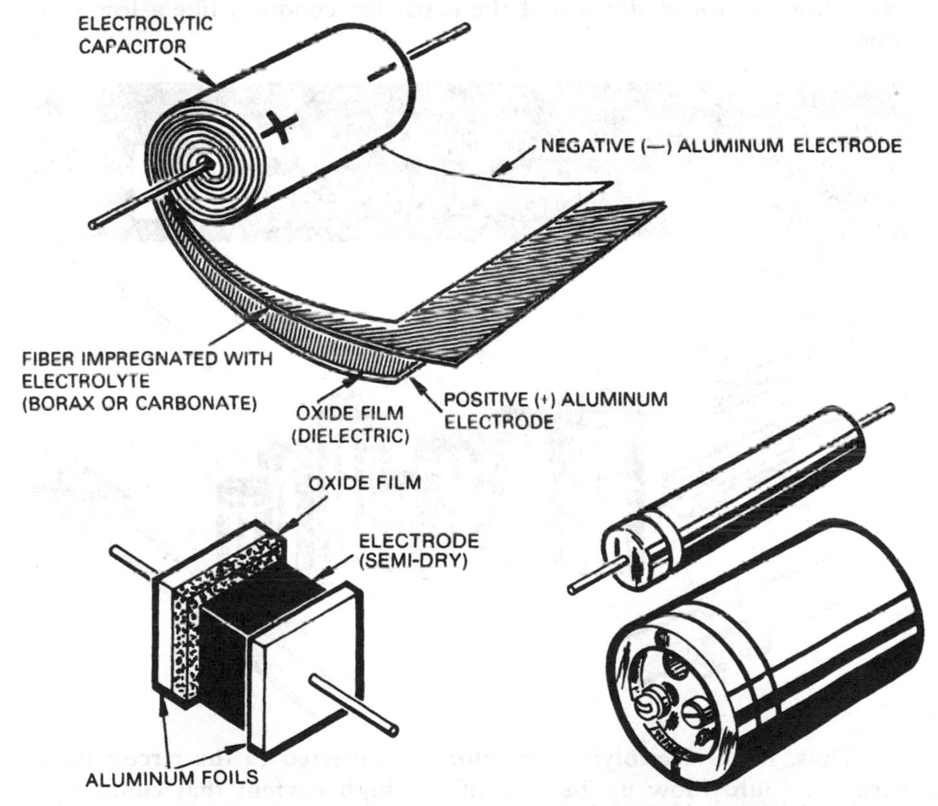

The electrolytic capacitor consists essentially of two foil sheets, separated by a layer of dielectric, and saturated with a chemical liquid called the *electrolyte*. These three sheets are rolled up together and sealed in a container. When the capacitor is first manufactured, a dc voltage is applied between the two foils. The resulting current causes a *thin oxide layer* to be formed on one foil sheet, and, as a result, the capacitor is *polarized* and must be operated so that the polarity is not reversed.

Types of Capacitors—
Electrolytic/Tantalytic Capacitors (continued)

The foil with the oxide layer is positive and serves as the positive (+) capacitor plate. The oxide layer serves as the dielectric for the capacitor. The electrolyte is the negative (−) capacitor plate, and the second foil sheet provides a connection between the electrolyte and external circuits.

When the electrolytic capacitor is *polarized*, it can only be used in circuits containing *fluctuating dc voltages*, where the polarity does *not* change. In addition, the positive and negative terminals of electrolytic capacitors must be connected in the circuit to point to the same polarity (negative to negative, and positive to positive). The reason for this is that the electrolytic capacitor acts as a capacitor for only *one* direction of applied voltage. When voltage is applied in the other direction, the oxide film is broken down and the capacitor conducts like a low-value resistor.

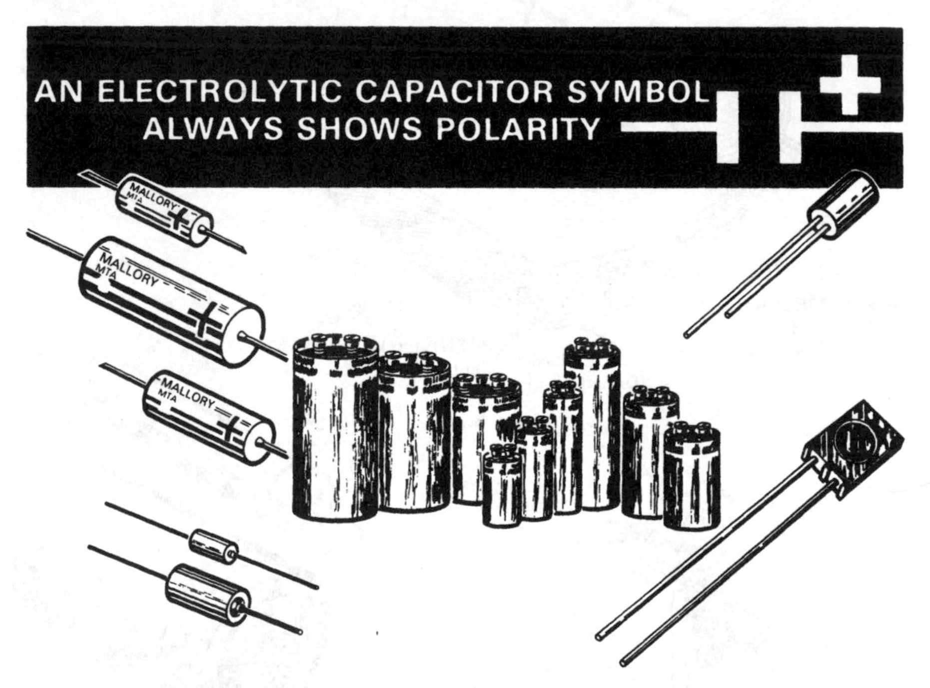

Thus, if an electrolytic capacitor is connected in the circuit backwards, it could blow up because of the high current that could flow through it. Special electrolytic capacitors are available for use in ac circuits. In these capacitors, two positive plates are used, thus permitting a reversal in polarity of the applied voltage. Essentially, this type of capacitor is made of two electrolytic capacitors connected back to back (− to +, + to −).

The electrolytic capacitor has much lower leakage resistance than the other capacitors and hence will not hold a charge for a long period of time. In addition, the oxide dielectric film is so thin that electrolytic capacitors are usually of limited voltage range.

Capacitor Color Codes

As in the case of resistors, capacitors are marked in such a way that their capacitance value and other important characteristics, such as voltage rating and tolerance, are indicated. Where possible, this information is printed directly on the capacitor as, for example, in the case of paper and electrolytic capacitors. In other cases where this is impractical, for example mica and ceramic capacitors, a system of color codes is used somewhat similar to the resistor color code that you learned about in Volume 2. The color code gives the capacitance in picofarads. You should remember, however, that the color codes for capacitors are not as well standardized as they are for resistors. For example, sometimes a 6-color system is used, and sometimes a 5-color system is used, as indicated by the following examples.

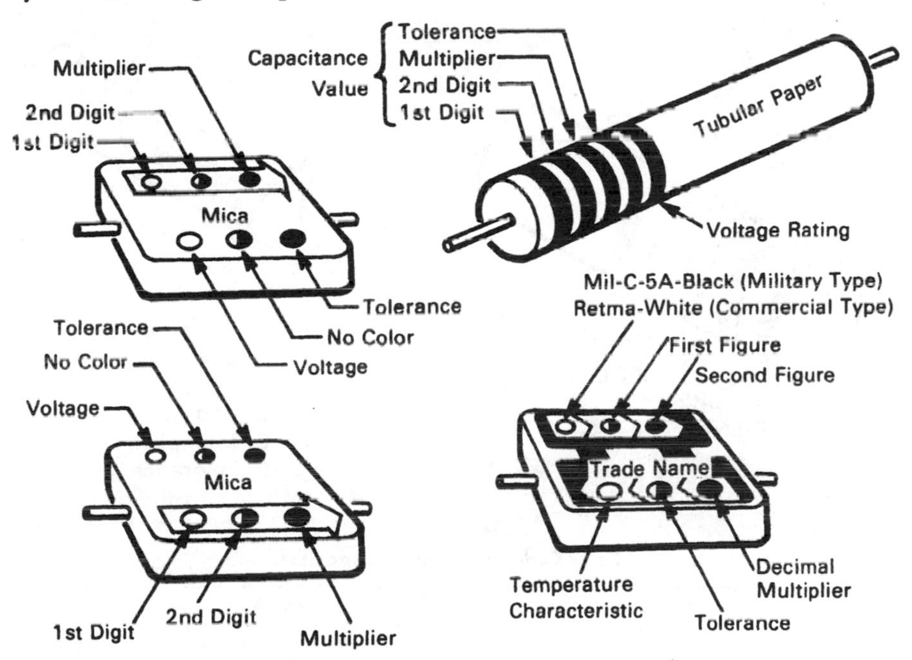

Capacitor Color Code

Color	*Significant Figure*	*Decimal Multiplier*	*Tolerance (%)*	*Voltage Rating*
Black	0	1	—	—
Brown	1	10	1	100
Red	2	100	2	200
Orange	3	1,000	3*	300
Yellow	4	10,000	4*	400
Green	5	100,000	5	500
Blue	6	1,000,000	6	600
Violet	7	10,000,000	7	700
Gray	8	100,000,000	8	800
White	9	1,000,000,000	9	900
Gold	—	0.1	5	1000
Silver	—	0.01	10	2000
No color	—	—	20	500

*Multiply by 10 for tubular paper capacitors.

Capacitor Color Codes (continued)

Ceramic capacitors are usually color coded with bands. In addition, the smaller ceramic capacitors may use dots. Temperature is very important in a ceramic capacitor. In some tubular ceramic capacitors, the temperature coefficient band is wider than the other identification bands. Sometimes this band is placed last. When it is, the rest of the code is the same. Sometimes the tolerance band is left out. No tolerance band indicates that the tolerance is ±20%.

The temperature coefficient indicates the change in temperature in parts per million that will take place for every degree change in the operating temperature above a nominal 20°C. A minus sign means that capacitance will decrease with increasing temperature, while a plus sign means the reverse.

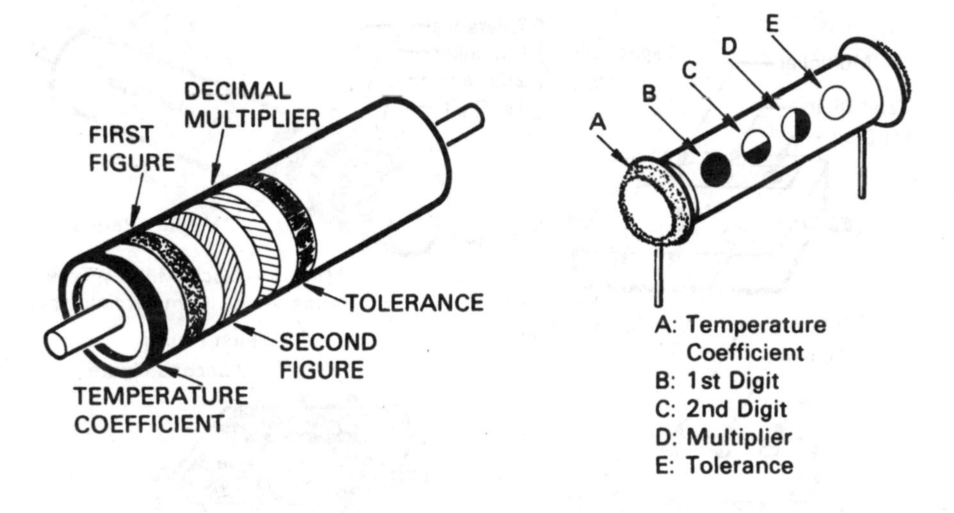

Color Code for Ceramic Capacitors

Color	First Digit	Second Digit	Multiplier	Tolerance: More than 10 pF (%)	Tolerance: Less than 10 pF (%)	Temperature Coefficient
Black	0	0	1.0	±20	±2	0
Brown	1	1	10	±1	–	−30
Red	2	2	100	±2	–	−80
Orange	3	3	1,000	±3	–	−150
Yellow	4	4	10,000	±4	–	−220
Green	5	5	–	±5	±0.5	−330
Blue	6	6	–	±6	–	−470
Violet	7	7	–	±7	–	−750
Gray	8	8	0.01	±8	±0.25	+30
White	9	9	0.1	±10	±1	+120 to −750
Gold	–	–	0.1	±5	–	–
Silver	–	–	0.01	±10	–	–

Capacitive Time Constant

When a voltage is applied across the terminals of a circuit containing capacitance, the voltage across the capacitance does not instantaneously equal the voltage applied to the terminals. You have already found that it takes time for the plates of a capacitor to reach their full charge, and that the voltage between the plates rises to equal the applied voltage in a curve similar to the current curve of an inductive circuit. The greater the circuit resistance, the longer the time that is required for the capacitor to reach its maximum voltage since the circuit resistance opposes the flow of current required to charge the capacitor.

The time required for the capacitor to become fully charged depends upon the product of circuit resistance and capacitance. This product RC—resistance × capacitance—is the *time constant* of a capacitive circuit. The RC time constant gives the time in seconds required for the voltage across the capacitor to reach 63.2% of its maximum value. Similarly, the RC time constant equals the time in seconds required for a discharging capacitor to lose 63.2% of its full charge. You can use the same curves as you would for inductance to calculate the percentage of full charge for more than one time constant. Thus, the voltage rises to 86.4% of maximum in two time constants, etc.

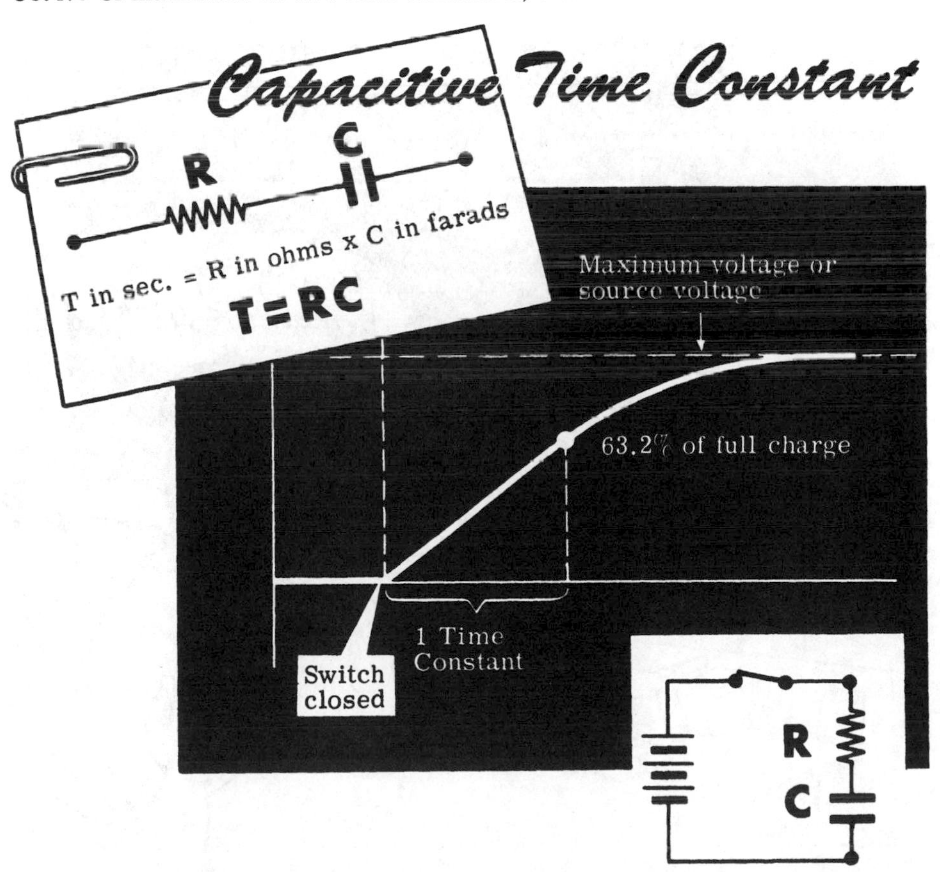

Capacitance—Charging and Discharging Currents

The definition of capacitance allows for calculation of the charge or discharge of a capacitor based on current flow and time. As you know, the farad is defined as the capacitance that results in 1-volt potential difference when 1 coulomb flows for 1 second. Based on this, the voltage charge for any capacitor can be calculated as

$$\Delta E = \frac{I \Delta T}{C}$$

where ΔE is the voltage charge that occurs in a capacitor of capacitance C (in farads) when a current of I amperes flows for a time ΔT. For example, a capacitor of 1 microfarad being charged with a current of 0.1 ampere for 1 millisecond (1/1000 second) can be calculated to have a voltage charge across its terminals as follows:

$$\Delta E = \frac{I \Delta T}{C} = \frac{0.1 \times 0.001}{0.000001} = 100 \text{ volts}$$

By the same calculation, you can show that if you take 0.1 ampere from a charged capacitor in 1 millisecond, then the voltage will drop by 100 volts.

Capacitance in AC Circuits

While a capacitor blocks the flow of dc current, it affects an ac circuit differently by allowing ac current to flow through the circuit. To see how this works, consider what happens in the dc circuit if a double-pole, double-throw switch is used with a dry cell so that the charge to each plate is reversed as the switch is closed—first in one position and then in the other.

When the switch is first closed, the capacitor charges, with each plate being charged in the same polarity as that of the cell terminal to which it is connected.

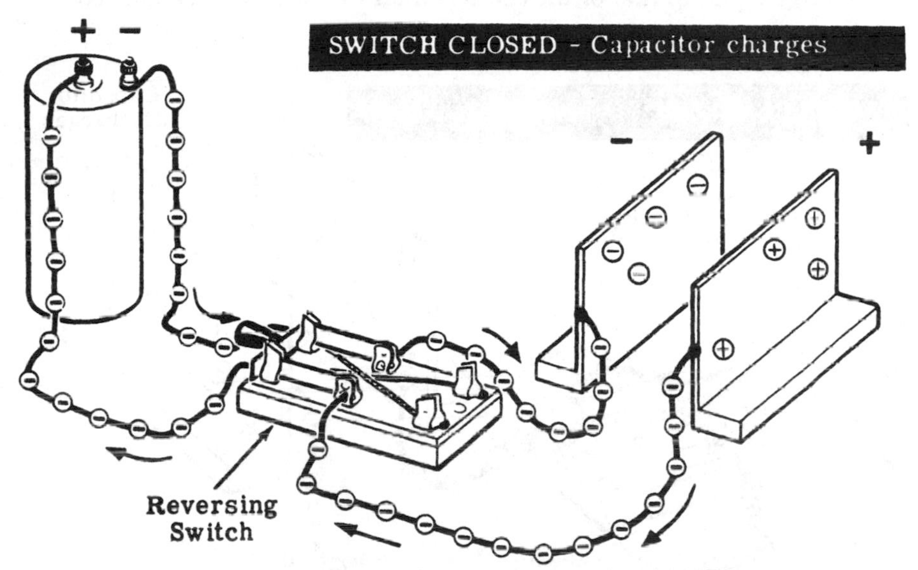

When the switch is opened, the capacitor retains the charges on its plates; the charges are equal to the cell voltage.

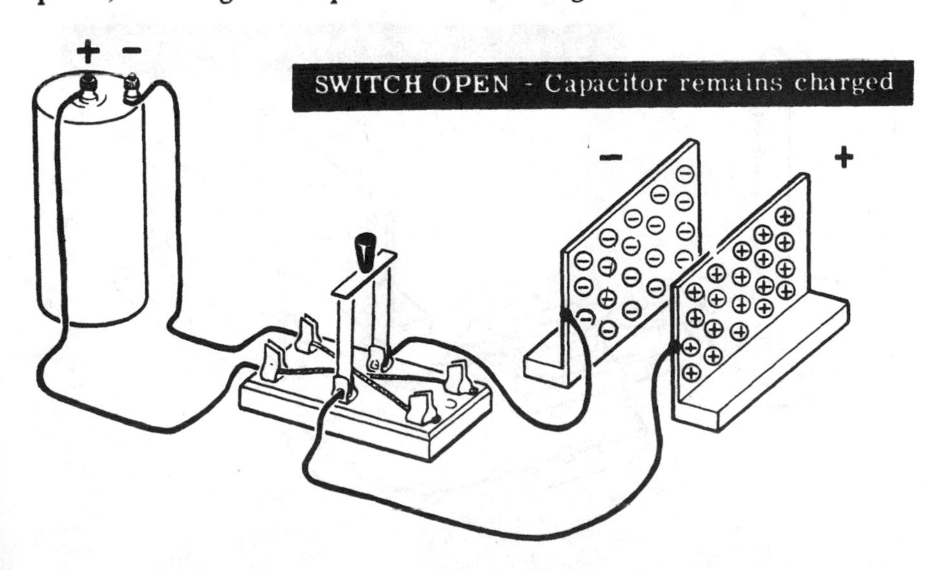

Capacitance in AC Circuits (continued)

If the switch is then closed in its original position, no current will flow, since the capacitor is charged in that polarity. However, if the switch is closed in the *opposite* direction, the capacitor plates will then be connected to cell terminals *opposite* to their charges in polarity. The positively charged plate is then connected to the negative cell terminal and will take electrons from the cell—first to neutralize the positive charge, then to become negatively charged, until the capacitor charge is in the same polarity and of equal voltage to that of the cell. The negatively charged plate gives up electrons to the cell, since it must take on a positive charge equal to that of the cell terminal to which it is connected.

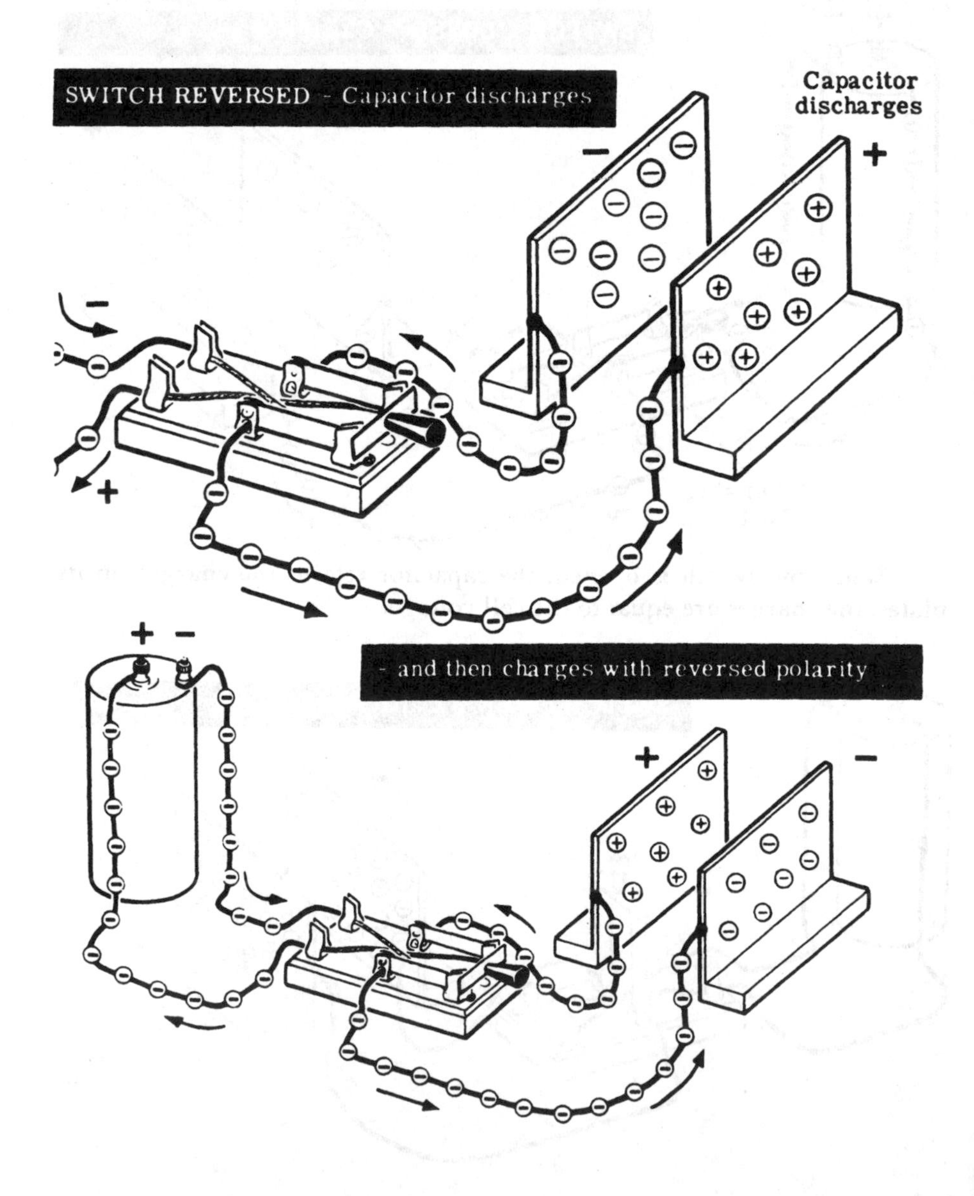
SWITCH REVERSED - Capacitor discharges
Capacitor discharges
- and then charges with reversed polarity

Capacitance in AC Circuits (continued)

If a zero-center microammeter, which can read current flow in both directions, is inserted in series with one of the capacitor plates, it will indicate a momentary current flow each time the plate is charged. When the reversing switch is first closed, it will show a current flow in the direction of the original charge. Then, when the cell polarity is reversed, it will show a momentary current flow in the opposite direction as the plate first discharges, then charges in the opposite polarity. The meter shows that current flows only momentarily each time the cell polarity is reversed and shows no current flow when the plates are fully charged.

CHARGE AND DISCHARGE CURRENT OF A CAPACITOR AS THE SOURCE VOLTAGE REVERSES

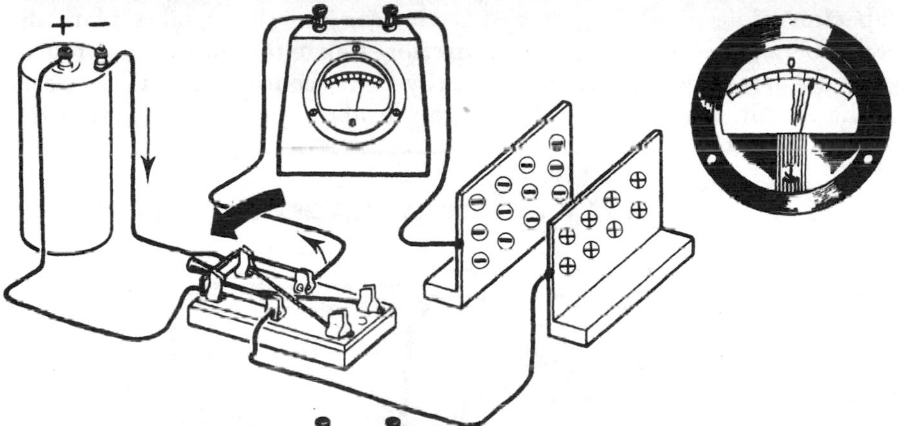

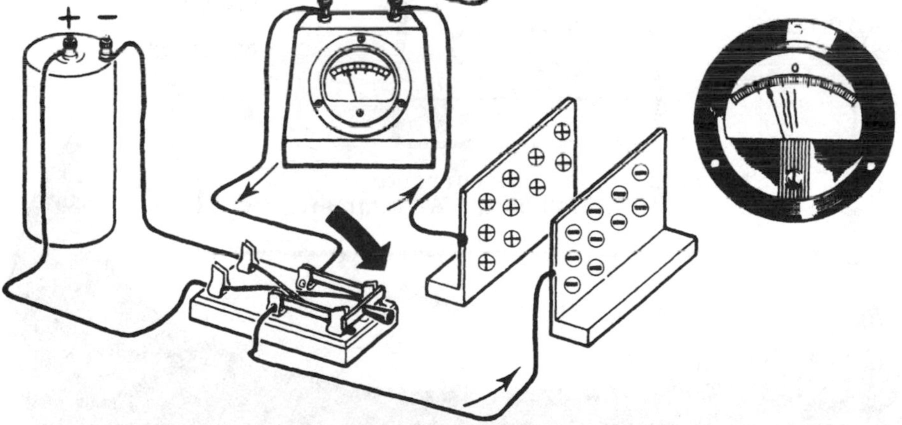

Suppose you switch the cell polarity fast enough so that, at the instant the capacitor plates become charged in one polarity, the cell polarity is reversed. The meter needle now moves continuously—first showing current flow in one direction, then in the opposite direction. While no electrons move through the air from one plate to the other, the meter shows that current is continuously flowing to and from the capacitor plates.

Capacitance in AC Circuits (continued)

If a source of ac voltage is used instead of the dc dry cell and reversing switch, the polarity of the voltage source will continually change each half cycle. If the frequency of the ac voltage is low enough, the ammeter will show current flow in both directions, changing each half cycle as the ac polarity reverses.

The standard power line frequency in North America is 60 hertz. Therefore, a zero-center ammeter will not show the current flow, since the meter pointer cannot move fast enough. However, an ac ammeter inserted in place of the zero-center ammeter will show a continuous current flow when the ac voltage source is used, indicating that there is a flow of ac current in the meter and in the circuit. Remember that this current flow represents the continuous charging and discharging of the capacitor plates, and that no actual electron movement takes place directly between the plates of the capacitor. Capacitors are considered to pass ac current because current actually flows continuously in all parts of the circuit, with the exception of the insulating material between the capacitor plates.

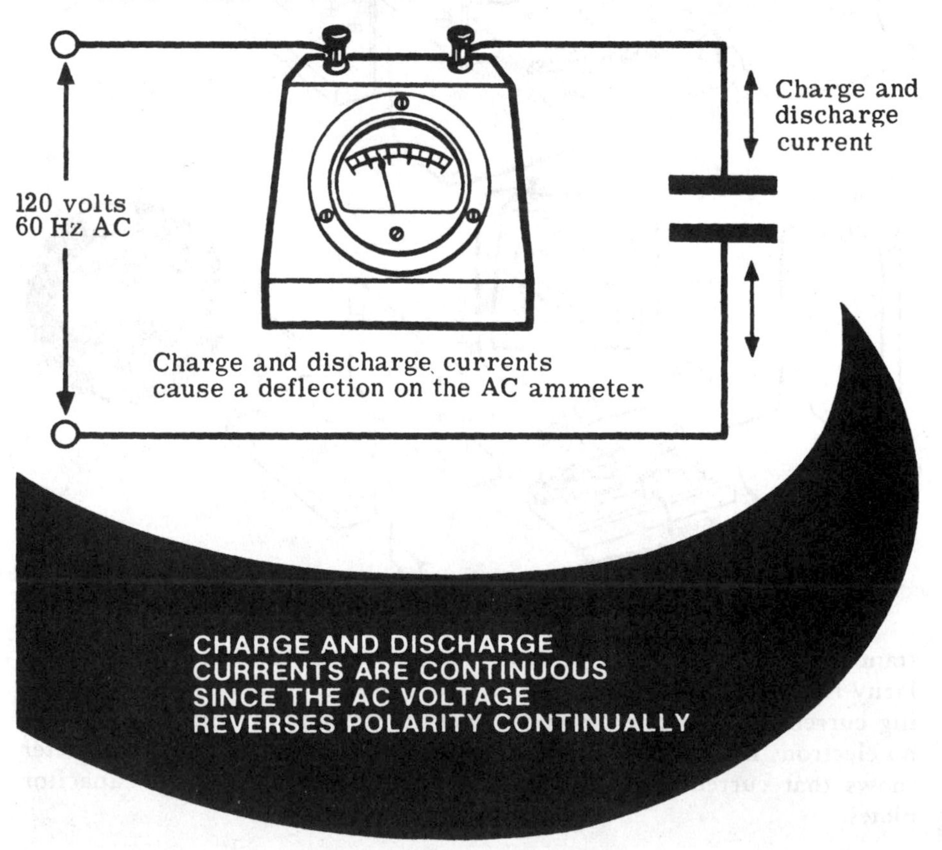
AC CURRENT in a capacitive circuit
Charge and discharge current
120 volts 60 Hz AC
Charge and discharge currents cause a deflection on the AC ammeter
CHARGE AND DISCHARGE CURRENTS ARE CONTINUOUS SINCE THE AC VOLTAGE REVERSES POLARITY CONTINUALLY

Capacitance in AC Circuits (continued)

Capacitive reactance is defined as the opposition to voltage change offered by the capacitance of a circuit. When a dc source is used, current flows only to charge or discharge the capacitor. Since there is no continuous flow of dc current in a capacitive circuit, the capacitive reactance to dc is considered *infinite.* Ac continuously varies in value and polarity. Therefore, the capacitor is continuously charging and discharging, resulting in a continuous circuit current flow and a *finite* value of capacitive reactance.

You know that the charge and discharge currents of a capacitor start at a maximum value and fall to zero as the capacitor becomes either fully charged or discharged. In the case of a charging capacitor, the uncharged plates offer little opposition to the charging current at first, but as they become charged, they offer more and more opposition, reducing the current flow. Similarly, the discharge current is high at the beginning of the discharge since the voltage of the charged capacitor is high, but as the capacitor discharges, its charge voltage drops, resulting in less current flow. Since the charging and discharging currents are highest at the beginning of the charge and discharge of a capacitor, the average current will become higher as the speed of charging and discharging becomes higher—keeping the current flowing at high values.

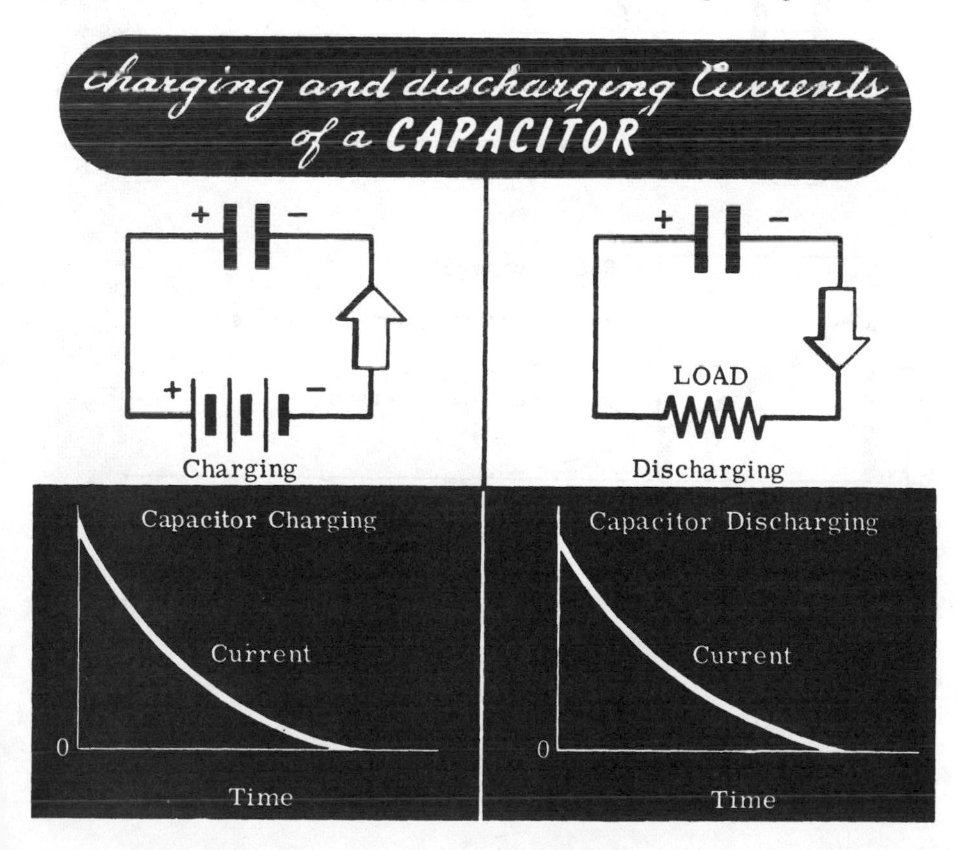

Capacitance in AC Circuits (continued)

The amount of current flow in an ac circuit depends on the *frequency* of the ac voltage for a given value of capacitance. The higher the frequency, the greater the current flow, since the charging current in each direction will be reversed before it has time to drop to a low value. If the source ac voltage is low in frequency, the current will drop to a lower value before the polarity reverses, resulting in a lower average value of current flow.

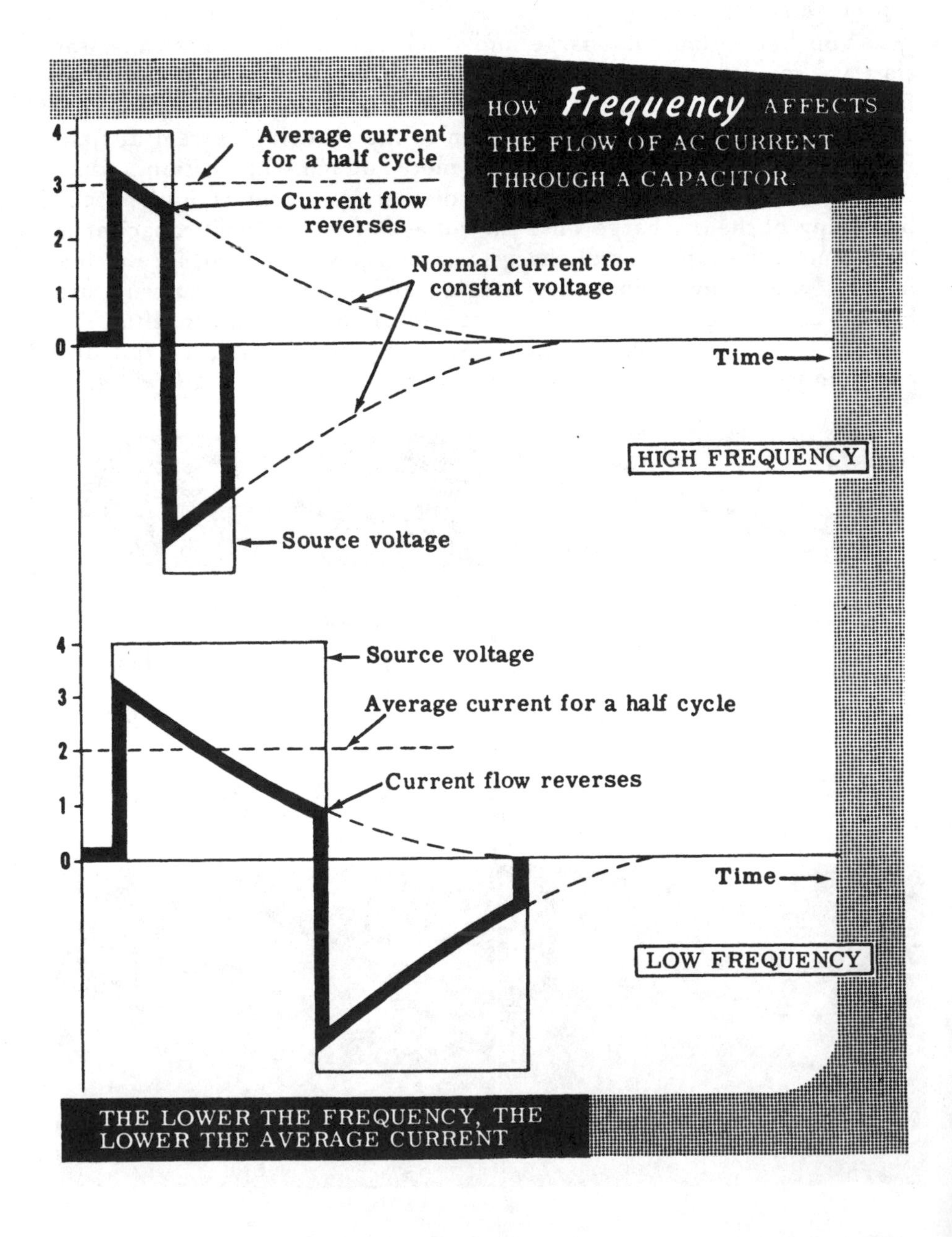

THE LOWER THE FREQUENCY, THE LOWER THE AVERAGE CURRENT

Capacitive Reactance in AC Circuits

When comparing the charging current for different values of capacitance, you see that the larger the capacitor, the longer the current remains at a high value. Thus, if the frequency is the same, a greater average current will flow through a larger capacitor than a smaller capacitor. This holds true only if the circuit resistances are equal, however, because the charge curve of a capacitor depends on the RC time constant of the circuit.

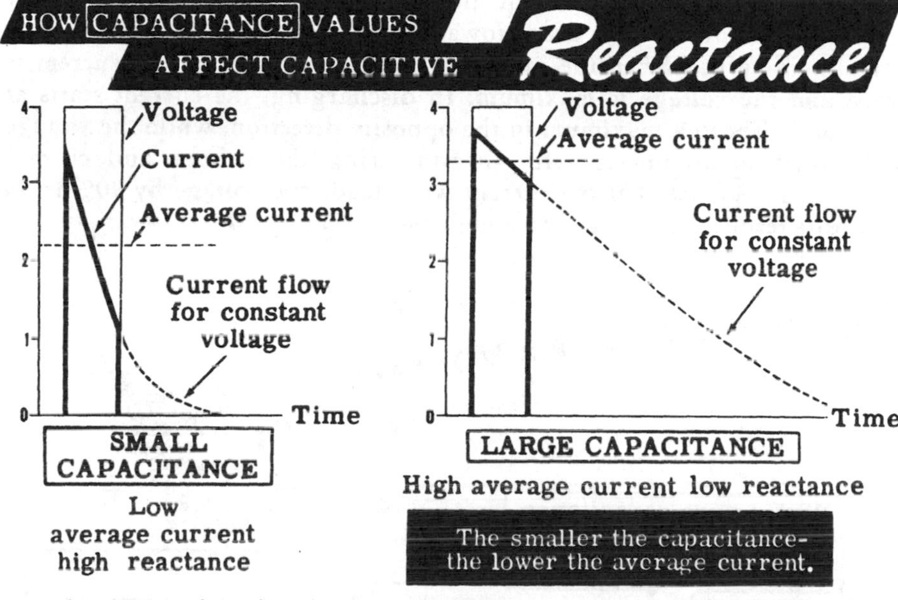

Assuming that there is no change in the resistance, the current flow in a capacitive circuit increases with an increase in either frequency or capacitance. Then capacitive reactance—the opposition to current flow through a capacitor—must decrease when the frequency or capacitance increases. The formula used to obtain capacitive reactance is

$$X_c = \frac{1}{2\pi fC}$$

In this formula, X_c is the capacitive reactance, f is the frequency in hertz, C is the capacitance in farads, and 2π is a constant (6.28). Since X_c, or capacitive reactance, represents opposition or resistance to current flow, it is expressed in ohms as with resistance and inductive reactance.

To illustrate how this equation is used, we can calculate the reactance of a 1-microfarad capacitor at 60 hertz.

$$X_c = \frac{1}{2\pi fC}$$

$$= \frac{1}{6.28 \times 60 \times 0.000001}$$

$$= 2653 \text{ ohms}$$

Phase Relationship in Capacitive Circuits

The phase relationship between current and voltage waves in a capacitive circuit is exactly the *opposite* to that of an inductive circuit. In a purely inductive circuit, the current wave *lags* the voltage by 90°, while in a purely capacitive circuit the current wave *leads* the voltage by 90°.

In a theoretical circuit of pure capacitance and no resistance, the voltage across the capacitor exists only after current flows to charge the plates. At the moment a capacitor starts to charge, the voltage across its plates is zero and the current flow is maximum. As the capacitor charges, the current flow drops toward zero while the voltage rises to its maximum value. When the capacitor reaches full charge, the current is zero and the voltage is maximum. In discharging, the current starts at zero and rises to a maximum in the opposite direction, while the voltage falls from maximum to zero. In comparing the voltage and current waves, you can see that the current wave leads the voltage by 90° or, in opposite terms, the voltage wave lags the current by 90°.

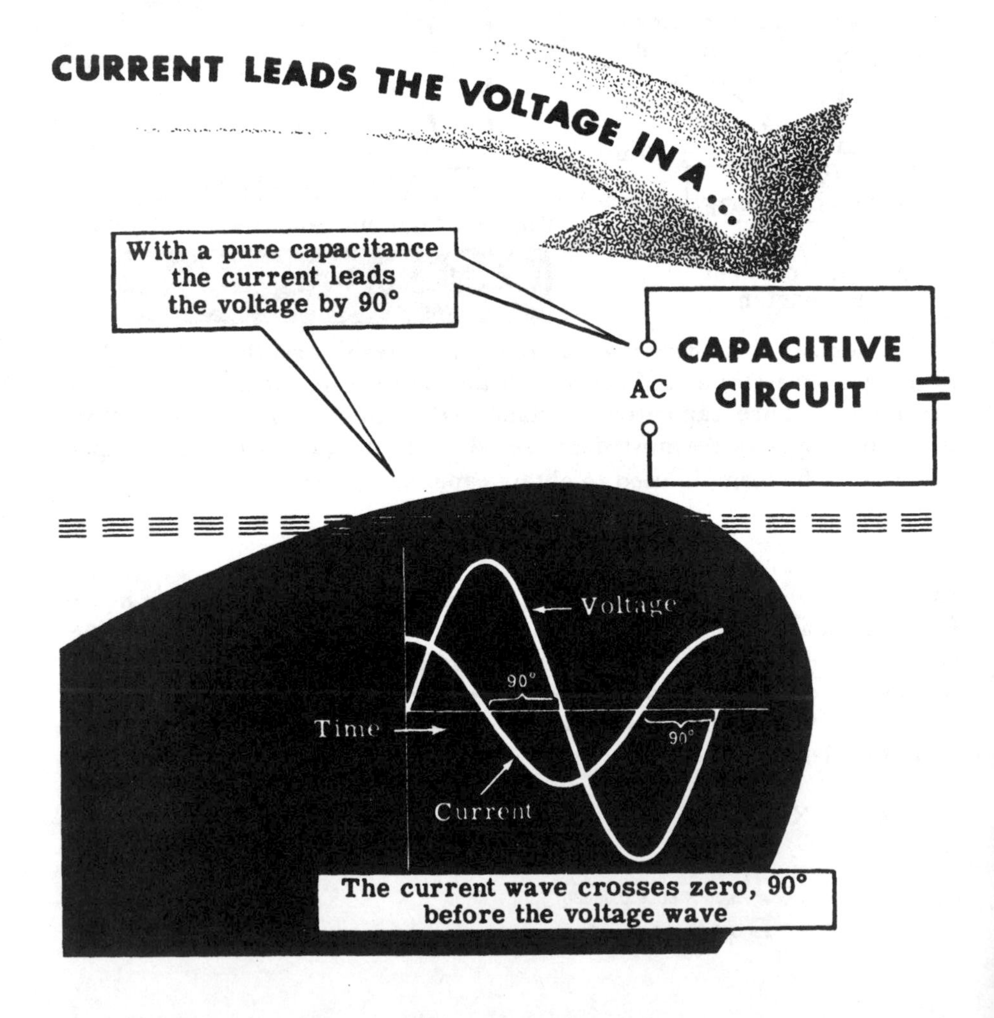
CURRENT LEADS THE VOLTAGE IN A...
With a pure capacitance the current leads the voltage by 90°
CAPACITIVE CIRCUIT
AC
Voltage
90°
Time
90°
Current
The current wave crosses zero, 90° before the voltage wave

Phase Relationship in Capacitive Circuits (continued)

Resistance affects capacitive circuits in almost the same way that it affects inductive circuits. Remember, in an inductive circuit containing both inductance and resistance, the current wave lags the voltage wave by an angle between 0° and 90°, depending on the ratio of inductive reactance to the resistance. For a purely capacitive circuit, current leads the voltage by 90°; but, with both resistance and capacitance in a circuit, the amount of lead—*phase angle*—is decreased depending on the ratio between the capacitive reactance and the resistance.

If the capacitive reactance and resistance are equal, they will have an equal effect on the angle of lead, resulting in a *phase angle* of 45° leading. As shown below, the current wave then leads the voltage wave by 45°.

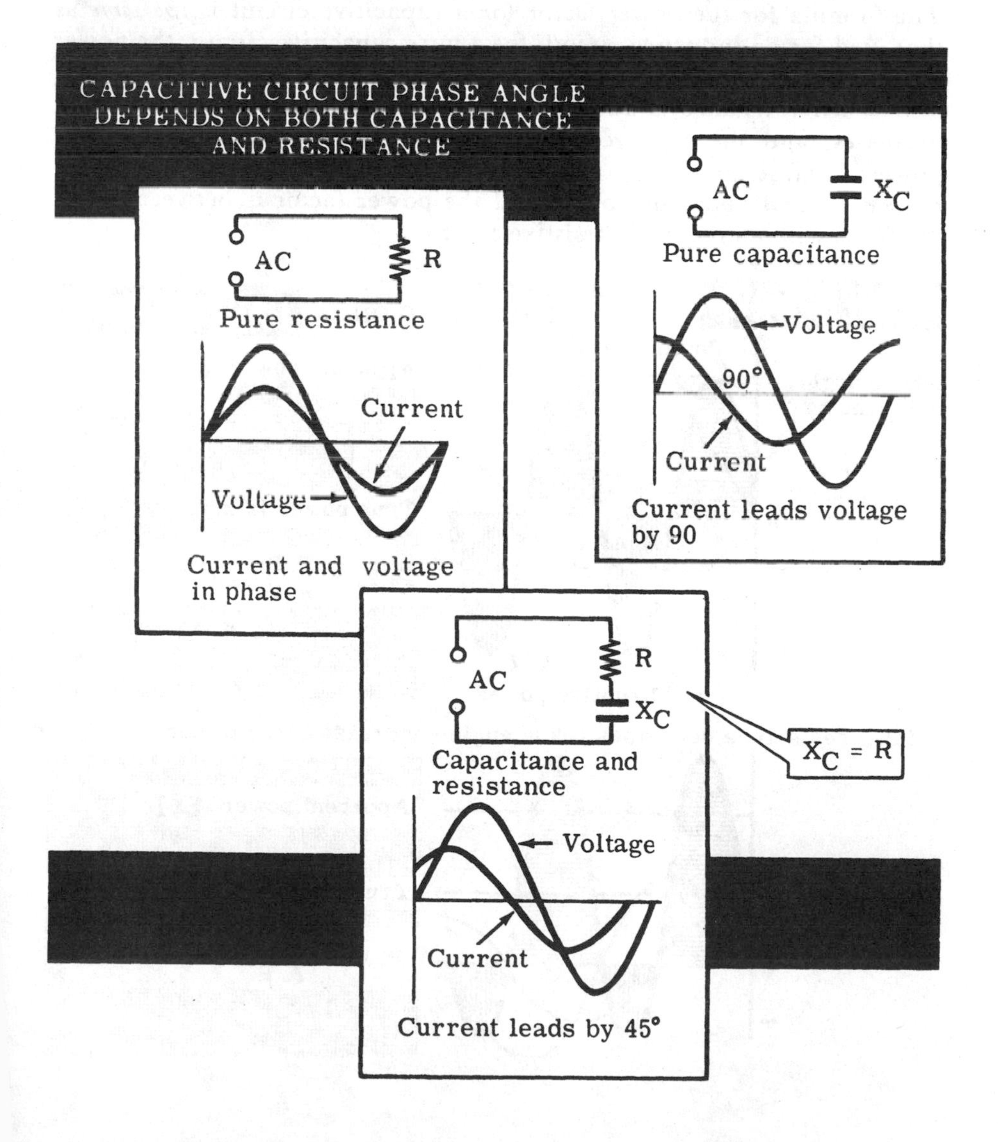

Power in a Capacitive Circuit

Since capacitive and inductive circuits behave with great similarity, it would be wise to review power in inductive circuits at this point. In a capacitive circuit, as in an inductive circuit, the true power used is less than the apparent power of the circuit. Current leads the voltage in a capacitive circuit. The power waveform is obtained by multiplying the corresponding values of voltage and current to obtain the instantaneous values of power. The power wave of an ac circuit consisting of pure capacitance is shown below and, like the power wave of a purely inductive circuit, its axis is the same as that of the voltage and current, while its frequency is twice that of the voltage and current. For this circuit, the phase angle between the current wave and voltage wave is 90°, and the negative power equals the positive power. Thus, the true power is zero. The formula for the power factor for a capacitive circuit is *the same* as that used for an inductive circuit; for a pure capacitive circuit, the power is also zero.

When resistance is added to a capacitive circuit, the phase angle decreases, and the positive power becomes greater than the negative power. Because the voltage and current are out of phase, power in watts does not equal apparent power, and the power factor is between 0 and 100%, as with an inductive-resistive circuit.

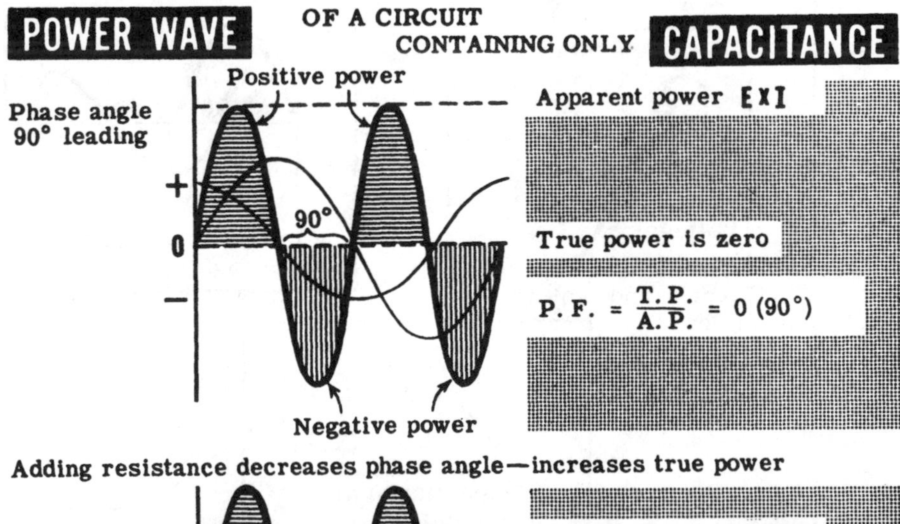

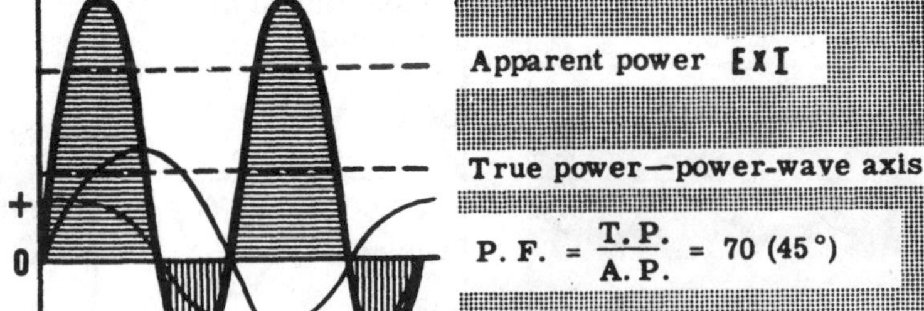

Review of Capacitors and Capacitance

You have now learned about capacitance and you have seen how it affects the flow of current in both dc and ac electric circuits. Now you are ready to perform Experiments/Applications on capacitance to find out more about it and its effects. Before performing the Experiments/Applications, suppose you recall what you have discovered about capacitance and capacitors.

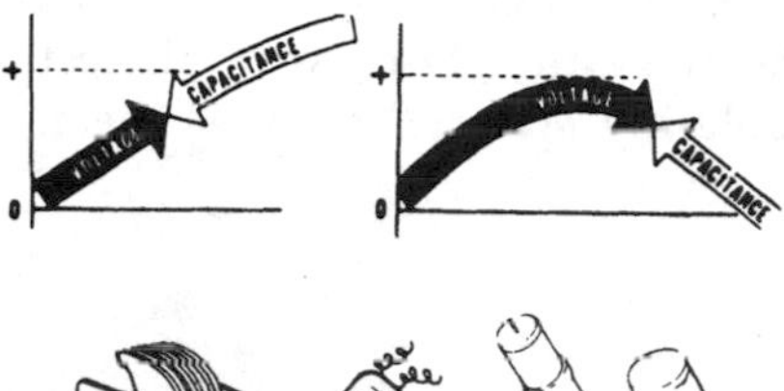

1. CAPACITANCE—The property of a circuit which opposes any change in the circuit voltage.

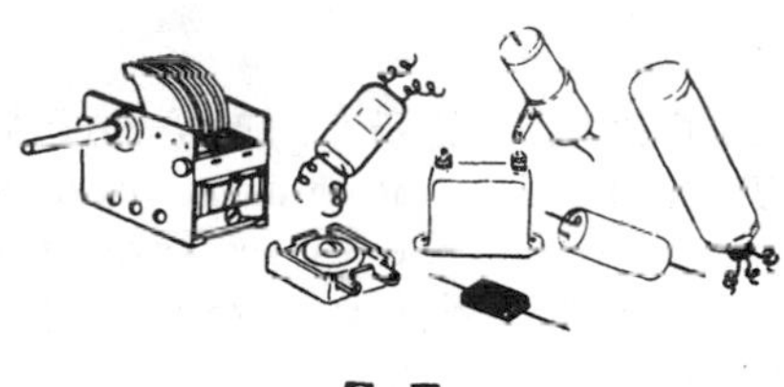

2. CAPACITOR—An electrical device used to supply capacitance in a circuit.

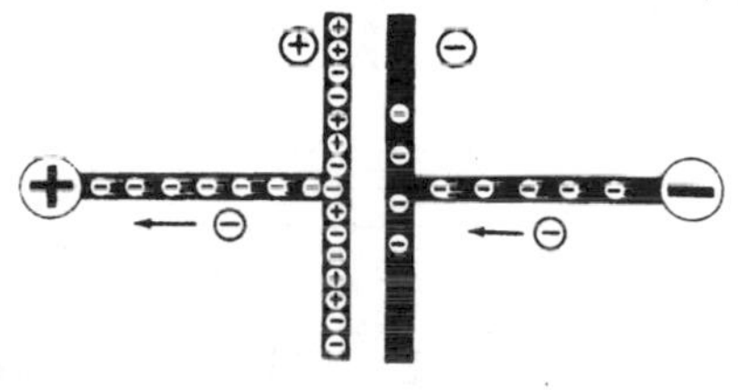

3. CAPACITOR CHARGE—The flow of electrons onto one plate and away from the other, resulting in a negative charge on one plate and a positive charge on the other.

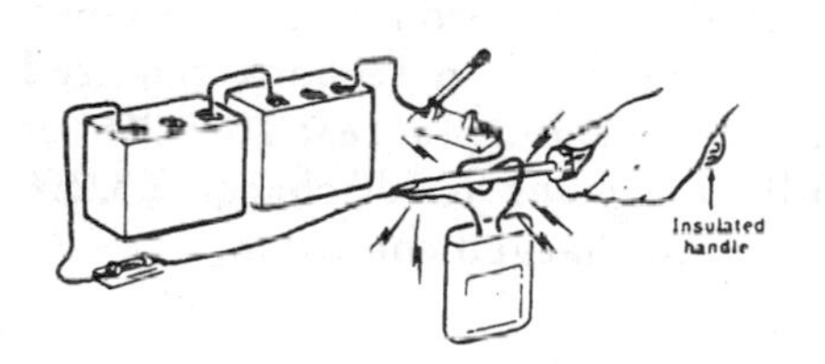

4. CAPACITOR DISCHARGE—The flow of electrons from the negatively charged plate of a capacitor to the positively charged plate, eliminating the charges on the plate.

5. FARAD—The basic unit of capacitance where 1 coulomb of charge for 1 second yields a potential difference of 1 volt.

$$1\ \mu F = \frac{1}{1{,}000{,}000}\ \text{farad}$$

$$1\ pF = \frac{1}{1{,}000{,}000{,}000{,}000}\ \text{farad}$$

6. PRACTICAL UNITS OF CAPACITANCE—The basic units are microfarad (one-millionth farad) and picofarad (one-millionth-millionth of a farad).

Review of Capacitors and Capacitance (continued)

Before performing the Experiments/Applications on capacitive circuit time constants, you should review some further facts that you have learned concerning capacitors, capacitive reactance, and RC time constant.

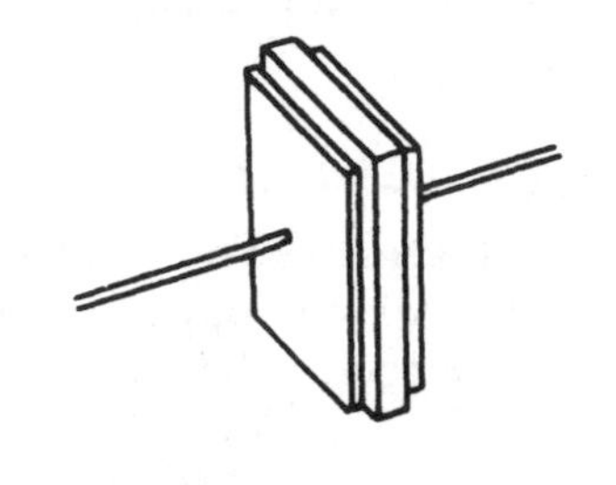

7. CAPACITOR COMPONENTS—The metallic plates with dielectric insulating material between them.

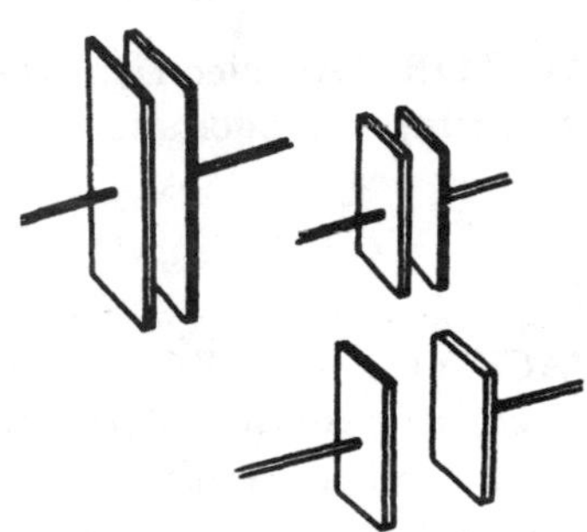

8. FACTORS AFFECTING CAPACITANCE—The plate area, the distance between the plates, and the dielectric material determine capacitance.

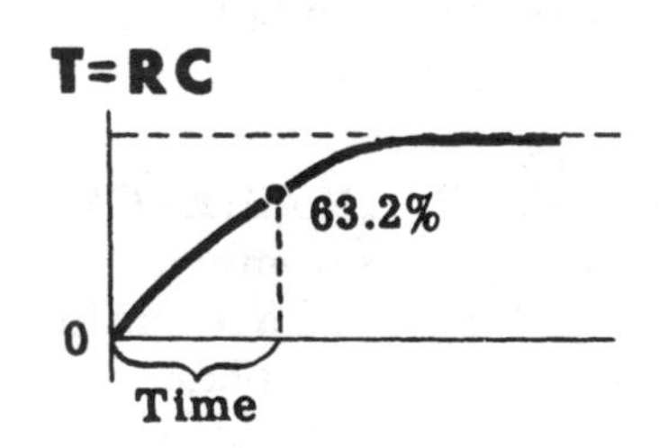

9. CAPACITIVE TIME CONSTANT—The product of RC which gives the time in seconds required for a capacitor to reach 63.2% of full charge or to discharge 63.2% from the charged condition.

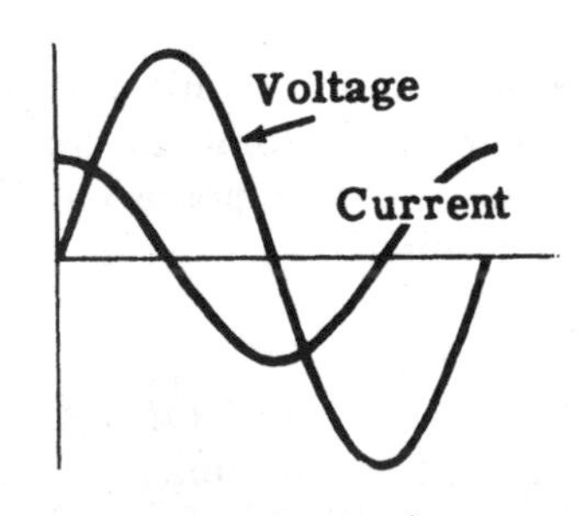

10. CAPACITIVE REACTANCE—The action of a pure capacitance in opposing the flow of ac current and in causing the current to lead the voltage by 90°. It is designated X_c and is measured in ohms.

APPARENT POWER = E x I

11. POWER IN PURE CAPACITIVE CIRCUIT—In a pure capacitive circuit, the true power is zero but the apparent power is E × I.

Self-Test—Review Questions

1. What is the property of capacitors that is analogous to inductors?
2. Sketch the components of a simple capacitor. How does the capacitance vary with plate size, plate spacing, and dielectric?
3. Define the farad. What are the practical units that are used?
4. What factor limits the voltage that can be applied to a capacitor?
5. Describe the fundamental differences between electrolytic capacitors and mica or paper capacitors.
6. What is the voltage across the capacitor at 3 seconds in the circuits below?

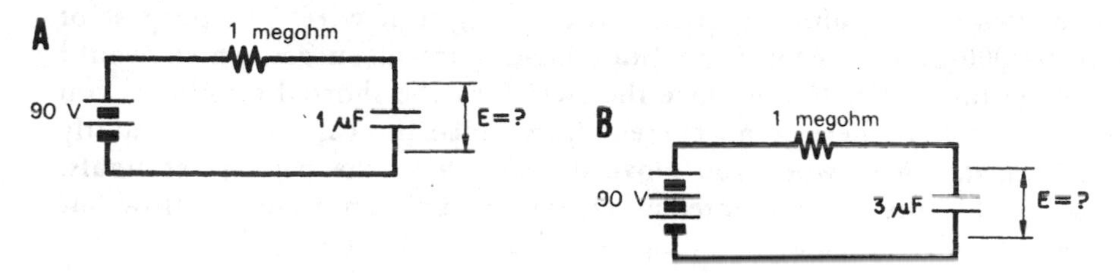

7. Explain why a capacitor can permit ac current flow but prohibits dc current flow.
8. Calculate the currents in the circuits below.

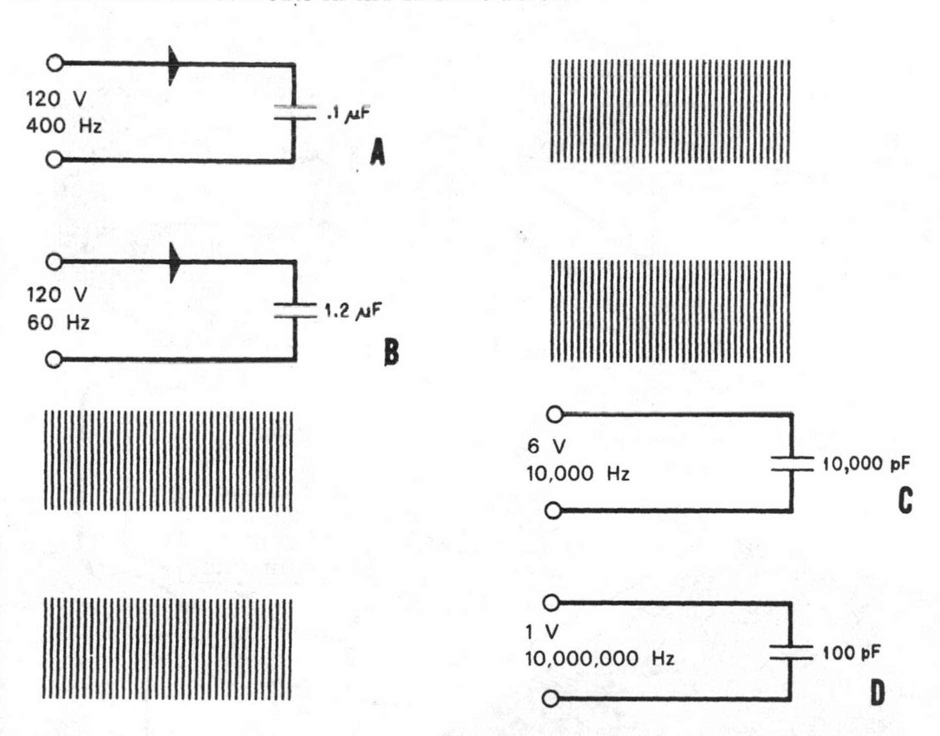

9. What is the power consumed in the circuits in question 8? Explain your answers.
0. In the circuits in question 8 what would the voltages be if the charging current were 1 milliampere applied for 1 millisecond?

Experiment/Application—Current Flow in a DC Capacitive Circuit

In circuits containing only capacitance, both the charge and discharge of a capacitor occur in a very short period of time. To show the circuit current flow during the charge and discharge of a capacitor, you could connect two 45-volt batteries in series to form a 90-volt battery. Next, connect the leads from this battery to two of the end terminals of a double-pole, double-throw switch. With the switch open, connect the 0-1-milliampere zero-center milliammeter and the 1-microfarad capacitor in series with the resistor to the switch as shown. Finally, connect the other two end terminals together with a length of wire. The purpose of the 91,000-ohm resistor is to limit large current surges which would damage the meter. If you move the switch to the shorted terminals, you would see that there is no current flow since the capacitor is initially uncharged. Then, when you move the switch to the battery terminals, you see that the meter pointer momentarily registers a current flow but drops quickly to zero again as the capacitor becomes charged.

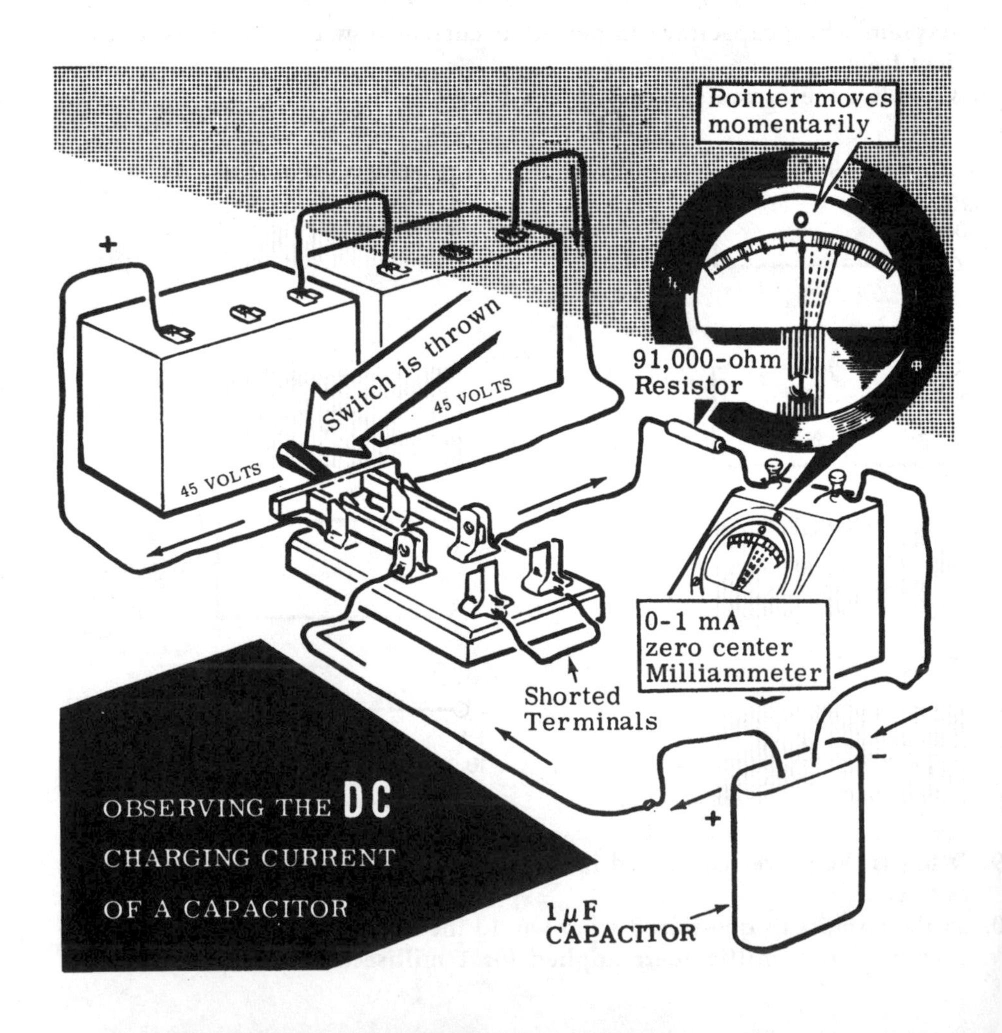

Experiment/Application—Current Flow in a DC Capacitive Circuit (continued)

If you then open the switch and move it to the shorted terminals, the meter pointer shows a momentary current flow in the opposite direction from that when the switch is closed, indicating the discharge of the capacitor.

You can then charge the capacitor as before, and you notice the instantaneous current flow. If you then open the switch and return it to its initial position, you notice no current flow, since the capacitor is already charged. When you move the switch to the shorted terminals, the current flows in the opposite direction again, showing the capacitor discharge.

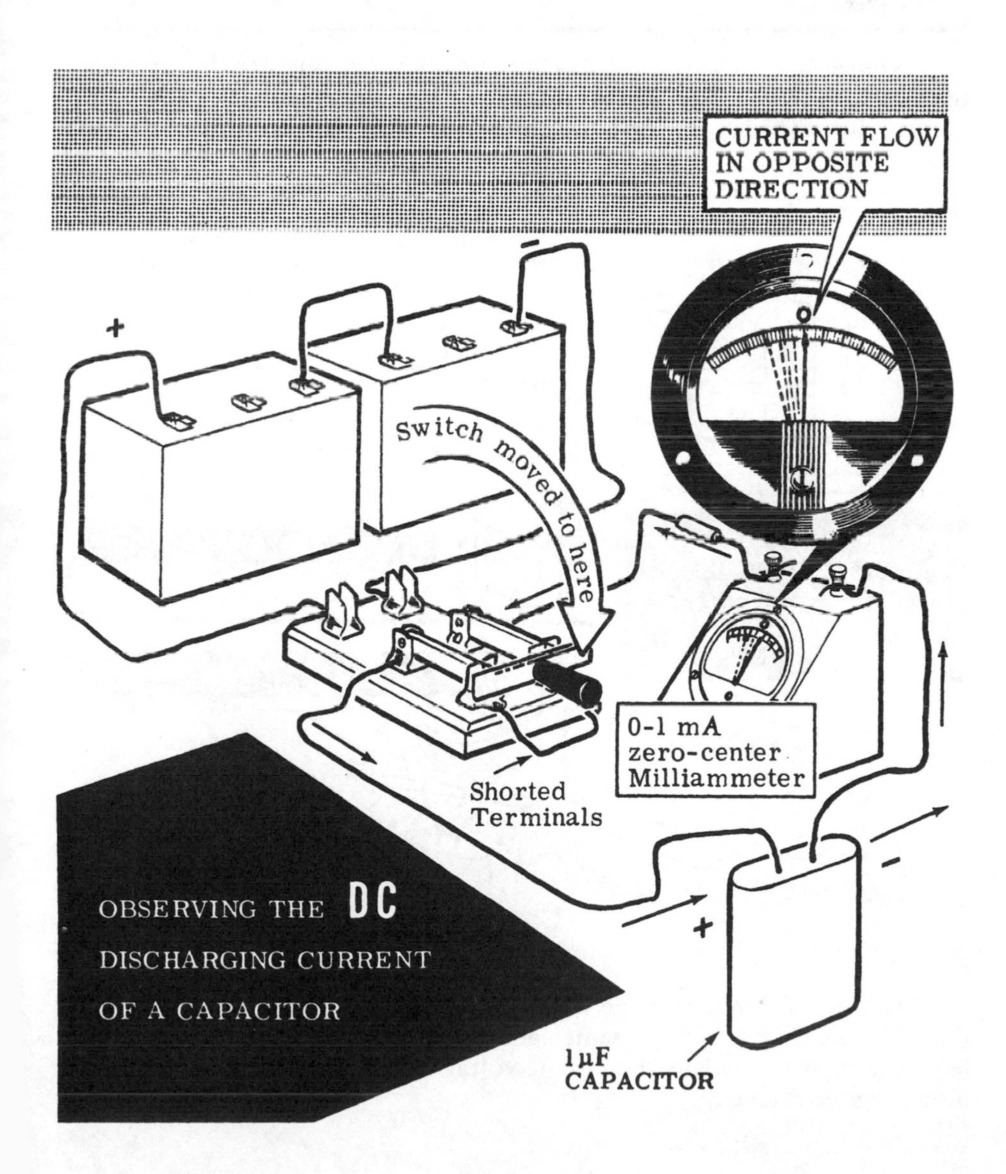

Experiment/Application—Current Flow in a DC Capacitive Circuit (continued)

Next connect the battery to the capacitor and switch in series. The capacitor is charged by closing the switch. Then open the switch and short across the capacitor terminals with a screwdriver blade, making *certain* to hold the screwdriver *only* by means of the *insulated* handle.

CAUTION:
NEVER DISCHARGE A CAPACITOR WHILE IT IS CONNECTED TO THE CIRCUIT VOLTAGE, WHETHER THE VOLTAGE SOURCE IS A BATTERY OR AC POWER LINE.

Notice that the capacitor discharge causes a strong arc. If you were to discharge the capacitor by touching the two terminals with your hands, the resulting electric shock—while not dangerous in itself—might cause a serious accident by making you jump.

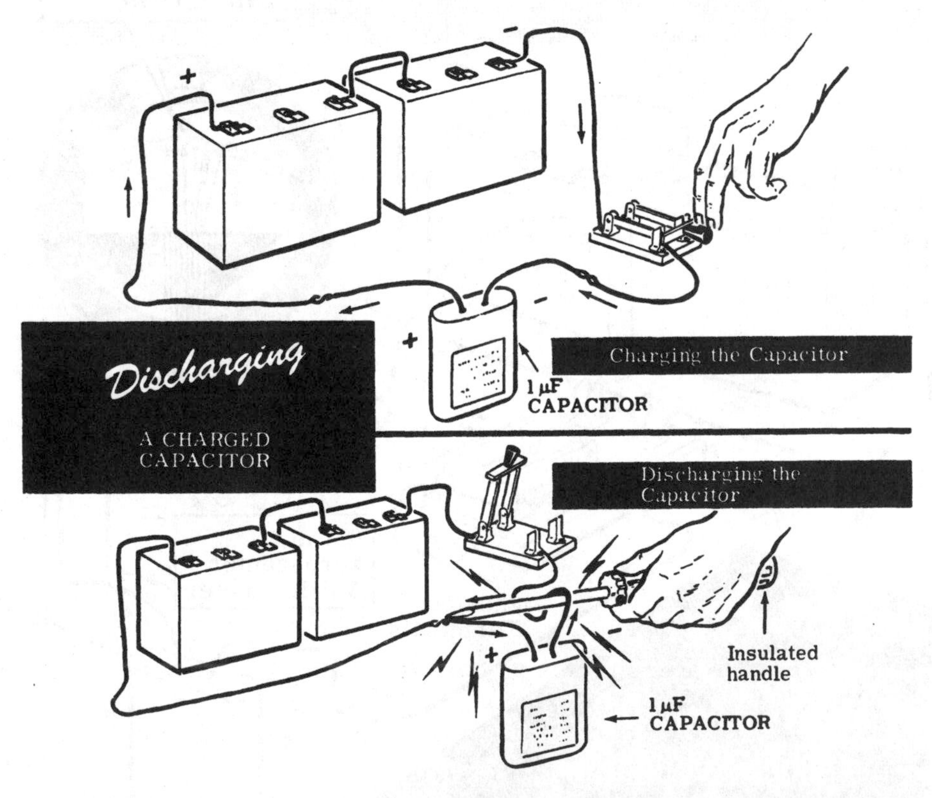

As you repeatedly charge and discharge the capacitor, you will see that the resulting arc is the same each time. This shows that the charge left in a capacitor when the circuit voltage is removed is always maximum in a dc circuit.

Experiment/Application—RC Time Constant

In an RC circuit, the charging current for the capacitor is limited by the resistance, and the voltage across the capacitor builds up at a rate determined by the RC time constant. If a voltmeter is used to measure the voltage across a capacitor to show the rise in voltage as the capacitor charges, the plates of the capacitor are connected together through the resistance of the voltmeter. This prevents the capacitor from reaching full charge and, since the meter completes the circuit across the capacitor, it will not show the rise in voltage across the capacitor.

A VOLTMETER CONNECTED ACROSS A CAPACITOR ACTS AS A RESISTOR

To show how voltage builds up across a capacitor, a device is needed which will indicate voltage but will not connect the capacitor plates together. A neon lamp may be used for this purpose, since it is an open circuit until the voltage across its terminals reaches a predetermined value. You should review how a neon lamp works by reference to the Experiment/Application on inductors. The lamp does not accurately measure voltage, nor does it show the actual building up of the voltage as the capacitor charges. However, if connected across a capacitor which is being charged, it does show that the build-up of voltage across a capacitance is delayed, since the neon lamp does not light *immediately* when the charging voltage is applied to the capacitor circuit. The lamp lights only after the voltage between the capacitor plates reaches the starting voltage of the lamp.

A NEON LAMP IS AN OPEN CIRCUIT UNTIL ITS STARTING VOLTAGE IS REACHED

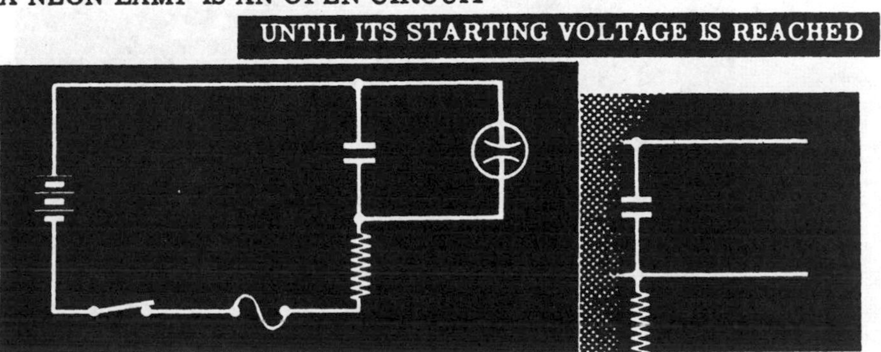

Experiment/Application—RC Time Constant (continued)

To show how a neon lamp may be used to indicate the delay in the rise of voltage across a capacitor, you could connect a 1-microfarad capacitor in series with a 2-megohm resistor. A neon lamp socket is connected across the capacitor and a neon lamp is inserted in the socket, thus completing the circuit except for a source of voltage. The unconnected end of the resistor is connected to the negative terminal of a 90-volt battery—two 45-volt dry cell batteries in series.

When you close the switch, you will observe that after a momentary delay the neon lamp flashes, indicating that the voltage across the capacitor terminals has reached the starting voltage of the lamp. Notice that the neon lamp continues to flash at intervals of about one second. Each time that the capacitor charges to a voltage equal to the starting voltage of the neon lamp, the lamp lights—providing a path for current flow between the capacitor plates and discharging the charge which has been built up. When the capacitor discharges through the lighted lamp, its voltage drops to a value which is too low to operate the lamp and the lamp ceases to conduct; it again becomes an open circuit. The capacitor then begins to charge again. You see that the lamp lights repeatedly, since each time the voltage across the capacitor reaches the lamp's starting voltage, the lamp lights, partially discharging the capacitor.

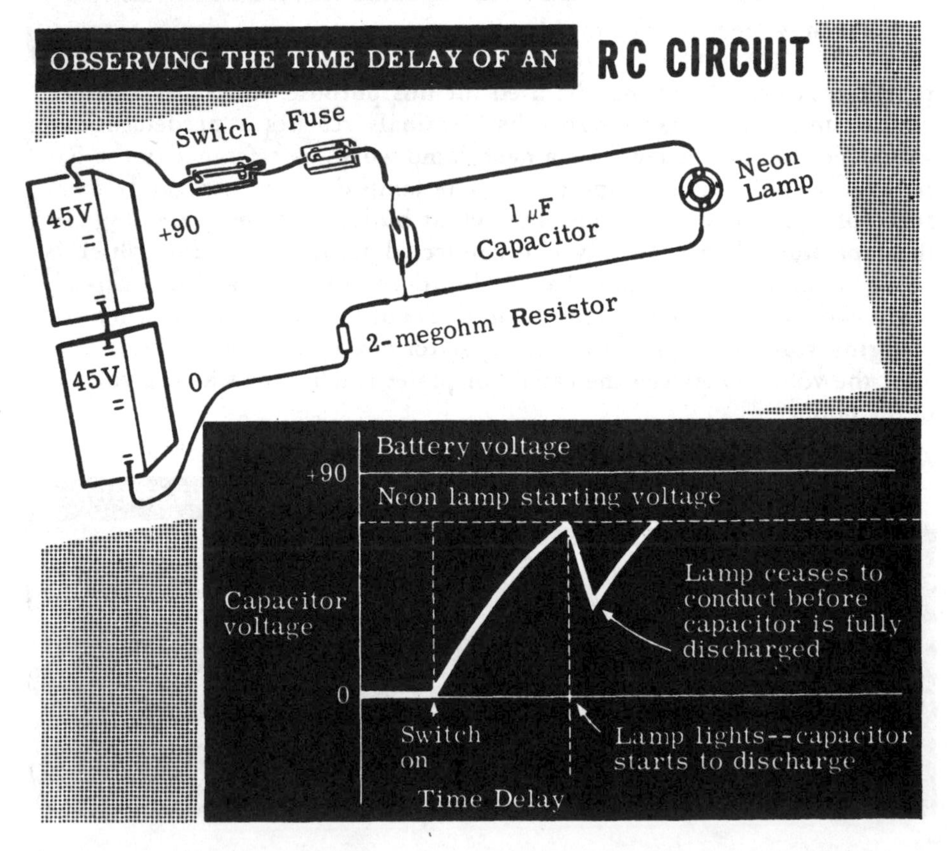

Experiment/Application—RC Time Constant (continued)

In order to show the effect of resistance on the time required for the capacitor to charge, additional resistors are added in series with the 2-megohm resistor. Observe that, when the resistance is doubled, the time required for the capacitor to charge is also doubled.

You already know that the RC time constant gives the time in seconds required for the voltage of a charging capacitor to reach 63.2% of the voltage applied to the circuit. The starting voltage of the neon lamp is between 65 and 70 volts, or about 75% of the total battery voltage used. However, there is a difference between the *computed* time constant and the *observed* time constant. If the circuit resistance is 4 megohms, according to the computed time constant, the neon lamp should flash 15 times per minute or 4 MΩ × 1 μF = 4 sec; but you actually see about 30 flashes per minute. The reason for this is that the lamp does not light at exactly 63.2% of charge and the capacitor does not fully discharge each time the lamp flashes. As various values of resistance are used, compare the time required to light the lamp with the computed time constant of the circuit.

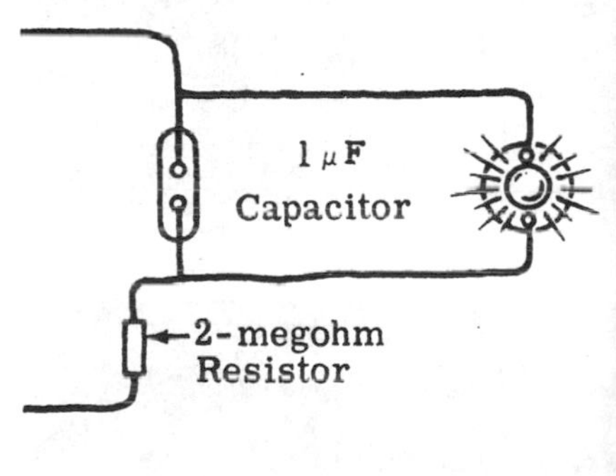

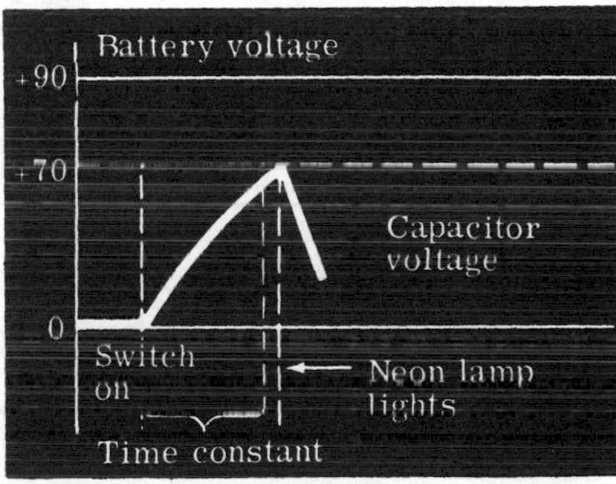

A CHANGE IN RESISTANCE CHANGES THE RC TIME CONSTANT

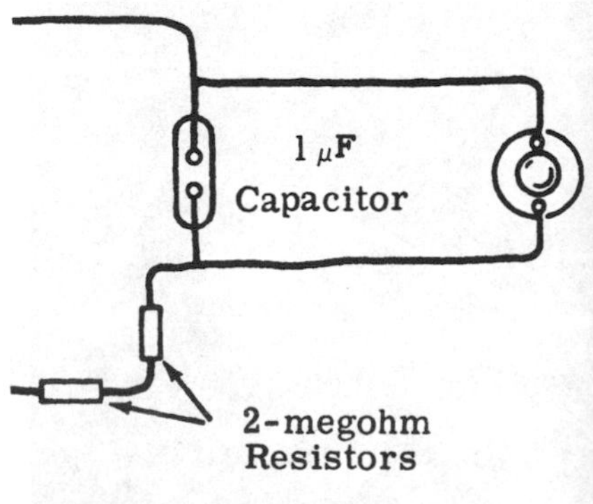

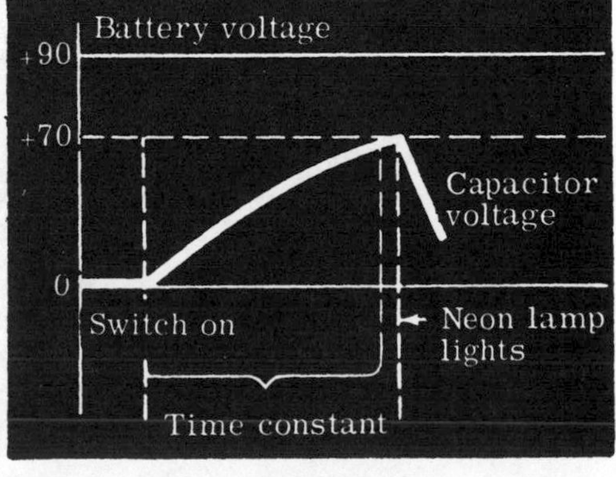

Experiment/Application—RC Time Constant (continued)

Next, all of the resistors are removed except the single 2-megohm resistor, and various values of capacitance are used. Notice that changes in the value of capacitance have the same effect on the circuit time constant as changes of resistance. When low values of capacitance are used, the time constant is shorter and the flashes occur so rapidly that the light appears to be steady, rather than flashing.

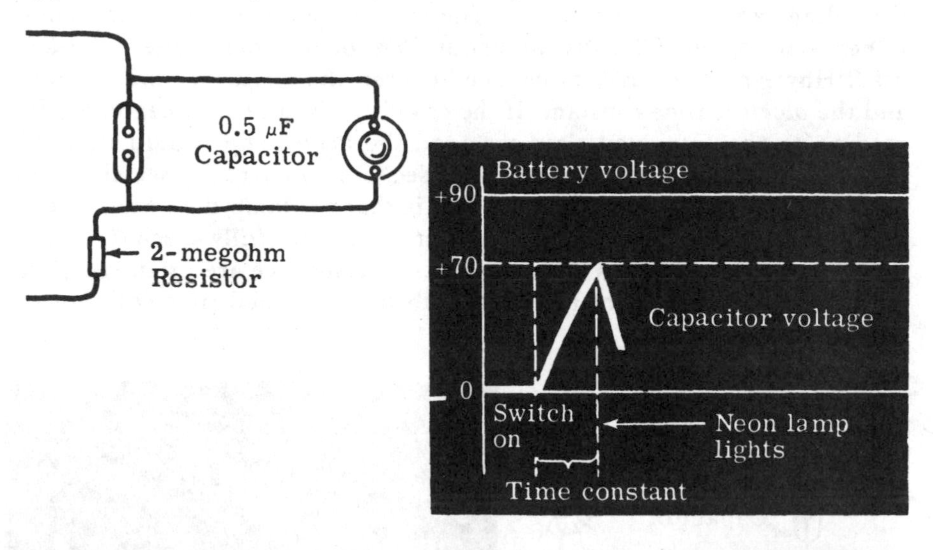

A CHANGE IN CAPACITANCE CHANGES THE RC TIME CONSTANT

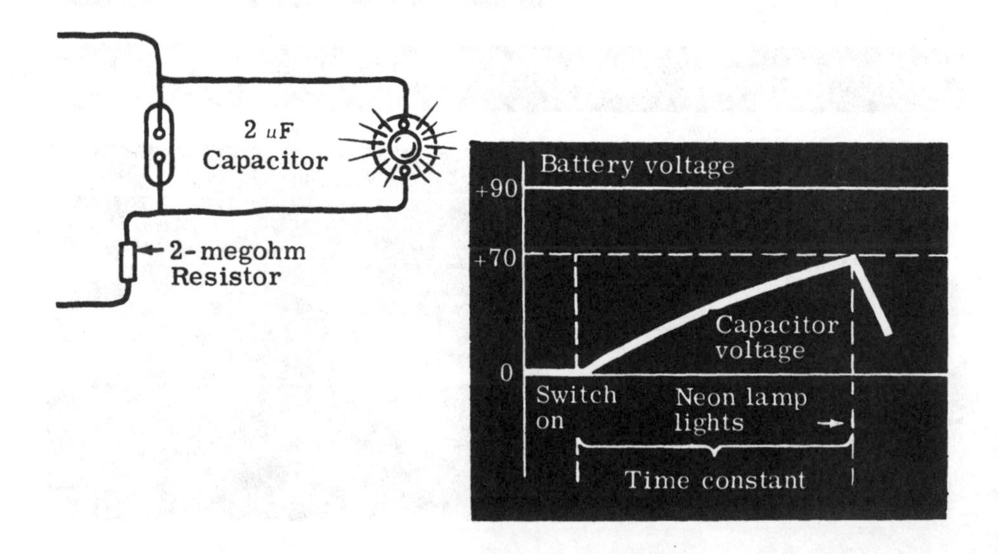

Experiment/Application—RC Time Constant (continued)

Current flow while charging a capacitor through a resistor can be easily demonstrated. A 0-1-mA zero-center milliammeter, a 200,000-Ω resistor and a 4-μF capacitor are series-connected in the circuit shown below.

At the moment the switch is closed, you see that the meter indicates a large current flow and, as the capacitor charges, the meter reading falls toward zero, reaching zero in about 4 seconds (5 time constants). Observe that once the current reaches zero (indicating a full charge on the capacitor plates), opening and closing the switch causes no further current flow. If you then open the switch and discharge the capacitor by shorting its terminals with a screwdriver, then, the switch is closed, and the meter will indicate a charging current again.

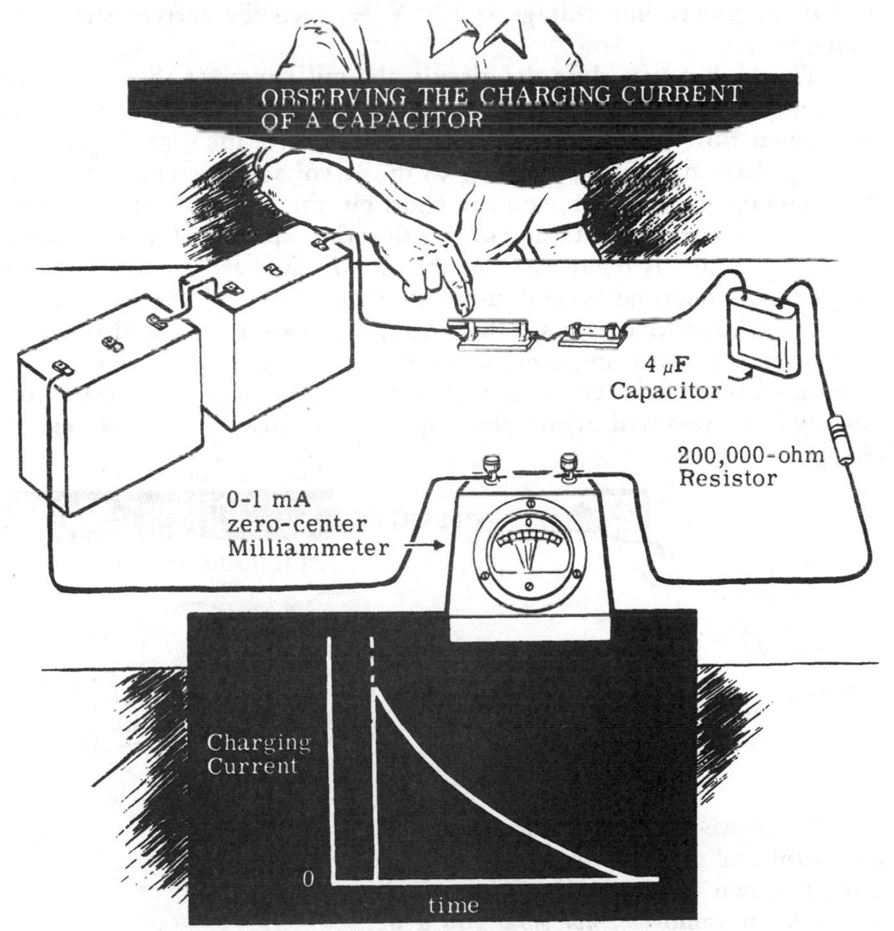

Substituting capacitors having various values, you can show that the current flow lasts longer when charging larger capacitors. Notice that, for small values of capacitance, the time required to charge them may be so short that it is difficult to read the current indicated on the meter.

Experiment/Application—Current Flow in an AC Capacitive Circuit

You can connect a capacitor to show that ac current flows continuously in an ac capacitive circuit. One lead of the line cord is connected to a terminal of the capacitor through the switch and fuses, while the other line cord lead is connected to the remaining capacitor terminal through the 0-50 mA ac milliammeter. When the circuit is plugged into the ac power line outlet, and the switch is closed, you see that a continuous flow of current is indicated on the milliammeter. You know that the capacitive reactance:

$$X_c = \frac{1}{2\pi fC} = \frac{1}{6.28 \times 60 \times 0.000001} = 2653 \text{ ohms}$$

and if the power line voltage is 120 V ac, then the current should be about 45 mA.

Thus, when you close the circuit, the milliammeter shows that approximately 45 milliamperes of ac current is flowing continuously. This continuous flow of circuit current is possible since the capacitor is continously charging and discharging as the ac voltage reverses its polarity. After you open the switch, you can short out the terminals of the capacitor with a screwdriver. Again you see that the capacitor retains a charge when the voltage is removed from the circuit. However, as the power is applied and removed several times in succession and the capacitor is discharged each time, you see that the sparks vary in size and intensity. This occurs because the amount of charge retained by the capacitor when used in an ac circuit is not always the same, since the circuit voltage may be removed while the capacitor is discharging or not yet charged.

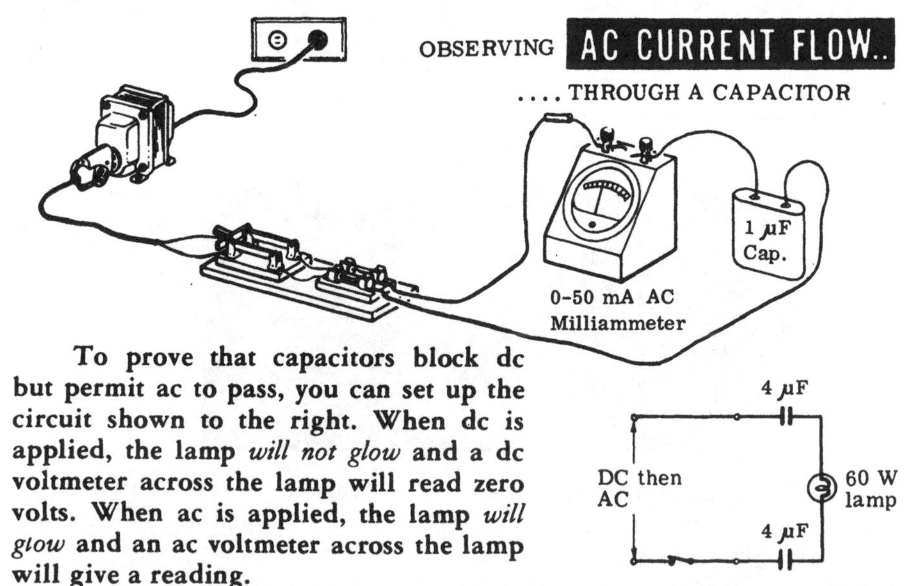

To prove that capacitors block dc but permit ac to pass, you can set up the circuit shown to the right. When dc is applied, the lamp *will not glow* and a dc voltmeter across the lamp will read zero volts. When ac is applied, the lamp *will glow* and an ac voltmeter across the lamp will give a reading.

Experiment/Application—Capacitive Reactance

You have already discovered that ac current can flow through a capacitive circuit and that the opposition to current flow in such a circuit depends on the capacitance, provided the frequency remains constant. Using the 120-V ac power line as a voltage source, you can connect a 0-100 mA ac milliammeter in series with a parallel-connected 1-μF and 0.5-μF paper capacitors. When the circuit is plugged in and the switch is closed, you will see that a current flow is indicated on the ac milliammeter.

To show the effect of changing the value of capacitance on the opposition offered to the flow of ac current, you could replace the 1.5-μF capacitor with first a 1-μF and then a 0.5-μF capacitor. Each time the capacitor is changed, you should first open the switch and then discharge the capacitor with a screwdriver. You see that *increasing* the value of capacitance *increases* the current flow, showing that the opposition or capacitive reactance *decreases* whenever the capacitance *increases*

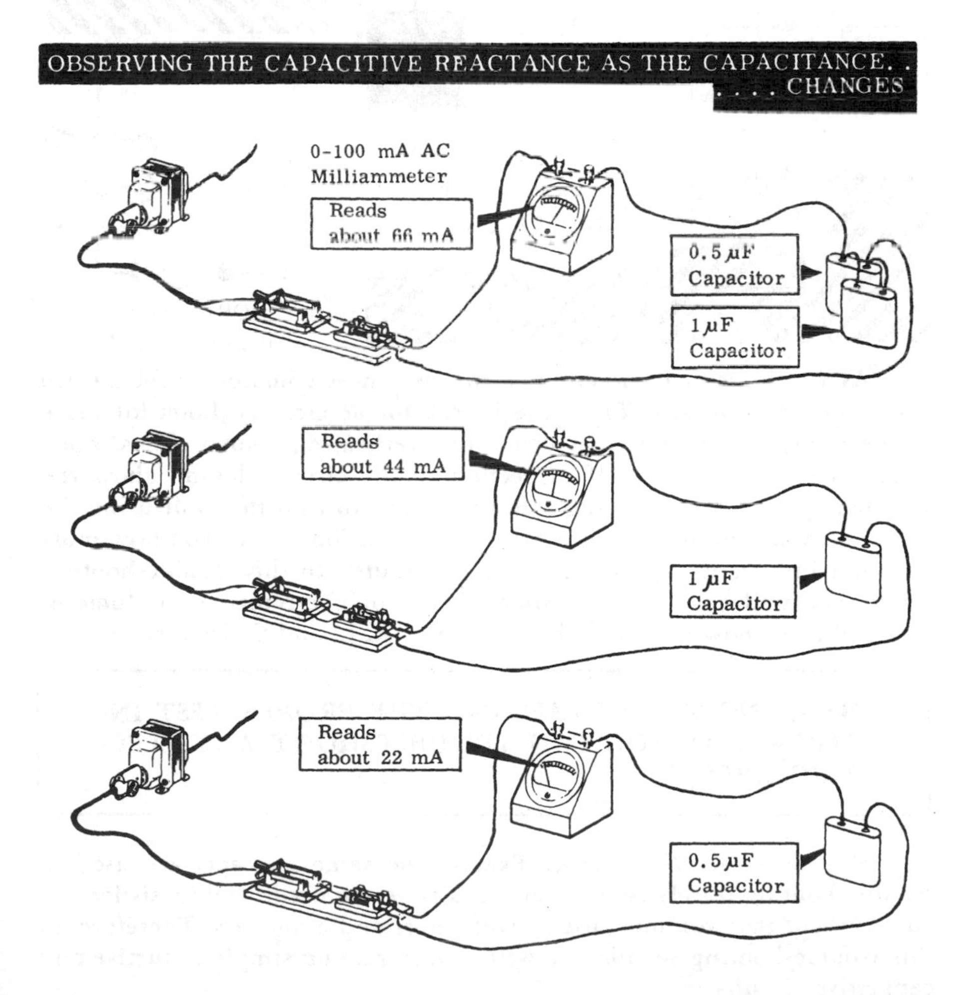

Basic Concepts— Resistive/Inductive/Capacitive Circuits (DC/AC)

As you learned in troubleshooting dc circuits in Volume 2, one of the most important things in acquiring troubleshooting skill is to *use your head and proceed logically* through a circuit. Also, you found that looking around in the circuit would often help to allow you to see the problem. The *same rules* and procedures that you learned in dc circuits can be applied to simple ac circuits. If you proceed logically, and remember the properties of resistance, inductance, and capacitance in circuits, you can troubleshoot ac circuits just as easily as dc circuits.

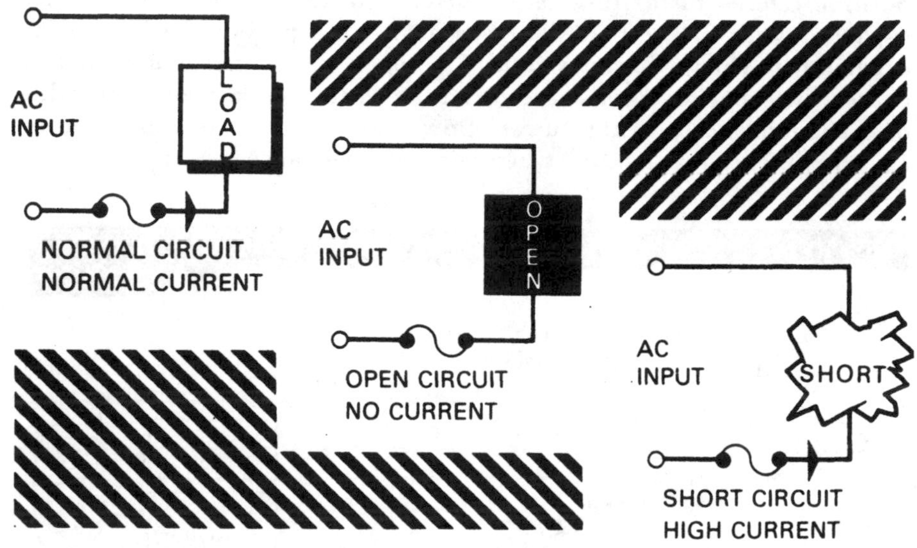

As you learned in dc circuits, the two most common problems are shorts or open circuits. The same is true for ac circuits. Look for loose connections, frayed wires, evidence of overheating, blown fuses, open switches, or loose plugs. If inspection of the circuit doesn't show the problem, you will have to dig into the circuit to find the fault using the same methods you used before with some additions, since you now must consider inductance and capacitance in circuits. In this troubleshooting section we will consider only simple R, L, or C circuits. In Volume 4, you will learn how to troubleshoot complex R, L, and C circuits.

BY USING YOUR HEAD AND THE PROPER TEST INSTRUMENTS YOU CAN TROUBLESHOOT ANY ELECTRIC CIRCUIT.

Note on ac resistive circuits: Exactly the same concepts are used in troubleshooting purely resistive ac circuits as you used for dc resistive circuits, except that you must use ac voltmeters and ammeters. Therefore, in this troubleshooting section, we will concentrate on simple inductive and capacitive circuits.

Troubleshooting Inductors (DC/AC Circuits)

Inductors or coils are connected into electric circuits to provide inductance. These inductors may develop faults which prevent them from doing their job. As coils are made in many shapes and sizes, it is not possible to specify precisely which faults will occur. There are, however, three basic faults you should always look for.

Open-circuit Winding: This most common fault may be caused by too high a current being passed through a coil, causing the winding to *burn out*; or it may be caused by a wire breaking at a soldering lug or wire connection. If the coil is made of fine wire, the wire may break where it passes through the coil former, or the break may be caused by corrosion of the wire.

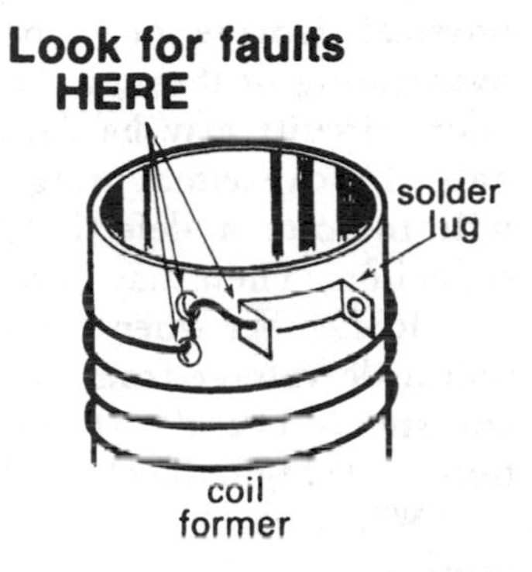

An open-circuit winding may be detected by visual inspection, or by checking the coil for continuity with an ohmmeter—that is, measuring the resistance of the coil winding. A very high resistance reading indicates an open-circuited winding.

Turns of Winding Shorting Together: If two or more turns of a coil are shorted, by reason of breakdown of the insulation between the turns, the inductance and resistance of the coil will be changed. However, this is usually very difficult to detect with an ohmmeter. Sometimes, this kind of fault may be detected by checking for continuity with an ohmmeter; a resistance reading appreciably lower than that specified by the manufacturer will indicate such a fault.

Usually, the coil must be removed from the circuit and a known good coil be substituted or the inductance checked as described below. If the symptoms which caused the first coil to be suspected are eliminated, the fault may have been cured. There is a danger, however, with this method that the new coil may also be made defective by a fault elsewhere in the circuit. Therefore, before the substitution is made, check the rest of the circuit for defects.

Shorting to Ground (Core): If the insulation between a coil winding and the core is defective, it may result in the coil being grounded, since the core of an inductor is often connected to ground. This type of defect can be easily detected by checking the resistance between the coil winding and the core (or ground). A zero or low value of resistance indicates defective insulation and requires that the coil be replaced.

Occasionally the short may only be apparent when the operating voltage is applied to the coil, in which case an *insulation tester* must be used, which applies a high voltage (100 or 500 V) to the insulation.

As you can tell from the above, it is usually necessary to disconnect at least one side of an inductor to make appropriate measurements. If the inductor can be removed, it is possible to measure the inductor in another known circuit or with an instrument called *the inductance bridge.*

Troubleshooting Capacitors (DC/AC Circuits)

Capacitors are connected into circuits for many purposes where use can be made of their special properties. When, for instance, a connection has to be made between two points in a circuit in such a way that ac will flow between the points, but not dc, a capacitor connected between the points will provide a suitable connecting link. As with inductors, there are three basic capacitor faults you should look for.

Short-circuited Plates: Short circuits in capacitors may be caused by physical damage, or by breakdown of the dielectric insulation through overheating or through the application of too high an operating voltage. Short circuits may be detected by testing continuity between the terminals of a capacitor, using an ohmmeter. A reading of less than 300 KΩ will indicate a defective capacitor, except in the case of electrolytic capacitors. These may have a *leakage resistance* of less than 500KΩ.

Remember when making such a test on small electrolytic capacitors that a dc voltage from the ohmmeter is applied across the capacitor, so you should check with the meter leads in both directions. If one direction reads high, the electrolytic capacitor is OK.

When checking variable capacitors for shorts, remember that the short will usually occur only when the spindle and its moving plates are in one particular position. The capacitor must, therefore, be tested over its *full range* of variation.

Open circuit Short circuit

Open Circuits: Since capacitors usually have a very high dc resistance (it may be *infinite*), it is impossible to test for open circuits with an ohmmeter. An exception is the case of electrolytic capacitors, where a measurable resistance indicates that the capacitor is probably good. With capacitors of larger capacity (say greater than 1 μF), it is possible to check for open circuits by removing at least one side of the capacitor from the circuit and charging the capacitor with the ohmmeter by putting it momentarily across the capacitor terminals. If the ohmmeter terminals are now reversed, the needle will jump down momentarily as the capacitor discharges into the ohmmeter circuit. As a last resort, the capacitor can be checked on a capacitance bridge or you can substitute a good capacitor in the circuit. *Be careful* because a circuit defect may have burned out the capacitor, so look the rest of the circuit over *before* you apply power. With electrolytic (polar) capacitors, make sure it is in the circuit in the *correct polarity*. Remember, electrolytic capacitors can *dry out* and lose their capacitance as they grow older, so a *working* capacitor may not be doing its job.

Short to Ground (Case): Sometimes, metal-cased capacitors develop a short between one of the plates and ground. This can easily be detected by use of an ohmmeter. Usually, both terminals of a capacitor have a very high impedance to the case. However, remember that some capacitors have one terminal connected to the case.

Troubleshooting Simple AC Inductive Circuits

Example

Suppose you had to troubleshoot the circuit shown below. Inspection reveals that the circuit wiring is in good condition.

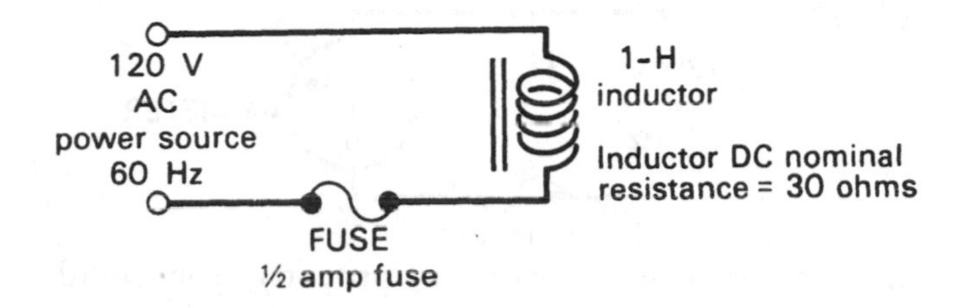

The symptom is that the fuse blows. Calculation shows that

$$X_L = 2\pi fL = 2 \times 3.14 \times 60 \times 1 = 377 \text{ ohms}$$

and the current should be

$$I = \frac{E}{X_L} = \frac{120}{377} = 0.32 \text{ ampere}$$

(actually somewhat less because of the dc resistance of the inductor).

Since there is only one component, we check it and find that the dc resistance is 2 ohms. Clearly, the inductor is shorted and must be replaced. On the other hand, if no current flowed and you measured the inductor resistance as *infinite*, it is obvious that the inductor has an open winding.

THE BASIS OF TROUBLESHOOTING LIES IN USING A LOGICAL PROCEDURE

By using your senses, your head, and test instruments, you can troubleshoot any electric circuit.

Troubleshooting Simple AC Capacitive Circuits

Troubleshooting capacitive circuits is just as easy as troubleshooting inductive circuits.

Suppose you had circuit A shown below and were told that the ammeter read 0.1 ampere.

Example 1

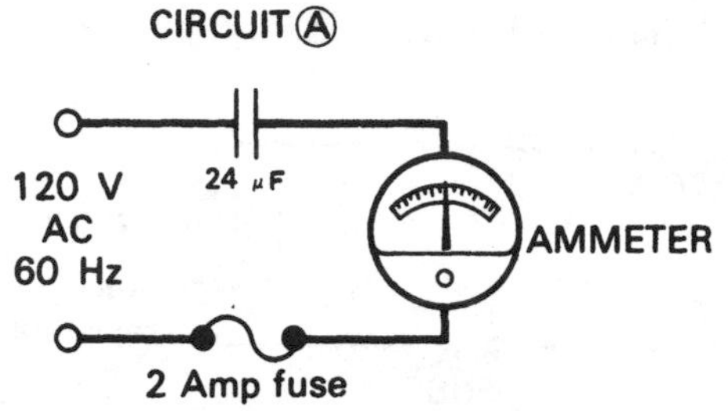

CIRCUIT Ⓐ

120 V AC 60 Hz — 24 μF — AMMETER — 2 Amp fuse

Assuming the ammeter is working properly, you could calculate that the proper current is

$$X_c = \frac{1}{2\pi fC} = \frac{1}{2 \times 3.14 \times 60 \times 0.000024} = 110 \text{ ohms}$$

$$I = \frac{E}{X_C} = \frac{120}{110} = 1.1 \text{ amperes}$$

Therefore, the capacitor is defective or has the wrong value. A check of the capacitor case showed that the capacity should be 24 μF. Therefore, the capacitor is defective.

Suppose you found that the fuse blew immediately when the power was applied in the same circuit. You would then suspect that the capacitor was shorted and you could confirm this by ohmmeter checks.

Example 2

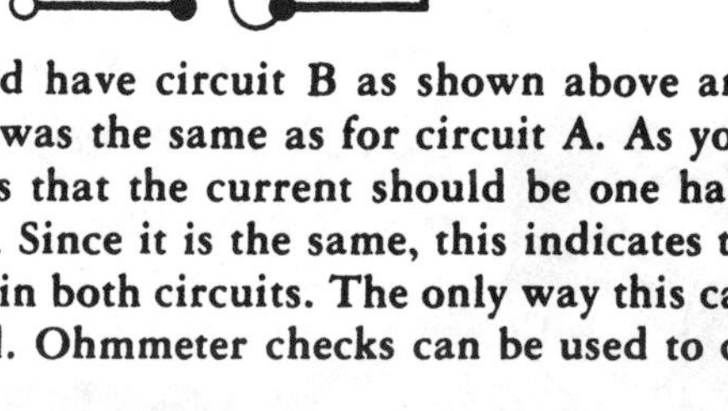

CIRCUIT Ⓑ

24 μF — 24 μF — AMMETER

Suppose you should have circuit B as shown above and the current shown on the ammeter was the same as for circuit A. As you know, analysis of the circuit shows that the current should be one half of the value calculated for circuit A. Since it is the same, this indicates that the circuit capacitance is the same in both circuits. The only way this can happen is if one capacitor is shorted. Ohmmeter checks can be used to confirm which capacitor is defective.

Note on dc capacitive circuits: While you will not learn specifically about troubleshooting capacitance in dc circuits, since this will be covered in Volume 4, it should be obvious that ohmmeter checks can also be used to discover the most common faults in dc capacitive circuits.

Drill in Troubleshooting Simple AC Circuits

1. You have been asked to troubleshoot the resistive circuit shown below. The symptom is that the fuse blows when the switch is closed. The wiring is correct and there are no obvious shorts.

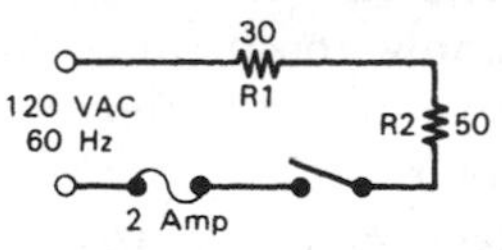

   (a) What current should flow?
   (b) How would you use an ohmmeter to determine what the problem is?
   (c) How would you correct the problem?
   (d) Voltage measurement across the resistors shows:

$$ER1 = 120\ V$$
$$ER2 = 0\ V$$

   What is the probable difficulty?
   (e) Can you explain now why the fuse blows?

2. An inductor has been used in series with a lamp as shown below to reduce the voltage to the lamp.

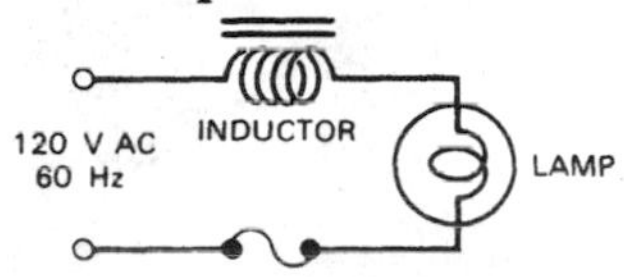

   (a) Suppose the lamp does not light, how would you proceed to find the defect? Suppose you had no ohmmeter but just an ac voltmeter, could you determine which component was defective? How?
   (b) Suppose the symptom was that the lamp was excessively bright and the inductor overheated badly, what would you suspect the difficulty to be? How would you verify this?

3. Suppose a capacitor had been used in the circuit for question 2 as shown below:

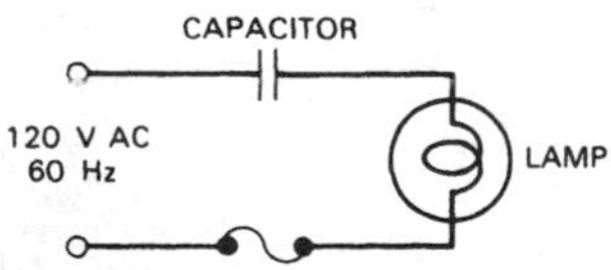

   (a) Suppose the lamp glows with excessive brightness. Inspection shows that the circuit is OK and there are no frayed nor loose wires. What do you suspect is wrong? How would you verify this? How would you correct this? Describe a simple test with an ohmmeter to determine whether the capacitor is OK.
   (b) Suppose the lamp glowed dimly or not at all. What do you suspect is wrong? Describe how you would test your theory on what is wrong.
   (c) If you had only a voltmeter, how would you use it to check for the defect in (a) or (b) above?

Learning Objectives—Next Volume

Overview—Now that you have learned about inductance and capacitance as well as resistance in basic ac and dc circuits, you are ready to use what you know about these three circuit elements. In Volume 4, you will study more complex circuits that use these circuit elements in combinations to control and influence current flow in electric and electronic circuits. Both dc and ac will be studied in conjunction with their typical applications in various types of machinery, electrical controls, etc. Also, you will learn the approach to troubleshooting complex ac (and dc) circuits and systems.

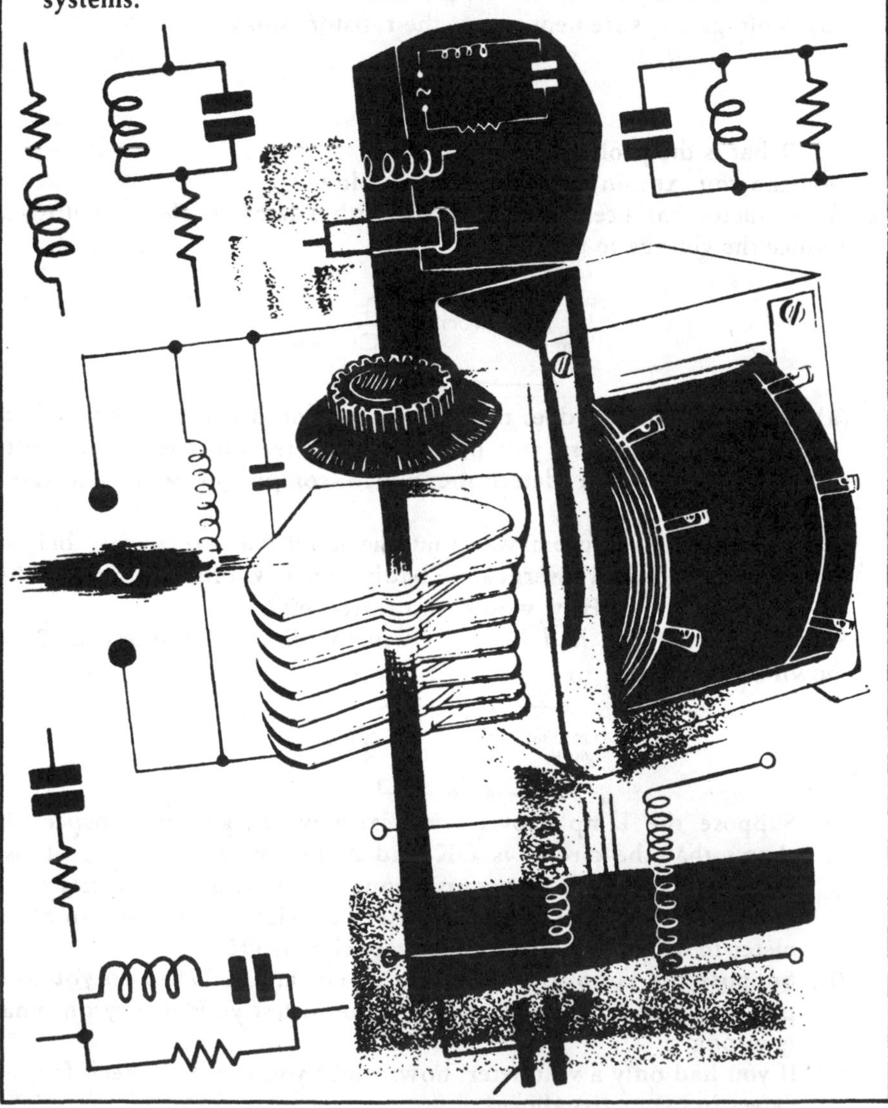

Basic Electricity

REVISED EDITION

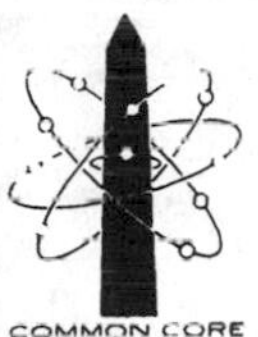

VAN VALKENBURGH,
NOOGER & NEVILLE, INC.

VOL. 4

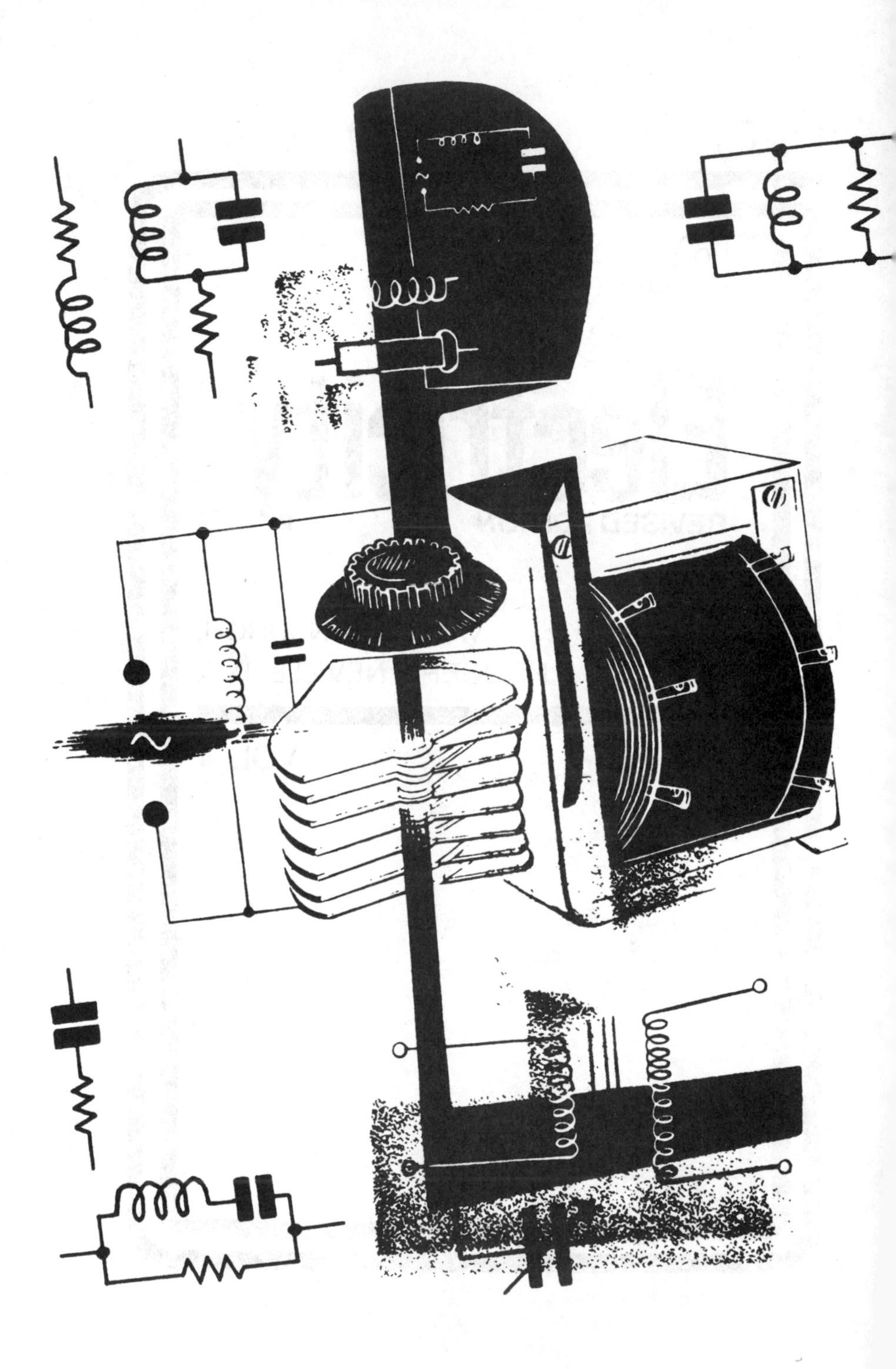

Introduction to AC Electric Circuits

Now that you know about direct current (dc) and alternating current (ac) and how they work from Volumes 2 and 3, and how resistance (R), inductance (L), and capacitance (C) behave in basic circuits carrying dc and ac, you are ready to learn how all three of these circuit elements can be used in combination to control and influence current flow in ac electric circuits. You also have learned how to analyze simple ac (and dc) circuits that contain *only one* of these circuit elements—resistance, inductance, or capacitance alone. You have found out how R, L, and C *individually affect* ac current flow, phase angle, and power in ac circuits. However, you have not as yet considered ac circuits having *two or more* of these circuit elements. So, here, in Volume 4, we now are going to put them *all together*!

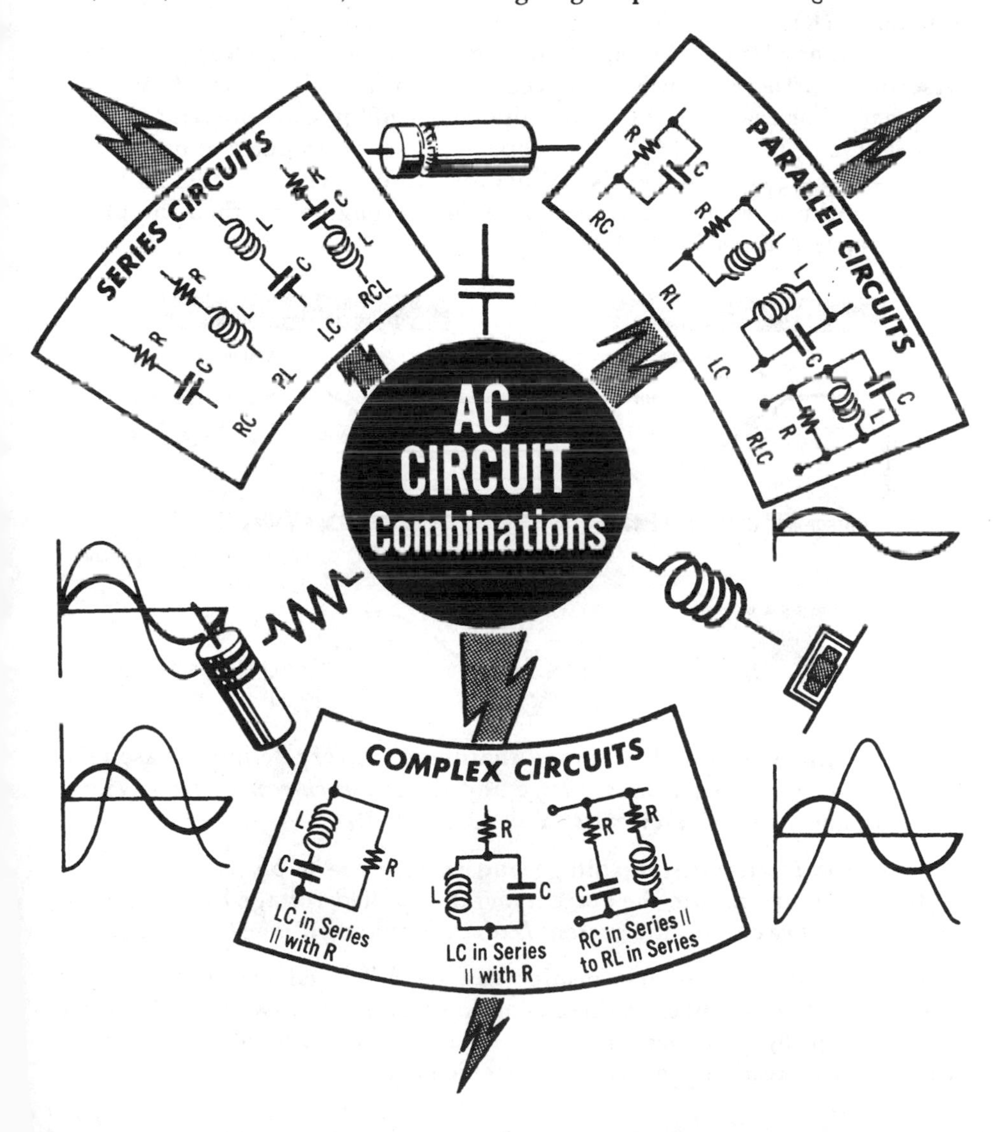

Introduction to AC Electric Circuits (continued)

In this volume you will learn how to analyze, solve, and troubleshoot ac circuits that contain both series and parallel combinations of the three circuit elements—resistance (R), inductance (L), and capacitance (C).

As you know, every electric circuit contains a certain amount of R, L, and C and inductive and capacitive reactance, X_L and X_C. Therefore, ac circuits can contain *three* factors which oppose current flow—R, X_L, and X_C. (In any given circuit, however, if any of these factors is negligible, it can be disregarded.) As you will learn in this volume, these factors must be combined in special ways to find the *net* opposition to the flow of current. This combined factor, called *impedance*, is represented by the symbol *Z*. The Ohm's law formula is used to calculate Z as you used it for finding resistance (R).

You know from Volume 3 that the term *phase angle* (θ) is used to describe the *time* relationship between ac voltages and currents of the same frequency. For example, if two ac voltages are of opposite polarity at *every instant* of *time*, they are 180 degrees out of phase, or, to put it another way, the phase angle *between them* is 180 degrees. If the current reaches its maximum amplitude when the voltage is going through zero, then the phase angle between them is 90 degrees.

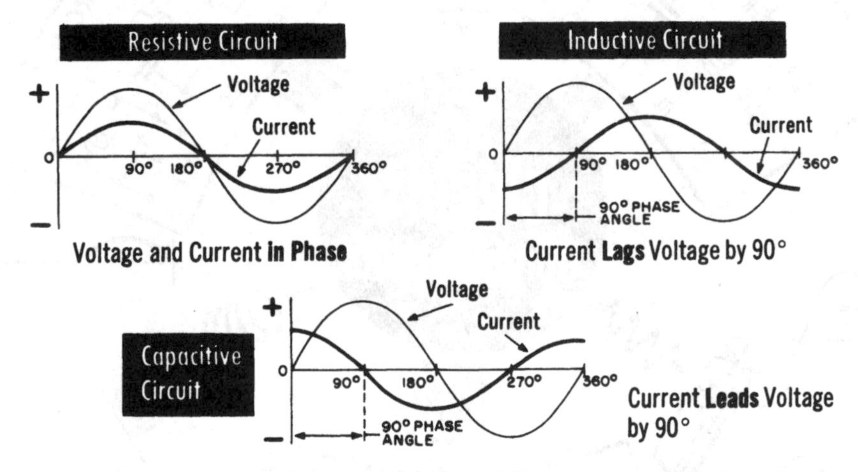

In Volume 3, you also learned that there are very definite phase relationships between the *applied voltage* and the *circuit current* in resistive, inductive, and capacitive circuits. You know that:

1. In a *resistive* circuit, the voltage and current are *in phase*.
2. In an *inductive* circuit, the current *lags* the applied voltage by 90 degrees.
3. In a *capacitive* circuit, the current *leads* the applied voltage by 90 degrees.

You know how these relationships can be described using *voltage* and *current waveforms*. However, there is another, and easier, way to show these relationships by using *vectors*. You must use vectors to solve ac electric circuits with more than one kind of circuit element.

Introduction to AC Electric Circuits (continued)

Since you will be thinking now about the interrelationship of all these various factors, you will need some additional tools to help you keep them all in mind and understand them. The additional tools that you are going to learn about as you continue to study ac electric circuits—starting with ac series circuits—are the following: (1) Use of *vectors* in the solution of ac circuits, and (2) solution of ac circuit problems either *graphically* or by *calculation*. As you will see, solving ac circuits by calculation will be made much easier if you have a hand calculator.

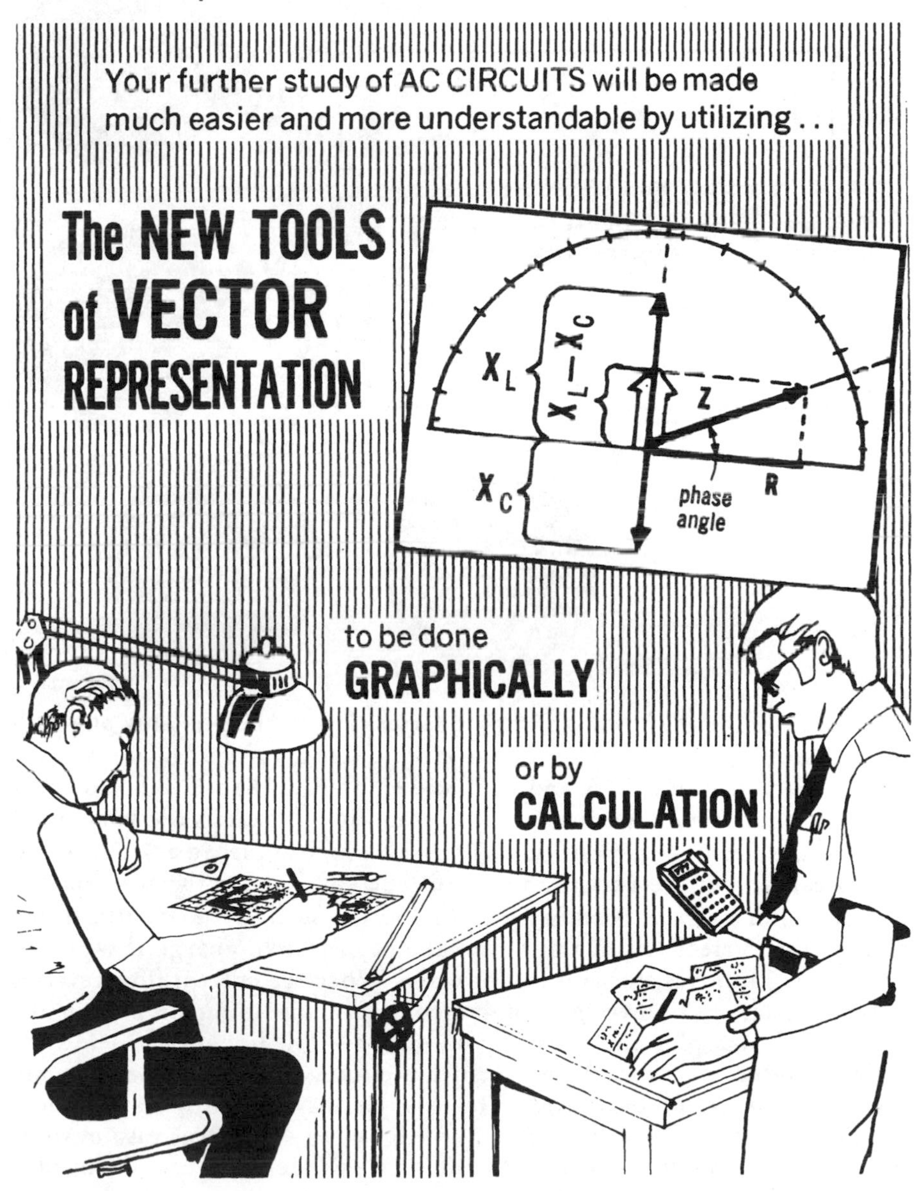

Your further study of AC CIRCUITS will be made much easier and more understandable by utilizing . . .
The NEW TOOLS of VECTOR REPRESENTATION
X_L
$X_L - X_C$
Z
X_C
phase angle
R
to be done GRAPHICALLY
or by CALCULATION

Current Flow in AC Series Circuits

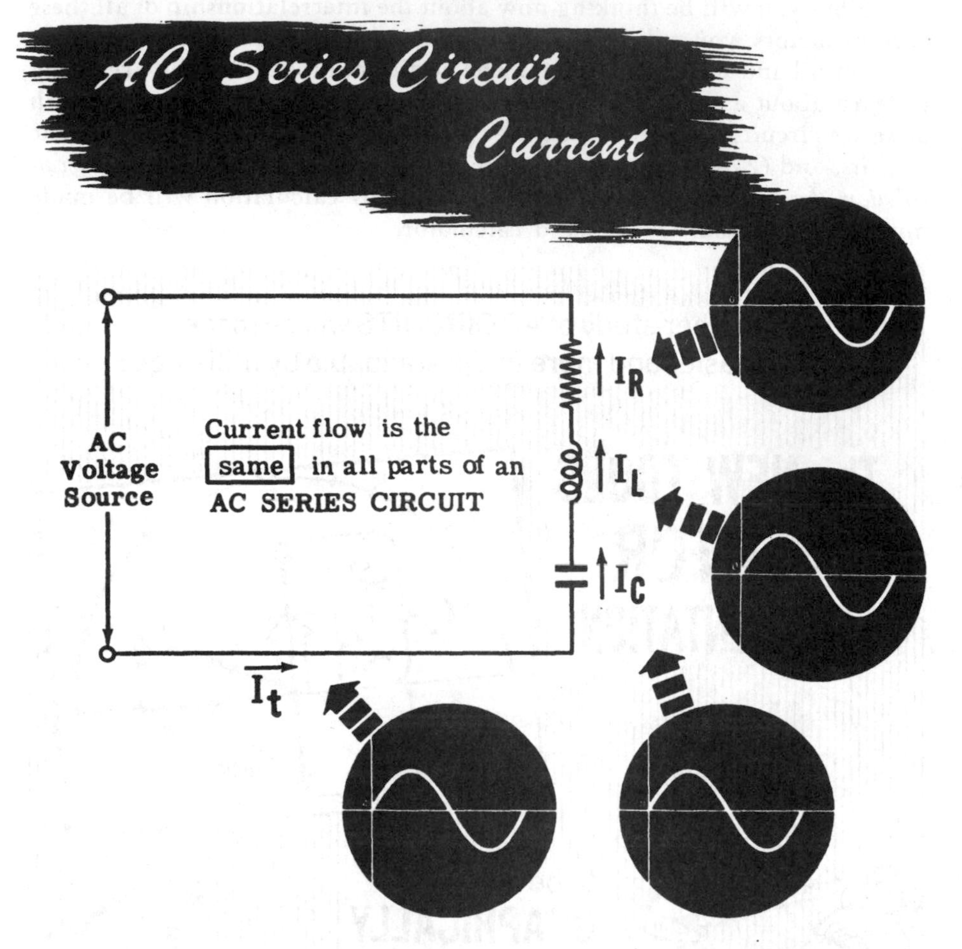

In an ac series circuit, as in a dc series circuit, there is only one path for current flow around the complete circuit. This is true regardless of the type of circuit: R and L, R and C, L and C, or R, L, and C. Since there is only one path around the circuit, the current flow is exactly the same in all parts of the circuit at any one time. Thus, all phase angles in a series circuit are measured with respect to the circuit current, unless otherwise stated.

Thus, in a circuit containing R, L, and C—such as the one illustrated above—the current which flows into the capacitor to charge it will also flow through the resistor and the inductor. When the current flow reverses in the capacitor, it reverses simultaneously in the inductor and in the resistor. If you plot the current waveforms— I_R, I_L, and I_C—for the resistor, inductor, and capacitor in such a circuit, the three waveforms are identical in *value* and *phase angle*. The total circuit current, I_t , is also identical to these three waveforms for $I_t = I_R = I_L = I_C$. It is also obvious that it doesn't matter in what order the circuit elements are arranged.

Voltages in AC Series Circuits

The total voltage, E_t, of an ac series circuit cannot be found by adding the individual voltages, E_R, E_L, and E_C, across the resistance, inductance, and capacitance of the circuit. Unlike dc circuits, these voltages cannot be added *directly* because the individual voltages across R, L, and C are *not in phase* with each other. Look below.

AC SERIES CIRCUIT VOLTAGE

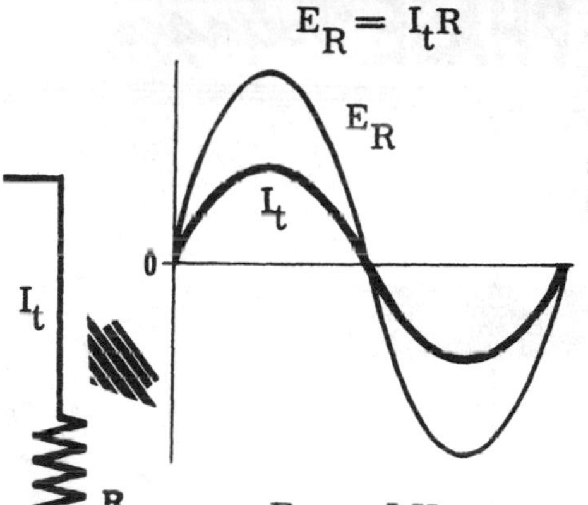

The voltage E_R across R is in phase with the circuit current, since current and voltage are in phase in pure resistive circuits.

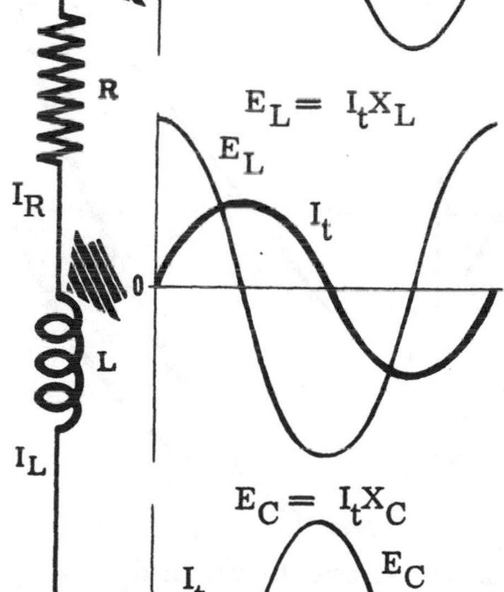

The voltage E_L across L leads the circuit current by 90 degrees, since current *lags* the voltage by 90 degrees in purely inductive circuits. Thus, E_L crosses the zero axis, going in the same direction, 90 degrees *before* the current wave.

$E_C = I_tX_C$

I_t E_C C I_C 0 I_t

$I = I_t$

$I_t = I_R = I_L = I_C$

The voltage E_C across C lags the circuit current by 90 degrees, since current *leads* the voltage by 90 degrees in purely capacitive circuits. Thus, E_C crosses the zero axis, going in the same direction, 90 degrees *after* the current wave.

Voltages in AC Series Circuits (continued)

To find the total voltage in a series circuit, the instantaneous values of the individual voltages must be added together to obtain the instantaneous values of the *total voltage waveform*. Positive values are added directly, and so are negative values. The difference between the total positive and negative values for a given moment is the instantaneous value of the total voltage waveform for that instant of time. After all possible instantaneous values have been obtained, the total voltage waveform is drawn by connecting together these instantaneous values.

Combining OUT-OF-PHASE *Voltages*

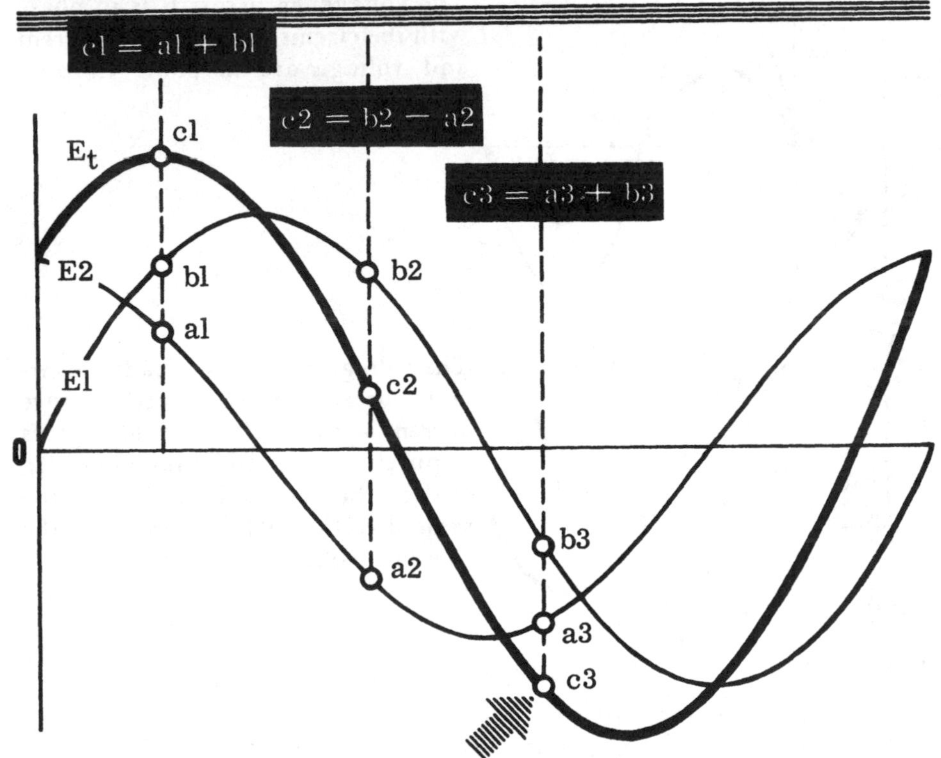

Combined instantaneous values are the result of combining the instantaneous values of E1 and E2

When combining out-of-phase voltages, the maximum value of the total voltage waveform is always less than the sum of the maximum values of the individual voltages. Also, the phase angle (which is the angle between one waveform and another chosen as a reference) of the total voltage wave differs from that of the individual voltages. It depends on the relative magnitude and phase angles of the individual voltages.

R and L Series Circuit Voltages

Suppose you consider an ac series circuit having negligible capacitance. The total circuit voltage depends on the voltage E_L across the circuit inductance and the voltage E_R across the circuit resistance. E_L leads the circuit current by 90 degrees, while E_R is in phase with the circuit current; thus E_L leads E_R by 90 degrees.

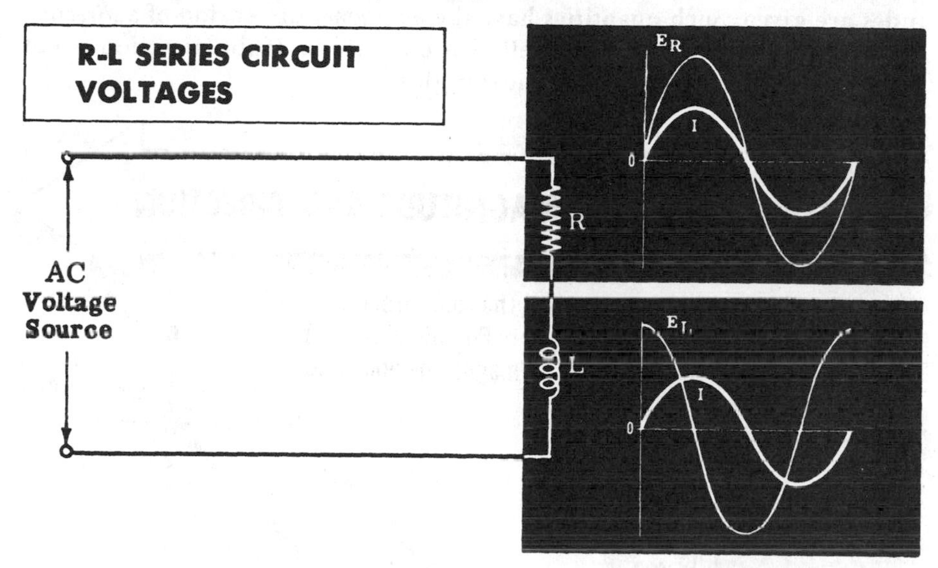

To add the voltages E_t and E_R, you can draw the two waveforms to scale and combine instantaneous values to plot the total voltage waveform. The total voltage waveform, E_t, then shows both the value and phase angle of E_t (the phase is measured with respect to the current).

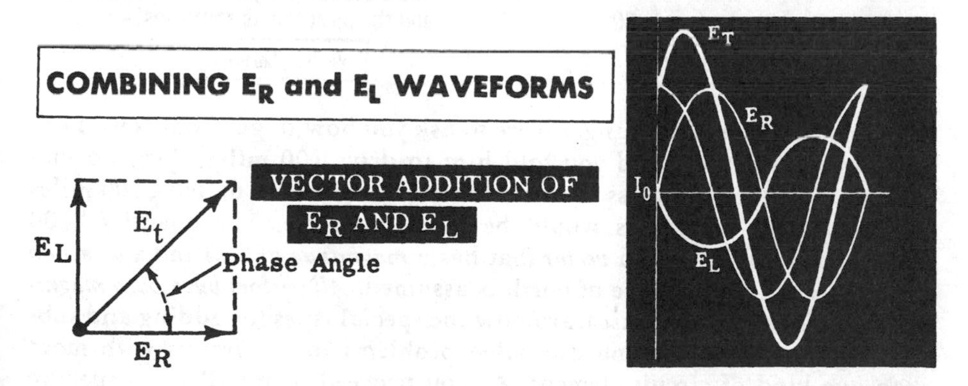

From the above, it should be clear that the resultant voltages are dependent on *both* magnitude and phase; when we add them, we must take *both* into account. To do this, we shall use *vectors* to make our work simpler with these two variables.

What a Vector Is

All physical quantities have a *basis* for specifying their size or amount. The terms "7 inches," "7 meters," "7 pounds," and "7 grams" all express the *magnitude* of physical quantities, and each is *completely* described by the number 7. Quantities that have *magnitude only* are called *scalar*. There are quantities, however, that are *not* completely described if only their magnitudes are given. Such quantities have the *additional* dimension of a *direction* with respect to a reference, as well as a magnitude. If the direction is not given, these quantities may be meaningless.

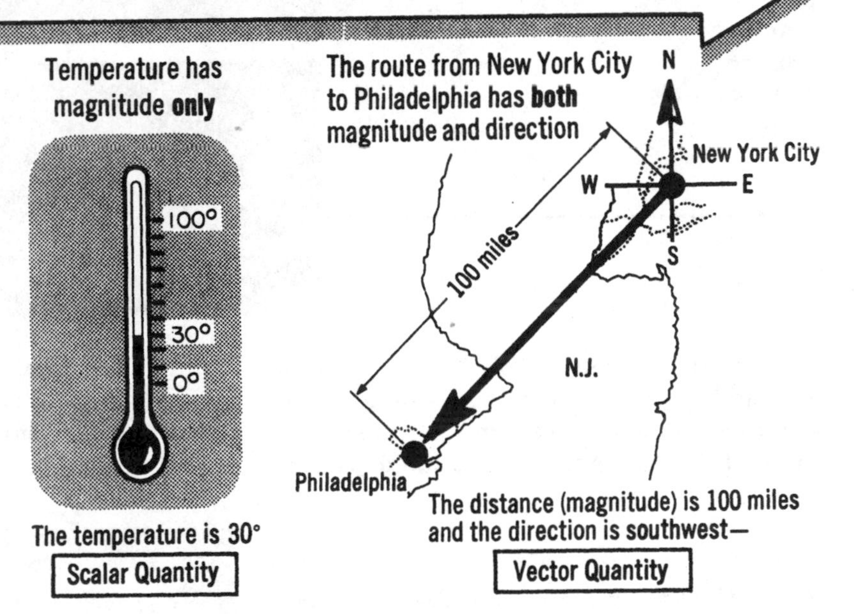
VECTORS HAVE BOTH MAGNITUDE AND DIRECTION
Temperature has magnitude only
100°
30°
0°
The temperature is 30°
Scalar Quantity
The route from New York City to Philadelphia has both magnitude and direction
N
New York City
W
E
S
100 miles
N.J.
Philadelphia
The distance (magnitude) is 100 miles and the direction is southwest—
Vector Quantity

For example, if a stranger were to ask you how to get from New York City to Philadelphia, and you told him to drive 100 miles, this information would be meaningless to him. However, if you said to drive 100 miles *southwest*, your directions would be more complete. The quantity "100 miles southwest" is thus a *vector* that has a *magnitude* of *100* and a *direction* of *southwest*. The reference of north is assumed. *All vectors have both magnitude and direction.* You must learn now the special rules for adding and subtracting vectors before you can solve problems in ac circuits with more than one kind of circuit element. As you proceed, you will learn specific points about vectors that you will need to know for solving the specific ac circuits either *graphically* or *by calculation.*

A *velocity* has both *direction* and *magnitude* and is therefore a *vector* quantity. You can see how to use vectors by adding two velocities.

What a Vector Is (continued)

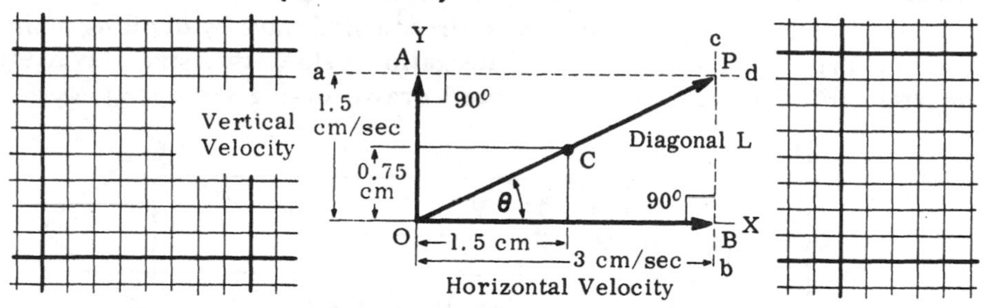

Suppose a point moves toward the top of this page from O on the line OY (OX is our reference) with a velocity of 1.5 cm/sec. One second after it starts, the point will arrive at A, which is 1.5 cm away from O. Suppose at the same time you started a point moving from O on the line OX toward B on the right-hand side of this page with a velocity of 3 cm/sec. One second after it starts, you know that it will be at B.

If the point starts at O and moves *diagonally* toward P, then it is in effect moving toward the top of the page and toward the right-hand side of the page *simultaneously*. Thus, this diagonal has *components* of velocity in *two* directions, toward A and toward B. Thus, at the end of 1 second, the point would be at P, which represents a distance of 1.5 cm vertically and 3 cm horizontally. And at the end of ½ second, the point would be at C, or 0.75 cm vertically and 1.5 cm horizontally away from O. If you had a point that had a velocity of 1.5 cm/sec toward the top of the page and simultaneously had a velocity of 3 cm/sec toward the right-hand side, you could ask, "Where will the point be at the end of 1 second?" You can represent these two velocity components as shown in the figure, by arrow OA and arrow OB. Using a scale so that OB is twice as long as OA, complete the parallelogram with lines AP and BP, drawn parallel to the axes as shown. The diagonal line OP, called the *resultant*, shows the path of the point during the 1-second interval, and P is where the point will be located at the end of 1 second. The vertical line OA and the horizontal line OB are the *vectors* representing the components or component velocities. Since the resultant diagonal line OP also has a magnitude and direction, it is thus a *resultant vector*.

The *angle* of the resultant with respect to the reference selected (line OX) gives the *direction* of the resultant. It is usually designated by the Greek letter θ (theta). Obviously, if someone gave you the location of point P, you could *resolve* the resultant into its component points by completing the parallelogram and then locating the vectors OA and OB on the drawing as the components that made up vector OP.

If you draw OA and OB carefully on graph paper and complete the parallelogram, you will find that the resultant OP is 3.35 cm long. The net velocity is thus 3.35 cm/sec. If you measure the angle θ with a protractor, you will find that the angle is about 27 degrees. As you will learn later, you can also solve vector problems analytically by using a hand calculator.

Vector Graphic Representation of AC Voltages and Currents

In Volume 3 you learned how to draw a sine wave by dividing a circle into many small arcs. In the illustration below, two sine waves 90 degrees out of phase have already been drawn by the same method.

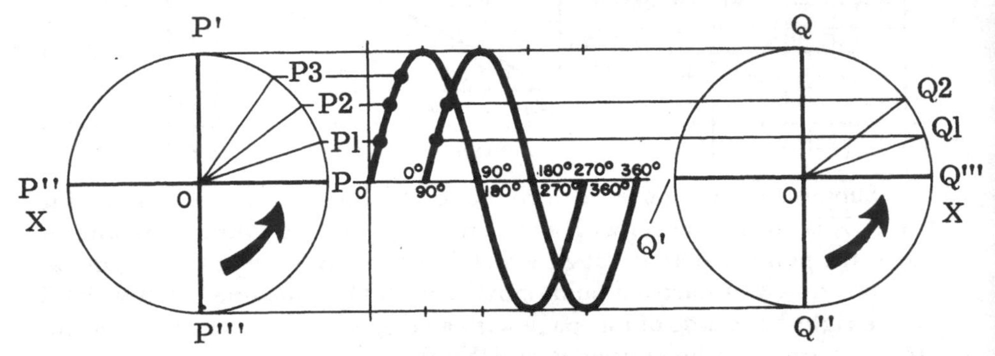

If you imagine the points P and Q to be rotating counterclockwise at the same speed on the two circles, you can see that the vertical distances of P and Q above the axis XX give the instantaneous values of the sine waves. As P and Q rotate through 360 degrees (one revolution), the sine waves complete one cycle. However, as the sine waves are 90 degrees out of phase, P reaches P′ when Q‴ has just arrived at the axis XX (the starting reference). Q starts to rotate and when Q reaches Q′, P′ has reached P″, and so on. Thus, to illustrate the relationship between the two sine waves (which could be, for example, the two voltages in a resistance and an inductor in series measured with respect to the current), you can use lines OP and OQ, provided that you draw them so that the angle between them is *equal to the phase angle between the sine waves.* Then OP and OQ are also vectors.

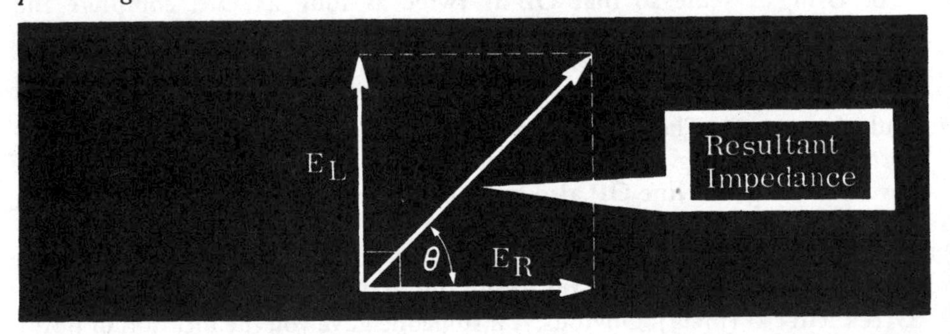

You have already learned that in an ac series circuit containing both inductance and resistance, the voltage across the inductance leads the voltage across the resistance by 90 degrees. You can, therefore, represent the voltages by vectors drawn at right angles to each other, as illustrated. The vectors are presumed to be rotating in a counterclockwise direction, so the fact that voltage across the inductor leads the voltage across the resistance is, in this case, represented by drawing the inductor voltage vector *vertically* and the resistance voltage vector *horizontally*.

Vector Graphic Representation of AC Voltages and Currents (continued)

If two out-of-phase voltages or currents are to be added together, you can represent them by vectors and find their resultant in the manner described on page 4-9. The lengths of the vectors are made proportional to the magnitude of the two quantities, and the angle between the vectors is made equal to the phase angle (θ) between them.

You can find the total voltage E_t in the R and L series circuit shown on page 4-7 by representing E_R and E_L by vectors and by finding the resultant using the method just described.

Draw a vector horizontally to represent E_R (the voltage across the resistance). E_R is in phase with the current, which is used as the reference vector. As the voltage E_L (the voltage across the inductor) leads the current by 90 degrees, draw the vector representing E_L at 90 degrees to E_R, ahead of E_R. (Remember that the vectors are rotating in a *counterclockwise* direction.) The resultant, E_t, is then obtained by completing the parallelogram. Draw lines parallel to E_L and E_R and join their crossing point at P (intersect at P).

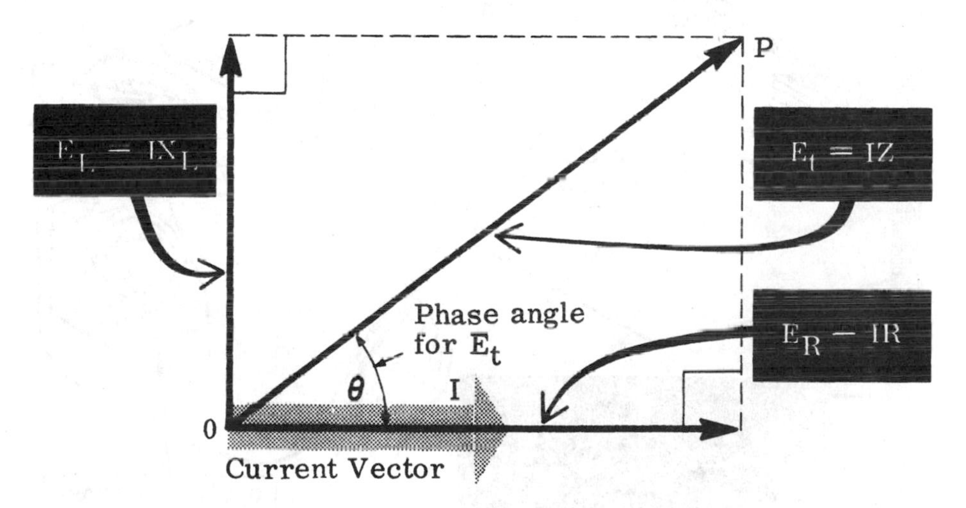

As E_R is in phase with the current through the circuit, the phase angle of the total voltage is the angle between E_R and E_t. In the case of an R and L series circuit, the total voltage leads the current.

If you now apply Ohm's law to the ac circuits you are considering, you will see that:

$$E_R = IR$$

$$E_L = IX_L \text{ (where } X_L \text{ is the inductive reactance)}$$

Z is the total opposition to current flow in the circuit and is called the *impedance.* You will learn about impedance on pages 4-13 to 4-22.

$$E_t = IZ$$

Experiment/Application—R and L Series Circuit Voltages

To verify the relationship of the various voltages in an R and L series circuit, you could connect a 1,500-ohm resistor and a 5-henry inductor to form an R and L series circuit (see illustration below). With the switch closed, individual voltage readings are taken across the inductor and resistor. Also, the total voltage across the series combination of the resistor and inductor is measured. Notice that if the measured voltages across the inductor and resistor are added directly, the result is greater than the total voltage measured across the two in series.

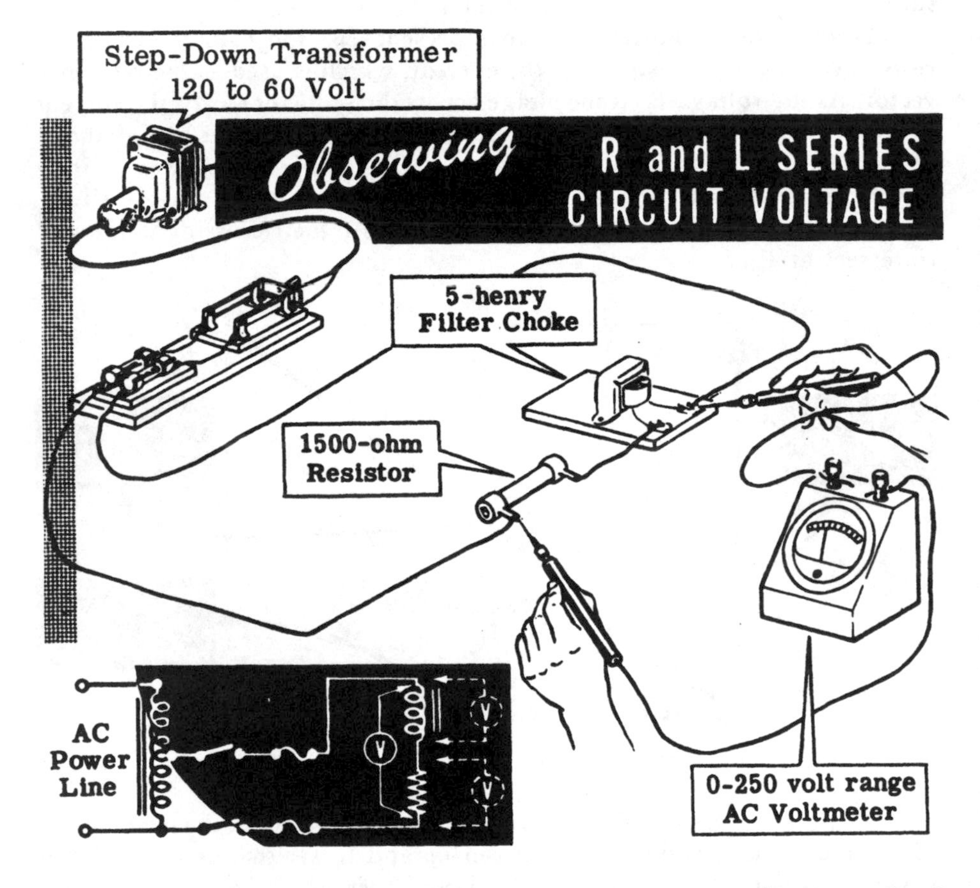

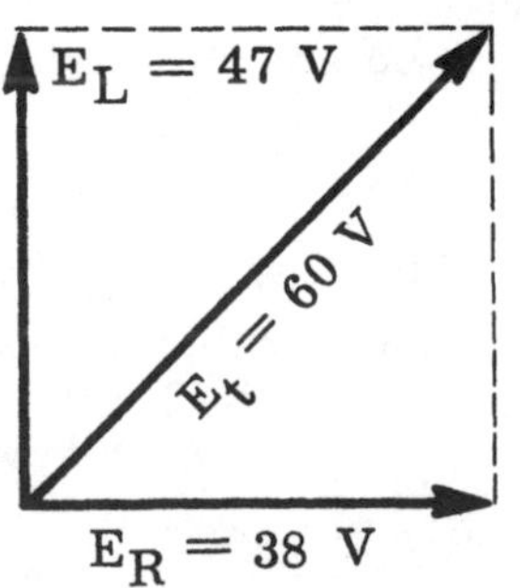

The voltage E_L measured across the 5-henry inductor is approximately 47 volts, and E_R measured across the 1,500-ohm resistor is about 38 volts. When added directly, the voltages E_L and E_R total approximately 85 volts, but the actual measured voltage across the resistor and inductor in series is only about 60 volts. Using vectors to combine E_R and E_L, you see that the result is about 60 volts—the actual total circuit voltage *as measured.*

R and L Series Circuit Impedance

As you know, when resistance and inductance are connected in series, the impedance (Z) is *not* found by adding these two values directly. Inductive reactance causes voltage to lead the current by 90 degrees, whereas for pure resistance the voltage and current are in phase. Thus, the effect of inductive reactance, as opposed to the effect of resistance, is shown by drawing two vectors to represent R and X_L at right angles to one another.

Look again at the voltage vector diagram for the R and L circuit on page 4-11. You will see that the lengths of the voltage vectors are proportional to the magnitudes of the voltages, E_L, E_R, and E_t.

Since $E_L = IX_L$, $E_R = IR$, and $E_t = IZ$, the vectors also represent, in their true proportions and relationship, X_L, R, and Z.

For example, suppose that the series circuit consists of 200 ohms of resistance in series with 200 ohms of inductive reactance at the frequency of the ac voltage source. The total impedance is *not* 400 ohms; it is approximately 283 ohms. *This value of impedance must be obtained by vector addition.*

To add 200 ohms of resistance and 200 ohms of inductive reactance, a horizontal line is drawn to represent the 200 ohms of resistance. The end of this line, which is to the left, is the reference point; the right end of the line is marked with an arrow to indicate its direction. To represent the inductive reactance, a vertical line is drawn upward from the reference point. Since X_L and R are each 200 ohms, the horizontal and vertical lines are equal in length.

In a series circuit, the resistance vector is usually drawn horizontally and used as a reference for other vectors in the same diagram.

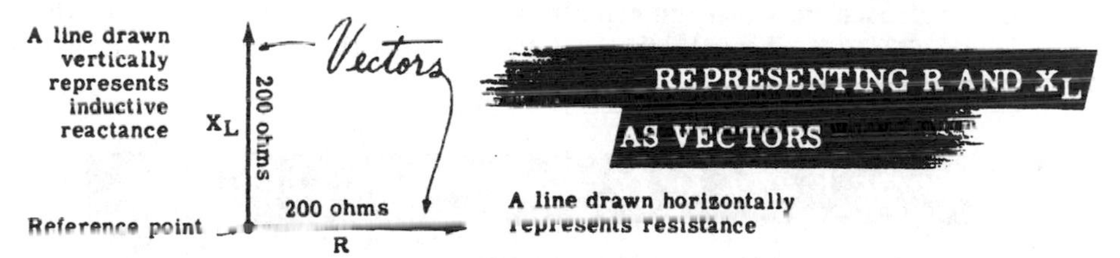
A line drawn vertically represents inductive reactance
X_L
200 ohms
Vectors
REPRESENTING R AND X_L AS VECTORS
200 ohms
A line drawn horizontally represents resistance
Reference point
R

A parallelogram is completed as shown in the right-hand diagram below. The diagonal of this parallelogram represents the impedance Z, the total opposition to current flow of the R and L combination. Using the same scale of measure, this value is found to be 283 ohms at a phase angle of 45 degrees.

COMBINING VECTORS R AND X_L TO FIND THE IMPEDANCE

X_L R + X_L R

X_L 200 ohms   283 ohms   45°   200 ohms   R

R and L Series Circuit Impedance (continued)

If the resistance and inductance values of a series circuit containing R and L are known, the impedance can be found by means of vectors. Suppose that the resistance is 180 ohms, the inductance is 400 mH, and the frequency of the applied ac voltage is 60 Hz. First, the inductive reactance, X_L, is computed in ohms by using the formula $X_L = 2\pi fL$. $X_L = 2 \times \pi \times 60 \times 0.4 = 150$ ohms. Then the vectors representing R and X_L are drawn to scale on graph paper. A common reference point is used with the resistance vector R drawn horizontally to the right, and the inductive reactance vector X_L is drawn upward, perpendicular to the resistance vector.

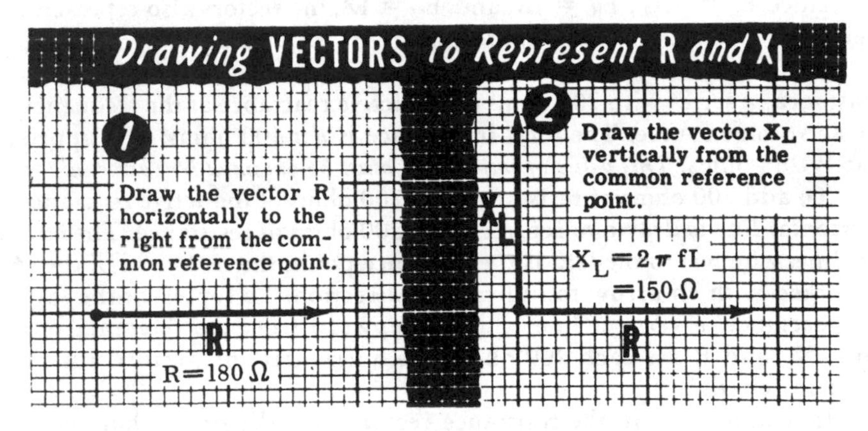

Next, the parallelogram is completed using dotted lines, and the diagonal is drawn between the reference point and the intersection of the dotted lines. The length of the diagonal represents the value of the impedance Z in ohms.

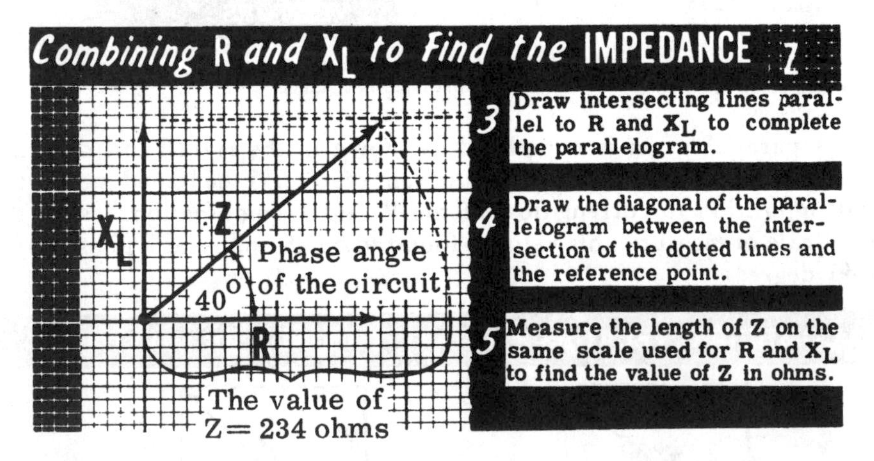

In addition to showing the value of the circuit impedance, the vector solution also shows the phase angle between the circuit current and voltage. The angle between the impedance vector Z and the resistance vector R

R and L Series Circuit Impedance (continued)

is the *phase angle* of the circuit in degrees. This is the angle between the circuit voltage and current and represents a current lag of 40 degrees (or voltage lead of 40 degrees).

Ohm's law for ac circuits may also be used to find the impedance Z for a series circuit. In applying Ohm's law to an ac circuit, Z is substituted for R in the formula. Thus, the impedance, Z, is equal to the applied voltage, E, divided by the circuit current, I. For example, if the circuit voltage is 120 volts and the current 0.5 ampere, the impedance Z is 240 ohms for a frequency of 60 Hz.

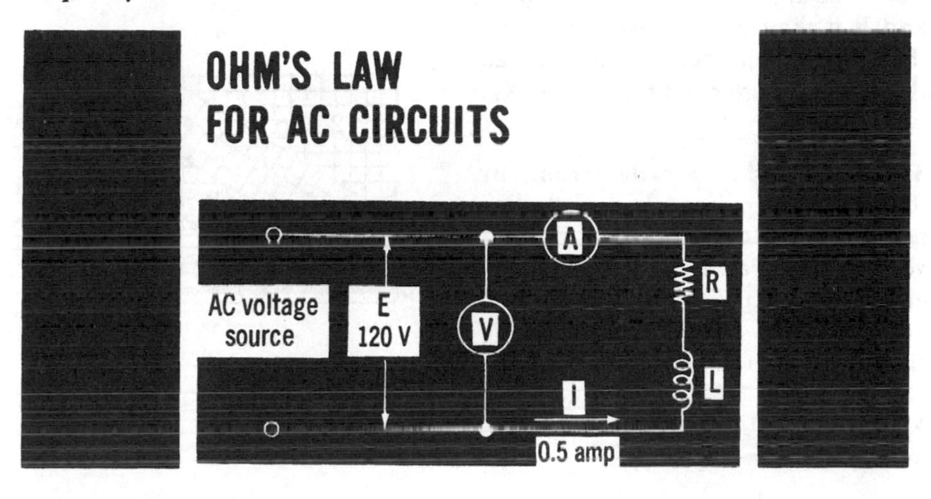

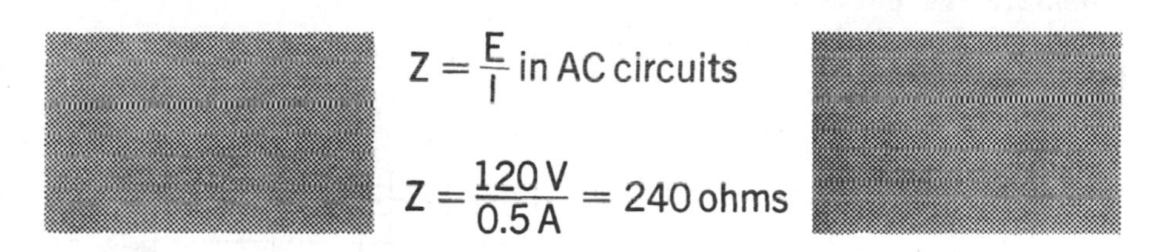

If the impedance of a circuit is found by applying Ohm's law for ac, and the value of R is known and X_L is unknown, the phase angle and the value of X_L may be determined graphically by using vectors. If the resistance in the circuit above is known to be 200 ohms, the vector solution is:

1. Since the resistance is known to be 200 ohms, the resistance vector is drawn horizontally from the reference point. At the end of the resistance vector, a dotted line is drawn perpendicular to it.

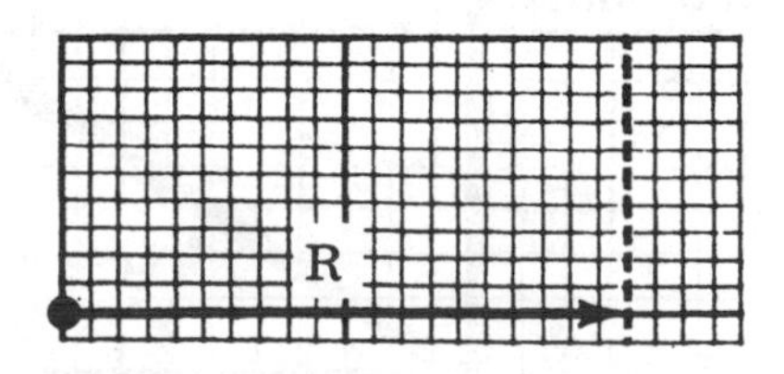

R and L Series Circuit Impedance (continued)

2. Using a straight edge (ruler) marked to indicate the length of the impedance vector, Z, find the point on the perpendicular dotted line which is exactly the length of the impedance vector from the reference point. Draw the impedance vector between that point and the zero position. The angle between the vectors Z and R is the circuit phase angle, and the length of the dotted line between the ends of the vectors represents X_L.

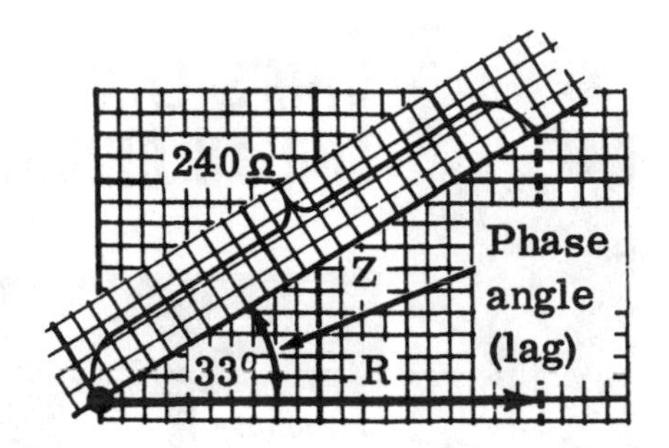

3. Complete the parallelogram by drawing a horizontal dotted line between the end of the vector Z and a vertical line drawn up from the reference point. X_L is this vertical line, and its length can be read by using the same scale. In the example shown, X_L = 133 ohms. And since $X_L = 2\pi fL$, L must equal $X_L/2\pi f$. Therefore, L = 0.353 H, or 353 mH.

4. Measure the phase angle with a protractor. It will be found to be about 33 degrees.

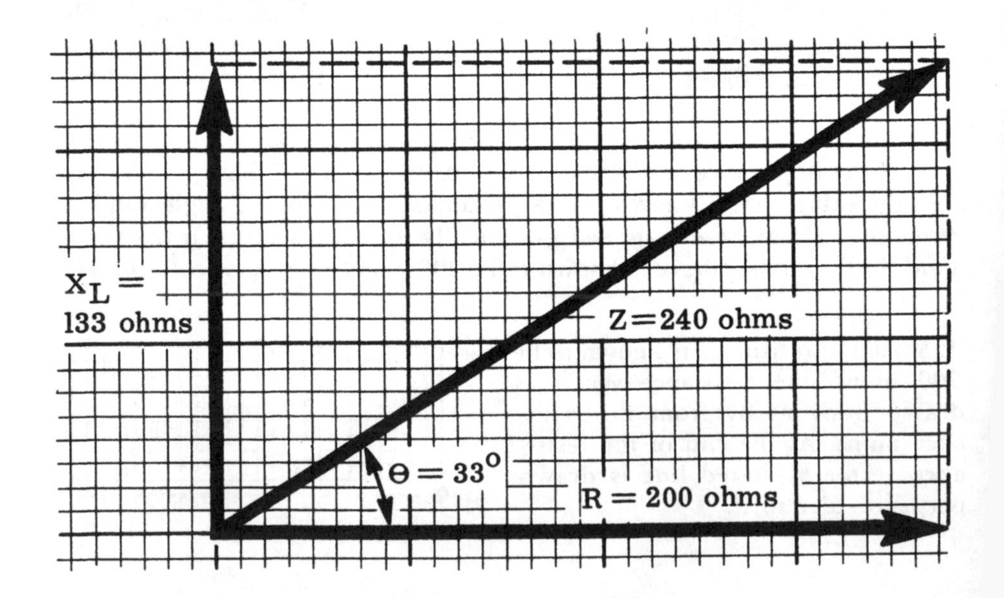

R and L Series Circuit Impedance (continued)

If the impedance and inductance are known, but the resistance is not, both the phase angle and resistance may be found by using vectors. For example, the impedance is calculated to be 300 ohms by measuring the current and voltage and applying Ohm's law for ac. If the circuit inductance is 0.5 henry, the vector solution is as follows:

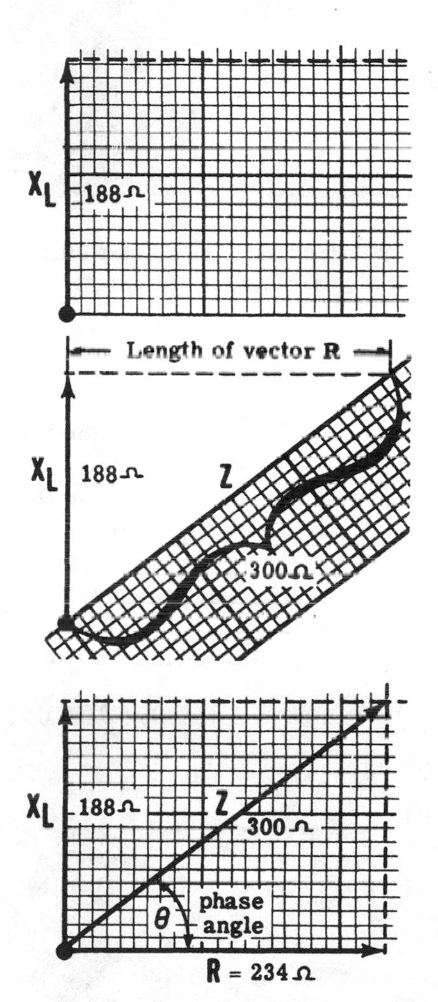

1. First the inductive reactance is computed by using the formula $X_L = 2\pi fL$. If the frequency is 60 Hz, then X_L is 188 ohms ($X_L = 6.28 \times 60 \times 0.5 = 188$ ohms). Draw the vector X_L vertically from the reference point. At the end of this vector draw a horizontal dotted line perpendicular to vector X_L.

2. Using a straight edge marked to indicate the length of the impedance vector, find the point on the horizontal dotted line which is exactly the length of the impedance from the reference point. Draw the impedance vector Z between that point and the reference point. The distance between the ends of the vectors X_L and Z represents the length of the resistance vector R.

3. Draw the vector R horizontally from the zero position and complete the parallelogram. The angle between R and Z is the phase angle of the circuit in degrees. The length of the vector R represents the resistance in ohms and is found to be 234 ohms.

4. Measure the phase angle with a protractor. It will be found to be 39 degrees.

You have already learned how to solve series circuits graphically and to calculate (by Ohm's law) the impedance of a series circuit which is composed of an inductor and a resistor connected to an ac voltage source. You will now learn how to calculate the impedance of such a circuit *without* the use of Ohm's law and *without* making measurements. A hand calculator is very useful for these calculations, and you will find that they are much more rapid and accurate than graphical methods.

R and L Series Circuit Impedance (continued)

To make use of them, you will have to know something about right triangles—that is, a triangle with a right angle. In any right-angled triangle, the lengths of the three sides bear a definite relationship to one another. Consider the triangle below:

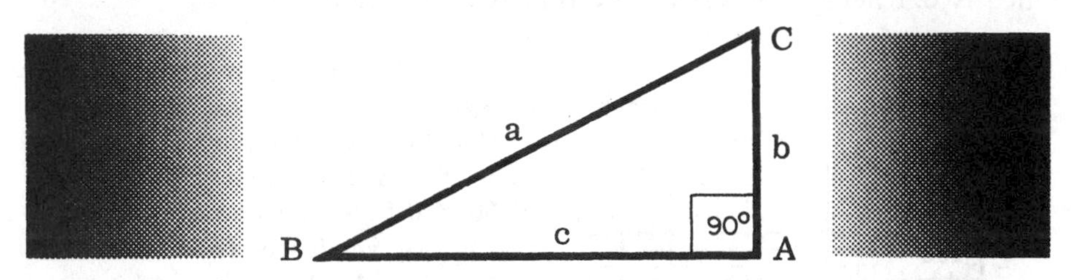

The angle BAC is a right angle (90 degrees). The relationship between the lengths of the sides of this triangle (a, b, and c) is expressed by the formula:

$$a^2 = b^2 + c^2$$

Side *a*, opposite the right angle, is called the *hypotenuse,* and the formula can be expressed in words as *the square of the hypotenuse is equal to the sum of the squares of the other two sides.* This is called the *Pythagorean theorem,* after the Greek philosopher, Pythagoras, who first propounded it.

If you look again at the vector diagram for the inductive reactance, resistance, and total impedance in an R and L ac series circuit, you will see that you can use the Pythagorean theorem to find either impedance, or reactance, or resistance, provided the other two quantities are known.

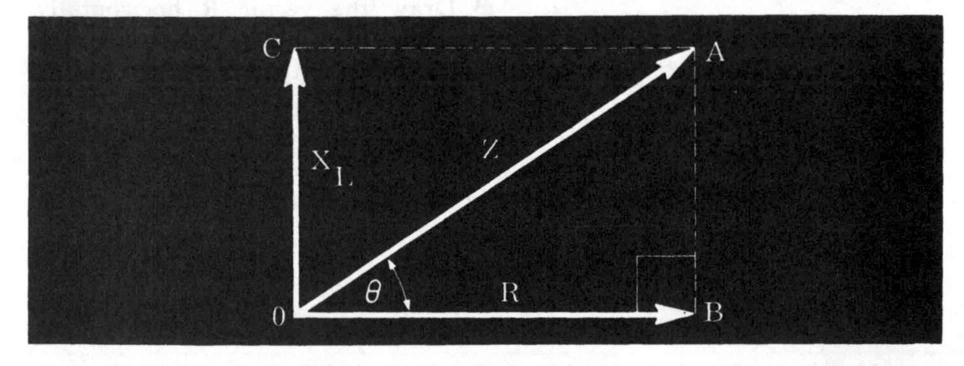

You can see that OAB (above) is a right-angled triangle in which Z is the hypotenuse, R (OB) one side, and X_L (AB= OC) the other. Therefore,

$$Z^2 = R^2 + X_L{}^2$$

and

$$Z = \sqrt{R^2 + X_L{}^2}$$

R and L Series Circuit Impedance (continued)

The two angles of a right triangle (other than the right angle itself) are related to the lengths of the sides by the trigonometric relationships called the *sine, cosine,* and *tangent*, abbreviated *sin, cos*, and *tan*, respectively. If the magnitude of an angle is known, the value of the sin, cos, or tan of the angle can be found in the tables of trigonometric functions at the back of this volume. Obviously, you can reverse the procedure and find the angle, if you know the value for the sin, cos, or tan, by looking up this value in the same tables.

$$\text{Sine of either angle} = \frac{\text{Length of side } \textit{opposite} \text{ angle}}{\text{Length of hypotenuse}}$$

$$\text{Cosine of either angle} = \frac{\text{Length of side } \textit{adjacent} \text{ to angle}}{\text{Length of hypotenuse}}$$

$$\text{Tangent of either angle} = \frac{\text{Length of side } \textit{opposite} \text{ angle}}{\text{Length of side } \textit{adjacent} \text{ to angle}}$$

Examples

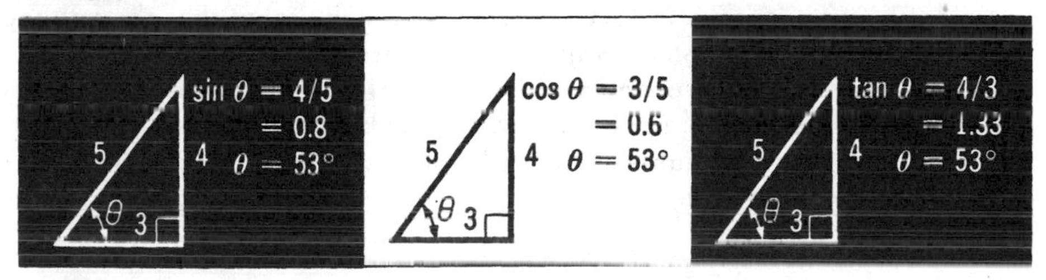

When the length of the hypotenuse and the angle between it and one of the other sides are known, the length of the other two sides can be found using the trigonometric relationships of sin or cos.

Example

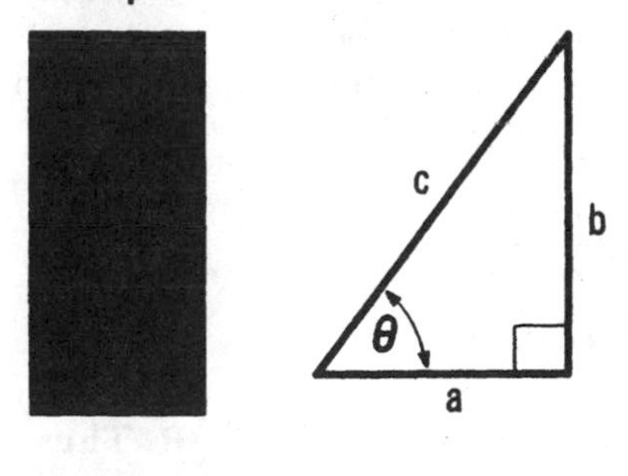

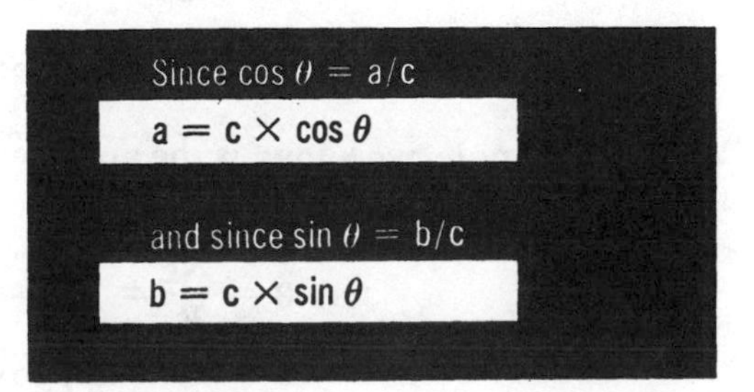

R and L Series Circuit Impedance (continued)

Suppose that in an ac series circuit the value of X_L and R are known to be 4 ohms and 3 ohms, respectively. Using the formula for the impedance Z from the previous pages:

$$Z = \sqrt{R^2 + X_L^2}$$
$$Z = \sqrt{3^2 + 4^2}$$
$$= \sqrt{9 + 16}$$
$$= \sqrt{25}$$
$$Z = 5 \text{ ohms}$$

Z = 5
X_L = 4
θ = 53°
R = 3

Phase angle θ is the angle between Z and R, and the tangent of that angle is X/R. Thus θ is the angle whose tangent is X/R, which is written algebraically as $\theta = \tan^{-1} X/R$ ($\tan^{-1}$ is read as "the angle whose tangent is"). Since we know that $X/R = 4/3 = 1.333$, $\theta = \tan^{-1} 1.333$.

The angle can now be found, by consulting the trigonometric tables at the back of this volume, to be 53 degrees, approximately.

As the circuit is an inductive circuit, we now know that the voltage leads the current by a phase angle of 53 degrees.

Consider again the problem solved on page 4-15 by drawing vectors to scale. In that problem, the impedance (Z) is known to be 240 ohms, and the resistance (R) is 200 ohms. The inductance and phase angle of the circuit remain to be found.

$$Z^2 = R^2 + X_L^2$$
$$X_L^2 = Z^2 - R^2$$
$$X_L = \sqrt{Z^2 - R^2} = \sqrt{240^2 - 200^2}$$
$$= \sqrt{57{,}600 - 40{,}000} = \sqrt{17{,}600}$$
$$X_L = 133 \text{ ohms, approximately}$$

(When doing calculations of this sort, you can most easily find the squares and square roots you want by looking them up in a set of mathematical tables or using the tables at the back of this book.)

The inductance can now be found by substituting known quantities in the formula for inductive reactance, $X_L = 2\pi fL$:

$$L = \frac{X_L}{2\pi f} = \frac{133}{6.28 \times 60}$$

$$L = 0.353 \text{ H or } 353 \text{ mH}$$

The phase angle, we know, is the angle whose tangent equals X_L/R. Thus,

$$\tan\theta = \frac{X_L}{R} = \frac{133}{200} = 0.665$$
$$\theta = 33.5° \text{ approximately}$$

Experiment/Application—Impedance in R and L Series Circuits

You can verify the methods of calculating impedance described in the preceding pages by first calculating the impedance of a given circuit, using both methods, and then measuring it in an actual circuit.

Consider an R and L circuit consisting of a 5-henry inductor in series with a 2,000-ohm resistor. First, calculate the inductive reactance, X_L:

$$X_L = 2\pi fL = 2\pi \times 60 \times 5 = 1{,}885 \text{ ohms}$$

Now, using the formula on page 4-18, calculate the impedance, Z:

$$Z = \sqrt{R^2 + X_L^2} = \sqrt{2{,}000^2 + 1{,}885^2} = 2{,}750 \text{ ohms, approximately}$$

The phase angle is calculated as $\theta = \tan^{-1}(X/R) = \tan^{-1}(1{,}885/2{,}000) = 43$ degrees.

A similar value for Z and θ will be found if a vector diagram is drawn to scale, and Z, the resultant, and θ, the phase angle, are measured.

Now connect a 2,000-ohm resistor, a 5-henry inductor, a switch, a fuse and a 0–50 mA ac milliammeter in series across the ac power source through a step-down transformer (120 to 60 volts) to form a series R and L circuit. A 0–150 volt ac voltmeter is connected across the transformer secondary to measure the circuit voltage. With the switch closed, you see that the voltmeter reads approximately 60 volts and the milliammeter reads about 22 mA. The Ohm's law value of Z is about 2,750 ohms ($60 \div 0.022 \approx 2{,}750$), and you see that the two methods of finding Z result in approximately equal values for Z. (The meter readings you observe will vary somewhat from those given because of variations in line voltage and meter accuracy and in the rating of the resistors and inductors used. Thus, the values of Z obtained in practice will always vary slightly from the values obtained by calculation.)

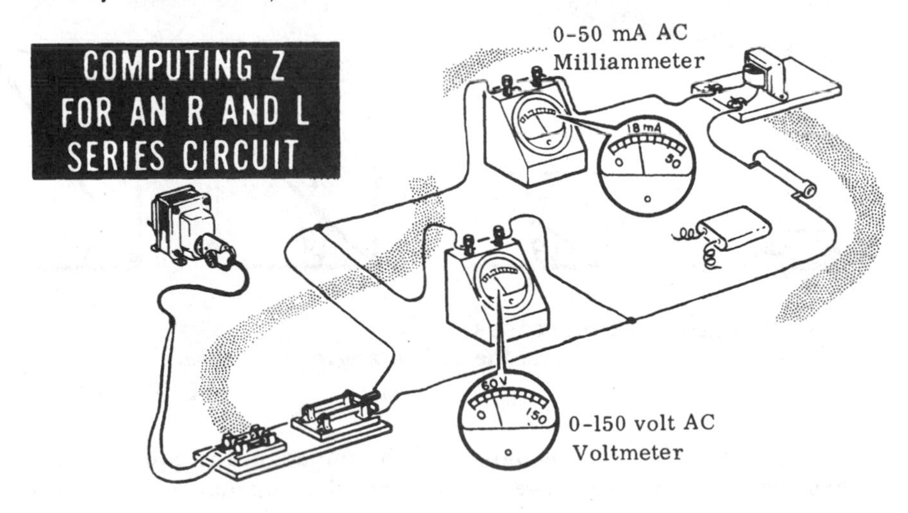

R and L Series Circuits Impedance Variation

The impedance of a series circuit containing only resistance and inductance is determined by the vector addition of the resistance and the inductive reactance. If a given value of inductive reactance is used, the impedance varies as shown below when the resistance value is changed.

How Impedance Varies

WHEN X_L IS A FIXED VALUE AND R IS VARIED

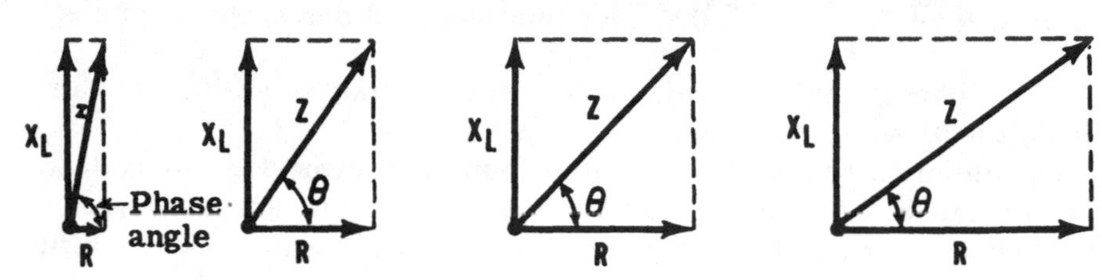

Impedance increases and the phase angle decreases as R increases

If a given value of resistance is used, the impedance, in its turn, varies as shown below when the inductive reactance is changed:

How Impedance Varies

WHEN R IS A FIXED VALUE AND X_L IS VARIED

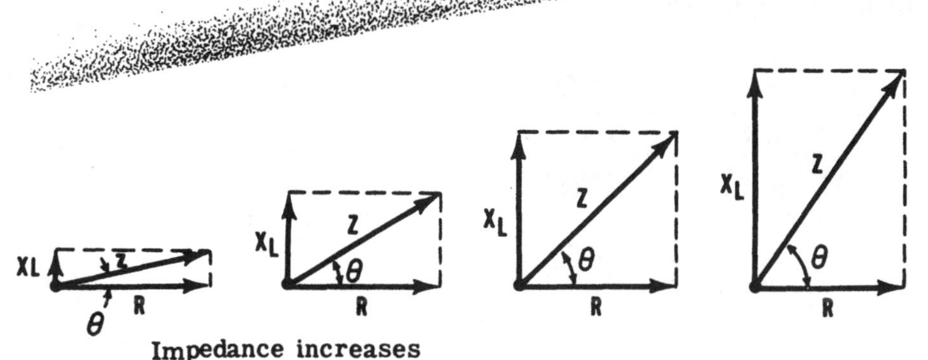

Impedance increases and the phase angle increases as X_L increases

When the resistance equals zero, the impedance is equal to X_L, and when the inductive reactance equals zero, the impedance is equal to R.

R and C Series Circuit Voltages

If an ac series circuit consists of only R and C, the total voltage is found by vectorially combining E_R, the voltage across the resistance, and E_C, the voltage across the capacitance. E_R is in phase with the circuit current, while E_C lags the circuit current by 90 degrees; thus, E_C lags by 90 degrees. The two voltages may be combined either graphically or by calculation as you did for R-L circuits.

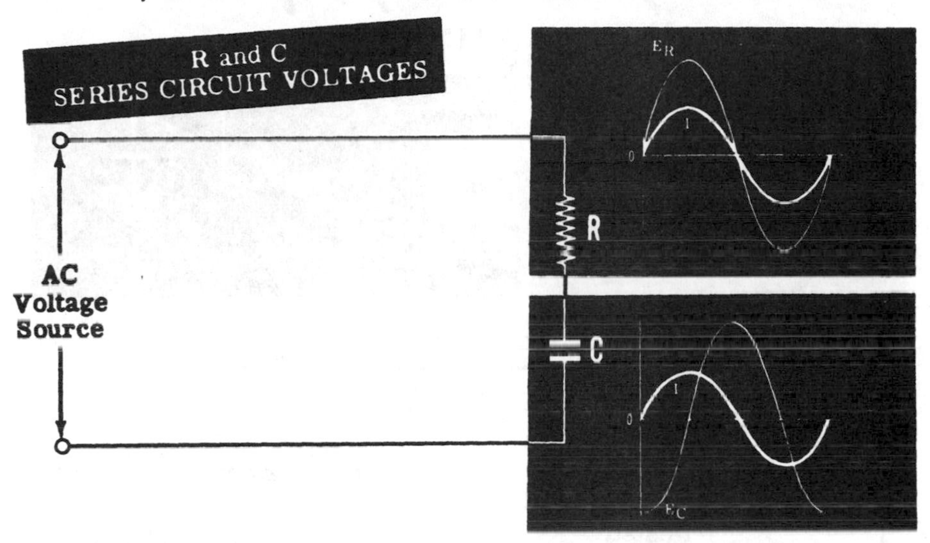

As the voltage across C lags the current in the circuit, the vector representing E_C is drawn downward from the horizontal vector representing E_R. Similarly, in the impedance vector diagram, X_C is drawn downward from R.

Ohm's Law

AND VECTOR RELATIONSHIP OF AN R AND C SERIES CIRCUIT

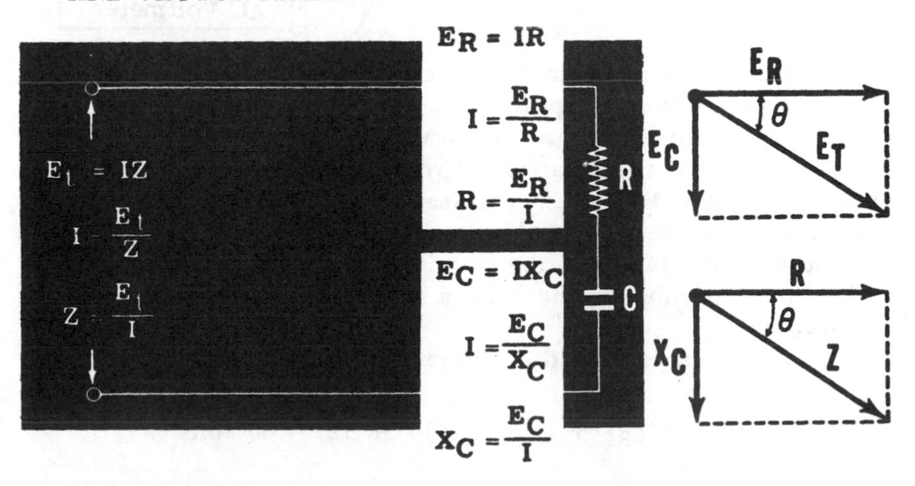

Experiment/Application—R and C Series Circuit Voltages

If you repeat the experiment/application described on page 4-12, using a 1-μF capacitor in series with a 1,500-ohm resistor, again you see that the sum of the voltages across the capacitor and resistor is greater than the actual measured total voltage.

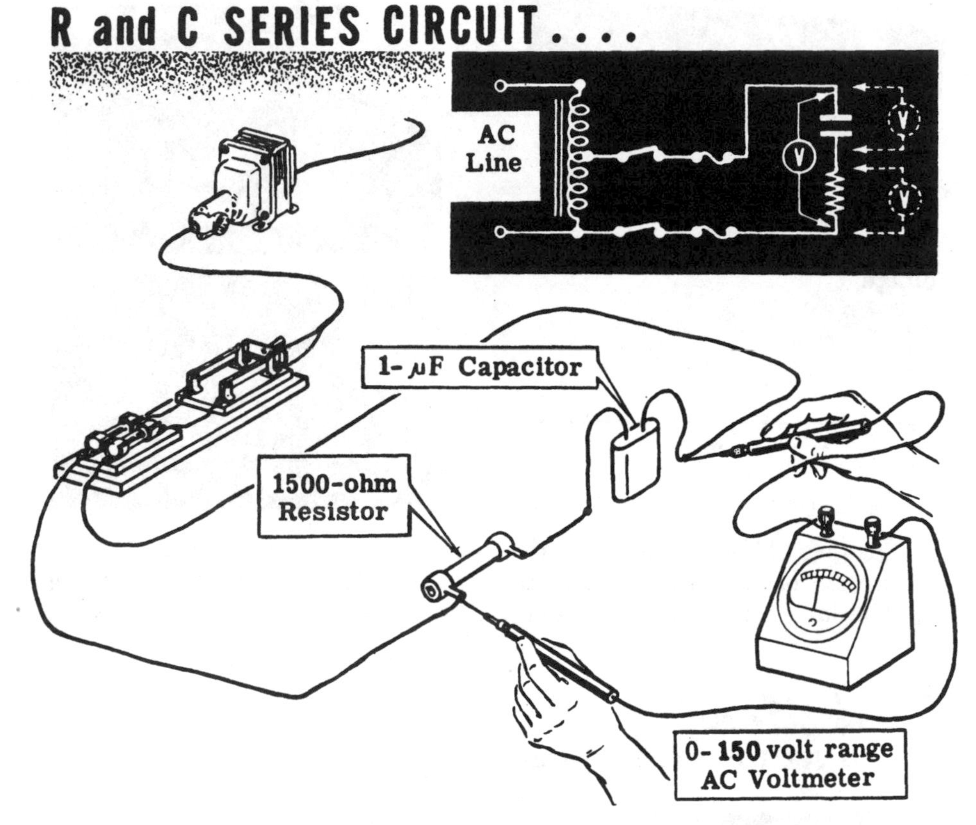

The measured voltage across the capacitor is about 52 volts, while that across the resistor is approximately 30 volts. When added, these voltages total 82 volts, but the actual measured voltage across the capacitor and resistor in series is only about 60 volts. Using vectors to combine the two circuit voltages, E_R and E_C, you see that the answer is about 60 volts, or about equal to the measured voltage of the circuit. (Remember that when you open the switch in the circuit, you must always short the terminals of the capacitor together with a screwdriver in order to discharge the capacitor.)

By calculation, you could combine the voltages as:

$$E_t = \sqrt{E_R^2 + E_C^2} = \sqrt{30^2 + 52^2} = 60 \text{ volts}$$

R and C Series Circuit Impedance

If an ac series circuit consists of resistance and capacitance in series, the total opposition to current flow (impedance) is due to two factors, resistance and capacitive reactance. The action of capacitive reactance causes the current in a capacitive circuit to lead the voltage (voltage lags current) by 90 degrees, so that the effect of capacitive reactance is at right angles to the effect of resistance. But, though the effects of inductive and capacitive reactance are both at right angles to the effect of resistance, their effects are exactly *opposite*—inductive reactance causing current to lag and capacitive reactance causing it to lead the voltage. Thus, the vector X_C, representing capacitive reactance, is still drawn perpendicular to the resistance vector, but drawn *down* rather than *up* from the zero position.

The impedance (Z) of a series circuit containing R and C is found in the same manner as the impedance of an R and L series circuit. Suppose in your R and C series circuit that R equals 200 ohms and X_C equals 200 ohms. To find the impedance, the resistance vector R is drawn horizontally from a reference point. Then a vector of equal length is drawn downward from the reference point at right angles to the vector R. This vector X_C, which represents the capacitive reactance, is equal in length to R since both R and X_C equal 200 ohms.

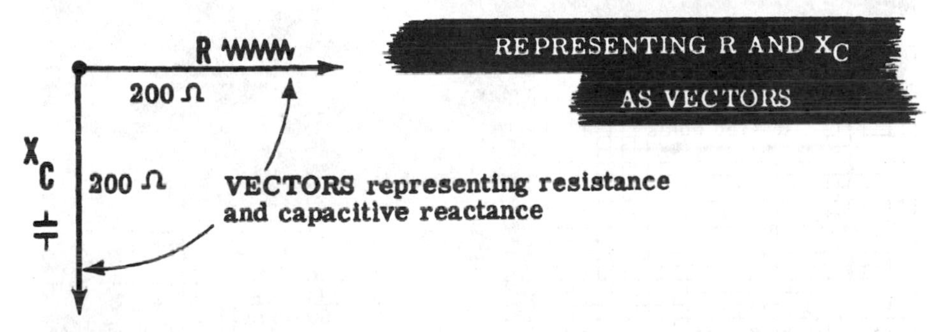

To complete the vector solution, the parallelogram is completed and a diagonal drawn from the reference point. This diagonal is the vector Z and represents the impedance in ohms (283 ohms). The angle between the vectors R and Z is the phase angle (θ) of the circuit, indicating the amount in degrees by which the current leads the voltage.

COMBINING VECTORS R AND X_C TO FIND THE IMPEDANCE

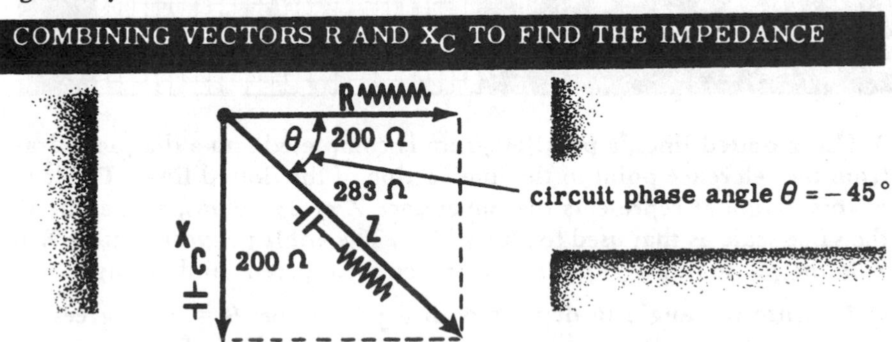

R and C Series Circuit Impedance (continued)

R and C series circuit impedances can be found by using the vector solution, by calculation, or by application of Ohm's law. To find the impedance Z (for example, in a series circuit containing 150 ohms resistance and 15 μF of capacitance) by using a vector solution, you should proceed as follows:

1. Compute the value of X_C by using the formula $X_C = 1/2\pi fC$. In this formula, 2π is a constant equal to 6.28, f is the frequency in Hz, and C is the capacitance in farads. For example, if C = 15 μF, $X_C = 1/6.28 \times 60 \times 0.000015 = 177$ ohms, or approximately 180.

2. Given that R = 150 ohms, draw vectors R and X_C to scale on graph paper, using a common reference point for the two vectors. R is drawn horizontally to the right from the reference point, and X_C is drawn downward from the reference point and perpendicular to the resistance vector R.

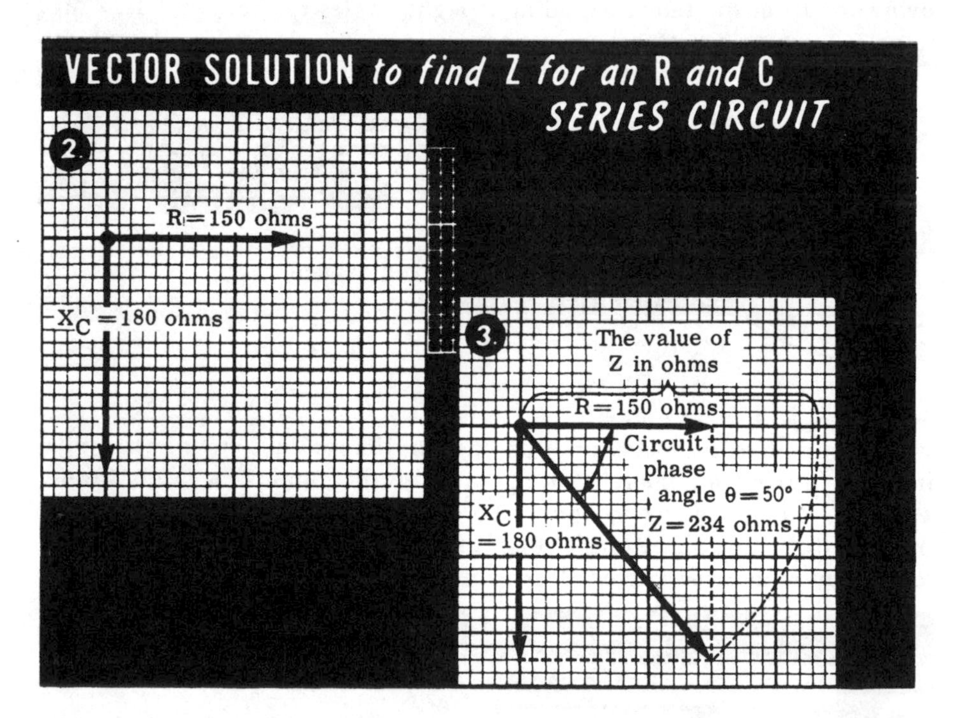

3. Using dotted lines, a parallelogram is completed and a diagonal drawn from the reference point to the intersection of the dotted lines. The length of this diagonal represents the impedance Z of 234 ohms, as measured to the same scale as that used for R and X_C. The angle between vectors R and Z is the phase angle (θ) between the circuit current and voltage.

4. Measure the angle in degrees with a protractor: θ = 50 degrees.

R and C Series Circuit Impedance (continued)

The impedance of an R and C series circuit may also easily be found by the application of the Pythagorean theorem.

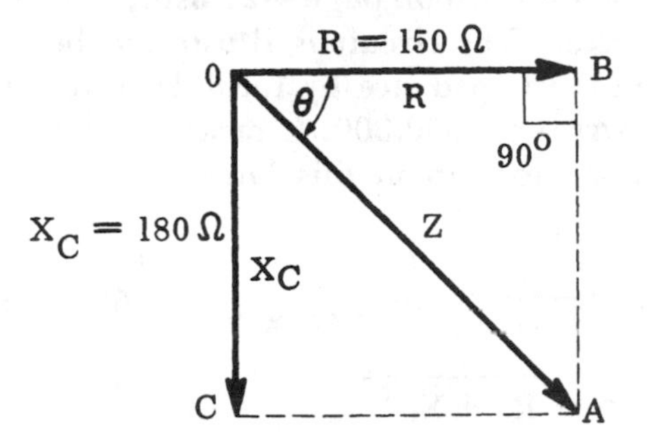

From the vector diagram above, you can see that X_C = AB and that

$$Z^2 = R^2 + X_C^2$$

$$Z = \sqrt{R^2 + X_C^2} = \sqrt{150^2 + 180^2} = \sqrt{22{,}500 + 32{,}400}$$

$$= \sqrt{54{,}900} = 234$$

$$\tan\theta = \frac{X_C}{R} = \frac{180}{150} = 1.2 \qquad \theta = \tan^{-1} 1.2 = -50°$$

A minus sign indicates that the phase angle is *below* the axis. Remember that in this case the effect of capacitive reactance (X_C) is to cause the voltage to lag the current by the phase angle (θ).

When the value of either R or X_C is unknown, but the values of the other and of Z are known, you can find the unknown value either by vector solution or by substituting known values in the formulas above.

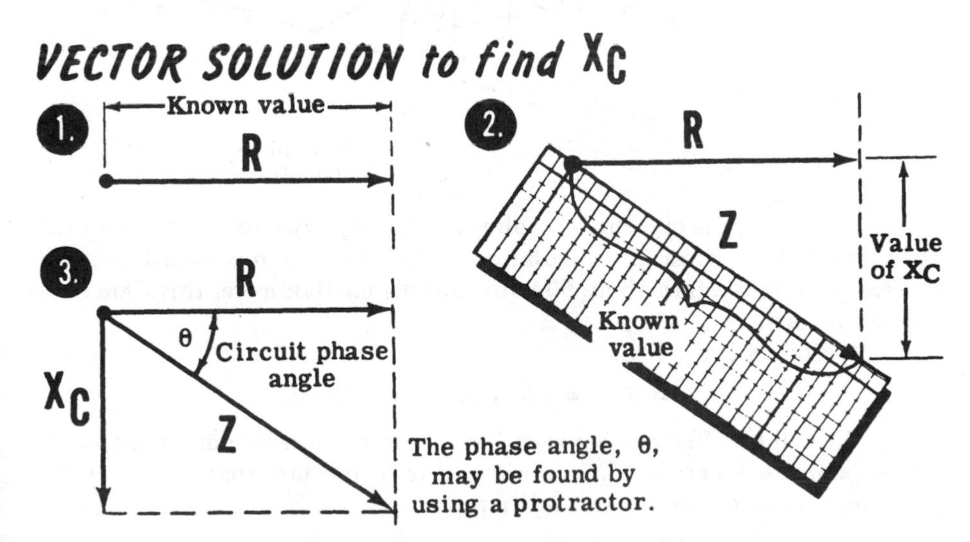

Experiment/Application—Impedance in R and C Series Circuits

You can verify that the methods of calculating impedance described on pages 4-25 and 4-26 are true for capacitive circuits by repeating the experiment/application outlined on page 4-21 using a 1-μF capacitor instead of the 5-henry inductor. The circuit is illustrated below.

First, calculate the impedance by formula. (Note that 10^6 is only a convenient way of writing 1,000,000. It means 10 multiplied by itself six times. You will learn more about this later.)

$$X_C = \frac{1}{2\pi fC} = \frac{10^6}{2\pi \times 60 \times 1} = 2{,}654 \text{ ohms}$$

$$Z = \sqrt{R^2 + X_C{}^2}$$

$$= \sqrt{2{,}000^2 + 2{,}654^2}$$

$$= \sqrt{4{,}000{,}000 + 7{,}043{,}716}$$

$$= \sqrt{11{,}043{,}716}$$

$$= 3{,}323 \text{ ohms}$$

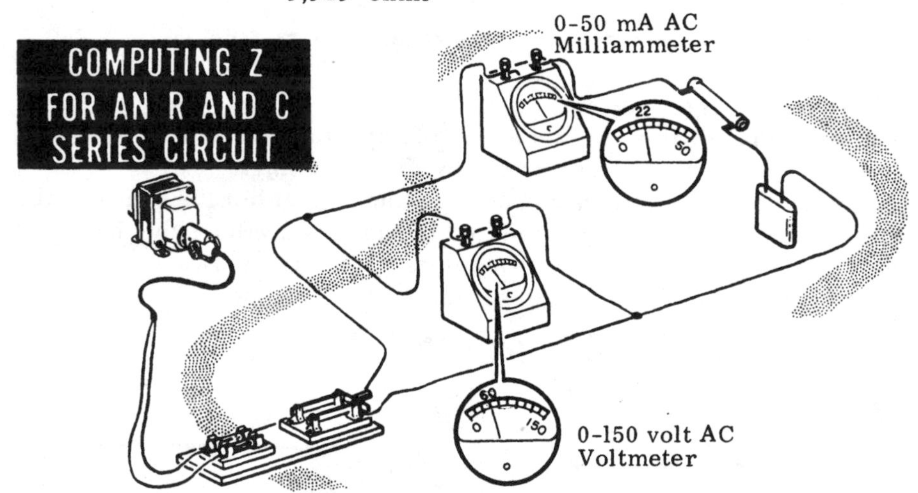

Now check the current and voltage values by experiment, and you will find that they confirm this calculation. When the switch is closed, the voltmeter reading is again about 60 volts, but the milliammeter this time reads about 18 mA.

By Ohm's law, then,

$$Z = E/I = 60/0.018 = 3{,}333 \text{ ohms}$$

The small difference between the computed and the measured values of Z (about 0.3 percent difference) is due to the fact that the current is measured only to the nearest milliampere.

R and C Series Circuit Impedance (continued)

The ratio of R to X_C determines both the amount of impedance and the phase angle in series circuits consisting only of resistance and capacitance. If the capacitive reactance is a fixed value and the resistance is varied, the impedance varies as shown below. When the resistance is near zero, the phase angle is nearly 90 degrees, and the impedance is almost entirely due to the capacitive reactance; but when R is much greater than X_C, the phase angle approaches zero degrees, and the impedance is affected more by R than by X_C.

How Impedance Varies

WHEN X_C IS A FIXED VALUE AND R IS VARIED

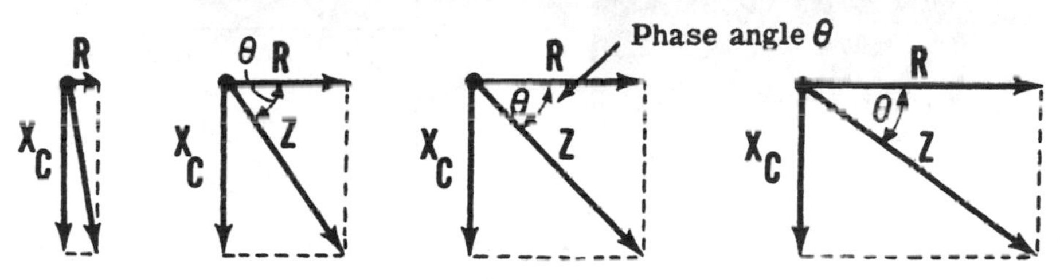

... Impedance increases and phase angle decreases as R increases

If the circuit consists of a fixed value of resistance and the capacitance is varied, the impedance varies as shown below. As the capacitive reactance is reduced toward zero, the phase angle approaches zero degrees, and the impedance is almost entirely due to the resistance. But as X_C is increased to a much greater value than R, the phase angle approaches 90 degrees, and the impedance is affected more by X_C than by R.

How Impedance Varies

WHEN R IS A FIXED VALUE AND X_C IS VARIED

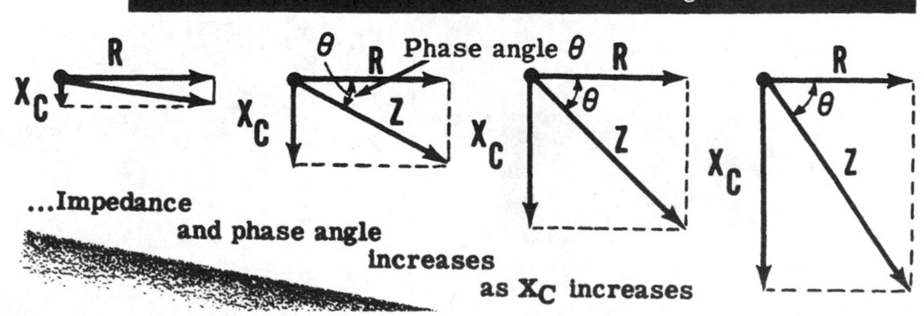

L and C Series Circuit Voltages

To find the total voltage of an L and C series circuit, you need only find the *difference* between E_L and E_C since they oppose each other directly. (Their vectors are 180 degrees apart or in exactly opposite directions.) E_L leads the circuit current by 90 degrees, while E_C lags it by 90 degrees. When the voltage waveforms are drawn, the total voltage is the arithmetic difference between the two individual values and is in phase with the larger of the two voltages—E_L or E_C. For such circuits, the total voltage can be found by subtracting the smaller voltage from the larger.

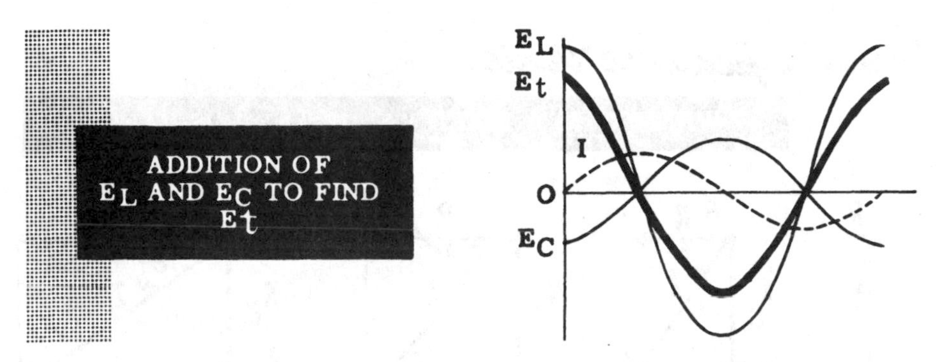

Either or both of the voltages E_L and E_C may be larger than the total circuit voltage in an ac series circuit consisting only of L and C.

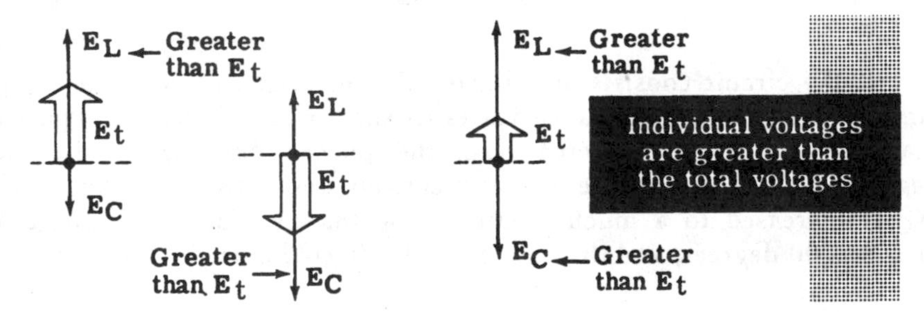

The voltage vectors and reactance vectors for the L and C circuit are similar to each other, except for the units by which they are measured. Ohm's law applies to each part, and to the total circuit outlined below.

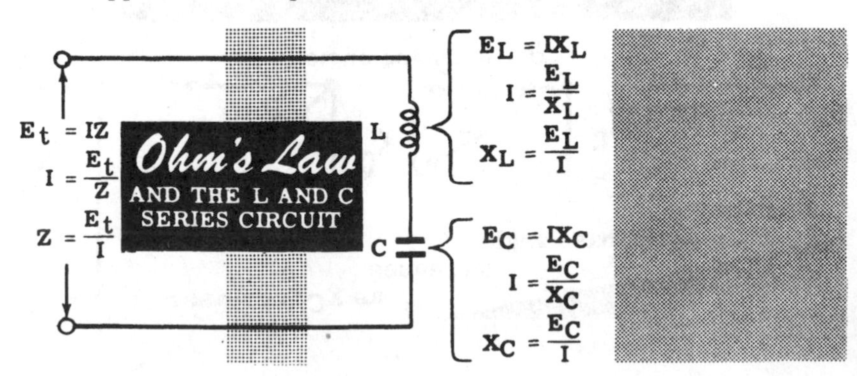

Experiment/Application—L and C Series Circuit Voltages

Suppose that you connected together a 1-μF capacitor and a 5-henry inductor to form an L and C series circuit having negligible resistance. With the power applied, if you measure individually the voltages across the inductor and the capacitor, and the total voltage across the series circuit, you will find that the voltage across the capacitor alone is about 207 volts—a value greater than the measured total voltage across the circuit. You will also find that the voltage across the inductor is about 147 volts. However, if you add the voltages vectorially (207 − 147), you will get the original supply voltage of 60 volts.

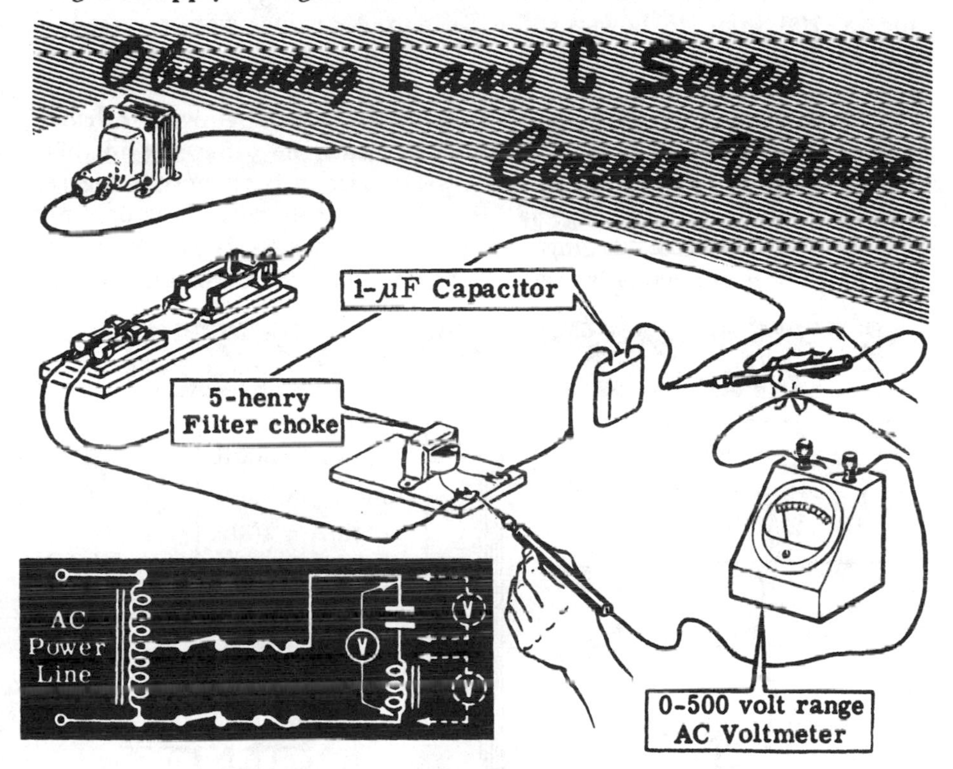

Thus, using vectors to combine the two voltages, you see that the result is approximately equal to the measured total voltage, or about 60 volts. (Although it is considered negligible, the resistance of the inductor wire does, in practice, cause a slight difference between the computed and the actual results.)

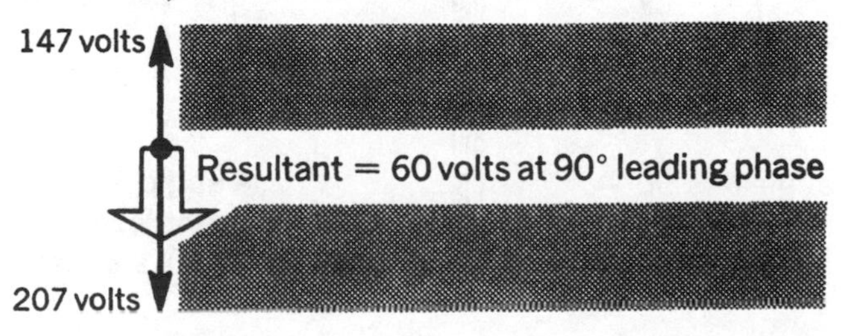

L and C Series Circuit Impedance

In ac series circuits consisting of inductance and capacitance, with only negligible resistance, the impedance is due only to inductive and capacitive reactance. Since inductive and capacitive reactances act in *opposite* directions, the total effect of the two is equal to the *difference* between them. For such circuits, Z can be found by subtracting the smaller value from the larger. The circuit will then act as an inductive or a capacitive reactance (depending on which is larger) that has an impedance equal to Z. For example, if X_L = 500 ohms and X_C = 300 ohms, the impedance Z is 200 ohms, and the circuit acts as an *inductance* having an inductive reactance of 200 ohms. If, however, the X_L and X_C values were reversed, Z would still equal 200 ohms, but the circuit would act as a *capacitance* having a capacitive reactance of 200 ohms.

The relationships of the above examples are shown below. Z is drawn on the same axis as X_L and X_C and represents the difference in their values. The phase angle of the L and C series circuit is always 90 degrees except when $X_L = X_C$ (a special case that is dealt with later on), but whether it is leading or lagging depends on whether X_L is greater or less than X_C. Phase angle θ is the angle between Z and the horizontal X axis.

COMBINING VECTORS X_L AND X_C

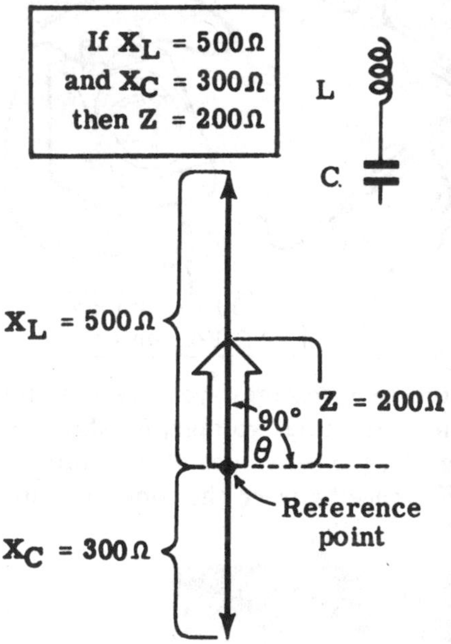

If $X_L = 500\Omega$
and $X_C = 300\Omega$
then $Z = 200\Omega$

L
C

$X_L = 500\Omega$
$Z = 200\Omega$
90°
θ
Reference point
$X_C = 300\Omega$

Phase angle θ is 90°—
current **lagging.**
Circuit acts as an **inductance.**

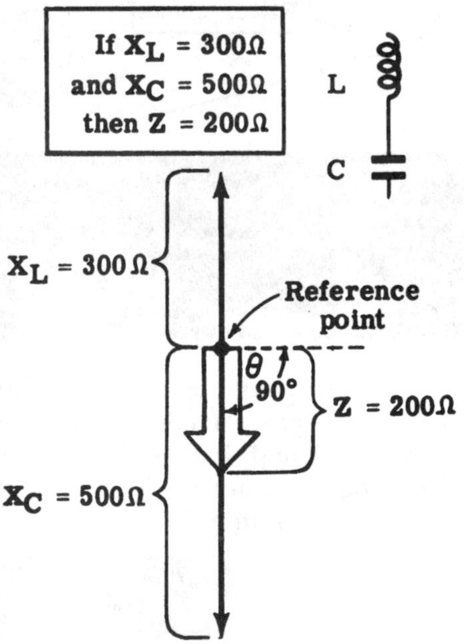

If $X_L = 300\Omega$
and $X_C = 500\Omega$
then $Z = 200\Omega$

L
C

$X_L = 300\Omega$
Reference point
θ
90°
$Z = 200\Omega$
$X_C = 500\Omega$

Phase angle θ is 90°—
current **leading.**
Circuit acts as a **capacitance.**

Experiment/Application—L and C Series Circuit Impedance

To see how inductive and capacitive reactance oppose each other, connect a 1-μF capacitor across the secondary of a 120-volt to 60-volt step-down transformer. (The circuit is illustrated below.) With the switch closed, you will see that the voltage is about 60 volts and that the current is about 22.6 mA. The Ohm's law value of the impedance is then approximately 2,655 ohms ($60 \div 0.0226 \approx 2.655$). Since only capacitance is used in the circuit, the value of Z is equal to X_C.

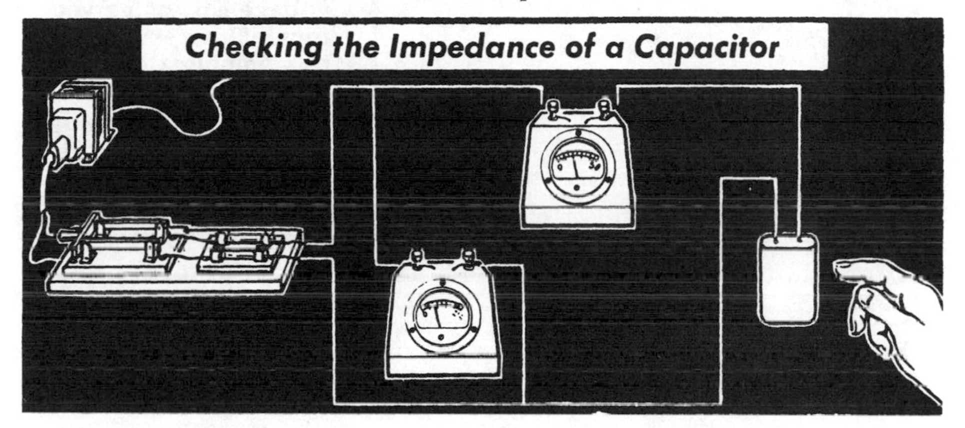

If you now replace the capacitor with a 5-henry inductor in the same circuit, you will find the circuit current to be about 31.8 mA, and the current impedance is then $60 \div 0.0318$, or approximately 1,887 ohms.

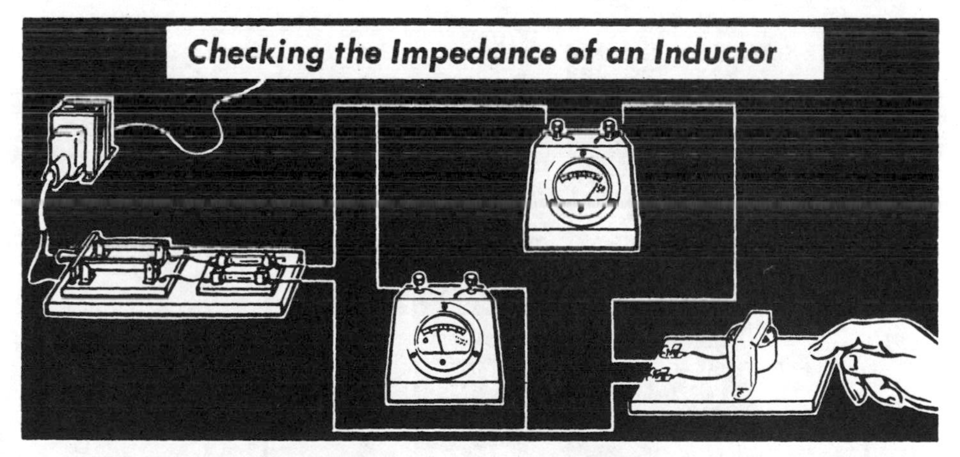

If you now redo the experiment on page 4-31 with an ammeter in the circuit where both the 5-henry inductance and the 1-μF capacitor are in series, you will find that the circuit current is 0.078 amp. The net circuit impedance is about 770 ohms ($Z = E/I = 60/0.078 = 770$). You will note that 770 ohms is approximately the difference between 2,655 ohms and 1,887 ohms. Thus, it is clear that the net circuit impedance is the difference between the inductive and capacitive reactances.

R, L, and C Series Circuit Voltages

To combine the three voltages of an R, L, and C series circuit by means of vectors, two steps are required:

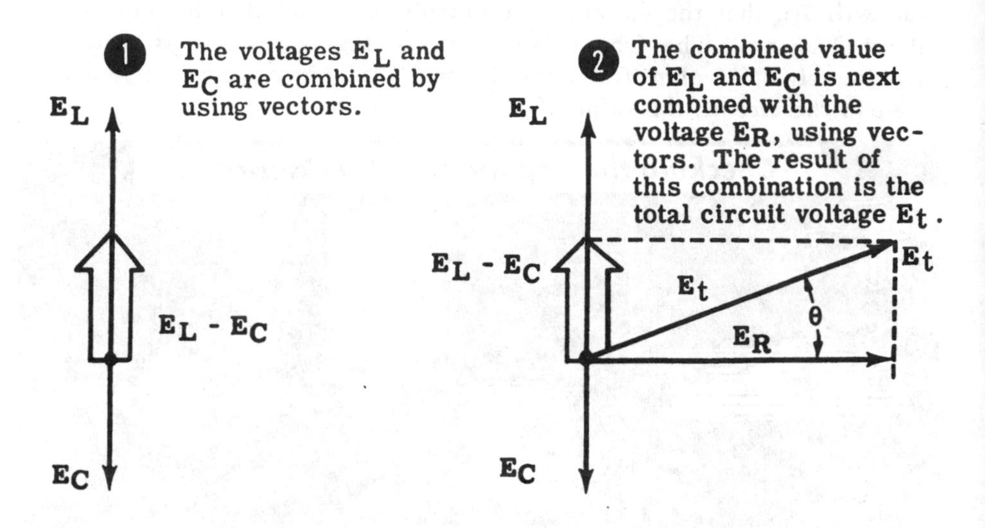
1
The voltages E_L and E_C are combined by using vectors.
E_L
$E_L - E_C$
E_C
2
The combined value of E_L and E_C is next combined with the voltage E_R, using vectors. The result of this combination is the total circuit voltage E_t.
E_L
$E_L - E_C$
E_t
E_t
θ
E_R
E_C

You can apply Ohm's law for any part of the circuit by using X_L, X_C, or R across inductors and capacitors or resistors, respectively. For the total circuit, Z replaces R as it is used in the original formula.

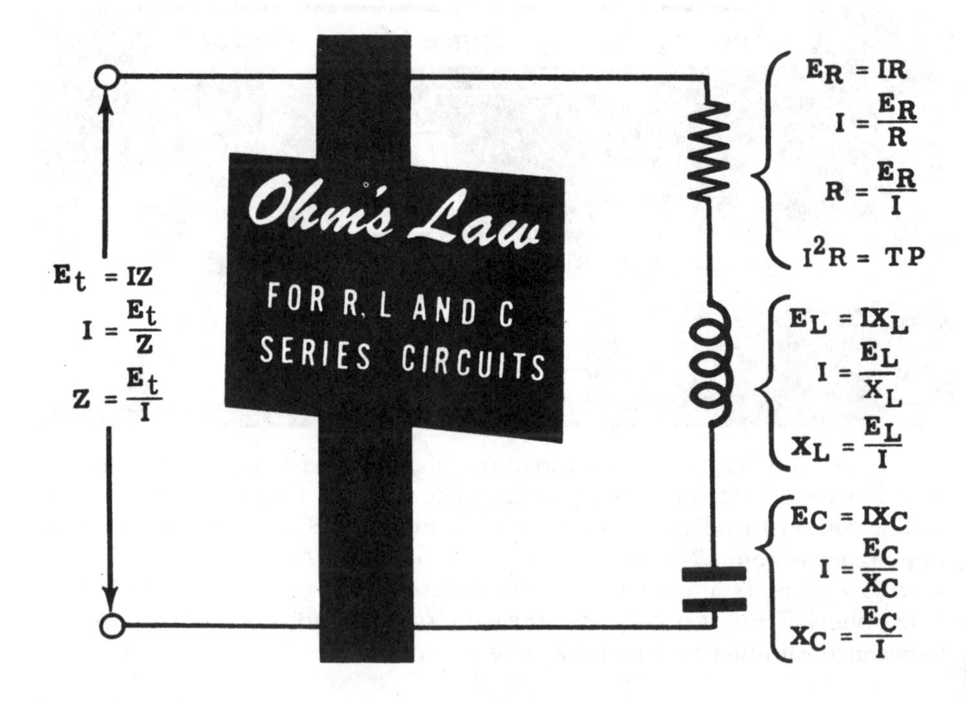
Ohm's Law
FOR R, L AND C
SERIES CIRCUITS
$E_t = IZ$
$I = \frac{E_t}{Z}$
$Z = \frac{E_t}{I}$
$E_R = IR$
$I = \frac{E_R}{R}$
$R = \frac{E_R}{I}$
$I^2R = TP$
$E_L = IX_L$
$I = \frac{E_L}{X_L}$
$X_L = \frac{E_L}{I}$
$E_C = IX_C$
$I = \frac{E_C}{X_C}$
$X_C = \frac{E_C}{I}$

R, L, and C Series Circuit Impedance—Graphic Solution

The impedance of a series circuit consisting of resistance, capacitance, and inductance in series depends on R, X_L, and X_C. If the values of all three factors are known, impedance Z may be found as follows:

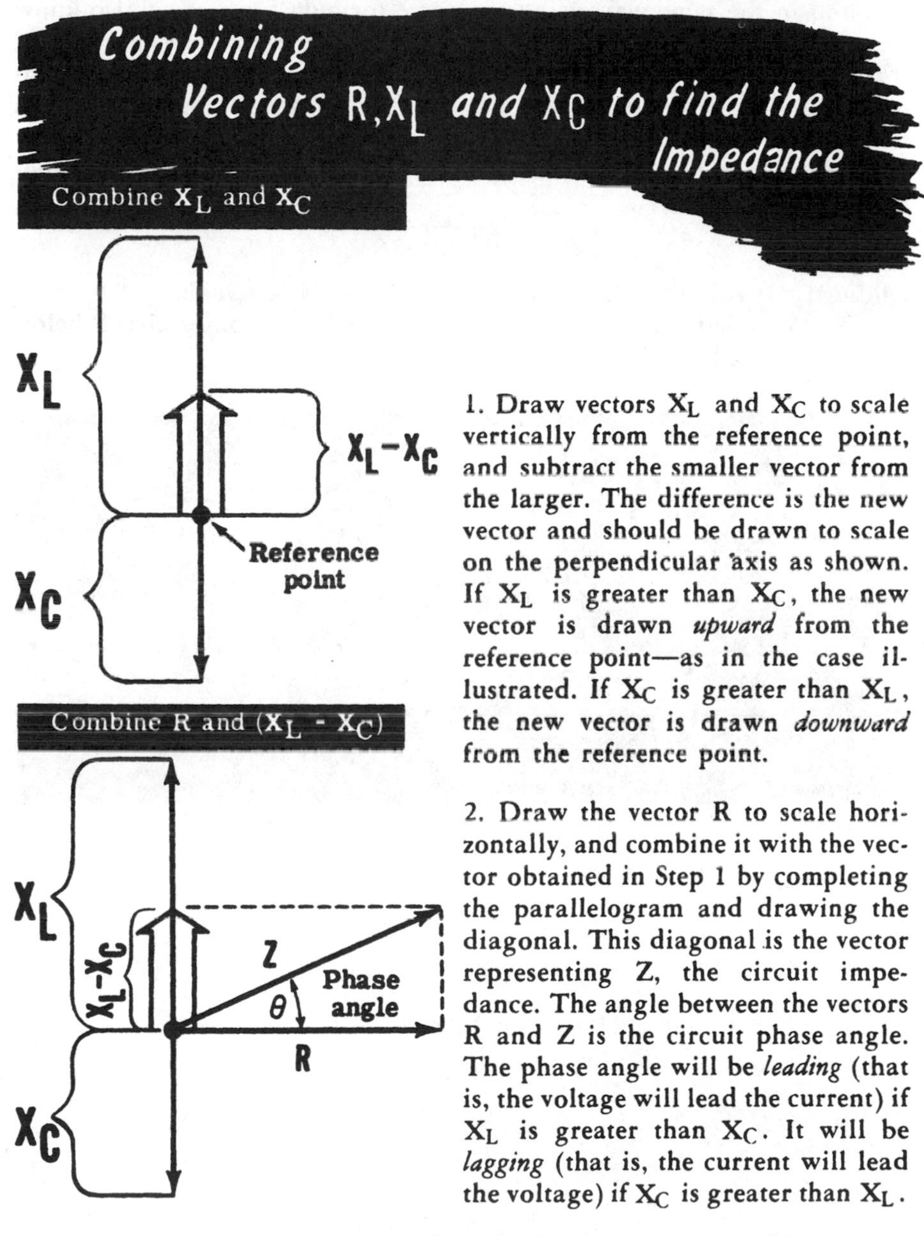

1. Draw vectors X_L and X_C to scale vertically from the reference point, and subtract the smaller vector from the larger. The difference is the new vector and should be drawn to scale on the perpendicular axis as shown. If X_L is greater than X_C, the new vector is drawn *upward* from the reference point—as in the case illustrated. If X_C is greater than X_L, the new vector is drawn *downward* from the reference point.

2. Draw the vector R to scale horizontally, and combine it with the vector obtained in Step 1 by completing the parallelogram and drawing the diagonal. This diagonal is the vector representing Z, the circuit impedance. The angle between the vectors R and Z is the circuit phase angle. The phase angle will be *leading* (that is, the voltage will lead the current) if X_L is greater than X_C. It will be *lagging* (that is, the current will lead the voltage) if X_C is greater than X_L.

You can always find the impedance of any circuit by applying Ohm's law for ac circuits, after measuring the circuit current and voltage.

R, L, and C Series Circuit Impedance—by Calculation

The impedance of a circuit which contains R, L, and C components may also be calculated by using a variation of the impedance formula, $Z = \sqrt{R^2 + X^2}$. You have learned that it makes no difference whether the reactivé component, X, is inductive or capacitive in nature; the impedance is found in the same way, using the same formula for Z. You also know that, when both inductive and capacitive reactances are present in a circuit, it is only necessary to subtract the smaller amount of reactance (either inductive or capacitive, as the case may be) from the larger amount and then to draw in the resultant diagonal vector Z. In calculating the value for impedance in a circuit containing both inductive and capacitive components, use the formula,

$$Z = \sqrt{R^2 + X_e^2}$$

where X_e is equal to $X_L - X_C$ or vice versa, as required.

Suppose that you wish to calculate the impedance of the circuit below when connected to a 60-Hz line:

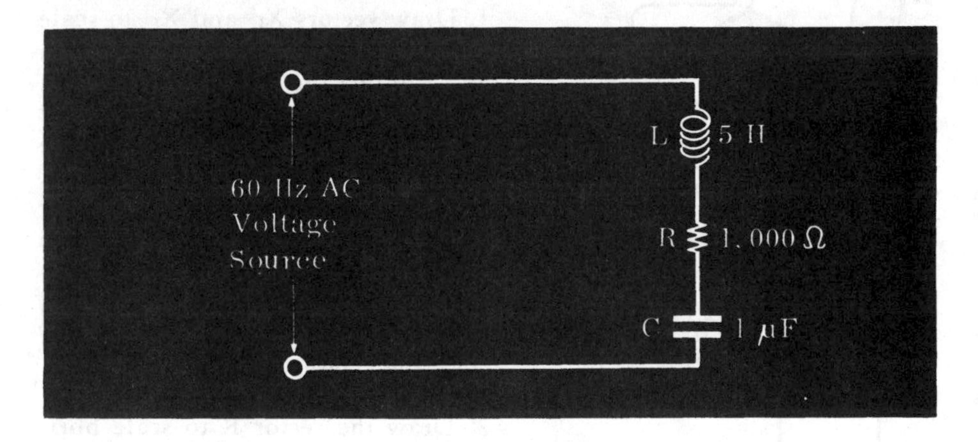

First calculate X_L and X_C:

$$X_L = 2\pi fL$$
$$= 6.28 \times 60 \times 5$$
$$= 1{,}884 \text{ ohms}$$

$$X_C = \frac{1}{2\pi fC}$$
$$= \frac{10^6}{6.28 \times 60 \times 1}$$
$$= 2{,}654 \text{ ohms}$$

R, L, and C Series Circuit Impedance—by Calculation (continued)

(Note that 10^6 is only a convenient way of writing 1,000,000. It means 10 multiplied by itself six times. You will learn more about this later.)

$$
\begin{aligned}
Z &= \sqrt{R^2 + (X_L - X_C)^2} \\
&= \sqrt{1{,}000^2 + (1{,}884 - 2{,}654)^2} \\
&= \sqrt{1{,}000^2 + (-770)^2} \\
&= \sqrt{10^6 + 592{,}900} \\
&= \sqrt{1{,}592{,}900} \\
&= 1{,}262 \text{ ohms}
\end{aligned}
$$

As X_C is greater than X_L, the current leads the total voltage across the circuit.

The phase angle can be calculated as before from

$$\tan \theta = \frac{X}{R}$$

where X is the net circuit reactance ($X_L - X_C$). Since $X_L - X_C = 1{,}885 - 2{,}654 = -770$,

$$\theta = \tan^{-1} \frac{X}{R} = \tan^{-1} \frac{-770}{1{,}000} = -37.6^\circ$$

and since X is negative, the phase angle is negative, indicating that the net circuit reactance is capacitive—that is, the voltage lags the current.

Series Circuit Resonance

As you may have suspected, in any series circuit containing both L and C, the current is greatest when the inductive reactance X_L equals the capacitive reactance X_C, since under those conditions the impedance is equal to R. Whenever X_L and X_C are unequal, the impedance Z is the vector combination of R and the difference between X_L and X_C. This vector is always greater than R, as shown below. When X_L and X_C are equal, Z is equal to R and is at its minimum value, allowing the greatest amount of current to flow. As you can see, if R is small, very large amounts of current can flow.

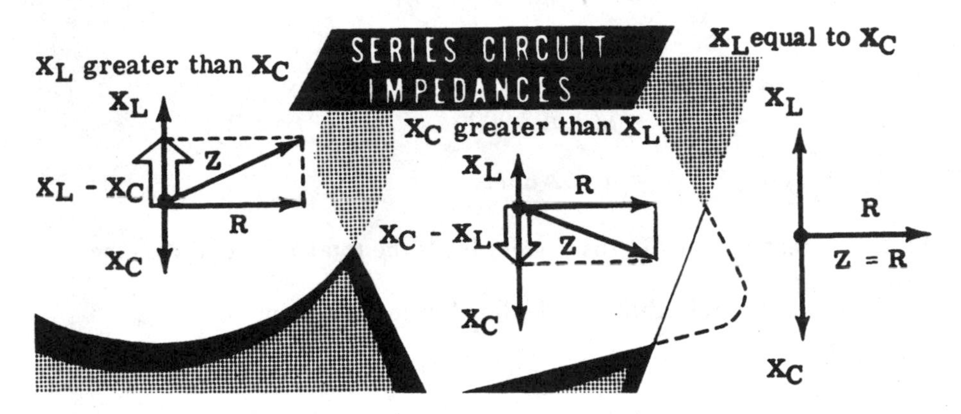

When X_L and X_C are equal, the voltages across them, E_L and E_C, are also equal, but of opposite phase, and the circuit is said to be *at resonance.* Such a circuit is called a *series resonant circuit.* (Note that both E_L and E_C may be much greater than E_t, which equals E_R at resonance.)

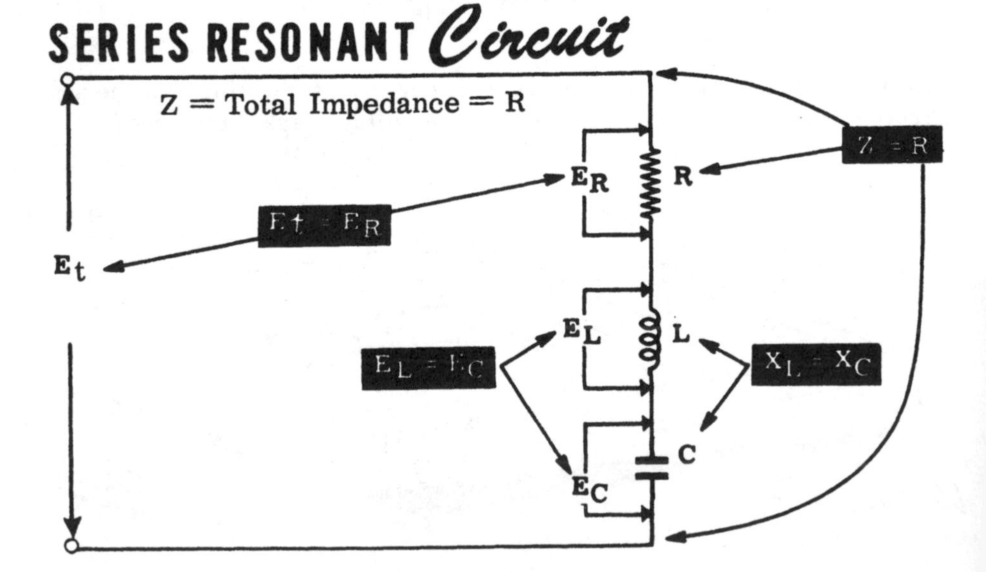

Series Circuit Resonance—Resonant Frequency

Suppose now that an ac voltage of variable frequency is applied to a circuit consisting of an inductor and a capacitor in series. The inductor is represented in the circuit diagram below by a pure inductance (L) in series with its internal resistance (R).

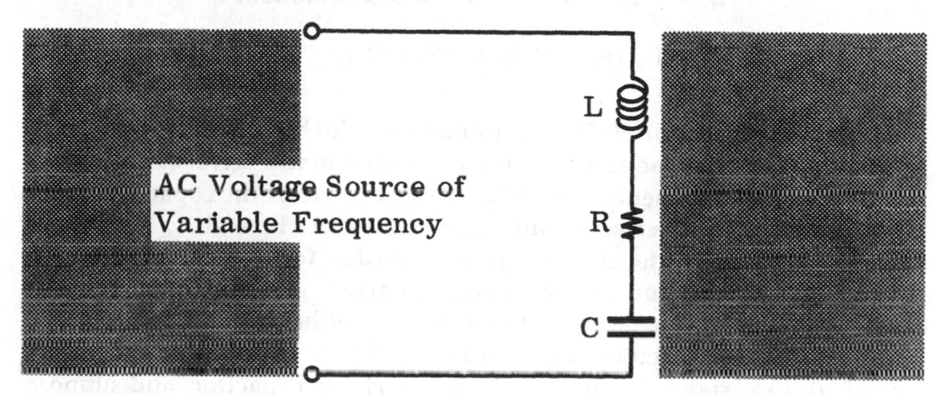

As the frequency of the applied voltage rises, the capacitive reactance X_C decreases, but the inductive reactance X_L increases. The graphs which follow show how X_C and X_L vary with changes in the frequency.

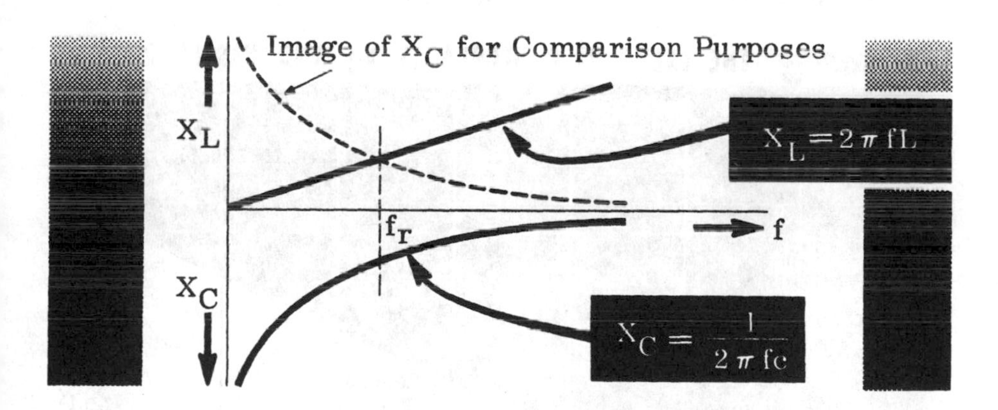

You can see that at only one frequency does X_C equal X_L. This is called the *resonant frequency, f_r*. The frequency at which a particular circuit will resonate can be calculated as follows:

$$X_C = X_L$$
$$2\pi f_r L = 1/2\pi f_r C$$
$$f_r^2 = 1/2^2\pi^2 LC$$
$$f_r = 1/2\pi\sqrt{LC}$$

As you can see, the resonant frequency is *not* affected by R.

Series Circuit Resonance—Resonant Frequency (continued)

For our circuit, with a 5-henry inductor and a 1-μF capacitor, we can calculate the resonant frequency as

$$f_r = 1/2\pi\sqrt{LC} = 1/2\pi\sqrt{5 \times 0.000001}$$

$$= 1 / (6.28 \times 0.0022) = 71 \text{ Hz}$$

This is not far from our 60-Hz line frequency, which accounts for the large circuit current even though X_L and X_C individually were quite high.

If either the frequency, the inductive reactance, or the capacitive reactance is varied in a series circuit consisting of R, L, and C, with other values kept constant, the circuit current variation forms a curve called the *resonance curve.* This curve shows a rise in current to a maximum value at exact resonance and a decrease in current on either side of resonance.

Consider, for instance, a circuit consisting of an inductor (of inductance L and resistance R) in series with a variable capacitor, and suppose that this circuit is connected across a suitable ac voltage source whose frequency is fixed. As the capacitor is varied from its minimum capacitance to its maximum capacitance, the impedance of the circuit will vary as illustrated below. You will notice that the current reaches its maximum, and the impedance, its minimum, when $X_C = X_L$.

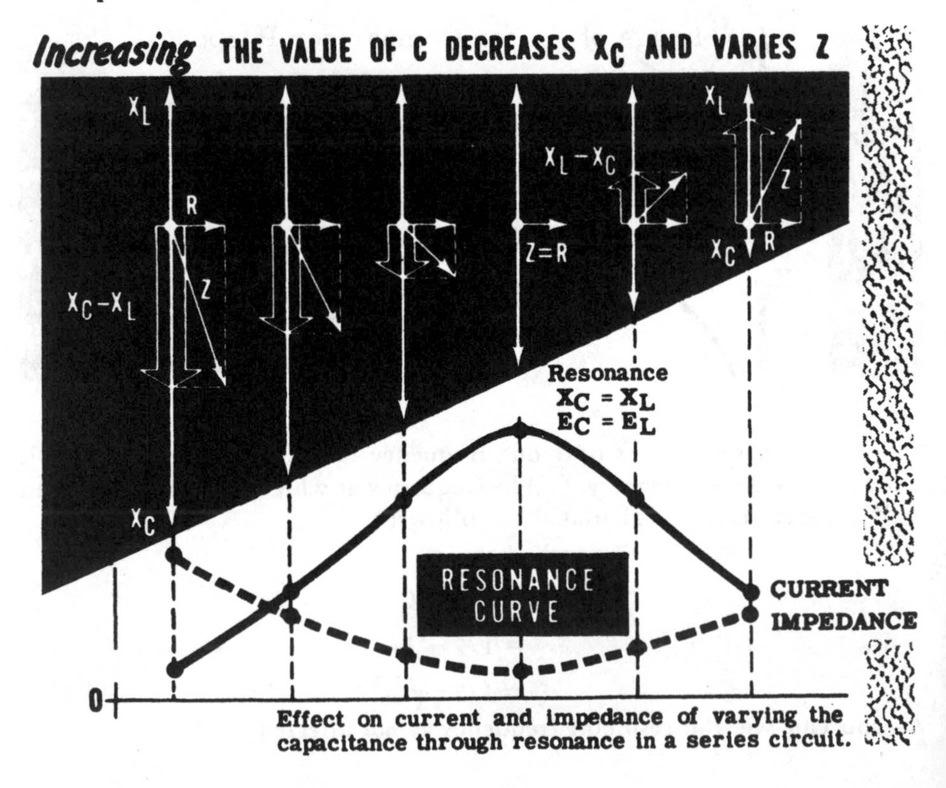

Effect on current and impedance of varying the capacitance through resonance in a series circuit.

Series Circuit Resonance—Resonant Frequency (continued)

Similar curves result if capacitance and frequency are held constant while the inductance is varied, and also if the inductance and capacitance are held constant while the frequency is varied.

Variation of the voltage input frequency to an R, L, and C circuit (over a suitable range) results in an output current curve which is similar to the resonance curve on the previous page. When operated below the resonant frequency, the current is low and the circuit impedance is high. Above resonance, the same condition occurs. At the resonant frequency, the impedance curve has its minimum value, equal to that of the circuit resistance; the current curve must, therefore, be at its maximum value (given, of course, by the equation $I = E/R$). The graphs of current and impedance below are typical of an R, L, C series circuit when only the applied voltage frequency is varied.

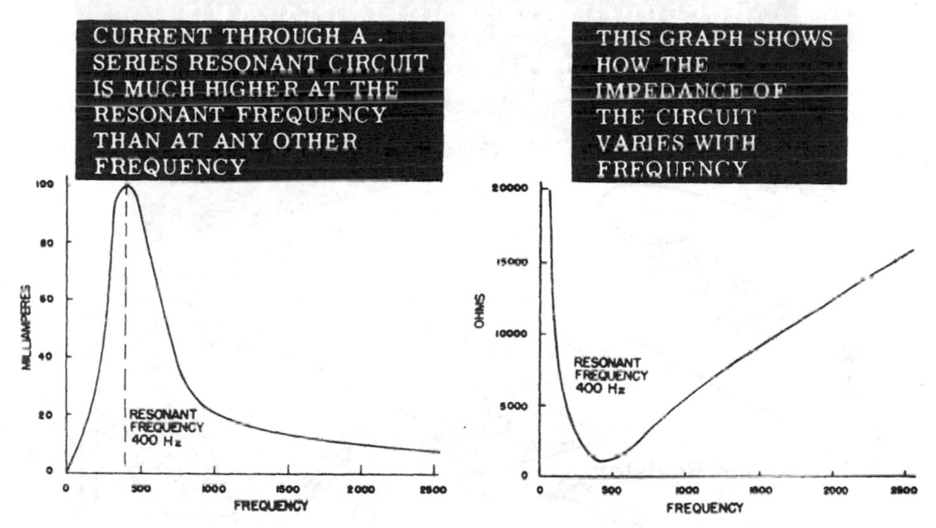

Note that below the resonant frequency the circuit acts as a capacitance because X_C is greater than X_L, and the current leads the applied voltage. Above the resonant frequency, the circuit acts as an inductance because X_L is greater than X_C, and the voltage leads the current. The variation of phase angle is illustrated below.

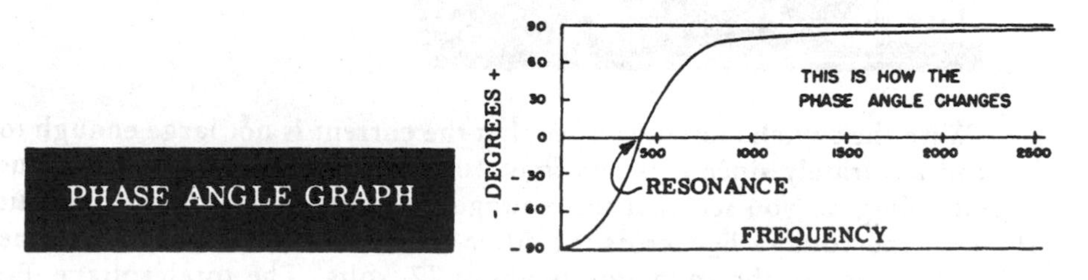

At resonance, the phase angle is zero.

Experiment/Application—Series Resonance

As you know from the earlier experiment/application, the series L-C circuit with a 5-henry inductor and 1-μF capacitor is resonant at about 71 Hz. To make it resonant at 60 Hz, you could increase either L or C. In this experiment/application, you will explore the effects of series resonance on the circuit impedance.

To demonstrate series resonance, suppose that you replaced the 1-μF capacitor with a 0.25-μF capacitor and inserted a 1,500-ohm, 10-watt resistor in series with the 5-henry filter choke and the capacitor. This forms an R, L, and C series circuit, in which C will be varied to show the effect of resonance on circuit voltage and current. A 0–50 mA ac milliammeter is connected in series with the circuit to measure the circuit current. A 0–250 volt ac voltmeter measures circuit voltages.

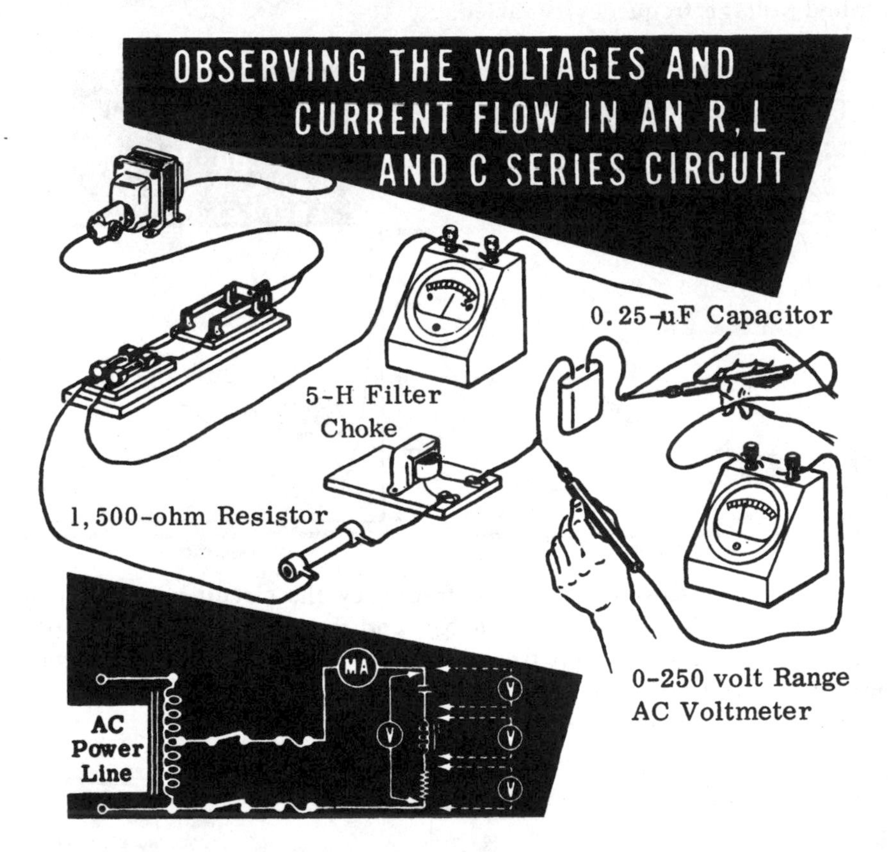

With the switch closed, you see that the current is not large enough to be read accurately since it is less than 10 mA. If you measure the various circuit voltages, you see that the voltage E_R across the resistor is about 10 volts, the voltage E_L across the filter choke is about 13 volts, and the voltage E_C across the capacitor is about 72 volts. The total voltage E_t across the entire circuit is approximately 60 volts.

Experiment/Application—Series Resonance (continued)

By using various parallel combinations of 0.25-μF, 0.5-μF, 1-μF, and 2-μF capacitors, you can vary the circuit capacitance from 0.25 μF to 3.75 μF in steps of 0.25 μF. It is important to remove the circuit power and discharge all capacitors used before removing or adding capacitors to the circuit. You will notice that as the capacitance is increased, the current rises to a maximum value at the point of resonance, then decreases as the capacitance is increased further.

OBSERVING THE CIRCUIT VOLTAGES AND CURRENT CHANGE AS THE CAPACITANCE VALUE IS CHANGED

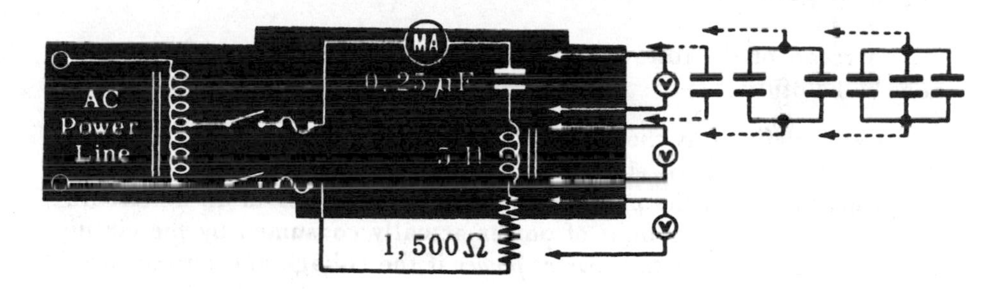

Except for the total circuit voltage, E_t, which is constant at 60 volts, the measured circuit voltages vary as the capacitance is changed. The voltage E_R across the resistor changes in the same manner as the circuit current. For capacitance values less than the resonance value, E_C is greater than E_L. At resonance, E_L equals E_C. As the capacitance is increased beyond the resonance point, E_L is greater than E_C when the circuit capacitance is greater than the value required for resonance.

You could, of course, have calculated the correct value from the resonance formula that you learned earlier, or, knowing that at resonance $X_L = X_C$, you can calculate as follows:

$$X_L = 2\pi fL$$

$$X_C = 1/2\pi fC$$

$$2\pi fL = 1/2\pi fC$$

$$2 \times 3.14 \times 60 \times 5 = 1/2 \times 3.14 \times 60 \times C$$

$$C = 1/(2 \times 3.14 \times 60 \times 5) \times (2 \times 3.14 \times 60)$$

$$C = 1/709{,}891 = 0.00000141 = 1.41\ \mu F$$

As you saw from the experiment, the capacitance value that gave the highest current in the circuit and the maximum voltage across the resistance was equal to 1.5 μF (close to 1.4).

Power Factor

The concept of power, particularly with regard to the calculation of power factors in ac circuits, will become a very important consideration as the circuits you work with are made more complex. In dc circuits, the expended power may be determined by multiplying the voltage by the current. A similar relationship exists for finding the amount of expended power in ac circuits, but certain other factors must be considered. For example, you know that pure inductive or capacitive ac circuits can draw current although the power consumed is zero. In pure resistive ac circuits, the power is equal to the product of the voltage and current *only* when the E and I are in phase or at resonance. If the voltage and current are not in phase, the power used by the circuit will be less than the product of E and I.

A recap of the following principles, which you already know, will now be helpful:

1. *Power* is defined as the rate of doing work.
2. A *watt* is the unit of electrical power.
3. *Apparent power* is the product of volts and amperes in an ac circuit.
4. *True power* is the amount of power actually consumed by the circuit.
5. *True power* is equal to *apparent power* if the voltage and current are in phase.
6. *True power* is equal to zero if the voltage and current are out of phase by 90 degrees.

Now consider again the vector diagram for the voltages in an R and L ac series circuit. E_L is the voltage across the inductance, E_R is the voltage across the resistance, and E_t is the total applied voltage.

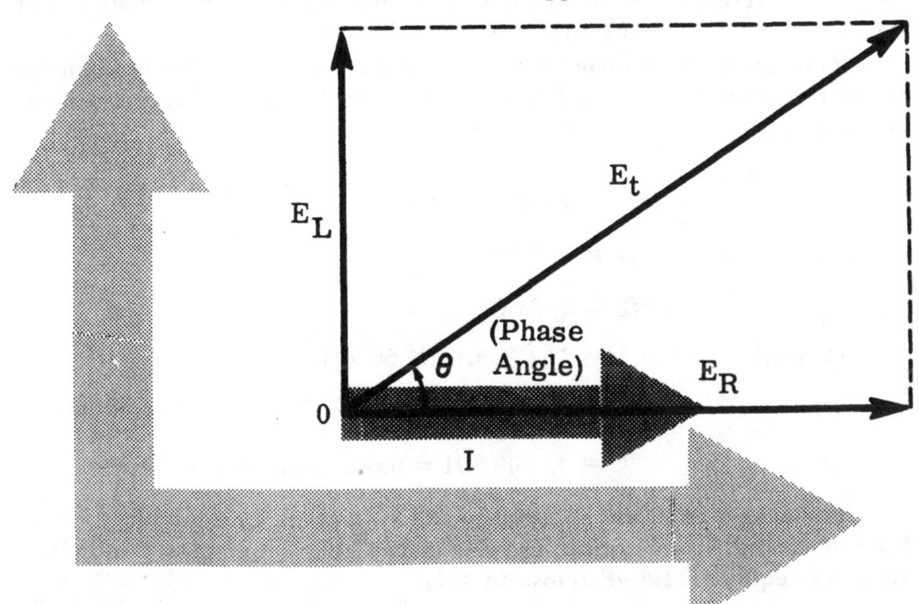

Power Factor (continued)

You can see that E_R is the component of E_t which is in phase with the current I, and E_L is the component of E_t which is 90 degrees out of phase with I.

From the definitions of true power and apparent power repeated above, we can see that:

$$\text{True power} = E_R I$$
$$\text{Apparent power} = E_t I$$

We define the "power factor" as the ratio of true to apparent power:

$$\text{Power factor} = \frac{\text{True power}}{\text{Apparent power}} = \frac{E_R I}{E_t I}$$

If we cancel the I in the numerator and denominator, we get:

$$\text{Power factor} = \frac{E_R}{I} \times \frac{I}{E_t} = \frac{E_R}{E_t} = \frac{R}{Z}$$

If you check back on your study of the right triangle, you will see that the ratio E_R/E_t is the cosine of angle θ, which is the phase angle. Thus, power factor = cos θ, and

$$\text{True power (P)} = \text{apparent power} \times \text{power factor}$$
$$P = E_t I \cos\theta$$

If you again compare the vector diagram of impedance, reactance, and resistance on page 4-13 with the right triangle on page 4-19, you will see that cos θ equals R/Z. You will find the quantity power factor expressed as a percentage calculated as (R/Z) × 100.

POWER FACTOR =

***TRUE* Power**

***APPARENT* Power**

Power

Suppose that you wish to find the amount of power expended (true power) in a circuit where the impedance is 5 ohms, the resistance is 3 ohms, and the inductive reactance is 4 ohms. The voltage is 10 volts ac, and the current is 2 amperes.

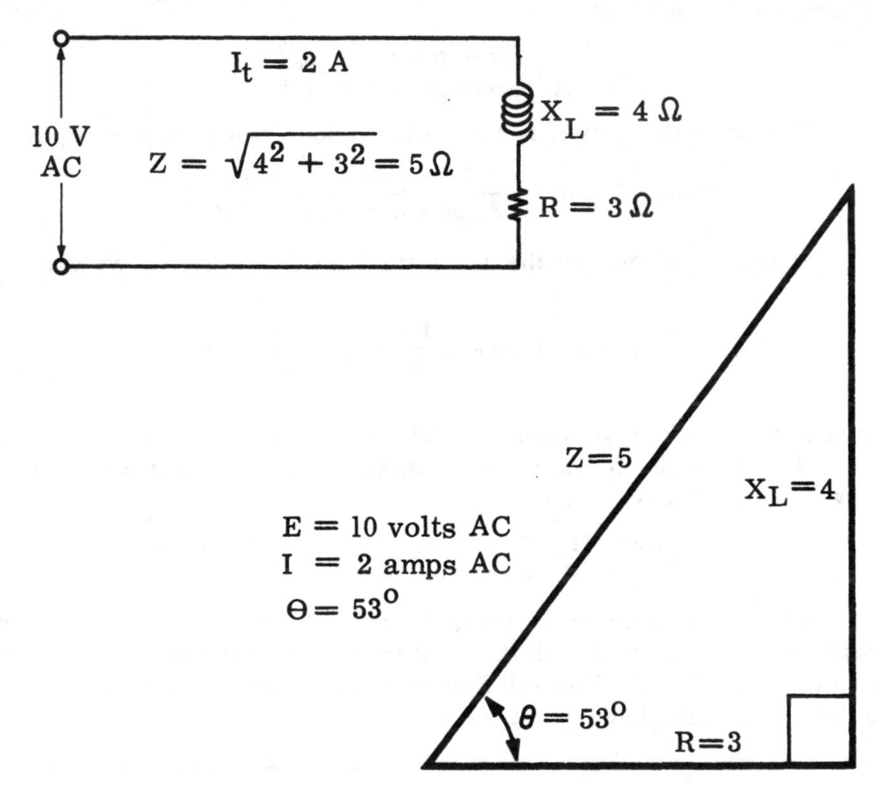

The apparent power is given by E × I, or 20 watts.

The formula for true power is $P = EI \cos \theta$. With reference to the impedance triangle, $\cos \theta$ is equal to the ratio of R divided by Z, or

$$\cos \theta = \frac{R}{Z} = \frac{3}{5} = 0.60 \text{ (60\% power factor)}$$

By looking up cos 53° in the back of this volume, we find it equal to 0.6 also. Substituting in the formula for true power,

$$\begin{aligned} \text{True power} &= EI \cos \theta \\ &= 10 \times 2 \times 0.6 \\ &= 12 \text{ watts} \end{aligned}$$

As you can see, the line current is 2 amperes but the power is only 12 watts. The line current would be only 1.2 ampere if the voltage and current were in phase. As you will learn later, this is a very important point since the power line has to carry the 2 amperes and any resistive loss is determined by the line resistance and the 2 amperes, not the 1.2 ampere.

Experiment/Application—Power and Power Factor

To demonstrate the difference between true and apparent power, you could use the series R, L, and C circuit with a wattmeter that you know measures true power. For this experiment/application, you will want to reduce the resistance to 500 ohms.

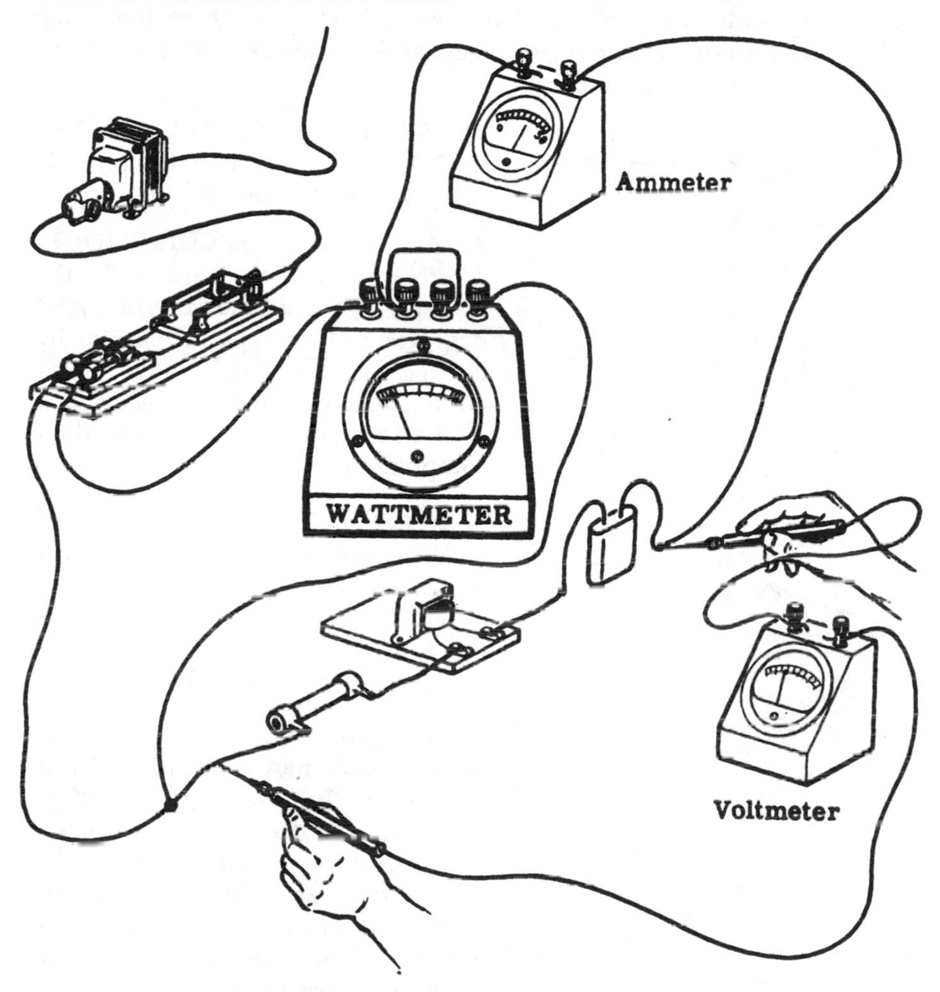

Vary the capacitance as before. You will find that the wattmeter (measuring true power) will read the true power as being equal to the current squared times the resistance (I^2R) whereas the apparent power, calculated as $E \times I$, will not be the same. If you calculate the power factor as true power divided by apparent power for each capacitor value, you will find that they will be equal at resonance (where there is no apparent reactance in the circuit). Thus, the power factor at resonance is unity (or 100 percent). You will see this important fact made use of later, where inductance and capacitance are added to power transmission lines to correct the power factor to unity so as to obtain the most efficient power transfer.

Review of Current, Voltage, Impedance, Resonance, Power, Power Factor

You have found that the rules for ac series circuit voltage and current are the same as those for dc series circuits, except that the various circuit voltages must be *added by means of vectors* because of the phase difference between the individual voltages. Now review what you have found out about ac series circuit current, voltages, and resonance, and how Ohm's law applies to an ac series circuit.

$$I_t = I_C = I_L = I_R$$

1. AC SERIES CIRCUIT CURRENT—The current is the same in all parts of a series circuit.

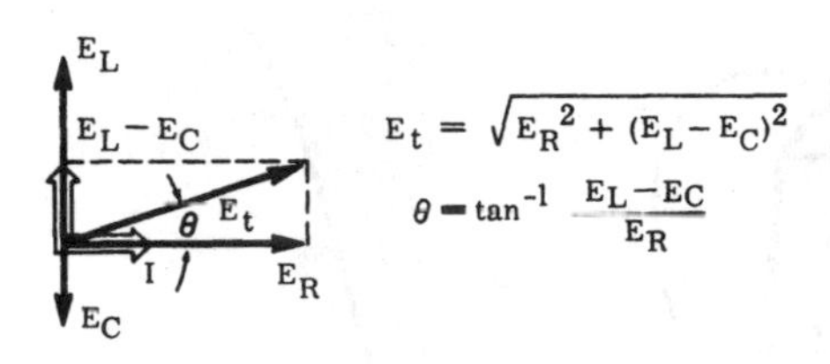

2. AC SERIES CIRCUIT VOLTAGES—E_R is in phase with the current, E_L leads the current by 90 degrees, and E_C lags the current by 90 degrees. The total voltage across an ac series circuit is the vector sum of the voltages across the individual components.

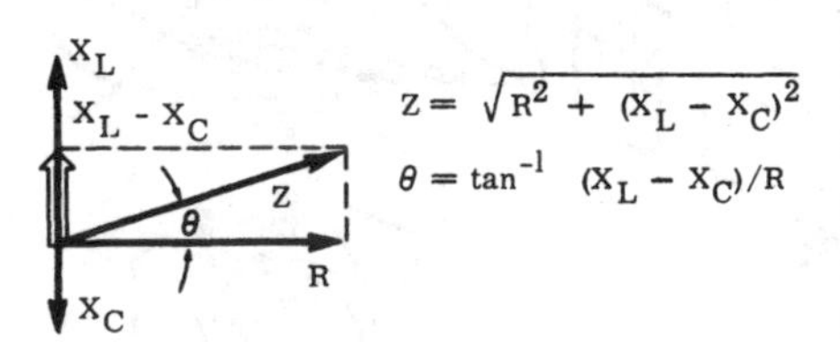

3. IMPEDANCE—Impedance, the total opposition to the flow of current in an ac circuit, is represented by Z and expressed in ohms. It is the vector sum of the resistance and reactance of the circuit.

$$f_r = 1/2\pi\sqrt{LC}$$

4. RESONANCE—An ac circuit containing R, L, and C is said to be at resonance when $X_L = X_C$ and $Z = R$. In the case of a series circuit, Z is a minimum at resonance, and the current through the circuit is a maximum at resonance. The frequency at which resonance occurs is called the *resonant frequency* (f_r).

$$\text{True power} = I^2R = EI\cos\theta$$
$$\text{Apparent power} = EI$$
$$\text{Power factor} = \frac{\text{True power}}{\text{Apparent power}}$$
$$\text{Power factor} = \cos\theta = R/Z$$

5. POWER/POWER FACTOR—The power in ac circuits containing reactances does not usually equal voltage times current ($E \times I$) as in dc circuits. This value is called the *apparent power*. *True power* must also take phase into account. The ratio of true to apparent power is called the *power factor*.

Self-Test—Review Questions

1. Draw a vector diagram to represent an R of 1000 ohms, an X_L for a 5-henry inductor, and an X_C for a 2-μF capacitor at 60 Hz.
2. For the circuits shown below (A through F), draw a vector diagram solution to find Z and θ (the frequency is 60 Hz).

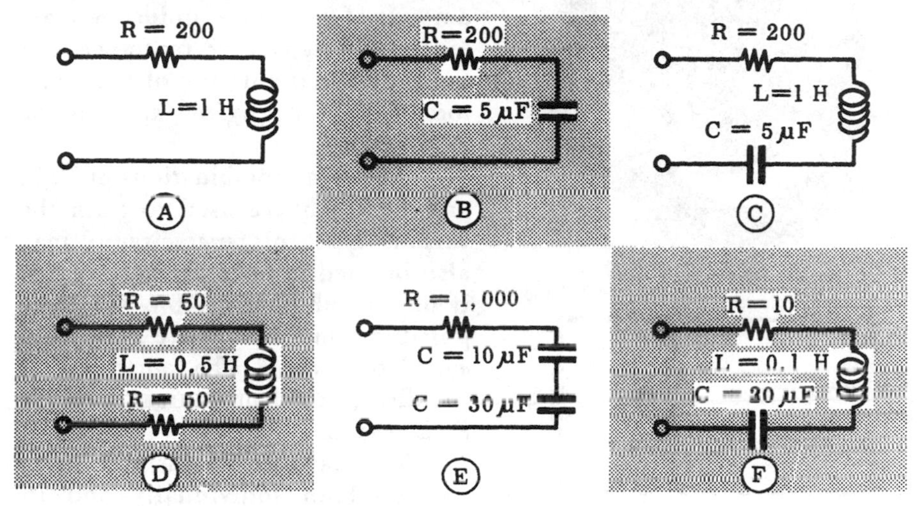

3. Solve the circuits above by calculation and compare your answers.
4. If the input voltage is 120 volts, 60 Hz, calculate the circuit currents and the voltage across each component in Question 2.
5. For each circuit in Question 2, calculate the true power, apparent power, and power factor.
6. Calculate the resonant frequencies for circuits (c) and (f). How much current will flow at resonance? What is the phase? What capacitance must be added to circuit (c) for it to be resonant at 60 Hz? What inductance must be added to circuit (c) for it to be resonant at 60 Hz?

Learning Objectives—Next Section

Overview—Now that you have learned about ac series circuits, you will learn in the next section about ac parallel circuits. They are most important since electric circuits are often made up of many devices connected in parallel across the ac line.

AC line — Lamp — TV — Washer — Motor

AC Parallel Circuit Combinations

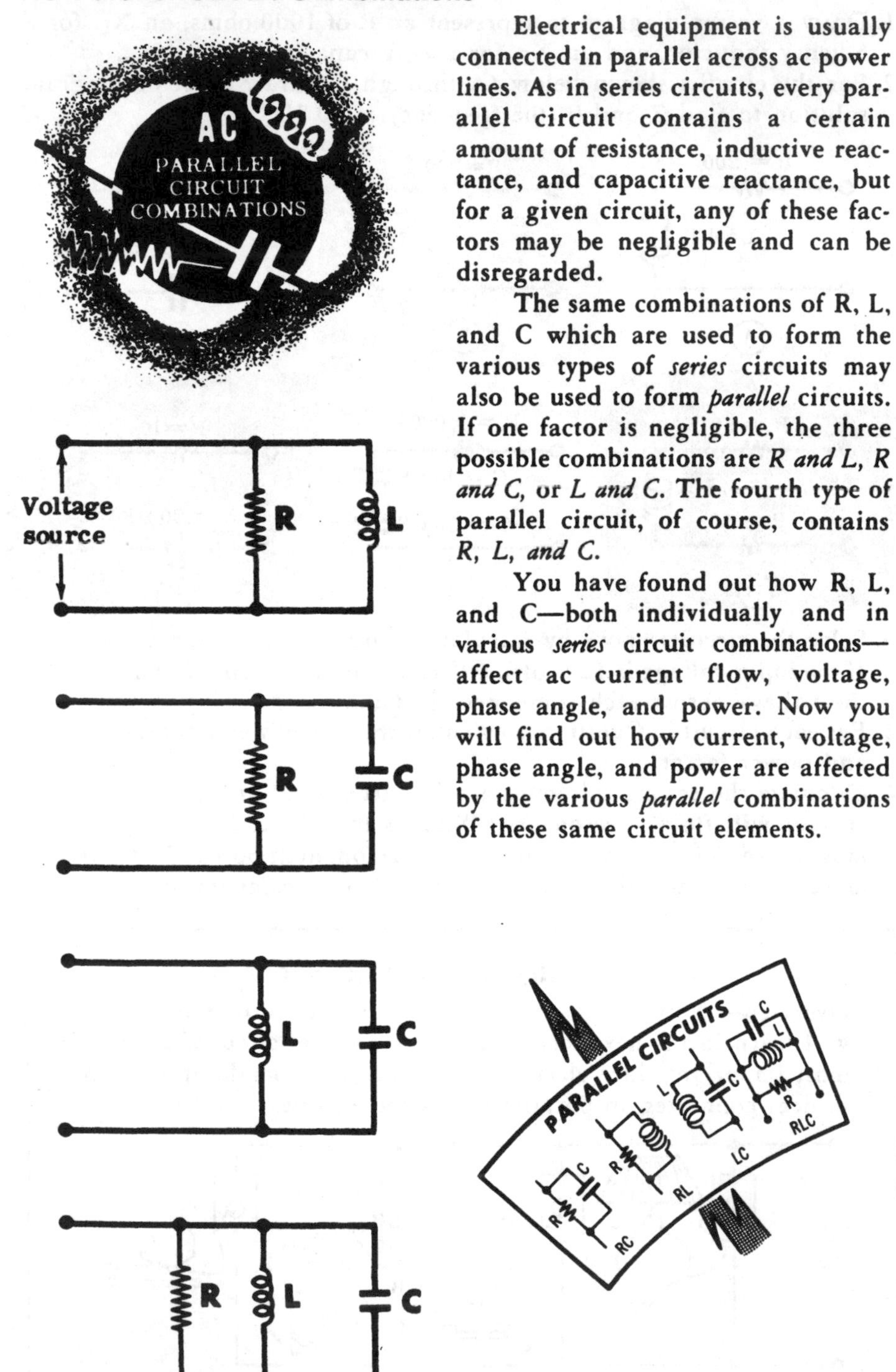

Electrical equipment is usually connected in parallel across ac power lines. As in series circuits, every parallel circuit contains a certain amount of resistance, inductive reactance, and capacitive reactance, but for a given circuit, any of these factors may be negligible and can be disregarded.

The same combinations of R, L, and C which are used to form the various types of *series* circuits may also be used to form *parallel* circuits. If one factor is negligible, the three possible combinations are *R and L, R and C,* or *L and C.* The fourth type of parallel circuit, of course, contains *R, L, and C.*

You have found out how R, L, and C—both individually and in various *series* circuit combinations—affect ac current flow, voltage, phase angle, and power. Now you will find out how current, voltage, phase angle, and power are affected by the various *parallel* combinations of these same circuit elements.

Voltages in AC Parallel Circuits

You will remember that in a parallel dc circuit the voltage across each of the parallel branches is *equal.* This is also true of ac parallel circuits; the voltages across each parallel branch are equal and also equal E_t, the total voltage of the parallel circuit. Not only are the voltages equal, but they are also *in phase.*

For example, if the various types of electrical equipment shown below—a lamp (resistance), a filter choke (inductance), and a capacitor (capacitance)—are connected in parallel, the voltage across each is *exactly the same.*

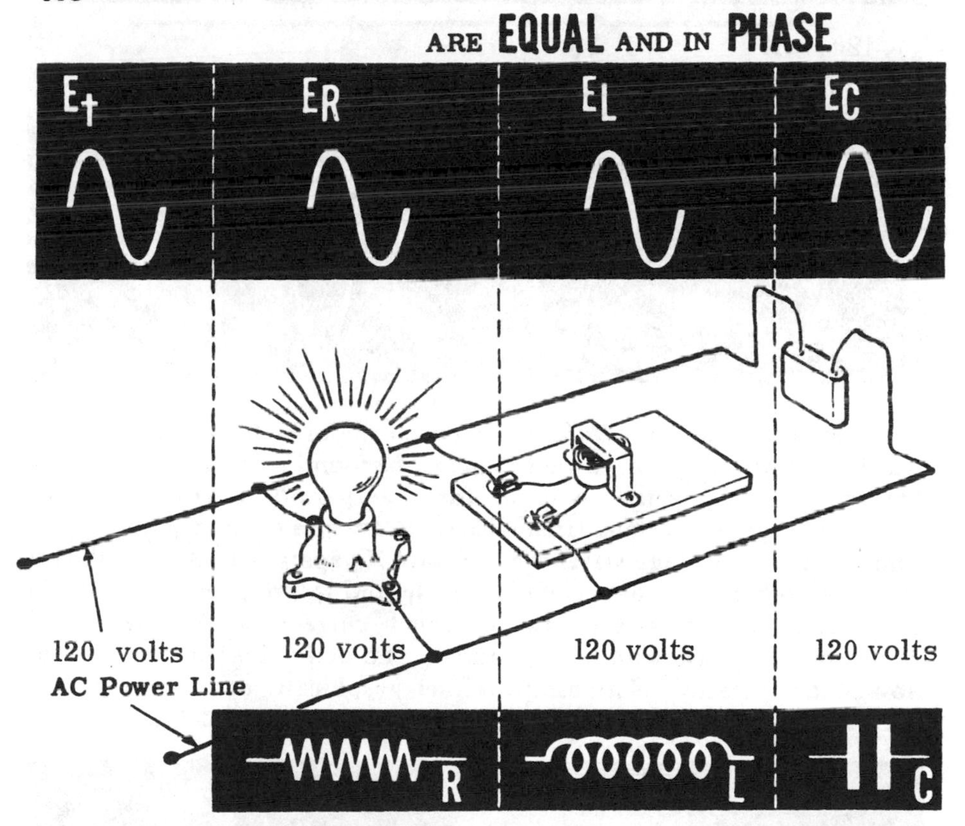

Regardless of the number of parallel branches, the value of the voltage across them is *equal* and *in phase.* All of the connections to one side of a parallel combination are considered to be one electrical point as long as the resistance of the connecting wire is neglected.

As you can see, in a parallel circuit the *voltages* across each element are the *same* while the *currents vary.* In the series circuits you studied earlier, on the other hand, the *current* through all elements is the *same,* but the *voltages* across the elements *vary.*

Currents in AC Parallel Circuits

The current flow through each individual branch is determined by the opposition offered by that branch. If your circuit consists of three branches—one a resistor, another an inductor, and the third a capacitor—the current through each branch depends on the resistance or reactance of that branch. The resistor branch current I_R is in phase with the circuit voltage E_t , while the inductor branch current I_L *lags* the circuit voltage by 90 degrees, and the capacitor branch current I_C *leads* the voltage by 90 degrees.

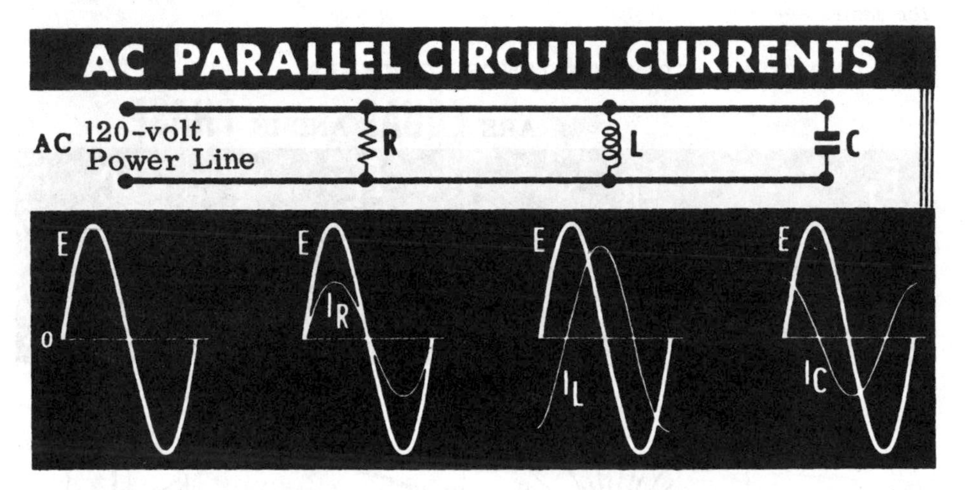

Because of the *phase difference* between the branch currents of an ac parallel circuit, the total current I_t *cannot* be found by adding the various branch currents directly—as it can for a dc parallel circuit. When the waveforms for the various circuit currents are drawn in relation to the common circuit voltage waveform, X_L and X_C again are seen to subtract from each other since the waveforms for I_L and I_C are exactly opposite in phase at all points. The resistance branch current I_R, however, is 90 degrees out of phase with both I_L and I_C. To determine the total current flow by using vector relationships (either graphically or by calculation), I_R must be combined with the difference between I_L and I_C.

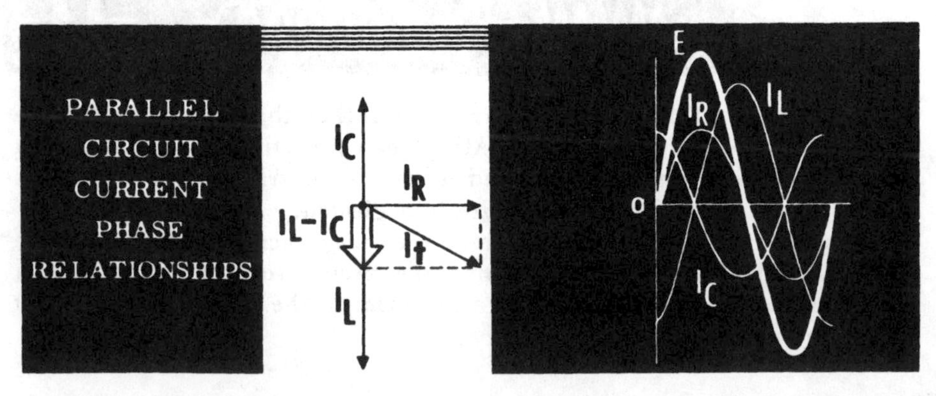

Currents in AC Parallel Circuits (continued)

To add the branch currents in an ac parallel circuit, the instantaneous values of current can be combined, as voltages were for series circuits, to obtain the instantaneous values of the total current waveform. After all the possible instantaneous values of current are obtained, the total current waveform is drawn by adding the instantaneous values at each point.

COMBINING PARALLEL CIRCUIT BRANCH CURRENTS

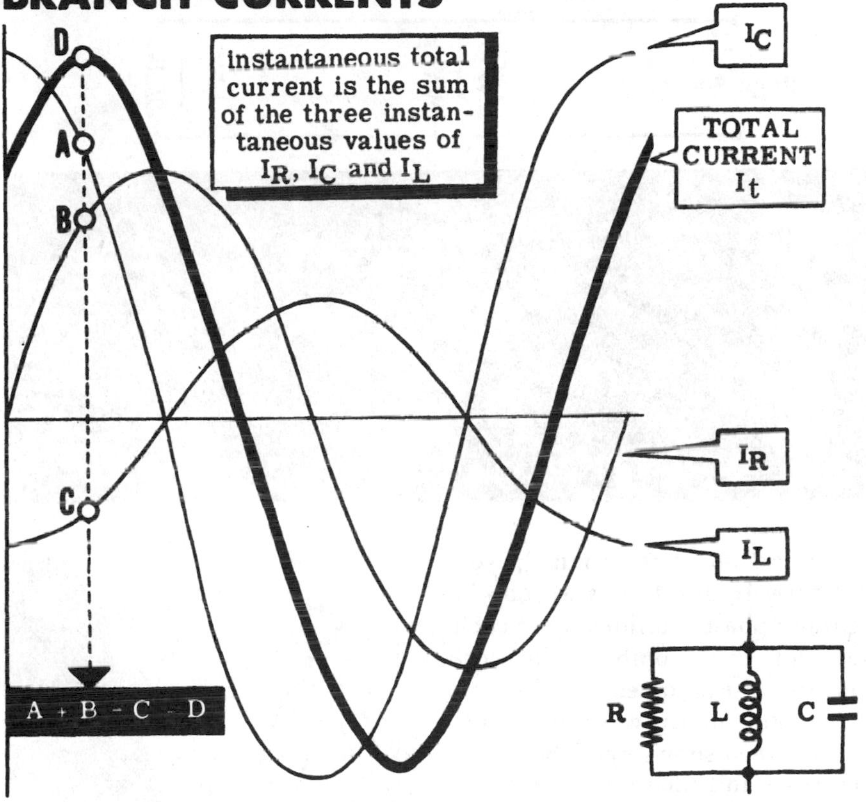

instantaneous total current is the sum of the three instantaneous values of I_R, I_C and I_L

I_C

TOTAL CURRENT I_t

I_R

I_L

D

A

B

C

A + B - C - D

R L C

The maximum value of I_t is equal to or less than the sum of the maximum values of the individual currents and is not usually in phase with the various branch currents. With respect to the circuit voltage, the total current either *leads* or *lags* by a phase angle between 0 and 90 degrees, depending on whether the inductive or capacitive reactance is greater.

A vector diagram for the various circuit currents and the circuit voltage of an ac parallel circuit is similar to the vector diagram for circuit current and voltages for an ac series circuit. In the series circuit, voltages are drawn with reference to total circuit *current,* while for parallel circuits the different currents are drawn with reference to the total circuit *voltage.*

R and L Parallel Circuit Currents

If an ac parallel circuit consists of a resistance and inductance connected in parallel, and the circuit capacitance is negligible, the total circuit current is a combination of I_R (the current through the resistance) and I_L (the current through the inductance). I_R is in phase with the circuit voltage E_t, while I_L lags the voltage by 90 degrees.

PHASE RELATIONSHIPS IN AN R AND L PARALLEL CIRCUIT

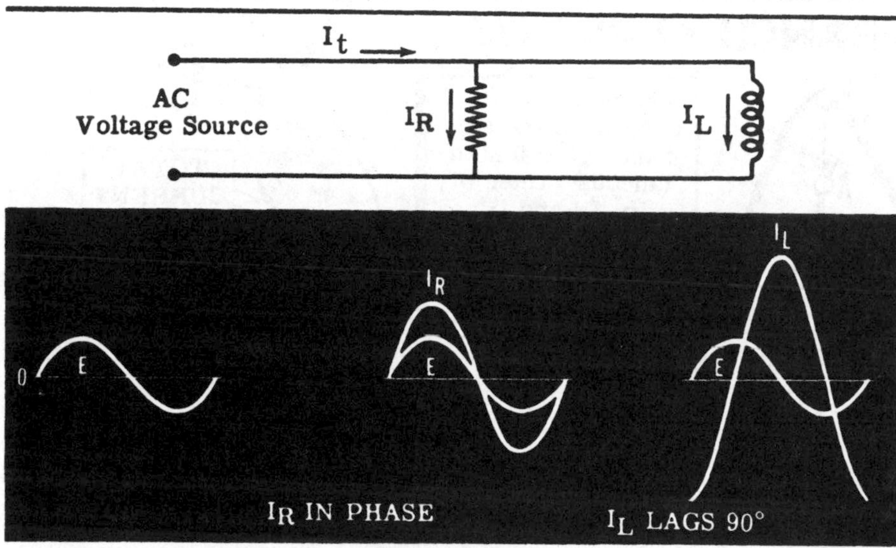

To find the total current I_t, you could draw I_R and I_L to scale and in the proper phase relationship to each other and then combine the corresponding instantaneous values to plot the total current waveform. This waveform then shows both the maximum value and phase angle of I_t.

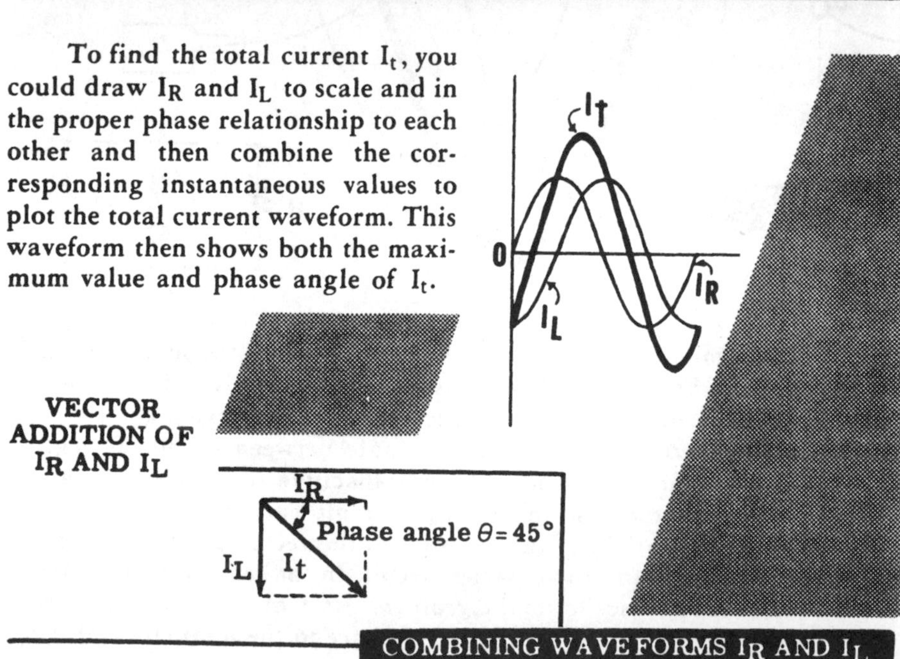

COMBINING WAVEFORMS IR AND IL

R and L Parallel Circuit Currents (continued)

You can use the much easier method of drawing vectors to scale to represent I_R and I_L, as you learned earlier, then combining the vectors by completing the parallelogram and drawing the diagonal, thus obtaining both the value and phase angle of I_t. The length of the diagonal represents the value of I_t, whereas the angle between I_t and I_R is the phase angle between total circuit voltage, E_t, and the total circuit current, I_t. As you know from right triangle relationships, the total current can be calculated from the Pythagorean theorem as

$$I_t = \sqrt{I_R^2 + I_L^2}$$

and the phase angle can be calculated as

$$\theta = \tan^{-1} \frac{I_L}{I_R}$$

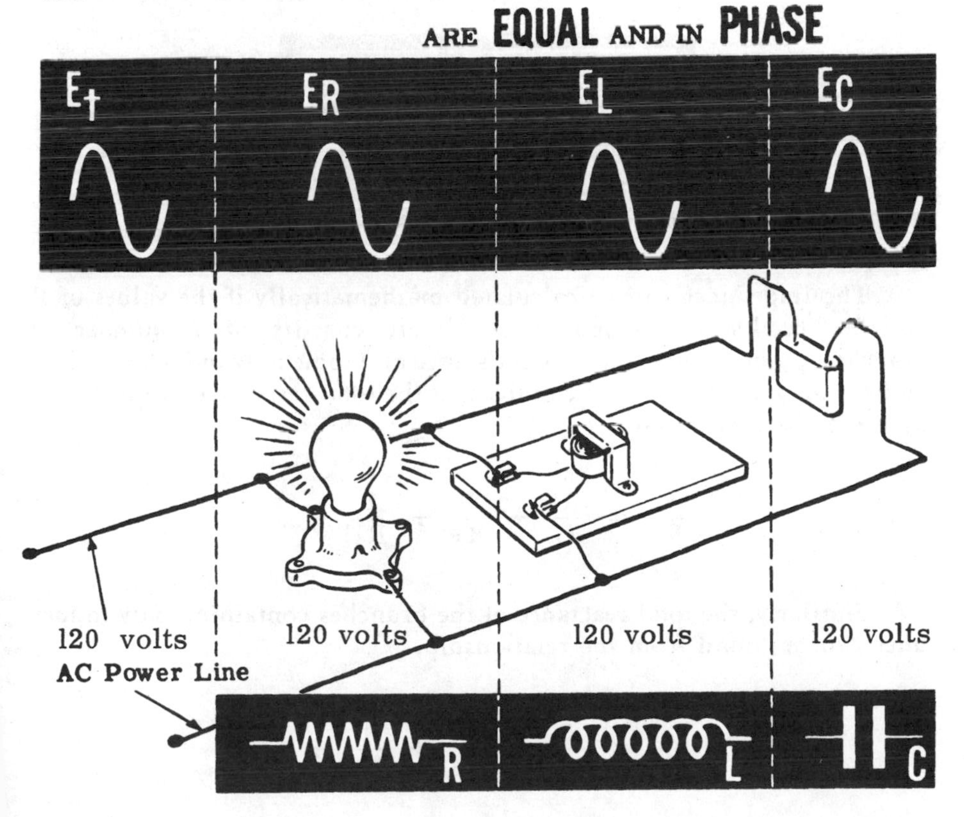

R and L Parallel Circuit Impedance

The impedance of a parallel circuit can be found by applying Ohm's law for ac current to the total circuit. Using Ohm's law, the impedance Z for all ac parallel circuits is found by dividing the circuit voltage by the total current, that is, $Z = E_t/I_t$.

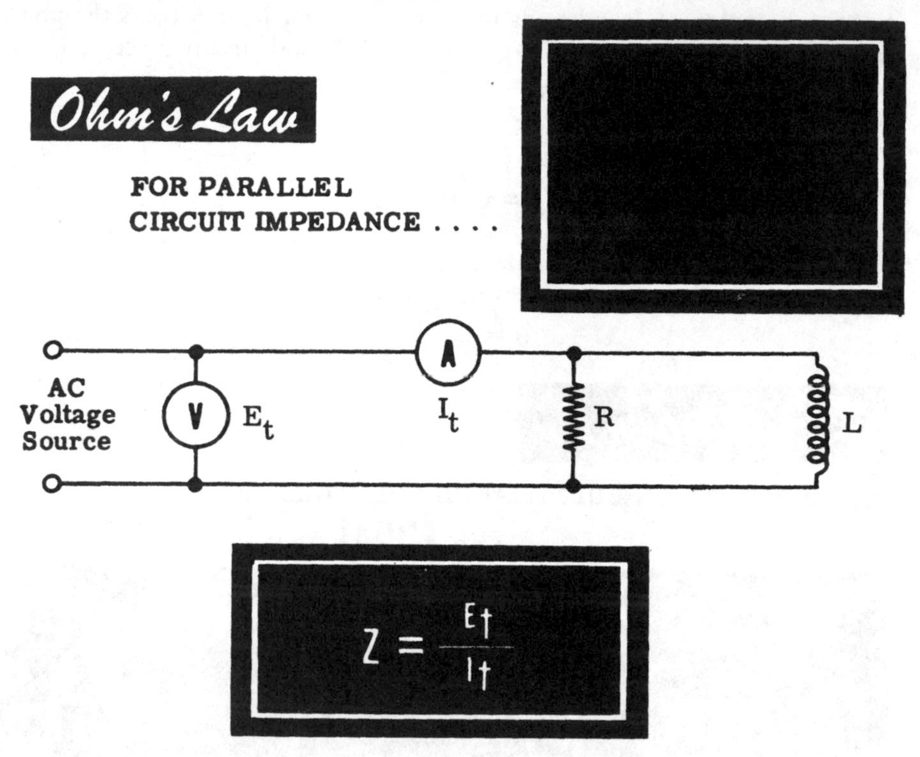

The impedance can be calculated mathematically if the values of R and L are known. Suppose the circuit consists of a number of branches—some containing only resistance and some only inductance. You know from your work on dc circuits that the total resistance can be found by using the relationship

$$\frac{1}{R_t} = \frac{1}{R_1} + \frac{1}{R_2} + \frac{1}{R_3} + \ldots$$

Similarly, the total reactance of the branches containing only inductance can be found from the relationship,

$$\frac{1}{X_t} = \frac{1}{X_1} + \frac{1}{X_2} + \frac{1}{X_3} + \ldots$$

R and L Parallel Circuit Impedance (continued)

The two quantities, R_t and X_t, can now be combined vectorially, as shown in the illustration below.

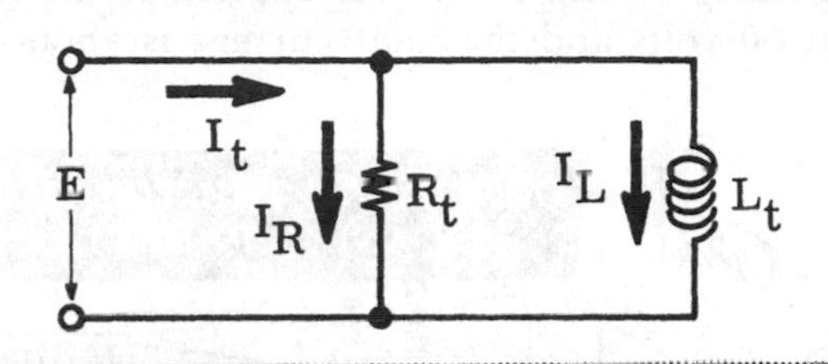

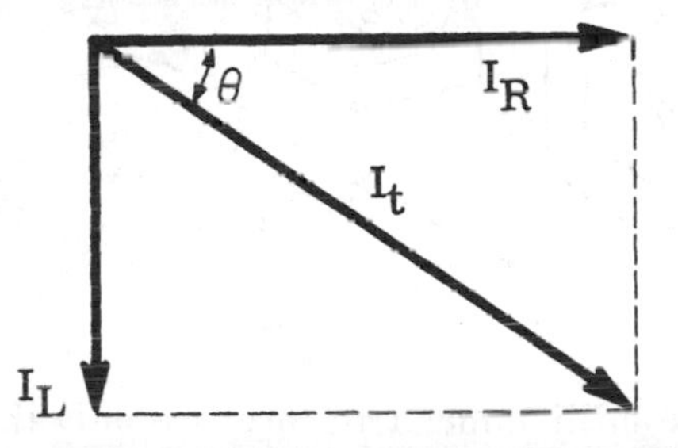

You will see that $I_R = E/R$, $I_L = E/X_L$, and $I_t = E/Z$, and since

$$I_t = \sqrt{I_R^2 + I_L^2}$$

we can say that

$$E/Z = \sqrt{(E/R)^2 + (E/X_L)^2}.$$

This can be rewritten

$$E/Z = E\sqrt{(1/R)^2 + (1/X_L)^2}.$$

Since E is on both sides of the equation, it can be deleted altogether, leaving

$$1/Z = \sqrt{(1/R)^2 + (1/X_L)^2}$$

so that

$$Z = 1/\sqrt{(1/R)^2 + (1/X_L)^2}$$

which can be simplified and expressed as

$$Z = RX_L/\sqrt{R^2 + X_L^2}$$

The phase angle can be calculated as always as $\theta = \tan^{-1}(I_L/I_R)$.

Experiment/Application—R-L Parallel Circuit Current and Impedance

To see for yourself the effects of connecting an inductor and a resistor in parallel, connect a 2,500-ohm, 20-watt resistor and a 5-henry inductor in parallel across the secondary of a step-down transformer to form an ac parallel circuit of R and L. A 0–50 mA ac milliammeter is connected to measure the total circuit current, and a 0–250 volt ac voltmeter is used to measure circuit voltage. With the power applied to the circuit, the circuit voltage is about 60 volts and the total current is about 40 mA.

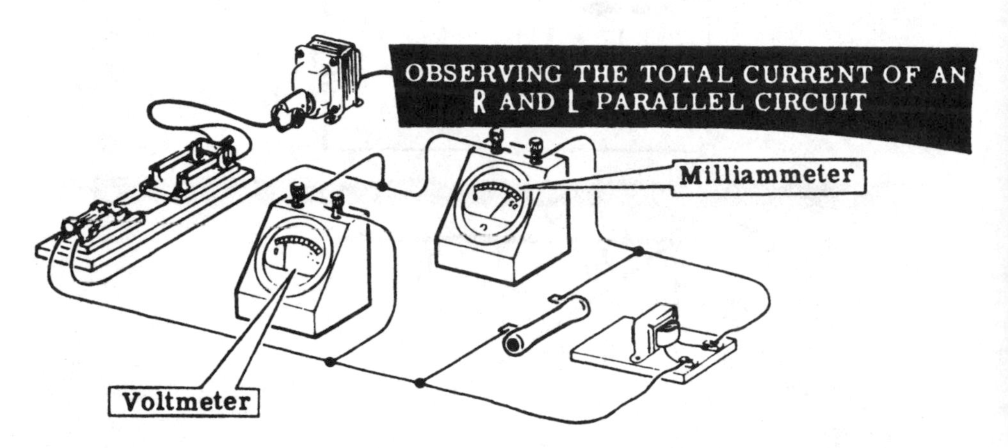

First connect the milliammeter to measure only the resistor current, then to measure only the inductor current. You see that the milliammeter reading for I_R is about 24 mA and that the current indicated for I_L is approximately 32 mA. The sum of these two branch currents I_R and I_L is thus 56 mA, while the actual measured total circuit current is about 40 mA. This difference shows that the branch currents must be added by means of vectors.

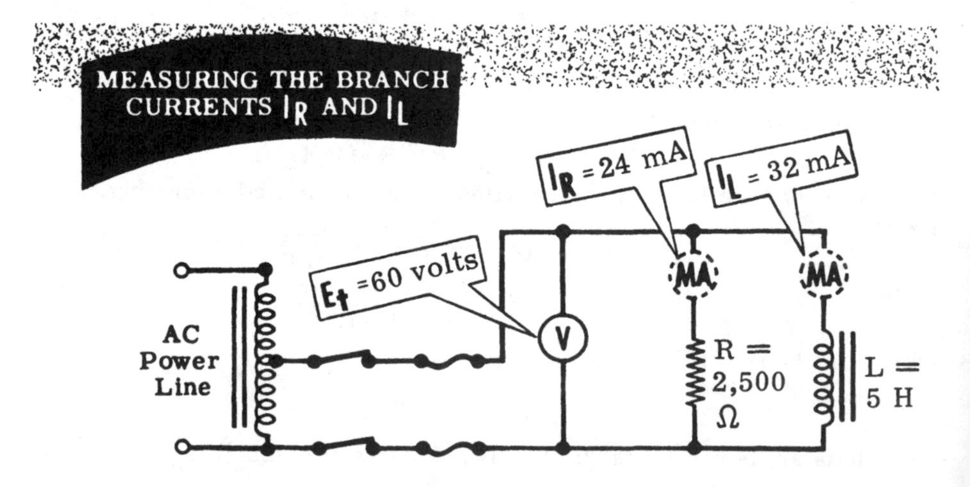

Experiment/Application—R-L Parallel Circuit Current and Impedance (continued)

Applying Ohm's law to the R and L circuit above to find the impedance, we get

$$Z = \frac{E}{I_t} = \frac{60}{0.040} = 1{,}500 \text{ ohms}$$

You can check the formula on page 4-57 by substituting the known values of R and X_L for this circuit and by comparing the value of Z found with that obtained above:

$$R = 2{,}500 \text{ ohms}; \quad X_L = 2\pi \times 60 \times 5 = 1{,}885 \text{ ohms}$$

$$Z = \frac{2{,}500 \times 1{,}885}{\sqrt{2{,}500^2 + 1{,}885^2}} = 1{,}500 \text{ ohms (approximately)}$$

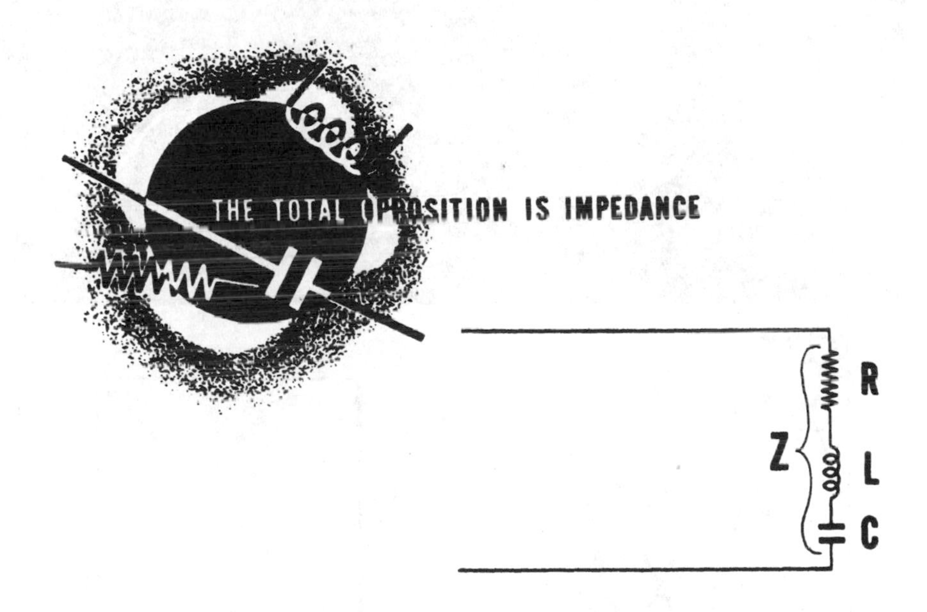

Ohm's Law for AC Circuits.

$$Z = \frac{E}{I_t}$$ **in AC circuits**

E
120 V
A
V
R
L
I

R and C Parallel Circuit Currents

The total current of an ac parallel circuit which consists only of R and C can be found by vectorially adding I_R (the resistive current) and I_C (the capacitive current). I_R is in phase with the circuit voltage E_t, while I_C leads the voltage by 90 degrees. To find the total current and its phase angle when I_R and I_C are known, you can draw either the waveforms of I_R and I_C or their vectors.

FINDING THE TOTAL CURRENT IN AN R AND C CIRCUIT. . .

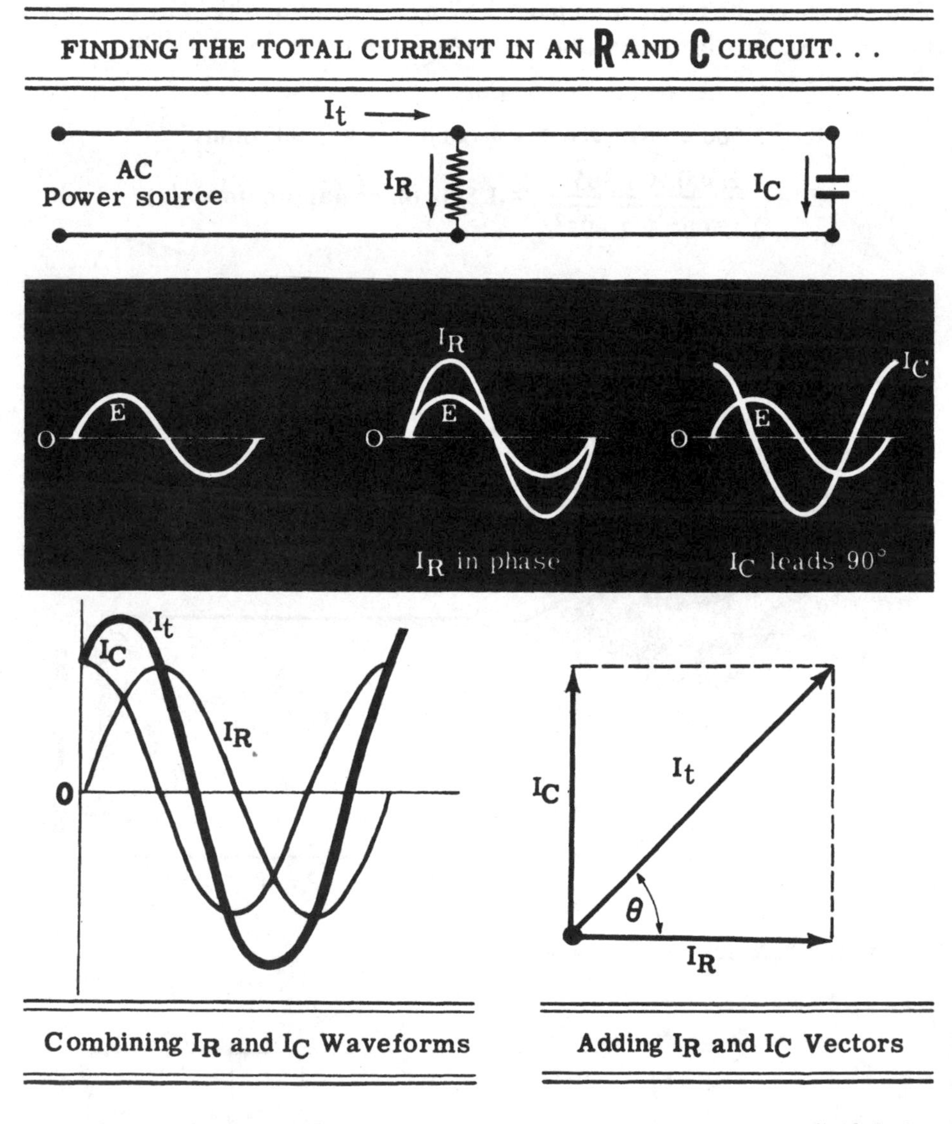

Combining I_R and I_C Waveforms

Adding I_R and I_C Vectors

As you did for R-L circuits, shown on page 4-54, you can find I_t by graphical addition or by calculation using the Pythagorean theorem with I_C in place of I_L.

R and C Parallel Circuit Impedance

The impedance in an R and C parallel circuit may be obtained by measuring the total current and voltage for the circuit and then finding Z by Ohm's law.

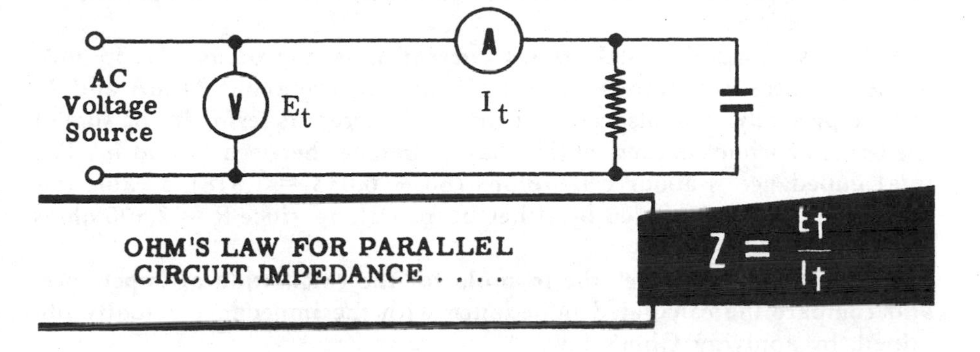

The impedance can also be calculated mathematically if the values R and C are known.

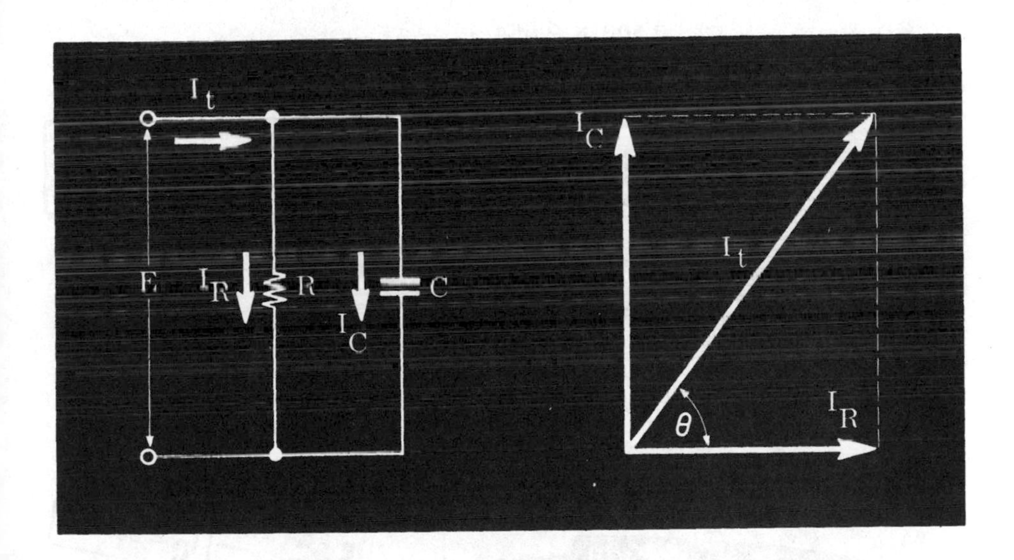

You will see that $I_R = E/R$, $I_C = E/X_C$ and $I_t = E/Z$ and since $I_t = \sqrt{I_R{}^2 + I_C{}^2}$, we can say that $E/Z = \sqrt{(E/R)^2 + (E/X_C)^2}$.

Now, if we eliminate E, as we did on page 4-57, we obtain $1/Z = \sqrt{(1/R)^2 + (1/X_C)^2}$ and $Z = RX_C/\sqrt{R^2 + X_C{}^2}$.

As always, the phase angle can be calculated as $\theta = \tan^{-1}(X_C/I_R)$.

Experiment/Application—R-C Parallel Circuit Current and Impedance

Repeat the experiment/application described on page 4-58 using a 1-μF capacitor instead of the 5-henry inductor. First the total circuit voltage and current (E_t and I_t) are measured, and then the branch currents I_R and I_C.

You see that the total circuit current I_t is approximately 33 mA, while the measured branch currents I_R and I_C are about 24 mA and 23 mA, respectively. You also see that the total current is less than the sum of the branch currents because of the phase difference between I_R and I_C. The total impedance is about 1,820 ohms (60 ÷ 0.033 = 1,818), a value less than the opposition offered by either branch alone, since R = 2,500 ohms and X_C = 2,654 ohms.

Again, you can check the formula for the calculation of impedance, and compare the calculated impedance with the impedance actually obtained, by applying Ohm's law:

$$R = 2{,}500; \qquad X_C = \frac{10^6}{2\pi \times 60 \times 1} = 2{,}654 \text{ ohms}$$

$$Z = \frac{2{,}500 \times 2{,}654}{\sqrt{2{,}500^2 + 2{,}654^2}} = 1{,}820 \text{ ohms}$$

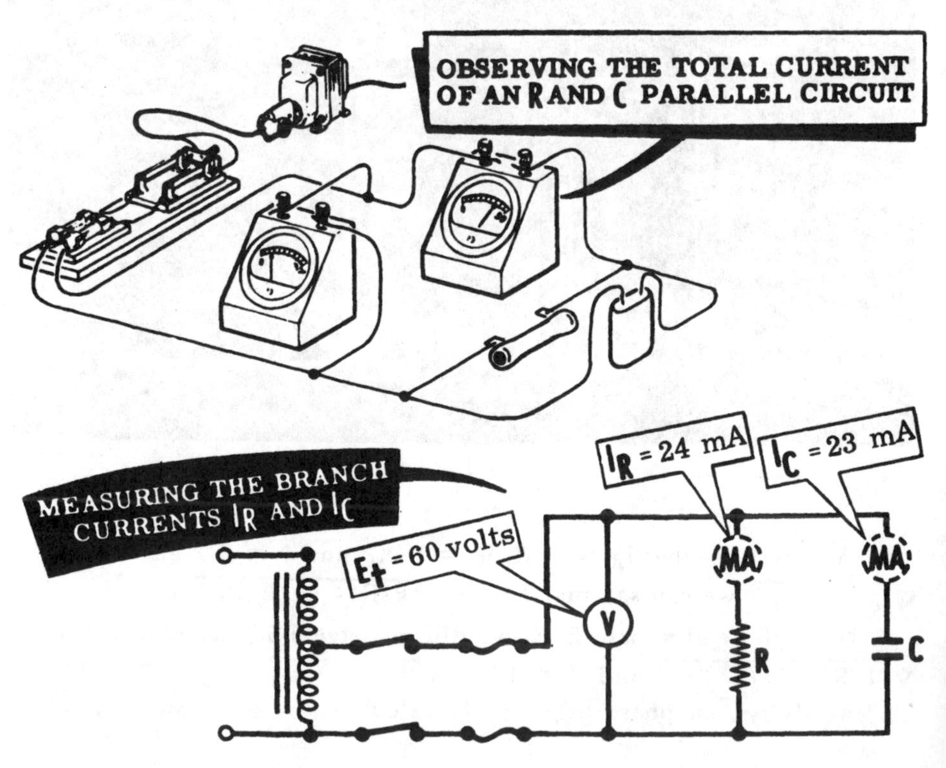

L and C Parallel Circuit Currents

When an ac parallel circuit consists only of L and C, the total current is equal to the *difference* between I_L and I_C, since they are *exactly opposite* in phase relationship. When the waveforms for I_L and I_C are drawn, you see that all the instantaneous values of I_L and I_C are of opposite polarity. If all the corresponding combined instantaneous values are plotted to form the waveform of I_t, the maximum value of this waveform is equal to the difference between I_L and I_C. For such circuits the total current can be found by subtracting the smaller current, I_L or I_C, from the larger.

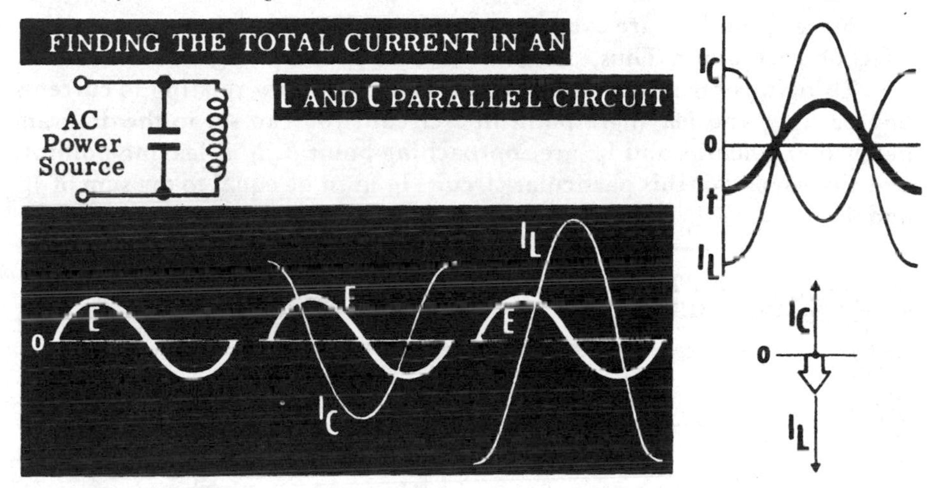

The relationships and paths of circuit currents in L and C circuits are shown below.

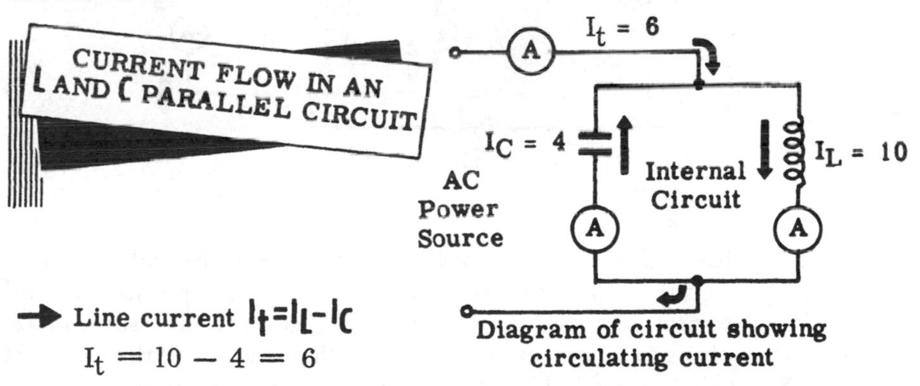

The parallel circuit can also be considered as consisting of an internal and an external circuit. Since the current flowing through the inductance is exactly opposite in polarity to that which is flowing through the capacitance at the same time, an internal circuit is formed. The amount of current flow around this internal circuit is equal to the smaller of the two currents I_L and I_C. The amount of current flowing through the external circuit (the voltage source) is equal to the difference between I_L and I_C.

L and C Parallel Circuit Currents and Impedance

The relationship between the various currents in a parallel circuit consisting of L and C is illustrated in the following example. A capacitor and an inductor are connected in parallel across a 60-Hz, 150-volt source so that $X_L = 50$ ohms and $X_C = 75$ ohms. The currents in the circuit are

$$I_L = \frac{E}{X_L} = \frac{150}{50} = 3\text{ A} \qquad I_C = \frac{E}{X_C} = \frac{150}{75} = 2\text{ A}$$

Since I_L and I_C are exactly opposite in phase, they have a cancelling effect on each other. Thus, the total current $I_t = I_L - I_C = 3 - 2 = 1$ A.

Using this phase relationship and Kirchhoff's law relating to currents approaching and leaving a point in a circuit, you can see in the diagram below that when I_C and I_t are approaching point A, I_L is leaving point A, and vice versa. For this particular circuit, I_L must be equal to the sum of I_t and I_C.

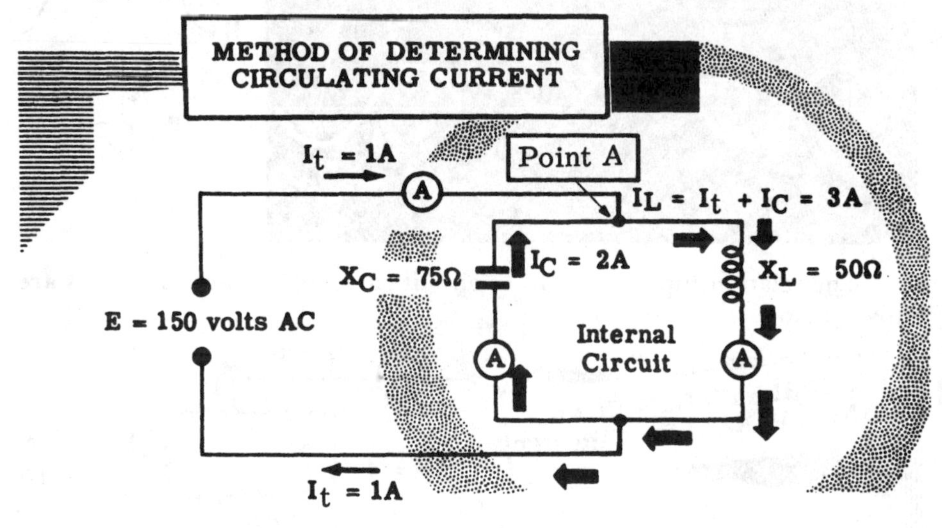

Since I_C flows through the capacitor and through the inductor, and then back through the capacitor, the result of the opposing phase of I_L and I_C is to form an internal circuit whose circulating current has a value equal to the smaller of the two currents I_L and I_C, in this case, I_C. If the values of X_L and X_C were reversed, I_L would be the circulating current.

The impedance of the circuit may be found by measuring the total current through the circuit and applied voltage, then by applying Ohm's law, or by calculation, if L and C are known. Since $I_t = I_L - I_C$,

$$\frac{E}{Z} = \frac{E}{X_L} - \frac{E}{X_C}$$

Hence,

$$\frac{1}{Z} = \frac{1}{X_L} - \frac{1}{X_C}$$

Experiment/Application—L-C Parallel Circuit Currents and Impedance

To observe the opposite effects of L and C in a parallel circuit, the 2,500-ohm resistor in the circuit of the experiment/application on page 4-62 is replaced by the 5-henry filter choke forming an L and C parallel circuit. Repeat each step of the experiment, first measuring the total circuit current, then that of each branch.

You see that the total circuit current is about 9 mA, while I_L is about 32 mA and I_C is about 23 mA. Thus, the total current is not only less than that of either branch but is actually the difference between I_L and I_C.

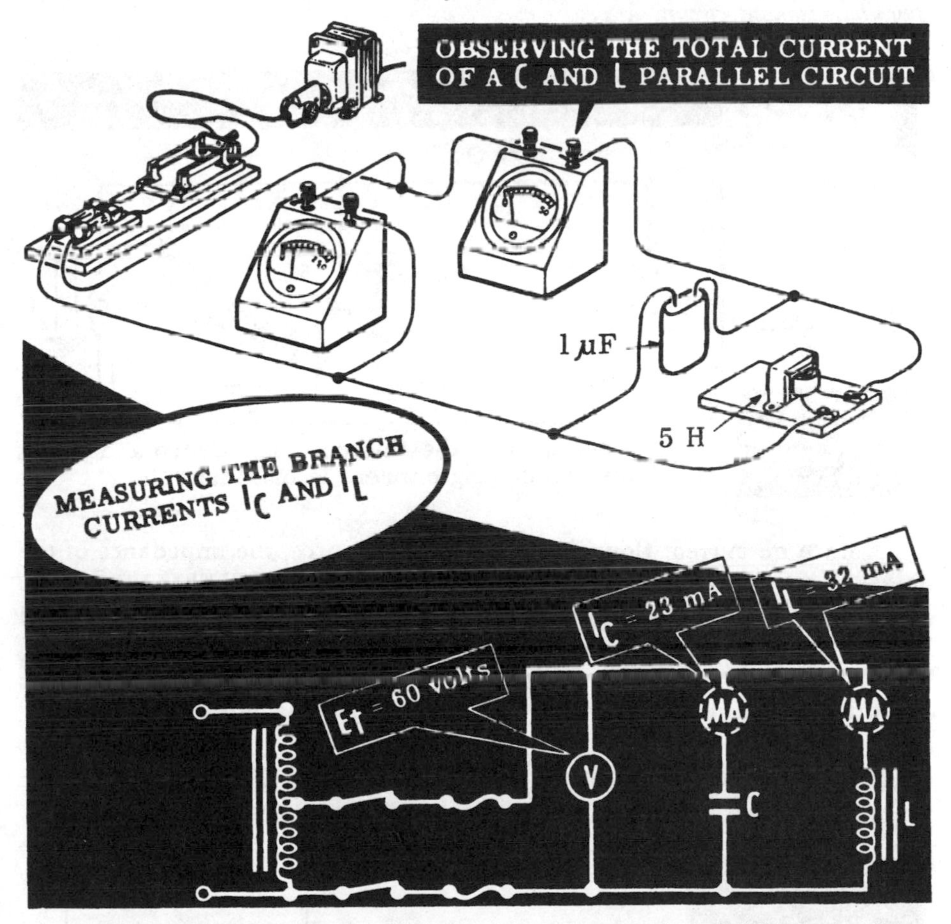

The total circuit impedance of the L and C circuit is 6,667 ohms (60 ÷ 0.009 ≈ 6,667), a value greater than the opposition of either the L or C branch of the circuit (X_L = 1,885 ohms, X_C = 2,650 ohms). Notice that when L and C are both present in a parallel circuit, the impedance increases, an effect *opposite* to that of a series circuit where combining L and C results in a lower impedance. As you will learn in the next pages, at parallel resonance, the line current is zero for a pure L-C parallel circuit.

Parallel Circuit Resonance

In an L and C parallel circuit containing equal X_L and X_C, the external circuit current is equal to that flowing through any parallel resistance. If the circuit contains no resistance, the external current is zero. However, within a theoretical circuit consisting only of L and C with $X_L = X_C$, a large current called the *circulating current* will flow, using no current from the power supply. This happens because the corresponding instantaneous values of the currents I_L and I_C are always in opposite directions, and since the line current is the difference between I_L and I_C, if these values are equal, no external circuit current will flow. Such a circuit is called a *parallel resonant circuit.*

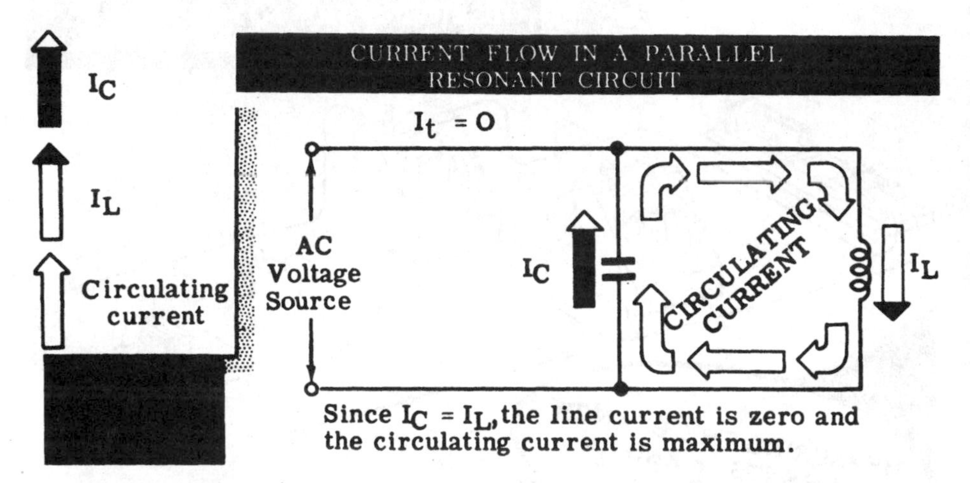

Since no current flows from the voltage source, the impedance of the ideal parallel resonant circuit is infinite. You will remember that in the case of the series circuit, the impedance at resonance was a minimum (and would be zero in a theoretical circuit).

Ohm's law for ac when applied to a parallel resonant circuit can be used to determine the value of the internal circulating current.

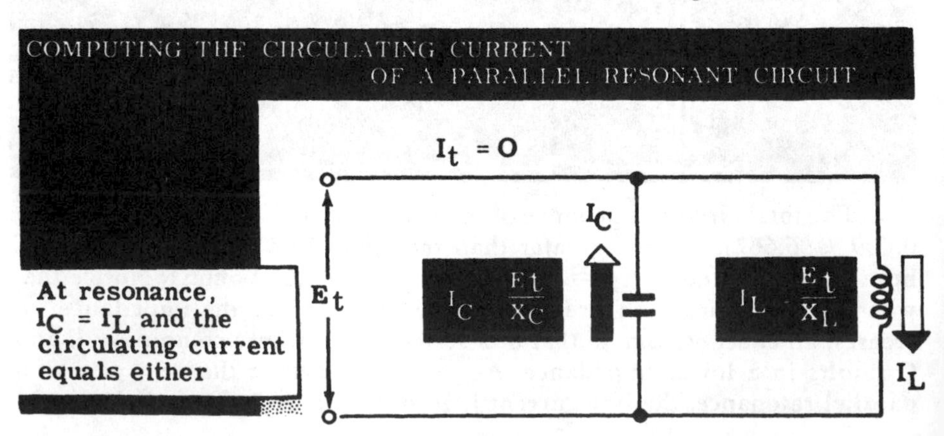

Parallel Circuit Resonance (continued)

Since the condition for parallel resonance is X_L being equal to X_C, the same as for series resonance, the same formula may be used to calculate parallel resonant conditions: $f_r = 1/2\pi\sqrt{LC}$.

As in the case of a series resonant circuit, if either the frequency, inductive reactance, or capacitive reactance of a circuit is varied and the two other values kept constant, the circuit current variation forms a resonance curve. However, the parallel resonance curve is the opposite of a series resonance curve. The series resonance current *increases* to a maximum at resonance, then *decreases* as resonance is passed, while the parallel resonance current *decreases* to a minimum at resonance, then *increases* as resonance is passed.

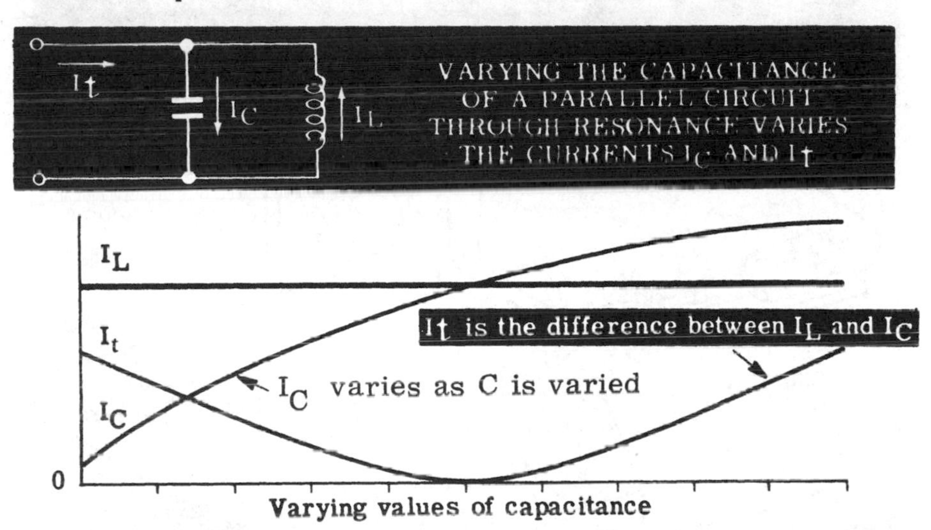

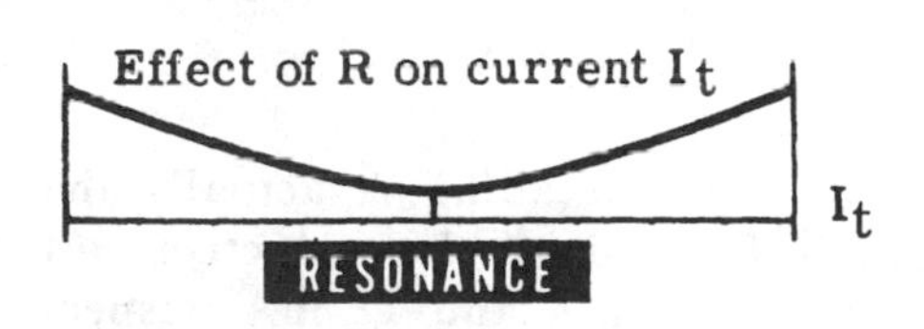

For a circuit of pure L and C, the curve would be as shown above. However, all actual capacitors and inductors have some resistance which prevents the current from becoming zero.

A comparison of circuit factors at resonance for series and parallel circuits, made in chart form, is shown below.

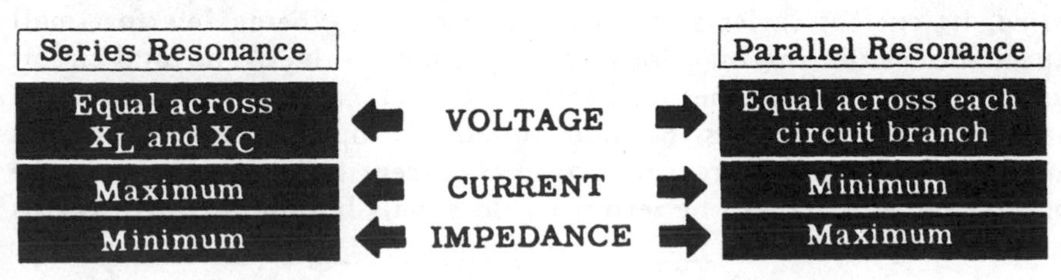

Series Resonance		Parallel Resonance
Equal across X_L and X_C	VOLTAGE	Equal across each circuit branch
Maximum	CURRENT	Minimum
Minimum	IMPEDANCE	Maximum

Experiment/Application—Parallel Circuit Resonance

To see the effect of parallel resonance, you could connect a 0.5-μF capacitor and a 5-henry inductor in parallel to form an L and C parallel circuit. A 0–50 mA ac milliammeter and a 0–250 volt ac voltmeter are connected to measure circuit current and voltage. This circuit is connected to the ac line through a switch, fuses, and step-down transformer. When the switch is closed, you observe that the current through the circuit is about 21 mA and that the voltage is approximately 60 volts.

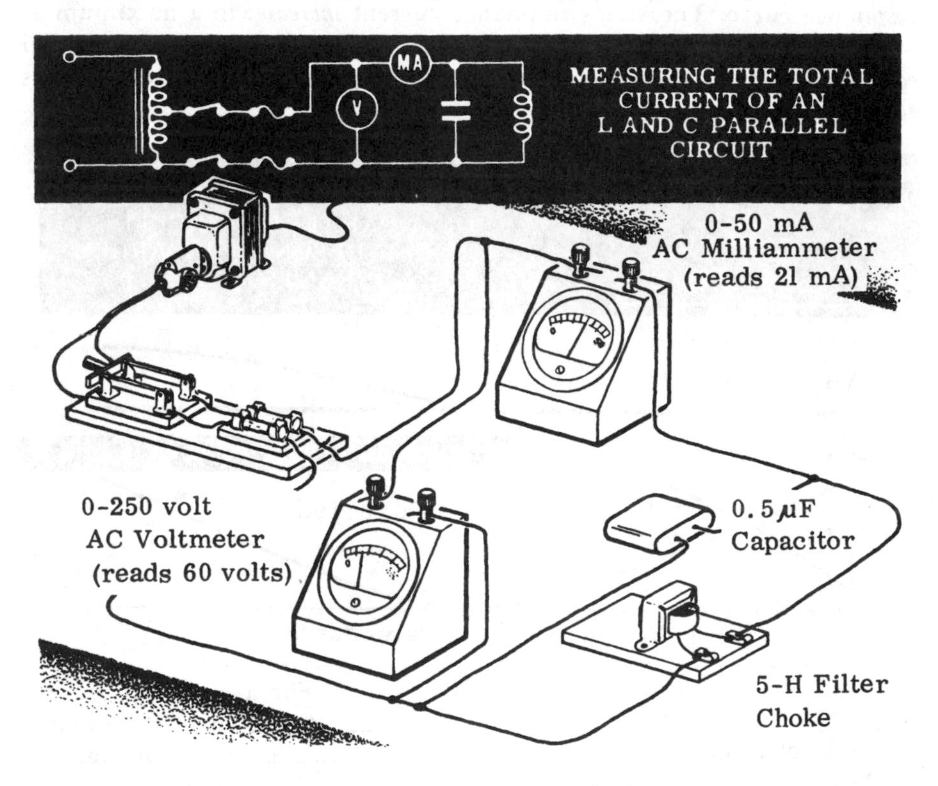

The total current indicated by the meter reading is actually the difference between the currents I_L and I_C through the inductive and capacitive branches of the parallel circuit (32 mA and 11 mA, respectively). The circuit voltage remains constant in parallel circuits so that, if a fixed value of inductance is used as one branch of the circuit, the current in that branch remains constant. If the capacitance of the other branch is varied, its current varies as the capacitance varies, being low for small capacitance values and high for large capacitance values. The total circuit current is the difference between the two branch currents; it is zero when the two branch currents become equal. If you increase the circuit capacitance, the total current will drop as current I_C increases toward the constant value of I_L, will be zero when I_C equals I_L, and then will rise as I_C becomes greater than I_L.

Experiment/Application—Parallel Circuit Resonance (continued)

Now vary the circuit capacitance in steps of 0.5 μF from 0.5 μF to 3.5 μF. Observe that the current decreases from approximately 21 mA to a minimum value less than 10 mA, then rises to a value beyond the range of the milliammeter. The current at resonance does not reach zero because the circuit branches are not purely capacitive and inductive and cannot be so in a practical parallel circuit. You will observe that the voltage does not change across either the branches or the total parallel circuit as the capacitance value is changed.

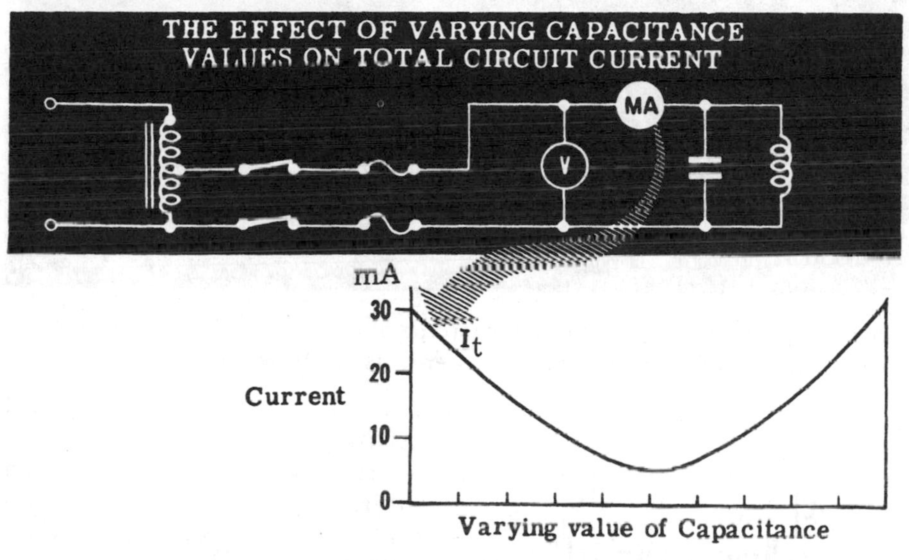

After the value of capacitance has been varied through the complete range of values, the value which indicates resonance—minimum current flow—is used to show that the circulating current exceeds the total current at resonance. Measure again the total current of the parallel resonant circuit; then connect the milliammeter to measure only the current in the inductive branch. You will see that the total current is less than 10 mA, yet the circulating current ($I_L = I_C$) is approximately 32 mA.

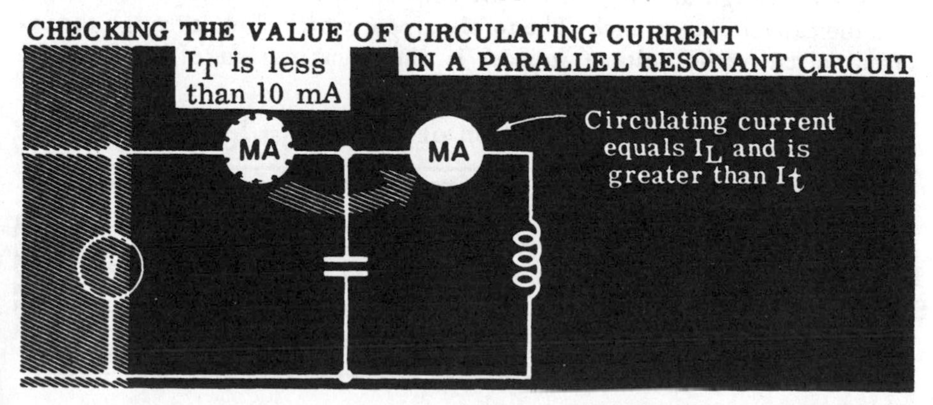

R, L, and C Parallel Circuit Currents

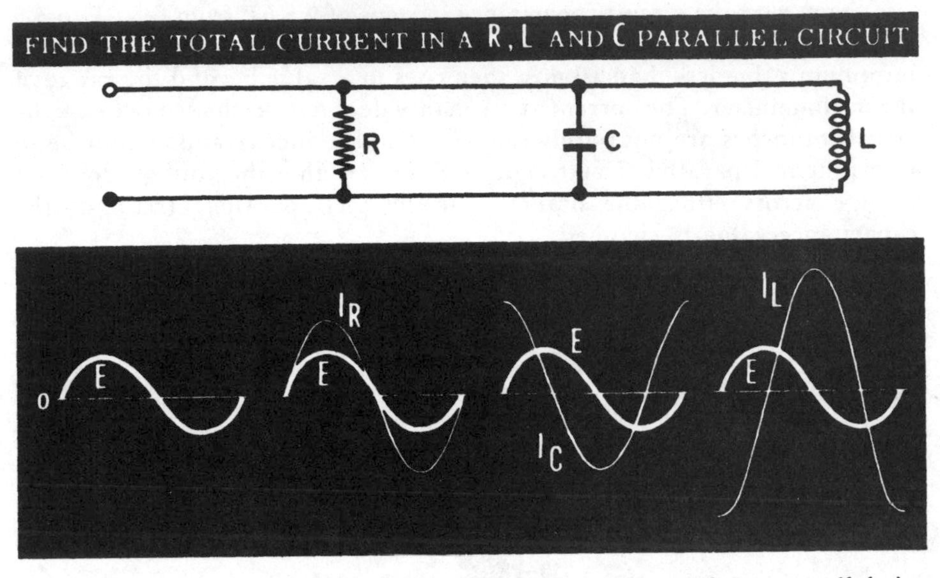

To combine the three branch currents of an R, L, and C ac parallel circuit by means of vectors takes two steps, as outlined below:

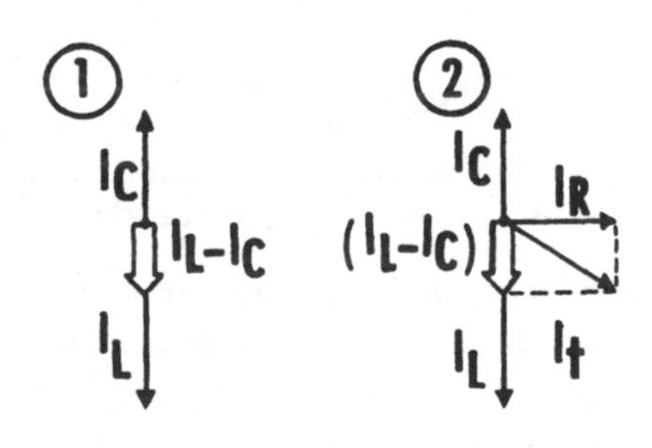

1. Currents I_L and I_C are combined by using vectors. (Both the value, which may be obtained by direct subtraction, and the phase angle of this combined current are required).
2. The combined value of I_L and I_C is then combined with I_R to obtain the total current.

In an R, L, and C parallel circuit—as in the parallel L and C circuit—a circulating current equal to the smaller of the two currents I_L and I_C flows through an internal circuit consisting of the inductance branch and the capacitance branch. The total current which flows through the external circuit (the voltage source) is the combination of I_R and the difference between currents I_L and I_C.

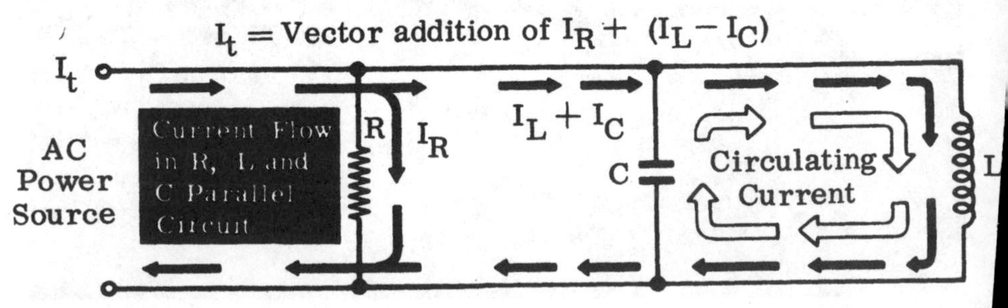

R, L, and C Parallel Circuit Impedance

The total impedance of an R, L, and C parallel ac circuit may be found by first measuring the current through the circuit and the applied voltage and then using Ohm's law.

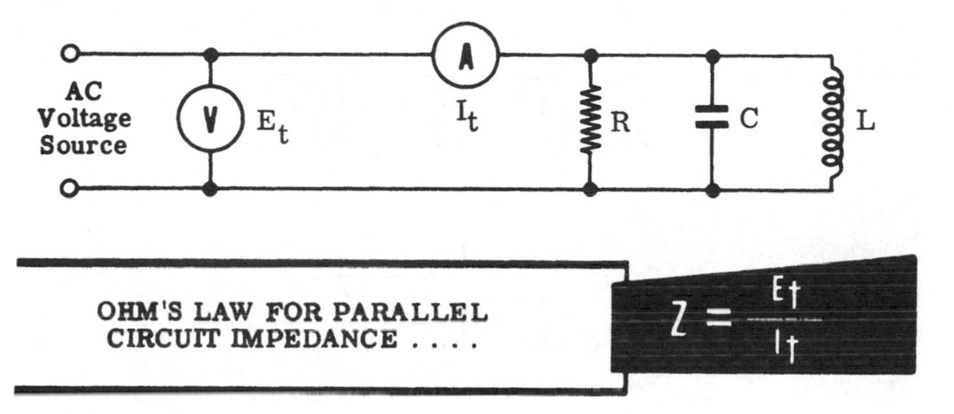

The impedance may be calculated mathematically (with the aid of a hand calculator) if the values of R, L, and C are known.

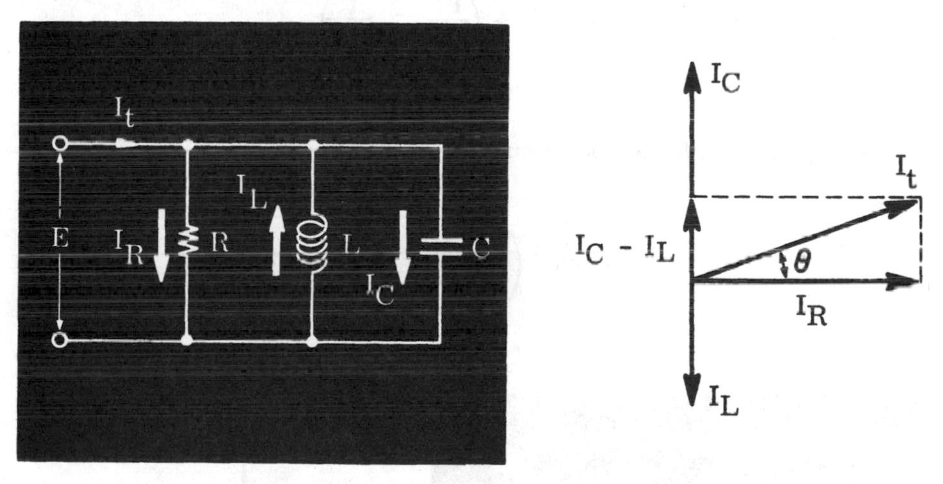

$$I_R = \frac{E}{R},\ I_L = \frac{E}{X_L},\ I_C = \frac{E}{X_C},\ \text{and } I_t = \frac{E}{Z}$$

$I_t = \sqrt{I_R^2 + (I_C - I_L)^2}$ in the example illustrated above, where the capacitive reactance (X_C) is smaller than the inductive reactance (X_L). But $I_t = \sqrt{I_R^2 + (I_L - I_C)^2}$ if the inductive reactance is smaller than the capacitive reactance.

The total current will lead the applied voltage if X_C is smaller than X_L and will lag the applied voltage if X_L is smaller than X_C, using the procedures of page 4-57.

R, L, and C Parallel Circuit Impedance (continued)

$$\frac{E}{Z} = \sqrt{\left(\frac{E}{R}\right)^2 + \left(\frac{E}{X_C} - \frac{E}{X_L}\right)^2}$$

and

$$\frac{1}{Z} = \sqrt{\left(\frac{1}{R}\right)^2 + \left(\frac{1}{X_C} - \frac{1}{X_L}\right)^2}$$

So

$$Z = \frac{1}{\sqrt{\left(\frac{1}{R}\right)^2 + \left(\frac{1}{X_C} - \frac{1}{X_L}\right)^2}}$$

or

$$\frac{1}{\sqrt{\left(\frac{1}{R}\right)^2 + \left(\frac{1}{X_L} - \frac{1}{X_C}\right)^2}}$$

The phase angle can be calculated as before as

$$\theta = \tan^{-1}\left(\frac{I_C - I_L}{I_R}\right)$$

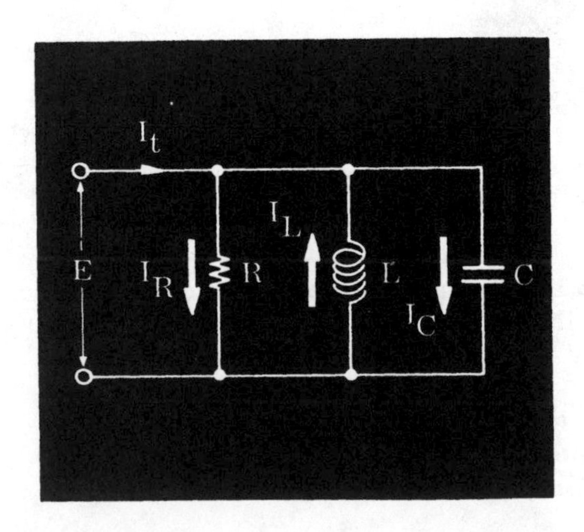

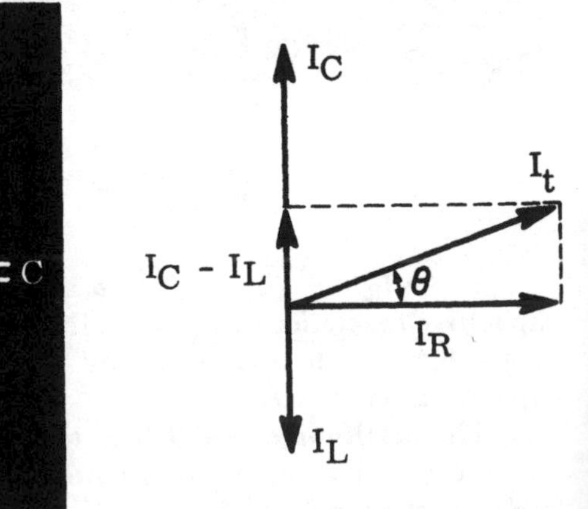

Experiment/Application—R, L, and C Parallel Circuit Currents and Impedance

By connecting a 2,500-ohm, 20-watt resistor in parallel with a 5-henry inductor and a 1-μF capacitor, you can make an R, L, and C parallel circuit. To check the various currents and find the total circuit impedance, measure the total circuit current and then the individual currents through the resistor, inductor, and capacitor in turn.

You will see that I_R is 24 mA, I_L is 31.8 mA, and I_C is 22.6 mA. Again you will see that the sum of the individual currents is much greater than the actual measured total current of 25.7 mA. The total circuit current is the sum of the resistive current I_R and the combined inductive and capacitive currents I_L and I_C added vectorially.

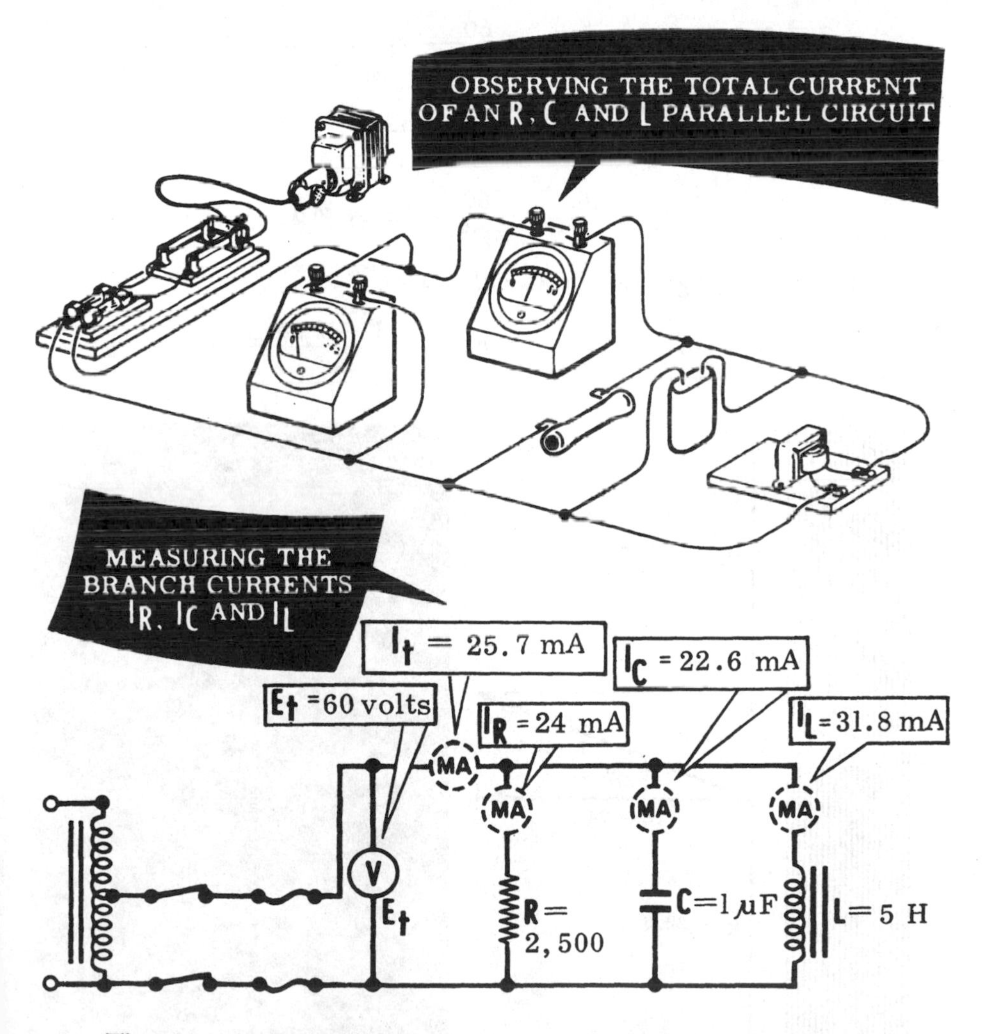

The total circuit impedance is about 2,335 ohms (60 ÷ 25.7 = 2,335).

Experiment/Application—R, L, and C Parallel Circuit Currents and Impedance (continued)

The impedance found by Ohm's law can be checked against the value found by substituting for R, X_L, and X_C in the formula on page 4-72:

$$Z = \frac{1}{\sqrt{\left(\frac{1}{R}\right)^2 + \left(\frac{1}{X_L} - \frac{1}{X_C}\right)^2}} = \frac{1}{\sqrt{\frac{1}{2{,}500^2} + \left(\frac{1}{1{,}885} - \frac{1}{2{,}650}\right)^2}}$$

$$= 2{,}335 \text{ ohms}$$

The circuit currents are:

$$I_L = \frac{E_t}{X_L} = \frac{60}{1{,}885} = 31.8 \text{ mA}$$

$$I_C = \frac{E_t}{X_C} = \frac{60}{2{,}650} = 22.6 \text{ mA}$$

$$I_R = \frac{E_t}{R} = \frac{60}{2{,}580} = 24 \text{ mA}$$

The phase angle can be calculated as:

$$\theta = \tan^{-1} \frac{I_L - I_C}{I_R} = \tan^{-1} \frac{31.8 - 22.6}{24} = 21.39^\circ$$

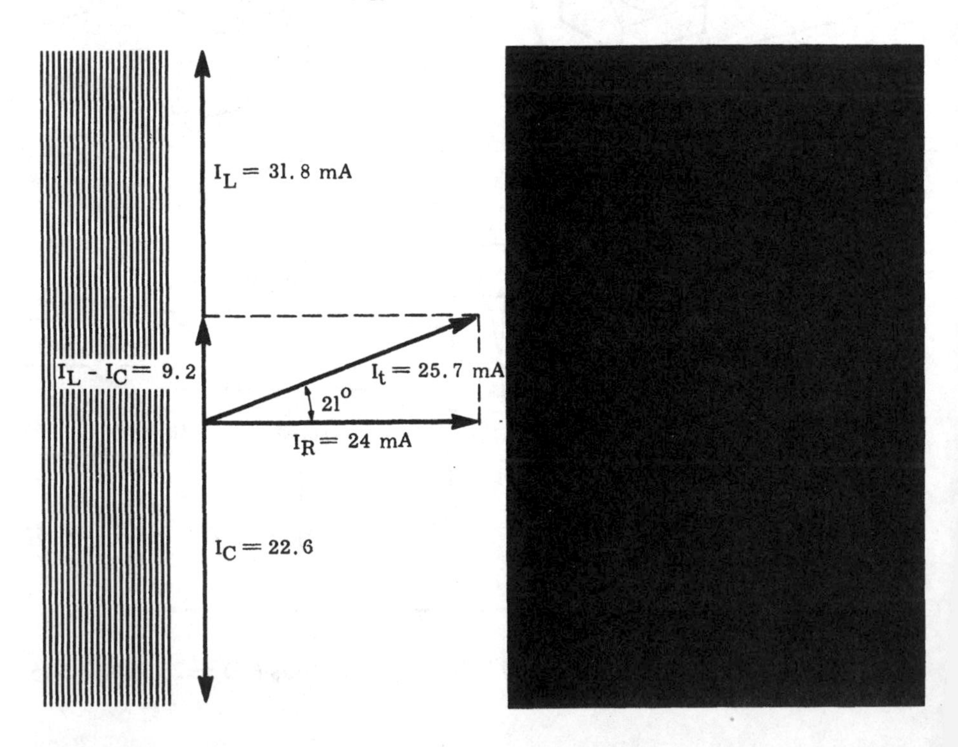

Power and Power Factor in Parallel Circuits

It would be wise at this point to review pages 4-44 to 4-46 on power and power factor in series circuits since the calculations are the *same* as for the parallel circuits we are now studying.

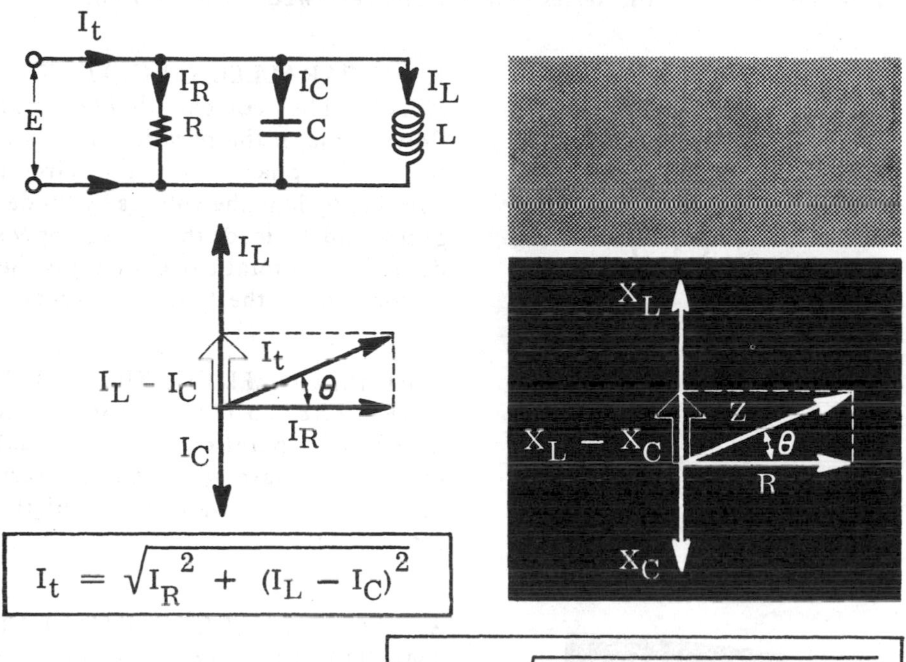

$$Z = 1/\sqrt{(1/R)^2 + (1/X_L - 1/X_C)^2}$$

$$\text{True power} = EI_R$$

$$\text{Apparent power} = EI_t$$

$$\text{Power factor} = \frac{\text{True power}}{\text{Apparent power}} = \frac{EI_R}{EI_t} = \frac{I_R}{I_t}$$

$$\text{Power factor} = \frac{R}{Z} = \frac{I_R}{I_t}$$

As before, the power factor is also defined as cos θ since

$$\cos\theta = \frac{R}{Z} \text{ or } \frac{I_R}{I_t}$$

$$\text{True power} = EI_t \cos\theta$$

As you can see, the relationships for power and power factor in parallel circuits are the *same* as those for series circuits.

Review of Current, Voltage, Impedance, Resonance, Power, Power Factor

Consider what you have found out so far about ac parallel circuits, and compare the effects of parallel connections of R, L, and C in ac circuits with those of the series connections reviewed on page 4-48.

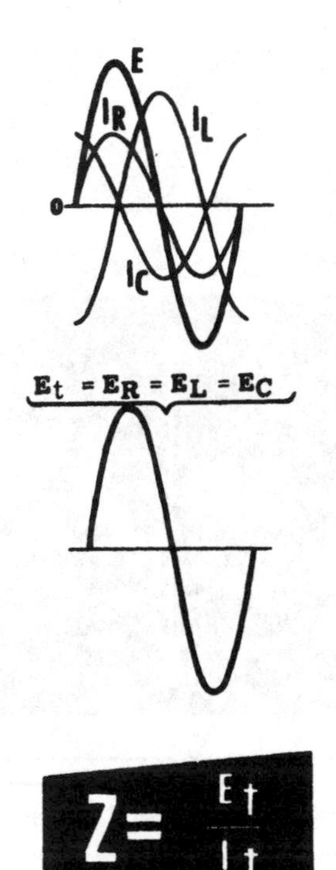

1. AC PARALLEL CIRCUIT CURRENT—The current divides and flows through the parallel branches. I_R is in phase with the circuit voltage, I_L lags the voltage by 90 degrees, and I_C leads the voltage by 90 degrees. The total current (I_t) is the vector sum of the branch currents.

2. AC PARALLEL CIRCUIT VOLTAGE—The voltage across each branch of a parallel circuit is equal to, and in phase with, that of every other branch as well as that of the total circuit.

3. AC PARALLEL CIRCUIT IMPEDANCE—The impedance of an ac parallel circuit is equal to the applied voltage divided by the total circuit current.

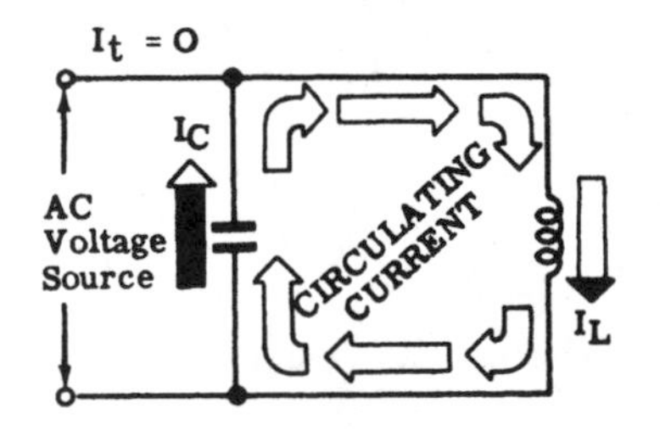

4. PARALLEL RESONANCE TOTAL CURRENT—The total current is minimum in a parallel resonant circuit. The circulating current is maximum in a parallel resonant circuit. At parallel resonance, the power factor is unity, and the phase angle is zero.

$$\text{True power} = I^2R = EI\cos\theta$$

$$\text{Apparent power} = EI$$

$$\text{Power factor} = \frac{\text{True power}}{\text{Apparent power}}$$

$$\text{Power factor} = \cos\theta = R/Z$$

5. POWER AND POWER FACTOR—As with all R, L, and C circuits, both true and apparent power are found in parallel circuits.

Self-Test—Review Questions

1. Draw a vector diagram to represent an R of 300 ohms, an X_L for a 1-henry inductor, and a 5-μF capacitor at 60 Hz.
2. Draw vector diagrams to solve the circuits shown below, finding both Z and θ (f = 60 Hz).

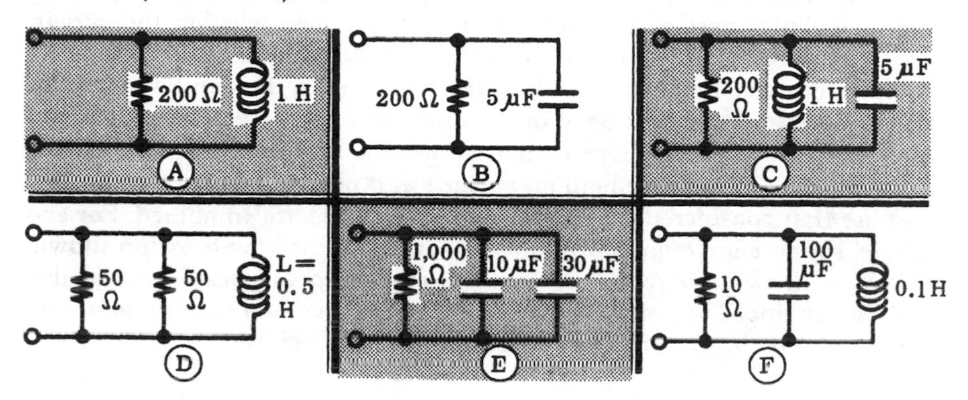

3. Solve the circuits above by calculation, and compare your answers.
4. If the input voltage is 120 volts, 60 Hz, calculate the circuit currents for each component and the total current in the six circuits of Question 2. Calculate the circuit currents where applicable.
5. For each circuit in Question 2, calculate true power, apparent power, and power factor.
6. Calculate the resonant frequency for circuits (e) and (f) of Question 2. How much current will flow at resonance? What is the phase? What capacitance must be added to circuit (c) to be resonant at 60 Hz? What inductance must be paralleled with the inductance of circuit (f) to make it resonant at 300 Hz?

Learning Objectives—Next Section

Overview—Now that you know about *series* and *parallel* ac circuits, you are ready to learn about *complex* ac circuits that have *both* series and parallel parts. You will find, as is true of dc complex circuits, that the learning trick is to reduce (simplify) the circuit by applying what you know about series and parallel circuit elements.

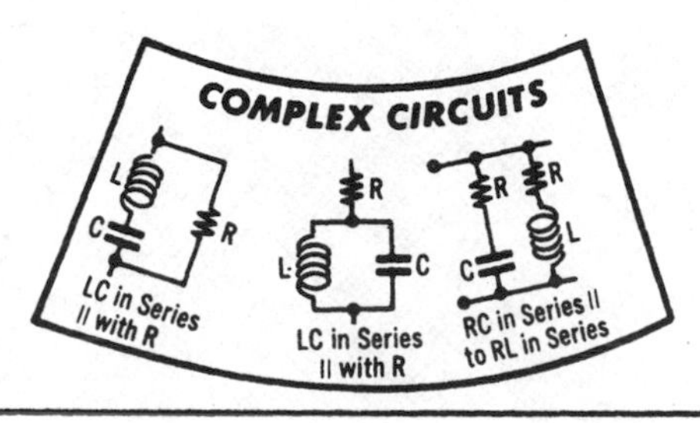

Series-Parallel Circuits

Many ac circuits are neither series circuits nor parallel circuits, but a *combination* of these two basic circuits. Such circuits are called *series-parallel* or *complex circuits.*

The values and phase relationships of the voltages and currents for each particular part of a complex circuit depend on whether the part is series or parallel. Any number of series-parallel combinations can form complex circuits. Regardless of the circuit variations, the step-by-step solution is similar to the solution of dc complex circuits that you learned about in Volume 2. The difference is that we must solve the circuits by vectors since we have to consider both magnitude and phase. The parts of the circuit are first considered separately; then the results are combined. For example, suppose a circuit consists of the series-parallel combination shown below, with two separate series circuits connected in parallel across the 240-volt ac line. The vector solution used to find the total current, total impedance, and the circuit phase angle is outlined below.

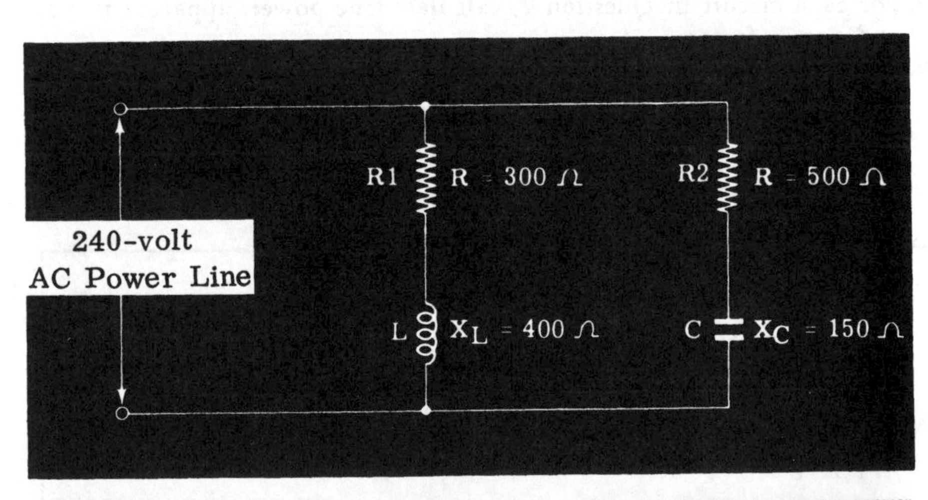

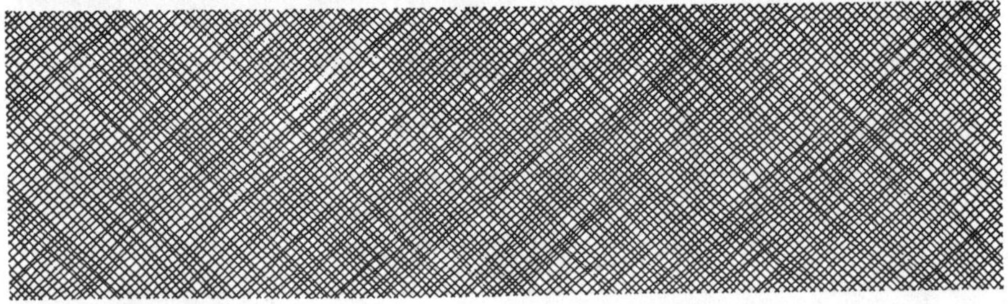

Series-Parallel Circuits (continued)

To find the values of branch currents I_1 and I_2, the impedance of each branch is found separately by using vectors. The current values are then determined by applying Ohm's law to the branches separately.

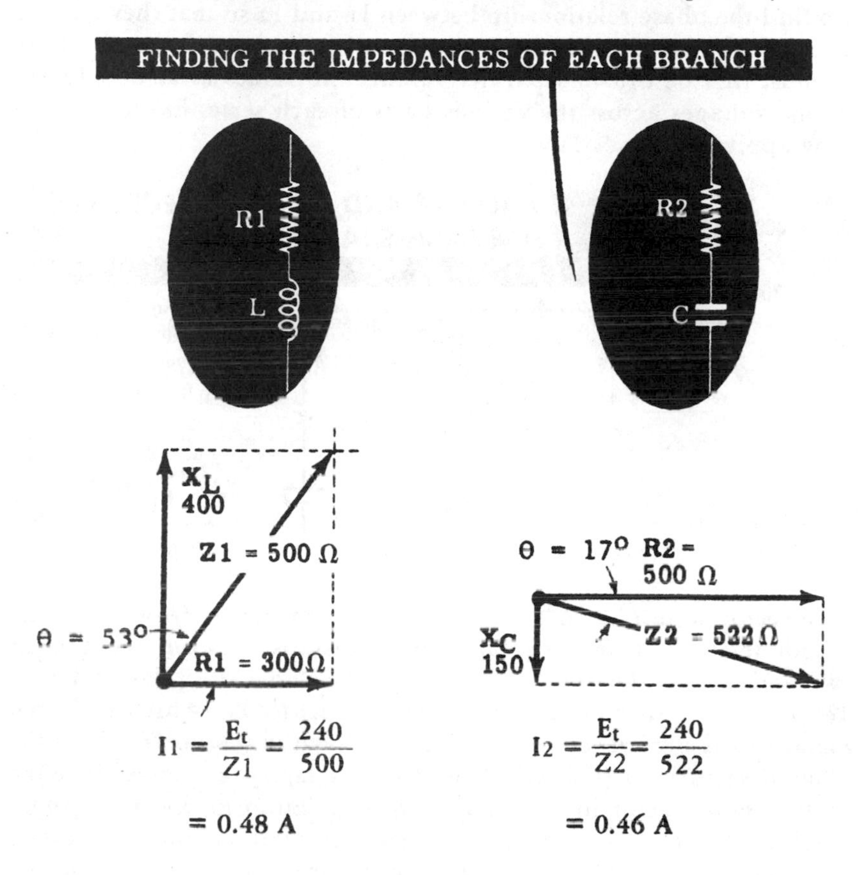

You can, of course, calculate these circuits analytically.

$$Z1 = \sqrt{300^2 + 400^2} = 500 \text{ ohms}$$

$$Z2 = \sqrt{500^2 + 150^2} = 522 \text{ ohms}$$

The phase angles are:

$$\theta = \tan^{-1}\frac{X_L}{R} = \tan^{-1}\frac{400}{300} = 53.1^\circ$$

$$\theta = \tan^{-1}\frac{X_C}{R} = \tan^{-1}\frac{150}{500} = 16.7^\circ$$

Series-Parallel Circuits—Vector Graphic Solution

Although you know branch currents I_1 and I_2, the total current I_t *cannot* be found by adding I_1 and I_2 *directly*. Since they are *out of phase*, the total current must be found by *vector addition*.

To find the phase relationship between I_1 and I_2 so that they may be added by using vectors, the voltage and current vectors for each series branch must first be drawn separately. (Since the values of I_1 and I_2 are known, the voltages across the various parts of each series branch can be found by applying Ohm's law.)

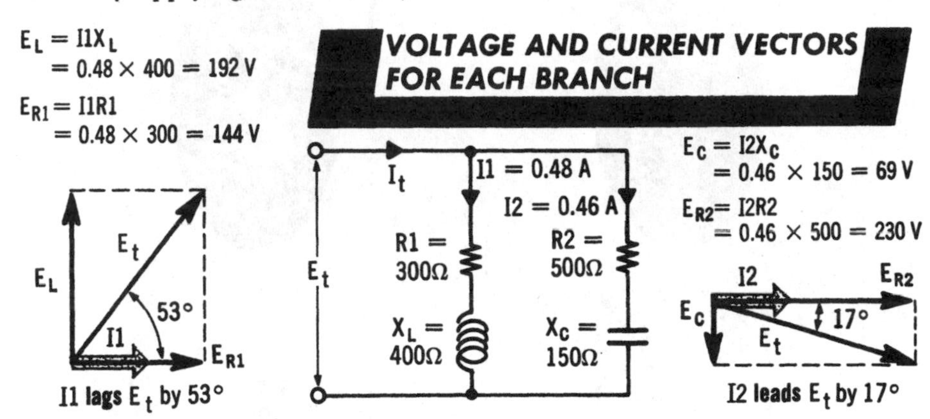

The vector solutions for each separate branch, when drawn to scale, show both the values and the phase relationships between the branch currents and the total circuit voltage, E_t. To show the phase relation between I_1 and I_2, they are redrawn with respect to E_t, which is drawn horizontally as the reference vector. Draw I_1 down in relation to E_t at the angle found vectorially (that is, 53 degrees). I_1 lags the voltage since the branch has the inductor in it. Draw I_2 up in relation to E_t at the angle found vectorially (that is, 17 degrees). I_2 leads the voltage since the branch has the capacitor in it. Complete the parallelogram by drawing a dotted line from the end of each vector parallel to the other. From the reference point, draw a line to where the two dotted lines cross. This vector represents the total current of the circuit. Measure with a protractor the angle between I_t and E_t, and this will be the phase angle of the circuit.

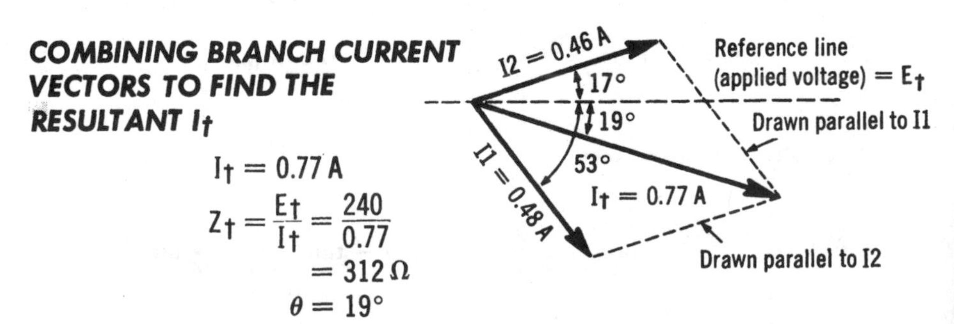

Series-Parallel Circuits—Solution by Calculation

Alternatively, the branch currents, total current, and total impedance of the circuit on page 4-80 can be calculated mathematically. Considering each branch separately,

$$Z1 = \sqrt{R1^2 + X_L^2} \quad \text{and} \quad Z2 = \sqrt{R2^2 + X_C^2}$$

$$Z1 = \sqrt{300^2 + 400^2} = 500 \text{ ohms} \quad \text{and}$$

$$Z2 = \sqrt{500^2 + 150^2} = 522 \text{ ohms}$$

The current through Z1, the inductive branch, lags behind the applied voltage by a phase angle $\theta 1$. $\theta 1 = \tan^{-1} X_L / R1 = 53$ degrees.

The current through the capacitance branch, Z2, leads the applied voltage by a phase angle $\theta 2$. $\theta 2 = \tan^{-1} X_C / R2 = 17$ degrees.

I_1, the current through the inductive branch, is E/Z1, or 240/500 = 0.48 ampere.

I_2, the current through the capacitive branch, is E/Z2, or 240/522 = 0.46 ampere.

As you know, both I_1 and I_2 can be resolved into two components—one component in phase with the applied voltage and the other 90 degrees out of phase with it.

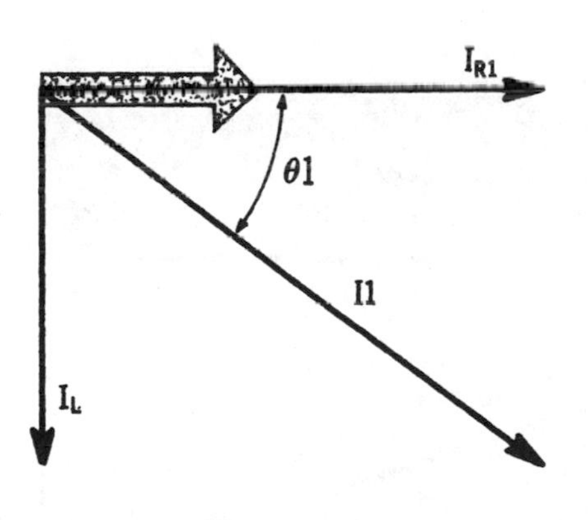

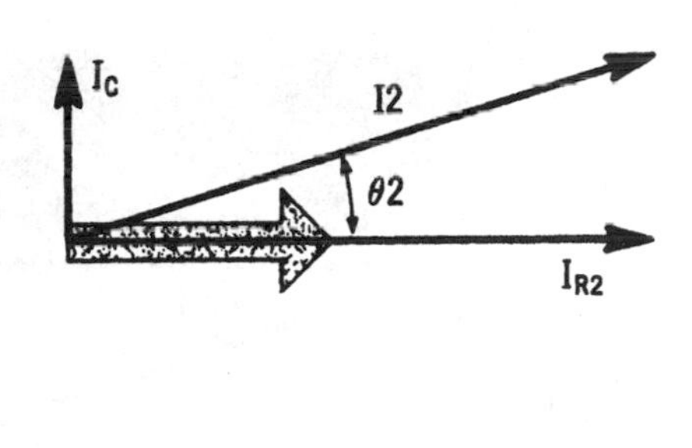

$I_{R1} = I1 \cos \theta 1 = 0.48 \cos 53° = 0.29$ A

$I_L = I1 \sin \theta 1$ (lagging) $= 0.48 \sin 53° = 0.38$ A

$I_{R2} = I2 \cos \theta 2 = 0.46 \cos 17° = 0.44$ A

$I_C = I2 \sin \theta 2$ (leading) $= 0.46 \sin 17° = 0.13$ A

Series-Parallel Circuits—Solution by Calculation (continued)

The total current is the vector sum of the two branch currents; it can be found by adding the in-phase components ($I_{in\text{-}phase}$) and the out-of-phase components ($I_{out\text{-}of\text{-}phase}$) and then combining them vectorially:

$$I_{in\text{-}phase} = I_{R1} + I_{R2} = 0.29\ A + 0.44\ A = 0.73\ A$$
$$I_{out\text{-}of\text{-}phase} = I_L - I_C \quad \text{or} \quad I_C - I_L = 0.38A - 0.13A = 0.25A\ \text{(lagging)}$$

Although the out-of-phase components are added, remember that they are in opposition; their algebraic sum is therefore the difference between them, and the direction of phase (leading or lagging) is the same as that of the larger one.

$$\text{Total current, } I_t = \sqrt{(I_{R1} + I_{R2})^2 + (I_L - I_C)^2}$$
$$= \sqrt{(0.73)^2 + (0.25)^2}$$
$$= 0.77\ A$$
$$\text{Phase angle } \theta = \tan^{-1}\frac{I_{out\text{-}of\text{-}phase}}{I_{in\text{-}phase}}$$
$$= \tan^{-1}\frac{0.25}{0.73} = 19^\circ$$
$$\text{Total impedance, } Z_t = E/I_t = 240/0.77 = 312\ \text{ohms}$$

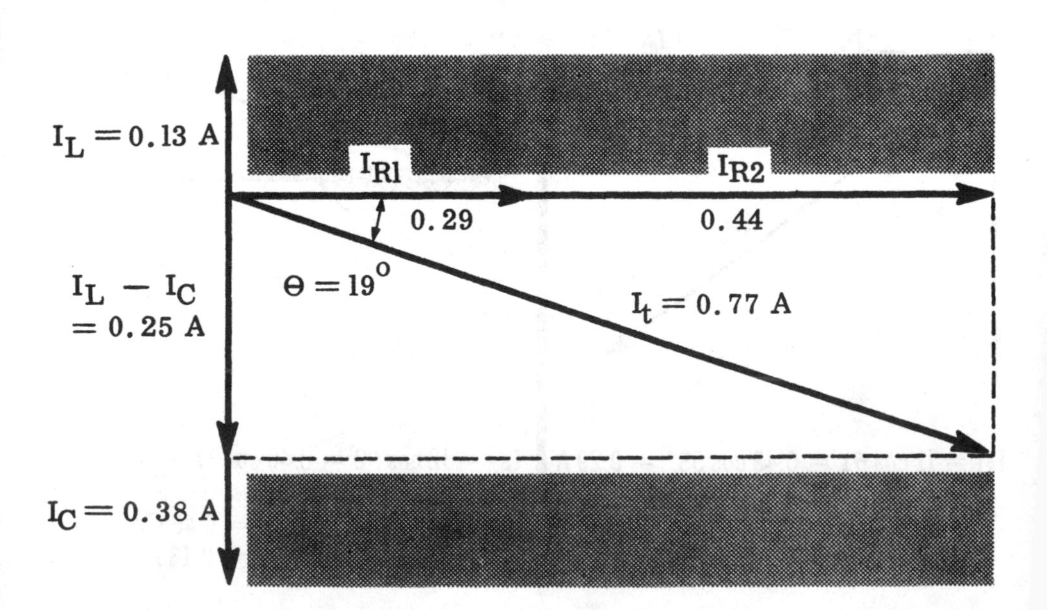

More Complex Series-Parallel Circuits—Vector Graphic Solution

Series-parallel circuits may be even more complex than the one just illustrated. For example, suppose that the series-parallel circuit under discussion is connected in series with an inductance and a resistance.

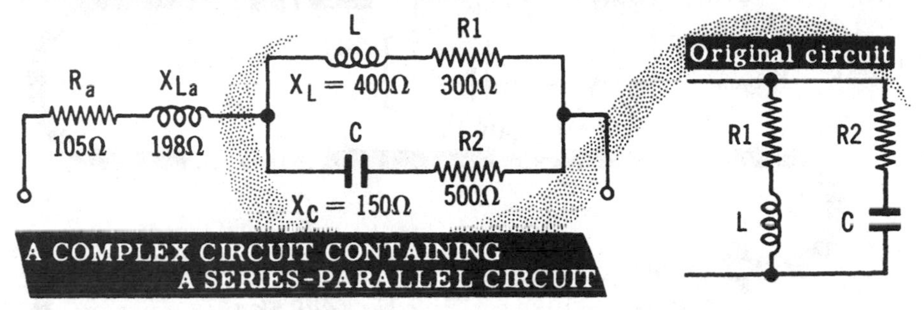

To calculate the impedance of the circuit, first resolve that part of the circuit considered earlier (see pages 4-78 through 4-81). In order to do this, you must assume an arbitrary voltage appearing across it.

This portion of the circuit, therefore, behaves as an inductor L_e and resistor R_e in series, having an effective impedance of 312 ohms and with the voltage leading the current by 19 degrees. Now draw the impedance sketch as shown below.

Finding X_{Le} and R_e

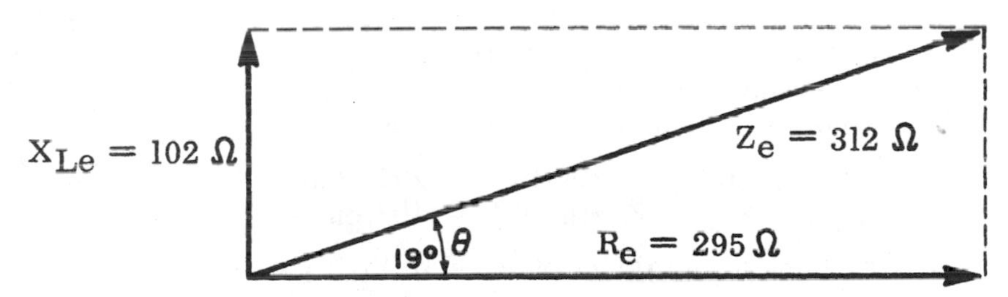

Z_e can be resolved into R_e and X_{Le} by vectors with values of 295 ohms and 102 ohms, respectively. These values are combined with those of R_a and X_{La}, giving resultants of 400 ohms (295 + 105) for R_t and 300 ohms (102 + 198) for X_{Lt}. Now draw the final impedance sketch.

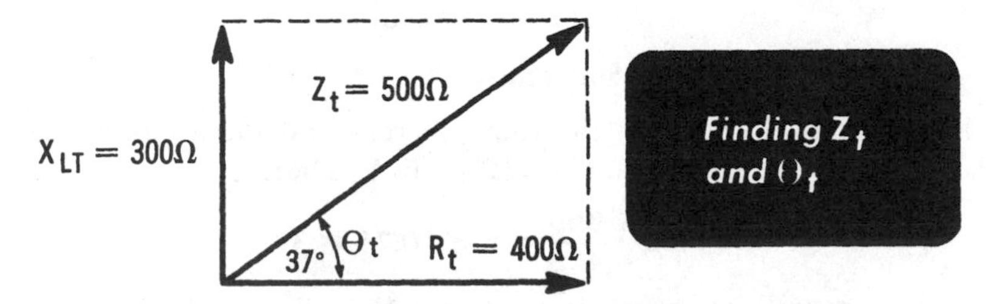

Thus, Z_t is 500 ohms, and the phase angle is 37 degrees, with the current lagging the applied voltage since the circuit is inductive.

More Complex Series-Parallel Circuits—Solution by Calculation

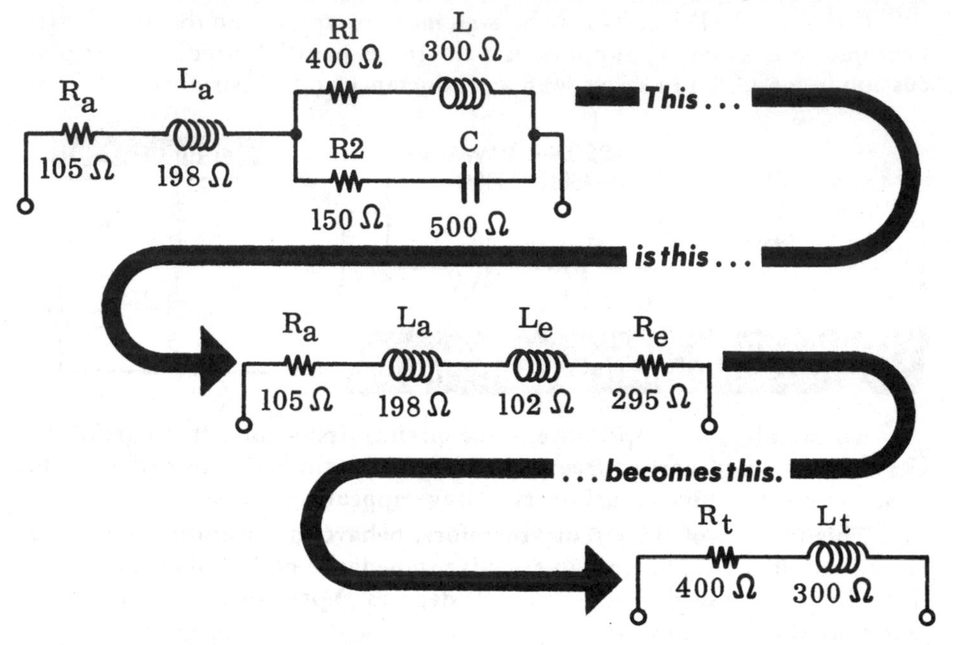

The same problem could be solved mathematically by using the method given on page 4-81.

The answers obtained for the parallel section would be $I_e = 0.77$ ampere, $Z_e = 312$ ohms, and $\theta_e = 19$ degrees—exactly the same values found by the vector method.

The values of R_e and X_{Le} are given by

$$R_e = Z_e \cos 19° = 295 \text{ ohms}$$
$$X_{Le} = Z_e \sin 19° = 102 \text{ ohms}$$

Adding these to the values of R_a and X_{La}, you get

$$R_t = 400 \text{ ohms} \quad \text{and} \quad X_{Lt} = 300 \text{ ohms}$$

Thus

$$Z_t = \sqrt{R_t^{\ 2} + X_{Lt}^{\ 2}} \text{ ohms}$$
$$= \sqrt{400^2 + 300^2} \text{ ohms}$$
$$= 500 \text{ ohms}$$

The phase angle between the total current and the applied voltage (θ_t) is the angle whose tangent is (X_{Lt} / R_t). Therefore,

$$\theta_t = \tan^{-1} \frac{300}{400} = \tan^{-1} 0.75 = 37°$$

The current will, of course, lag the voltage by this angle since the circuit is inductive.

Experiment/Application—Graphic and Calculated Solution in Complex Circuits

In this experiment/application, you will be using both of the methods described on pages 4-78 through 4-81 to find—in a series-parallel circuit containing resistance, inductance, and capacitance—the total circuit current, the branch currents, and the impedance. Your graphic and calculated results will then be checked with *actual* voltage and current measurements, and you will see how those values compare.

Because pure inductances and pure capacitances are only *theoretical* possibilities, there will be a difference between the *measured* and the *calculated* results, but the measured results will show that the calculations are accurate enough for practical use in electric circuits.

Connect a 500-ohm resistor, a 1,000-ohm resistor, a 1-μF capacitor, and a 5-henry inductor to form the complex circuit shown below. Because the inductor has a dc resistance of approximately 50 ohms, the total resistance of the R and L branch of the circuit is 1,050 ohms, and R2 is shown as a 1,050-ohm resistor rather than as a 1,000-ohm resistor. The transformer is a 120-volt to 60-volt step-down transformer, the circuit being connected across the secondary.

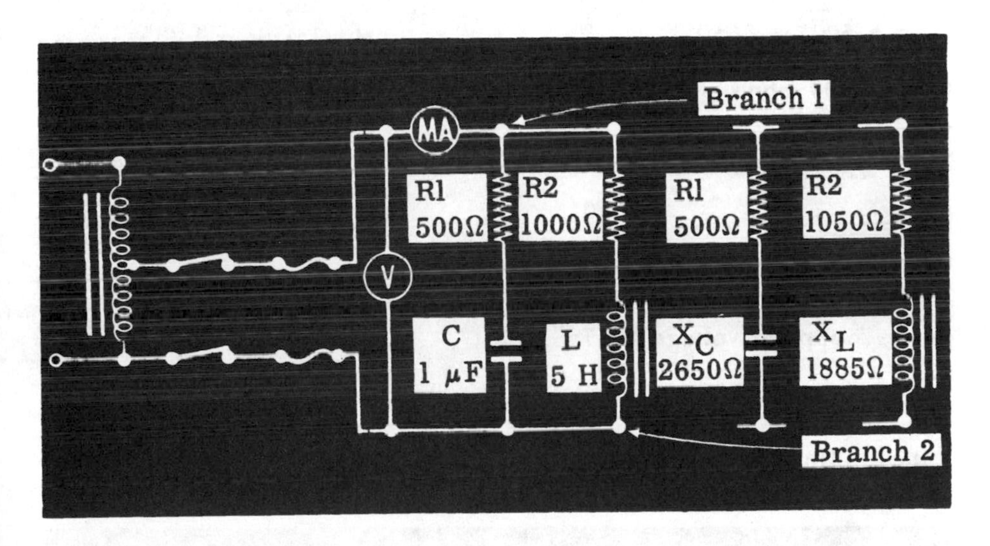

Before switching on, calculate the impedance of, and the current through, the circuit. First the values of X_L and X_C are computed, using 60 Hz as the line frequency. Rounded off to the nearest 10 ohms, these values are 1,885 ohms for X_L and 2.650 ohms for X_C.

Using the known values of R1 and R2 together with the computed values of X_L and X_C, the impedances of each series branch are found separately by using vectors or by calculation. From these values of impedance and a source voltage of 60 volts, the values of I_1 and I_2 are found.

Experiment/Application—Graphic and Calculated Solution in Complex Circuits (continued)

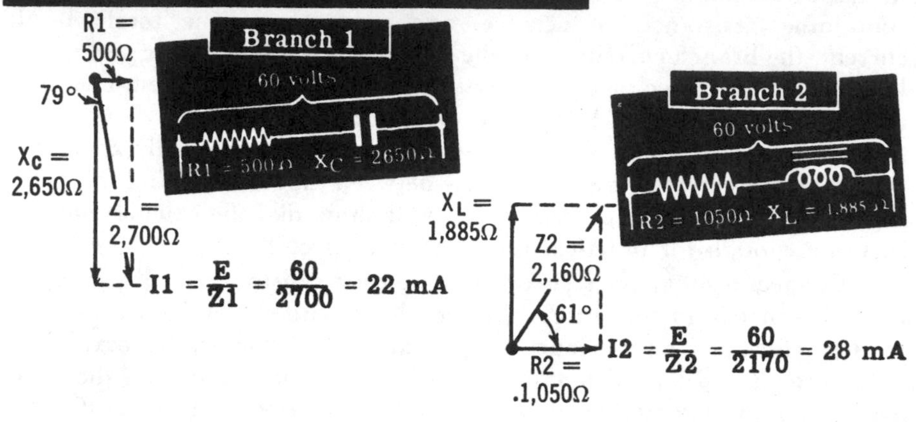

$$Z1 = \sqrt{R1^2 + X_C^2}$$
$$= \sqrt{500^2 + 2{,}650^2}$$
$$= 2{,}700 \text{ ohms approx.}$$
$$I1 = \frac{E}{Z1} = \frac{60}{2{,}700} \cong 22 \text{ mA}$$
$$\theta 1 = \tan^{-1}\frac{X_C}{R1} = \tan^{-1}\frac{2{,}650}{500}$$
$$= \tan^{-1} 5.3 = 79^\circ$$

I1 *leads* applied voltage by 79°

$$Z2 = \sqrt{R2^2 + X_L^2}$$
$$= \sqrt{1{,}050^2 + 1{,}885^2}$$
$$= 2{,}160 \text{ ohms approx.}$$
$$I2 = \frac{E}{Z2} = \frac{60}{2{,}160} = 28 \text{ mA}$$
$$\theta 2 = \tan^{-1}\frac{X_L}{R2} = \tan^{-1}\frac{1{,}885}{1{,}050}$$
$$= \tan^{-1} 1.8 = 61^\circ$$

I2 *lags* applied voltage by 61°

The branch current vectors are now drawn with reference to the common total voltage vector, E, and the two branch current vectors are combined to find the total circuit current. From this value of the total current and the given value of voltage, 60 volts, the total impedance is computed.

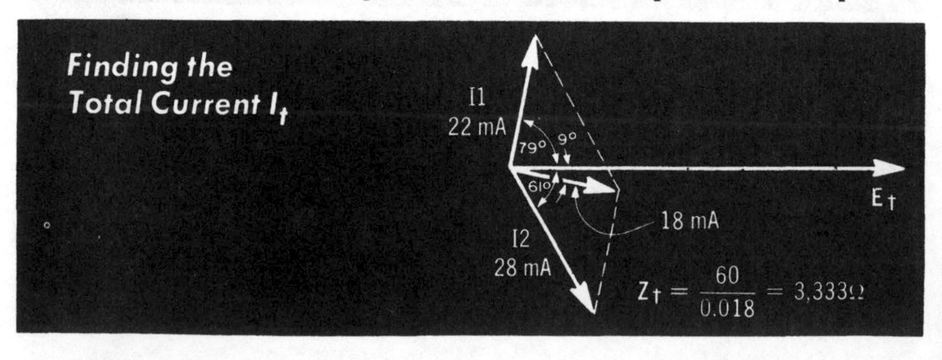

Experiment/Application—Graphic and Calculated Solution in Complex Circuits (continued)

By calculation, you can solve for the total current by resolving each current (I_t and I_2) into their respective components, adding these components, and solving for the total current:

$$I_{R1} = I_1 \cos \theta 1 = 22 \cos 79^\circ = 4.2 \text{ mA}$$

$$I_{R2} = I_2 \cos \theta 2 = 28 \cos 61^\circ = 13.6 \text{ mA}$$

$$I_C = I_1 \sin \theta 1 = 22 \sin 79^\circ = 21.6 \text{ mA}$$

$$I_L = I_2 \cos \theta 2 = 28 \sin 61^\circ = 24.5 \text{ mA}$$

$$I_{R1} + I_{R2} = 4.2 + 13.6 = 17.8 \text{ mA}$$

$$I_L - I_C = 24.5 - 21.6 = 2.9 \text{ mA}$$ (inductive; therefore total current lags voltage)

$$I_t = \sqrt{(I_{R1} + I_{R2})^2 + (I_L - I_C)^2} = \sqrt{(17.8)^2 + (2.9)^2}$$

$$= 18 \text{ mA} = 0.018 \text{ A}$$

$$Z_t = \frac{60}{0.018} = 3{,}333 \text{ ohms}$$

The θ angle of the current is given by

$$\theta = \tan^{-1} \frac{(I_L - I_C)}{I_R}$$

$$\approx 9^\circ$$

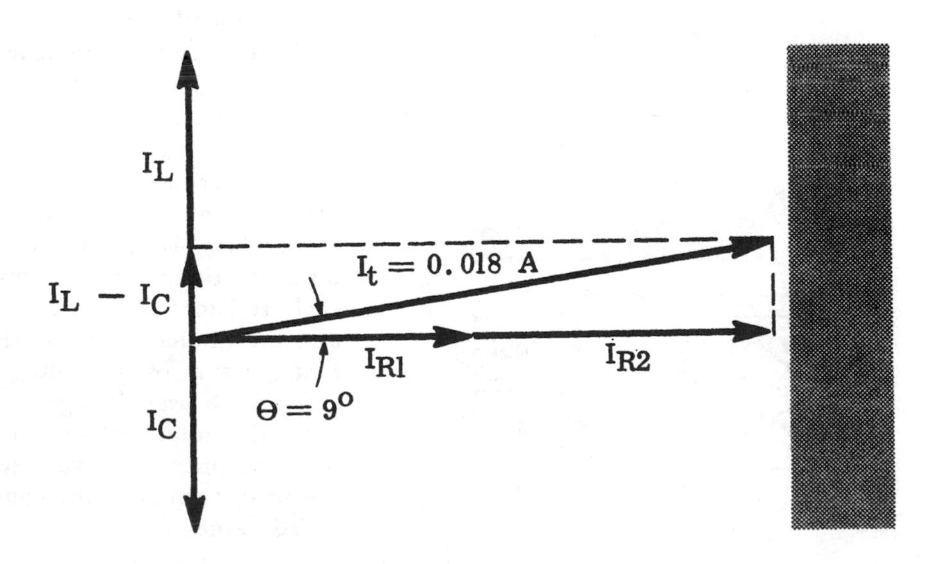

Experiment/Application—Verification of Calculated Current and Impedance

To check the computed values of the total current and circuit impedance, close the switch and measure the total circuit current and voltage. You will see that the measured total current is approximately the same as the computed value. From these measured values, the impedance Z_t is determined and compared with the value obtained by calculation.

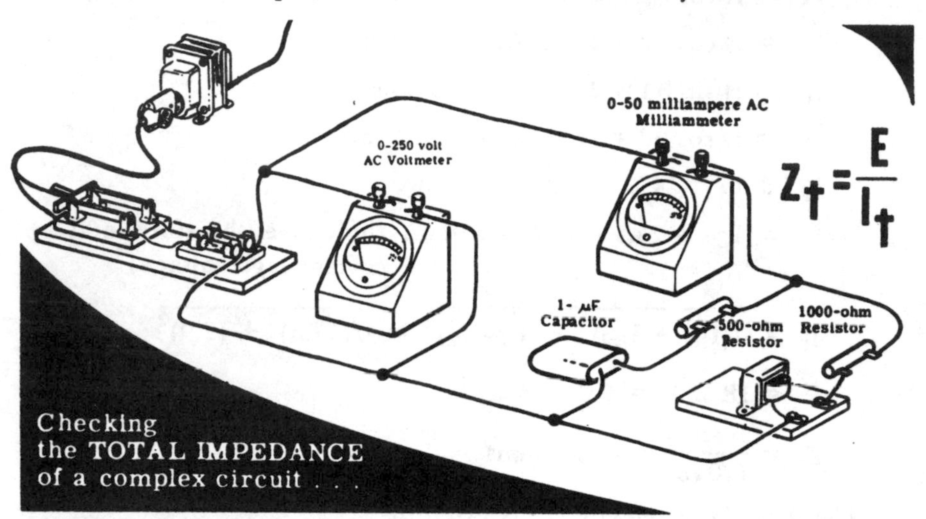

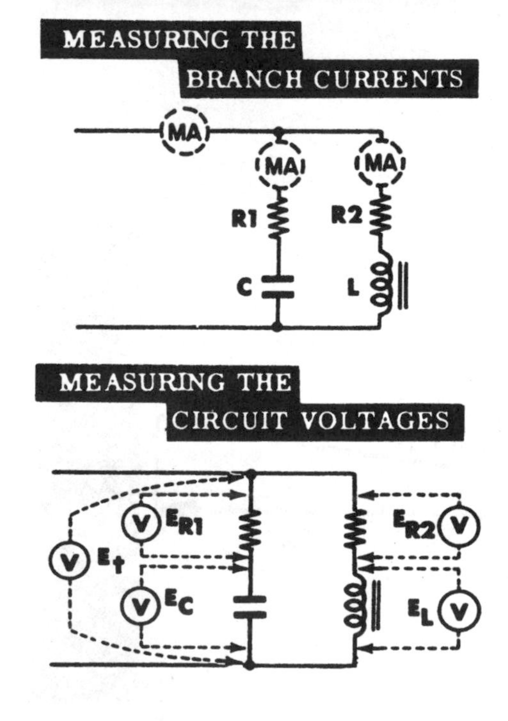

Next connect the milliammeter to measure each branch current in turn. You see that the measured values are about the same as the computed values and that the total circuit current is less than the sum of the two branch currents.

Now calculate the voltage across each component in the circuit (I_1R1, I_2R2, I_1X_C, I_2X_L), and then measure the voltage across each resistor, the capacitor, and the inductance. You see that the sum of the voltages across each branch is greater than the total voltage and that the measured voltages are nearly equal to the computed values.

Review—AC Complex Circuits

The solution of an ac complex circuit requires a step-by-step use of vectors and Ohm's law to find unknown quantities of voltage, current, and impedance. Suppose you review the vector solution of a typical complex circuit either graphically or by calculation.

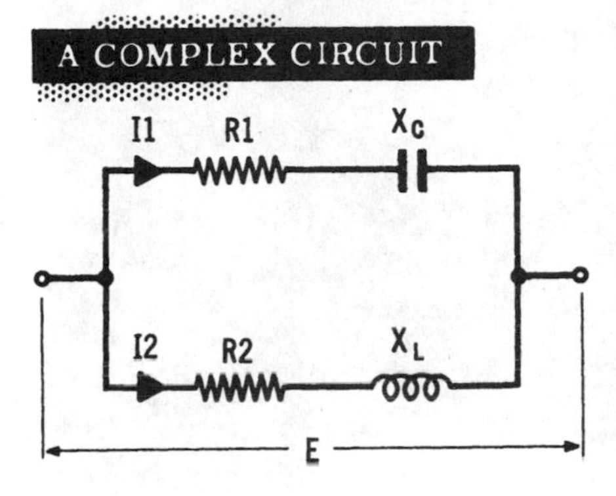

1. Calculate the reactance values of the circuit capacitances and/or inductances.

2. Using vectors, or by calculation, find the impedance of each series branch separately, and compute the Ohm's law values of the branch currents.

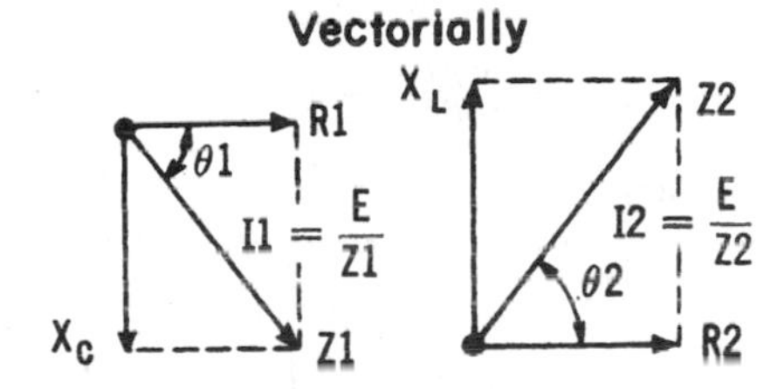

3. You can find the individual voltages of each branch, with respect to their particular branch current. Complete the voltage parallelograms to find the phase relationship between each branch current and the total circuit voltage, or use the formulas for calculation.

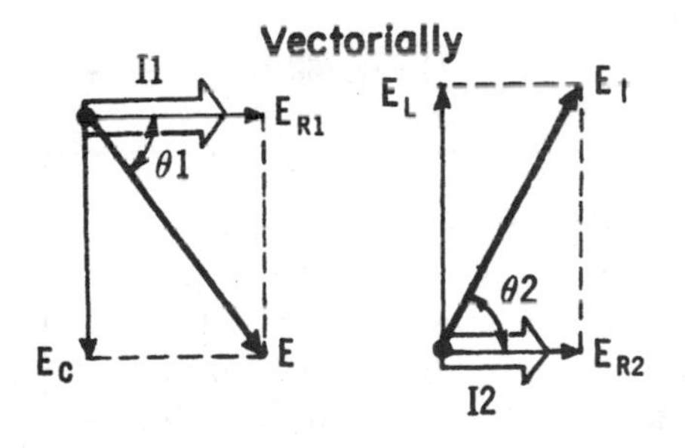

4. Now find the branch current vectors with respect to a common total circuit voltage vector; then combine the currents to find the total circuit current. Using this total current, you can find the total circuit impedance.

Vectorially

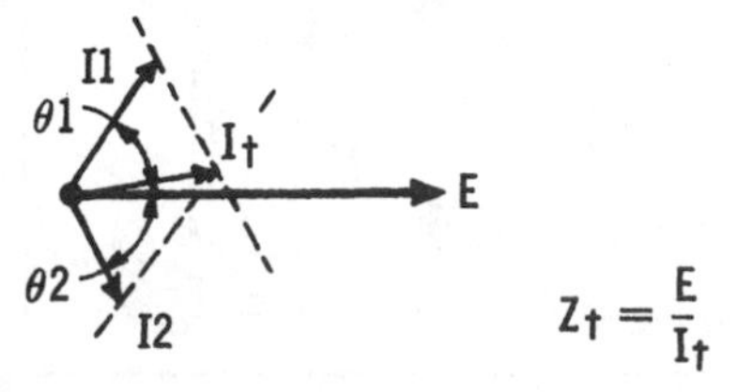

$$Z_t = \frac{E}{I_t}$$

5. You can now calculate the true power, apparent power, and power factor.

Self-Test—Review Questions

1. For the circuits shown below, determine graphically and by calculation the line current and its phase. Also find the true power, apparent power, and power factor.

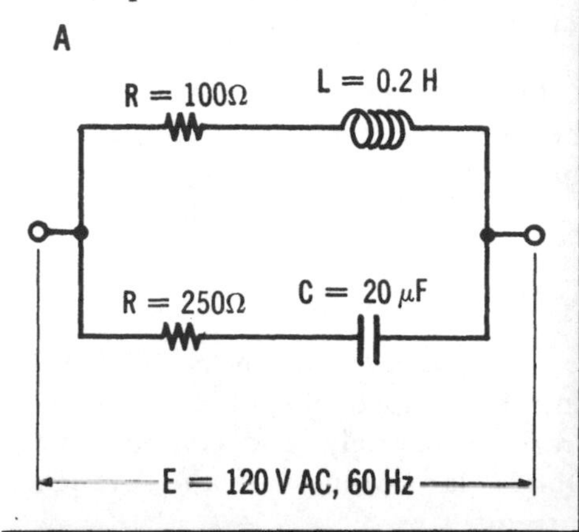

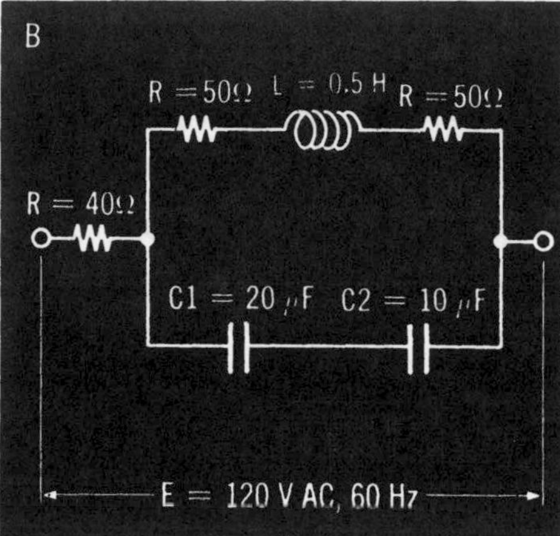

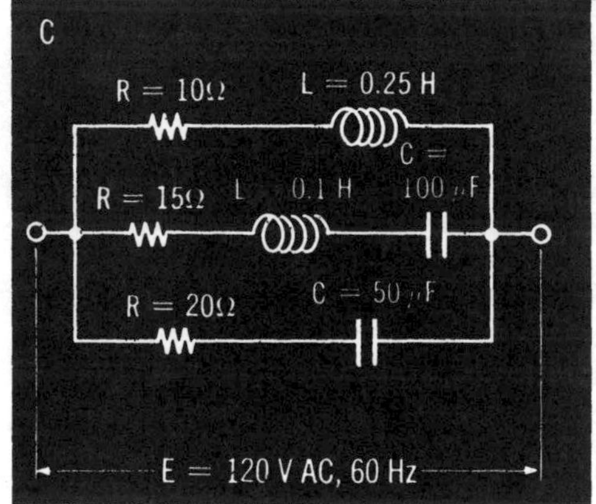

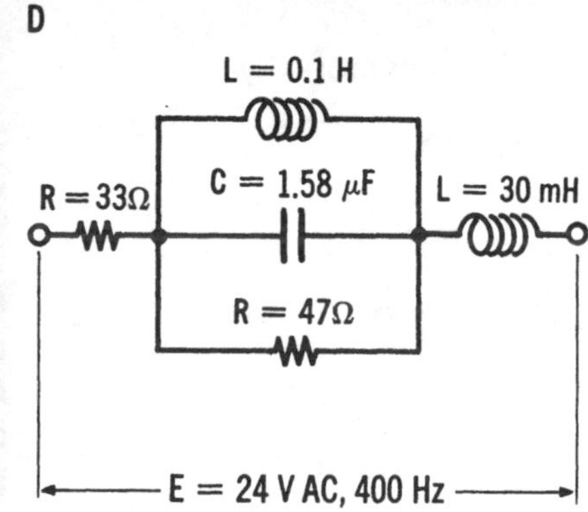

Learning Objectives—Next Section

Overview—You have been using transformers in your ac experiments/applications as a means for obtaining a desired voltage from a line source. You are now ready to learn about these important devices that are indispensable in modern electrical and electronic circuits and equipment.

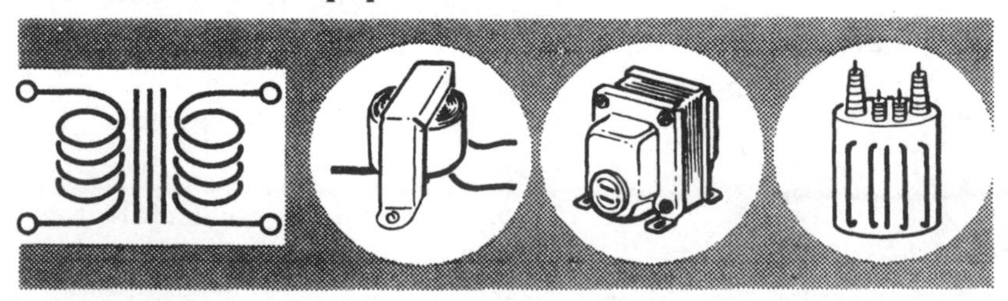

Mutual Induction—Faraday's Law

As you know, ac current is used because it can be readily stepped (transformed) upward or downward in voltage by means of *transformers.* You will now learn that transformers are inductances *coupled* together by their mutual magnetic fields, or *mutual induction.* To start our study of transformers, we will need to learn about mutual induction.

Michael Faraday, an English scientist, did a great deal of important work in the field of electromagnetism, and his work on mutual induction eventually led to the development of the transformer. Faraday found that if the total magnetic flux linking a circuit changes with time, an emf is induced in the circuit. He also found that if the rate of flux change is increased, the magnitude of the induced emf is increased as well. Stated in other terms, Faraday found that the character of an emf induced in a circuit depends upon (a) the amount of flux, and (b) the rate of change of flux which links a circuit. These effects are what is meant by mutual induction.

You have seen in Volume 1 an illustration of the mutual induction principle just stated. When a conductor is made to move with respect to a magnetic field, an emf is induced in the conductor which is directly proportional to the velocity of the conductor's movement with respect to the field. Moreover, the voltage induced in a coil is proportional to the number of turns of the coil, the magnitude of the inducing flux, and the rate of change of this flux.

An example of mutual induction (inducing an emf in a neighboring conductor or inductor) is shown below. Current flows in the direction indicated in primary coil A. This current produces a magnetic field, and if the current remains constant, the number of flux lines produced is fixed. If, however, the current is reduced by opening the switch, the number of flux lines in coil A is decreased, and consequently the flux linking secondary coil B is decreased also. This changing flux induces an emf in coil B, and a current I_B flows, as shown by the movement of the indicator pointer. Thus, it is seen that energy can be transferred from one circuit to another by the principle of *electromagnetic induction*—mutual induction.

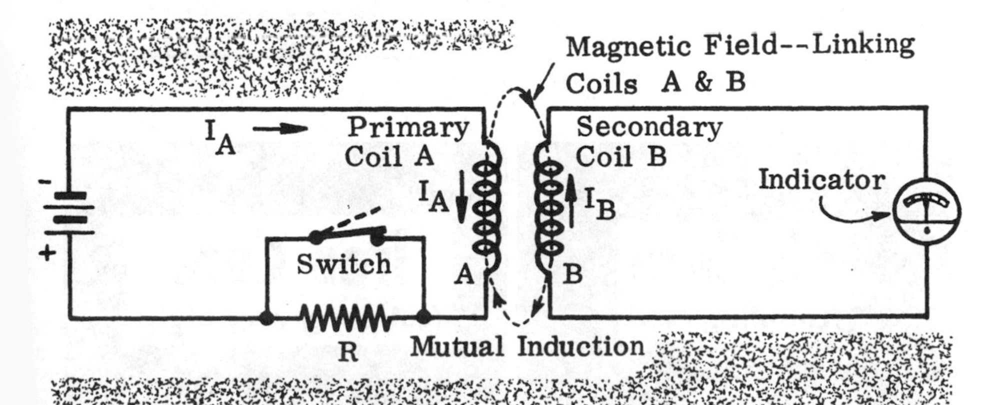

Mutual Induction—Faraday's Law (continued)

A battery was used in the experiment on the previous page as a source of emf. The only way current, and thus magnetic field, variations can be developed is by opening or closing the switch. If an ac voltage source with an extremely low frequency (1 Hz) is used to replace the battery, the indicator will show continuous variations in current. The indicator needle moves to the left (or right) first, and then reverses its position, thus showing the reversal in ac current flow.

If the battery from the circuit on the previous page were replaced with 60-Hz ac source in the *primary* coil and the indicator meter with an ac voltmeter in the *secondary* coil, you would find that an ac voltage existed across the secondary terminals. It is obvious that the mutual inductance or coupling between two coils depends on their *flux linkages*. Maximum coupling is when all the flux lines from the primary cut the secondary winding. When this happens, the degree of coupling (called the *coefficient of coupling*) is unity. While coupling factors of less than unity are desirable in some electronic circuits you may learn about later, in transformers it is desirable to get the coupling between the two windings as *high* as possible. Thus, both coils are usually wound on an iron core so that the path of the flux lines can be controlled and kept where desired. A special alloy of silicon steel is commonly used for transformer cores.

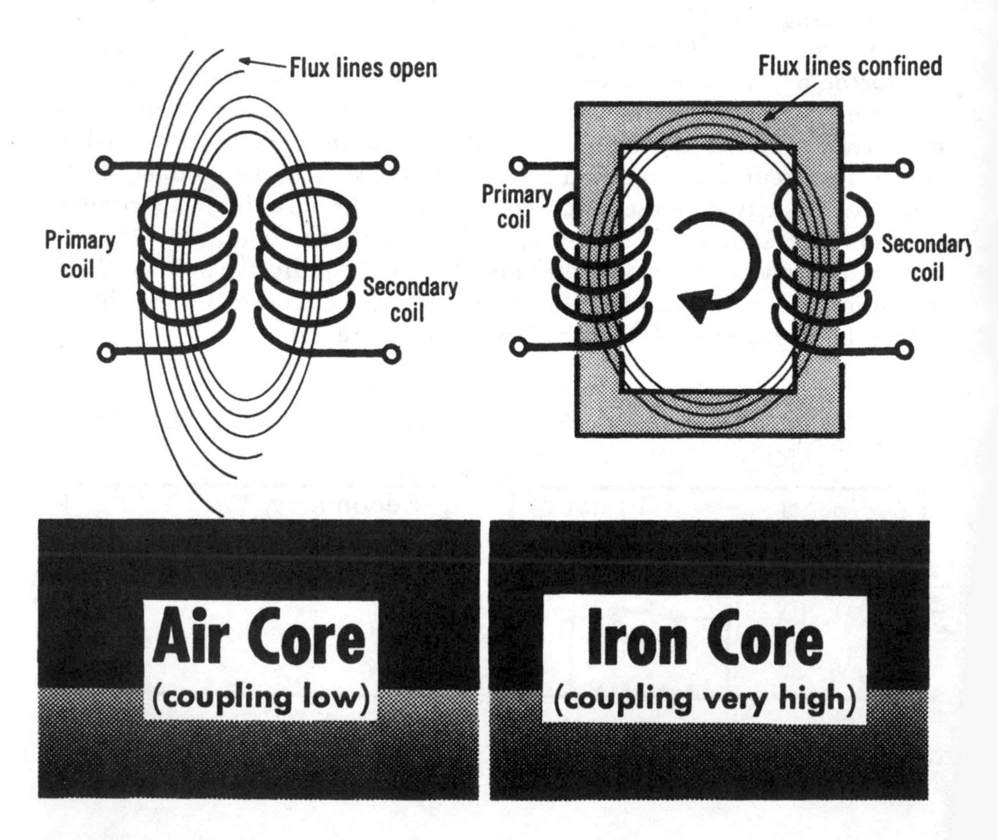

How a Transformer Works

A simple transformer consists of two windings very tightly coupled together, usually with an iron core, but electrically insulated from each other. The winding to which an ac voltage source is applied is called the *primary.* It generates a magnetic field which cuts through the turns of the other coil, called the *secondary,* and generates a voltage in it. The windings are not physically connected to each other. They are, however, *magnetically coupled* to each other. Thus, a transformer transfers electrical power from one coil to another by means of an *alternating magnetic field.*

Assuming that all the magnetic lines of force from the primary cut through all the turns of the secondary, the voltage induced in the secondary will depend on the ratio of the number of turns in the secondary to the number of turns in the primary. This is mathematically expressed as $E_s = (N_s/N_P)\ E_P$. For example, if there are 1,000 turns in the secondary and only 100 turns in the primary, the voltage induced in the secondary will be *10 times* the voltage applied to the primary (1,000/100 = 10). Since there are more turns in the secondary than there are in the primary, the transformer is called a *step-up transformer.* If, on the other hand, the secondary has 10 turns and the primary has 100 turns, the voltage induced in the secondary will be *one-tenth* of the voltage applied to the primary (10/100 = 1/10). Since there are less turns in the secondary than there are in the primary, this transformer is called a *step-down transformer.* The symbol for a transformer is shown below.

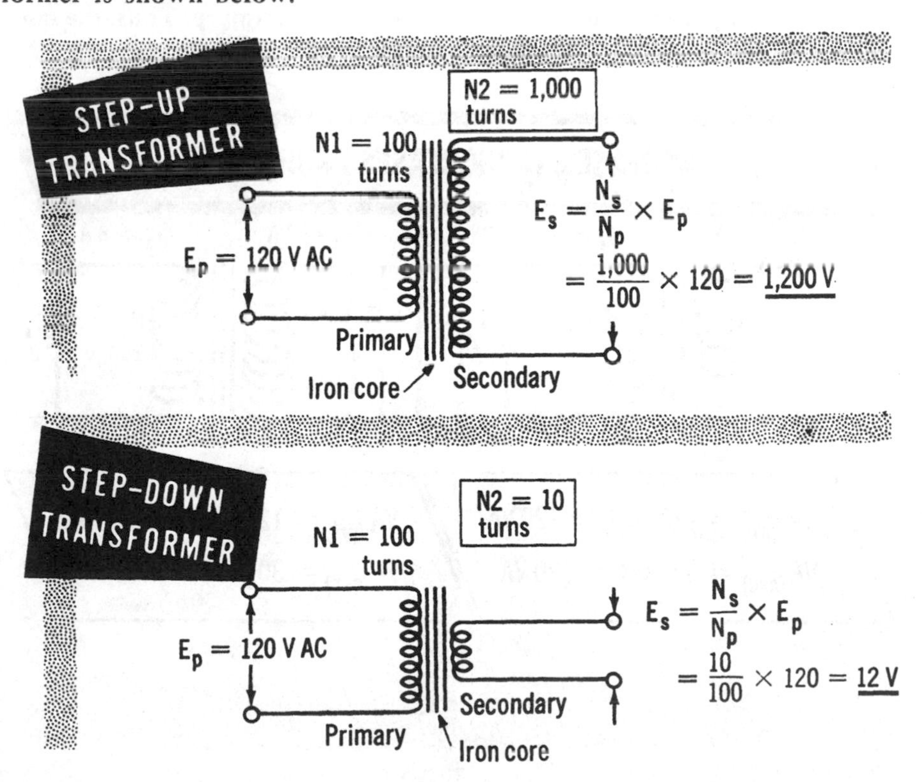

How a Transformer Works (continued)

To find any unknowns in a transformer, use the formula $E_P/E_S = I_s/I_P = N_P/N_s$ and cross-multiply to find the required information.

The current in the secondary of a transformer flows in a direction opposite to that which flows in the primary because of the emf of mutual induction. An emf of self-induction is also set up in the primary which is in opposition to the applied emf.

When no load is present at the output of the secondary, the primary current is very small because the emf of self-induction is almost as large as the applied emf. If no load is present at the secondary, there is no secondary current, but there is a small primary current flow to account for transformer losses. Since there are no fields caused by secondary current flow, the magnetic field of the primary will develop to its maximum strength. When the primary field is at its maximum strength, it produces the strongest possible emf of self-induction, and this opposes the applied voltage. The difference between the emf of self-induction and the applied emf causes a small current to flow in the primary, and this is the exciting or magnetizing current that accounts for transformer losses.

Since any current made to flow in the secondary is opposite to the current in the primary, the lines of flux are opposite, too. As a load is applied to the secondary, causing current to flow, it causes a reduction in the total flux (primary flux–secondary flux) which reduces the flux linking the primary. The reduction in flux lines reduces the emf of self-induction and permits more current to flow in the primary. Thus, the more current in the secondary, the more current in the primary.

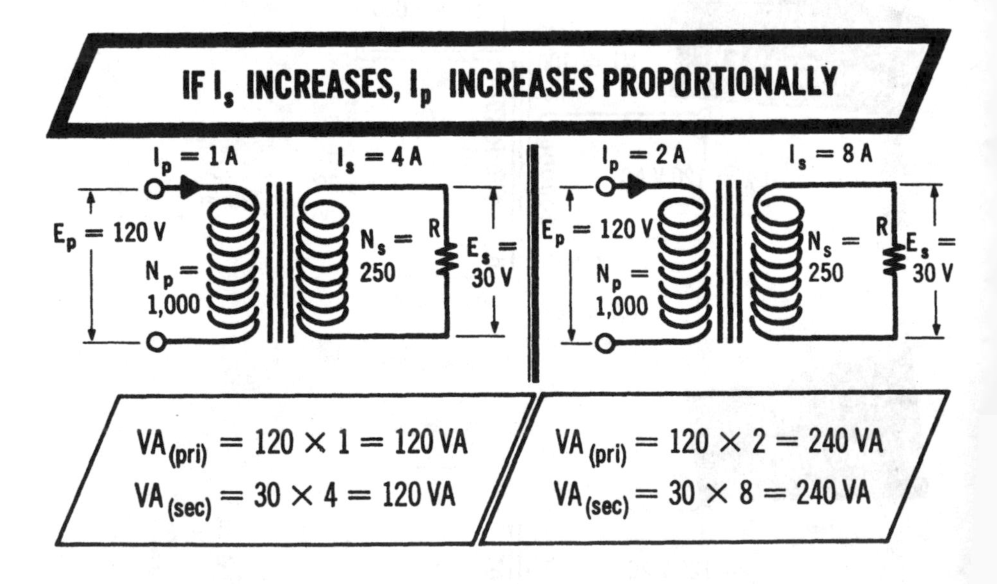

How a Transformer Works (continued)

A transformer *does not* generate electrical power. It simply *transfers* electrical power from one coil to another by magnetic induction. Although transformers are not 100 percent efficient (lossless), they are very nearly so. (You will find out about some of the losses in transformers a little bit later.) Therefore, a transformer can be defined as *a device that transfers power from its primary circuit to its secondary circuit with little loss.*

Transformers are usually rated in volt-amperes rather than watts since they must handle the total current no matter what the phase is. Since power equals voltage times current, if $E_p I_p$ represents the primary power and $E_s I_s$ represents the secondary power, then $E_p I_p = E_s I_s$. If the primary and secondary voltages are equal, the primary and secondary currents must also be equal.

Suppose E_p is twice as large as E_s. Then, for $E_p I_p$ to equal $E_s I_s$, I_p must be one half of I_s. It follows that a transformer which *steps voltage down* must *step current up.*

Similarly, if E_p is only half as large as E_s, I_p must be twice as large as I_s. So a transformer which *steps voltage up must step current down.*

Transformers are classified as step-down or step-up by reference to their effect on *voltage only.*

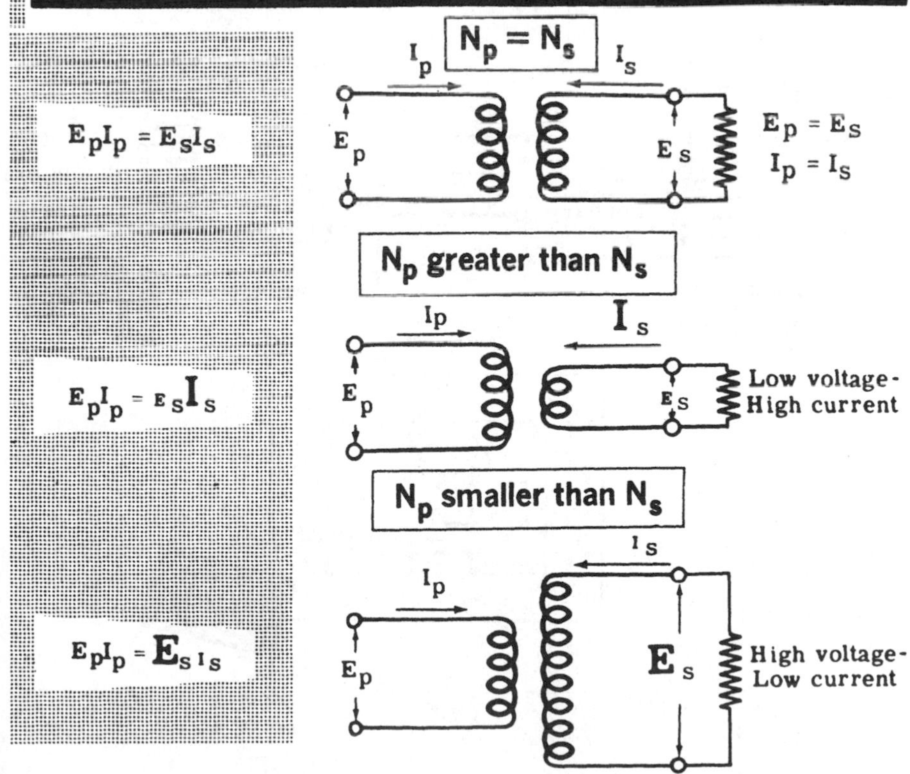

Transformer Construction

Transformers designed to operate on low frequencies (power transformers) have their coils, called *windings*, wound on iron cores. Since iron offers little resistance to magnetic lines, nearly all the magnetic field of the primary flows through the iron core and cuts the secondary.

Iron cores are constructed in three main types—the open core, the closed core, and the shell core. The open core is the least expensive to build since the primary and the secondary are wound on one cylindrical core. The magnetic path is partly through the core, partly through the air. Since the air path opposes the magnetic field, the magnetic interaction or *linkage* is weakened. The open core transformer is therefore inefficient and is never used for power transmission.

The closed core improves the transformer efficiency by offering more iron paths and less air path for the magnetic field, thus increasing the magnetic linkage or *coupling*. The shell core further increases the magnetic coupling, and therefore the transformer efficiency, because it provides two parallel magnetic paths for the magnetic field. It thus permits maximum coupling to be attained between the primary and the secondary.

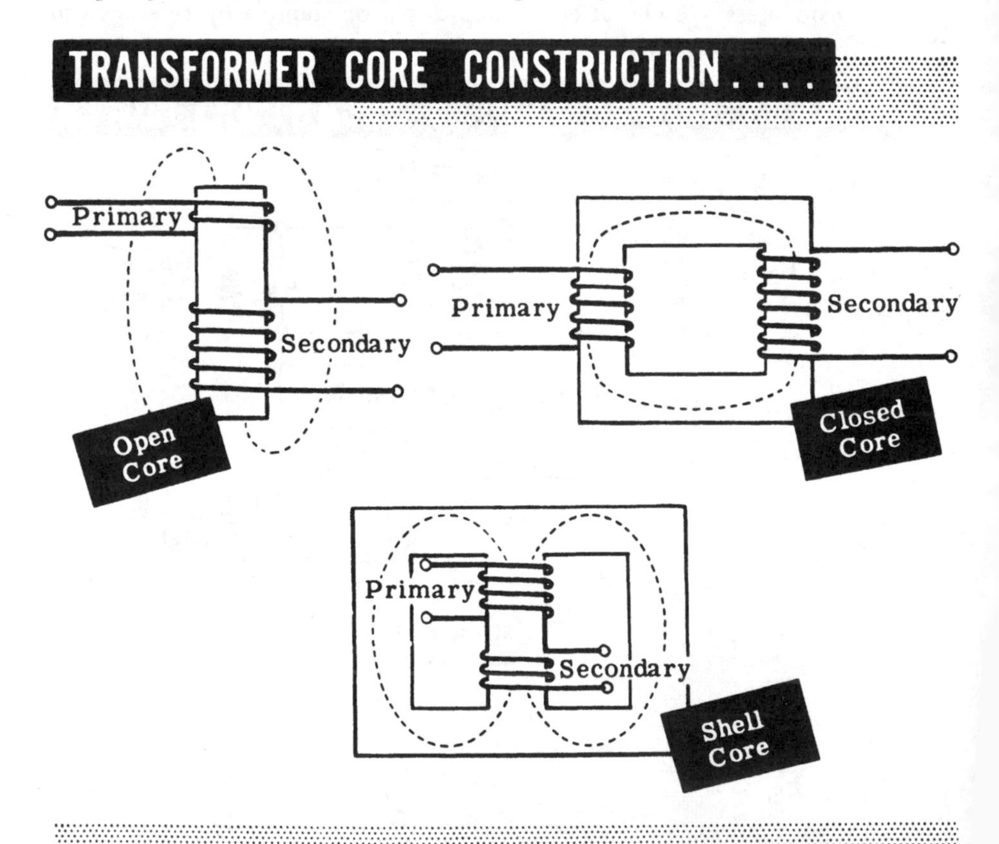

Transformer Losses

Up to now, you have assumed that transformers are perfect and have no internal losses. While this is usually almost true, there are some losses present. Most transformers are between 90 and 99 percent efficient. The major transformer losses are copper loss (resistance loss), flux leakage loss, hysteresis loss, eddy current loss, and saturation loss.

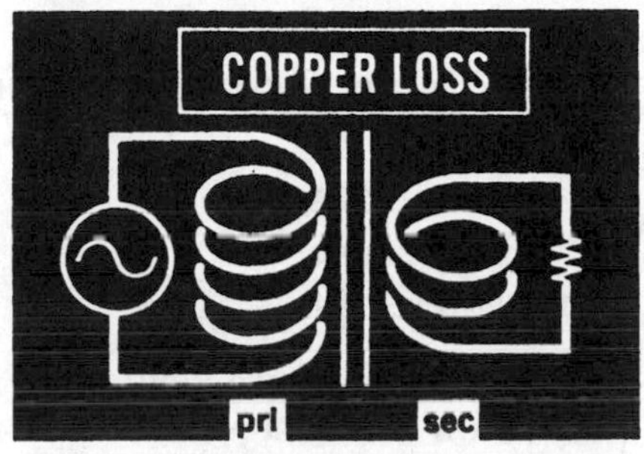

Loss from resistance of (copper) wire in windings of transformer = I^2R, where R is winding resistance.

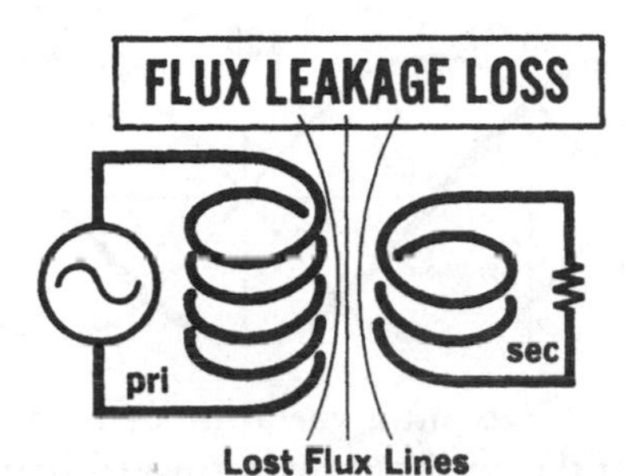

Loss from flux lines that leak from windings or core so that they do not link the flux of primary and secondary and, therefore, represent lost energy.

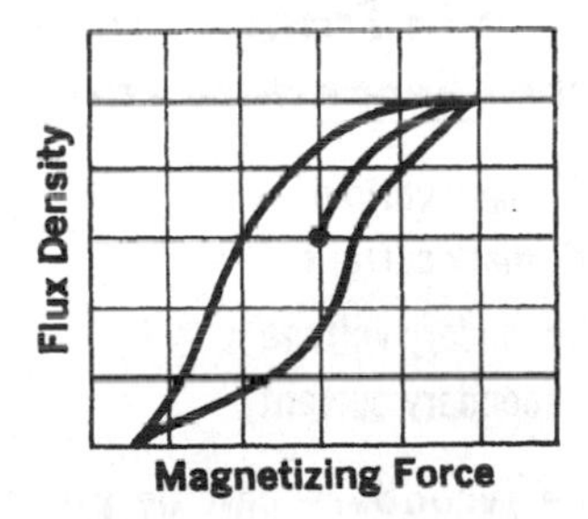

When the current reverses, magnetic alignment of the core also reverses, but the magnetic domains are a little behind. Energy is needed to get the magnetic domains aligned and to turn them around. This energy, which is not available in the secondary, is called **hysteresis loss.** Some transformers employ a powdered iron core to reduce these losses.

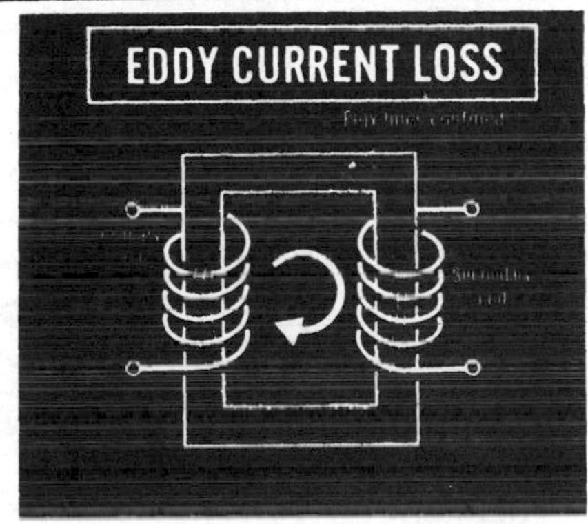

The core of the transformer conducts electricity and acts like a single-turn shorted secondary. The current that flows in the core is called **eddy current.** Eddy currents are kept at a minimum by using a core made up of many flat sections, called **laminations,** stacked together.

SATURATION LOSS

As primary current increases, the number of flux lines in the core increases. Finally, a limit is reached, and additional current results in no more flux lines. When this point is reached, the core is said to be **saturated,** and to increase the number of flux lines would require that the core be made larger.

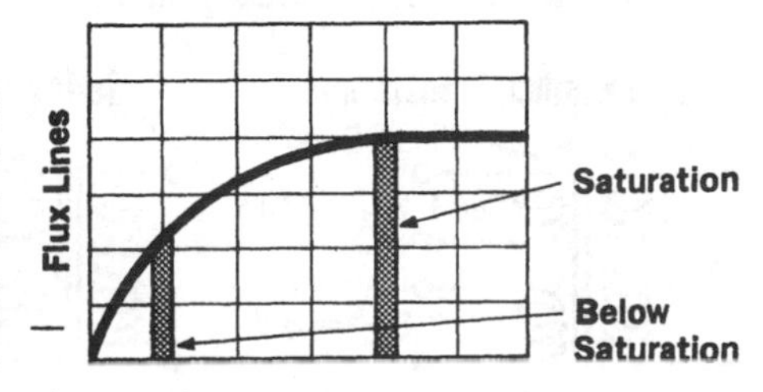

Phase Relationships in Transformers—Multiple Winding Transformers

A transformer with no load on the secondary (open circuit) acts like a simple inductor. Since the losses are usually small, as you might suspect, the current lags the voltage by 90 degrees. The secondary voltage is 180 degrees out of phase with the applied voltage.

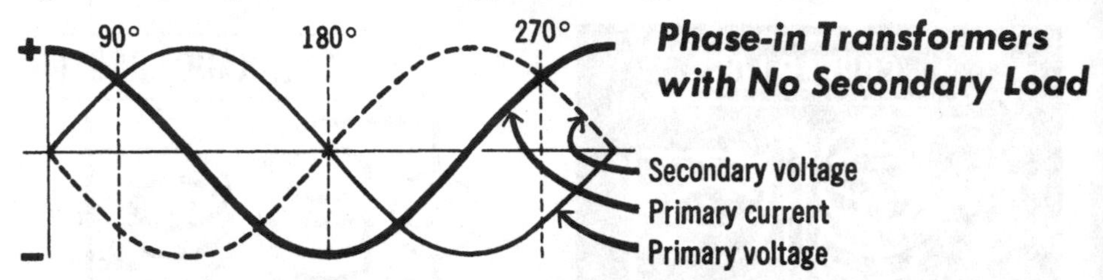

Phase-in Transformers with No Secondary Load

If we put a variable load resistance on the secondary, we would find that the current moves toward being in phase with the voltage as the load resistance increases. When the load current increases so that it is much larger than the no-load current (the way transformers normally operate), the voltage and current in the primary and the secondary have the same relative phase relationship, even though their magnitudes may be different. Thus, a load with a less than unity power factor on the secondary will be reflected as a load with the *same* power factor in the primary.

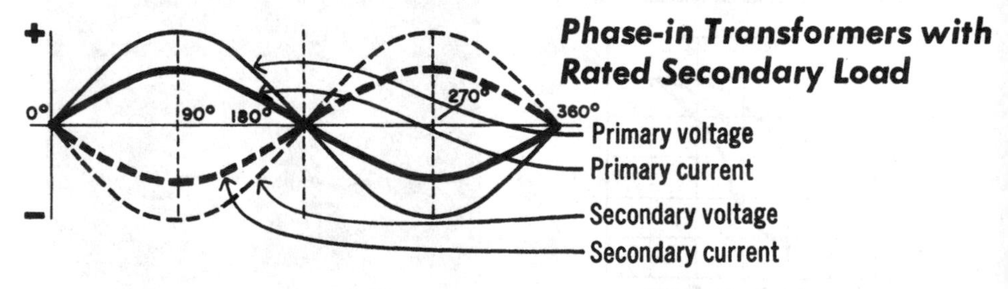

Phase-in Transformers with Rated Secondary Load

It should be obvious that more than one secondary can be put on a transformer. When this is done, the phase relationship between all secondaries will be the same if the loads are similar. Sometimes, it is desirable to connect secondary windings in series so that the voltage will be increased or decreased. When the windings are connected in *series aiding*, the voltage is the *sum* of the voltages for both windings. When the windings are connected in *series opposing*, the voltage is the *difference* of the voltages for both windings. Sometimes, a dot or similar marking is used to indicate the terminals that have the same phase.

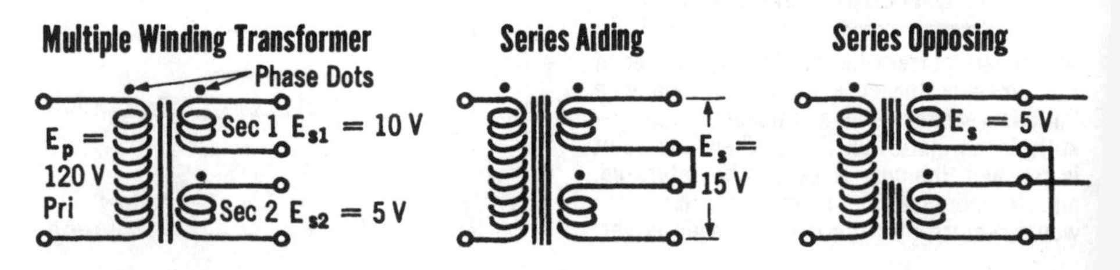

Autotransformers—Tapped Transformers

Autotransformers make common use of part of a winding for both the primary and secondary. They have a tap (a wire brought out from a point on the winding) that is necessary for their operation. As shown below, the operation of an autotransformer is exactly like that of a conventional transformer where one lead of the primary and one of the secondary are tied together in the right way.

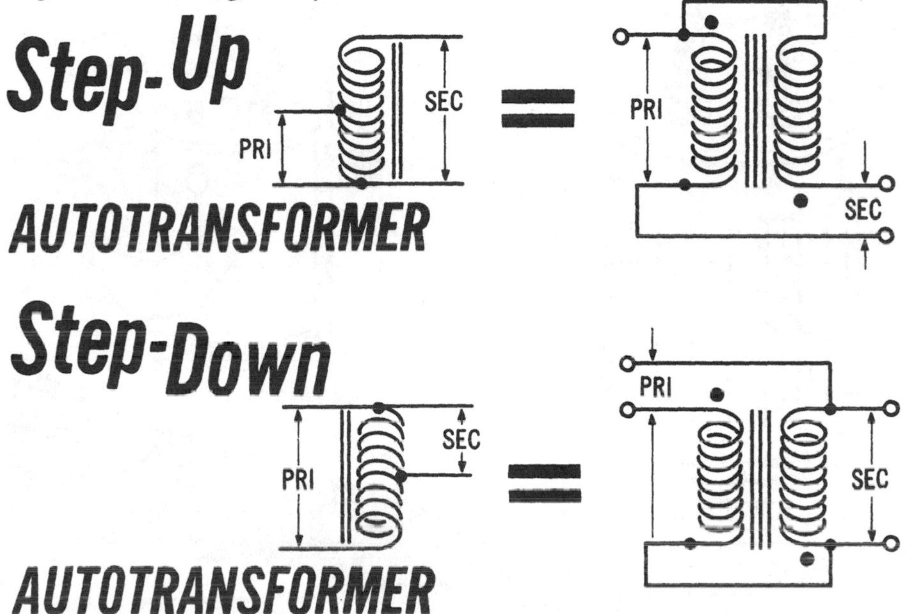

A common type of autotransformer is arranged so that the tap is continuously adjustable in order to provide a voltage range from 0 to about 130 percent of the nominal voltage available. The variable transformer is useful in circuits where it is necessary to set the voltage precisely to a known value.

In some cases, when continuous adjustment is not necessary, fixed taps are used to change the turns ratio. In such cases, the taps can be used in an autotransformer configuration or as taps on the primary or secondary of a conventional transformer.

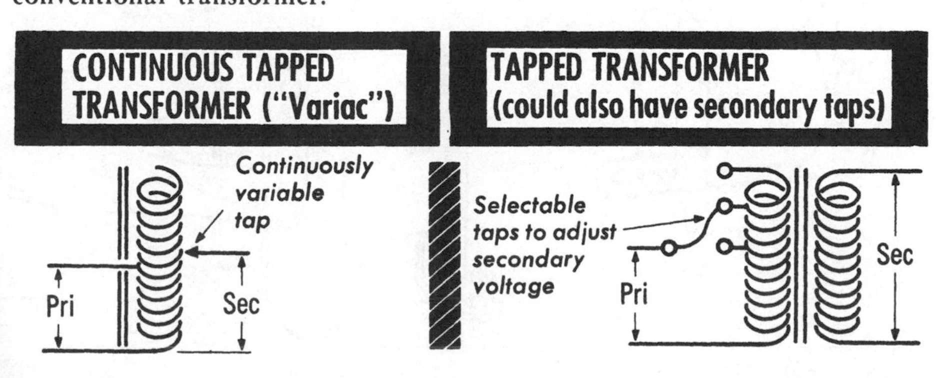

Transformer Types

Transformers come in all sizes and shapes. Very large, high-voltage, high-current transformers are used for power distribution; subminiature transformers are used in highly portable, low-power equipment. Large transformers are sometimes immersed in oil for cooling and improved insulation.

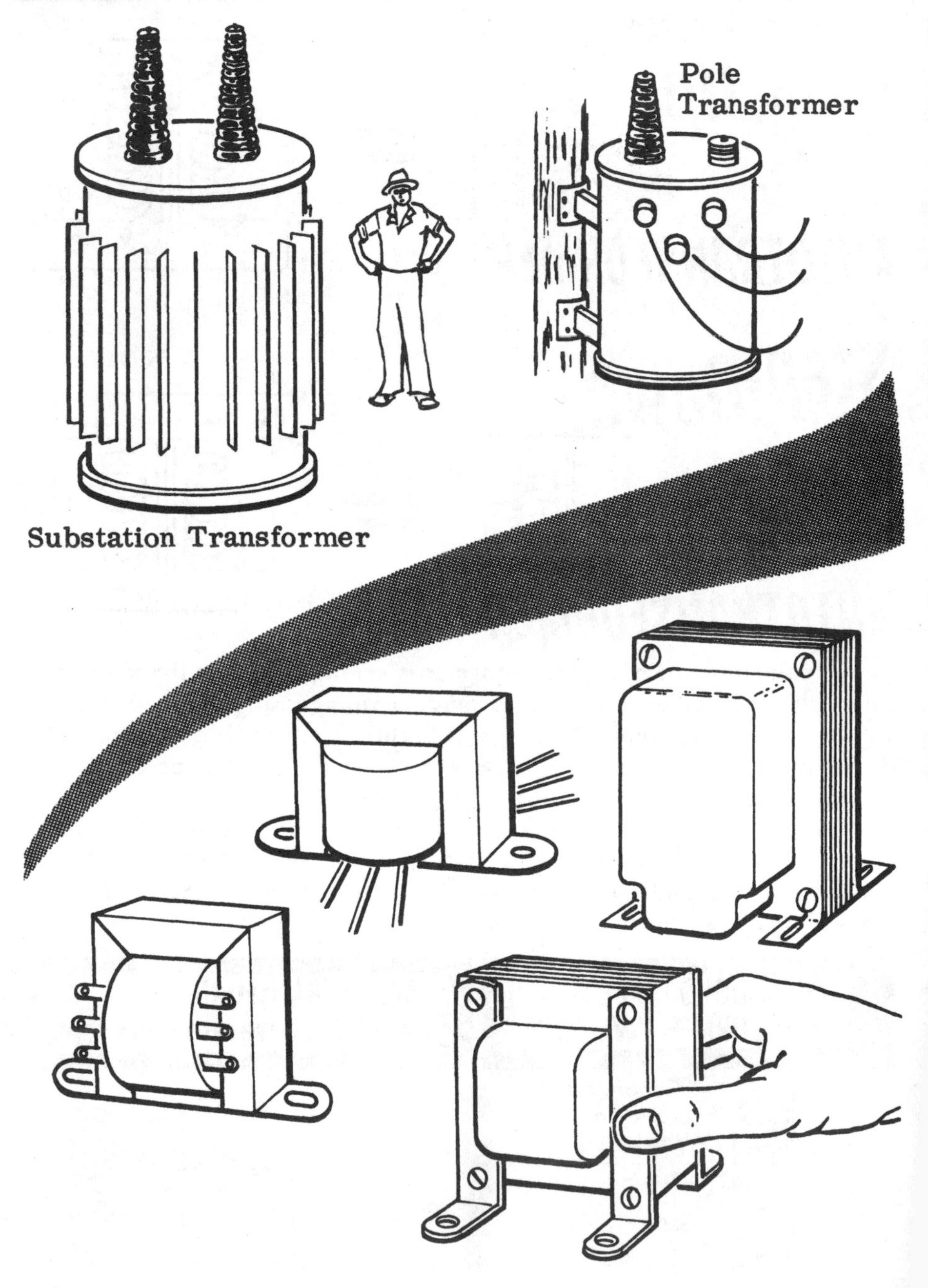

Experiment/Application—Transformer Action

You can observe a transformer in action by making a very simple one and seeing how it works. Suppose you had an iron bar (a bundle of soft iron wires would be better) and you wound a layer of thin insulating tape on this. You could then take some No. 22 or No. 24 enameled wire and wind three coils of it as follows: Winding 1 = 500 turns, Winding 2 = 100 turns, Winding 3 = 250 turns.

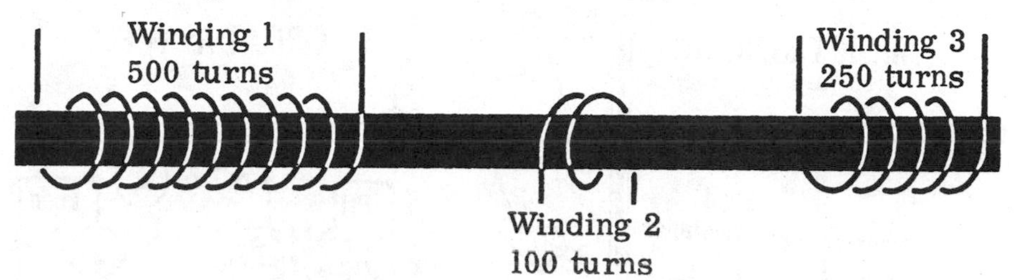

Suppose you then connected up the transformer, using Winding 1 as the primary and Windings 2 and 3 as the secondary, with a 7.5-volt ac source. Using a voltmeter to measure the voltages, you will see that they essentially agree with basic transformer theory.

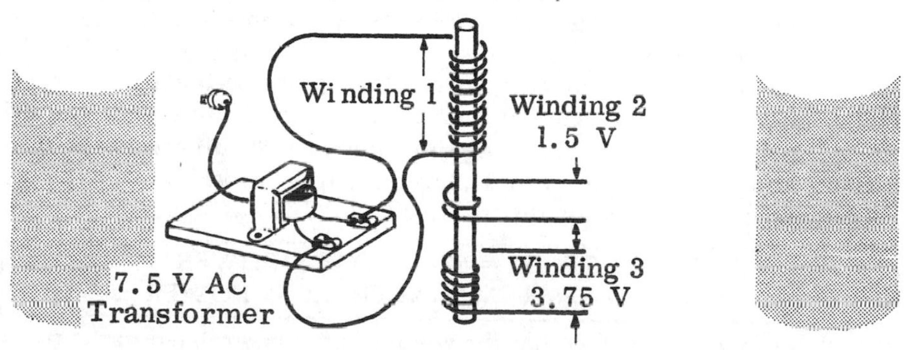

Reverse the setup by using Winding 3 as the primary, and measure the voltages across Winding 1 and Winding 2.

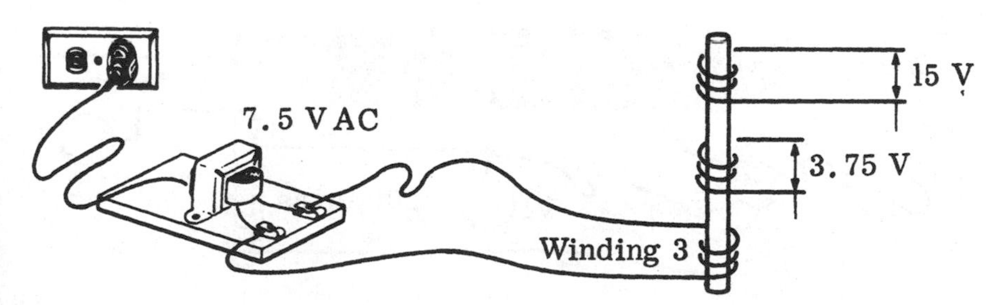

If you have an ammeter available, measure the primary current with no load and calculate the transformer losses. Since this transformer is not very efficient, it will not transfer enough power for you to be able to test the power relationship.

Experiment/Application—Transformer Action (continued)

To demonstrate further how transformers work, you could also take a filament transformer (120 volts to 6.3 volts) and connect it to the line as shown below. If a resistive load is put on the secondary, and primary power and secondary power are measured, the efficiency of the transformer can be found.

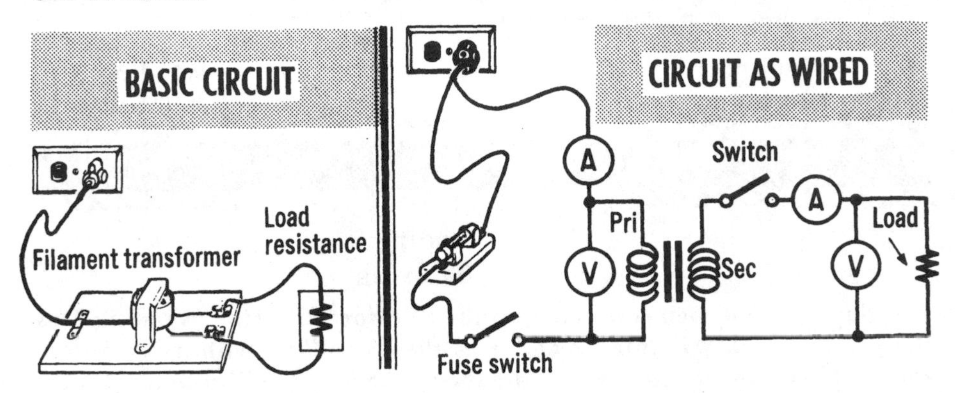

With the primary switch closed and the secondary switch open, you can determine the losses due to hysteresis, eddy currents, and flux leakage, which are often lumped together as *magnetic loss*. This is represented by the power (VA) with zero secondary current. When the secondary switch is closed, the load will draw current from the secondary. If we assume that the losses listed above do not change significantly, then the difference between primary and secondary power (VA), less the magnetic loss power, will give the copper, or I^2R, loss for this particular load. Thus,

PRIMARY POWER = SECONDARY POWER + COPPER LOSSES + MAGNETIC LOSS POWER

You can verify the fact that the voltage is directly related to the turns ratio by using an open winding transformer with a known turns ratio and operating at low voltage.

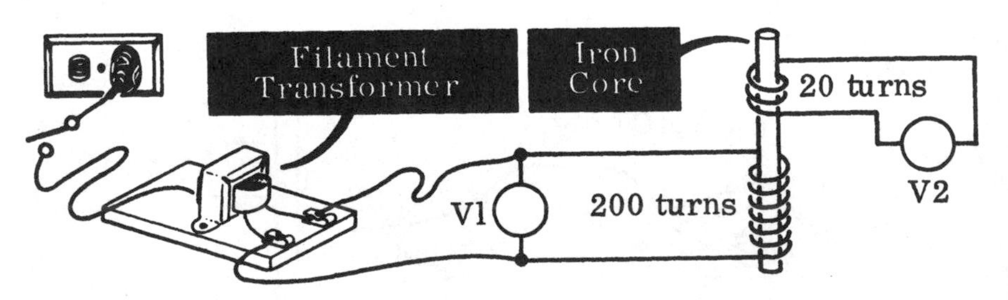

You will see that V1 equals about 6 volts and that V2 is one-tenth of V1, or about 0.6 volt.

$$V2 = \frac{V1 \times N2}{N1} = \frac{6 \text{ volts} \times 20 \text{ turns}}{200 \text{ turns}} = 0.6 \text{ volt}$$

Review of Transformers

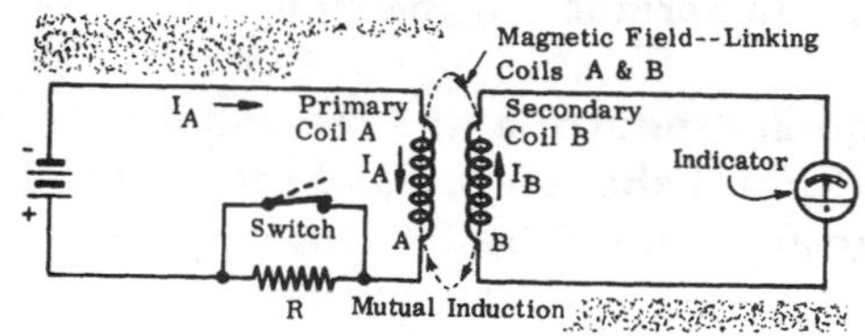

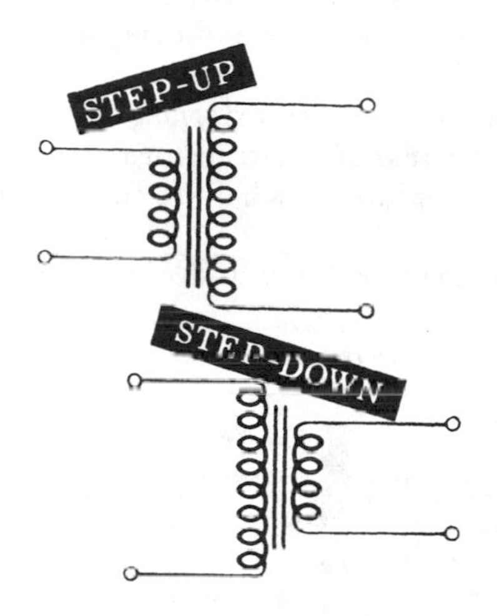

$$E_pI_p = E_sI_s$$

SATURATION LOSS

COPPER LOSS

HYSTERESIS LOSS

EDDY CURRENT LOSS

FLUX LEAKAGE LOSS

Phase-in Transformers with Rated Secondary Load

0° 90° 180° 270° 360°

Primary voltage

Primary current

Secondary voltage

Secondary current

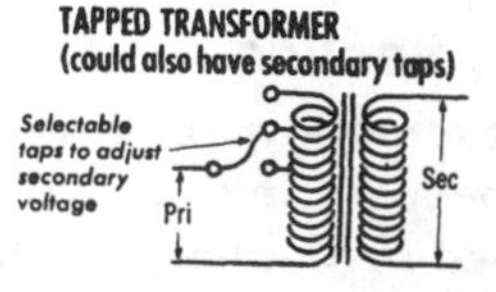

1. MUTUAL INDUCTION—Mutual induction is the induction of an emf in one coil as a result of a current change in a neighboring coil or conductor.

2. TRANSFORMER—A transformer consists of two or more windings coupled together, usually by an iron core, so that almost all flux lines interact with the windings.

3. STEP-UP TRANSFORMER/STEP-DOWN TRANSFORMER—The voltage ratio in a transformer is proportional to the turns ratio. When the number of secondary turns is *greater* than the number of primary turns, the transformer is a *step-up* transformer. When the number of secondary turns is *less* than the number of primary turns, the transformer is a *step-down* transformer.

4. PRIMARY POWER = SECONDARY POWER—As the turns ratio of a transformer changes, the voltage and current change to keep the primary power equal to the secondary power.

5. TRANSFORMER LOSSES—The losses in transformers, which are usually small, include copper loss, flux leakage loss, hysteresis loss, eddy current loss, and saturation loss.

6. PHASE RELATIONSHIP—In a resistively loaded transformer, the primary voltage and current are essentially in phase, and the secondary voltage and current are also in phase but opposite in polarity to the primary.

7. ADJUSTABLE-VARIABLE TRANSFORMERS—These allow for the adjustment of the turns ratio to permit voltage changes.

Self-Test—Review Questions

1. State Faraday's law and explain its importance in the study of transformers.
2. What is the difference between mutual induction and self-induction?
3. Draw a transformer in schematic form. Label all parts of the circuit.
4. Why is an iron core used in transformers? Why is it usually laminated?
5. List and briefly describe the losses that occur in transformers.
6. Draw the phase relationships between the voltages and currents in (a) an unloaded transformer, and (b) a loaded transformer.
7. You have a load of 50 ohms connected to a transformer secondary which produces an ac voltage of 30 volts. If the transformer has no losses and has a turns ratio of 4-to-1 step-down, what are the primary voltage and current?
8. For the same load, calculate the primary voltage and current if the transformer is a 3-to-1 step-up.
9. For the transformer circuits shown below, calculate the unknown quantities indicated.

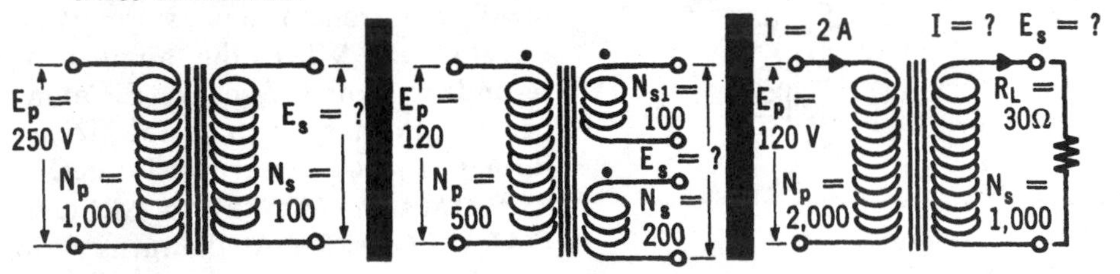

10. Show how the transformer below can be used to obtain the following voltages: 55, 165, 22, 77, and 98 volts.

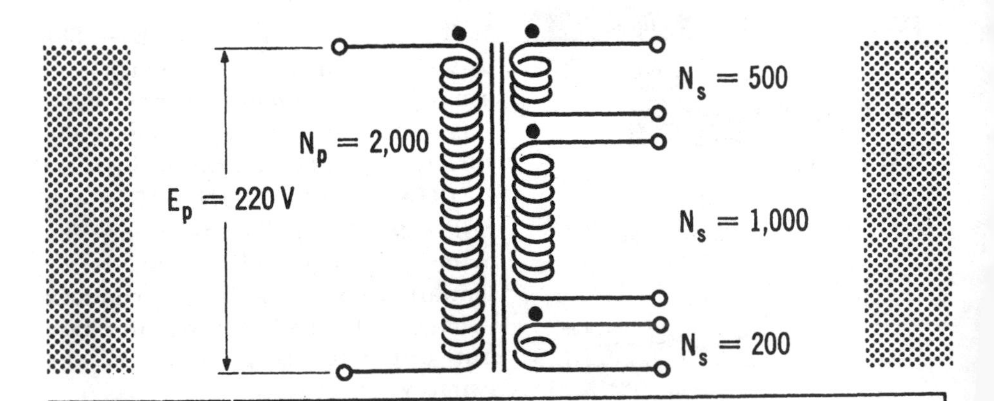

Learning Objectives—Next Section

Overview—One of the most important things to know about is how to troubleshoot ac and dc circuits. In this next section, we will expand on the troubleshooting you have learned in Volumes 2 and 3.

AC Power Systems

As you know, in the U.S. and most North and Central American countries, the power line frequency is 60 Hz and the nominal line voltages for household and industrial use are specified as 120/240 single-phase or 120/208 three-phase. Some older houses have only 120-volt ac single-phase. As you will learn in Volume 5, there are many advantages to the use of three-phase power for some industrial applications and for power transmission. For most homes and very small industrial plants, a single-phase 120/240-volt system is used. This is obtained from the main line that is at a voltage between 600 and 1,000 volts and stepped down by a transformer.

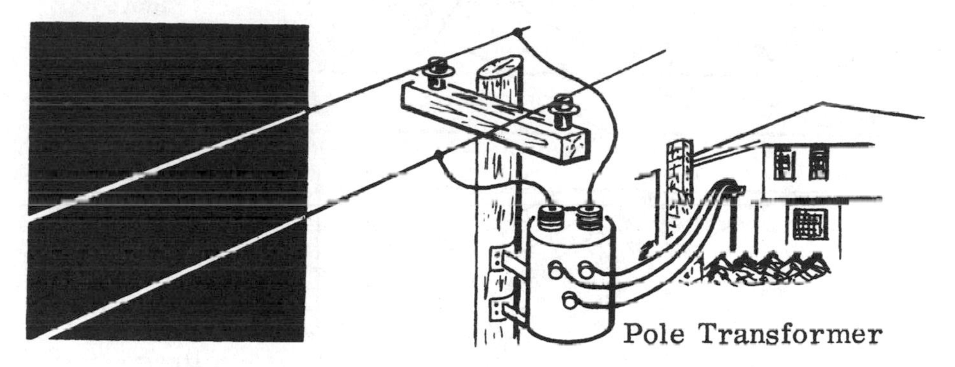

The transformer has a center-tapped secondary so that from each side of the center tap—often called the *neutral*—the line voltage is 120 volts and the total voltage is 240 volts.

Schematically, the system looks as shown below:

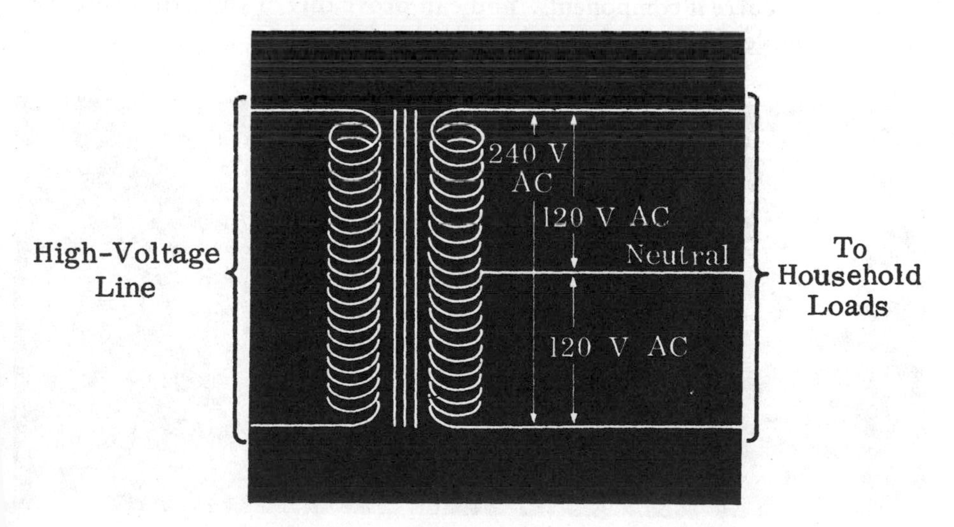

The neutral line is grounded at the pole and at each subscriber's power input box.

Local AC Power Distribution

Usually, the line goes through a recording wattmeter to record the amount of electricity used. How this meter works will be discussed in Volume 5. The power at the household level is distributed via protective devices to the household circuits. For most appliances and lighting, etc., the circuit consists of one side and the neutral so the line voltage is 120 volts ac. For heavy loads (stoves, heaters, dryers, etc.) the appliance is put across the 240-volt lines to reduce the current required.

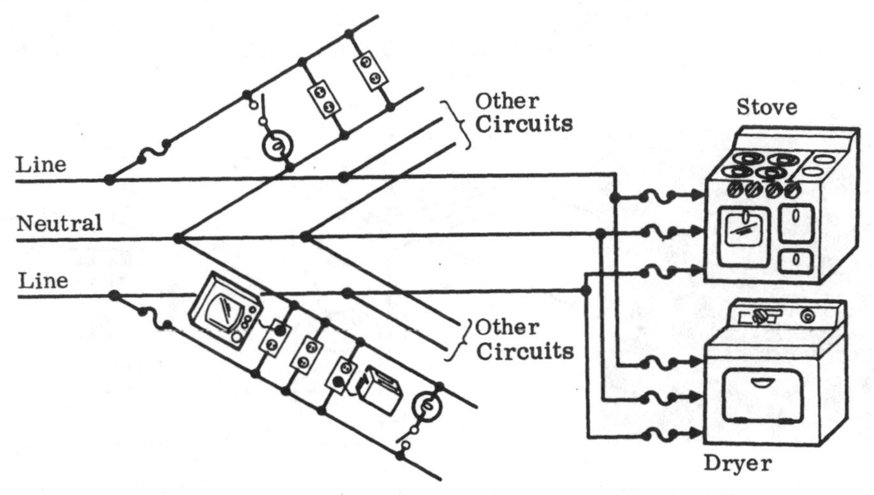

As you can see, the 120-volt circuits are normally loaded evenly so that the line currents are fairly well balanced. If the loads are perfectly balanced there is no current in the neutral. Thus, the neutral only carries the unbalanced current components. You can prove this to yourself by analyzing the simple diagram shown below:

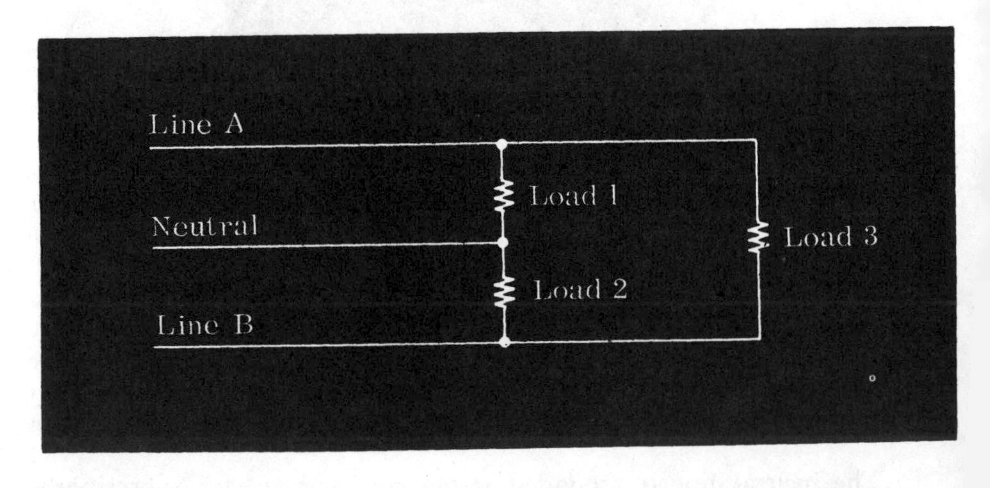

Remember that line A and line B are 180 degrees out of phase!

Troubleshooting Local Power Distribution Systems

Because of the way the circuits are connected to the lines, and because there are protective devices (fuses or circuit breakers about which you will learn more in Volume 5), it is relatively easy to isolate a circuit that has a fault. A blown fuse or open circuit breaker indicates an overload on a circuit. *You should never increase the size of the fuse or circuit breaker beyond the rated value.* Instead, if the line is simply overloaded from too many devices, you can move some of the loads to other circuits. If the overload persists, then either one of the devices on the circuit is defective or the line is shorted somewhere. This can be checked by removing the devices from the line and checking to see whether the overload still exists. A clamp-on ammeter is very useful in troubleshooting to allow for current measurement without disconnecting circuits.

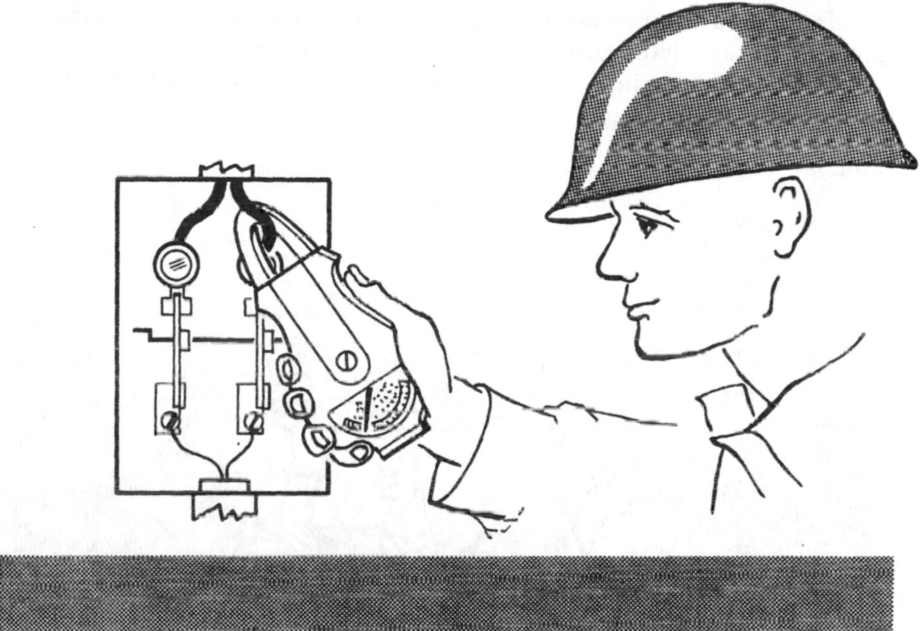

An open circuit can also easily be found because the voltage will be zero at all outlets or other convenient test points. If you know the way the wiring is arranged, you can trace the circuit to the point where the open circuit is and, more than likely, visually find the defect.

The importance of keeping the neutral line intact cannot be overemphasized! You could see this if you were to take the diagram on the previous page and put a very large load on one (low resistance) leg of the circuit and a very small load (high resistance) on the other leg. As long as the neutral is intact, the voltage across each load will be 120 volts and neutral current will flow. If you were to open the neutral, however, you would find that most of the voltage from the 240-volt line would appear across the smaller load, which, obviously, would burn out this device.

Introduction

You have already learned how to troubleshoot dc circuits in Volume 2 and also how to troubleshoot simple ac circuit elements (inductance and capacitance) in Volume 3. In this section, you will learn how to troubleshoot the ac circuits you have studied in this volume. As you learned in Volume 2, the important thing in troubleshooting is to use your head—that is, adopt a logical procedure and apply what you know. While hot components are not necessarily defective, the appearance of excessive heating can be a sure sign of difficulty. On the other hand, if components that should be at least warm are actually cold, then this condition may be equally suspicious. At this point, it is strongly recommended that you review the troubleshooting procedures described in Volumes 2 and 3. Use the ideas developed in Volume 2 to determine whether you are possibly dealing with a short or open circuit. As with dc circuits, such troubles are easier to locate than those arising from a partial failure.

Since you have just learned about transformers, we will learn how to troubleshoot them first.

Troubleshooting Transformer Circuits—Open Circuits

Since transformers are important parts of ac electrical equipment, you must know how to test them and how to locate faults that develop in them.

The three things that can cause transformer failures are (1) open circuit windings, (2) shorted windings, and (3) shorts to ground.

When one of the windings in a transformer develops an open circuit, no output current can flow and the transformer will not deliver any output. The symptom of an open-circuited transformer is, therefore, that the circuits which derive power from the transformer go dead. A check with an ac voltmeter across the transformer output terminals will show a reading of zero volts, although a voltmeter check across the transformer input terminals shows that voltage is present. Before you go further, this measurement is essential to make sure that the problem is not in the circuits feeding the transformer. Since there is voltage at the input and no voltage at the output, one of the windings must be open circuited. To find the fault, disconnect the transformer and check the windings for *continuity*, using an ohmmeter. Continuity (a continuous circuit) is indicated by a fairly low resistance reading, whereas an open winding will indicate an infinite resistance. In the majority of cases, the transformer will have to be replaced, unless of course the break is accessible and can be repaired.

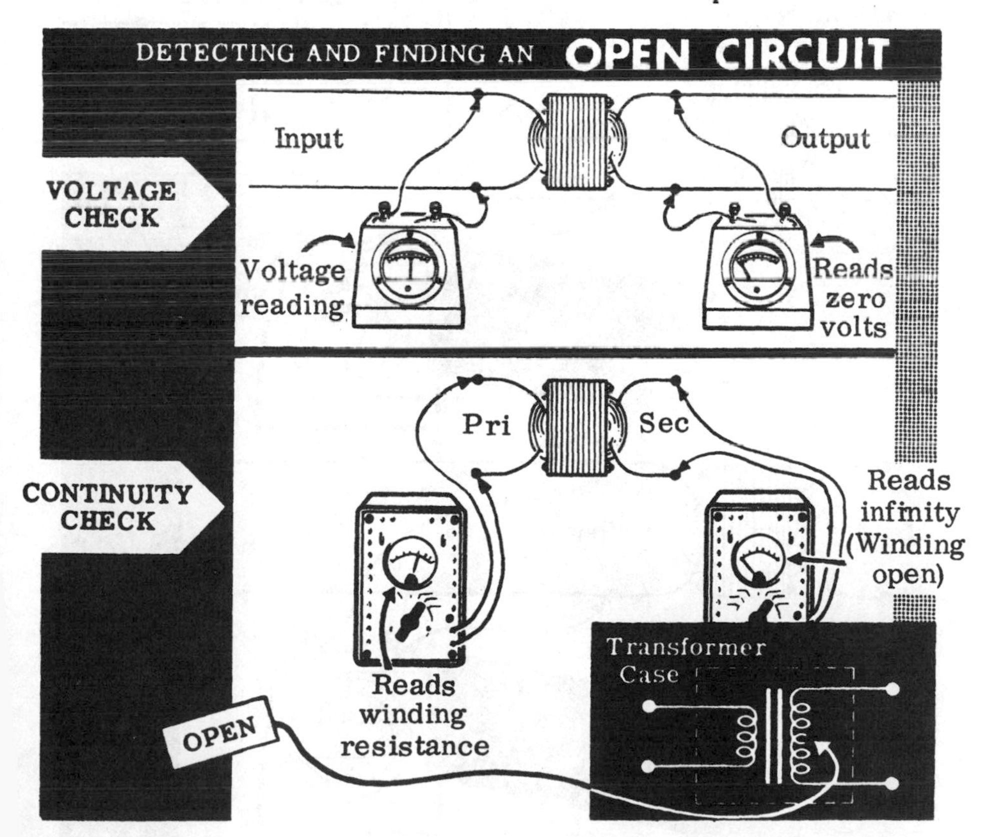

Troubleshooting Transformer Circuits—Short Circuits

When a few turns of a secondary winding of a transformer are shorted, the output voltage drops, and the transformer may overheat because of the increase in current in the secondary. The winding with the short may give a lower reading than normal on the ohmmeter. Usually the normal resistance reading is so low that a few shorted turns cannot be detected by using an ordinary ohmmeter. It is important to make sure that the problem is not in the load, which could be the reason why the transformer is drawing excessive current. One way to check this is to disconnect the load. If the transformer still overheats or the open circuit voltage is low, then it may be that the transformer has a shorted winding. In this case, a sure way to tell if the transformer is bad is to replace it with another equivalent transformer. If the replacement transformer operates satisfactorily, it should be used; the original transformer should be either repaired or discarded, depending on its size and type.

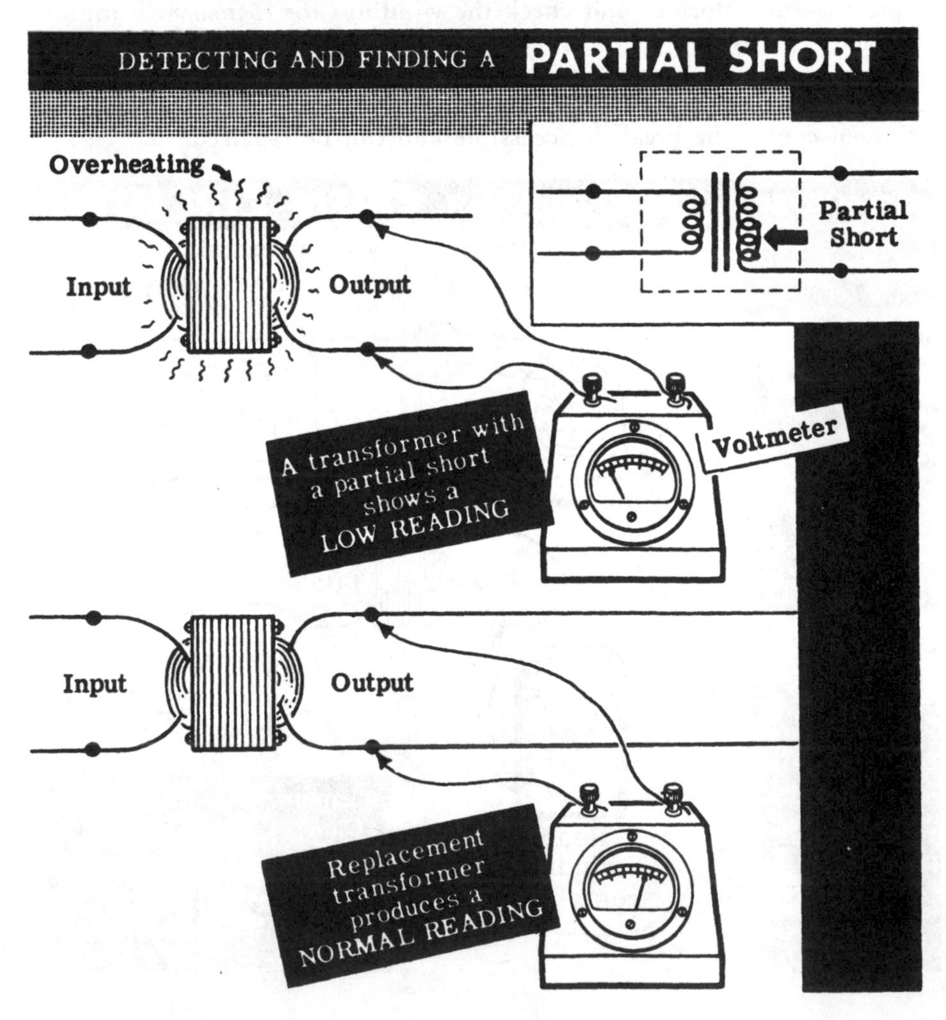
DETECTING AND FINDING A PARTIAL SHORT
Overheating
Input
Output
Partial Short
Voltmeter
A transformer with a partial short shows a LOW READING
Input
Output
Replacement transformer produces a NORMAL READING

Troubleshooting Transformer Circuits—Short Circuits (continued)

Sometimes a winding has a complete short across it. The short may be in the external circuit connected to the winding or in the winding itself. Again, one of the symptoms is excessive overheating of the transformer because of the very large circulating current. The heat often melts the insulating materials inside the transformer, a fact which you can quickly detect by the smell. Also, there will be no voltage output across the shorted winding, and the circuit across the winding will be dead. In equipment which is fused, the heavy current flow will blow the fuse before the transformer is damaged completely. But if the fuse does not blow, the shorted winding may burn out.

The way to isolate the short is to disconnect the external circuit from the winding. If the voltage is normal with the external circuit disconnected, the short is in the external circuit. But if the voltage across the winding is still zero, it means the short is in the transformer, which will have to be repaired or replaced.

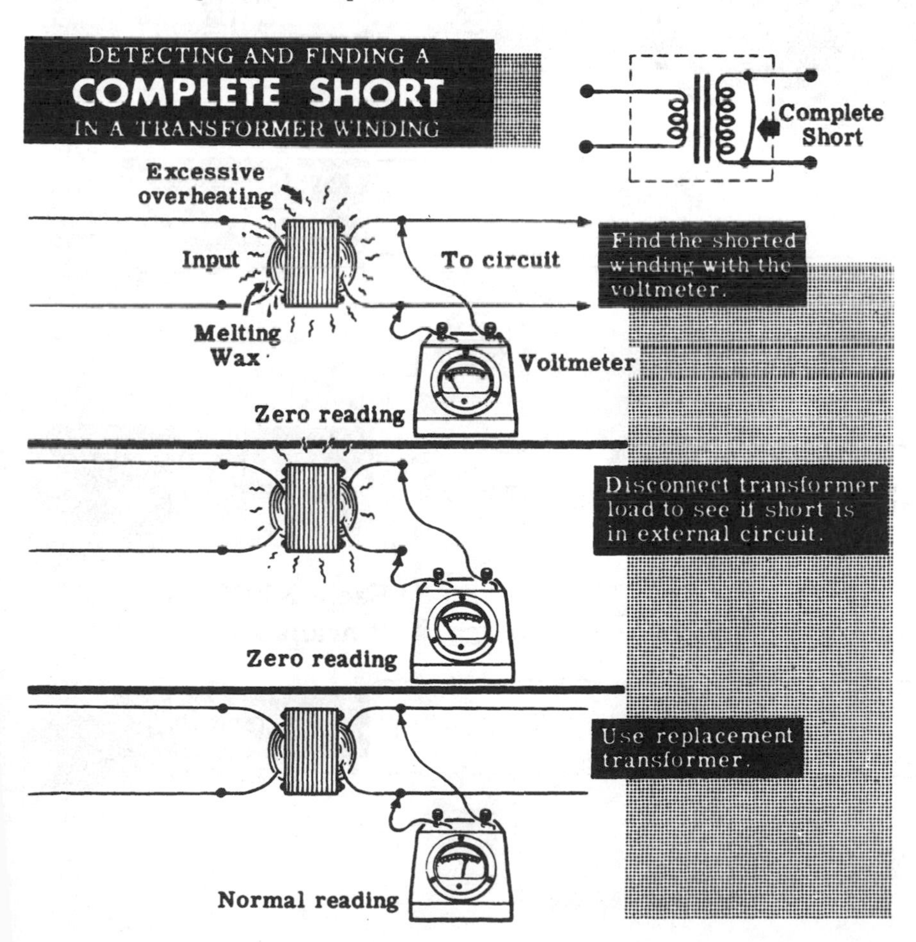
DETECTING AND FINDING A
COMPLETE SHORT
IN A TRANSFORMER WINDING
Complete Short
Excessive overheating
Input
To circuit
Find the shorted winding with the voltmeter.
Melting Wax
Voltmeter
Zero reading
Disconnect transformer load to see if short is in external circuit.
Zero reading
Use replacement transformer.
Normal reading

Troubleshooting Transformer Circuits—Short Circuits (continued)

Sometimes the insulation at some point in the winding breaks down, and the wire becomes exposed. It may touch the inside of the transformer case, shorting the wire to the case and thus grounding the winding.

If a winding develops a short to ground, and a point in the external circuit connected to this winding is also grounded, part of the winding will be shorted out. The symptoms will be the same as those for a shorted winding, and the transformer will have to be replaced or repaired. You can check whether a transformer winding is shorted to ground by connecting an ohmmeter between one side of the winding in question and the transformer case, but only after all the transformer leads have been disconnected from the circuit. A zero or low reading on the ohmmeter indicates that the winding is grounded.

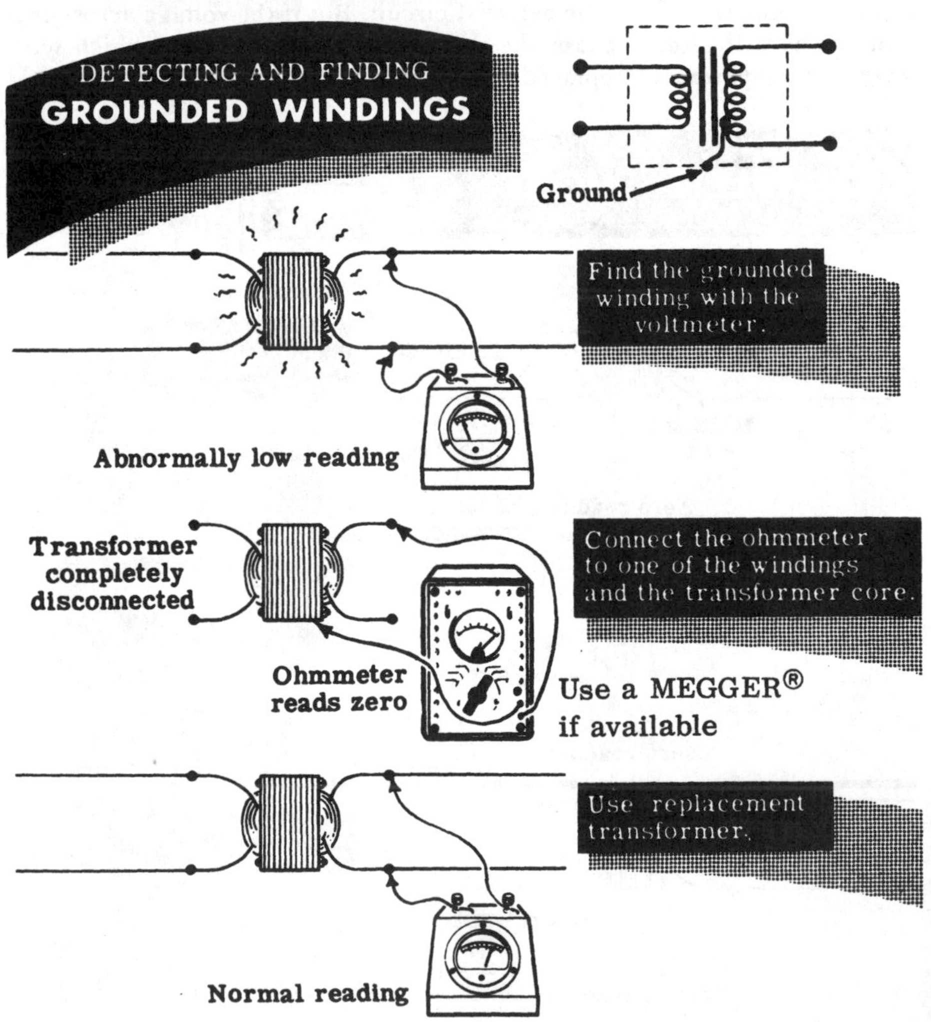
DETECTING AND FINDING
GROUNDED WINDINGS
Ground
Find the grounded winding with the voltmeter.
Abnormally low reading
Transformer completely disconnected
Connect the ohmmeter to one of the windings and the transformer core.
Ohmmeter reads zero
Use a MEGGER® if available
Use replacement transformer.
Normal reading

Troubleshooting AC Circuits

As for any troubleshooting, make sure that the input voltage is present and correct. In ac circuit work, you can use a clamp-on ammeter (see Volume 3) to measure circuit current without opening the circuit. You will find it a very convenient test instrument to have available. In addition, most of these meters have an auxiliary ac voltmeter that is very useful for checking circuit voltage.

Except for control applications, which you will learn about in Volume 5, series circuit connections are rarely used in ac circuits. In addition, you will find that complex ac circuits are found mainly in electronic systems. Usually, ac circuits are parallel connected. As you know, in parallel circuit connections the circuit voltage is common to all components (line voltage) but the currents depend on the component's impedance.

Suppose you had a line that was fused for 15 amperes (120 volts ac, 60 Hz) and, so far as you could tell, the normal load consisted of eight 100-watt lamps plus a small appliance that presented a load approximating that of a 50-ohm resistor in parallel with a 0.1-henry inductor.

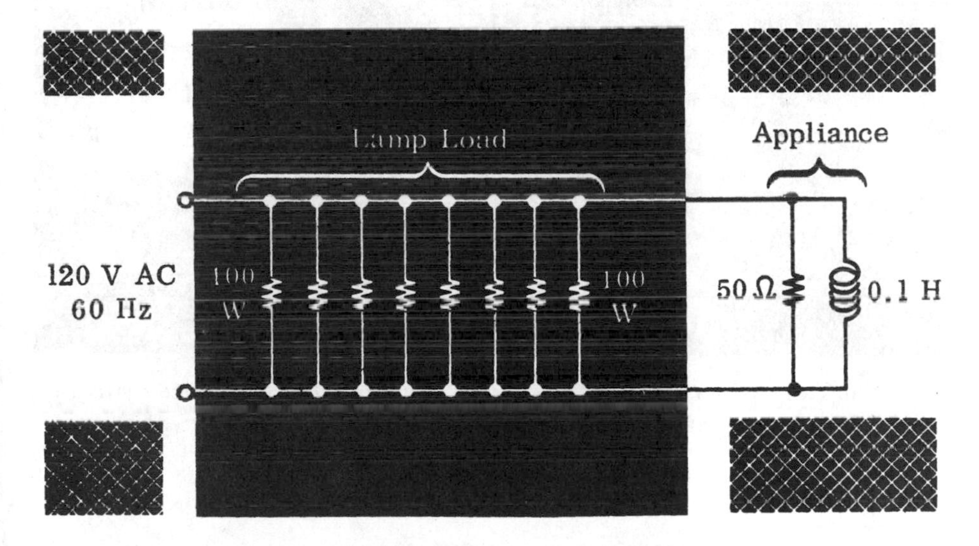

The fuse keeps blowing, indicating a circuit fault. With all the lamps and the appliance disconnected, the circuit is all right.

As you might suspect, a good way to find out which element is causing the difficulty would be to plug in each component one at a time to isolate the one at fault. You would, of course, leave the elements that function in the circuit, since the defective element alone might not draw enough current to cause the fuse to blow. As you return each circuit element to the line, check the current drawn by using the clamp-on ammeter.

You can use your knowledge of parallel circuits to figure out how much current each circuit component should draw.

Troubleshooting AC Circuits (continued)

Current Drawn by 100-Watt Lamp (Resistive)

$$P = EI$$

$$I = \frac{P}{E} = \frac{100}{120} = 0.8 \text{ A/lamp}$$

For 8 lamps, the load is $0.8 \times 8 = 6.4$ A

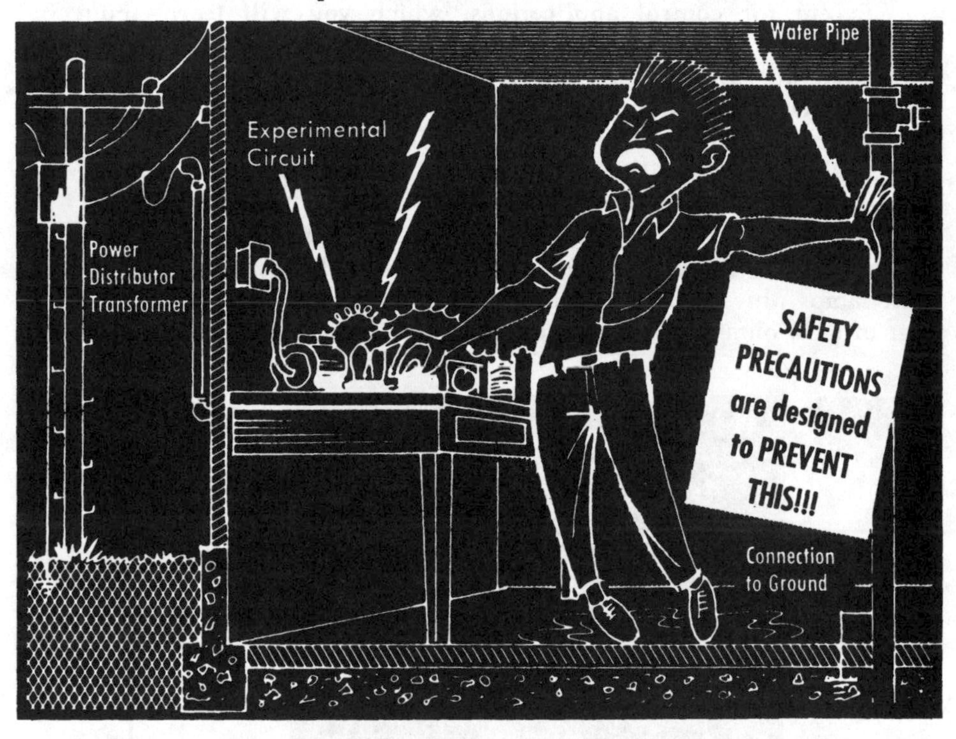

You can easily calculate the current drawn by the appliance by using Ohm's law and what you know about vectors. The impedance of the circuit is calculated as follows:

$$X_L = 2\pi fL = 2 \times \pi \times 60 \times 0.1 = 37.7 \text{ ohms}$$

$$Z = \frac{37.7 \times 50}{\sqrt{50^2 + 37.7^2}} = \frac{1{,}885}{62.6} = 30.11 \text{ ohms}$$

The current drawn by the appliance circuit is normally:

$$I = \frac{120}{30.11} = 3.99 \text{ A} \cong 4 \text{ A}$$

This, in combination with the lamp load, gives a total reading of

$$6.4 + 4 = 10.4 \text{ A}$$

Troubleshooting AC Circuits (continued)

A check with a clamp-on ammeter, or a series ammeter, shows that the appliance actually draws 10 amperes, which explains why the fuse blows.

The appliance can be checked out individually with an ohmmeter. In this case, the ohmmeter showed a partial short so that the inductor had a very low resistance. The problem was corrected by replacement of this part.

Remember that the fuse has to carry the *total* current and will blow at over 15 amperes, regardless of the phase angle. Thus, it is *volt-amperes* that need to be considered when a fuse or circuit breaker is involved. (However, only the in-phase component is involved with the real power used by the appliance.)

In later volumes, you will learn how to service the various electrical devices that are discussed. This special information coupled with what you know about electric circuits, and how to calculate them, will allow you to troubleshoot any electric circuit. More important than any particular item of test equipment, however, is *learning how to apply what you know.* Use your head and you will find that troubleshooting is not too difficult, even if you have just a few rudimentary tools. An essential part of troubleshooting is to know your electricity and, just as important, to know how the device or system you are troubleshooting works. The combination of this knowledge and skill will be the secret of your success in troubleshooting.

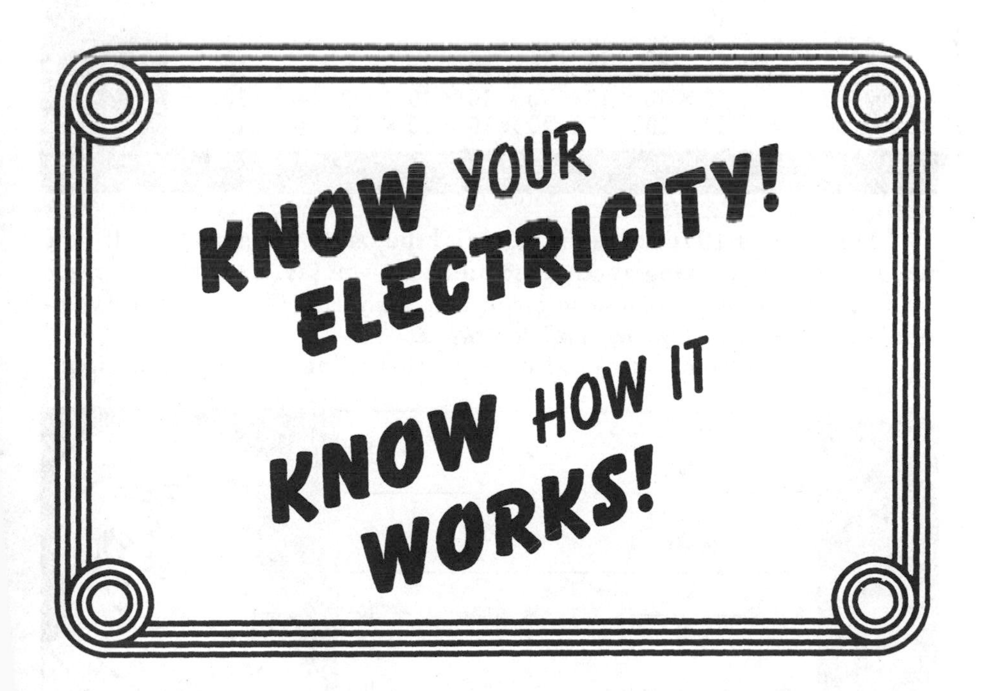

Scientific Notation

Your study so far has shown you that in working with electricity you will often use quantities as small as thousandths or millionths; in future work you may meet quantities as large as hundreds of thousands and even millions. While you have learned the rules for moving decimal points around to convert from one unit of measurement to another, there is a much simpler method available. That method is based upon understanding and using *scientific notation.*

Scientific notation is a shorthand method of writing down very large and very small numbers. Its use greatly simplifies making calculations with those numbers. This shorthand method makes use of the number 10 as the basis for writing down all large and small numbers.

The number of tens that have to be multiplied together to equal the desired number is written as a small numeral above and to the right of 10:

Scientific Notation

Desired Number		Number of Tens		Scientific Notation
1	=	Zero	=	10^0
10	=	10	=	10^1
100	=	10×10	=	10^2
1,000	=	$10 \times 10 \times 10$	=	10^3
10,000	=	$10 \times 10 \times 10 \times 10$	=	10^4
100,000	=	$10 \times 10 \times 10 \times 10 \times 10$	=	10^5
1,000,000	=	$10 \times 10 \times 10 \times 10 \times 10 \times 10$	=	10^6
10,000,000	=	$10 \times 10 \times 10 \times 10 \times 10 \times 10 \times 10$	=	10^7

The number 10 is called the *base*, and the smaller number is called the *exponent*. When speaking about these numbers, you say, *ten to the first, ten to the fourth, ten to the sixth*, and so forth. The numbers 10^2 and 10^3 are commonly known as *ten squared* and *ten cubed.*

Very small numbers can also be written using scientific notation:

$\frac{1}{10}$ =	$\frac{1}{10}$	= 0.1	=	10^{-1}
$\frac{1}{100}$ =	$\frac{1}{10 \times 10}$	= 0.01	=	10^{-2}
$\frac{1}{1000}$ =	$\frac{1}{10 \times 10 \times 10}$	= 0.001	=	10^{-3}
$\frac{1}{1,000,000}$ =	$\frac{1}{10 \times 10 \times 10 \times 10 \times 10 \times 10}$	= 0.000001	=	10^{-6}

Scientific Notation (continued)

You can use scientific notation to write down numbers smaller than 1 (one) by making the (negative) exponent equal to the count of zeros below the fraction line. For decimal numbers, the exponent is equal to one more than the number of places that the decimal point is located to the left of the first digit higher than zero. When speaking about these numbers you say *ten to the minus one, ten to the minus two, ten to the minus three,* and so forth. (You do not use the term *square* and *cube* when speaking of numbers smaller than one.)

Simply by following the concepts already explained, scientific notation can be used to indicate any number at all. For example:

The number to the left of the multiplication sign is the *multiplier.*

Although the examples that have been given should make the procedure for writing in scientific notation quite clear, there is a standard procedure that can be followed to resolve any questionable cases:

1. Write the desired quantity in terms of a decimal number.
2. Move the decimal point to the right or left depending on whether you are dealing with very large or very small numbers.
3. Count the number of places the decimal point was moved. This number is the exponent of the base 10.
4. If the decimal point was moved to the left, the exponent is positive (but no plus sign is required). If the decimal point was moved to the right, place a minus sign in front of the exponent.

Scientific Notation (continued)

In addition to simplifying the writing of large and small numbers, scientific notation makes it very easy to multiply and divide them. The rules are as follows:

Multiplication

1. *Multiply* the two *multipliers*. This gives you the new multiplier.
2. *Add* the two *exponents*. This gives you the new exponent.
3. Write the results in terms of scientific notation.
4. If necessary, move the decimal point to the right of the first digit larger than zero and adjust the value of the exponent. Refer to rules 3 and 4 on the previous page.

Example of Multiplication

Multiply 4×10^3 by 5.5×10^6
1. $4 \times 5.5 = 22.0$ (multiply)
2. $10^3 \times 10^6 = 10^{(3+6)} = 10^9$ (add exponents)
3. 22.0×10^9 (answer)
4. 2.2×10^{10} (answer adjusted)

Division

1. *Divide* the *first* multiplier (numerator) by the *second* (denominator). This gives you the new multiplier.
2. *Subtract* the denominator exponent from the numerator exponent. This gives you the new exponent.
3. Write the results in terms of scientific notation.
4. If necessary, adjust the value of the exponent. Refer to rules 3 and 4 on adjusting numbers.

Example of Division

Divide 3.6×10^9 by 6.0×10^3
1. $3.6 \div 6.0 = 0.6$
2. $10^9 \div 10^3 = 10^{(9-3)} = 10^6$
3. 0.6×10^6 (answer)
4. 6.0×10^5 (answer adjusted)

Note that $1/10^6$ is the same as 10^{-6}. You can prove this as follows:

$$\frac{1}{10^6} \times \frac{10^{-6}}{10^{-6}} = \frac{10^{-6}}{10^0} = \frac{10^{-6}}{1} = 10^{-6}$$

Scientific Notation (continued)

Conversely, $1/10^{-6} = 10^{6}$. Therefore, you can move factors-of-10 quantities from numerator to denominator or vice versa just by changing the sign of the exponent. You *cannot* do this with the multiplier; other rules of arithmetic hold in these cases.

You can also add *and* subtract numbers in scientific notation. However, *the value of the exponent must be the same before either operation:*

$$3 \times 10^{4} + 4.7 \times 10^{4} = 7.7 \times 10^{4}$$
$$8.2 \times 10^{-5} + 2 \times 10^{-5} = 10.2 \times 10^{-5} = 1.02 \times 10^{-4}$$

If you want to add or subtract numbers with different exponents, you must make the exponents *the same* before addition (or subtraction). For example, addition would be performed as follows:

$$5.1 \times 10^{3} + 3.2 \times 10^{4} = 5.1 \times 10^{3} + 32 \times 10^{3} = 37.1 \times 10^{3} = 3.71 \times 10^{-4}$$

$$3.5 \times 10^{-4} + 2.2 \times 10^{-3} = 3.5 \times 10^{-4} + 22 \times 10^{-4} = 25.5 \times 10^{-4} = 2.55 \times 10^{-3}$$

or you could do it as:

$$3.5 \times 10^{-4} + 2.2 \times 10^{-3} = 0.35 \times 10^{-3} + 2.2 \times 10^{-3} = 2.55 \times 10^{-3}$$

or, using subtraction:

$$6.3 \times 10^{3} - 2.1 \times 10^{2} = 6.3 \times 10^{3} - 0.21 \times 10^{3} = 6.09 \times 10^{3}$$
$$5.1 \times 10^{3} - 3.2 \times 10^{4} = 5.1 \times 10^{3} - 32 \times 10^{3} = -26.9 \times 10^{3} = -2.69 \times 10^{4}$$

To find the square of a number in scientific notation, multiply the multiplier by itself and double the exponent.

Example

$$(5 \times 10^{3})^{2} = 25 \times 10^{6}$$
$$(6.3 \times 10^{6}) = 39.69 \times 10^{12}$$

To take the square root of a number in scientific notation, adjust the exponent so that it is an even number. Take the square root of the multiplier and halve the exponent.

Example

$$\sqrt{5 \times 10^{3}} = \sqrt{50 \times 10^{2}} = 7.07 \times 10^{1} = 70.7$$
$$\sqrt{6.3 \times 10^{6}} = 2.51 \times 10^{3}$$

Measurement Conversion

Earlier you learned how to convert units of current measurement by moving the decimal point. These procedures and your knowledge of scientific notation will make it a simple matter to convert other units.

By learning the simple mathematical relationships that exist between all units of electrical and electronic measures, you will not only be able to convert current measurements, but you will also be able to convert the measurements of volts, ohms, watts, farads, and henrys.

Measurement Conversion (continued)

All electrical, electronic, and many other types of scientific measurement make use of standard prefixes which are attached to the front of the word that is used as the standard unit of measure. These prefixes indicate the precise multiplier or fraction of that standard unit. You already know that the prefix "milli" means one-thousandth and the prefix "micro" means one-millionth. The range of prefixes in common use is as follows:

Prefix	Abbrev.	Meaning	Mathematical Equivalent
pico (micromicro)	p ($\mu\mu$)	1 millionth of 1 millionth part of	10^{-12}
nano (millimicro)	n (mμ)	1 thousandth of 1 millionth part of	10^{-9}
micro	μ	1 millionth part of	10^{-6}
milli	m	1 thousandth part of	10^{-3}
centi	c	1 hundredth part of	10^{-2}
—	Unit	unit standard of measurement	10^{0}
kilo	k	1 thousand times	10^{3}
mega	M	1 million times	10^{6}
giga	G	1 thousand million times	10^{9}

For example, a thousandth of a volt is known as a *millivolt* and a million watts is known as a *megawatt*.

Measurement Conversion: Units of Current

Scientific notation makes it a simple matter to convert from one unit of measure to another, since the exponent tells you how the decimal point should be moved:

1. To change from any unit of measure to a *larger* one, move the decimal point to the *left* by a number of places equal to the difference between their exponents (as shown on the preceding page).
2. To change from any unit of measure to a *smaller* one, move the decimal point to the *right* by a number of places equal to the difference between their exponents (as shown on the preceding page).

To keep the explanation on familiar grounds, consider changing units of current measurement in the following examples.

Measurement Conversion: Units of Current (continued)

Example

Convert 3.5 milliampere (3.5 mA) to amperes.

Solution:
1. Milli = 10^{-3}; ampere (unit of measure) = 10^{0}
2. According to the rules, move the decimal point three places to the left.
3. 3.5 milliamperes = 0.003.5 ampere = 0.0035 ampere

Example

Convert 0.001 ampere to microamperes.

Solution:
1. Micro = 10^{-6}; ampere (unit of measure) = 10^{0}
2. According to the rules, move the decimal point six places to the right.
3. 0.001 ampere = .001000. microamperes = 1,000 μA

Example

Convert 2 milliamperes to microamperes.

Solution:
1. Micro = 10^{-6}; milli = 10^{-3}
2. According to the rules, move the decimal point three places to the right.
3. 2 milliamperes = 2.000. microamperes = 2,000 μA.

Measurement Conversion: Units of Voltage and Power

Units of voltage measurement are converted in exactly the same manner as units of current.

The exponents for the common units of voltage are as follows:

1 microvolt (μV) = 10^{-6}
1 millivolt (mV) = 10^{-3}
1 volt (V) = 10^{0}
1 kilovolt (kV) = 10^{3}
1 megavolt (MV) = 10^{6}.

Example

Convert 6,600 volts to kilovolts.

Solution:
1. Kilo = 10^{3}; volt (unit of measure) = 10^{0}
2. According to the rules, move the decimal point three places to the left.
3. 6,600 volts = 6.600. = 6.6 kilovolts

Measurement Conversion: Units of Voltage and Power (continued)

Likewise, similar conversions can be made with power units. The exponents for common units of power are:

1 microwatt (μW) = 10^{-6}
1 milliwatt (mW) = 10^{-3}
1 watt (W) = 10^{0}
1 kilowatt (kW) = 10^{3}
1 megawatt (MW) = 10^{6}.

Example

Convert 2.75 kilowatts to watts.

Solution:
1. Kilo = 10^{3}; watt (unit of measure) = 10^{0}
2. According to the rules, move the decimal point three places to the right.
3. 2.75 kilowatts = 2.750. = 2,750 watts

Measurement Conversion: Units of Resistance

Resistance is normally expressed in ohms. However, in some special applications, the resistance used may be as small as a fraction of an ohm or as large as many millions of ohms. This means that you may find resistance values expressed in microhms, milliohms, ohms, kilohms, and megohms and that you may have to convert from one unit of resistance measurement to another.

Units of resistance are converted in the same manner as units of current or voltage, by moving the decimal point as indicated by the exponent in scientific notation, and according to the table and rules of measurement conversion.

The exponents for the common units of resistance measurements are:

1 microhm (μΩ) = 10^{-6}
1 milliohm (mΩ) = 10^{-3}
1 ohm (Ω) = 10^{0}
1 kilohm (K or KΩ) = 10^{3}
1 megohm (M or MΩ) = 10^{6}

Example

Convert 47 K to ohms.

Solution:
1. Kilo = 10^{3}; ohm (unit of measure) = 10^{0}
2. According to the rules, move the decimal point three places to the right.
3. 47 K = 47.000. = 47,000 ohms

Measurement Conversion: Units of Resistance (continued)

Example

Convert 1.2 megohms to kilohms.

Solution: 1. Mega = 10^6; kilohm = 10^3
2. According to the rules, move the decimal point three places to the right.
3. 1.2 megohms = 1.200. kilohms = 1,200 K

Applying Scientific Notation

Example A

Calculate the impedance of the following circuit at 1 MHz.

0.1 mH 1,000 pF 100 Ω

$$X_L = 2\pi fL = 2 \times 3.14 \times 10^6 \times 0.1 \times 10^{-3} = 628 \text{ ohms}$$
$$X_C = 1/2\pi fC = 1/2 \times \pi \times 10^6 \times 10^{-9} = 159 \text{ ohms}$$
$$Z = \sqrt{(100)^2 + (628 - 159)^2} = \sqrt{10^4 + 2.76 \times 10^5}$$
$$= \sqrt{10^4 + 27.6 \times 10^4} = \sqrt{28.6 \times 10^4} = 5.35 \times 10^2 = 535 \text{ ohms}$$

Example B

Calculate the resonant frequency of the parallel circuit below:

0.01 μF 0.05 H (50 mH)

$$f_r = \frac{1}{2\pi\sqrt{LC}}$$
$$= \frac{1}{2\pi\sqrt{50 \times 10^{-3} \times 0.01 \times 10^{-6}}}$$
$$= \frac{1}{2\pi\sqrt{50 \times 10^{-3} \times 10^{-8}}} = \frac{1}{2\pi\sqrt{50 \times 10^{-11}}}$$
$$= \frac{1}{6.28\sqrt{5 \times 10^{-10}}} = \frac{1}{6.28 \times 2.24 \times 10^{-5}}$$
$$= \frac{1}{14.067 \times 10^{-5}} = \frac{10^5}{14.067} = 7{,}108 \text{ Hz} = 7.108 \text{ kHz}$$

Example

Example

Example C

What are the apparent and true powers for a 235-kV transmission line carrying 200 amperes at a phase angle of 15 degrees?

Apparent power $= E \times I = 235 \text{ kV} \times 200 = 235 \times 10^3 \times 2 \times 10^2$
$= 470 \times 10^5$ watts = 47 MW

True power = apparent power $\times \cos\theta$
$= 47 \text{ MW} \times \cos\theta = 47 \text{ MW} \times 0.966 = 45.4 \text{ MW}$

Table of Squares and Square Roots

To use this table for values of N greater than 100, move the decimal point to the left two places at a time until a number less than 100 is obtained. Look up the square or square root as required for this number, then adjust your answer as follows: If the number is being squared, add *two* zeros to the left of the decimal point for each place you moved the decimal point originally. If the square root is being taken, move the decimal point one place to the right for each two places you originally moved it to the left.

Examples: **What is the square of 900?**

$900^2 = 9^2$ (moved decimal two places to the left) = 81 (move decimal *back* two places to the right for *each place* moved to the left originally) = 81 00 00 = 810,000.

What is the square root of 900?

$\sqrt{900} = \sqrt{9}$ (moved decimal two places to the left) = 3 (move decimal *back* one place to the right for *each two places* moved to the left originally) = 3 0 = 30.

N	N^2	$\sqrt{N}$	N	N^2	$\sqrt{N}$	N	N^2	$\sqrt{N}$
1	1	1.000	36	1,296	6.000	71	5,041	8.426
2	4	1.414	37	1,369	6.083	72	5,184	8.485
3	9	1.732	38	1,444	6.164	73	5,329	8.544
4	16	2.000	39	1,521	6.245	74	5,476	8.602
5	25	2.236	40	1,600	6.325	75	5,625	8.660
6	36	2.449	41	1,681	6.403	76	5,776	8.718
7	49	2.646	42	1,764	6.481	77	5,929	8.775
8	64	2.828	43	1,849	6.557	78	6,084	8.832
9	81	3.000	44	1,936	6.633	79	6,241	8.888
10	100	3.162	45	2,025	6.708	80	6,400	8.944
11	121	3.317	46	2,116	6.782	81	6,561	9.000
12	144	3.464	47	2,209	6.856	82	6,724	9.055
13	169	3.606	48	2,304	6.928	83	6,889	9.110
14	196	3.742	49	2,401	7.000	84	7,056	9.165
15	225	3.873	50	2,500	7.071	85	7,225	9.220
16	256	4.000	51	2,601	7.141	86	7,396	9.274
17	289	4.123	52	2,704	7.211	87	7,569	9.327
18	324	4.243	53	2,809	7.280	88	7,744	9.381
19	361	4.359	54	2,916	7.348	89	7,921	9.434
20	400	4.472	55	3,025	7.416	90	8,100	9.487
21	441	4.583	56	3,136	7.483	91	8,281	9.539
22	484	4.690	57	3,249	7.550	92	8,464	9.592
23	529	4.796	58	3,364	7.616	93	8,649	9.644
24	576	4.899	59	3,481	7.681	94	8,836	9.695
25	625	5.000	60	3,600	7.746	95	9,025	9.747
26	676	5.099	61	3,721	7.810	96	9,216	9.798
27	729	5.196	62	3,844	7.874	97	9,409	9.849
28	784	5.292	63	3,969	7.937	98	9,604	9.899
29	841	5.385	64	4,096	8.000	99	9,801	9.950
30	900	5.477	65	4,225	8.062	100	10,000	10.000
31	961	5.568	66	4,356	8.124			
32	1,024	5.657	67	4,489	8.185			
33	1,089	5.745	68	4,624	8.246			
34	1,156	5.831	69	4,761	8.307			
35	1,225	5.916	70	4,900	8.367			

Table of Sines (Sin), Cosines (Cos), and Tangents (Tan)

To find sin, cos, or tan, look up the nearest angle and read across to the right to the appropriate value. To find $\sin^{-1}$, $\cos^{-1}$, or $\tan^{-1}$, look for the closest value and read across to the left and find the angle. For more accuracy, you can estimate between values.

Angle	*Sin*	*Cos*	*Tan*
0	0	1	0
2.5	0.044	0.999	0.044
5	0.087	0.996	0.087
7.5	0.131	0.991	0.132
10	0.174	0.985	0.176
12.5	0.216	0.976	0.222
15	0.259	0.966	0.268
17.5	0.301	0.954	0.315
20	0.342	0.940	0.364
22.5	0.383	0.924	0.414
25	0.423	0.906	0.466
27.5	0.462	0.887	0.521
30	0.500	0.866	0.577
32.5	0.537	0.843	0.637
35	0.574	0.819	0.7
37.5	0.609	0.793	0.767
40	0.643	0.766	0.839
42.5	0.676	0.737	0.916
45	0.707	0.707	1.000
47.5	0.737	0.676	1.091
50	0.766	0.643	1.192
52.5	0.793	0.609	1.303
55	0.819	0.574	1.428
57.5	0.843	0.537	1.570
60	0.866	0.500	1.732
62.5	0.857	0.462	1.921
65	0.906	0.423	2.145
67.5	0.924	0.383	2.414
70	0.940	0.342	2.747
72.5	0.954	0.301	3.172
75	0.966	0.259	3.732
77.5	0.976	0.216	4.511
80	0.985	0.174	5.671
82.5	0.991	0.131	7.596
85	0.996	0.087	11.430
87.5	0.999	0.044	22.904
90	1.000	0	∞ (all reactance)

Learning Objectives—Next Volume

Overview—In the first four volumes of this course on basic electricity you have dealt with the fundamental principles of electricity and dc and ac electric circuits. To complete your study of electricity, you will learn how electrical machines make use of the properties of electricity and magnetism to generate electricity—generators—and to do mechanical work—motors. The construction and operation principles of generators, alternators, electric motors, and other dc and ac electrical devices and equipment will be described. Then you will see how these dc and ac generators and motors are connected to form a system and how various devices are used to control the system. In addition, you will learn how electric power is controlled and how it is used in its many applications. Also, you will learn something about solid-state devices that are used in electric power conversion, control, and distribution. Finally, you will be introduced to the maintenance and troubleshooting of dc and ac machinery and systems.

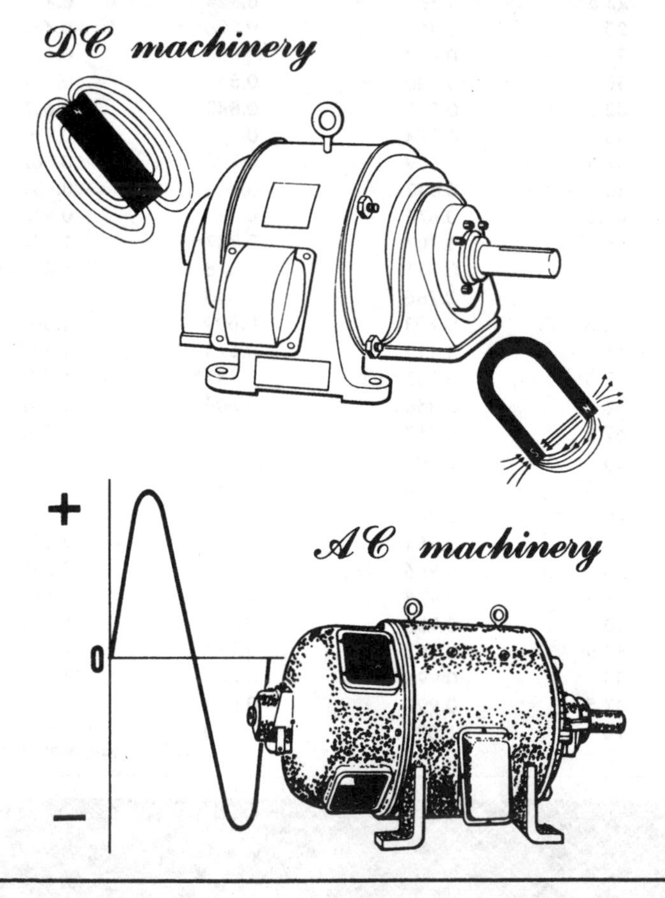

Basic Electricity

REVISED EDITION

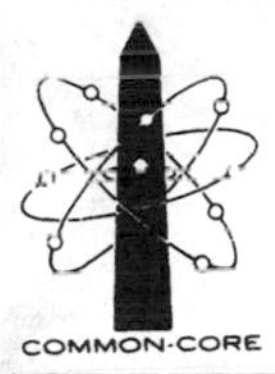

VAN VALKENBURGH,
NOOGER & NEVILLE, INC.

VOL. 5

DC machinery

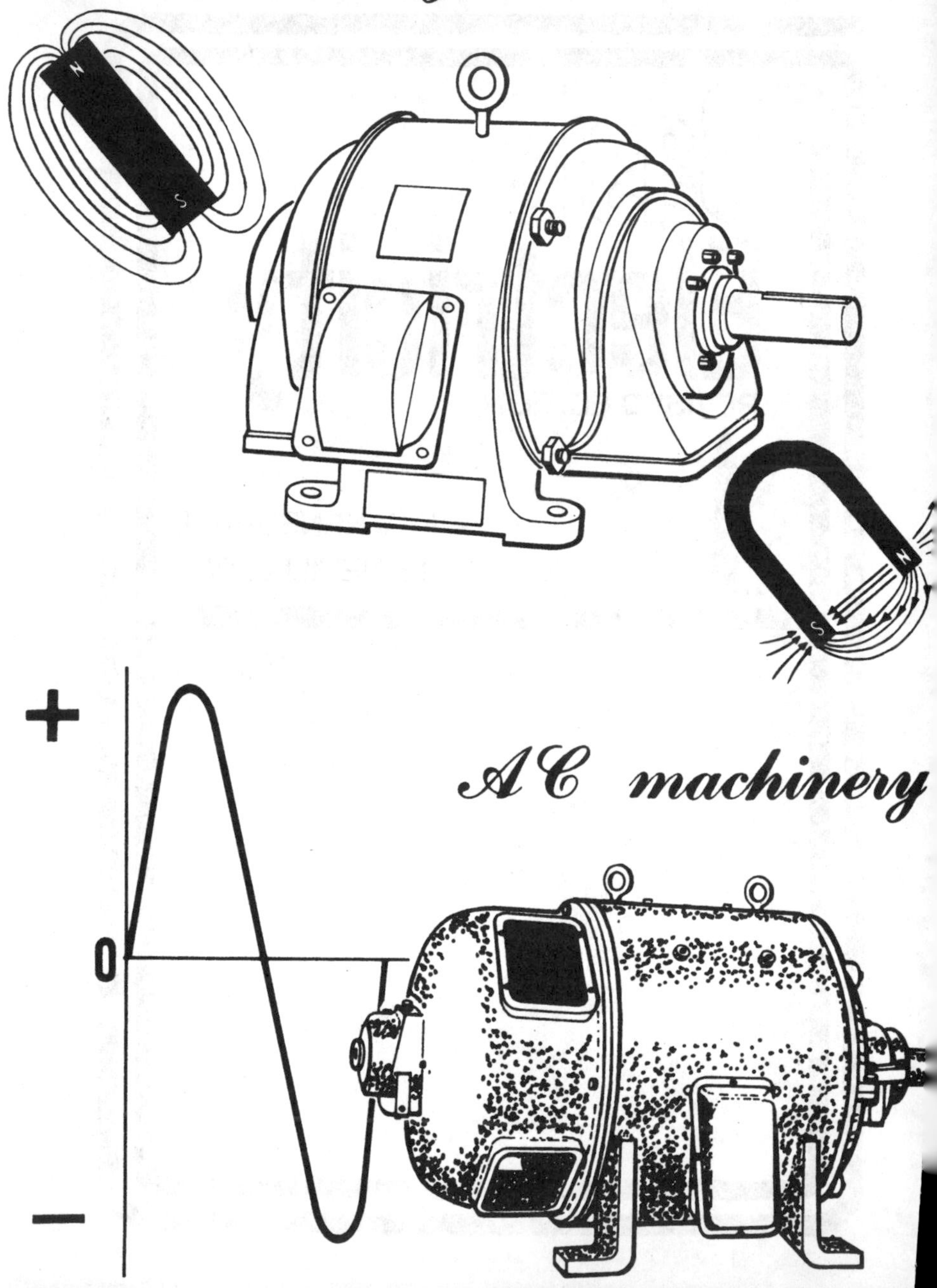

Basic Generators and Motors

Most of us live and work at the end of an electric wire. You will remember from "How Electricity Is Produced and Used" in Volume 1 of *Basic Electricity* that different forms of energy can be converted to electricity (electrical energy) and that, likewise, electricity (electrical energy) can be converted to different forms of energy. The electric generator is the device that converts *mechanical* energy to *electrical* energy—the electric motor, which is essentially a generator used differently, converts the *electrical* energy *back* to *mechanical* energy. As you may remember, the generator is used to provide almost all of the electrical energy that we use today. One of our major problems is finding energy sources to run these generators. Because of this, we may find it necessary increasingly to use new and alternative sources of energy in the future. But for now, we depend almost entirely on the electric generator for the electrical energy we use. Of course, we use electrical energy for many purposes other than providing mechanical energy via motors.

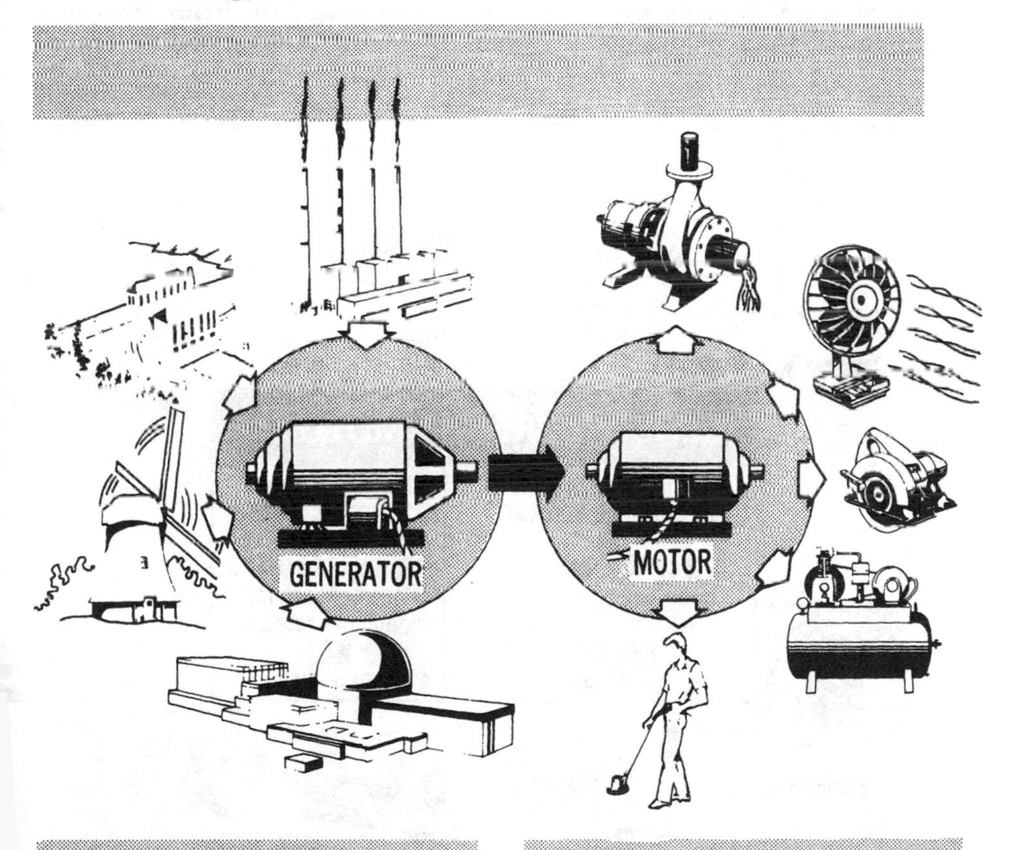

MECHANICAL→ ELECTRICAL ENERGY

ELECTRICAL→ MECHANICAL ENERGY (and other uses)

Basic Generators and Motors (continued)

You already know the principles necessary to understand how generators and motors work. In this volume, you will find out how generators and motors work, how they are controlled, and how to troubleshoot them.

Although there are many variations in generators and motors, you will find them all basically very similar. All electric generators and motors use the interaction between moving conductors and magnetic fields (or vice versa). It is the difference in the *way* these conductors and magnetic fields are *arranged* and in their *outputs* (electrical or mechanical) that results in the differences in the devices that we will learn about here. It would be wise at this point to review the information on magnetism and magnetic effects in Volumes 1, 3, and 4 before proceeding with your study of generators and motors.

Although you will learn about dc generators and motors and ac generators and motors, it is important to realize now that the operation of any one of these depends simply on interacting magnetic fields involving current-carrying conductors.

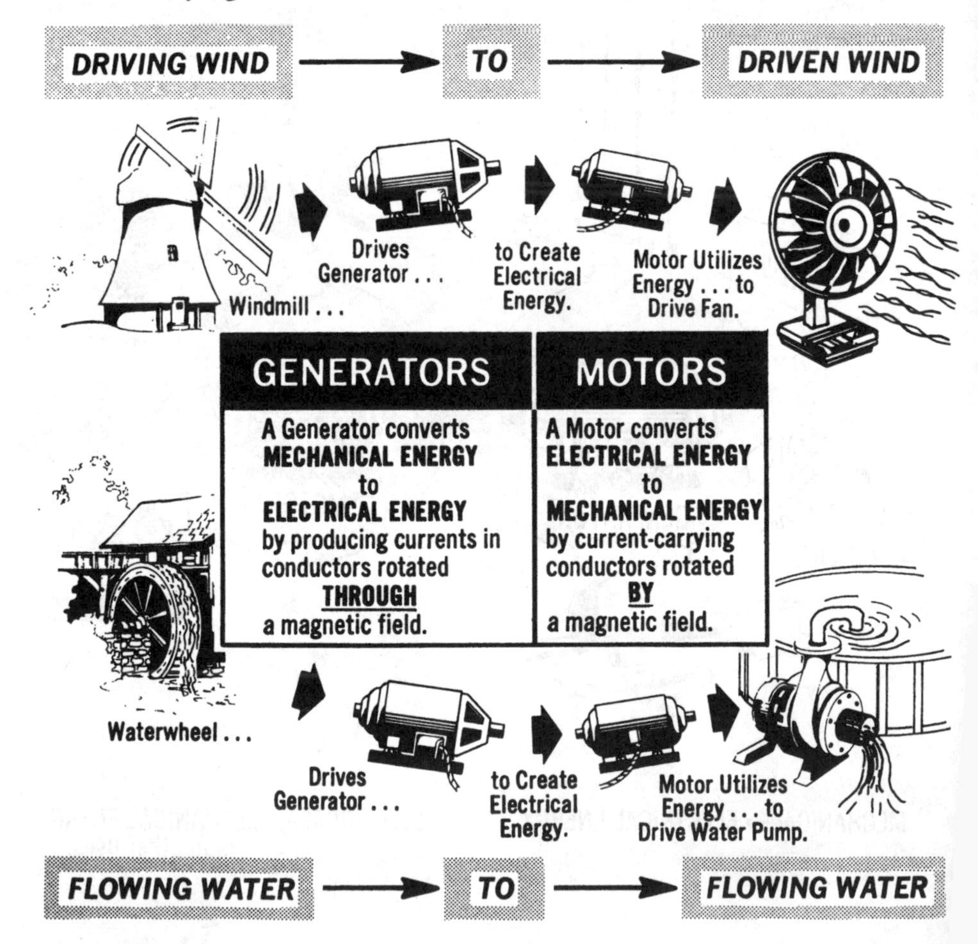
DRIVING WIND
TO
DRIVEN WIND
Windmill . . .
Drives Generator . . .
to Create Electrical Energy.
Motor Utilizes Energy . . . to Drive Fan.
GENERATORS
A Generator converts MECHANICAL ENERGY to ELECTRICAL ENERGY by producing currents in conductors rotated THROUGH a magnetic field.
MOTORS
A Motor converts ELECTRICAL ENERGY to MECHANICAL ENERGY by current-carrying conductors rotated BY a magnetic field.
Waterwheel . . .
Drives Generator . . .
to Create Electrical Energy.
Motor Utilizes Energy . . . to Drive Water Pump.
FLOWING WATER
TO
FLOWING WATER

Generator Energy Sources

As you learned in Volume 1, voltage, or an emf, is induced in a conductor moving through a magnetic field. All of the power stations supplying almost all of the electric power consumed in the world today make use of this simple principle to convert some source of energy into electrical energy.

Most power stations use the heat energy produced by burning fossil fuels such as coal, oil, or natural gas to produce steam. The steam is then used to drive a turbine coupled to a generator. Several changes in the form of the energy thus take place: the *chemical* energy of the fuel is first converted into *heat* energy; the heat energy is converted into *mechanical* energy of motion in the turbine; and finally, the mechanical energy of motion is converted into *electrical* energy in the generator. Whatever the original source of energy—coal, oil, gas, plutonium, uranium, a head of water, the sun, the wind—the final step is always the conversion of the *mechanical* energy of rotation into *electrical* energy in a generator.

This is also true of all the small transportable generating sets that you find in ships and motor vehicles or in sources of emergency power.

Review of Electricity from Magnetism

You will recall that electricity can be generated by moving a wire through a magnetic field. As long as there is relative motion between the conductor and the magnetic field, electricity is generated. The generated voltage is called an *induced voltage*, or *induced emf*, and the method of generating it by cutting a magnetic field with a conductor is called *induction*.

You know that the amount of voltage induced in the wire cutting through the magnetic field depends on a number of factors. First, if the speed of the relative cutting action between the conductor and the magnetic field increases, the induced emf or voltage increases. Second, if the strength of the magnetic field is increased, the induced voltage or emf also increases. Third, if the number of turns cutting through the magnetic field is increased, the induced emf (voltage) is increased again.

The polarity of this induced voltage or emf will be such that the resulting current flow will build up a field to react with the field of the magnet and to oppose the movement of the coil. This phenomenon illustrates *Lenz's law*, which states that *in electromagnetic induction, the direction of the induced emf, and hence current flow, is such that the magnetic field setup opposes the motion which produces it.*

FACTORS WHICH DETERMINE INDUCED EMF STRENGTH

. . . the SPEED of conductor through magnetic field

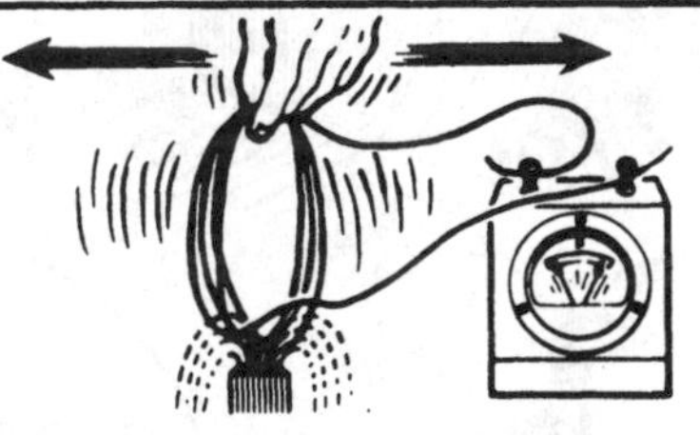

. . . the STRENGTH of magnetic field

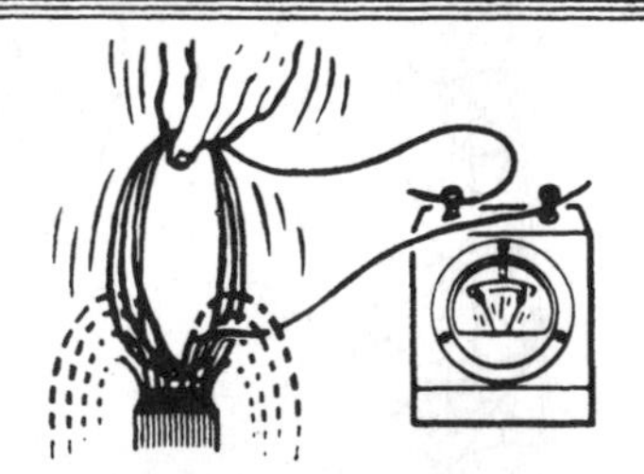

. . . the NUMBER of turns

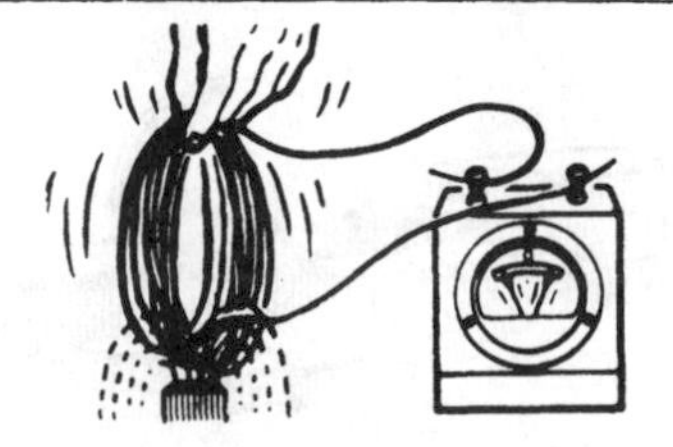

Review of Electricity from Magnetism (continued)

The induced voltage (E) in each conductor is proportional to the strength of the magnetic field times the speed of the conductor through the field:

$$E \approx \text{flux} \times \text{speed}$$

You also know that the polarity of the induced voltage, and hence the direction of the generated current flow, is determined by the *direction* of the *relative motion* between the magnetic field and the cutting conductor.

DIRECTION OF RELATIVE MOTION DETERMINES DIRECTION OF CURRENT FLOW

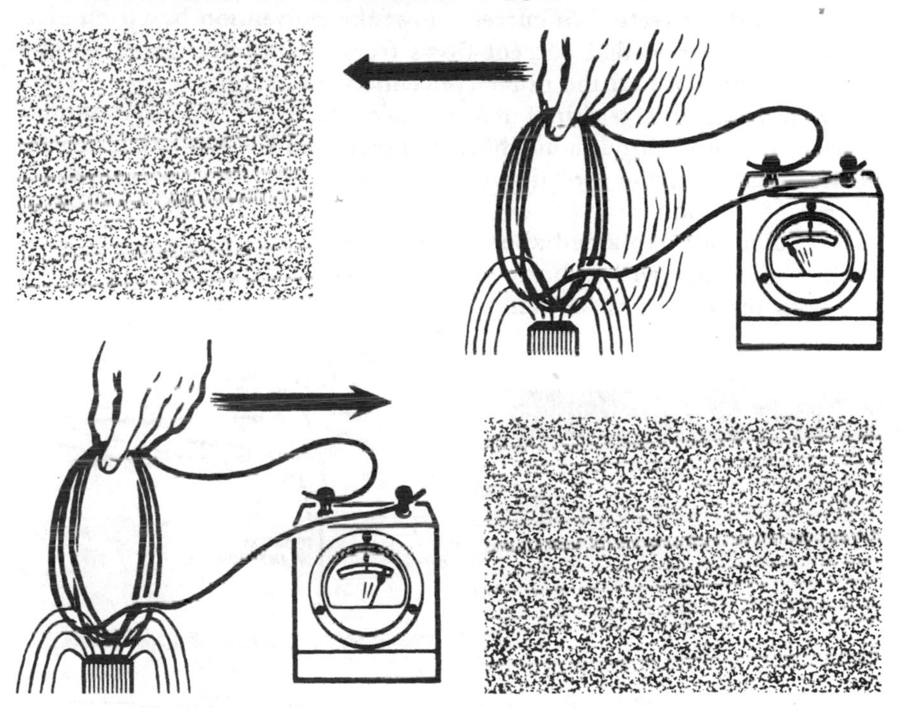

So to sum up what you already know about electricity from magnetism:

1. Moving a conductor through a magnetic field generates an emf or voltage that produces a current flow.
2. The faster the conductor cuts through the field, the more turns there are, and the stronger the magnetic field—the greater the induced emf or voltage, and the greater the current flow.
3. Reversing the direction of movement of the conductor reverses the polarity of the induced emf (voltage), and therefore the direction of current flow is reversed.
4. It doesn't matter whether the conductor or the magnetic field changes or moves, the *result* is the *same*.

The Left-Hand Rule

You have seen how an emf or voltage is generated in the coil of the elementary generator. There is a simple method for remembering the direction of the emf induced in a conductor moving through a magnetic field; it is called the *left-hand rule for generators.* This rule states that if you hold the thumb and the first and middle fingers of the left hand at right angles to one another, with the first finger pointing in the flux direction and the thumb pointing in the direction of motion of the conductor, the middle finger will point in the direction of the induced emf. Direction of induced emf means the direction in which current will flow as a result of this induced emf.

You will remember from Volume 1 that there are two conventions (views) as to the direction of current flow: the convention based on electron theory that states that current flows from the negative terminal of a source of electricity, and the older convention that supposes that current flows from the positive terminal of a source of electricity. The first of these conventions is used throughout these volumes, and the left-hand rule for generators indicates the direction of electron current flow in accordance with that convention.

The same rule explained above, but applied to the *right* hand instead of to the left hand, will indicate the direction of current flow in accordance with the older convention.

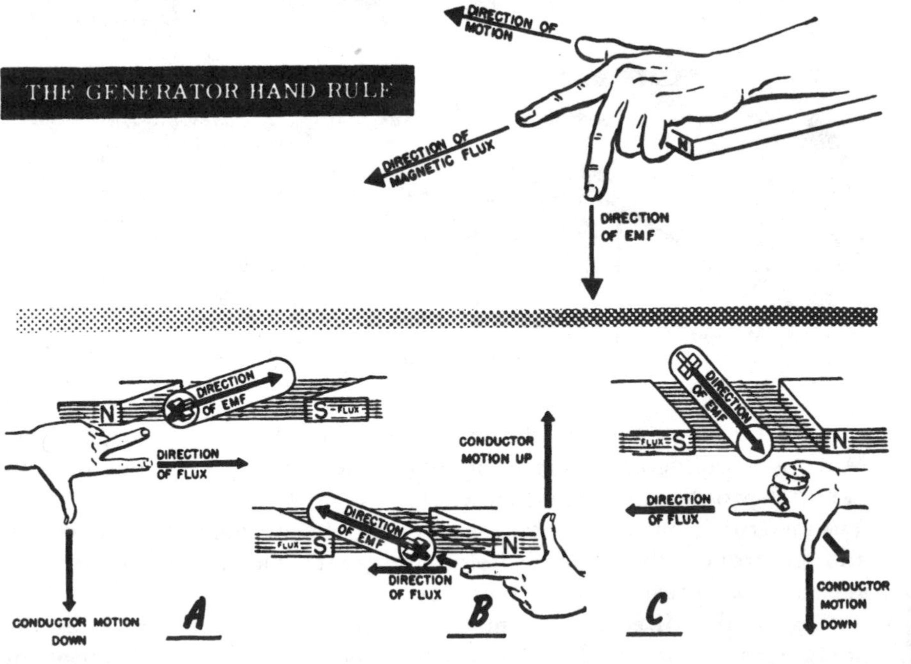

We will also use the convention that a conductor shown as ⊕ means that the current is flowing *away* from the observer, whereas one shown as ⊙ means that the current is flowing *toward* the observer.

The Parts of an Elementary Generator

An elementary generator consists of a loop of wire so placed that it can be rotated in a uniform magnetic field to cause an induced emf in the loop. A pair of sliding contacts are employed to connect the loop to an external circuit in order to use the induced emf.

The *pole pieces* are the north and south poles of the magnet that supplies the magnetic field. The loop of wire that rotates through the field is called the *armature.* The ends of the armature loop are connected to rings, called *slip rings,* that rotate with the armature. *Brushes* ride up against the slip rings to connect the armature to the external circuit. You will recall that this is the same elementary generator used to generate an ac voltage as described at the beginning of Volume 3.

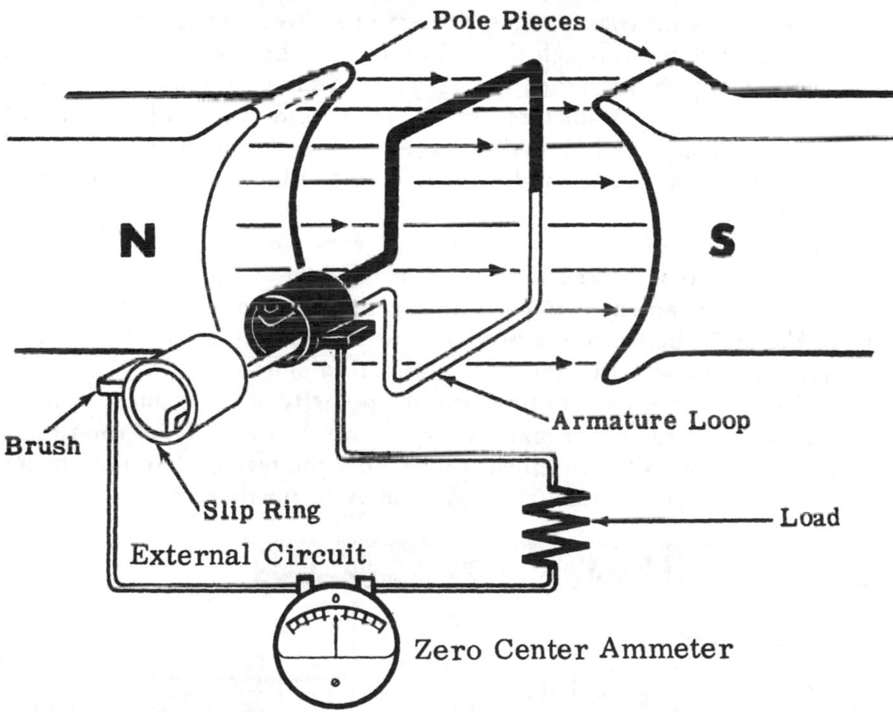

In the description of the generator action given in the following pages, you should visualize the loop rotating through the magnetic field. (However, please remember that you could just as easily rotate the *magnet assembly.*) As the sides of the loop cut through the magnetic field, an emf is induced in them which causes a current to flow through the loop, slip rings, brushes, ammeter, and load resistor—all connected in series. The magnitude of the induced emf generated in the loop, and therefore of the current that flows, depends on the instantaneous position of the loop in the magnetic field.

Elementary Generator Operation

As you remember from Volume 3, the elementary generator works like this: Assume that the armature loop is rotating in a clockwise direction and that its initial position is at A (0°) (see diagram below). In position A, the loop is perpendicular to the magnetic field, and the black and white conductors of the loop are moving parallel to the magnetic field. If a conductor is moving parallel to a magnetic field, its relative motion is zero, it does not cut through any lines of force, and no emf is generated in the conductor. This applies to the conductors of the loop at the instant they go through position A. Thus, no emf is induced in the loop, and no current flows through the circuit. The ammeter reads zero.

As the loop rotates from position A to position B, the conductors are cutting through the lines of force at a faster and faster rate, until at 90° (position B) they are cutting through lines of force at the maximum rate. In other words, between 0° and 90°, the induced emf in the conductors builds up from zero to a maximum value. Observe that from 0° to 90° the black conductor cuts down through the field while at the same time the white conductor cuts up through the field. As you can show by Lenz's law, the induced emfs in both conductors are in series; and the resultant voltage across the brushes (the terminal voltage) is the sum of the two induced emfs, or double that of one conductor, since the induced voltages are equal to each other.

The current through the circuit will vary just as the induced emf varies, being zero at 0° and rising to a maximum at 90°. If the ammeter could follow the variations in current, it would show an increasing deflection to the right between positions A and B, indicating that the current through the load was flowing in the direction shown.

The direction of current flow and the polarity of the induced emf depend on the direction of the magnetic field and the direction of rotation of the armature loop. The waveform shows how the terminal voltage of the elementary generator varies from position A to position B.

HOW THE ELEMENTARY GENERATOR WORKS

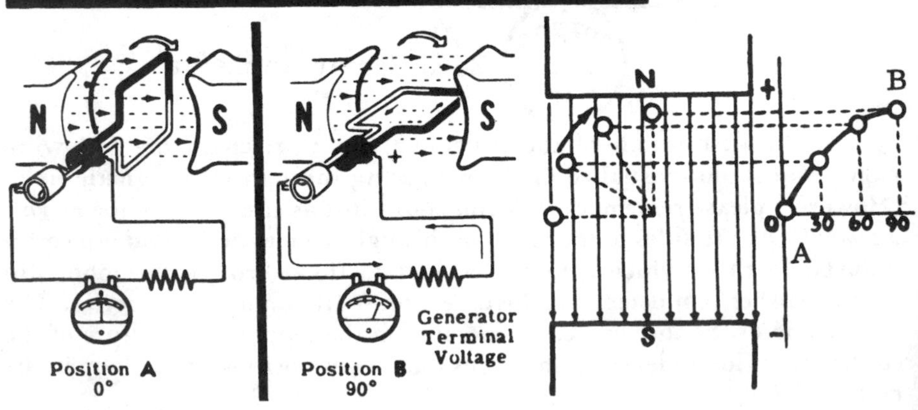

Elementary Generator Operation (continued)

As the loop continues rotating from position B (90°) to position C (180°), the conductors, which are cutting through the lines of force at a maximum relative rate at position B, begin to cut through lines more and more slowly, until at position C they are moving parallel to the magnetic field, and there is no relative velocity between the field and the conductor. The induced emf will therefore decrease as the loop moves from 90° to 180°, and the current flow will also vary as the voltage varies.

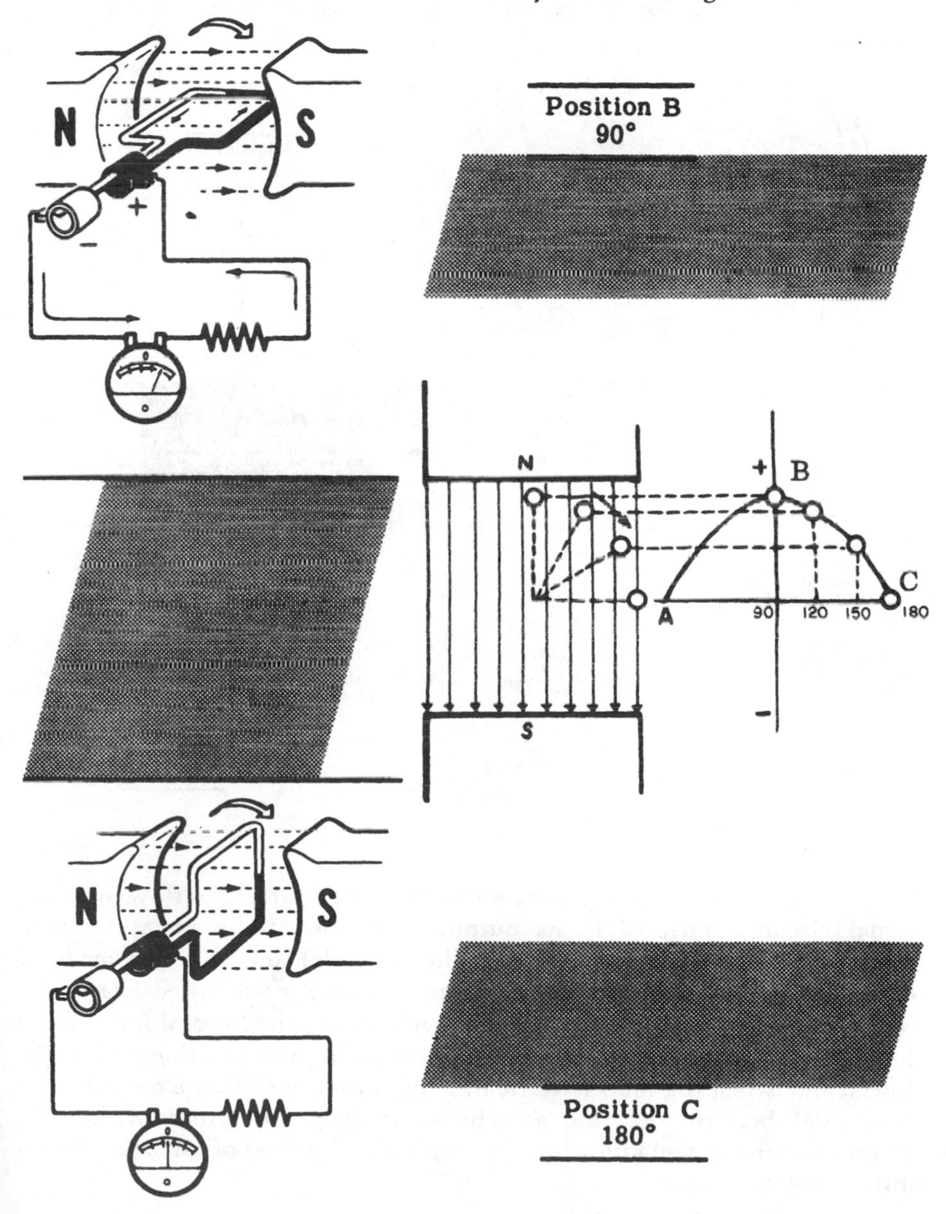

Elementary Generator Operation (continued)

From 0° to 180° the conductors of the loop have been moving in the same direction through the magnetic field, and the polarity of the induced emf has therefore remained the same. As the loop starts rotating beyond 180° back to position A, however, the direction of the cutting action of the conductors through the magnetic field reverses. Now the black conductor cuts *up* through the field, and the white conductor cuts *down* through the field. As a result, the polarity of induced emf and the current flow, will reverse. The voltage output waveform for the complete revolution of the loop is shown below.

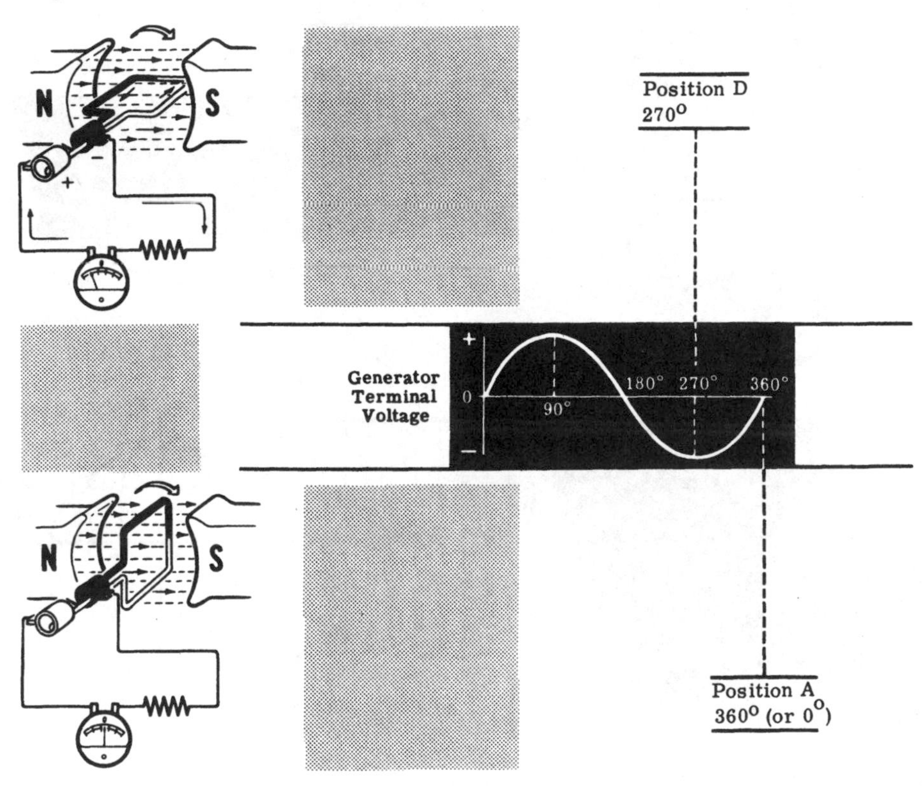

As you know, the emf generated will cause ac current to flow in an external circuit connected to its output terminals. If the armature were rotated 60 times (cycles) per second (hertz or Hz), then the frequency of the ac output would be 60 Hz. Ac generators are often called *alternators*. When the loop is rotated at constant speed, it cuts the lines of force at the highest rate when the loop moves from the horizontal position and at the lowest rate when the loop is vertical. This causes the voltage output to be sinusoidal because the rate at which the loop cuts lines of force is sinusoidal, and as you know it is the rate of cutting lines of force that determines output voltage.

The Commutator

As you have seen, the elementary generator is an ac generator. If an ac output is desired, then the generator is complete. You will study practical ac generators later in this volume.

In the elementary generator, the ac voltage induced in the loop reverses its polarity every time the loop goes through the 0° and 180° points. At these points, the conductors of the loop reverse their direction through the magnetic field. As you already know, the polarity of the induced emf depends on the direction in which a conductor moves through a magnetic field; and if the direction reverses, the polarity of the induced emf reverses. Since the loop continues rotating through the field, the emf induced in the conductors of the loop will always be alternating. Therefore, the only way that dc can be obtained from the generator is to convert the ac output to dc.

One way to do this is to have a switch connected across the generator output in such a way that it will reverse the connections to the load every time the polarity of the induced emf changes inside the generator. The switch, illustrated in the diagram below, must be operated manually every time the polarity of the voltage changes. If this is done, the voltage applied to the load will always have the *same* polarity, and the current flow through the resistor will not reverse direction, although it will rise and fall in value as the loop rotates.

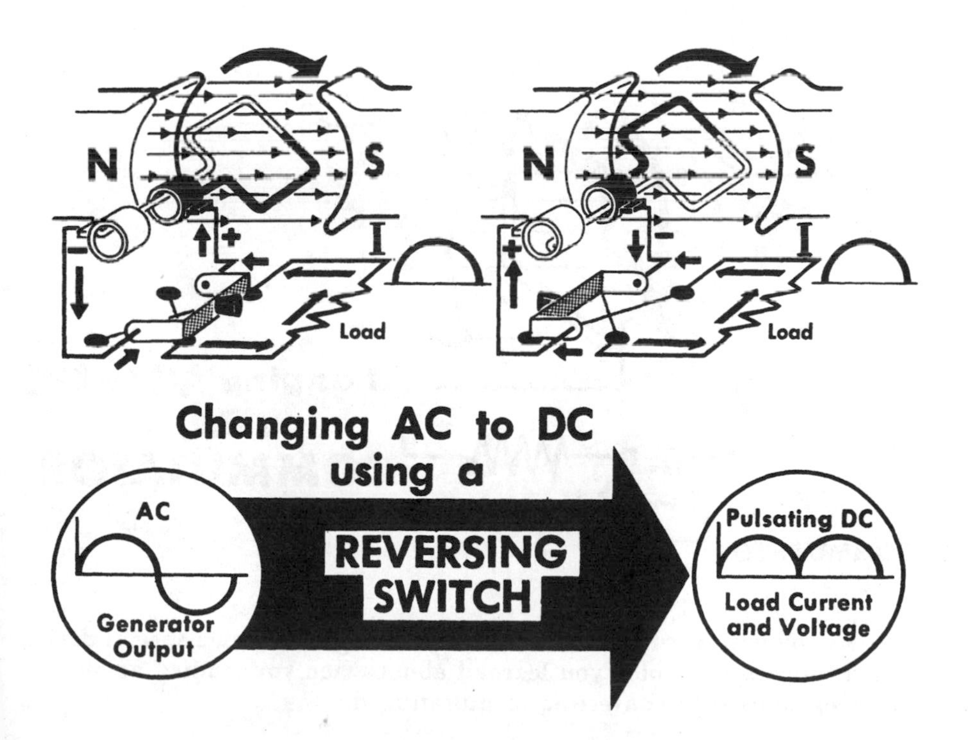

The Commutator (continued)

To convert the generated ac voltage into a pulsating voltage the switch must be operated twice for every cycle. If the generator output is alternating at 60 Hz, the switch must be operated 120 times per second to convert the ac to dc. Obviously, it would be impossible to operate a switch manually at such a high speed.

The problem has been solved simply by mounting the switch on the rotating shaft. This is done by changing the slip rings, so that they give the same result as the mechanical switch. Essentially, one of the slip rings is eliminated, and the other is split along its axis. The ends of the coil are then connected, one to each of the segments of the slip ring. The segments are insulated so that there is no electrical contact between them, the shaft, or any other part of the armature. The entire split ring is known as the *commutator*, and its action in converting the ac into dc is known as *commutation.*

The brushes are positioned opposite each other, and the commutator segments are mounted so that they are short-circuited by the brushes as the loop passes through the zero voltage points; thus, no current flow in the short circuit. Notice also that, as the loop rotates, each conductor will be connected by means of the commutator, first to the positive brush and then to the negative brush.

When the armature loop is rotated, the commutator automatically switches each end of the loop from one brush to the other, every time the loop completes a half revolution. The action is *exactly* the *same* as that of the manual reversing switch.

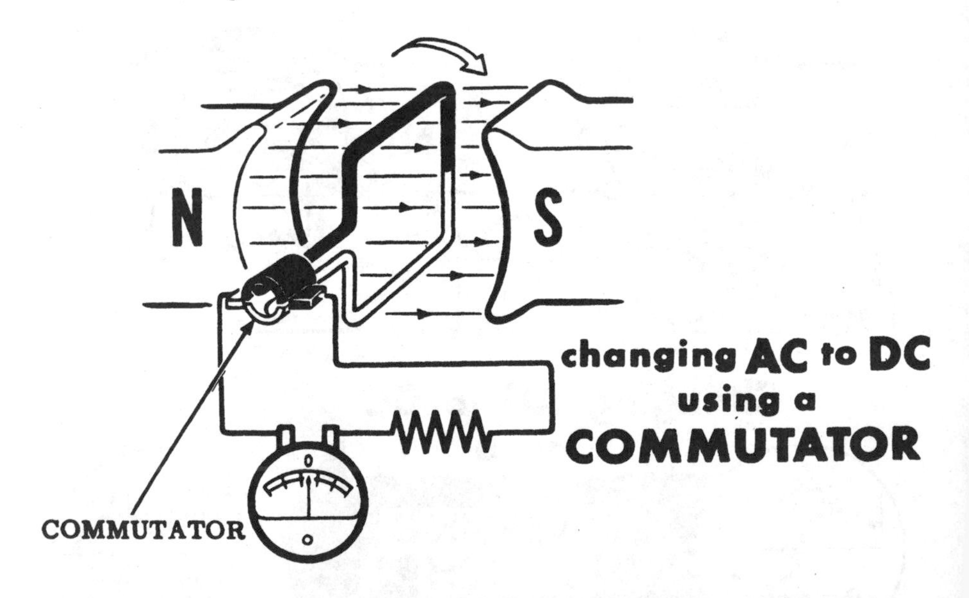

As you will learn later when you study dc generators, a special device called the *rectifier*, which you learned about when you studied ac meters, can also be used to convert ac to pulsating dc.

Converting AC to DC by Use of the Commutator

Suppose you now analyze the action of the commutator in converting the generated ac into dc. In A position (0°), the loop is perpendicular to the magnetic field, no emf is generated in the conductors of the loop, and there is no current flow. Notice that the brushes are in contact with both segments of the commutator, effectively short-circuiting the loop. This short circuit does not create any problem, since there is no current flow.

The moment the loop moves slightly beyond A position, however, the short circuit no longer exists. The black brush is in contact with the black segment, and the white brush is in contact with the white segment.

As the loop rotates clockwise from A position to B position, the induced emf starts building up from zero, until at B position (90°) the induced emf is at a maximum. Since the current varies with the induced emf, the current flow will also be at a maximum 90°. As the loop continues rotating clockwise, from B position to C, the induced emf decreases, until at C position (180°), it is zero once again.

The waveform pictured below shows how the terminal voltage of the generator varies from 0° to 180°.

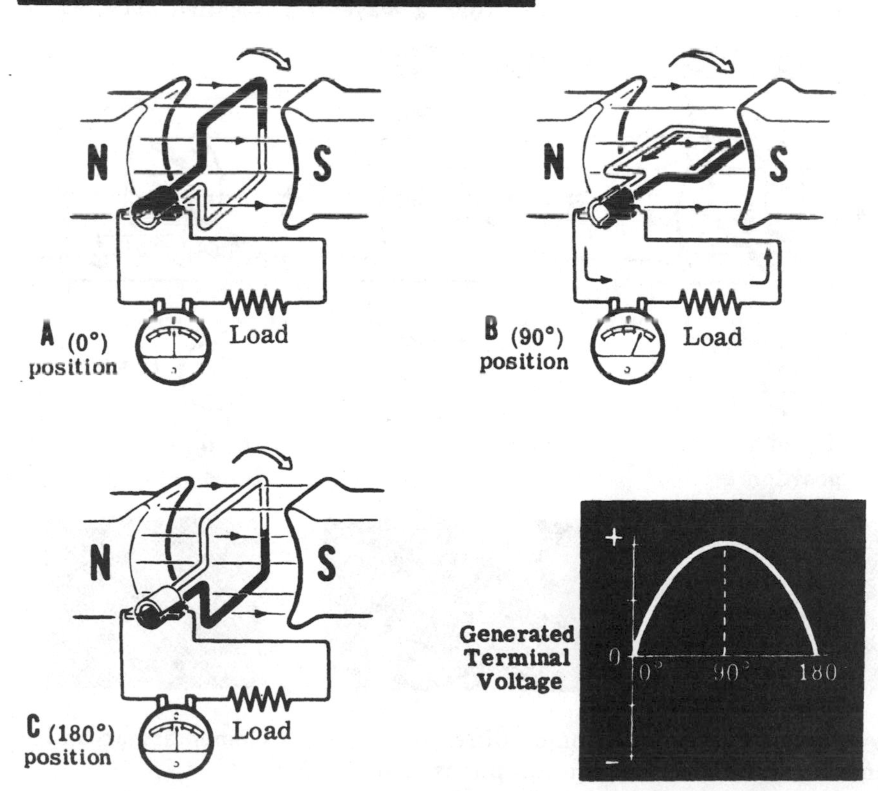

Converting AC to DC by Use of the Commutator (continued)

Notice that in C position the black brush is slipping off the black segment onto the white segment while at the same time the white brush is slipping off the white segment onto the black segment. In this way, the black brush is always in contact with the conductor of the loop moving downward, and the white brush is always in contact with the conductor moving upward. Since the upward-moving conductor has a current flow toward the brush, the white brush is the negative terminal and the black brush is the positive terminal of the dc generator. You can easily verify this by the left-hand rule.

As the loop continued rotating from C position (180°) through D position (270°) and back to A position (360° or 0°), the black brush is connected to the white wire, which is moving down, and the white brush is connected to the black wire, which is moving up. As a result, the same polarity voltage waveform is generated across the brushes from 180° to 360° as was generated from 0° to 180°. Notice that the current flows in the same direction through the ammeter after commutation even though it reverses in direction every half cycle in the loop itself.

The voltage output then has the same polarity at all times; but it varies in value, rising from zero to maximum, falling to zero, then rising to maximum and falling to zero again for each complete revolution of the armature loop. As you will recall, such a waveform is called *pulsating dc.*

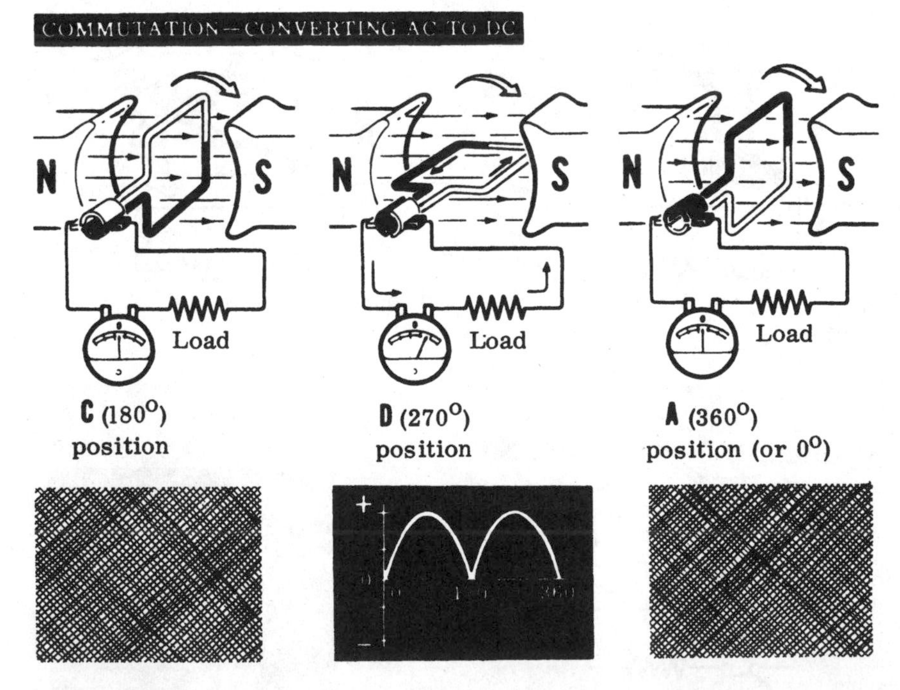

As you can see, the only major difference between elementary ac and dc generators is in the way the output is manipulated.

Improving the DC Output

Before you learned about generators, the only dc voltage you were familiar with was the smooth and unvarying voltage produced, for example, by a battery. Now you find that the dc output of an elementary dc generator is very uneven—a pulsating dc voltage varying periodically from zero to maximum. Although this pulsating voltage is dc, it is not constant enough to operate many dc appliances and equipment. Therefore, the elementary dc generator must be modified so that it will produce a smoother output. This is accomplished by adding more coils of wire to the armature.

The illustration shows a generator with a two-coil armature, the two coils (A and B) positioned at right angles to each other. Notice that the commutator is broken up into four *segments*, with opposite segments connected to the ends of a coil. In the position shown, the brushes connect to the white coil (A) in which a maximum voltage is generated, since it is moving at right angles to the field. As the armature rotates clockwise, the output from coil A starts dropping off. After an eighth of a revolution (45°) the brushes slide over to the black commutator segments in whose coil (B) the induced emf is increasing. The output voltage starts to pick up again, reaches a peak at 90°, and starts dropping off as coil B cuts through fewer lines of force. At 135°, commutation takes place again and the brushes are once more in contact with coil A.

In the illustration below, the voltages from both coils are shown superimposed on the single coil voltage. *Notice that the output never drops below point Y.* You will also notice that commutation occurs at the points where the voltages in the coils are equal. Since there is a momentary short developed on the commutator at this instant, the fact that the points are of equal potential means that no current will flow across the two commutator segments. The rise and fall in voltage is therefore now limited to the distance between Y and the maximum, rather than to that between zero and the maximum. This reduced variation in the output voltage of a dc generator is known as *generator ripple*. It is apparent that the output of the two-coil armature is much closer to steady dc than the output of the one-coil armature.

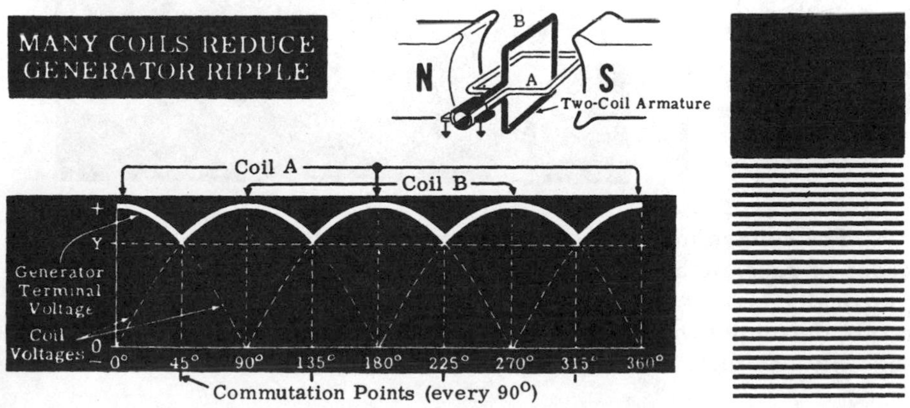

MANY COILS REDUCE GENERATOR RIPPLE

Improving the DC Output (continued)

Although the output of the two-coil generator is much closer to constant dc than the output of the one-coil generator, there is still too much ripple in the output to make it useful for some electrical equipment. Simple generators, however, can be very useful in applications where ripple is not important (or can be filtered), for example, in battery chargers, welding equipment, etc. To make the output really smooth, the armature is made with a large number of coils, and the commutator is similarly divided into a large number of segments. The coils are so arranged around the armature that at every instant there are some turns cutting through the magnetic field at right angles. As a result, the generator output contains very little ripple and is for all practical purposes a constant, or *pure*, dc.

MANY-TURN COILS INCREASE VOLTAGE OUTPUT

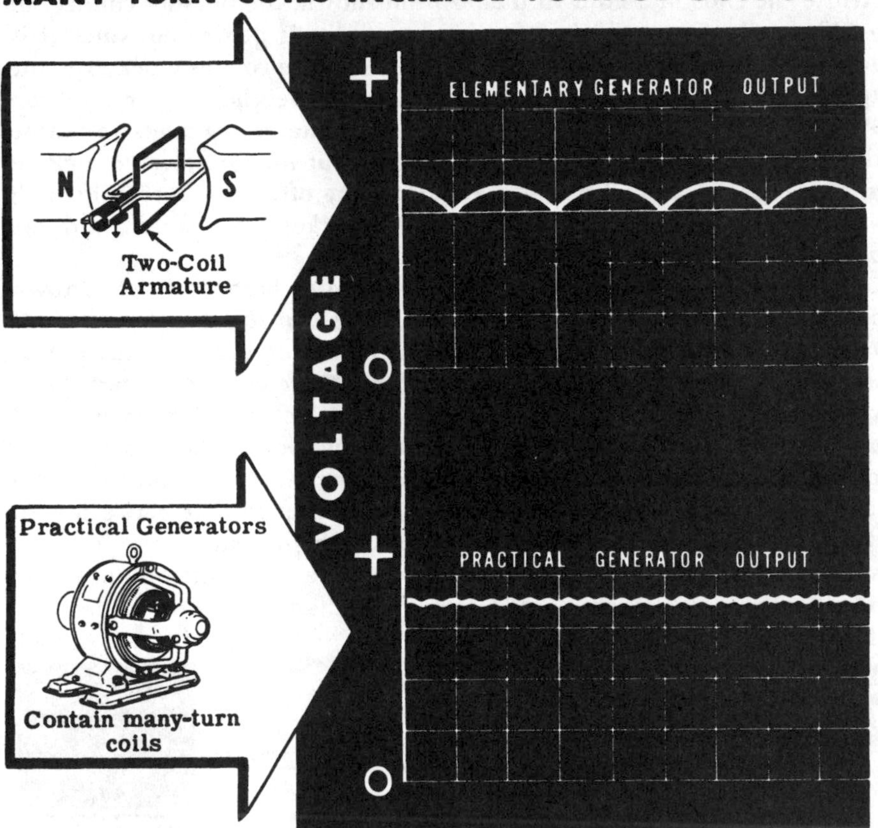

The voltage induced in a one-turn coil or loop is not very large. In order to generate a large voltage output, each coil on the armature of a practical generator consists of many turns of wire connected in series. As a result, the output voltage is much greater than that generated in a coil having only one turn.

Improving the Generator Output

When you considered the way an elementary generator works, you assumed that the coil was rotated in a uniform magnetic field. To obtain this uniform field in practical generators, concave pole pieces are used, and the armature coils are carried by an iron rotor so as to confine the magnetic field to desired areas. It should be apparent that the generator greatly resembles the transformer in principle and that what you learned about transformers earlier will help you understand generators and motors.

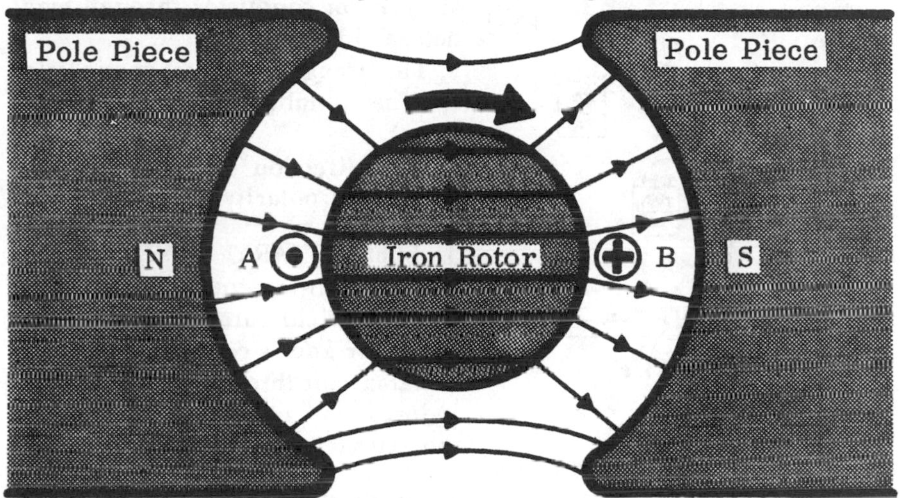

The flux can be made greater and even more uniform by using more than one pair of poles. Two pairs of poles are common, but more may be used in large machines.

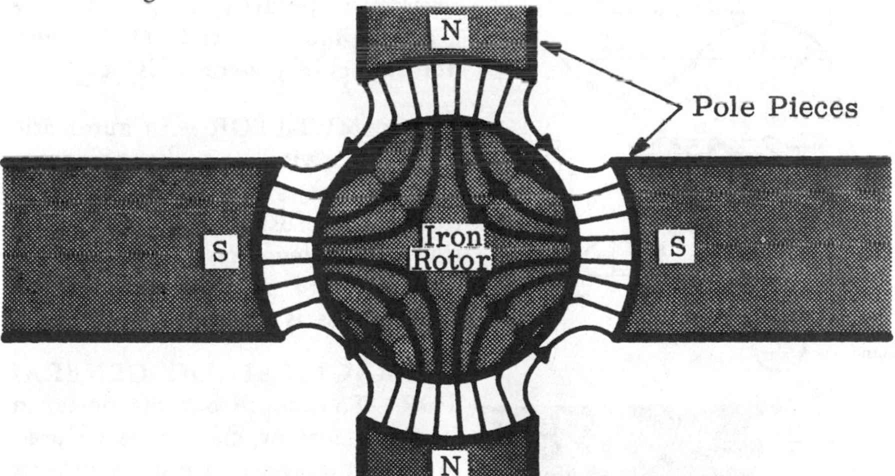

It should also be apparent that the generator can be arranged in several configurations, just so long as the principle of conductors moving through magnetic fields is preserved. Later in this volume we will study some of these different configurations.

Review of the Elementary Generator

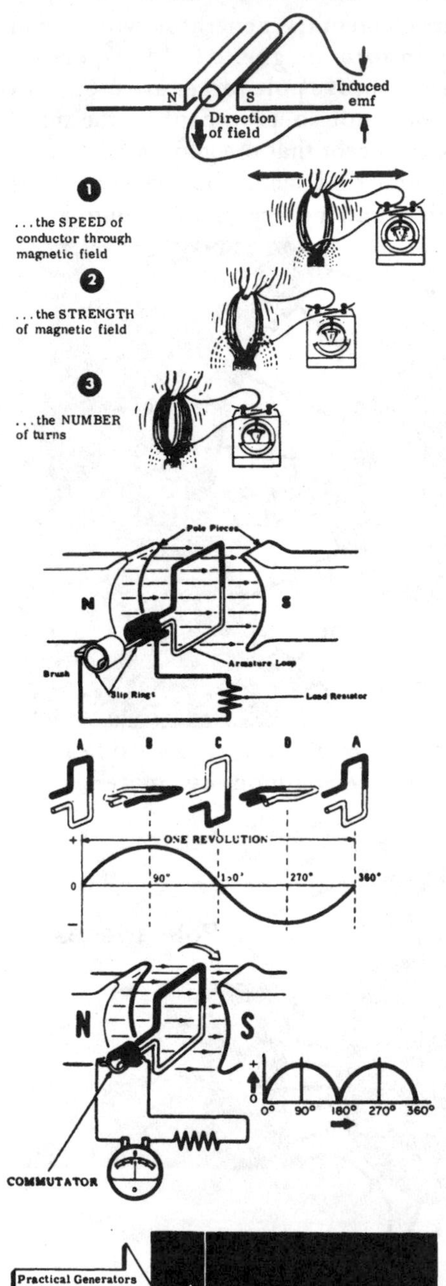

1. INDUCED EMF—The induced emf is the result of a conductor moving through a magnetic field and cutting the lines of force.

2. FACTORS AFFECTING INDUCED EMF—
a. Speed of conductor through magnetic field.
b. The strength of the magnetic field.
c. The number of inductors (or turns).
d. The direction of relative motion determines polarity.

3. ELEMENTARY GENERATOR—A loop of wire rotating through a magnetic field forms an elementary generator and is connected to an external circuit through slip rings. The induced emf causes current flow in the external circuit.

4. ELEMENTARY GENERATOR OUTPUT—The emf and current flow of an elementary generator reverse in polarity every time the armature loop rotates 180°. The output of such a generator is ac.

5. COMMUTATOR—An automatic reversing switch on the generator shaft, which switches coil connections to the brushes every half revolution of an elementary generator. Its purpose is to provide a dc output. The process is called *commutation.*

6. PRACTICAL DC GENERATOR—To smooth out the dc taken from a generator, many coils are used in the armature, and more segments are used to form the commutator. A practical dc generator has a voltage output that is near maximum at all times and has only a light ripple.

Self-Test—Review Questions

1. List five practical uses for generators.
2. What factors control generator output?
3. If you doubled the speed of a generator, what would happen to the voltage?
4. If you doubled the number of turns on the coil of an elementary generator, what would happen to the output?
5. How would the output voltage change if the strength of the magnetic field were increased? Decreased?
6. Draw an elementary generator and show, using the left-hand rule, how a commutator works.
7. State Lenz's law.
8. Check back into Volume 3 on Faraday's law and restate it here.
9. Draw an elementary generator. Label and define the functions of all of the parts.
10. What is the essential difference between an ac and a dc elementary generator?

Learning Objectives—Next Section

Overview—Now that you know something about the principle of the elementary generator, you are ready to learn about the practical dc generator in the next section.

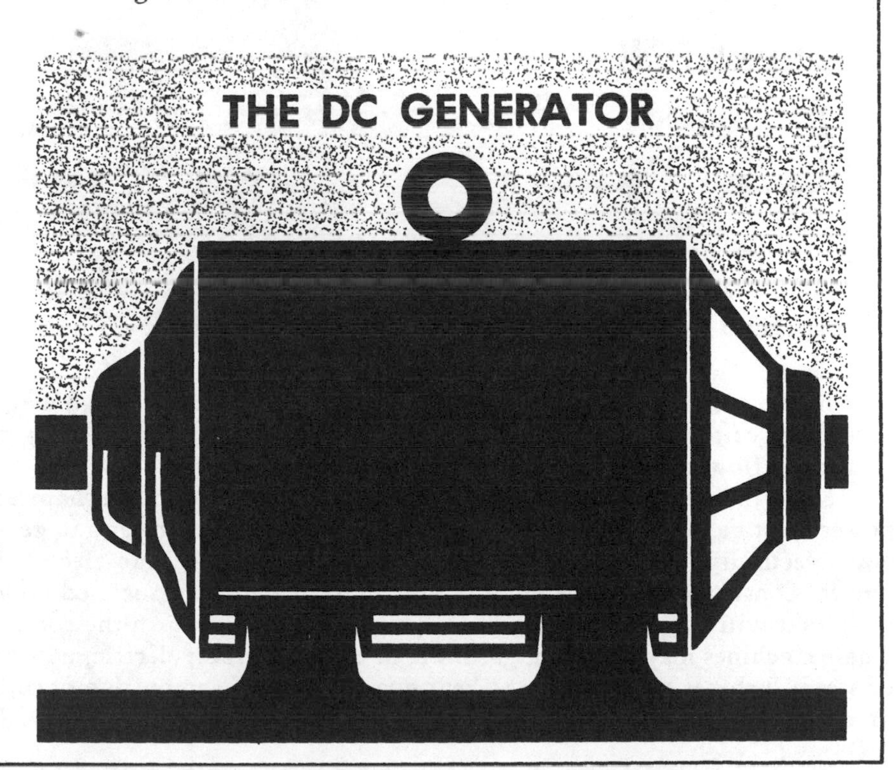

Generator Construction

So far you have learned the fundamentals of generator action and the theory of operation of elementary ac and dc generators. Now you are ready to learn about actual generators and how they are constructed. In fact, dc generators have become relatively unimportant because of the ease with which ac can be converted to dc. However, an understanding of dc generators provides a basis for understanding all the generators and motors that you will study later in this volume.

There are various components essential to the operation of a complete generator. Once you learn to recognize these components and become familiar with their function, you will find it much easier to do fault-finding and maintenance work on generators.

All generators—whether ac or dc—consist of a rotating part and a stationary part. In most dc generators the coil that the output is taken from is mounted on the rotating part, which is referred to as the *armature*. The coils that generate the magnetic field are mounted on the stationary part, which is referred to as the *field*. In most ac generators the opposite is true, that is, the field is mounted on the rotating part—the *rotor* ; and the armature is wound on the stationary part—the *stator*. In modern generators there are exceptions to the above cases where a permanent magnet or a simple iron core is the rotor, and both the field and the output coils are mounted on the stator. You will learn about these later in this volume.

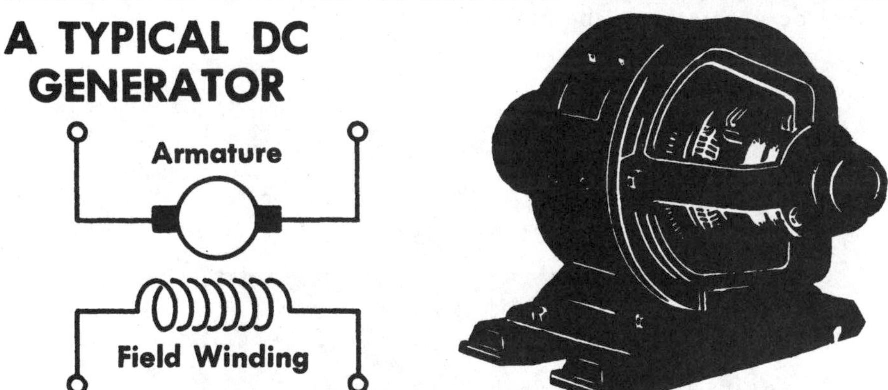

In all cases, there is relative motion such that a coil cuts through magnetic lines of force. As a result, an emf is induced in the coil, causing a current to flow through the external load.

Since the generator supplies electrical power to a load, mechanical power must be put into the generator to cause the rotor to turn and to generate electricity. The generator converts mechanical power into electrical power. Consequently, all generators must have machines associated with them that will supply the mechanical power necessary to turn the rotors. These machines may be steam, gasoline, or diesel engines; electric motors; or steam turbines actuated by the heat given off in the combustion of coal or oil, or nuclear fission; or turbines driven by water power.

DC Generator Construction

The relationship of the various components making up a dc generator is illustrated below. In the generator shown the field coils are the stator, and one end housing (not illustrated) is bolted to the generator frame. The armature is inserted between the field poles and the housing end, with the brush assemblies mounted last.

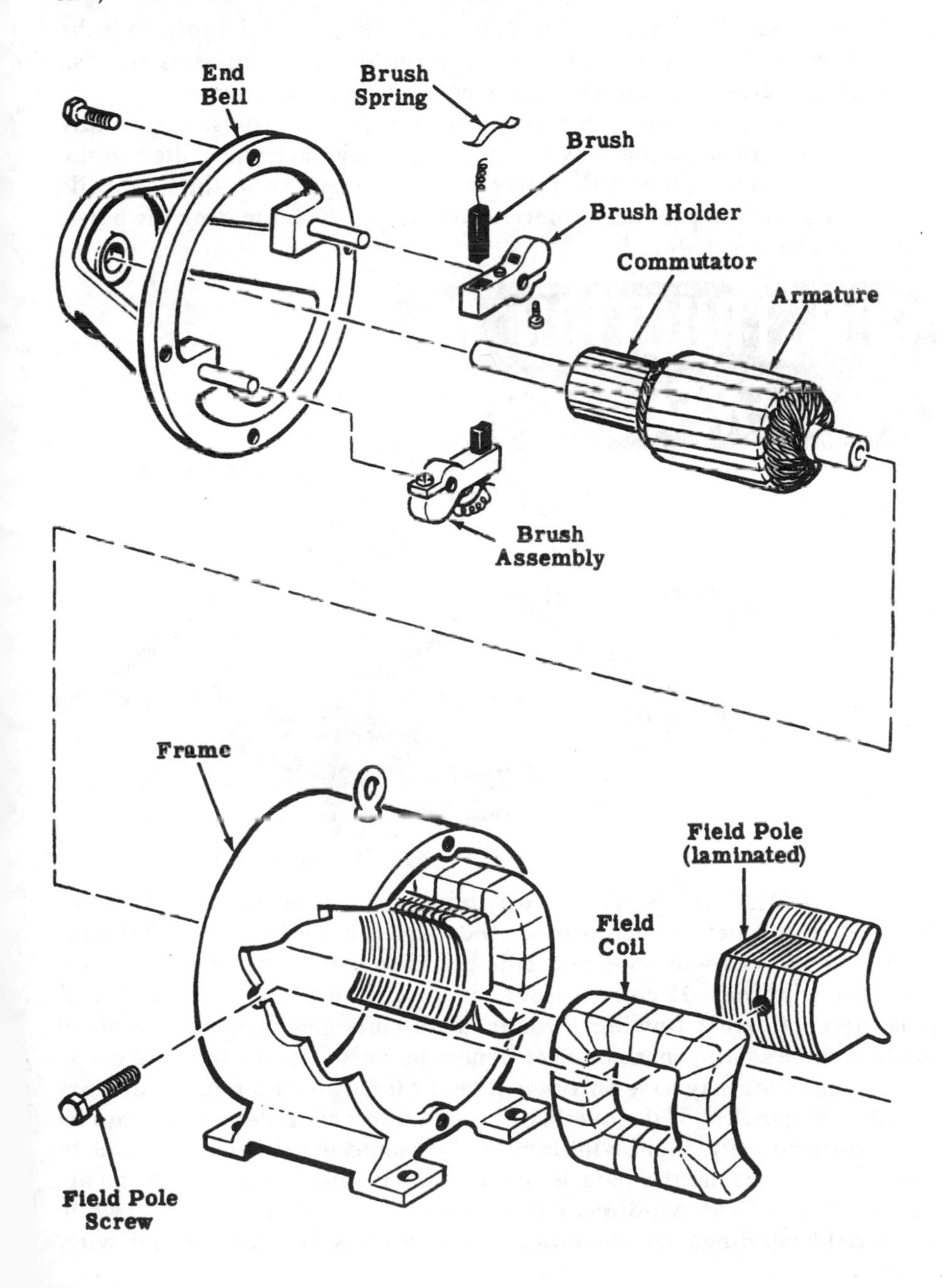

DC Generator Construction (continued)

Generator design varies depending on the size, type, and manufacturer; but the general arrangement of parts is as illustrated above. The major parts of the dc generator are described on the following pages. Compare each part and its function with the elementary generator described earlier.

MAIN FRAME: The main frame is sometimes called the *yoke*. It is the foundation of the machine and supports the other components. It also serves to complete the magnetic field between the pole pieces.

POLE PIECES: The pole pieces, like transformer cores, are usually made of many thin laminations of iron, joined together and bolted to the inside of the frame. These pole pieces provide a support for the field coils and are designed to produce a concentrated field. By laminating the poles, eddy currents are reduced.

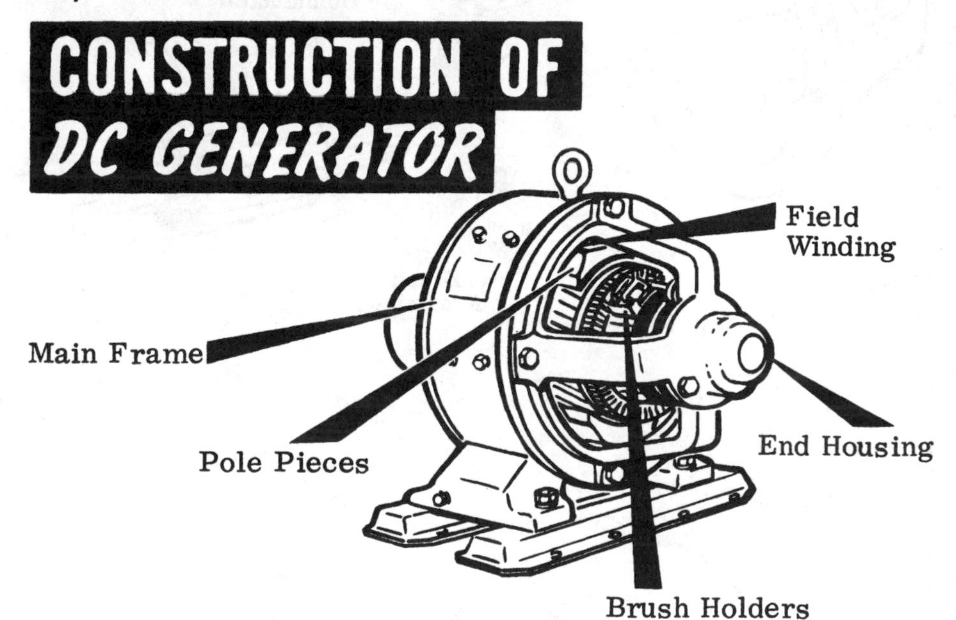

FIELD WINDINGS: The field windings mounted on the pole pieces form electromagnets, which provide the magnetic field necessary for generator action. The windings and pole pieces together are often called the *field*. The windings are coils of insulated wire wound to fit closely around pole pieces. Current flowing through these coils generates the magnetic field. In some small generators, permanent magnets replace the field coils.

A generator may have only two poles, or it may have a number of pairs of poles. Regardless of the number of poles, alternate poles will always be of opposite polarity. Field windings can be connected either in series or in parallel (or *shunt*, as the parallel connection is often called) with the armature. Shunt field windings consist of many turns of fine wire, while series field windings are composed of fewer turns of fairly heavier wire.

DC Generator Construction (continued)

END HOUSINGS: These are attached to the ends of the main frame and contain the bearings for the armature. The rear housing usually supports only the bearing, whereas the front housing also supports the brush assemblies.

BRUSH HOLDER: This component supports the brushes and their connecting wires. The brush holders are secured to the front end housing with clamps. On some generators, the brush holders can be rotated around the shaft for adjustment.

ARMATURE ASSEMBLY: In practically all dc generators, the armature rotates between the poles of the field. The armature assembly is made up of a shaft, an armature core, armature windings, and a commutator. The armature core is laminated to reduce eddy current losses and is slotted to receive the armature windings. The armature windings are usually wound in forms and then placed in the slots of the core.

The commutator is made up of copper segments insulated from one another and from the shaft by mica or heat-resistant plastic. These segments are secured by retainer rings to prevent them from slipping out under the force of rotation. Small slots are provided in the ends of the segments to which the armature windings are soldered. The shaft supports the entire armature assembly and rotates in the end bearings.

There is a small air gap between the armature and pole pieces to prevent rubbing between the armature and pole pieces during rotation. This gap is kept to a minimum in order to keep the field strength at a maximum.

BRUSHES: The brushes ride on the commutator and carry the generated voltage to the load. They are usually made of a high grade of carbon, or a carbon-copper mixture, and are held in place by brush holders. The brushes can slide up and down in their holders so that they are able to follow irregularities in the surface of the commutator. A flexible braided conductor, called a *pigtail,* connects each brush to the external circuit.

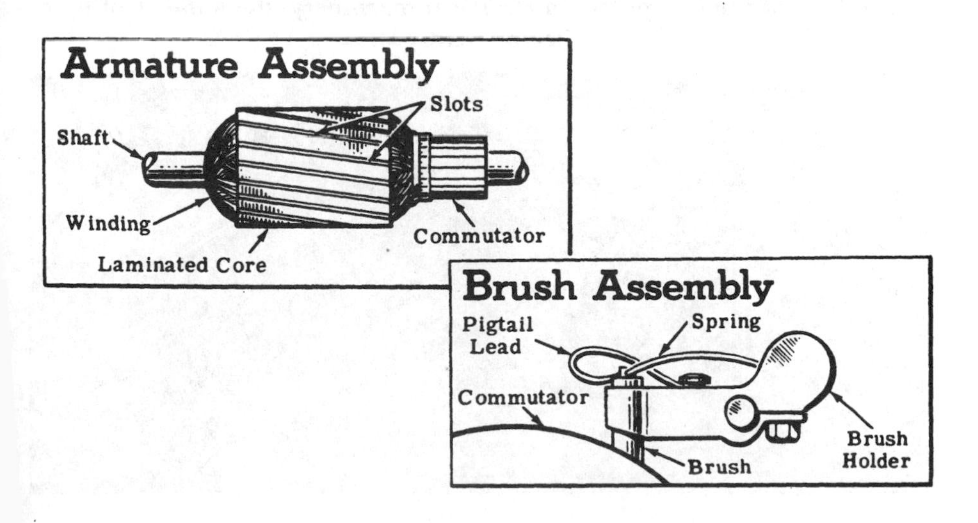

Types of Armatures

Armatures used in dc generators are divided into two general types. These are the *ring*-type armature and the *drum*-type armature. In the ring-type armature, the insulated armature coils are wrapped around an iron ring, with taps taken off at regular intervals to form connections to the commutator segments. The ring-type armature was used in early designs for rotating electrical machinery but is not used today.

The drum-type armature is the modern standard armature construction. The insulated coils are inserted into slots in the cylindrical armature core. The ends of the coils are then connected to each other.

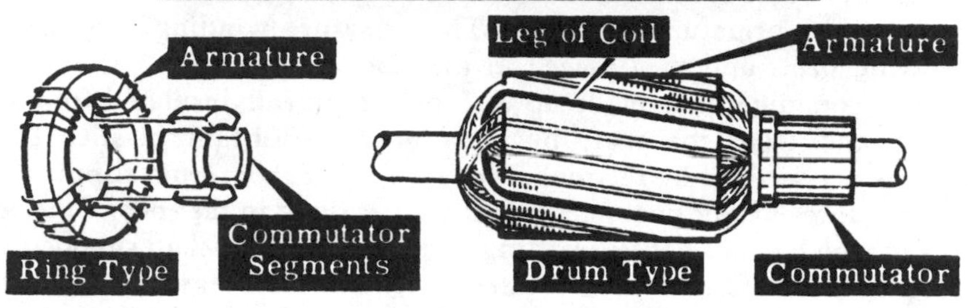

As a rule, most dc armatures use form-made coils. These coils are wound by machine with the proper number of turns and in the proper shape. The entire coil is then wrapped in tape, or otherwise insulated, and inserted into the armature slots as one unit.

The coils are so inserted that the two legs of each coil are under unlike poles. In a two-pole machine, the legs of each coil are situated on opposite sides of the core and therefore automatically come under opposite poles. In a four-pole machine, the legs of the coils are placed in slots about one-quarter the distance around the armature, thus again keeping the legs of the coils under unlike poles. In electrical machinery, the number of poles specified is actually the number of field poles.

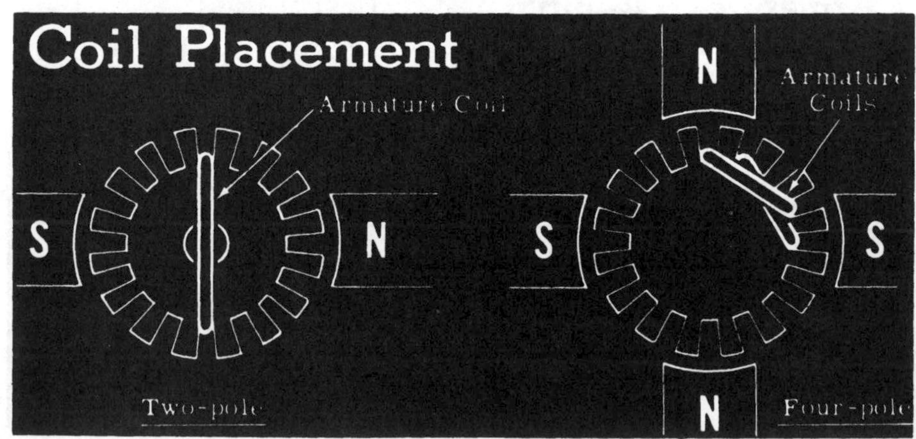

Types of Armature Winding

The windings on a drum-type armature may be one of two kinds, a *lap winding* or a *wave winding*.

In a lap winding the two ends of each coil are connected to adjacent commutator segments, and the winding forms the pattern illustrated below. The drawing is made as if the armature were laid out flat, so the extreme right-hand side actually connects to the extreme left-hand side. The illustration is for a four-pole generator.

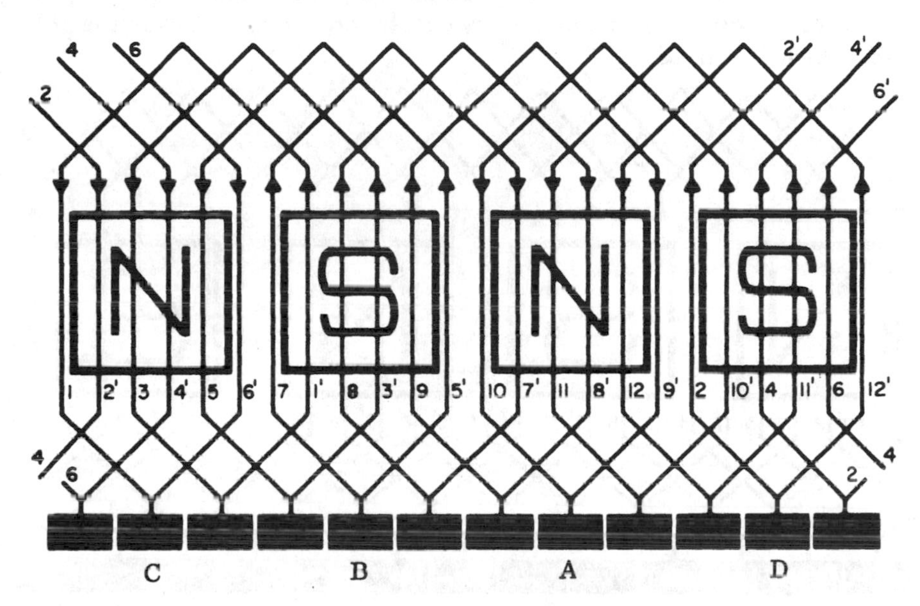

The direction in which the emf in each conductor is induced is indicated in the diagram. The way in which the coils in the lap winding are connected can be seen more easily in the simplified diagram below.

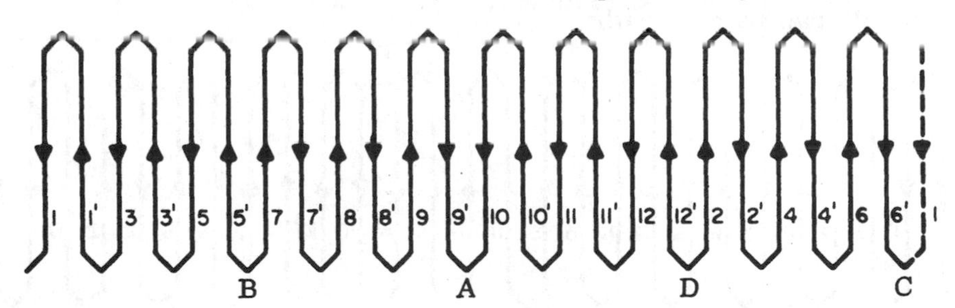

You can see that there are two points where the emfs in adjacent conductors meet: A and C; and two points where the emfs in adjacent conductors are diverging: B and D. If brushes are placed at these points, current will flow from the armature winding at A and C and into the winding at B and D. Brushes having the same polarity can be connected together, and the armature is thus effectively divided into four parallel paths.

Types of Armature Winding (continued)

In general, the number of parallel paths through a lap winding is equal to the number of poles. The terminal emf is equal to the emf induced in one path; the current delivered to the external circuit is equal to the sum of the currents in each of the parallel paths. For this reason the lap winding is used for high-current applications.

The other type of winding used on drum-type armatures is the wave winding, which is illustrated below. In this case the connections are made so that the winding passes under every pole before it comes back to the pole from which it started.

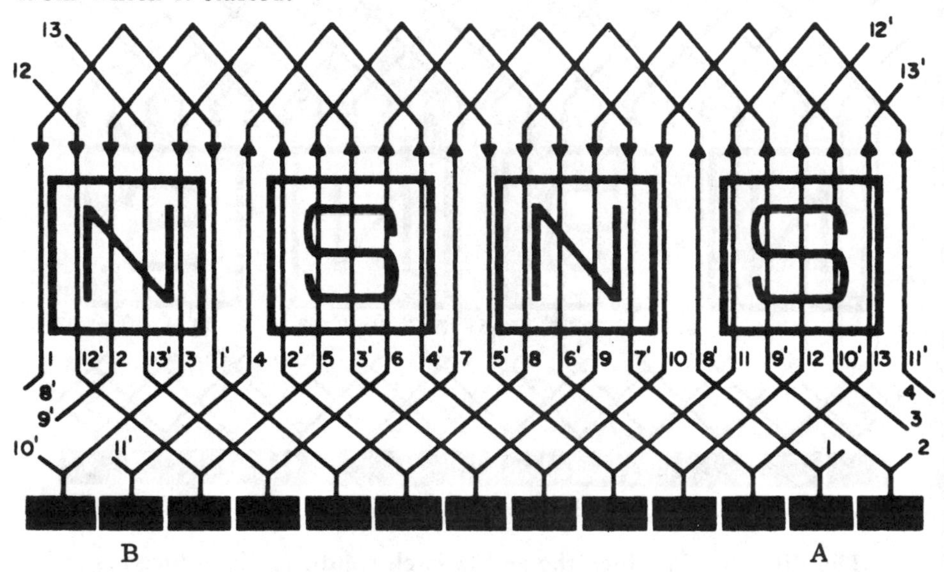

The connections between coils are shown in the simplified drawing of the winding below. You will notice that in this case the winding is divided into only two parallel paths.

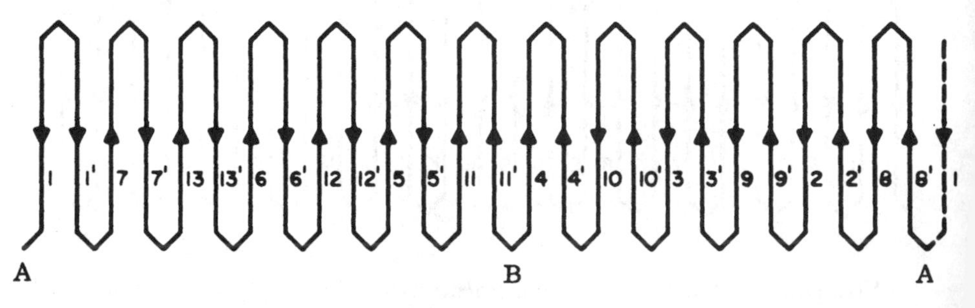

If brushes are placed at points A and B and you trace the two paths through the winding from A to B, you will see that most of the emfs in each path are acting in the direction A to B. A small number of emfs (small arrows) in each path are in opposition to the others, but if you check the position of these conductors, you will see that they are in the spaces

Types of Armature Winding (continued)

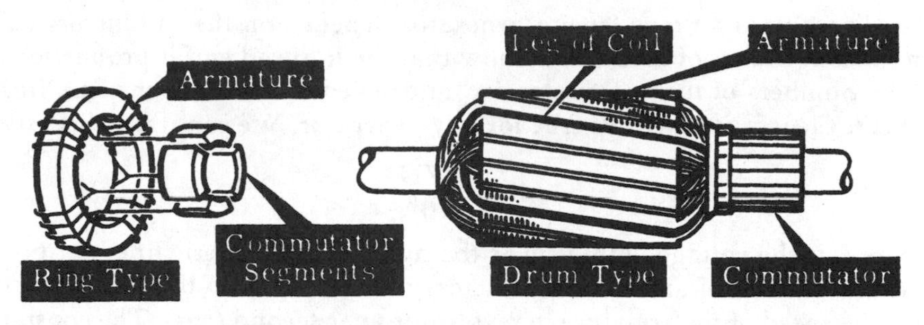

between poles, so that the emfs induced in them are small (at the instant shown).

In general, the wave winding has only two parallel current paths and uses only two brushes regardless of the number of poles. The emf developed in a wave winding is equal to that induced in one half of the total number of armature conductors. For this reason, the wave winding is used for high-voltage applications.

The current delivered to the external circuit is twice the current in an individual armature conductor because both halves of the armature winding are effectively in parallel.

Generator Voltage Output

The output voltage from a generator depends on the conductors cutting through lines of force. You know that the induced emf is proportional to the numbers of turns, flux density, and speed with which the flux lines are cut. Clearly, this will be true for any generator. We can write this as:

$$E = \frac{\Phi Z N}{10^8}$$

where E is the voltage output, Φ is the number of magnetic lines per pole, Z is the number of armature conductors interacting with the field Φ, and N is the speed of the armature in revolutions per second (rps). The constant 10^8 (100,000,000) is used so that all the factors come out as volts. Remember that a single turn has two conductors interacting, so Z is not just the number of turns between two commutator segments.

Example

Suppose you had a generator with 4 poles and 440 armature conductors and 10^6 (1,000,000) lines of magnetic flux that were rotating at 3,000 rpm (50 rps); the output voltage would be:

$$E = \frac{10^6 \times 440 \times 50}{10^8} = 220 \text{ volts}$$

Since each armature conductor that is effective is in parallel under a pole piece, and there are four poles, the load current/armature conductor is one-fourth of the total current. If the load current were 1,000 amperes, the armature conductor current would be:

$$\frac{1{,}000}{4} = 250 \text{ amperes}$$

GENERATOR OUTPUT DEPENDS ON

1. The Magnetic Field Strength
2. The Number of Conductors
3. The Speed at which it is Driven

Types of DC Generators

Most large dc generators have electromagnetic fields. Permanent magnet fields are also used in small generators. To produce a constant electromagnetic field for use in a generator, the field coils must be connected across a dc voltage source. The dc in the field coils is called the *excitation* and may be supplied either from a separate dc voltage source or by using the dc output of the generator itself.

If the field is supplied with current from an external source, the generator is said to be *separately excited*; but if some of the generator output is used to supply the field current, it is said to be *self-excited*. In a self-excited generator the field coils may be connected either in series with the armature coils (*series*); in parallel with the armature coils (*shunt*); or with two windings, with one in series and the other in parallel with the armature coils (*compound*). The symbols illustrated below are used to represent the armature and field coils in the various generator circuits.

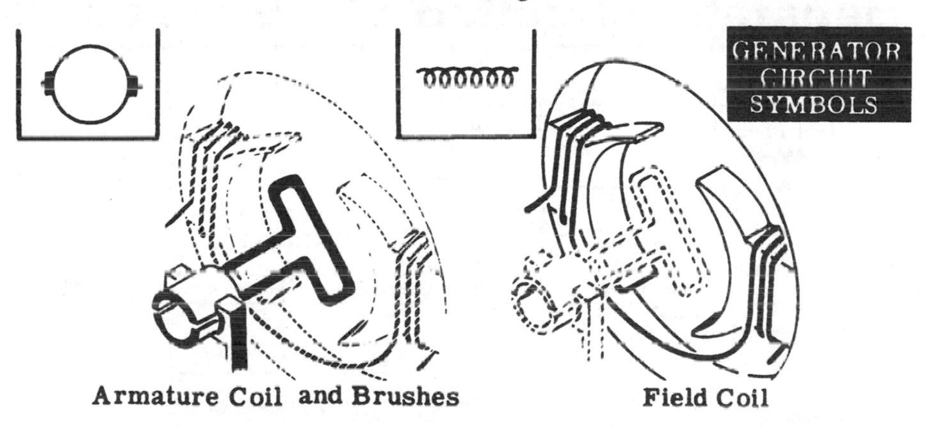

Separately excited dc generators have two external circuits: the field circuit, consisting of the field coils connected across a separate dc source, and the armature circuit, consisting of the armature coil and the load resistance. (Two or more field or armature coils connected in series with one another are represented by a single symbol.) The two circuits of a separately excited generator are illustrated below, showing the current flow through the various parts of the circuit.

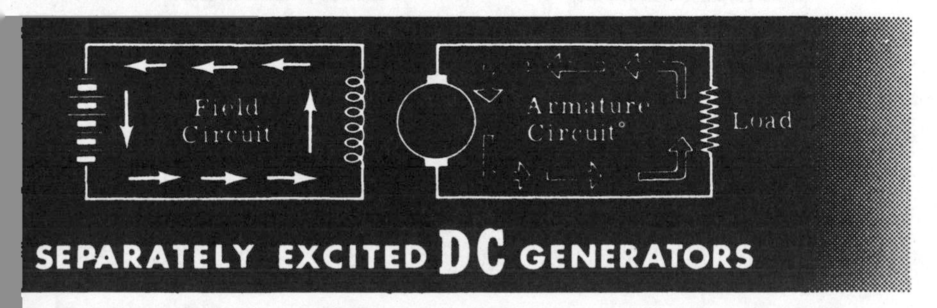

Separately Excited DC Generators

In a separately excited dc generator, the field is independent of the armature, since it is supplied with current from another generator (exciter), an amplifier, or a battery. The separately excited field provides a means of controlling the voltage output of the generator because a change in the field strength changes the magnitude of the induced voltage. Thus, with small changes in the field current, a large change in the load voltage (and current) will result.

The separately excited generator is frequently used in automatic motor control systems. In these systems the field current is controlled by an amplifier, and the output of the generator supplies the current that drives a dc motor. The motor may be used to drive a machine at constant speed, rotate a radar antenna, or power any other heavy load.

Separately Excited DC Generators

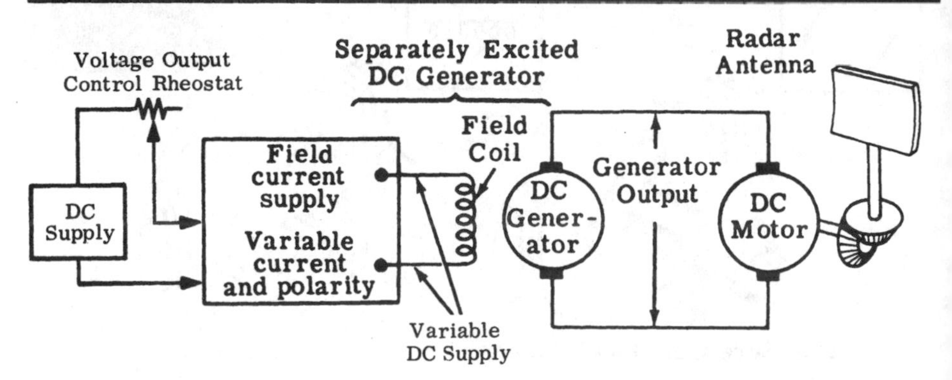

Sometimes, information about the motor speed is fed back by means of a small generator (called a *tachometer*), connected to the motor shaft, that generates a voltage proportional to the speed. The output from the tachometer is arranged so that, if the speed is to increase, the field current is decreased—thus reducing the generator output and lowering the motor's speed. In this way, the motor speed can be held to a constant value.

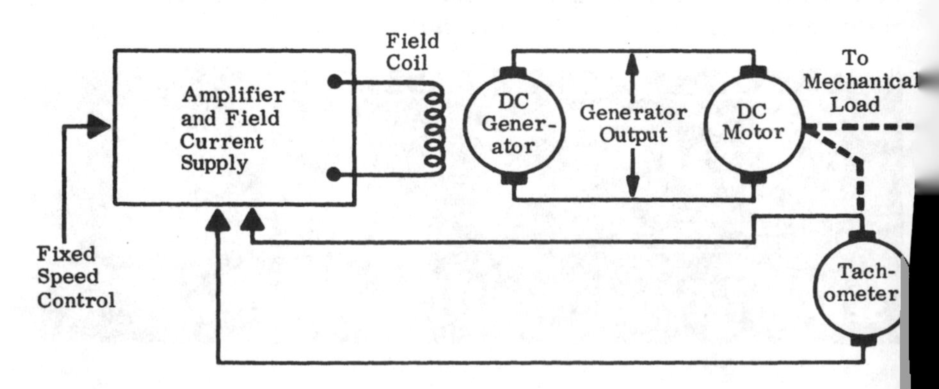

Self-Excited DC Generators

Self-excited generators use part of the generator's output to supply excitation current to the field. These generators are classified according to the type of field connection used.

In a *series generator*, the field coils are connected in series with the armature winding, so that the entire armature current flows through both the field and the load. If the generator is not connected across a load, the circuit is incomplete, and no current will flow to excite the field. The series field contains relatively few turns of heavy wire.

Shunt generator field coils are connected across the armature circuit, forming a *parallel*, or *shunt*, circuit. Only a small part of the armature current flows through the field coils; the rest flows through the load. Since the shunt field and the armature form a closed circuit independent of the load, the generator is excited even with no load connected across the armature. The shunt field contains many turns of finer wire.

A *compound* generator has both a series and a shunt field, forming a series-parallel circuit. Two coils are mounted on each pole piece, one coil series connected and the other shunt connected. The shunt field coils are excited by only a part of the armature current; however, the entire load current flows through the series field. Therefore, as the load current increases, the strength of the series field also increases.

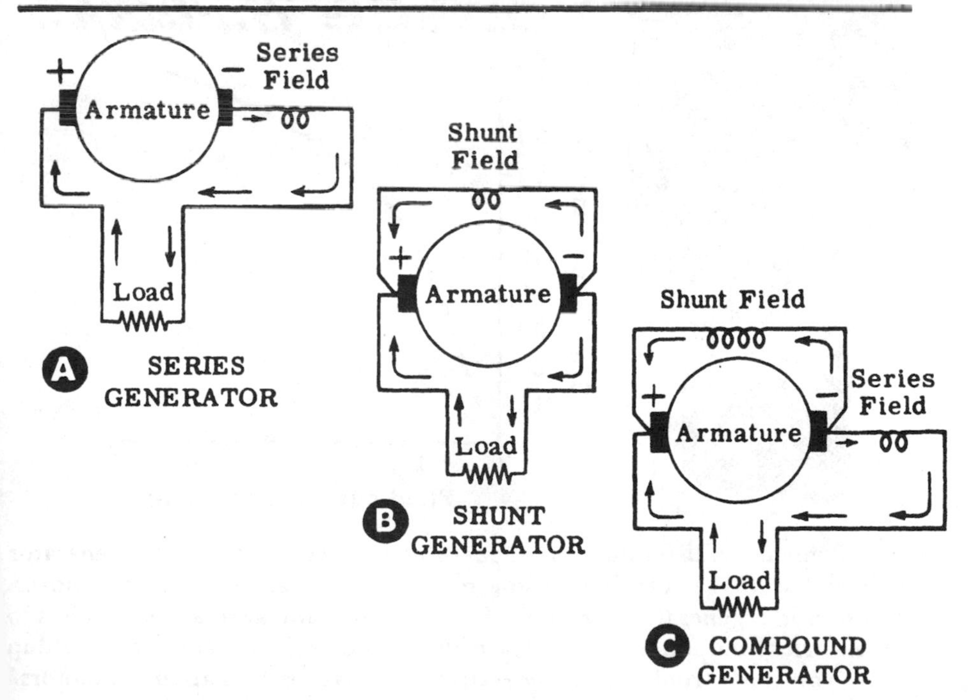

Self-Excited DC Generators (continued)

Almost all of the dc generators in use are either controlled by feedback or are directly of the self-excited type. However, if the original field excitation depends on the armature output, and if no voltage is induced in the armature coil unless it moves through a magnetic field, you may wonder how the generator output can build up. In other words, if there is no field to start with (since no current is flowing through the field), how can the generator produce an emf?

What happens is that the field poles retain a certain amount of magnetism, called *residual magnetism*, from a previous generator use. When the generator starts turning, a field does exist, which, although very weak, is still generally enough to induce an emf in the armature.

This induced emf forces current through the field coils, reinforcing the original magnetic field and strengthening the total magnetism. This increased flux, in turn, generates a greater emf, which again increases the current through the field coils—and so on until the machine attains its normal field strength.

All self-excited generators build up in this manner. The buildup time is normally less than 30 seconds. The graph shows how generator voltage and field current build up in a shunt generator.

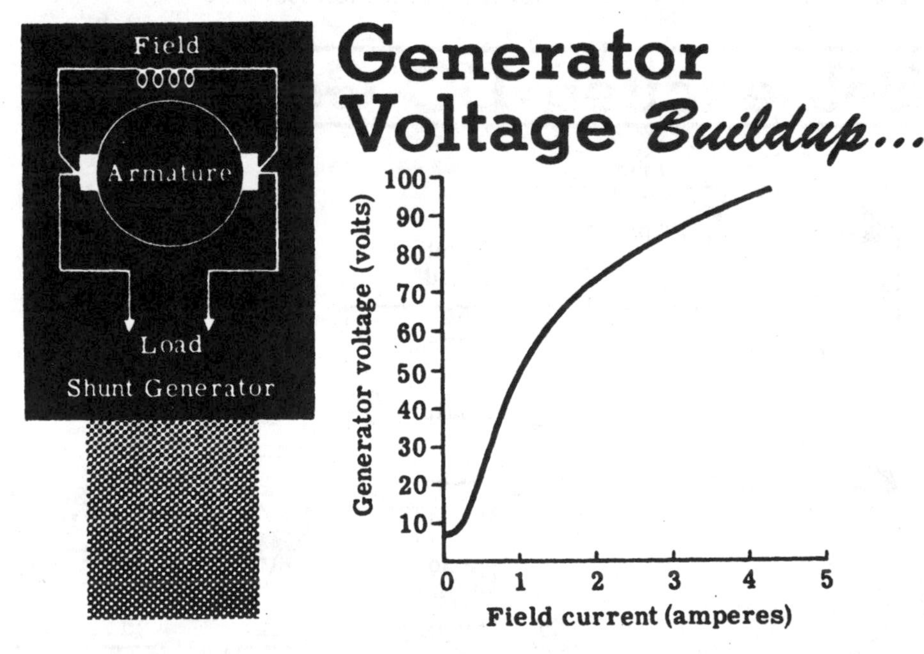

Remember, the output of a generator is electrical power. A generator always has to be turned by some mechanical means—the prime mover. Buildup in a generator does not refer to its mechanical rotation; it refers to its electrical output. You will learn what to do when there is no buildup when you study troubleshooting—after you have learned about dc motors.

The Self-Excited Series DC Generator

In a self-excited series dc generator, the armature, the field coils, and the external circuit are all in series. This means that the same current that flows through the armature and external circuit also flows through the field coils. Since the field current, which is also the load current, is large, the required strength of magnetic flux is obtained with a relatively small number of turns in the field windings.

The illustration shows the schematic of a typical dc series generator. With no load, no current can flow, and therefore very little emf will be induced in the armature—the amount actually produced depends on the strength of the residual magnetism. If a load is connected, current will flow, the field strength will build up, and the terminal voltage will increase. As the load draws more current from the generator, this additional current increases the field strength, generating more voltage in the armature winding and further increasing the current in the load. A point, A, is soon reached where further increase in load current does not result in greater voltage because the magnetic field has reached saturation and can increase no more.

The Series Generator

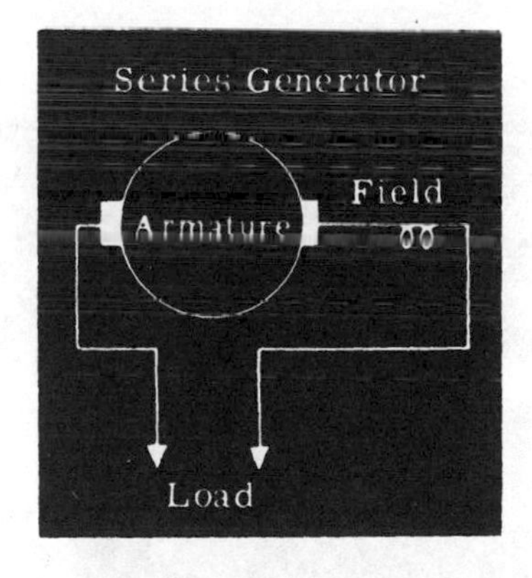

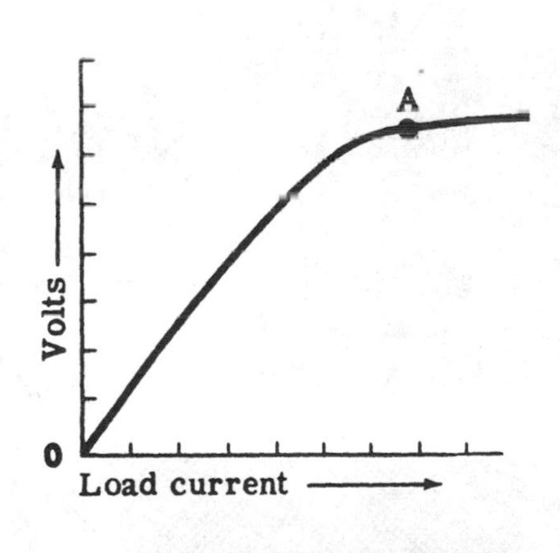

CHARACTERISTIC CURVE

The series generator is not used as an ordinary generator, but the fact that its output voltage is proportional to its armature current makes it useful in some special applications.

The Self-Excited Shunt DC Generator

A self-excited shunt dc generator has its field winding connected in shunt (or parallel) with the armature. Therefore, the current through the field coils is determined by the terminal voltage and the resistance of the field. The shunt field windings have a large number of turns, and therefore require a relatively small current to produce the necessary field flux.

When a shunt generator is started, the buildup time for rated terminal voltage at the brushes is rapid, since field current flows even though the external circuit is open. As the load draws more current from the armature, the terminal voltage decreases because the increased drop across the armature subtracts from the generated voltage and hence reduces the field strength.

The illustration shows the schematic diagram and characteristic curve for the shunt generator. Observe that over the normal operating region of no load to full load (A to B), the drop in terminal voltage as the load current increases is relatively small (typically 5% to 10%). The shunt generator is therefore used where a relatively constant voltage is desired, regardless of load changes. If the load current drawn from the generator increases beyond point B, the full-load point, the output may start to drop off sharply due to saturation and other effects.

The terminal voltage of a self-excited shunt generator can be controlled by varying the resistance of a rheostat in series with the field coils.

The Shunt Generator

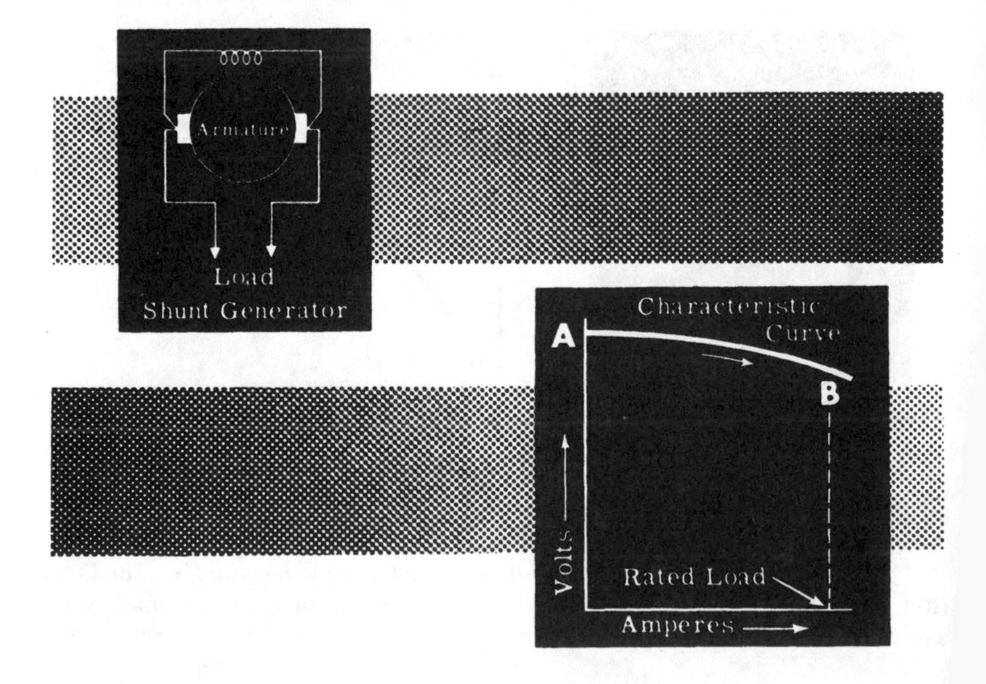

The Self-Excited Compound Generator

A self-excited compound generator is a combined series and shunt generator. There are two sets of field coils—one in series with the armature and one in parallel with the armature. One shunt coil and one series coil are always mounted on a common pole piece and are sometimes enclosed in a common covering.

If the series field is connected so that it aids the shunt field, the generator is called *cumulatively compound*. If the series field opposes the shunt field, the generator is called *differentially compound*. The field may also be connected either *short shunt* or *long shunt*, depending on whether the shunt field is in parallel with both the series field and the armature, or only with the armature. The operating characteristics of both types of shunt connection are practically the same.

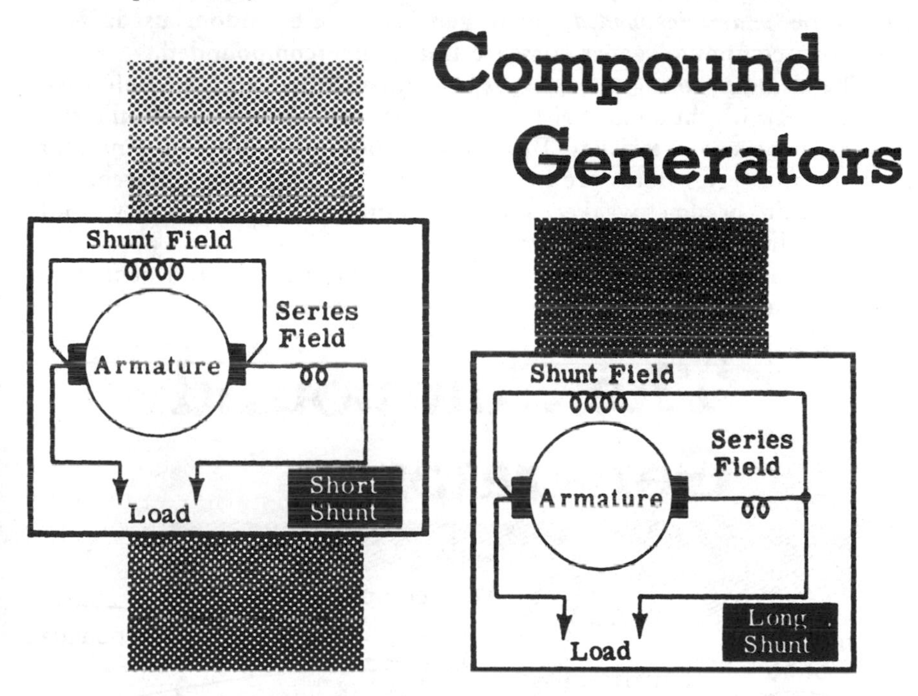

Cumulatively compound generators are designed to overcome the drop in terminal voltage that occurs in a shunt generator when the load is increased. This voltage drop is undesirable when constant voltage is required by the load. By adding the series field, which increases the strength of the total magnetic field when the load current is increased, the voltage drop due to the added current flowing through the armature resistance is overcome, and constant voltage output can be obtained.

The voltage characteristics of the cumulative compound generator depend on the ratio of the turns in the shunt and series field windings.

The Self-Excited Compound Generator (continued)

If the series windings are so proportioned that the terminal voltage is practically constant at all loads within its range, it is *flat-compounded.* Usually in these machines the full-load voltage is the same as the no-load voltage, and the voltage at intermediate points is somewhat higher. Flat-compounded generators are used to provide a constant voltage to loads a short distance from the generator where live voltage drops are small.

An *overcompounded* generator has a series coil so designed that the full-load voltage is greater than the no-load voltage. This generator is used when the load is some distance away. The increase in terminal voltage compensates for the drop in the long feeder lines, thus maintaining a constant voltage at the load.

When the rated voltage is less than the no-load voltage, the machine is said to be *undercompounded.* Such generators are seldom used. Most cumulative compound generators are flat or overcompounded.

The terminal voltage can always be controlled by varying a field rheostat in series with the shunt field; however, this can interfere with the voltage regulation characteristics. In a differentially compounded generator, the shunt and series fields are in opposition. Therefore, the difference, or resultant field, becomes weaker, and the terminal voltage drops very rapidly with an increase in load current.

The characteristic curves for the four types of compound generators are shown below.

The Compound Generator

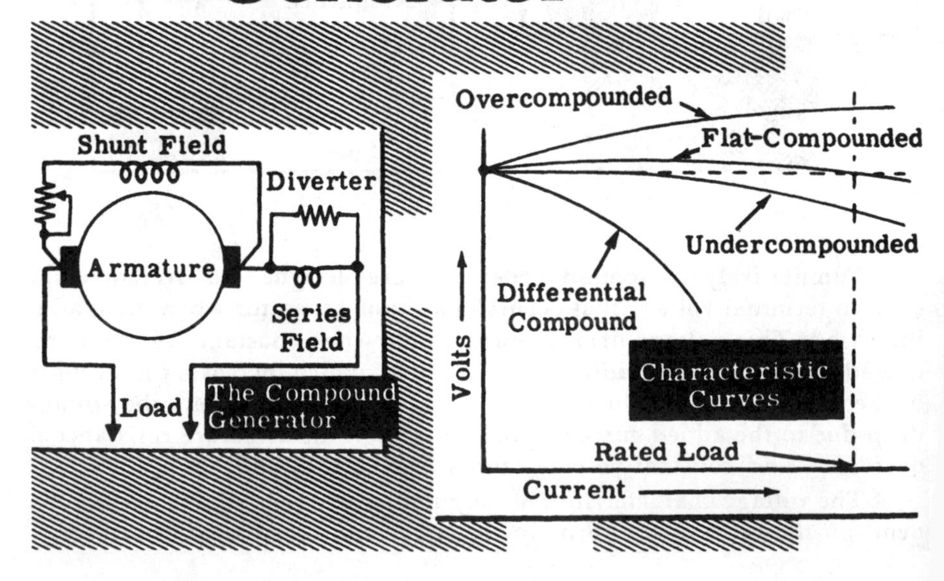

Voltage Regulation

As you have learned, the output voltage from a generator is not usually constant. The curve for a generator that shows the value of the terminal voltage as the load current varies is called *the regulation curve.* The voltage regulation is the difference between the no-load and full-load terminal voltage of a generator divided by the full-load voltage, where the full-load voltage is the rated output voltage. The voltage regulation is usually expressed as percent regulation.

$$\text{Percent regulation} = \frac{\text{no-load voltage} - \text{full-load voltage}}{\text{full-load voltage}} \times 100$$

Example

Suppose you had a compound wound generator that had a terminal voltage of 120 volts at 0 current, and 107 volts at full load. The percent regulation is

$$\frac{120 - 107}{107} \times 100 = 12\% \text{ regulation}$$

The voltage regulation characteristics of generators are very important since they often determine whether generators can be used for a specific application. As you continue your study of electricity, you will need to calculate voltage regulation, so remember this formula.

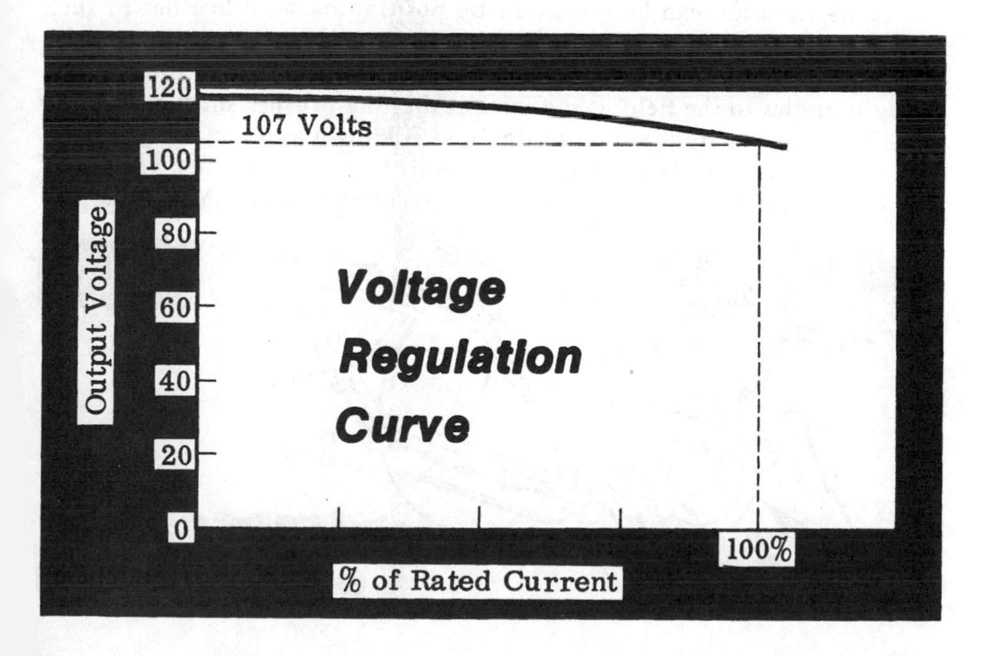

Commutation in DC Generators

When you studied the elementary dc generator, you learned that the brushes are positioned so that they short-circuit the armature coil when it is not cutting through the magnetic field. At that instant no current flows, and there is no sparking at the brushes (which are in the process of slipping from one segment of the commutator to the next).

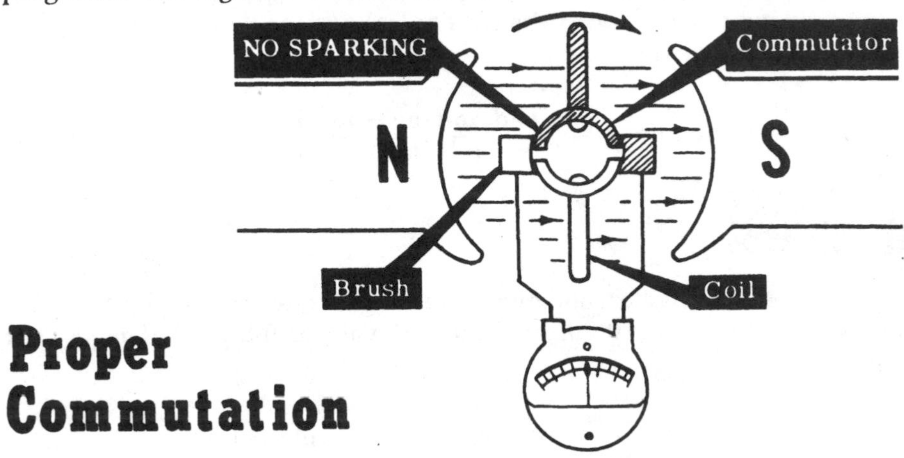

Proper Commutation

Should the brushes be moved a few degrees from their correct position, however, they will short-circuit the armature coil when it is cutting through the field. As a result, a voltage will be induced in the short-circuited coil and a short-circuit current will flow and will cause sparking at the brushes. This short-circuit current may seriously damage the coils and burn the commutator.

This situation can be remedied by positioning both brushes in such a way that commutation takes place when the coil is moving at right angles to the field. Dc generators operate efficiently when the plane of the coil is at right angles to the field at the instant that the brushes short-circuit the coil. The plane at right angles to the field is known as the *plane of commutation* or *neutral plane.* When the coil is in this plane, the brushes will short-circuit the coil at a moment when no current is flowing through it.

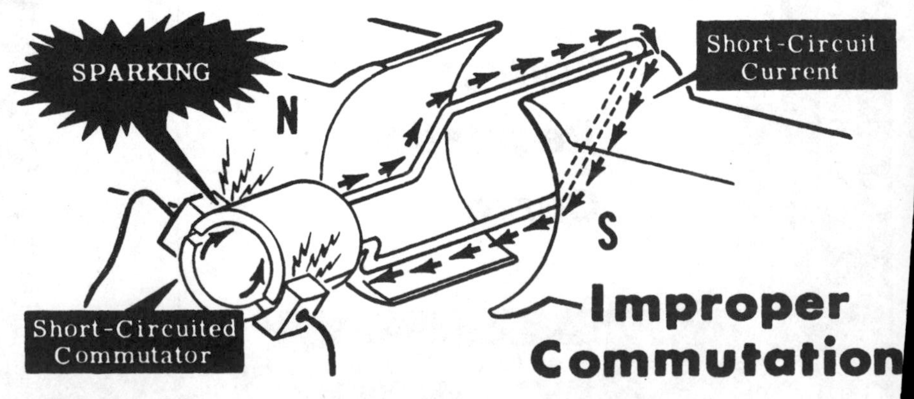

Improper Commutation

Armature Reaction in DC Generators

Suppose you consider the operation of a simple two-pole dc generator. The armature is shown in a simplified view below, with the cross section of one of its conductors represented as little circles. When the armature rotates clockwise, the sides of the conductor to the left will have current flowing out of the page, and the sides of the conductor to the right will have current flowing into the page. The field generated around each leg of the conductor is also shown.

Now you have two fields in existence—the main field and the field around each leg. The diagram shows how the armature field distorts the main field and how the neutral plane is shifted in the direction of rotation.

But you learned on the previous page that the brushes must short-out the coil *when it lies in the neutral plane.* If the brushes are allowed to remain in the old neutral plane, they will be short-circuiting coils that have voltage induced in them. Consequently, there will be arcing between the brushes and commutator.

To prevent this situation, the brushes must be shifted to the new neutral plane. The reaction of the armature in displacing the neutral plane is known as *armature reaction.*

Armature Reaction

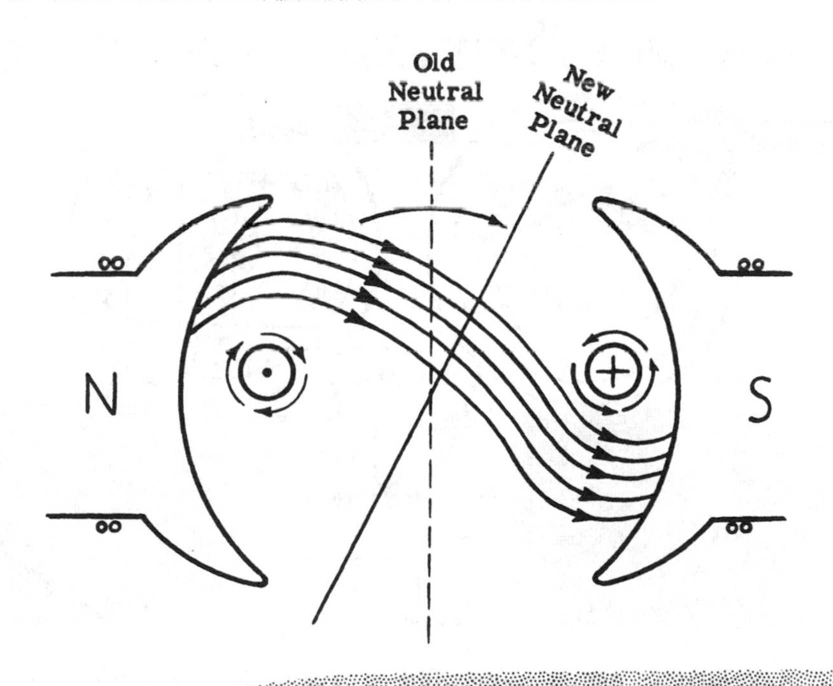

Compensating Windings and Interpoles

A mere shifting of the brushes to the advanced position of the neutral plane does not, however, completely solve the problem of armature reaction. For the effect of armature reaction varies with the load current; and every time the load current varies, the neutral plane shifts, meaning that the brush position will have to be changed.

In small machines the effects of armature reaction are minimized by mechanically shifting the position of the brushes to a compromise position where the small amount of arcing is accepted. In larger machines, more elaborate means are taken to eliminate armature reaction, such as using compensating windings or interpoles.

The *compensating windings* consist of series of coils embedded in slots in the pole faces. The coils are connected in series with the armature in such a way that the field they generate will just cancel the effects of armature reaction for all values of armature current. As a result, the neutral plane remains stationary; and once the brushes have been set correctly, they do not have to be moved again.

Another way to minimize the effects of armature reaction is to place small auxiliary poles, called *interpoles,* between the main field poles. The interpoles have a few turns of large wire connected in series with the armature. The field generated by the interpoles also just cancels the armature reaction for all values of load current and improves commutation.

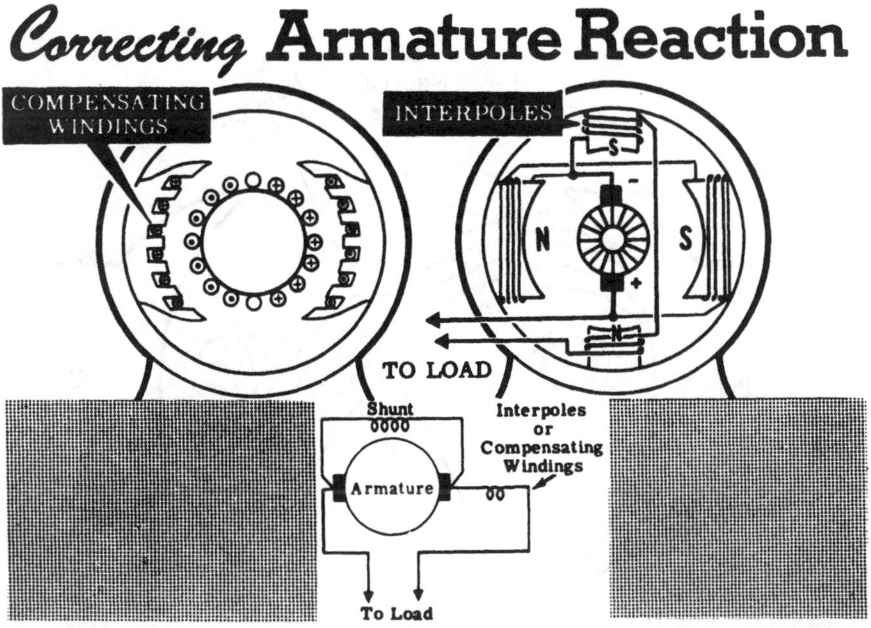

Correcting Armature Reaction

Interpoles or compensating windings are put in series with the load so that the cancellation is correct at all loads.

Review of DC Generators

The major parts of a dc generator are the armature, field windings, brushes, and commutator and the frame and bearing assembly. You should review each of these in this section so that you are thoroughly familiar with each and can identify them on an actual generator. You should also study the drawings for lap and wave windings so that you understand how and why they are used.

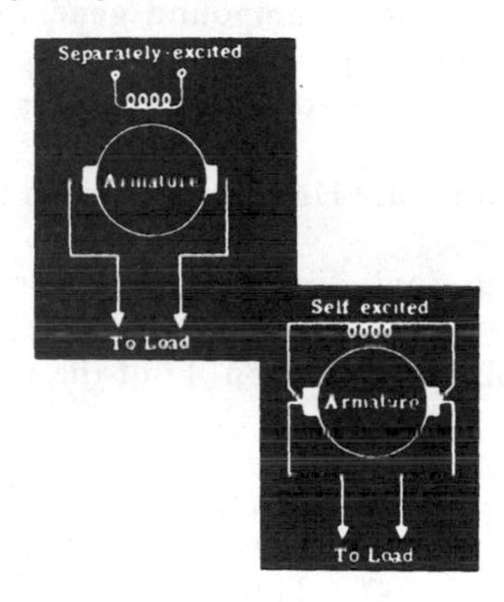

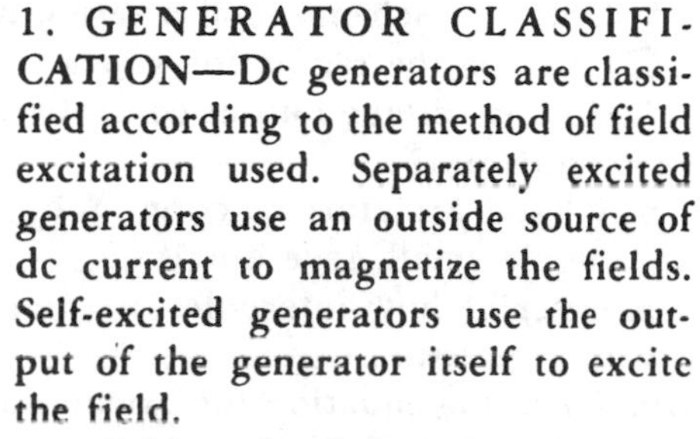

1. GENERATOR CLASSIFICATION—Dc generators are classified according to the method of field excitation used. Separately excited generators use an outside source of dc current to magnetize the fields. Self-excited generators use the output of the generator itself to excite the field.

Self-excited generators are further divided into classifications according to their field winding connections.

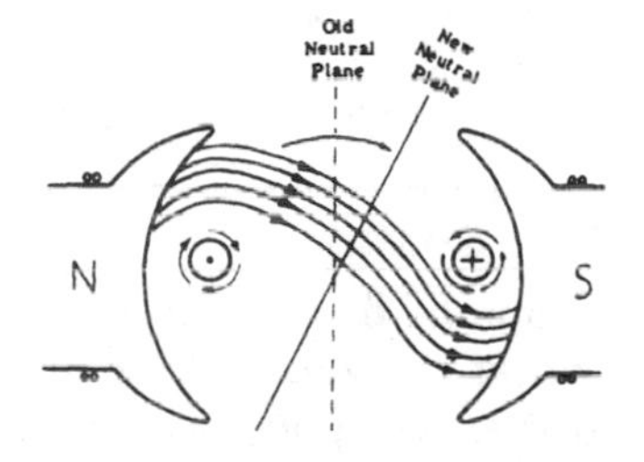

2. ARMATURE REACTION—Current flow in the armature coil generates a magnetic field at right angles to that of the generator field poles. The resultant total field shifts the neutral plane.

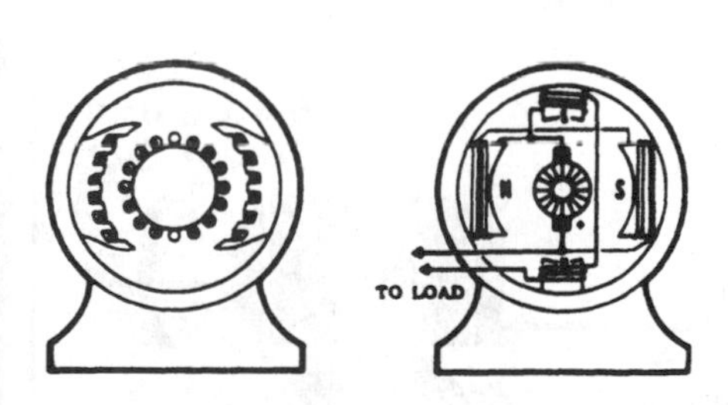

3. COMPENSATING WINDINGS AND INTERPOLES—Compensating windings are additional windings in the pole face carrying the armature current but in opposite polarity in order to counteract the armature field. Interpoles are small poles placed between the main field windings to generate a field opposite to the armature field. Both are designed to correct for armature reaction.

$$\text{Percent regulation} = \frac{\text{no-load voltage} - \text{full-load voltage}}{\text{full-load voltage}} \times 100$$

4. REGULATION—A measure of how steady the output voltage is when the load changes.

Self-Test—Review Questions

1. Name the major parts of a practical dc generator.
2. What are the significant differences between an elementary and a practical dc generator? What is the importance of these differences?
3. Draw the symbols that identify dc generators.
4. Draw the schematic diagram for a self-excited and a separately excited dc generator.
5. Draw the schematic of a series, a shunt, and a compound generator.
6. Describe the regulation characteristics of each.
7. How is voltage output controlled in a dc generator? What factors are significant?
8. What is armature reaction? Why does it occur? How is it corrected in a simple small generator?
9. Describe how interpoles or compensating windings counteract armature reaction.
10. Write the equation for calculating voltage regulation. Find the regulation of the following generators:

	No-Load Voltage	*Full-Load Voltage*
(a)	130.0	110.0
(b)	28.0	24.0
(c)	9.3	6.3
(d)	15.0	13.4
(e)	32.0	35.0

Learning Objectives—Next Section

Overview—Now that you have learned about the dc generator, you are ready to find out how the dc motor works. As you will see, the motor operates very much like a generator except that, instead of the armature being driven across a magnetic field by an external mechanical source, the armature is driven across a magnetic field generated by an external source of electricity.

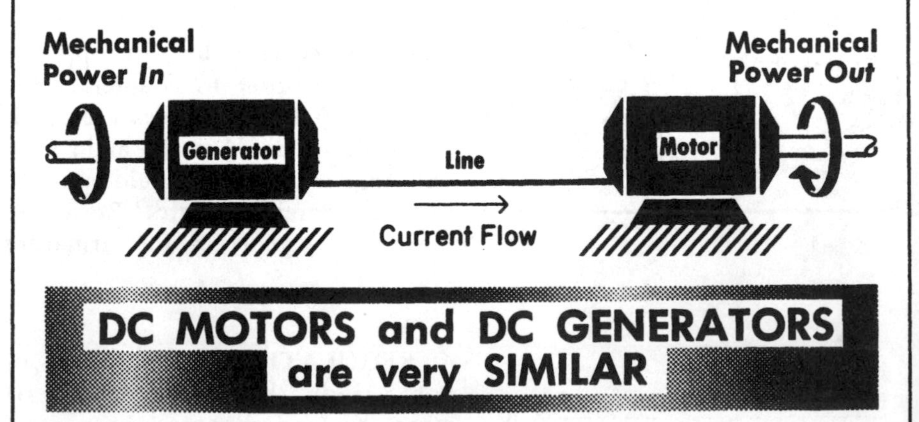

Converting Electrical Power to Mechanical Power

Dc motors and dc generators have essentially the same components and are very similar in outward appearance. They differ only in the way that they are used. In a generator, mechanical power turns the armature, forcing the winding of the armature through a field, and the moving armature generates electrical power. In a motor, electrical power generates a field that forces the conductors of the armature to turn, and the moving armature, by means of a mechanical system of belts or gears, turns a mechanical load.

A generator converts mechanical energy to electrical energy. A motor converts electrical energy into mechanical energy.

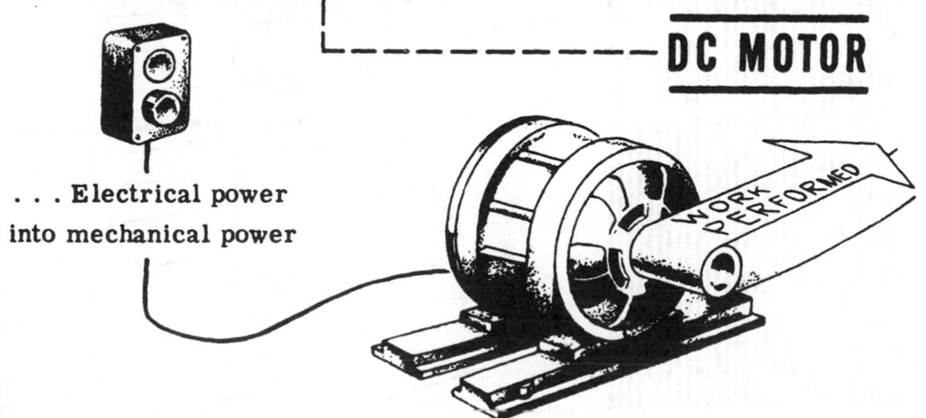

Principles of DC Motor Operation

Two nineteenth-century scientists, Heinrich Lenz of Germany and Sir John Ambrose Fleming of Britain, made discoveries that will help you to understand the principle of the dc motor.

While a conductor is moving so as to cut across the flux of a magnetic field, there is an emf induced in it; and the direction of the emf—and of the current if the conductor is part of a complete circuit—is such that it opposes the change producing it. This direction rule, which is true in every case where an induced current flows, is known as *Lenz's law*, as you learned earlier.

Fleming discovered the method for determining the direction of rotation of a motor if the direction of the current is known. He found that there is a definite relationship between the direction of the magnetic field, the direction of current in the conductor, and the direction in which the conductor tends to move. This relationship is called the *right-hand rule for motors*.

If the thumb, index finger, and third finger of the right hand are extended at right angles to each other, and if the hand is so placed that the index finger points in the direction taken by the flux lines of the magnetic field, then the thumb will point in the direction of motion of the conductor, and the third finger will point in the direction taken by the current through the conductor. Obviously, if the direction of the magnetic field is not known, but the motion of the conductor and the direction of the current through the conductor are known, the index finger must point in the direction of the magnetic field, provided that the right hand is placed in the proper position.

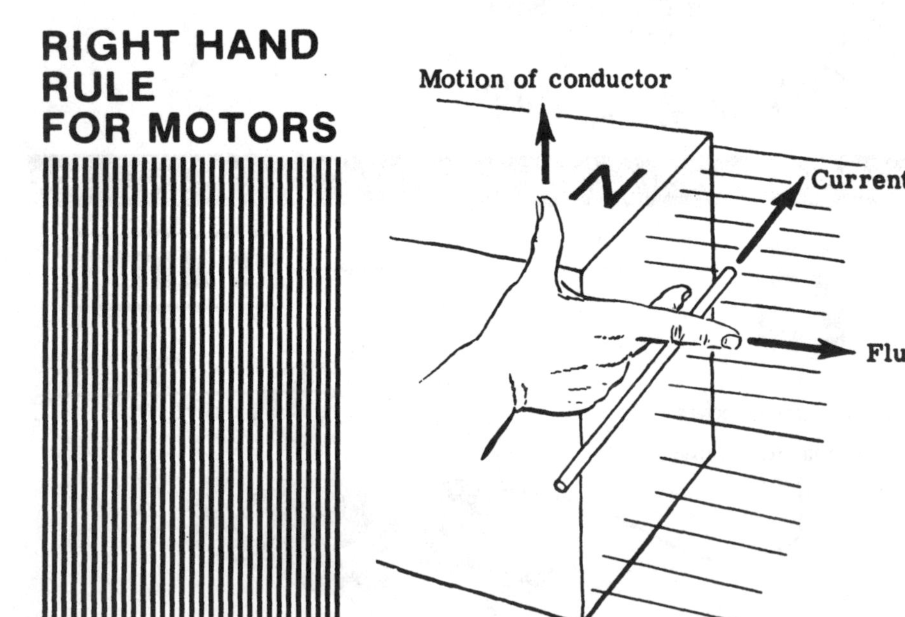

Principles of DC Motor Operation (continued)

Fleming was also first to propound the rule for remembering the direction of the emf induced in a generator (page 5-6). When he first stated these rules, the older convention for direction of current was in use; you will therefore sometimes come across textbooks that continue to use the older convention, and that refer to these rules as *Fleming's right-hand rule for generators* and *Fleming's left-hand rule for motors.*

As you know, a conductor that carries a current is surrounded by a magnetic field. If the conductor lies in another magnetic field, the two fields interact.

Since magnetic fields never cross, the lines of the two fields crowd together on one side and cancel each other out on the other side, producing either strong or weak fields, respectively.

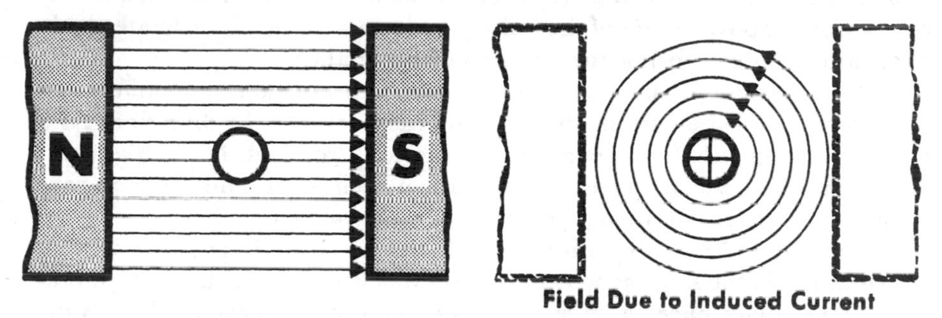

INTERACTION Between MAGNETIC FIELDS

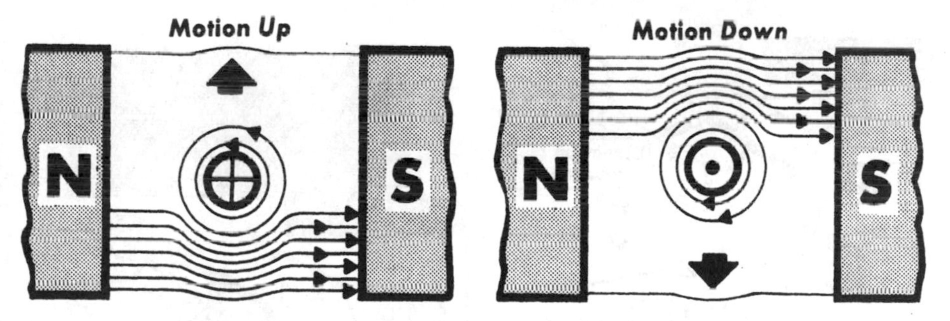

Remember that flux lines tend to push each other away. Thus the flux lines under the conductor shown above, in pushing each other away, tend to move the conductor upward or, if the direction of current in the conductor is reversed, to push the conductor downward.

This movement of the conductor causes it to cut through the magnetic field of the magnet. Hence, an emf is induced in the conductor, which according to Lenz's law, will be such that it opposes the motion producing it. That is to say, the induced emf will be opposite in polarity to the emf applied to the conductor externally. It is therefore called the *back*, or *counter, emf.*

Principles of DC Motor Operation (continued)

The back emf is never as great as the applied emf; the difference between the applied emf and the back emf is always such that current can flow in the conductor and produce motion.

The elementary dc motor is similar to the elementary dc generator. It consists of a loop of wire positioned between the poles of a magnet. The ends of the loop connect to commutator segments, which in turn make contact with the brushes. The brushes have connecting wires leading to a source of dc voltage.

Keep in mind the action of a conductor in a moving field, and compare it with that of the elementary dc motor. With the loop in position 1, the current flowing through the loop makes the top of the loop a north pole and the underside a south pole, according to the left-hand rule. The magnetic poles of the loop will be repelled by the like and attracted by the corresponding opposite poles of the field. The loop will therefore rotate clockwise, attempting to bring the unlike poles together.

When the loop has rotated through 90° to position 2, commutation takes place, and the current through the loop reverses its direction. As a result, the magnetic field generated by the loop reverses. Now, like poles face each other, which means that they repel each other, and the loop continues to rotate in an attempt to bring unlike poles together.

Rotating 180° past position 2, the loop finds itself in position 3. Now the situation is the same as when the loop was back in position 2. Commutation takes place once again, and the loop continues to rotate. This is the fundamental operation of the dc motor.

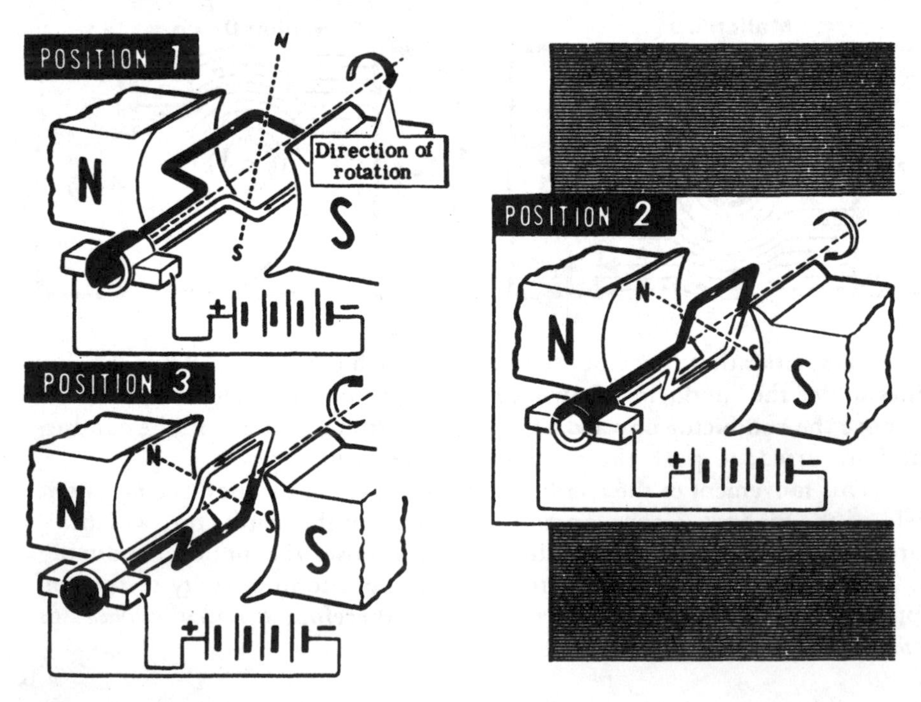

Commutator Action in a DC Motor

The commutator thus plays a very important part in the operation of the dc motor. It causes the current through the loop to reverse at the instant unlike poles are facing each other. This causes a reversal in the polarity of the field; repulsion exists instead of attraction; and the loop continues to rotate.

In a multicoil armature, the armature winding acts like a coil whose axis is perpendicular to the main magnetic field, and which has the polarity shown below. The north pole of the armature field is attracted to the south pole of the main field. This attraction exerts a turning force on the armature, which moves in a clockwise direction. Thus, a smooth and continuous torque, or turning force, is maintained on the armature by reason of the large number of coils. Since there are so many coils close to one another, a resultant armature field is produced which appears to be fixed.

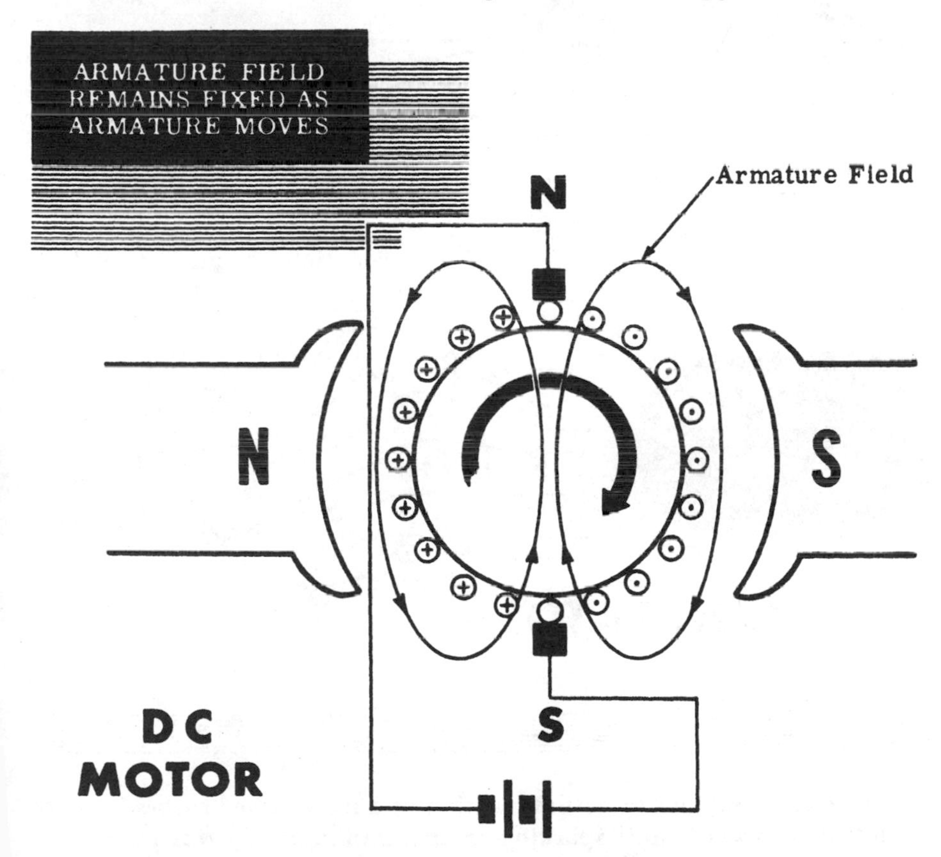

You will notice that a generator commutates when the armature coil is lined up with the field, that is, where the induced emf is zero. In a motor, commutation is at right angles to the field, at the point where a reversal of the field causes no net torque change because the armature coil fields are symmetrical in the external fields.

Armature Reaction

Since the motor armature has current flowing through it, a magnetic field will be generated around the armature coils as a result of this current. This armature field will distort the main magnetic field—in other words, the motor has *armature reaction*, just as the generator has. However, the direction of distortion due to armature reaction in a motor is just the opposite of what it is in a generator. In a motor, armature reaction shifts the neutral commutating plane *against* the direction of rotation.

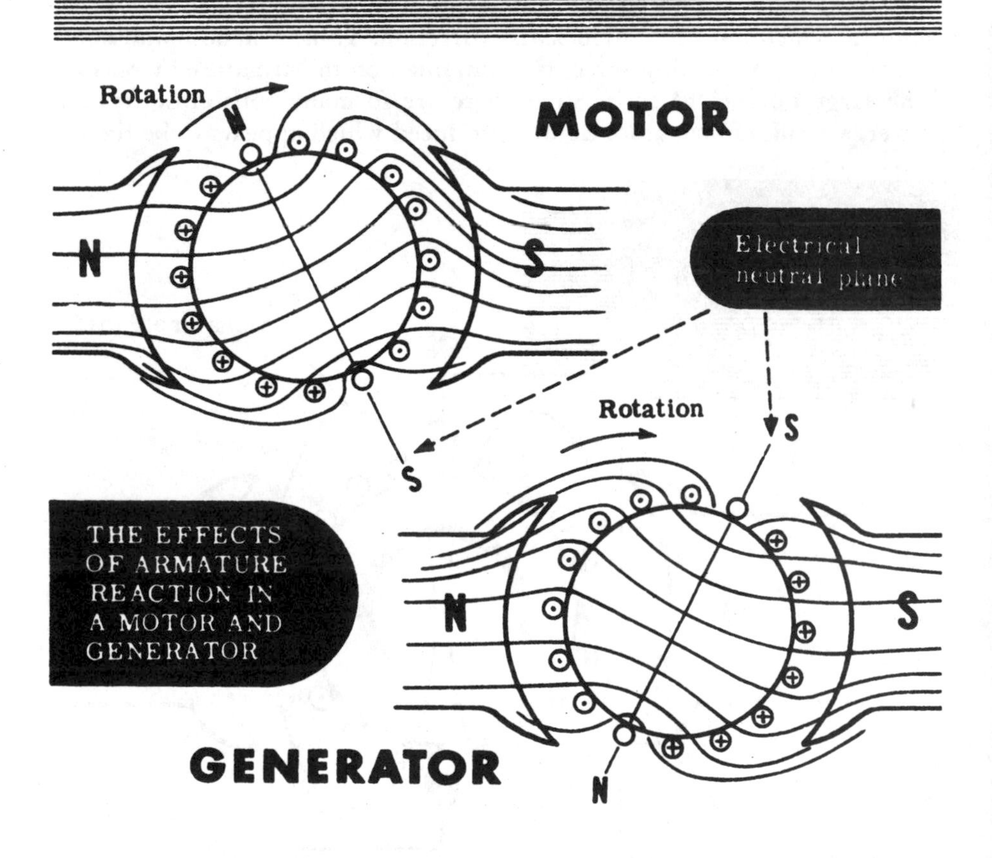

To compensate for armature reaction in a motor, the brushes have to be shifted backward until sparking is at a minimum. At this point, the coil, being short-circuited by the brushes, is in the neutral plane, and no emf is induced in it. Armature reaction can also be corrected by means of compensating windings and interpoles, just as in a generator, so that the neutral plane is always exactly between the main poles and the brushes do not have to be moved once they are properly adjusted.

DC Motor Torque

As you know, the force on a conductor carrying current is proportional to the current flowing, the field strength, and the active length of the conductor. The force, F, can be calculated from the equation:

$$F = \frac{8.85 \times BLI_a}{10^8}$$

where the force is in pounds; B, the flux density, is in lines per square inch; and the current, I_a, is armature current in amperes. The purpose of the constant 8.85 is to make everything come out in proper units. You can see that, for a constant field and a fixed conductor length, the force is proportional to the current in the armature winding.

Example

If you have a conductor 10 inches long, with $I_a = 30$ amperes, and B = 37,700 lines per square inch, what is the force on the conductor?

$$F = \frac{8.85 \times BLI_a}{10^8} = \frac{8.85 \times 37{,}700 \times 10 \times 30}{10^8} = 1 \text{ pound}$$

Since the conductor is wound around an armature, there is an additional force of 1 pound on the other part of the coil. Thus, the net force would be 2 pounds.

Torque, T, is a measure of the twisting capability of a motor and is proportional to the force times the number of conductors, N, times the distance from the center of the rotations of the conductor, R. It is measured in pound-feet. Thus, if the armature had a radius of 1.5 feet, for the example above with a single conductor (N = 1)

$$T = N \times F \times R = 1 \times 1 \times 1.5 = 1.5 \text{ pound-feet}$$

For the conductor pair making up a coil, the total torque would be 3 pound-feet. Actually the conductors are not all centered in the magnetic field, so for such calculation we use the average radius (or moment arm), which is 0.637 times the actual (or maximum) radius. Thus, for an actual motor the torque is proportional to the magnetic field strength, the armature current, and the effective radius of the armature, or $T = K\Phi I_a$ where K is a constant that includes the number of conductors, the number of paths, the armature radius, etc., and I_a and Φ are defined as before.

Metric example: Using lines of force in lines/square centimeters and conductor length in centimeters

$$F = \frac{10.4 \text{ BLIa}}{10^8}$$

and for B = 5,731 lines/square centimeter (same as 37,700/square inch) and a conductor length of 25.4 cm (10 in.) we get

$$F = \frac{10.4 \text{ BLI}_a}{10^8} = \frac{(10.4)(5{,}731)(25.4)(30)}{10^8} = 0.45 \text{ kg}$$

which is the same as the answer in English units, since 1 pound is 0.45 kg.

DC Motor Horsepower and Efficiency

As you know, work is done when force acts through a distance, and a unit of work is the foot-pound. You also know that power is work per unit time. The unit of mechanical power is the *horsepower* (hp), which is equal to 33,000 foot-pounds per minute. Thus, 1 horsepower is developed when the product of distance and pounds equals 33,000 and this is done in 1 minute. Thus, 1 horsepower is developed if 33,000 pounds are lifted 1 foot in 1 minute or 1 pound is lifted 33,000 feet in 1 minute. In electrical terms, 1 horsepower is equal to 746 watts (1 hp = 746 W).

A point on the armature (rotor) of a motor travels a distance equal to the circumference of the rotor with each revolution. Therefore, its velocity, V, is equal to

$$V = 2\pi RN \text{ (feet/minute)}$$

where R is the circumference of the rotor and N is the armature speed in revolutions per minute (rpm). Horsepower (hp) is therefore equal to

$$hp = \frac{F \times V}{33{,}000} = \frac{\text{foot-pounds/minute}}{33{,}000}$$

or substituting

$$hp = \frac{2\pi RN \times F}{33{,}000}$$

and as you know, T = RF. Therefore

$$hp = \frac{2\pi TN}{33{,}000} = \frac{TN}{5{,}252}$$

If you had a motor that rotated at 100 rpm, and the torque was calculated as before as 1.5 pound-feet/conductor, and the motor had 200 conductors, then the total torque would be 300 pound-feet. The horsepower is calculated as

$$hp = \frac{TN}{5{,}252} = \frac{(300)\ (100)}{5{,}252} = 5.7$$

Thus, the horsepower depends on speed and torque. Large, slow-speed motors develop a large torque while high-speed, smaller motors develop reduced torque.

Motor efficiency is the ratio of output power (P_{out}) to input power (P_{in}).

$$\text{Efficiency} = \frac{P_{out}}{P_{in}} \times 100$$

Example

For example, if a motor had an input of 10 amperes at 120 volts and delivered 1 horsepower, its efficiency would be

$$\text{Efficiency} = \frac{P_{out}}{P_{in}} \times 100 = \frac{746 \times 1}{10 \times 120} \times 100 = \frac{746}{1{,}200} \times 100 = 62\%$$

Back EMF (Counter EMF)

Remember that in a dc motor, as the armature rotates, the armature coils cut the magnetic field and thereby induce a voltage, or electromotive force (emf), in these coils. Since this induced voltage opposes the applied terminal voltage, it is called the *back emf*, sometimes referred to as the *counter emf*. This back emf depends on the same factors as the generated emf in the generator—the speed and direction of rotation and the field strength. The stronger the field and the faster the rotating speed, the larger will be the back emf. But the back emf will always be less than the applied voltage because of the internal voltage drop caused by the resistance of the armature coils. The illustration represents the back emf as an imaginary battery opposing the applied voltage, with the total armature resistance shown symbolically as a single resistor.

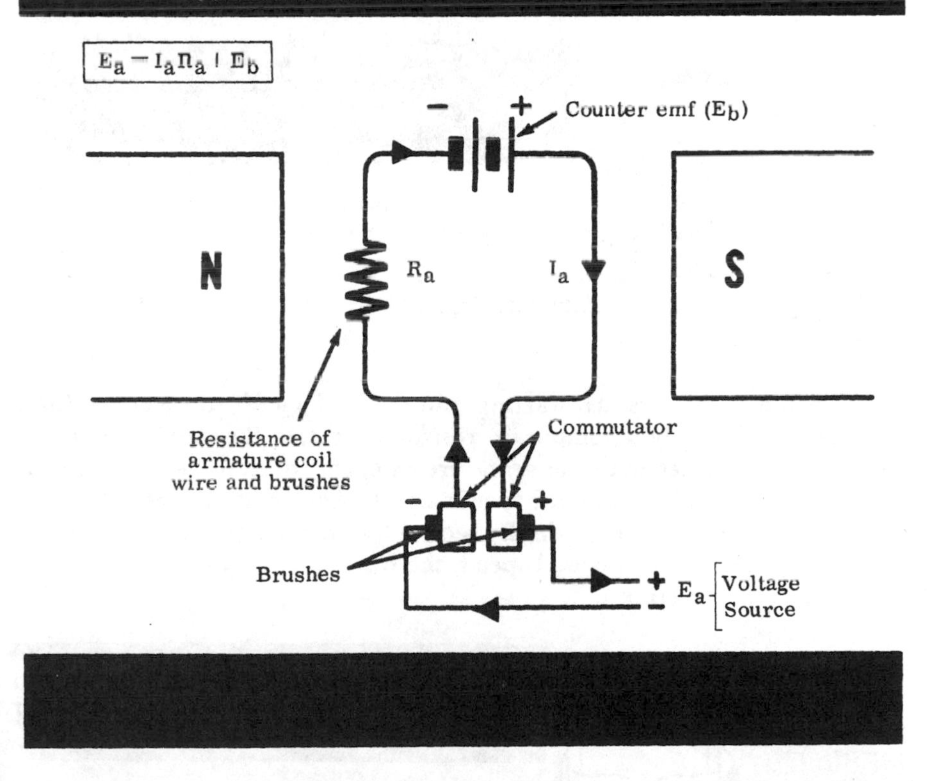

The emf that actually moves the armature current through the armature coils is the difference between the voltage applied to the motor, E_a, and the back emf, E_b. Thus, $E_a - E_b$ is the actual voltage effective in the armature, and it is this effective voltage which determines the value of the armature current. Since $I = E/R$ (Ohm's law), in the case of the dc motor, $I_a = (E_a - E_b)/R_a$. Also, since (according to Kirchhoff's second law) the

Back EMF (Counter EMF) (continued)

sum of the voltage drops around any closed circuit must equal the sum of the applied voltages, $E_a = E_b + I_a R_a$.

The internal resistance of the armature of a dc motor is usually very low, sometimes less than 1 ohm. If this resistance were all that limited the armature current, the current would be very high. For example, if the armature resistance is 1 ohm and the applied voltage is 230 volts, the resulting armature current, according to Ohm's law, would be: $I_a = E_a/R_a = 230/1 = 230$ amperes. This excessive current could burn out the armature.

However, the back emf is in opposition to the applied voltage and limits the amount of armature current that can flow. For example, if the back emf is 220 volts, then the effective voltage acting on the armature is the difference between the terminal voltage and the back emf: 230 − 220 = 10 volts. The armature current is then only 10 amperes:

$$I_a = (E_a - E_b)/R_a = 10/1 = 10 \text{ amperes}$$

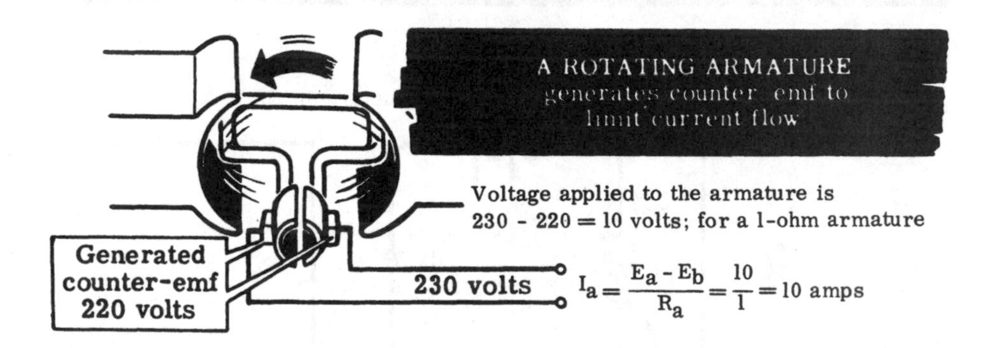
A ROTATING ARMATURE
generates counter emf to
limit current flow
Voltage applied to the armature is
230 - 220 = 10 volts; for a 1-ohm armature
Generated
counter-emf
220 volts
230 volts
$I_a = \frac{E_a - E_b}{R_a} = \frac{10}{1} = 10$ amps

When the motor is just starting and the back emf is too small to limit the current effectively, a temporary resistance, called the *starting resistance*, must be put in series with the armature in order to keep the current flow within safe limits. As the motor speeds up, the back emf increases and this starting resistance can be gradually reduced, allowing a further increase in speed and back emf. At normal speed, the starting resistance is completely shorted out of the circuit.

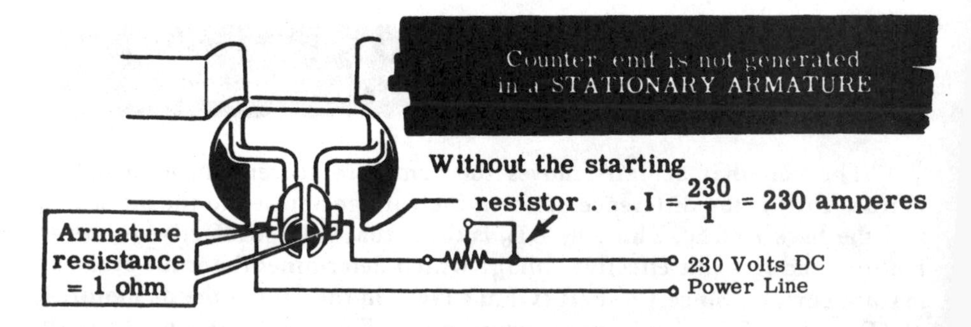
Counter emf is not generated
in a STATIONARY ARMATURE
Without the starting
resistor . . . $I = \frac{230}{1}$ = 230 amperes
Armature
resistance
= 1 ohm
230 Volts DC
Power Line

Shunt Motors

As with generators, the field winding of motors can be arranged in shunt, series, and compound. Also, the turning moment, or torque, developed by a motor is caused by the force arising from interaction of the magnetic field around the armature coils with the main field. The amount of torque developed, therefore, varies with the strength of the main field and of the armature current.

In a shunt-connected motor, the field is connected directly across the voltage source, and is therefore independent of variations in load and armature current. So the torque developed varies directly with the armature current. If the load on the motor increases, the motor slows down, reducing the back emf (which depends on motor speed as well as on constant field strength). The reduced back emf allows the armature current to increase, thereby furnishing the greater torque needed to drive the increased load. If the load is decreased, the motor speeds up, increasing the back emf and thereby decreasing the armature current and the torque developed. Thus, whenever the load changes, the speed also changes until the motor is again in electrical balance, that is, until $E_b + I_a R_a = I_a$ again. In a shunt motor, the variation of speed from no load to normal, or *full* load, is usually only about 10% of the no-load speed. For this reason, shunt motors are considered relatively constant-speed motors.

When a shunt motor is started, the starting current is low because of the added starting resistance, so the starting torque will also be low. Shunt motors are normally used when constant speed under varying load is desired and when it is possible for the motor to start under light or no-load conditions. Controlling motor speed will be discussed later.

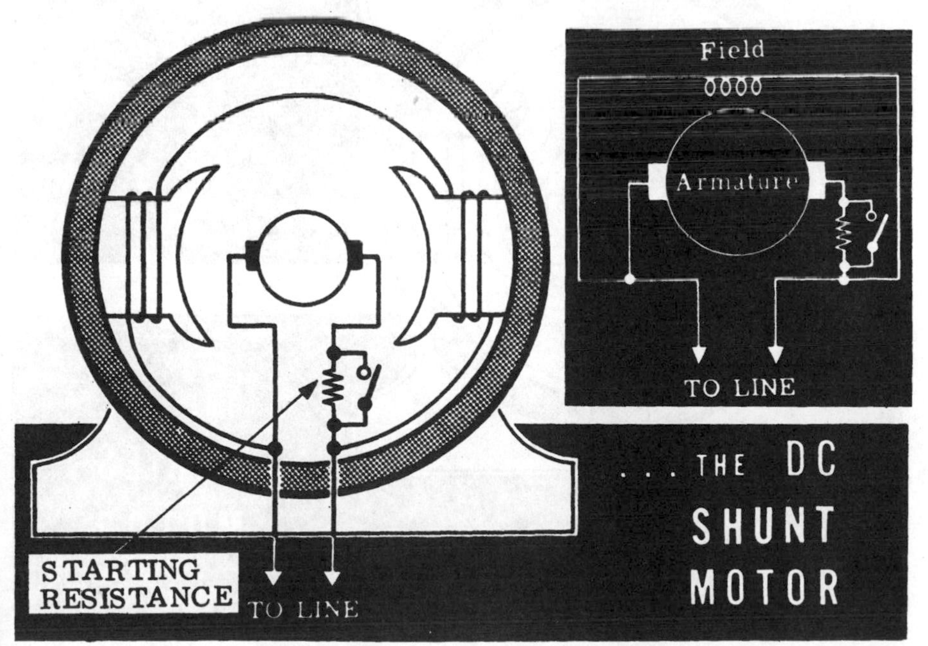

Series Motors

The series motor has its field connected in series with the armature and with the load, as shown below. The field coil consists of a few turns of heavy wire and since the entire armature current flows through it, the field strength varies directly with the armature current. As the load increases, the motor slows down and the back emf decreases, which allows the current to increase and supply the heavier torque needed by both increasing the field strength and the armature current.

The series motor runs very slowly with heavy loads and very rapidly with light loads. If the load is completely removed, the series motor will speed dangerously and may fly apart, for the current required is very small and the field very weak, so that the motor cannot turn fast enough to generate the amount of back emf needed to restore the balance. Series motors must never be run under no-load conditions, and they are therefore seldom used with belt drives from which the load can be removed.

You can see, also, that series motors are variable-speed motors, that is, their speed changes a great deal when the load is changed. For this reason series motors are not used when a constant operating speed is needed. And they are not used when the load is intermittent—in other words, when the load is put on and removed while the motor is running.

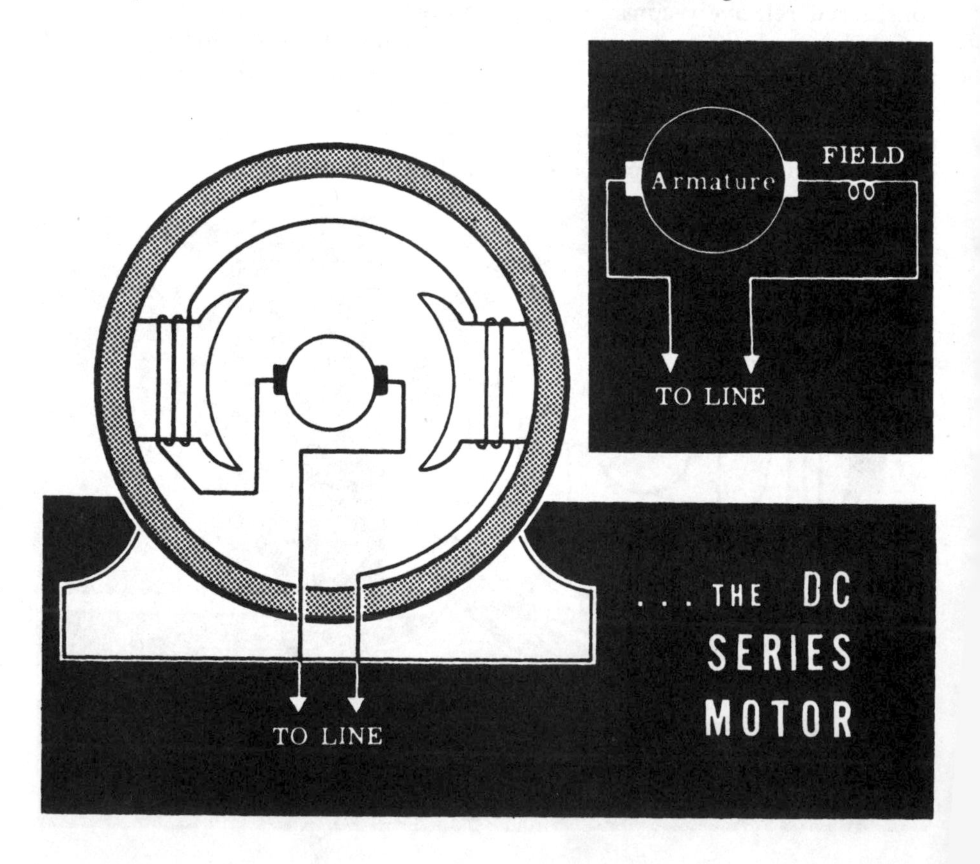

Series Motors (continued)

The torque—the turning force—developed by any dc motor depends on the armature current and the field strength. In the series motor, the field strength itself depends on the armature current, so the amount of torque developed depends doubly on the amount of armature current flowing. When the motor speed is low, the back emf is, of course, low and the armature current is high. This means that the torque will be very high when the motor speed is low or zero, for example, when the motor is starting.

Thus the series motor has a high starting torque. Because of its high starting torque, the dc series motor must not be started unloaded. For if there is no opposing torque on starting, the motor will accelerate furiously and race to a dangerously high speed. The armature may disintegrate, with the coils and commutator segments flying out and causing damage.

There are special jobs that require a heavy starting torque and the high rate of acceleration that this heavy torque allows. Some machines that have these features are cranes, electric hoists, electrically powered trains and trolleys, cars, and buses. The motors used in these machines are normally series motors because the loads that they handle are very heavy upon starting but become lighter once the machine is in motion.

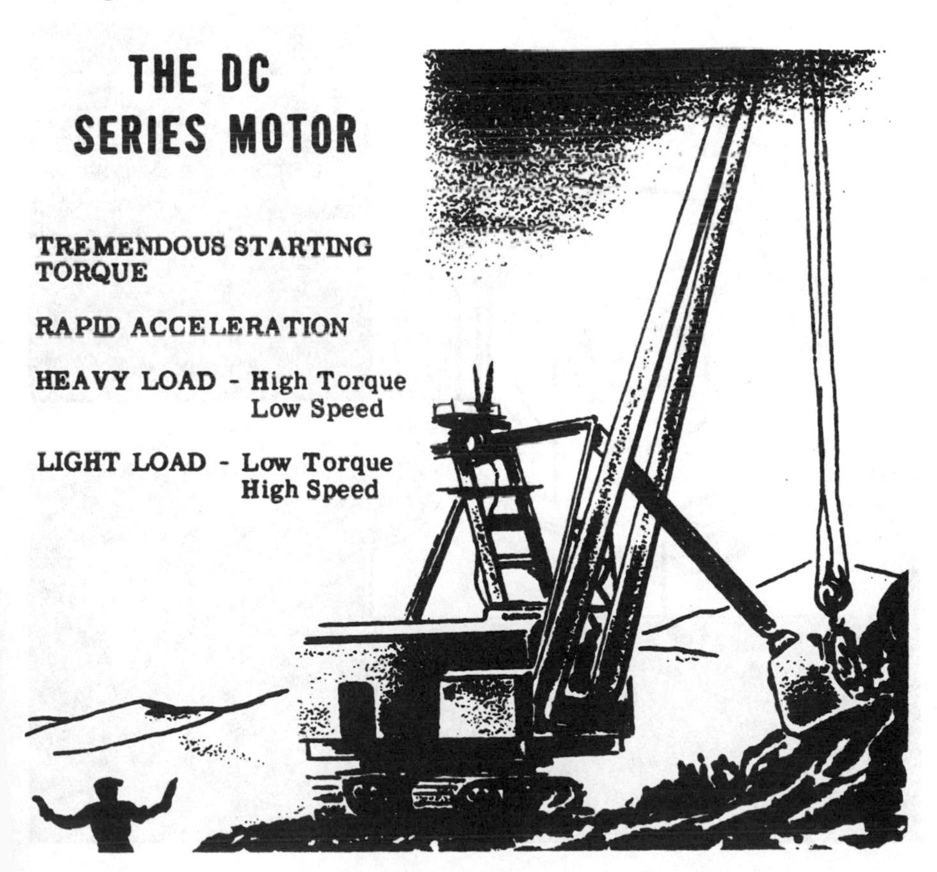

Compound Motors

A compound motor is a combination series and shunt motor. The field consists of two separate sets of coils. One set, with coils wound with many turns of fine wire, is connected across the armature as a shunt field. The other set, with coils wound with few turns of heavy wire, is connected in series with the armature as a series field.

The characteristics of the compound motor combine the features of the series and shunt motors. *Cumulatively compound* motors, whose series and shunt fields are connected to aid each other, are the most common. In a cumulatively compound motor, an increase in load decreases the speed and greatly increases the developed torque. The starting torque is also large.

Thus the cumulatively compound motor is a fairly constant-speed motor, with excellent pulling power for heavy loads and good starting torque.

In a *differentially compound* motor, the series field opposes the shunt field, and the total field is weakened when the load increases. This allows the speed to increase with increased load up to a safe operating point. As starting torque is very low, the differentially compound motor is rarely used.

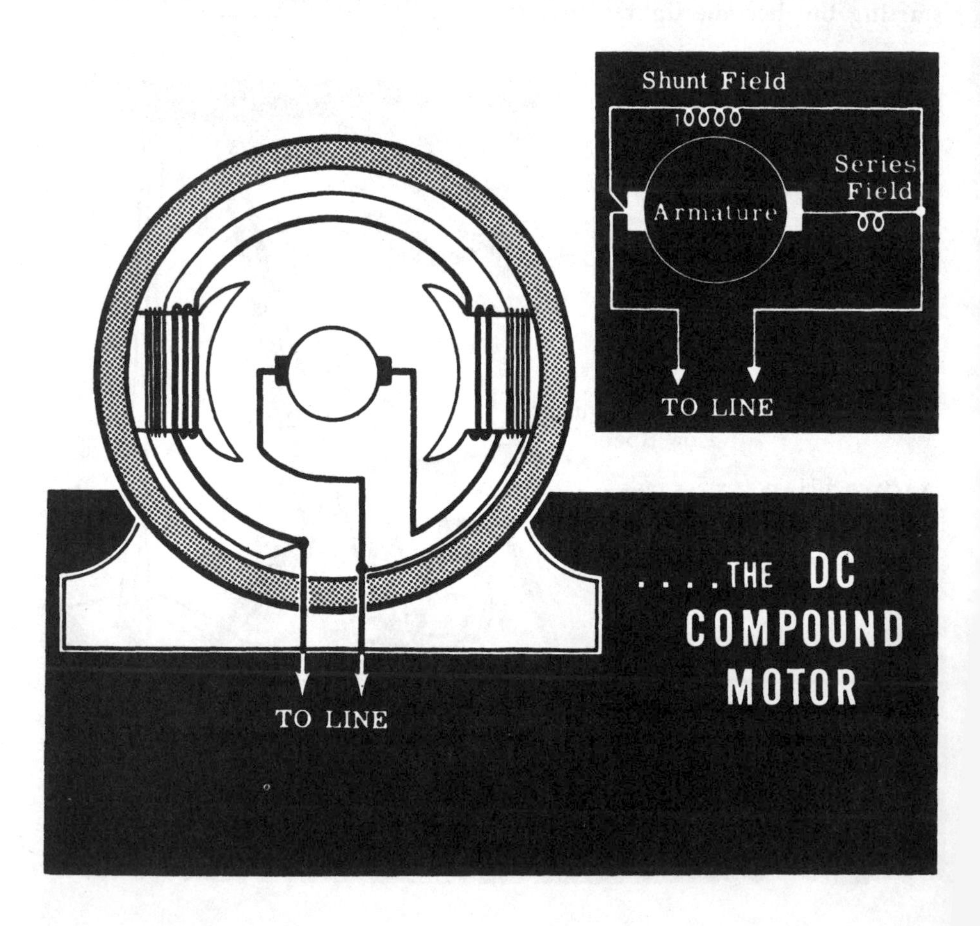

Comparison of DC Motor Characteristics—Speed Regulation

The operating characteristics of the different types of dc motors can be summarized by drawing a graph that shows how the speed varies with the torque, or load, on the motor. Notice that the speed of the shunt motor varies the least as the torque requirements of the load increase. On the other hand, the series motor speed greatly drops as the torque requirements increase. The cumulatively wound compound motor has speed characteristics lying somewhere between the series and shunt machines. Observe that the more heavily compounded it is (i.e., the greater its percentage of series turns compared with shunt turns), the more the motor acts like a series motor.

The second graph shows how the torque developed varies with armature current for the different motors of the same horsepower rating. The torque curve for the shunt motor is a straight line because the field remains constant and the torque varies directly with the armature current. The curves for the series and compound motors show that, above the full load

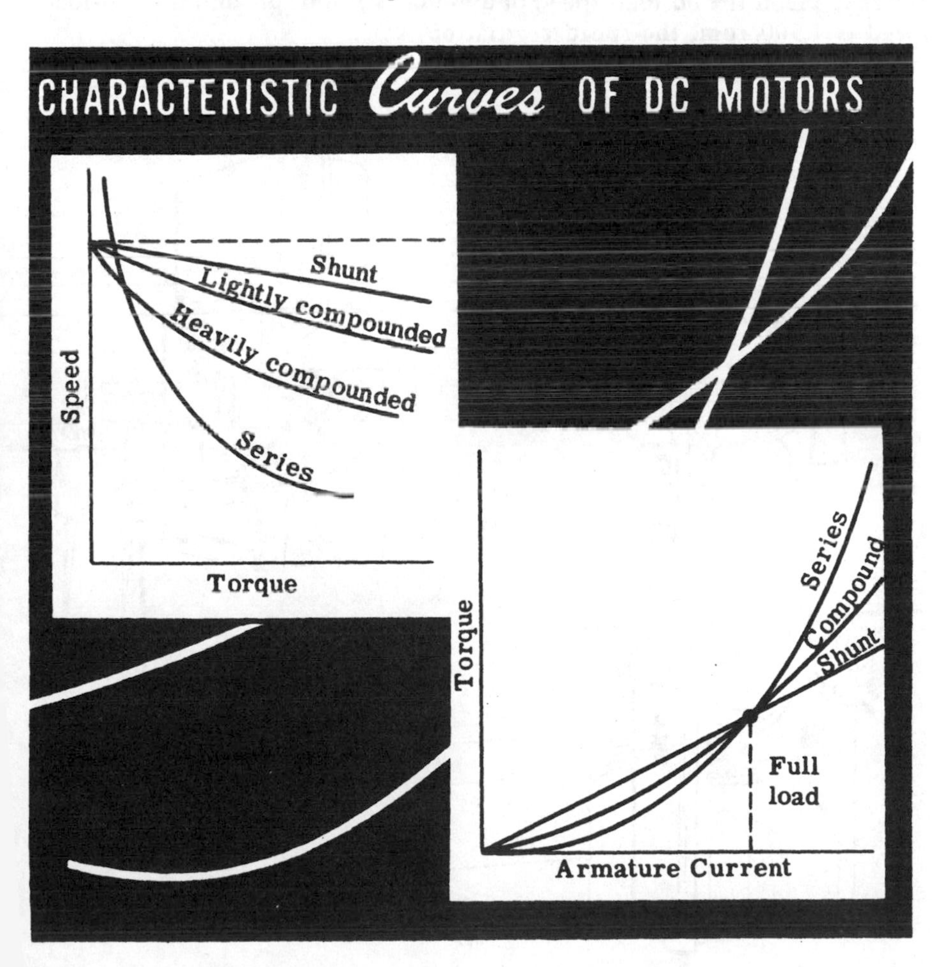

Comparison of DC Motor Characteristics— Speed Regulation (continued)

or normal operating current, the developed torque is much greater than for the shunt motor. Below the full-load current, the field strength of the series and compound machines has not reached its full value; therefore the developed torque is less than in the shunt machine.

Speed regulation can be calculated, like voltage regulation, as a percentage of full-load speed. Thus

$$\text{Percentage of speed regulation} = \frac{\text{no-load speed} - \text{full-load speed}}{\text{full-load speed}} \times 100$$

Example

For example, if the no-load speed of a motor is 1,600 rpm and the full-load speed is 1,500 rpm, the speed regulation is

$$\frac{1{,}600 - 1{,}500}{1{,}500} \times 100 = 6.6\%$$

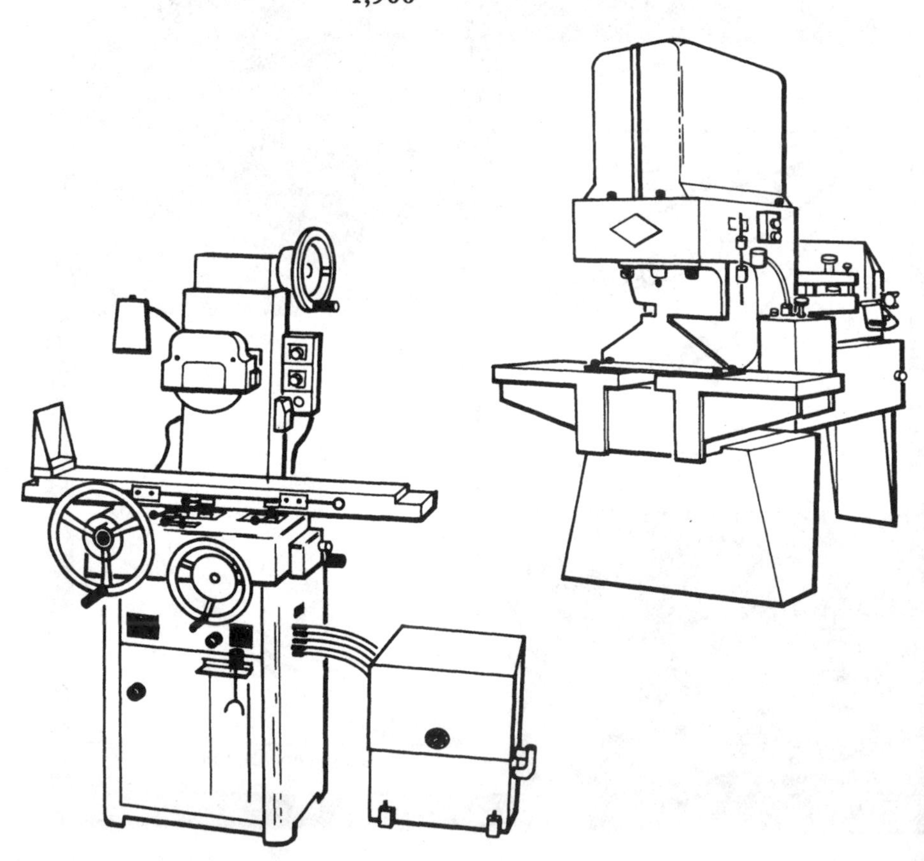

Reversing the Direction of Motor Rotation

The direction of rotation of a motor depends on the direction of the field and the direction of current flow in the armature. Current flowing through a conductor will set up a magnetic field around this conductor. The direction of this magnetic field is determined by the direction of current flow. If the conductor is placed in a magnetic field, force will be exerted on the conductor by reason of the interaction of its magnetic field with the main magnetic field. This force causes the armature to rotate in a certain direction between the field poles.

If either the direction of the field or the direction of current flow through the armature is reversed, the rotation of the motor will reverse. However, if both of the above factors are reversed at the same time, the motor will continue rotating in the same direction.

Ordinarily a motor is set up to do a particular job, which requires a fixed direction of rotation; but there are times when you may find it necessary to change the direction of rotation. Remember that to do this you must reverse the connections of *either* the armature *or* the field, *but not both*.

On larger machines, manufacturers usually provide some means of easily reversing the field connections.

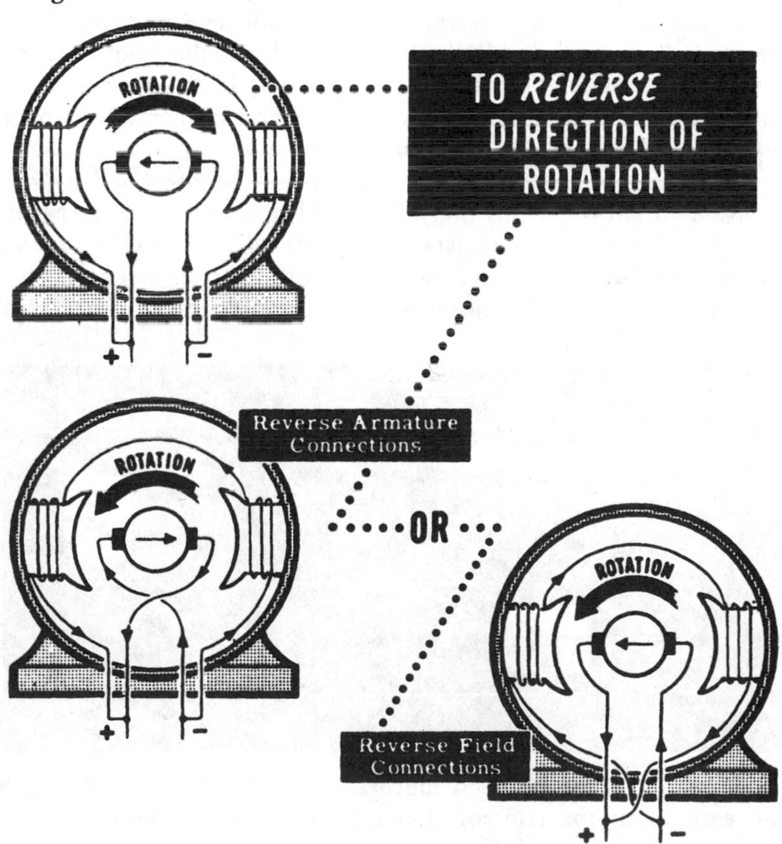

Controlling the Speed of a DC Motor

The speed of a dc motor depends on the strength of the magnetic field and the voltage applied to the armature, as well as on the load. The speed may, therefore, be controlled either by varying the field current or by varying the voltage applied to the armature.

You could decrease the voltage applied to the armature (and so decrease the speed of the motor) by increasing the resistance in the armature circuit. But this method is seldom used because a very large rheostat would be necessary and also because the starting torque is reduced.

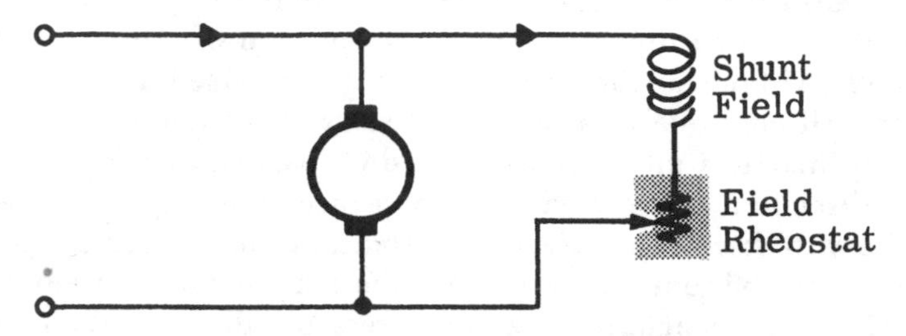

In shunt motors, however, the control of speed may be provided by connecting a rheostat in series with the shunt field winding. An increase in the resistance in series with the field reduces the field current and thus weakens the magnetic field. A decreased field strength means the motor must turn faster so as to maintain the back emf needed to ensure that the equation $E_a = I_a R_a + E_b$ remains in balance.

In series motors, control of speed may be provided by connecting a rheostat in parallel with the series field winding; such a rheostat is often called a *diverter*. As the resistance in parallel with the field is increased, so the current through the field winding, and the field strength, increase also, with the result that the motor must turn more slowly in order to maintain the same back emf.

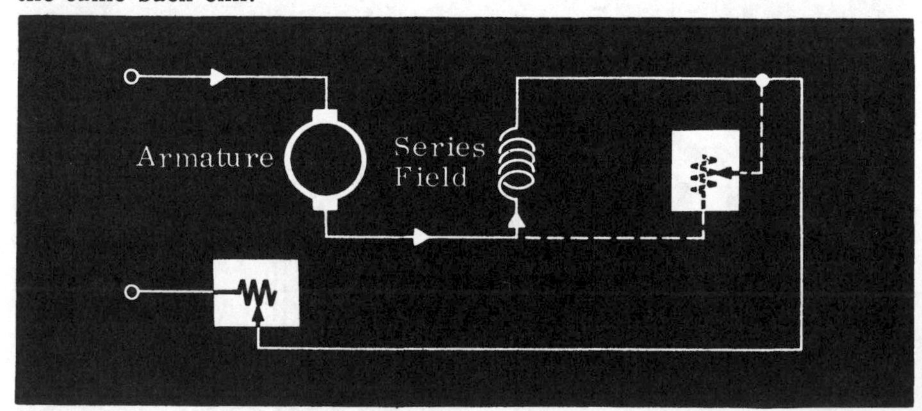

Usually variable-speed motors are of the shunt field type because of the ease of motor control. Later in this volume you will learn about automatic speed regulators.

Experiment/Application—DC Motor Speed Control—Back EMF

You can readily observe the effect of a rheostat in a dc motor field circuit by setting up the following experiment:

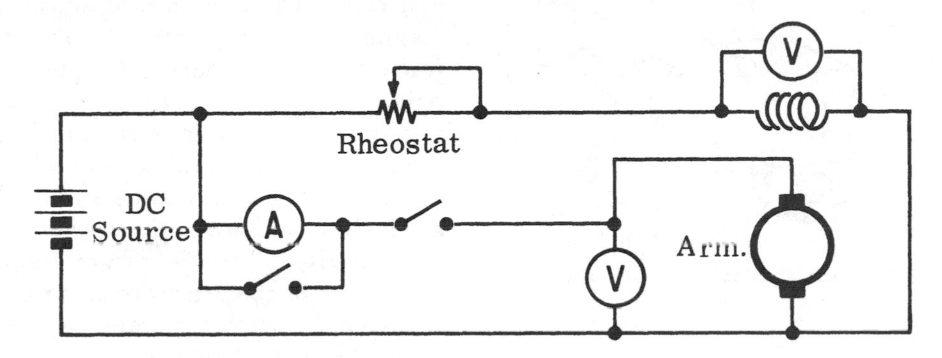

The switch across the ammeter should be closed when the motor is started to avoid meter damage from the large input current at that time. The switch in series with the armature is normally closed.

As you vary the rheostat, notice that the field voltage changes and the motor speed varies—increasing as the field voltage decreases. Also notice that the armature current decreases as the speed increases and vice versa (for a constant motor load).

You can estimate the back emf by opening the switch momentarily and rapidly observing the voltmeter across the armature winding. The voltmeter provides a measure of the back emf. You will observe that the back emf increases as the speed increases. Since the armature current depends on the differences between the input voltage and the back emf, you can see that the armature current goes down as the back emf goes up.

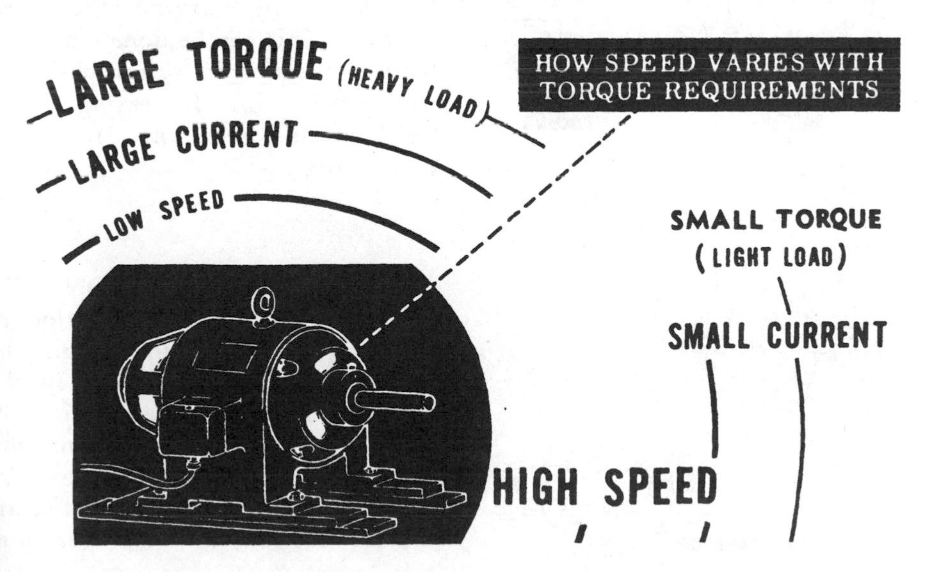

Review of DC Motors

1. DC MOTOR PRINCIPLES—Current flow through the armature coil causes the armature to act as a magnet. The armature poles are attracted to field poles of opposite polarity, causing the armature to rotate. The commutator reverses the armature current at the moment when unlike poles of the armature and field are facing each other, thus reversing the polarity of the armature field. Like poles of the armature and field then repel each other, causing continuous armature rotation.

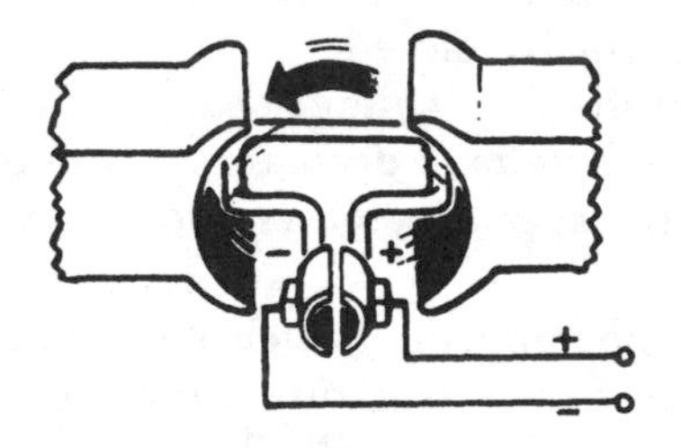

2. DC MOTOR BACK EMF—The rotating armature coil of a dc motor generates an electromotive force that opposes the applied voltage. This generated back emf limits the flow of armature current.

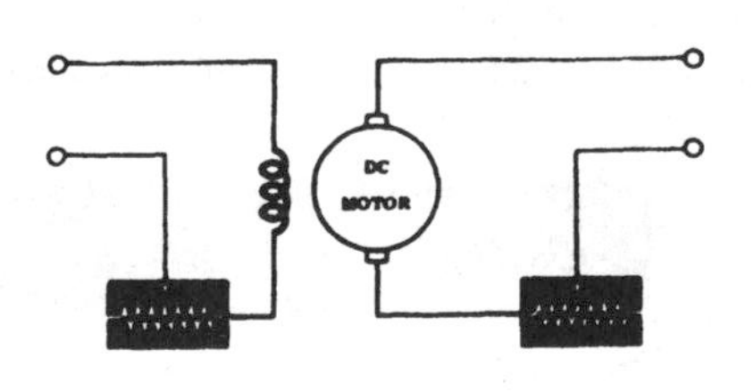

3. DC MOTOR SPEED CONTROL—The speed of a dc motor can be varied by varying the field strength. This can be done manually by a resistance in series with the shunt field coil. Increasing shunt field circuit resistance increases motor speed.

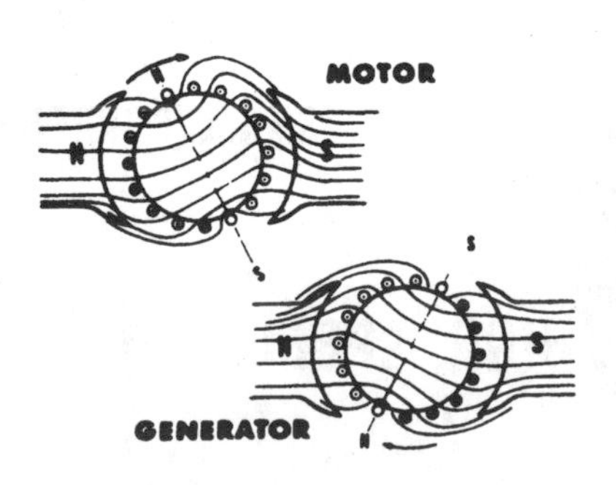

4. ARMATURE REACTION—The armature field causes distortion of the main field in a motor, causing the neutral plane to be shifted in the direction opposite to that of armature rotation. Interpoles, compensating windings, and slotted pole pieces are all used to minimize the effect of armature reaction on motor operation.

Review of DC Motors (continued)

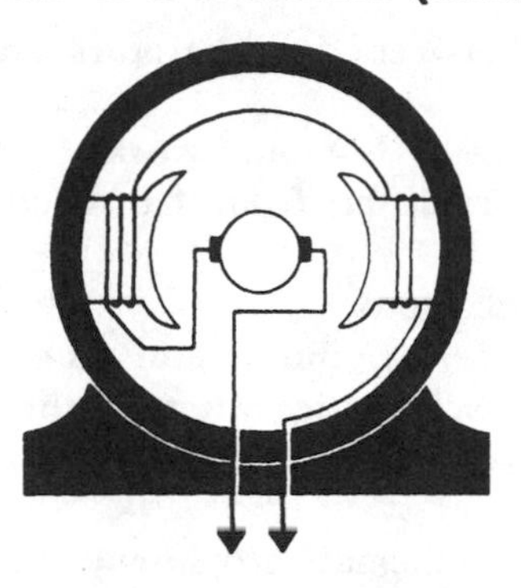

5. SERIES MOTORS—The field windings are connected in series with the armature coil, and the field strength varies with changes in armature current. When its speed is reduced by a load, the series motor starts to develop greater torque. Its starting torque is greater than that of other types of dc motors.

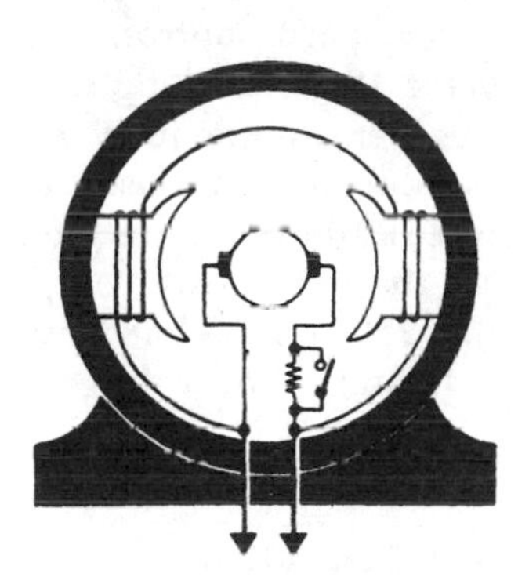

6. SHUNT MOTORS—The field windings are connected in parallel across the armature coil, and the field strength is independent of the armature current. Shunt motor speed varies only slightly with changes in load; the starting torque is less than that of other types of dc motors.

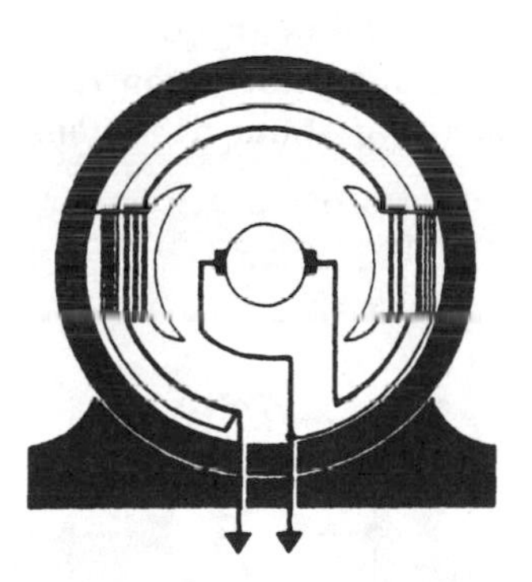

7. COMPOUND MOTORS—One set of field windings is connected in series with the armature, and one set is parallel connected. The speed and load characteristics can be changed by connecting the two sets of fields so that they either aid or oppose each other.

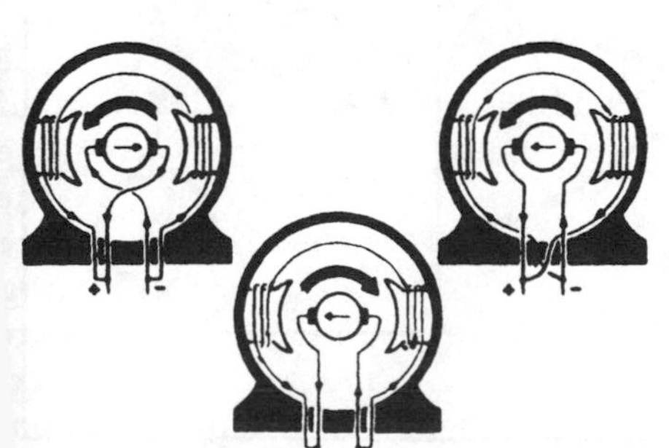

8. MOTOR REVERSAL—The direction of rotation of a dc motor can be reversed either by reversing the field connection or by reversing the armature connections, but not both.

Self-Test—Review Questions

1. List the similarities and differences between dc generators and dc motors.
2. Show by diagrams and explanation how a dc motor works.
3. Explain how armature reaction in motors differs from that reaction in generators.
4. What factors limit current flow in a dc motor?
5. Draw simple schematic diagrams for series, shunt, and compound motors. Show separately excited motor schematics when applicable.
6. For the motors listed above, describe the characteristics of each, and list possible applications.
7. For the motors listed in question 5, show schematically how you would reverse their direction of rotation.
8. Describe the principle involved in dc motor speed control.
9. Calculate the force generated by a conductor 18 inches long in a field with a flux density equal to 50,000 lines per square inch where a current of 60 amperes flows through the conductor. If the armature has a diameter of 12 inches, what is the torque if 100 turns are effective?
10. For the motor conductor above, calculate the horsepower output (at 100% efficiency) if the motor is rotating at 250 rpm. What would the horsepower output be at 75% efficiency (assume line voltage is 120 volts dc)?

Learning Objectives—Next Section

Overview—Now that you know how dc generators and dc motors work, you will learn about dc generator and dc motor control in the next section. You will also learn something about dc systems.

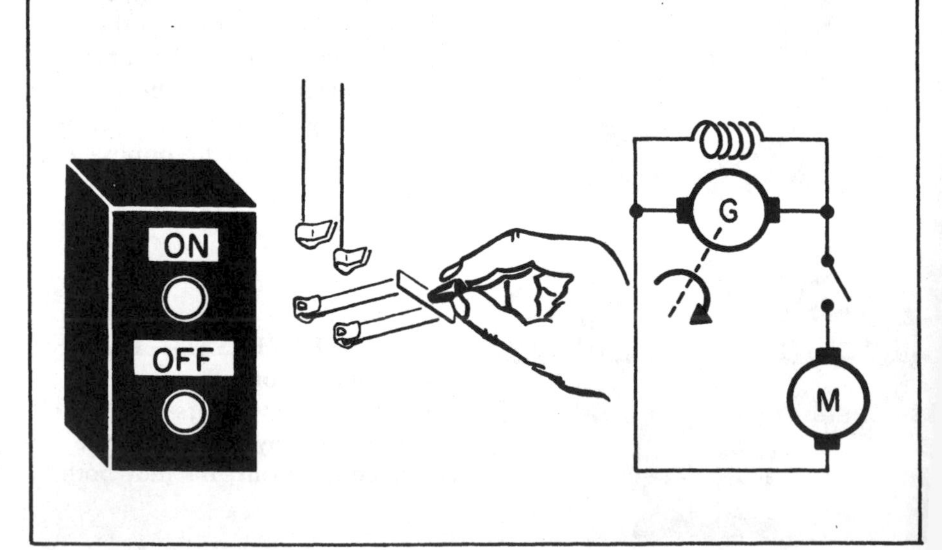

Circuit Control Devices—Relays

Earlier you learned about switches, which are the basic control device. A relay is an electrically operated switch. Relays are classed as *power controls* or *sensing relays*, depending on their function. Power relays can be used to control large amounts of power from a remote point or for operation of a circuit that has little power in it. Sometimes these power relays are called *contactors*.

The diagram on the left below illustrates a very simple magnetic relay, the essential parts of which are an electromagnet and a movable arm called an *armature*. When current flows through the coil of the magnet, a magnetic field is set up that attracts the iron arm of the armature to the core of the magnet. The set of contacts on the armature and relay frame then close, completing a circuit across terminals A and B. When the magnet is deenergized, the return spring returns the armature to the open position and the contacts open, breaking the circuit across terminals A and B.

The diagram shows only one set of contacts, but there can be any number of sets, depending on the requirements of the circuit.

The relay shown on the left is a normally open relay because the contacts are open when the relay is deenergized. The relay shown on the right is a normally closed relay because the contacts are closed when the relay is deenergized. When the relay is energized, the armature is pulled to the magnet, the contacts open, and the circuit across terminals A and B breaks.

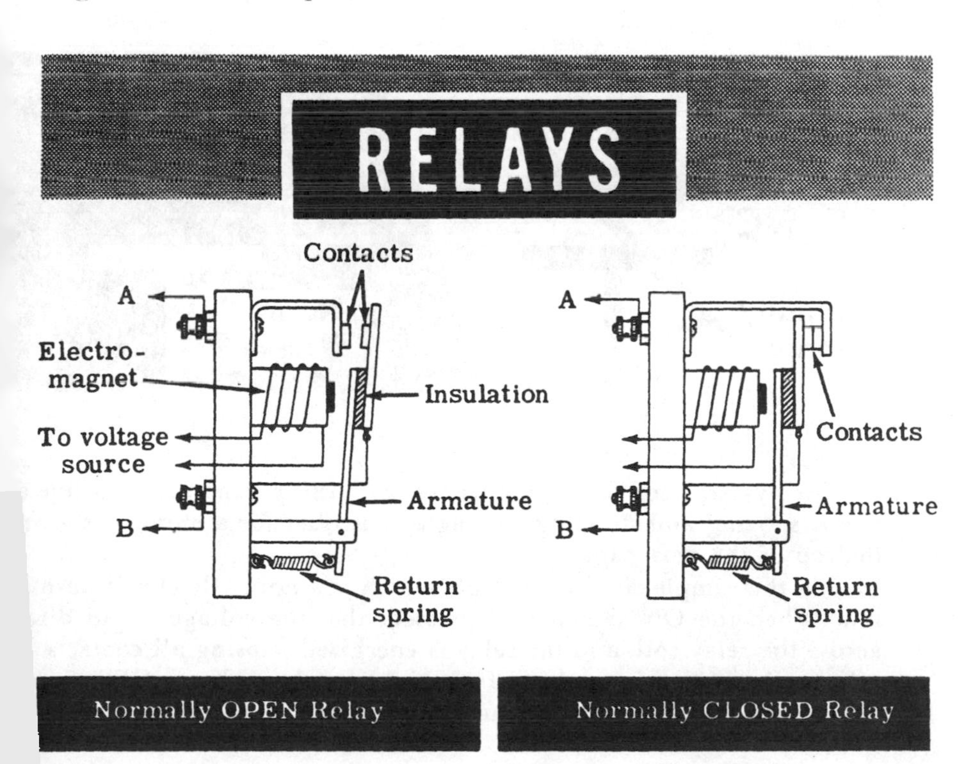

Circuit Control Devices—Relays (continued)

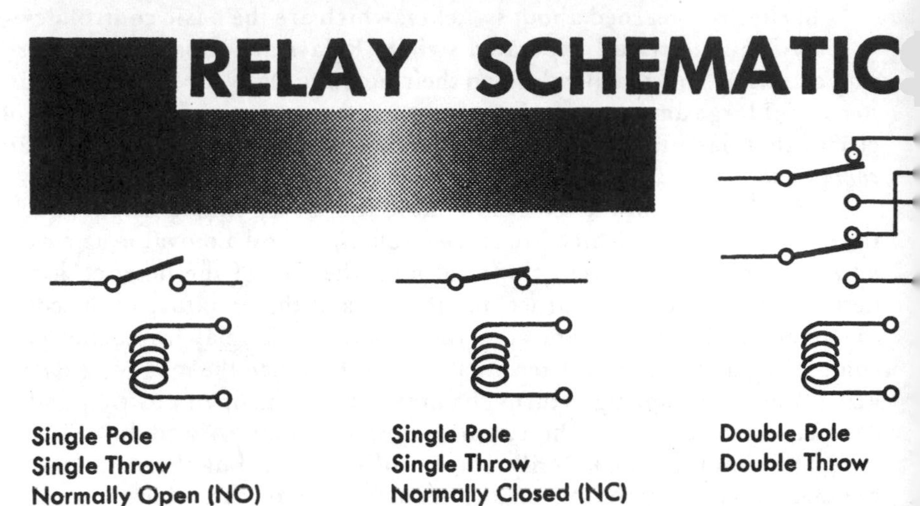

Some relays have several sets of different contacts for performing var ous functions, such as energizing a holding coil. You will learn more abo this on the next page. Sometimes relay contacts are shown schematically below.

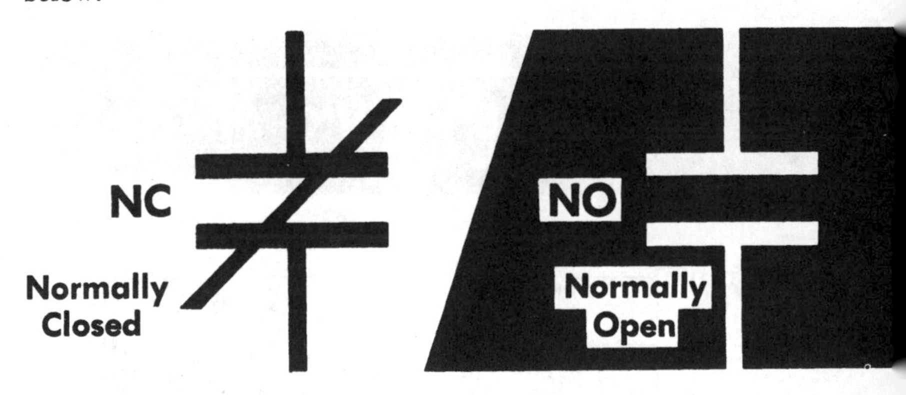

Relays are used for such purposes as putting generators on line a motor starting and stopping. A single controller for a motor is showr the top of the next page.

In this simple circuit, the OFF switch is a normally closed push b ton. When the ON switch is depressed, the line voltage is put dire across the relay coil, and the relay is energized, closing all contacts. relay coil circuit is completed from the line, through the OFF swit normally closed (NC)—to the auxiliary contacts on the relay, so as to

Circuit Control Devices—Relays (continued)

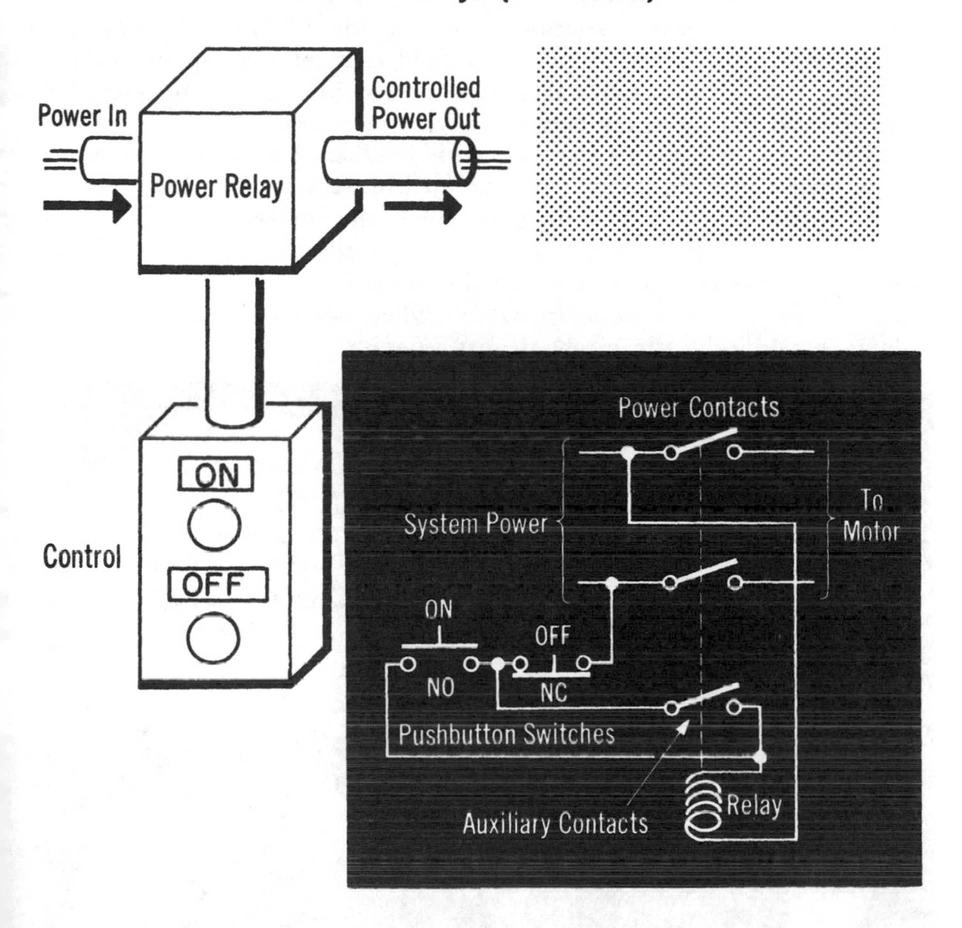

the relay closed. To turn the motor off, the OFF push button is depressed. This opens the line holding the relay closed, and the relay circuit opens, stopping the motor. Such arrangements are very common, and a relay of this type is often called a *latching relay.* In some motor starters and generator controllers that we will study later, the relay is held closed by a latching coil, called a *holding coil,* through which part of the circuit current is directed. If the circuit is interrupted, this holding coil is deenergized and the relay opens.

You will probably encounter *thermal overload relays* if you work with power equipment. These are relays which consist of a bimetallic strip that bends and opens a set of contacts when heated. They are often used to protect electrical equipment from excessive temperatures that would usually destroy it. The symbol for a thermal relay is the same as for a pair of NC contacts except that a *T* is written alongside them to designate it as a thermal relay (TNC).

Circuit Protection Devices—Circuit Breakers

A circuit breaker is a variation of the relay that is designed to open the circuit if the current exceeds a predetermined value. It is commonly used in place of a fuse and sometimes also serves as a switch. Circuit breakers *trip* (open the circuit) when the current flow is excessive.

There are two main types of circuit breakers based on the current sensing mechanism. In the *magnetic* circuit breakers, the current is sensed by a coil that forms an electromagnet. When the current is excessive, the electromagnet actuates a small armature that pulls the trip mechanism—thus opening the circuit breaker. In the *thermal-type* circuit breaker, the current heats a bimetallic strip, which when heated sufficiently bends enough to allow the trip mechanism to operate.

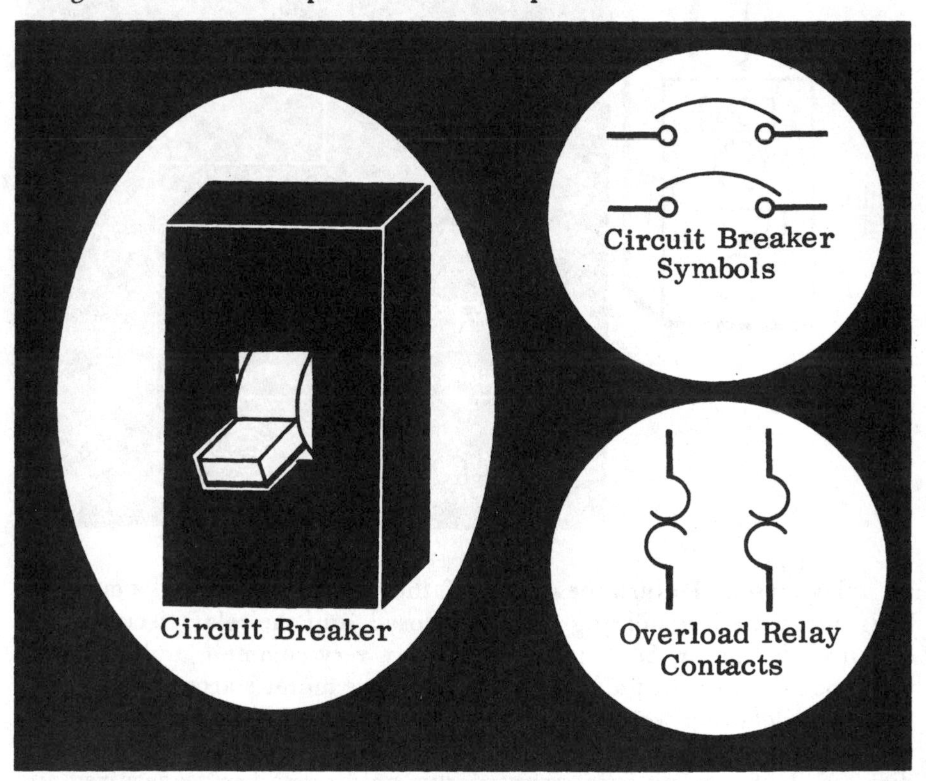

Most circuit breakers require a manual reset, that is, when tripped, they must be manually closed again. Some circuit breakers are electrically resettable by operating a *reset button*. As you will learn, some circuit breakers are combined with motor starting equipment so that all control and protective devices are in a single unit. In many circuits you will see overload relay contacts in series with the line. These contacts are normally closed, but if the temperature of the device becomes excessive, the thermal relays will close, causing the overload relay to become energized and opening the line.

DC Generator Voltage Regulators—Manual and Automatic

As you learned earlier, the voltage from a dc generator can be varied by changing the speed of the prime mover and by changing the field current in a shunt (or compound) wound generator. The internal resistance of the generator windings also causes the output voltage to vary with the load. Under many circumstances, these voltage changes are undesirable, and therefore a *voltage regulator* is used.

The simplest system consists of a rheostat in the field winding with a voltmeter on line to read the generator output. As conditions change, the rheostat is adjusted to keep the voltage constant.

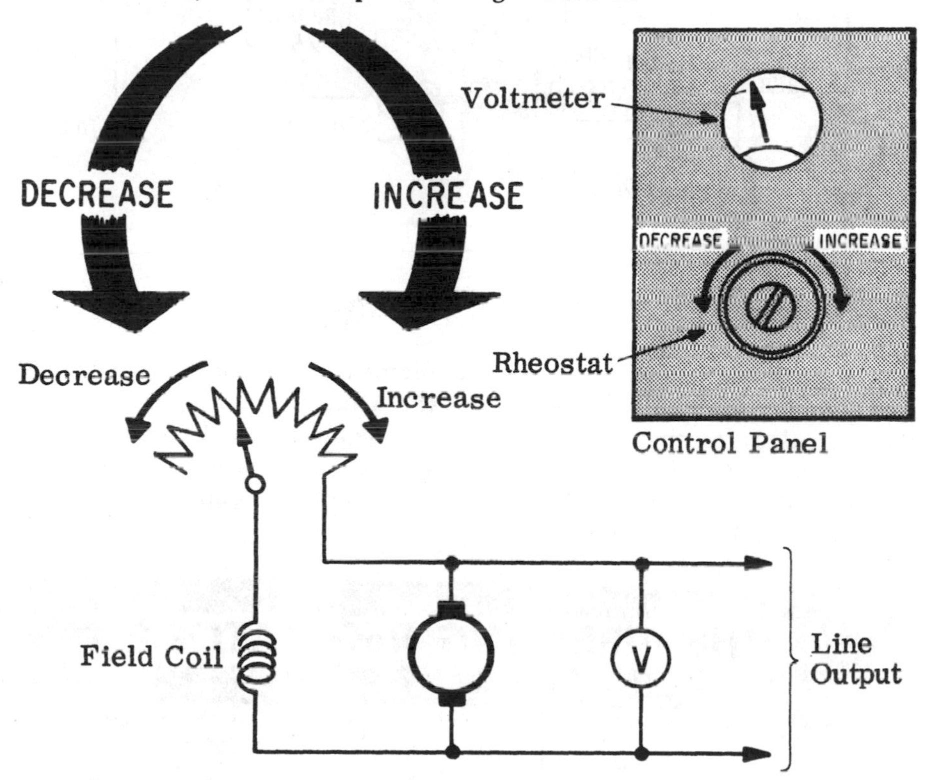

There are many types of automatic voltage regulators that are used to control the output voltage of a dc generator. An early type of automatic regulator was the *carbon pile regulator*. Although it is not commonly used today, its basic principle is common to many regulators. Fundamentally, it consists of a stack of carbon washers loose in a holder. The resistance of the stack is proportional to the pressure applied. A spring applies a pressure to the pile establishing a minimum resistance. An electromagnet with a sliding core (called a *solenoid*) is arranged to pull against the spring and thus increase the resistance of the pile as the pressure decreases. As shown on the next page, the carbon pile rheostat is in series with the field and essentially makes fine adjustments in the field current.

DC Generator Voltage Regulators—Automatic

It is apparent from the drawing below that, as the voltage output from the generator starts to increase, the pressure on the pile is reduced. This increases the field resistance, thus decreasing the current flow in the field. As a result, the voltage output is reduced.

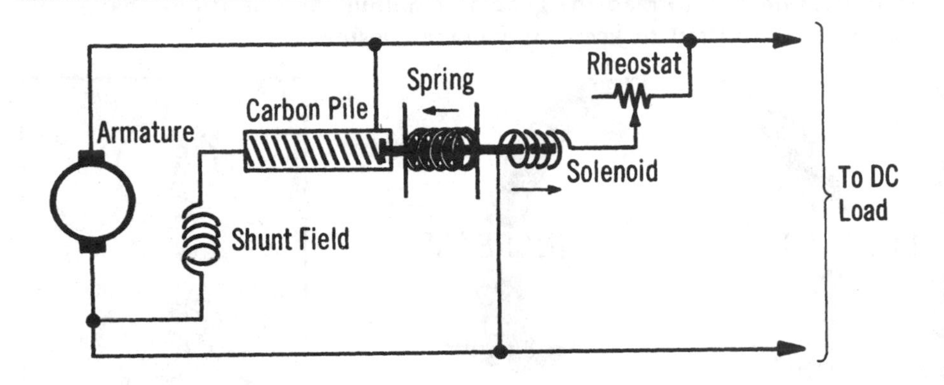

A more common type of voltage regulator is the *vibrating regulator*. It is an automatic regulator that works on the principle that an intermittent short circuit across the field rheostat will cause the terminal voltage to pulsate within narrow limits and thus maintain an average steady voltage. A schematic diagram is shown below. Operation of the vibrating regulator is described on the next page.

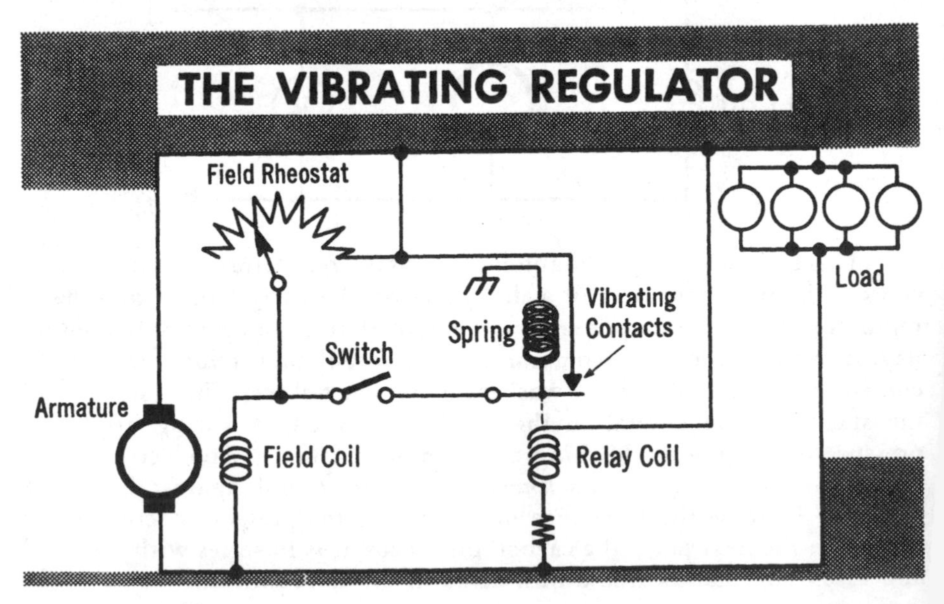

DC Generator Voltage Regulators—Automatic (continued)

The vibrating regulator, shown schematically on the previous page, operates by varying the average field current of the generator. With the generator running at normal speed, the switch is opened and the generator output set at about 70% of normal output by means of the field rheostat. If the relay contacts are closed when the switch is closed, the field rheostat is shorted out and the output voltage will start to rise. When it reaches a high enough value, the relay coil will exert sufficient force on the moving contact to pull it away, thus inserting the field rheostat into the field circuit. This will start to reduce the voltage. This cycle repeats rapidly and continuously—hence the name vibrating regulator. As the load increases, the field rheostat is shorted for a longer period of time, and thus the relay contacts vibrate more slowly and spend more time in the shorted position. As the load decreases, the contacts vibrate more rapidly and spend more time in the open position.

In some applications, for example, in aircraft, boats, automobiles, etc., where speed varies *widely*, the generator is driven by an engine. While most modern electrical systems in these vehicles use *alternators*, which you will study later, it is also important to know about the dc regulator for variable-speed generators because this regulator in modified form is also used with alternators.

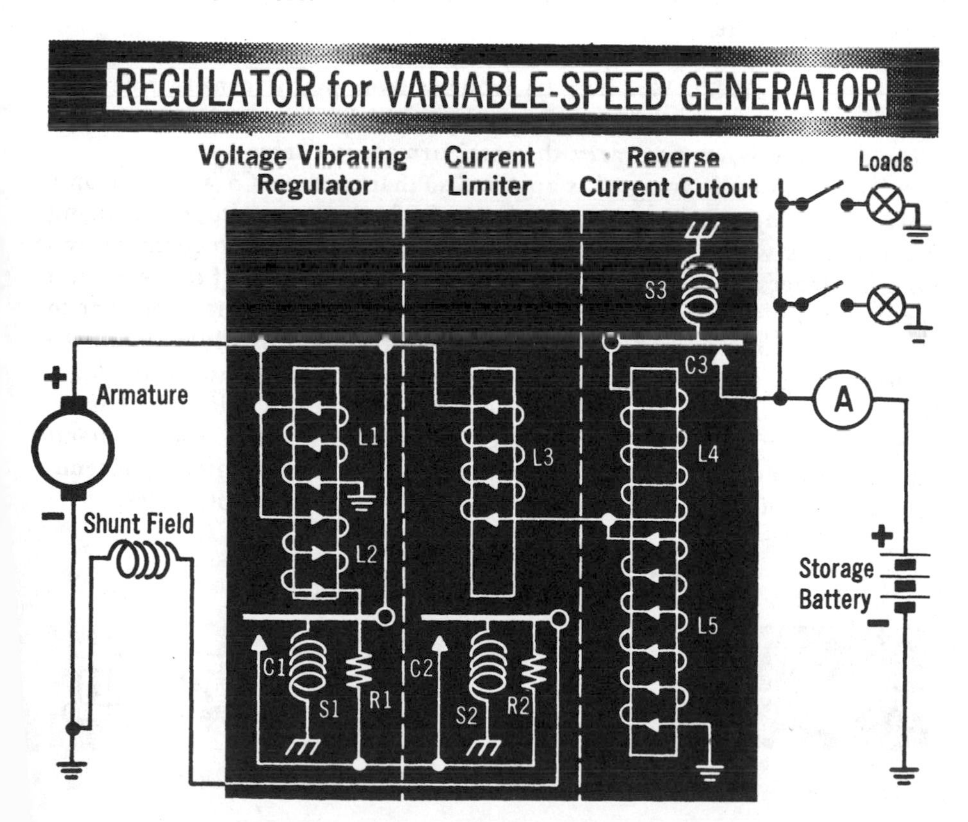

DC Generator Voltage Regulators—Automatic (continued)

The regulator shown on the previous page is a three-unit regulator used to charge a battery as well as to operate system loads. The first unit is a vibrating voltage regulator as described earlier. It operates in conjunction with the second unit, which functions as a current limiter to avoid damage to the generator if the load becomes excessive. As you can see from the diagram, the voltage regulator contacts C1 add R1 alternately to the field circuit—or short it out when contacts C2 are closed. Winding L3 on the current regulator consists of very few turns of heavy conductor, since it carries the *full* generator output. Normally contacts C2 are closed, shorting out additional field resistance R2. However, if the current becomes excessive, the contacts C2 are opened against spring tension from S2. When R2 is in the field circuit, the generator voltage is reduced, so the current limiter keeps the current down by reducing the voltage. Like the voltage regulator, the current limiter contacts vibrate to limit the current (and reduce output voltage).

Since the generator shown is used to charge a battery as well as to drive loads directly, some means are necessary to make sure that the current only flows from the generator *to* the battery. If the reverse were true, the generator would operate as a motor load (motorize) and *discharge* the battery. The third unit of the regulator, the reverse current cutout, serves this purpose.

Contacts C3 are held open by spring S3. Contacts C3 are operated by windings L4 and L5. Winding L5 is made from many turns of fine wire and connected across the generator output, and L4 is made from a few turns of heavy wire that carry the total armature current.

The tension of spring S3 is adjusted so that contacts C3 will close only when the voltage of the generator is above the battery voltage—assuring that charging will take place. Coil L4 is wound so that, when current flows from the generator to the load, its field aids that of L5, and if the reverse, it opposes. With this arrangement, contacts C3, connecting the generator to the load, are only closed when the generator voltage is higher than the battery voltage or when the current flow is from the generator to the load. It is apparent that, if the generator voltage should drop below the battery voltage, the momentary reverse current through L4 would open contacts C3. In more modern systems, reverse current flow is prevented by a semiconductor device called a *diode,* which you will learn about later.

SEMICONDUCTOR DIODES

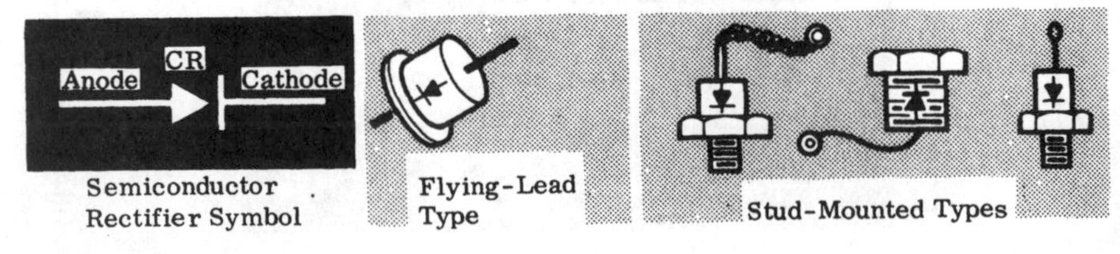

Semiconductor Rectifier Symbol

Flying-Lead Type

Stud-Mounted Types

DC Starters and Controllers

In studying dc motors, you learned that the armature resistance is usually very low. If this resistance were the only opposition to current flow, the armature current would be excessively high. When the motor is running, however, the back emf generated in the rotating armature opposes the applied voltage and limits the armature current.

But when the motor is just starting, the back emf is zero or very low, and the starting current would be very high if it was not limited in some way. To prevent this high starting current (which would damage the armature windings and commutator), a resistance called the *starting resistor* is put in series with the armature at starting. As the motor speed and the back emf increase, the starting resistor is gradually removed from the circuit.

The complete starting resistor assembly is called a *dc starter.* In addition to limiting the value of the starting current, the dc starter usually contains devices intended to protect the motor in case the field circuit becomes open-circuited, or in case the applied voltage becomes too low. It is also necessary to ensure that the starting resistance is automatically reconnected into the circuit every time the motor stops.

When the starter is so constructed that it can also *control* the operating speed of the motor, it is called a *controller.*

There are various types of starters, some manually controlled and some automatic. Usually the starting current is limited to about one and one half times the normal full-load current.

There are some small dc motors whose armatures contain many turns of fine wire, offering enough resistance to the current flow to make a starter unnecessary. But *all large* dc motors require some type of starter or controller.

A DC Starter.....

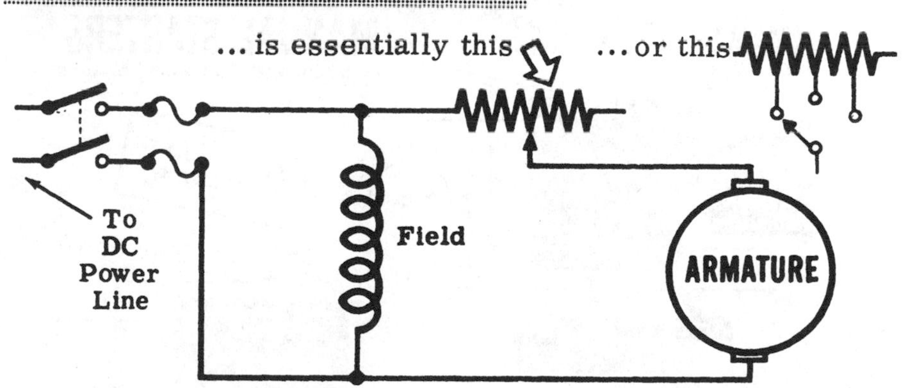

Depending on the physical design, starters are of the *faceplate* or *drum type.* In a faceplate starter the contacts are on a flat surface. In a drum-type starter the contacts are arranged on a rotary drum.

Shunt Motor Starters—Manual

One common type of manual shunt motor starter has three terminals and is called the *three-point starter.* As illustrated on the left below, point L goes to one side of the supply, point A to the armature, and point F to the field. When starting the motor, the arm is moved to the first contact, and the entire resistance is placed in series with the armature circuit. The field coil is connected in series with an electromagnet across the supply. As the motor builds up speed and the back emf increases, the arm is moved to each of the contacts in turn, decreasing the resistance in steps. As the arm moves across the starter, some of the resistance is also in series with the field and the electromagnet coil. When the arm is all the way to the right, in the *run position*, the armature is directly across the supply, and the motor is operating at full speed.

In the run position, a small piece of iron on the arm is held by the electromagnet through whose coil the field current flows. If the supply voltage fails or is switched off, or if the field circuit is broken, the magnet no longer attracts the iron, and a return spring pulls the arm back to the OFF position—thus, disconnecting the motor from the supply, and preventing it from being started without any starting resistance when the supply voltage is applied once again.

The return spring can also be set to return the arm if the voltage drops by a certain amount. This is called *low-voltage* protection.

Overload protection is provided by a second electromagnet connected so that the armature current flows through its coil. It is adjusted so that when the current exceeds a predetermined safe value, a piece of iron attached to a spring-loaded arm is attracted to the magnet. At the end of this arm is a link, which then connects two terminals and shorts out the coil of the electromagnet that is holding the starter arm in the run position. The starter arm is thus released and disconnects the motor from the supply.

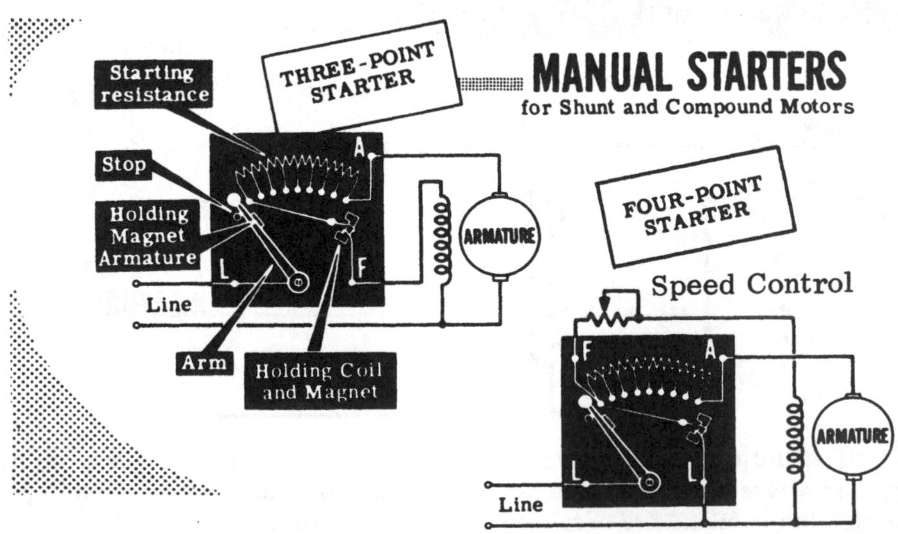

Shunt Motor Starters—Manual (continued)

When it is desirable to obtain variable speed control of the motor by varying the field current, the electromagnet that holds the starter arm in the run position may be connected directly across the supply (as illustrated in the right-hand diagram on the previous page). This is called a *four-point starter* because four wires are brought out. This type of connection allows variation of field current without altering the pull of the electromagnet. But it does not give protection against an open-circuited field. The speed control rheostat is in the field circuit for adjusting the motor speed.

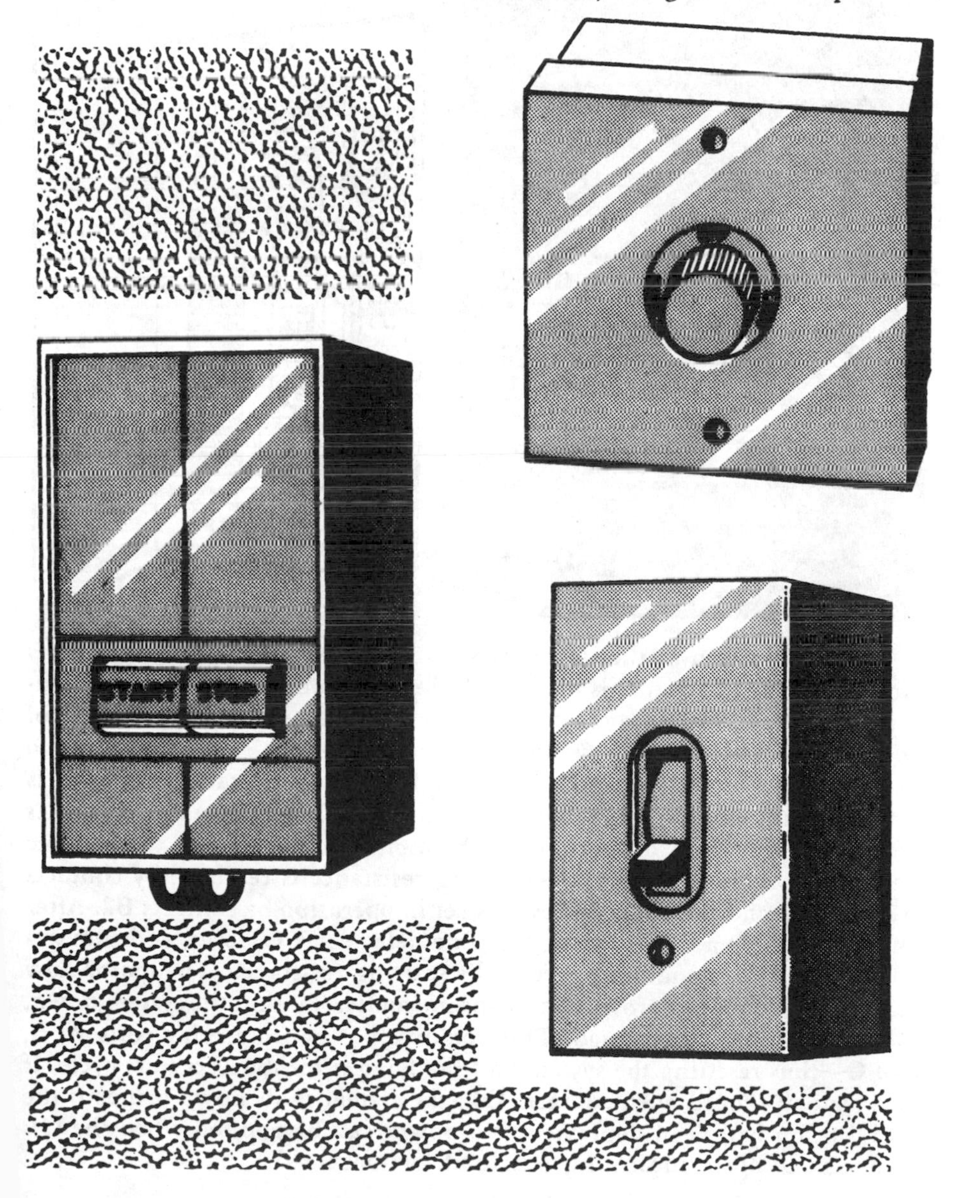

Shunt Motor Starters—Automatic

There are many types of automatic starters available for dc motors. These have some automatically timed way of removing the resistance from the motor input. Relays and protective circuits are also often built into these devices. A typical automatic starter for a compound generator is shown below.

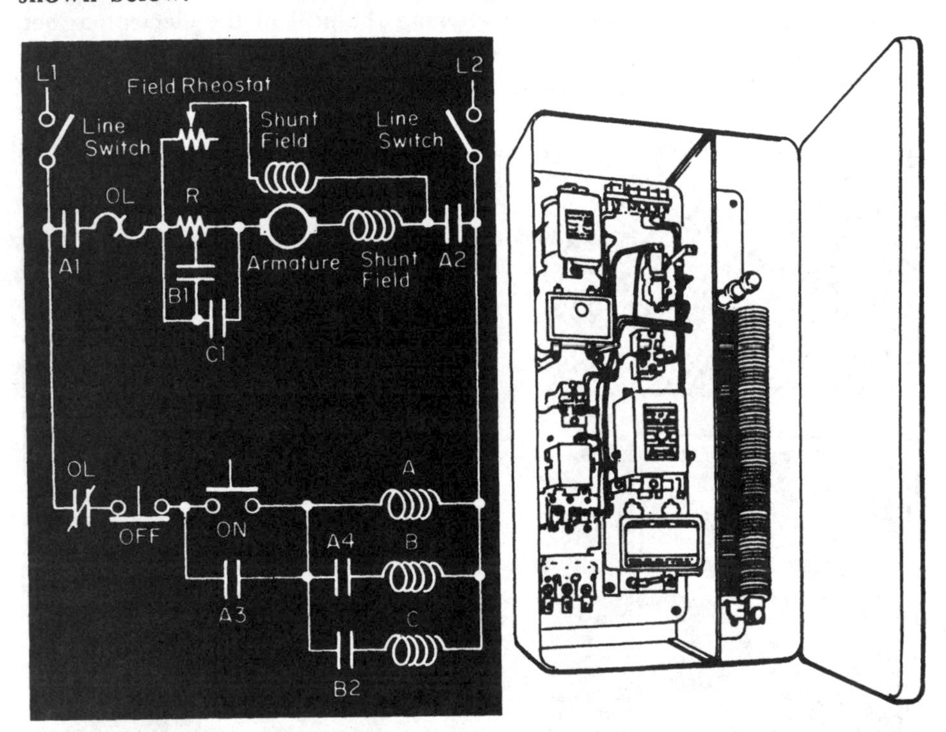

The operation of the starter is straightforward. If the line switch is closed and the ON button is depressed, the main contactor A closes, assuming that the overload relay is closed (as it would be normally). The coil of contactor A is held closed by closed relay contacts A3. Power is applied to the motor via contacts A1 and A2 through starting resistance R. Power is also applied to timing relay B via contact set A4. Time delay relay B closes after a suitable time, depending on the motor characteristics. When time delay relay B closes, part of the starting resistance is removed by contacts B1 and a second time delay relay C is set in operation by contacts B2. After another suitable period of time, the time delay relay C closes, and contacts C1 short out the remaining part of the starting resistance. The motor now has full-line voltage across it and runs normally. To shut the system down, the OFF button is depressed. This removes power from relay coils A, B, and C—thus resetting the system and readying it for the next start cycle. If an overload should open contacts OL, the system shuts down just as if the OFF button had been depressed. No power can be applied when the OL contacts are open, indicating overload has occurred.

DC Motor Speed Control

The basic method for controlling dc motor speed is by changing the shunt field current. Later when we learn about universal (ac or dc) motors, we will learn about other methods of speed control. In general, only shunt or compound wound dc motors are used in variable-speed applications. Dc motor speed controllers can be of either the faceplate or drum type. The faceplate type is typically used in applications where speed adjustment is infrequent, whereas the drum-type controller is used where speed changes and operation in both directions are more frequently required.

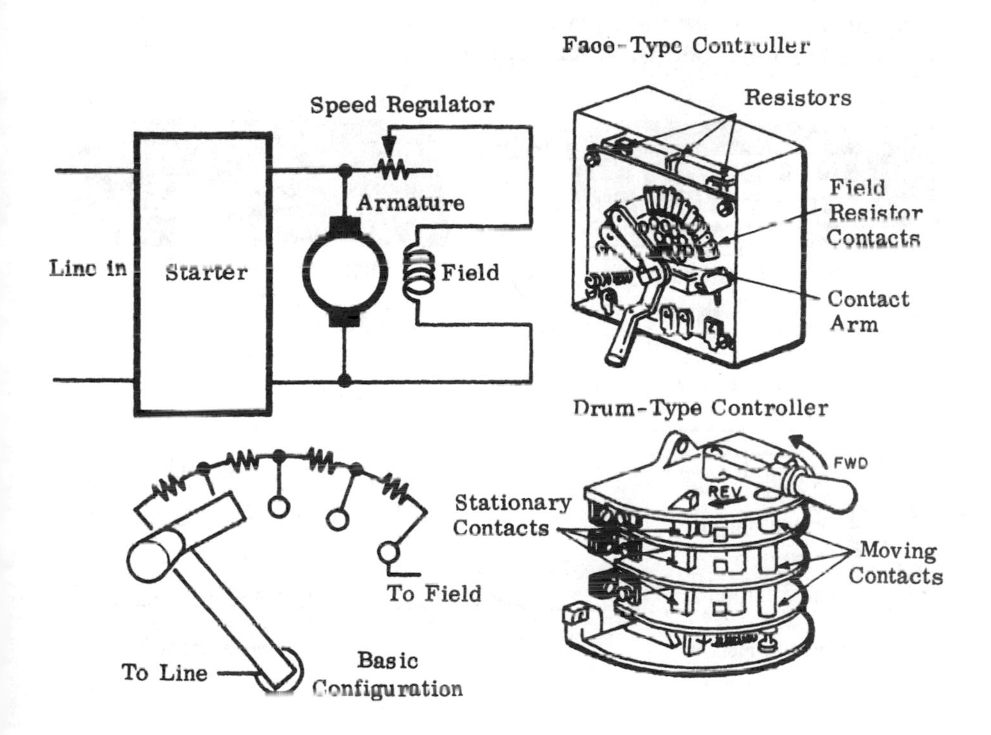

In either case, the control of speed is by control of the motor shunt field current.

Sometimes the speed controller is made part of the starting circuit. In this case, operation of a control lever is involved with starting the motor on its initial travel, and is used for motor speed control during the remainder of its travel.

For constant-speed operation where accurate speed control is essential, electronic control systems are used. These generally use a small generator to measure motor speed and feed this voltage back to the field current of the motor via electronic amplifiers to vary the field current of the motor. Such systems are beyond the present scope of your study of basic electricity.

DC Aircraft Electrical System

Earlier in this section you studied a control system for dc generators driven at variable speed, such as the generator for an automobile or an aircraft. You will now see how an aircraft dc system (usually 28 volts dc) functions with two on-line generators paralleled for reliability and greater output.

The system is shown schematically below. In addition to the usual shunt field, each generator has an auxiliary winding called the *compensating field.* To make sure that the generators carry equal loads, an automatic load balancing circuit is built into the regulator system. As shown, there is a current flow through the equalizing coils whenever the potentials at the compensating field points, A, are unequal.

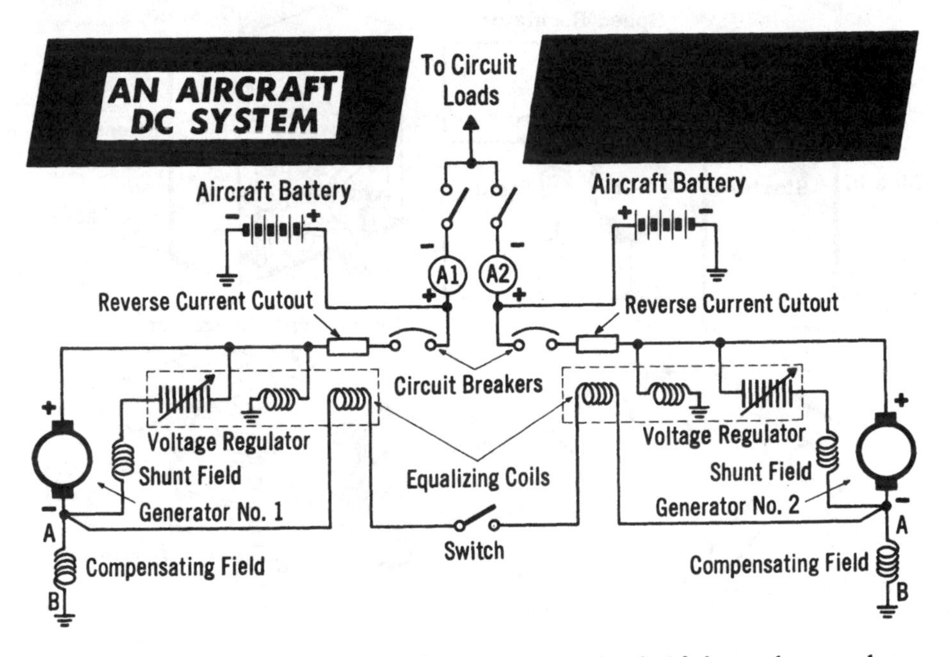

This current flow either aids or opposes the field from the regulator potential coils, depending on the direction of current flow. The load between the two generators is thus equalized. The switch is opened when only one generator is used.

The generator outputs go through reverse current cutouts to avoid discharging the batteries when the generators are not on. The circuit breakers protect the generators against overload, and the protected generator output is used to charge the batteries or to supply power to the electrical circuits via ammeters in each line.

While not all systems will have the configurations described above, they will in general be similar, with similar elements, so that if you understand how one system operates, you can extend your understanding to other systems.

Experiment/Application—Operation of DC Voltage Regulator

You can show the operation of the dc voltage regulator if you have a small dc generator driven by a variable-speed motor. To see how a vibrating reed-type regulator works, you can connect up a circuit as below.

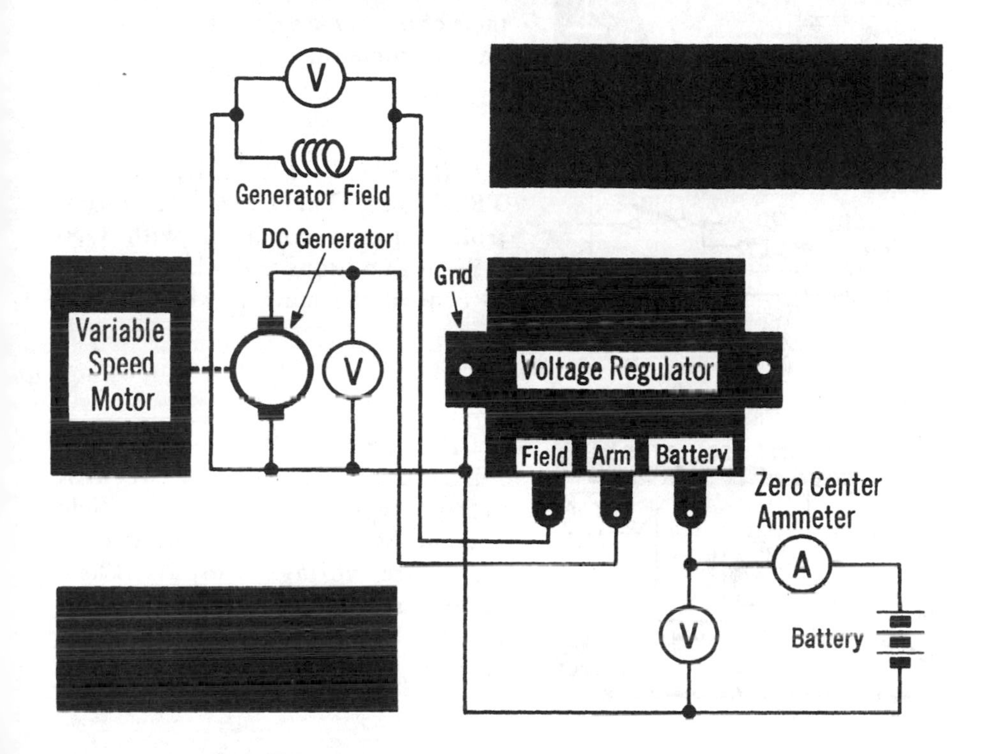

The voltmeters measure the generator output, the load voltage, and the field voltage. Remove the cover of the voltage regulator unit. With the generator operating at very low speed, you will notice that the voltage of the field is about the same as the generator output (voltage regulator relay closed most of time). You will also notice that the current limiting relay is closed and the reverse current relay is open. No current flows into the battery. As the speed of the generator is increased, the generator output voltage will increase until, finally, the reverse current relay closes and the battery is connected to the generator. You will notice at this point that the ammeter shows current flowing *into* the battery. You will also notice that the voltage across the field has increased.

As the generator speed is increased further, you will notice that the field voltage will drop—indicating that the voltage regulator relay is open a greater part of the time. Eventually, as the speed of the generator is increased further, the current limiting relay starts to operate—further reducing the field voltage and keeping the load current near a constant value. The relay vibration can be sensed with a finger to verify its action.

Review of DC Systems and Controls

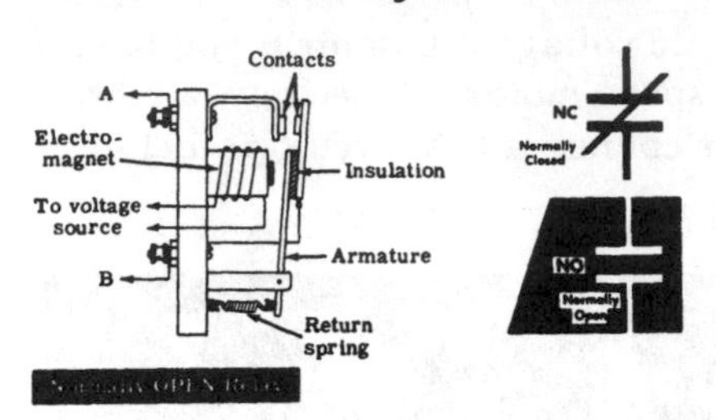

1. RELAY-CONTACTOR—The relay or contactor is a device that allows a low-power circuit to control a high-power circuit. The switch contacts of a relay are open and closed by an electromagnet.

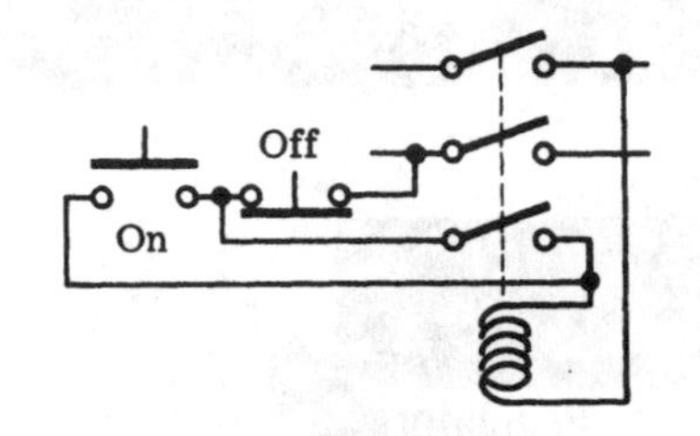

2. BASIC CIRCUIT CONTROLLER—The basic circuit controller uses a contactor with ON-OFF push buttons and a latching coil to control the main power to a machine.

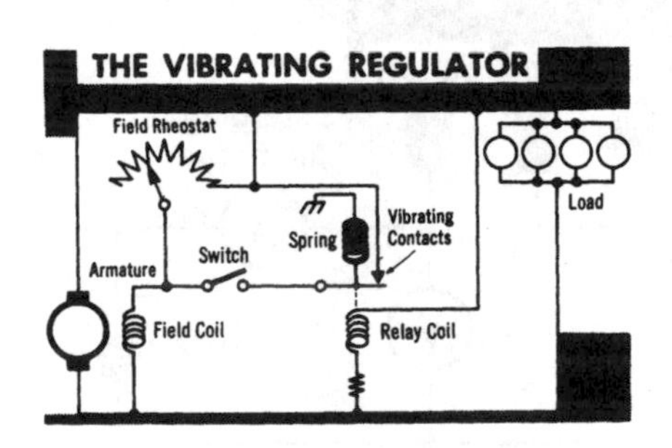

3. DC GENERATOR VOLTAGE REGULATORS—Dc generator voltage regulators control the field current by increasing field current for more voltage output. These voltage regulators vary from simple manual rheostats to elaborate vibrating regulators.

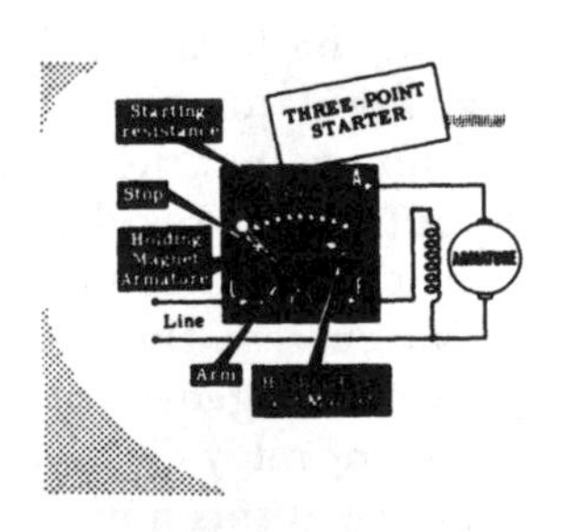

4. DC MOTOR STARTERS—Dc motors in all but the smallest sizes need to be started with some resistance in series with the line in order to limit the initial current. Motor starters vary from simple manually operated starters to more complicated automatic starters that automatically sequence the starting resistance out of the circuit.

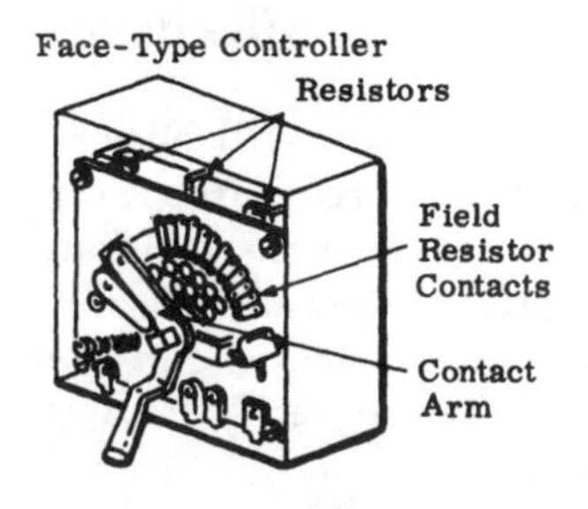

5. DC MOTOR SPEED CONTROL—Dc motor speed is controlled by adjustment of field current. Reversal is done by reversing the field ends.

Self-Test—Review Questions

1. Sketch the essential parts of a contactor and label all of the parts. Show the schematic representation.
2. Draw a circuit for motor (or generator) control using push buttons and a contactor. Describe how the circuit works.
3. Add circuit breakers and an overload circuit to the circuit in question 2.
4. Sketch and describe the operation of the simple rheostat voltage control for generators.
5. Draw the circuit for the vibrating voltage regulator and tell how it works. Compare it with the carbon pile regulator.
6. Compare the control system for a variable-speed generator with the simple vibrating regulator.
7. Draw the circuit for the three-point faceplate motor starter. How does it work? Why is a dc motor starter necessary?
8. What are the differences between a three-point motor starter and an automatic dc motor starter? Draw the circuit for the automatic starter and describe the function of each component.
9. Discuss control of dc motor speed. Compare it with control of the output voltage from a dc generator.
10. What is the function of the equalizing coils in a parallel generator system?

Learning Objectives—Next Section

Overview—Now you know the operating principles for dc generators and motors and how these devices are controlled. In the next section you will learn about the maintenance and troubleshooting of dc machinery.

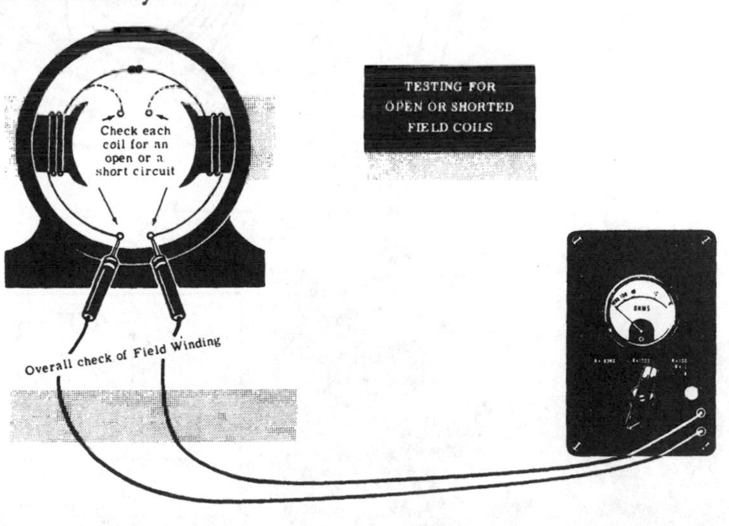

Generator Maintenance

Once the prime mover has been coupled to a generator shaft, the only maintenance necessary should be to oil bearings and to ensure good contact between the brushes and the commutator.

It is seldom desirable to tamper with the generator leads. For if they are wrongly reconnected, the field may be reversed. In the case of a self-excited generator, this reversal of the field will cancel the residual magnetism of the field, and the generator will not build up even when the leads are properly reconnected. It may, very occasionally, be desirable to reverse the polarity of a dc generator; however, this may be done by reversing the output leads only. *The field connections should never be reversed.* The field coils are only connected to the terminal board so as to make their replacement easier in case of damage. And once the field wires have been properly connected in the initial installation, they should *never* be changed.

Generator Maintenance (continued)

When a generator does *not* build up, one of several things may be wrong. There may be too little, or no, residual magnetism. In this case, to provide the initial field necessary, the generator must be momentarily excited by an external dc source. This is called *flashing the field.* When flashing the field, it is important that the externally produced field be of the *same* polarity as the residual magnetism. If these polarities are opposed, the initial field will be further weakened, and the generator will still not build up.

Often a rheostat is connected in series with the shunt field in order to control the field current. If this rheostat adds too much resistance to the circuit at first, the field current will be too small for a proper buildup.

Finally, if the field coil circuit has become *open-circuited,* the generator will not build up. The break or open circuit must be found and repaired.

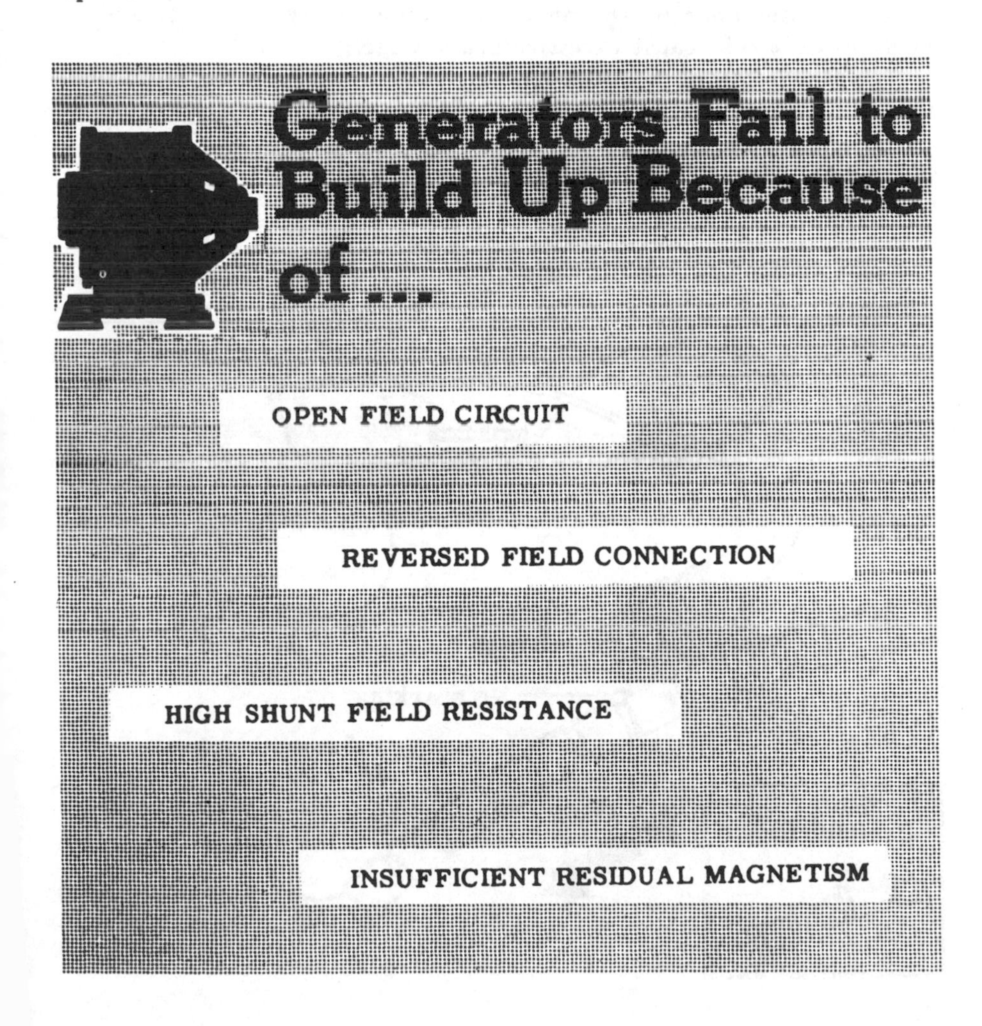

Commutators and Brushes

Commutators and brush assemblies are a source of trouble in dc rotating machinery. The continual sliding of the brushes against the commutator wears the brushes down and tends to push them out of alignment, causing bad contact between the commutator and brushes. When something does go wrong in commutation, it may be accompanied by excessive sparking, which aggravates the original trouble at once. When correct commutation is taking place, there will be very little sparking.

For satisfactory commutation of dc machines, a continuous contact must be maintained between commutator and brushes. The commutator must be mechanically true, the unit in good balance, and the brushes in good shape and well adjusted. These brushes are free to slide up and down in the brush holders and are made to bear on the commutator by a spring that produces a pressure of one and one-half to two pounds per square inch, or 0.5 to 0.7 kilograms per square centimeter, of brush surface. Too little pressure causes poor brush contact and unnecessary sparking, and too much pressure will cause excessive brush wear.

Note that the mica insulation between the commutator segments is usually cut below the surface of the segments to allow for wear.

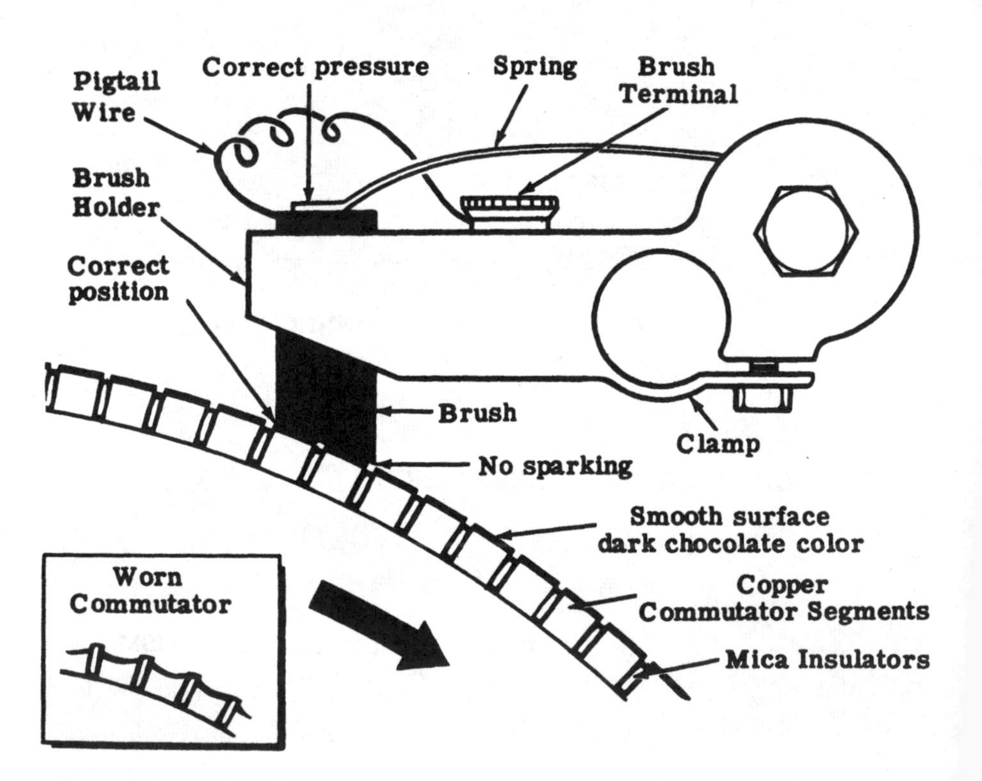

Commutators and Brushes (continued)

Excessive sparking can be caused by a number of external factors, such as shorted or open armature windings, overload, and improper field strength (see Troubleshooting Chart, page 5-89).

When there is excessive sparking at the commutator and good commutation cannot be obtained, the commutator and brush assembly must be checked, and any defect corrected as soon as possible. The inspection procedure and the steps taken to eliminate troubles are as follows:

1. Watch the machine operating to see if you can spot any arcing or excessive sparking elsewhere that might indicate a loose connection.
2. Check the brush positions to be sure that the brushes are commutating at the proper point (neutral plane).
3. Turn off the machine, making sure that all power is disconnected *before* proceeding with your check.
4. Inspect all connections and make sure that none are loose.
5. Check the relative position of the brushes on the commutator. If they are unequally spaced, look for a bent brush holder.
6. Check the condition of the brushes. If they are badly worn, they should be replaced. When removing a brush, first lift the spring lever to release the pressure, then remove brush. Insert a new brush, making sure that it can move freely in the holder. The end of the brush must then be fitted to the commutator by sanding it as illustrated below. Adjust the brush spring pressure. Check the pigtail wire and its terminal for tightness. The pigtail wire must not touch any metal except the brush to which it is attached.
7. Check the commutator for dirt, pitting, irregularities, etc. Dirt can be removed with a piece of light canvas. Fine sandpaper will remove slight roughness. *Never* use emery cloth on a commutator.
8. A badly worn commutator, such as the one shown on page 5-84, should be *skimmed* on a lathe. This refers to machining away the irregularities of the surface.

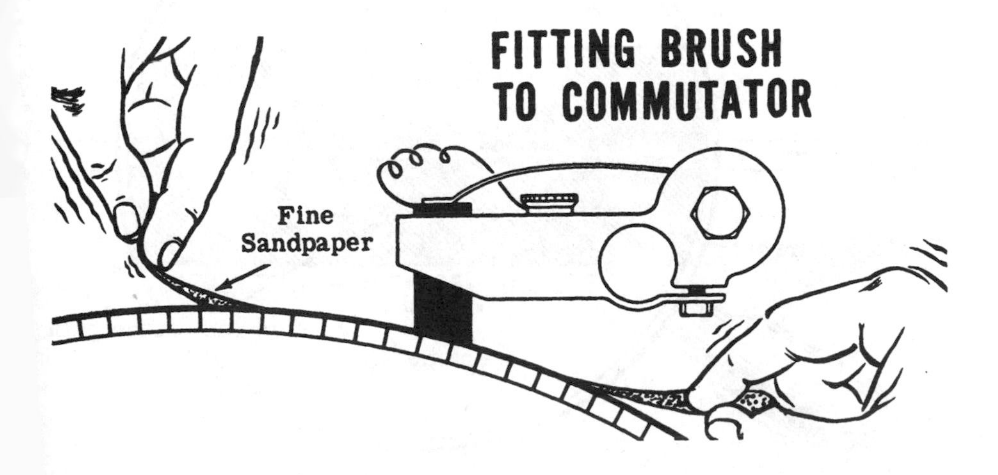

Insulation Breakdown—Shorts and Open Circuits

Under normal operating conditions, the field and armature windings of generators and motors are completely insulated from the frame of the machine, which is bolted to the bench or floor. A resistance measurement from the frame to the armature or field should read infinity, or several million ohms.

Sometimes, however, because of excessive heat generated by overloading or because of high moisture content in the air, the resistance of the insulation decreases and some of the current leaks through to the frame. This leakage current adds to the breakdown of the insulation; and if the leakage is not found in time, the breakdown will be complete and the winding will be shorted to the frame. This will cause it to overheat or burn out. The armature and field windings should therefore be checked at regular intervals to detect *leaks* and *shorts* to ground before they cause serious damage.

An ordinary ohmmeter cannot be used for insulation testing in large practical machines, since the leakage will often show itself only when a high voltage is applied to it, and an ohmmeter cannot be used when a high voltage is applied. An instrument called a MEGGER® is used instead. It supplies the necessary high voltage and is calibrated to read very high resistance values.

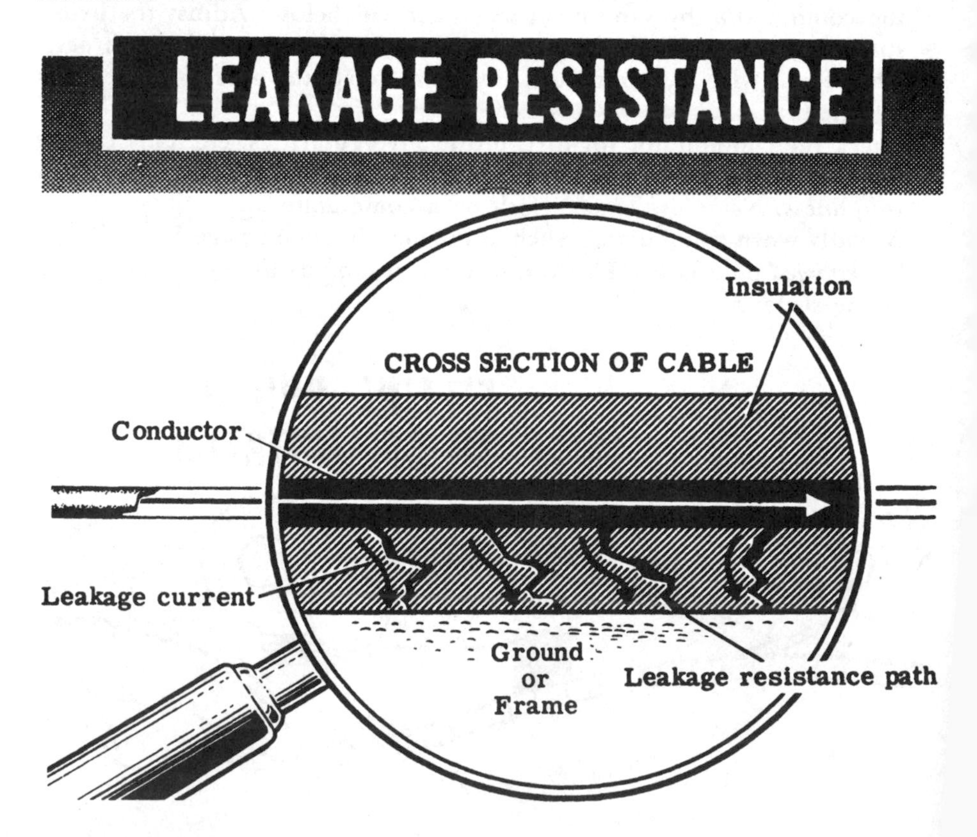

Insulation Breakdown— Shorts and Open Circuits (continued)

Use an ohmmeter for testing for open and shorted (internally and externally) field coils. Disconnect the field coil leads from the armature, to avoid parallel circuits in testing, and place the ohmmeter across the field leads as shown in the simplified illustration below. If the ohmmeter reads a very high resistance, there is an open circuit somewhere in the field winding. The open-circuited coil can be detected by testing each coil individually. The coil with the open circuit should be disconnected from the other coils and replaced. The armature resistance of a dc machine is normally so low that an ordinary ohmmeter will not be able to measure it. If the armature has an open-circuited turn, the ohmmeter will also read zero because of the numerous parallel paths. Therefore, special equipment must be used to test armatures.

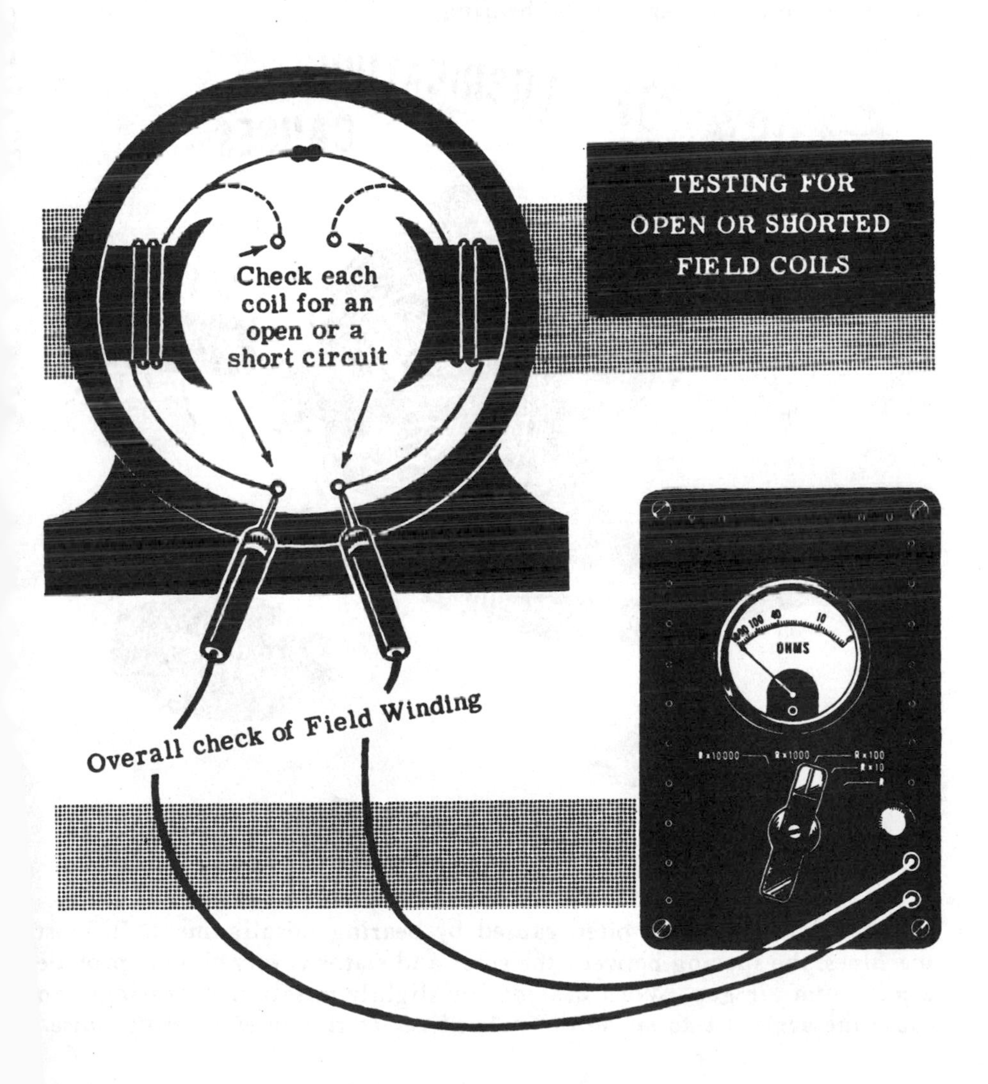

Bearing Lubrication and Maintenance

Excessive lubrication can cause faults in generator or motor operation; the *lack* of lubrication is more serious. A bearing which is not properly lubricated will overheat immediately, causing an expansion of the shaft and bearing assembly that may be sufficient to stop shaft rotation altogether. Lack of lubrication also results in noisy operation because of the direct contact between the shaft and bearing.

Bearing housings should be checked periodically for overheating and noisy operation. In normal operation, the temperature of a generator or motor will rise, so that the bearing housings will heat up a certain amount. But if they heat up too much, do *not* add or change lubrication without *first inspecting* the bearing to make certain that lack of lubrication is the cause. Shafts may be forced out of line by a coupling unit, or the lubrication may not be reaching all parts of the bearing.

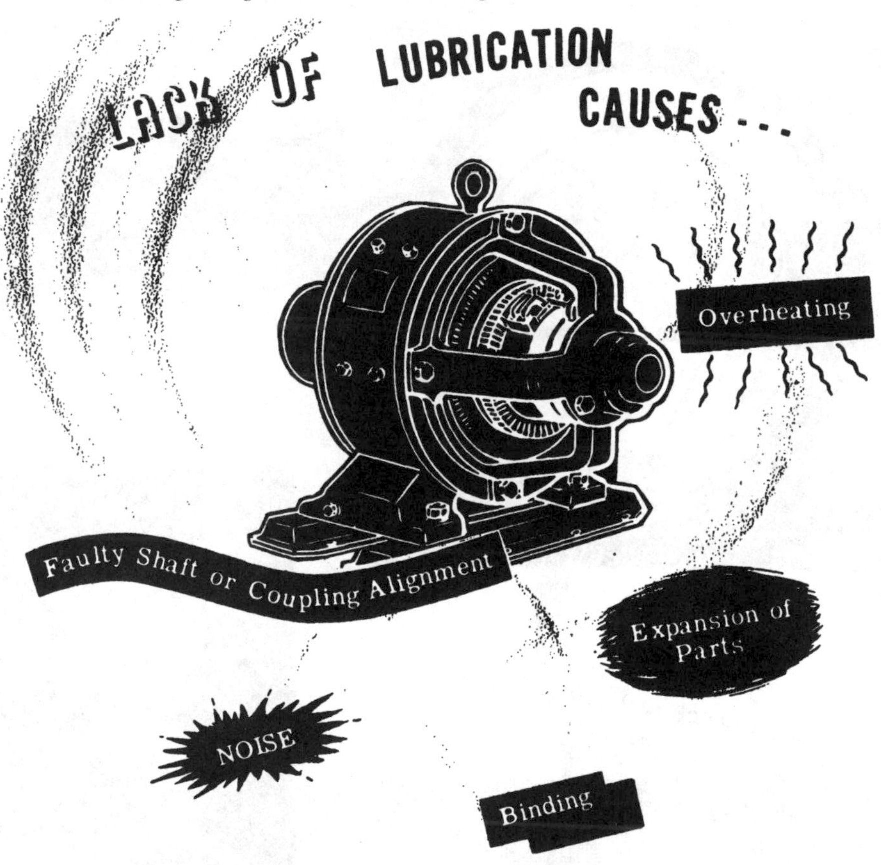

Noisy operation is often caused by bearing misalignment. In most machines, the spacing between the rotor and stator is very close to provide a minimum air gap. Worn bearings or slightly misaligned bearings can cause the armature to strike the pole pieces as it rotates, causing noise.

Troubleshooting Chart—DC Electrical Machines

Symptom	*Probable Causes*
1. Generator fails to build up	a. Speed too low or reversed b. Field circuit open c. Insufficient residual magnetism d. Output short-circuited
2. Generator voltage too high (or low)	a. Field current high (low) b. Prime mover too fast (slow) c. Regulator defective
3. Motor does not start	a. No power (circuit open) b. Improper connections c. Heavy electrical overload; mechanical bearing or commutator jammed d. Starter defective
4. Motor runs too fast (or too slow)	a. Field current low (high) b. Open field (partially shorted) c. Improper connections d. Short-circuited armature winding e. Improper load—too small (or excessive)
5. Commutator sparking of brushes	a. Overload b. Poor brush contact, commutator rough c. Brushes at wrong angle or commutation windings defective d. Armature windings open or shorted e. Improper field current
6. Noise, excessive vibration	a. Bad bearings—excessive end play b. Armature striking pole pieces c. Unbalanced armature
7. Hot bearings	a. Out of lubricant or dirty bearings b. Defective bearings c. Poor alignment or bearings too tight d. Crooked or unbalanced shaft
8. Electrical parts overheating	a. Overload (all parts hot) b. Excessive field current (field hot) c. Brushes and commutator misaligned or worn (brushes and commutator hot) d. Armature coils shorted (armature hot)

Introduction to AC Generators

The ac generator is the most important means for the production of electrical power in use today. As you learned earlier ac is used for most power applications because of the ease with which it can be transformed.

Ac generators, or *alternators*, vary greatly in size depending on the power they are required to supply. For example, one single alternator used in a generating station in New York City generates 1,000 megawatts. On the other hand, the alternators used in modern automobiles usually generate less than 500 watts.

Regardless of size, however, all electrical generators, whether dc or ac, depend for their operation on the action of a coil cutting through a magnetic field or of a magnetic field cutting through a coil. As long as there is *relative* motion between a conductor and a magnetic field, a voltage will be generated.

As you know, so that relative motion may take plaçe between a conductor and a magnetic field, all generators are made up of two mechanical parts—a rotor and a stator.

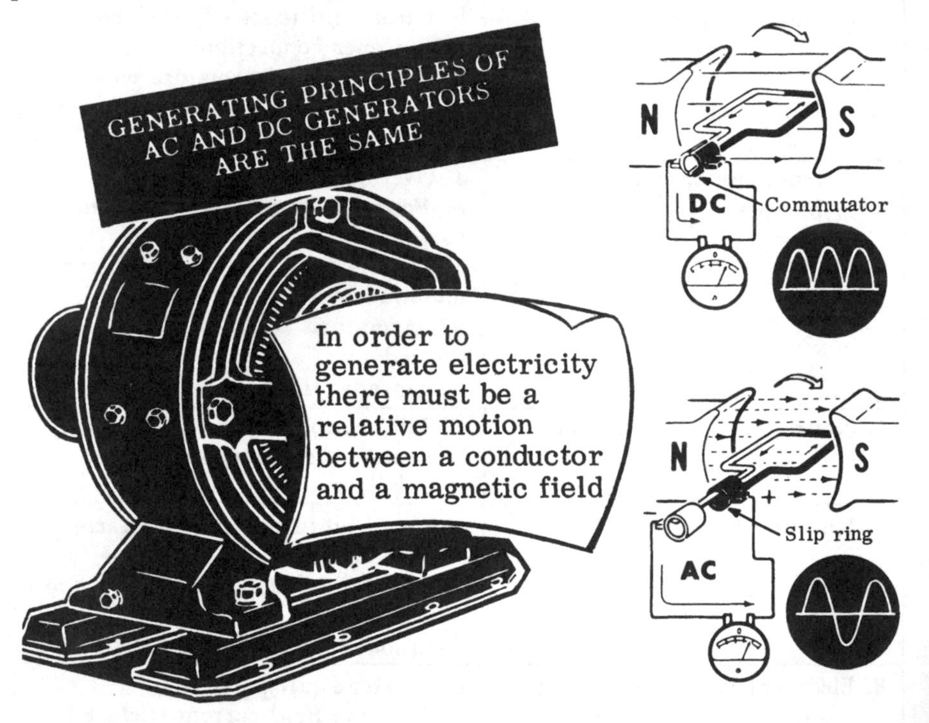

As you learned of dc generators, the output of the generator is proportional to the field strength and the speed at which the coils and flux field interact. Since ac generators are usually operated at constant speed to keep the frequency constant, the control of voltage output is done by varying the field strength. You will learn more about this when you study ac generator and motor control later in this volume.

Types of Alternators

You learned earlier that in a dc generator the revolving part is always the armature. In an ac generator, however, this is not usually true.

There are two basic types of alternators—the *revolving armature* type and the *revolving field* type. The revolving armature type is similar in construction to the dc generator in that the armature rotates through a stationary magnetic field. But in the dc generator, the generated ac in the armature windings is converted into dc by means of the commutator, whereas in the alternator, the generated ac is brought to the load as ac by means of slip rings. The revolving armature alternator is only found in alternators with small power ratings. A variation of the revolving armature ac generator is often used for very low power (e.g., an automobile alternator) where both windings are interlaced on a stator and the fields are coupled by a rotating soft iron (electromagnetic) armature without windings. This arrangement will be described later.

The revolving field type of alternator has a stationary armature winding and a rotating field winding. The advantage of having a stationary armature winding is that the generated voltage can be connected directly to the load without slip rings. Fixed connections are much more easily insulated than are slip rings at very high voltages, so high-voltage, high-power alternators are usually of the rotating field type. Since the voltage applied to the rotating field is low-voltage dc, the problem of arc-over at the slip rings is not encountered.

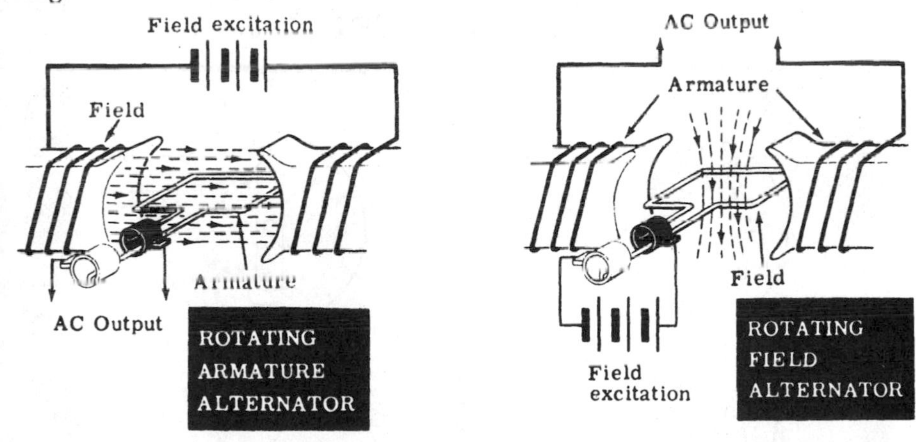

The maximum current that an alternator can supply depends on the maximum heating loss that can be sustained in the armature. This heating loss (which is an I^2R power loss) heats the conductors, and if excessive, can destroy the insulation. Alternators are rated in terms of this current as well as in terms of voltage output. That is to say, every alternator rating is expressed in volt-amperes (or in the case of large machines, in kilovolt-amperes) of the apparent power that the alternator can supply. Remember that even though the load may be drawing reactive (wattless) power, the I^2R loss in the generator and feed line are real.

Frequency of Alternator Output

The frequency of the ac generated by an alternator depends on the number of poles and on the speed of the rotor. When a rotor has rotated through a sufficiently wide angle for two opposite poles—a north and a south—to have passed one stator winding, the voltage induced in the winding will have passed through a complete cycle of 360° (electrical degrees). So a single-phase, two-pole alternator rotating at 3,600 rpm will generate a 60-Hz output.

The more poles there are in the rotating field, the lower the speed of rotation will need to be for a given frequency. An eight-pole alternator, for example, will only have to rotate at 900 rpm to generate a 60-Hz output.

The relationship between the generated frequency, f, expressed in Hz (cycles per second); the speed of the rotor (N), expressed in rpm; and the number of poles, P, is given in the formula:

$$f = \frac{NP}{120}$$

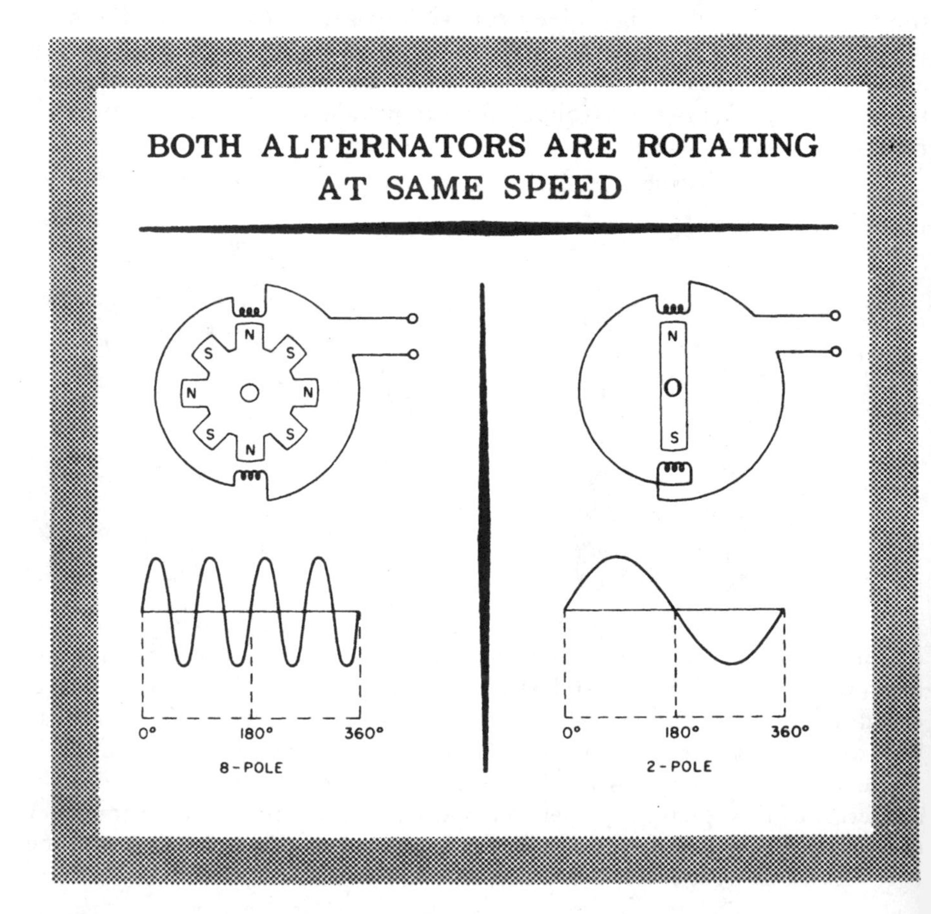

Alternator Construction

Alternators having high kilovolt-ampere ratings are usually of the turbine-driven, high-speed type. The prime mover for this type of alternator is a high-speed steam turbine driven by steam under high pressure. The steam is generated by nuclear power or the burning of oil, coal, or gas. The rotor of the turbine-driven alternator is cylindrical, small in diameter, and has windings firmly imbedded in slots in its face. The windings are arranged to form two or four distinct field poles. Only with this type of construction can the rotor withstand the terrific centrifugal force developed at high speeds without flying apart.

In slower speed alternators, which are driven by variable-speed gasoline engines, water power, geared turbines, or electric motors, a *salient pole rotor* is used. In this type of rotor, a number of separately wound pole pieces are bolted to the frame of the rotor. The field windings are either connected in series, or in series groups connected in parallel. In either case, the ends of the windings usually connect to slip rings mounted on the rotor shaft. Regardless of the type of rotor field use, its windings are separately excited, usually by a dc generator called an *exciter*. In large ac generators, the exciter is connected to the same shaft as the main generator. However, in some very small ac generators, the rotor field is generated by a permanent magnet.

The stator, or stationary armature of an alternator, holds the windings which are cut by the rotating magnetic field. The voltage generated in the stator as a result of this cutting action is applied to the load.

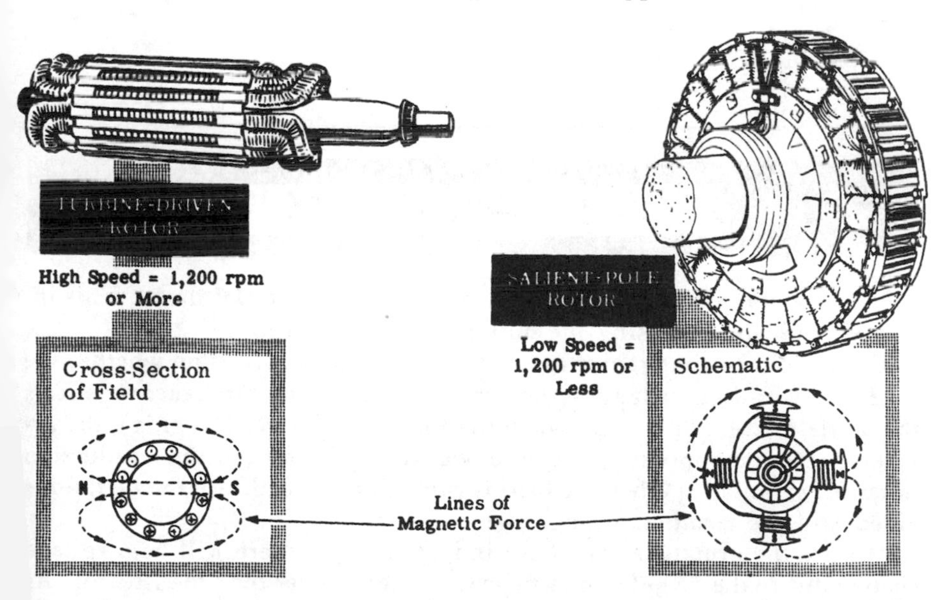

The stators of all alternators are essentially the same. They consist of a laminated iron core, with the stator windings embedded in this core. The core is secured to the stator frame.

AC Generator Characteristics

The voltage output from an ac generator is similar to that from a dc generator except that we must deal in rms voltage in ac circuits. The rms voltage, E per phase, is given by

$$E = 2.22 \Phi Zf$$

where Φ is the number of flux lines per pole; Z is the number of conductors in series, per phase; and f is the frequency in Hz. Thus, as for any generator, the output depends on the flux density, the number of conductors in the field, and the speed with which the conductor moves through the field.

When the load on a generator is changed, the terminal voltage will vary. The causes of this change are armature resistance, armature reaction, and armature reactance.

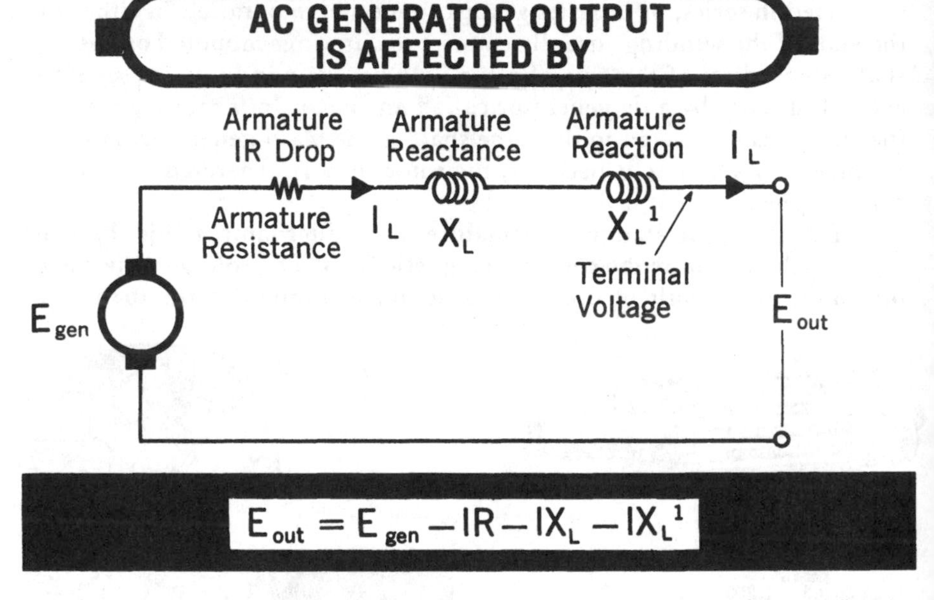

The voltage drops because of armature resistance (IR drop). This increases with load, causing the output voltage to drop.

The voltage change due to armature reaction depends on whether the load was a leading or lagging power factor. The armature reaction affects the dc field strength so that, when the load is inductive (lagging), the armature reaction opposes the dc field, weakening it and causing a reduction in terminal voltage. When the load is capacitative (leading), the dc field is aided and the terminal voltage rises.

Since the armature coils have inductance (L), there will be a voltage change due to the IX_L that may be many times greater than the IR drop. As with armature reaction, the voltage may decrease or increase depending on the power factor of the generator load. Voltage regulation for ac generators is calculated the same as for dc generators.

Single-Phase Alternator

A single-phase alternator has all the armature conductors connected in series, as one winding, across which an output voltage is generated. If you understand the principle of the single-phase generator, you will easily understand multiphase generators.

The schematic diagram below illustrates a two-pole, single-phase alternator. The stator winding is in two distinct coils, both being wound in the same direction around the stator frame. The rotor consists of two poles of opposite polarity.

As the rotor turns, its poles induce ac voltages in the stator windings. The two coils of the stator winding are connected to each other in such a way that the ac voltages induced in them are in phase; therefore, they add directly.

SINGLE-PHASE ALTERNATOR

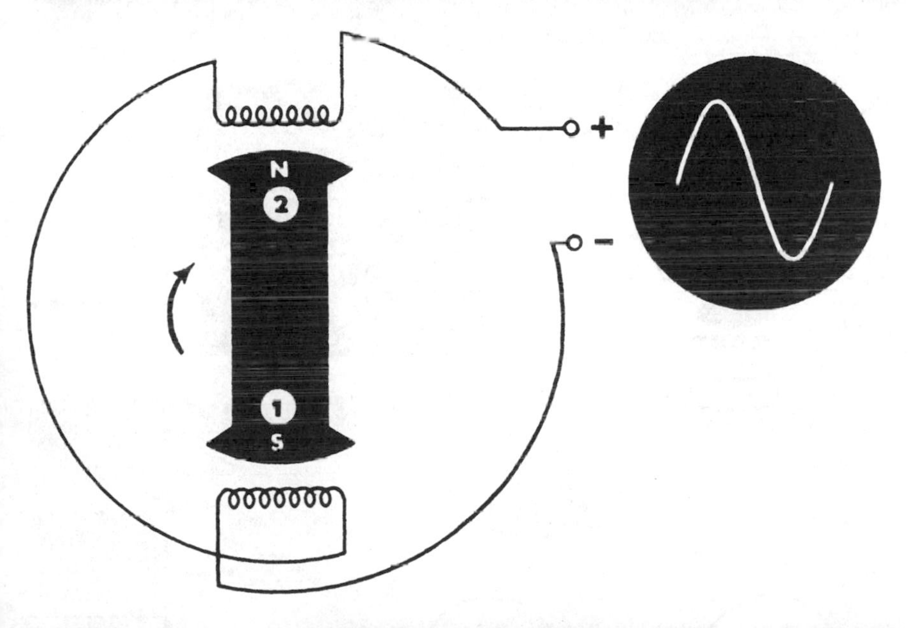

Obviously, a single-phase alternator can be built with many pairs of poles for lower speed operation. Single-phase alternators are often used for low-power generators; and in small sizes, a permanent magnet is often used for the rotor.

Two-Phase Alternators

Multiphase (or *polyphase*) alternators have two or more single-phase windings symmetrically spaced around the stator. In a two-phase alternator there are two single-phase windings, physically spaced so that the ac voltage induced in one is 90° out of phase with the voltage induced in the other. The windings are electrically separate from each other. The only way to get a 90° phase difference is to space the two windings so that, when one is being cut by maximum flux, the other is being cut by minimum flux.

The schematic diagram below illustrates a two-pole, two-phase alternator. The stator consists of two single-phase windings completely separated from one another. Each winding is made up of two parts, which are so connected that their voltages add. The rotor is identical to that used in the single-phase alternator.

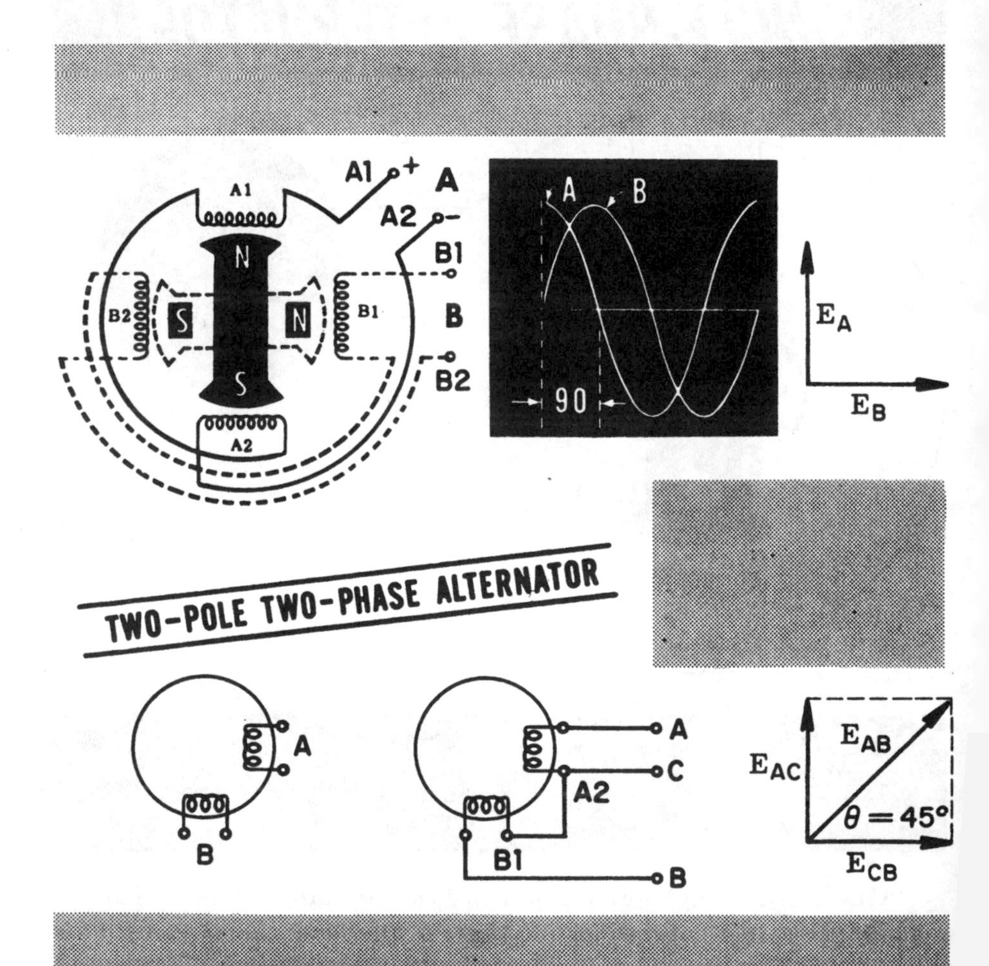

Two-Phase Alternators (continued)

In the schematic, the rotor poles are opposite the windings of phase A. Therefore, the voltage induced in phase A is maximum, and the voltage induced in phase B is zero. As the rotor continues rotating, it moves away from the A windings and approaches the B windings. As a result, the voltage induced in phase A decreases from its maximum value, and the voltage induced in phase B increases from zero, and so on, for a complete rotation of the rotor. Notice that a 90° rotation of the rotor corresponds to one quarter of a cycle, or 90°.

The waveform picture shows the voltages induced in phase A and phase B for one cycle. The two voltages are 90° out of phase.

In some cases, you will find the two-phase generator connected with a three-wire output. As shown in the vector diagram, the voltage from terminals A to B (E_{AB}) is equal to the vector sum of the voltage from A to C and from C to B. Thus, if the voltage from each winding is 100 volts, the total output E_{AB} is 141 volts. As you know E_{AB} can also be obtained by solving the right triangle analytically. Thus

$$E_{AB} = \sqrt{(E_{AC})^2 + (E_{CB})^2}$$

$$\theta = \text{arc tan } \frac{E_{AC}}{E_{CB}}$$

$$= \text{arc tan } 1$$

$$= 45°$$

TWO-POLE TWO-PHASE ALTERNATOR

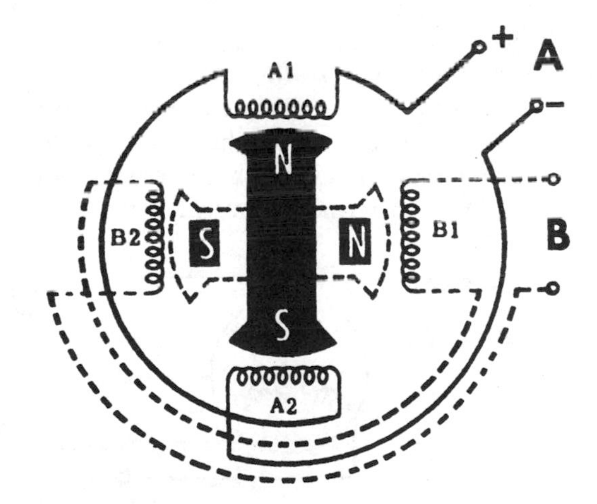

$$E_{AB} = \sqrt{100^2 + 100^2}$$
$$= \sqrt{10,000 + 10,000}$$
$$= \sqrt{20,000}$$
$$= 141 \text{ volts}$$

Three-Phase Alternators

The three-phase alternator, as the name implies, has three single-phase windings so spaced that the voltage induced in any one is phase-displaced by 120° from the other two. A schematic diagram of a three-phase stator, showing all the coils, becomes complex, and it is difficult to see what is actually happening. So the simplified schematic below shows the windings of each single phase lumped together as one winding. The rotor is omitted for simplicity. The voltage waveforms generated across each phase are drawn on a graph, phase-displaced 120° from one another.

The three-phase alternator shown in this schematic is essentially three single-phase alternators whose generated voltages are out of phase by 120°. The three phases are independent of each other.

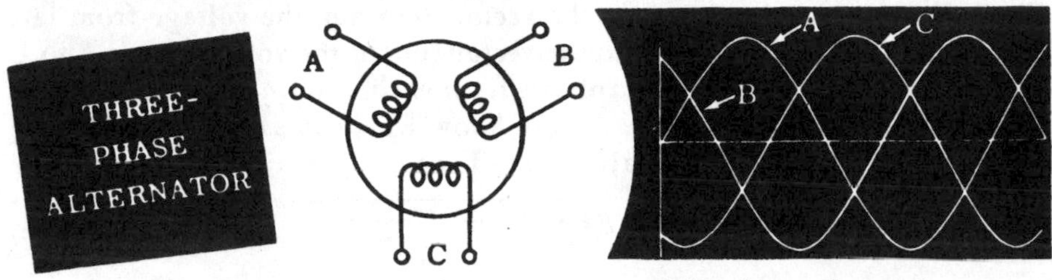

Instead of having six leads coming out of the three-phase alternator, three leads—one from each phase—are connected together to form what is called *Y* (usually designated a *wye*, or a *star, connection*). The point of connection is called the *neutral*, and the voltage from this point to any one of the line leads will be the phase voltage. The total voltage, or line voltage, across any two line leads is the vector sum of the individual phase voltages. Since the windings form only one path for current flow between phases, the line and phase currents are equal.

A three-phase stator can also be connected so that the phases are connected end to end; it is then *delta connected*. In the delta connection the line voltages are equal to the phase voltage, but the line currents will be equal to the vector sum of the phase currents. Wye and delta connections are used not only for alternators but also for power transmission and ac motors.

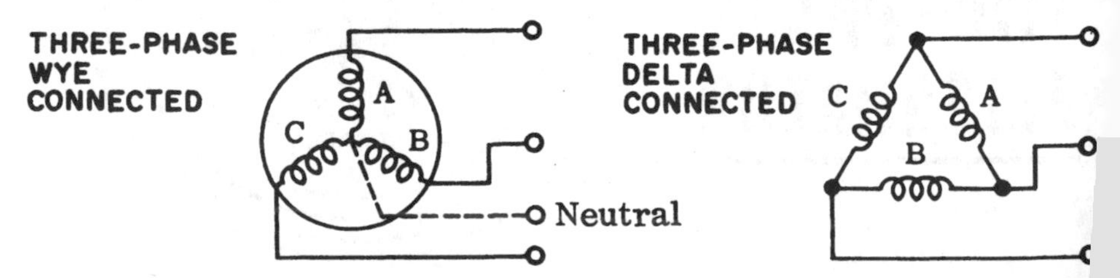

Very large ac generators and power systems use six or more phases; however, the principles remain the same. As you will learn shortly, it is possible to go from wye to delta and vice versa by using transformers. Multiphase circuits are used to reduce the number of lines necessary fo power transmission.

Wye Connection

As you learned on the previous page, the wye-connected generator may have a neutral lead brought out. Each load has two phases in series. Thus, the current and voltage must be determined by vectors. Three-phase wye-connected systems are the most common power available; however, it is usually obtained from a three-phase transmission line via transformers.

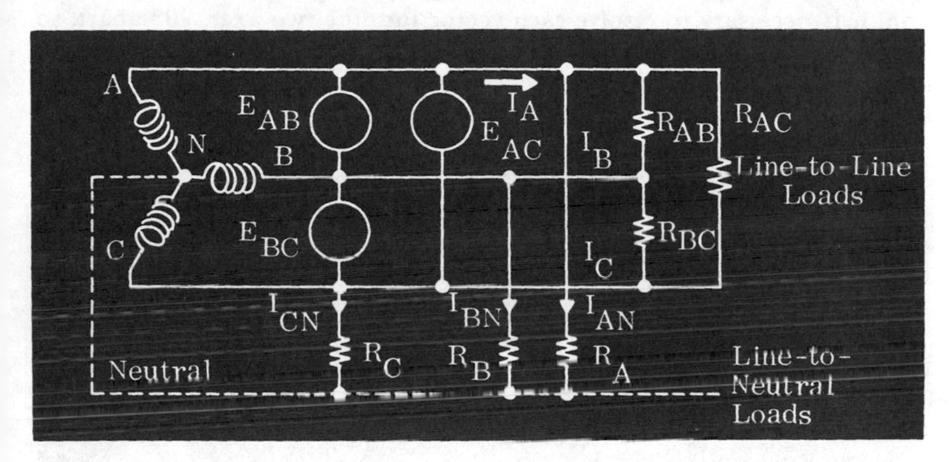

In the wye-connected system, the voltage and current from line to line (E_{AB}, E_{BC}, E_{AC}) are the vector sums of the voltage from two of the windings. Thus:

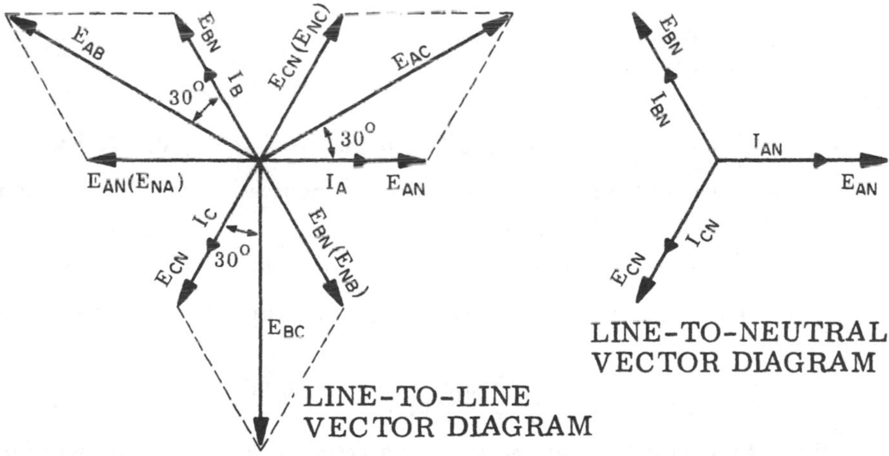

For the line-to-line loads, the resultant voltages are still 120° apart in phase, but the voltages are equal to the vector sum of the output from two windings. Tracing the circuit, for example, from A to B, and using Kirchhoff's voltage law, it is apparent that $E_{AB} = E_{AN} + E_{NB}$ (vectorially)! Since E_{NB} is the negative of E_{BN} (other direction), it is necessary to reverse the direction of E_{BN} before addition, as shown in the vector diagram above.

Wye Connection (continued)

The voltage can be obtained graphically or analytically. If you obtained it graphically, you would find that the resultant voltage, line to line, would be about 1.7 times the voltage from line to neutral (each separate winding). You can calculate the resultant voltage analytically, but you must do it in a slightly more complicated way. To perform the vector addition, it is necessary to resolve each vector into the two axes, 90° apart, so we can use the Pythagorean theorem to find the resultant voltage. This is easy for E_{AN}, since it is already on one of the axes. For E_{NC}, we have to resolve it into its two components, 90° apart.

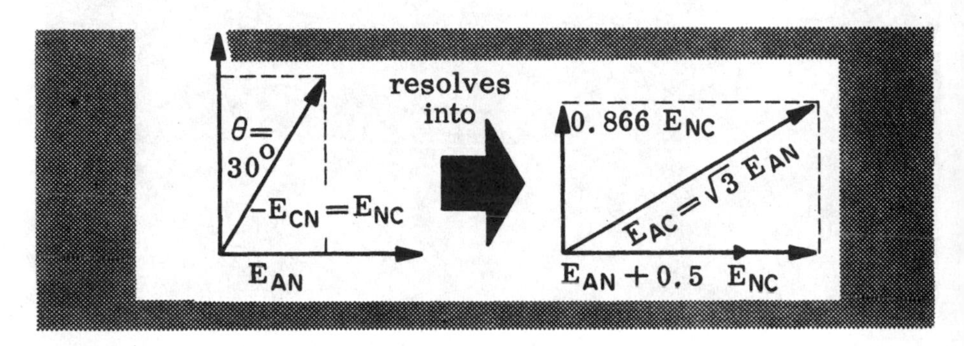

As you remember we can use what we know about right triangles to do this. Thus, the projections of E_{NC} into the vertical and horizontal axes are obtained from the relationships

$$\text{Cos } \theta = \frac{\text{vertical}}{\text{hypotenuse}}; \text{ vertical} = \cos 30° \times E_{NC} = 0.866 \times E_{NC}$$

and

$$\text{Sin } \theta = \frac{\text{horizontal}}{\text{hypotenuse}}; \text{ horizontal} = \sin 30° \times E_{NC} = 0.5 \times E_{NC}$$

If the individual phase voltages (line to neutral) are 120 volts each, then by the Pythagorean theorem

$$E_{AC} = \sqrt{(0.866 \times 120)^2 + (1.5 \times 120)^2} = 208 \text{ volts } (\sqrt{3} \times 120)$$

The phase angle (relative to E_{AN}) can be obtained from the relationship

$$\theta = \text{arc tan} \frac{0.866\ E_{NC}}{1.5\ E_{AN}} \text{ or arc tan} \frac{0.866}{1.5} = 30°$$

Since E_{NC} and E_{AN} are equal for a resistive load (unity power factor), the voltages, line to line, are $\sqrt{3}$ (1.73) times (not twice) the individual winding voltage and are shifted (lagging) by 30° with respect to the individual winding phase. The line currents are 30° out of phase with the line voltage.

When a neutral is used, the line currents and line voltages are equal to the individual phases from the generator. No current flows in the neutral when the loads are equal across each phase.

Delta Connection

A three-phase stator can also be connected in a delta configuration. In the delta connection, the end of the first winding is connected to the start of the second, the end of the second to the start of the third, etc.

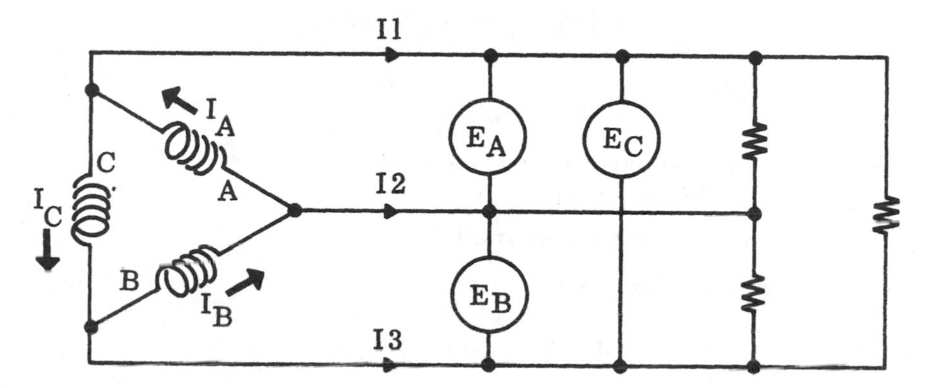

The delta connection is similar to the wye except that the line currents have two components contributed from two windings (e.g., I1 has a contribution from I_a and I_c). As for the wye connection, a vector diagram can be drawn based on these line currents.

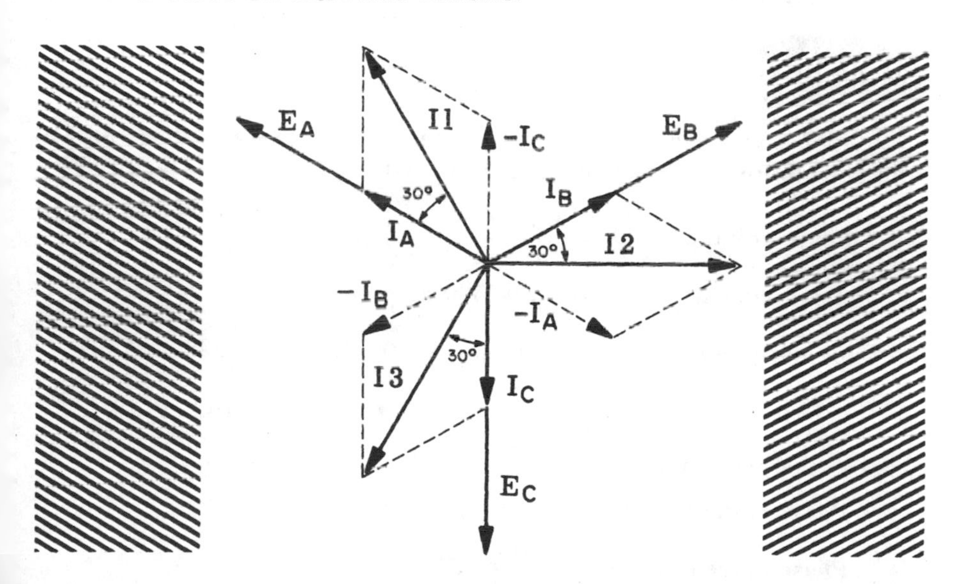

The vector diagram for the wye connection can be solved both graphically and analytically. By analogy to the wye circuit solution, it can be shown that when the power factor is unity and the phases are uniformly loaded, the line currents are equal and are 120° apart. The line voltages are equal to the phase voltages and are 120° apart. The line currents are 30° out of phase with the line voltages. And the line currents are $\sqrt{3}$ (1.73) times the phase current.

Power in Three-Phase Circuits

The total power (P_t) delivered by a balanced three-phase wye-connected system is equal to three times the power delivered by each phase (I_p).

$$P_t \text{ (phase)} = 3\ I_p E_p \cos \Theta$$

since

$$E_p = E_L/\sqrt{3} \text{ and } I_p = I_L$$

where E_p = phase voltage, E_L is the line voltage, and I_L is also the line voltage. The total true power is

$$P_t \text{ (line)} = \sqrt{3}\ E_L I_L \cos \Theta$$

For a three-phase delta-connected system

$$E_p = E_L \text{ and } I_p = \frac{I_L}{\sqrt{3}}$$

Therefore, the total true power

$$P_t = 3E_L \frac{I_L}{\sqrt{3}} \cos \Theta = \sqrt{3}\ E_L I_L \cos \Theta$$

which is the same as for a wye-connected system.

Example

Suppose you had a wye-connected generator that was delivering 440 volts (line to line) at 300 amperes per line with an 80% power factor (PF):

$$\text{Phase voltage} = \frac{E_L}{\sqrt{3}} = \frac{440}{\sqrt{3}} = 254 \text{ volts}$$

$$\text{Load current} = I_p = I_L = 300 \text{ amperes}$$

$$\text{Apparent power} = \sqrt{3}\ E_L I_L = \sqrt{3} \times 400 \times 300 = 228.36 \text{ kVA}$$

$$\text{True power} = \text{kVA} \times \text{PF} = 228.36 \times 0.8 = 182.69 \text{ kW}$$

Suppose the generator above were delta connected:

$$\text{Phase voltage} = E_p = E_l = 440 \text{ volts}$$

$$\text{Phase current} = \frac{I_L}{\sqrt{3}} = \frac{300}{\sqrt{3}} = 173 \text{ amperes}$$

$$\text{Apparent power} = \sqrt{3}\ E_L I_L = \sqrt{3} \times 440 \times 300 = 228.36 \text{ kVA}$$

$$\text{True power} = \text{kVA} \times \text{PF} = 228.36 \times 0.8 = 182.69 \text{ kW}$$

Phase Sequence—Synchronizing Generators

The vector diagrams that you studied earlier in connection with wye and delta connections can be viewed either clockwise (sequence ACB) or counterclockwise (sequence ABC). Either sequence is obtained by interchanging *one* pair of leads. The actual sequence becomes important in determining direction of rotation of three-phase motors, which you will study later in this volume and in the parallel operation of generators.

When ac generators are to be paralleled, they must not only be at the correct voltage but they must also be in phase and have the correct phase sequence. Large power plants may have elaborate equipment to correct the phasing before putting a generator on line. In fact, it is often done automatically to avoid the possible disaster that paralleling of generators with incorrect phasing can cause. There is a very simple method (shown below) for determining when two sources are properly phased.

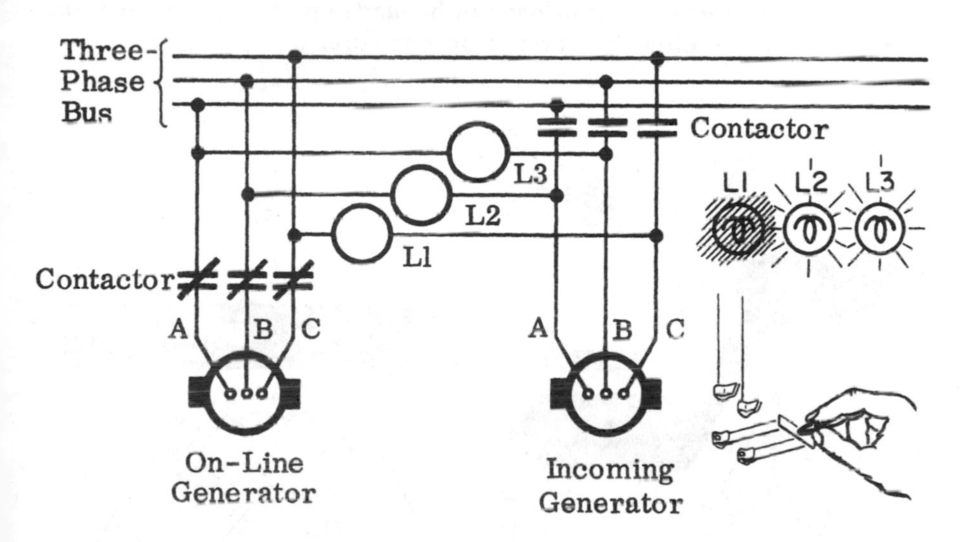

By connecting three lamps, as shown above, you can make sure that the generators are synchronized before closing the contactor for the incoming generator. You can see from the schematic that lamps L2 and L3 are cross-connected, but lamp L1 is not. At the instant when lamps L2 and L3 are both brightest (equally) and L1 is dark, the two generators are synchronized. When the generators are out of sync, the lamps will flicker in sequence. And the closer to synchronism, the slower the flicker rate will be. Lamp L1 will be dark when phase C is either 0° or 180° (zero voltage points). Therefore it is necessary to have lamps L2 and L3 cross-connected so that when the voltage between buses A and B is correct, they will glow with equal brightness, assuring us that lamp L1 is sensing the in-phase condition.

Multiphase Transformers

As you learned earlier, transformers can be used to step ac voltages up and down with very high efficiency. Transformers are used in multiphase circuits to do this stepping up and down and are also used to convert from wye to delta connections and vice versa. What we discuss here will be applicable to all multiphase circuits; however, we will address the three-phase circuit in particular, since it is most commonly encountered. To ensure proper connection in multiphase circuits, the polarity of the windings must be known.

As you learned earlier, smaller transformers use colored dots to indicate points of the same polarity. Large power transformers use many different designations that are too involved to go into here. In this discussion, we shall use the dot reference. The transformer arrangement is usually described in terms of the primary and secondary arrangement shown below. The multiphase transformer can be made up of single-phase transformers, or all windings can be put on one core.

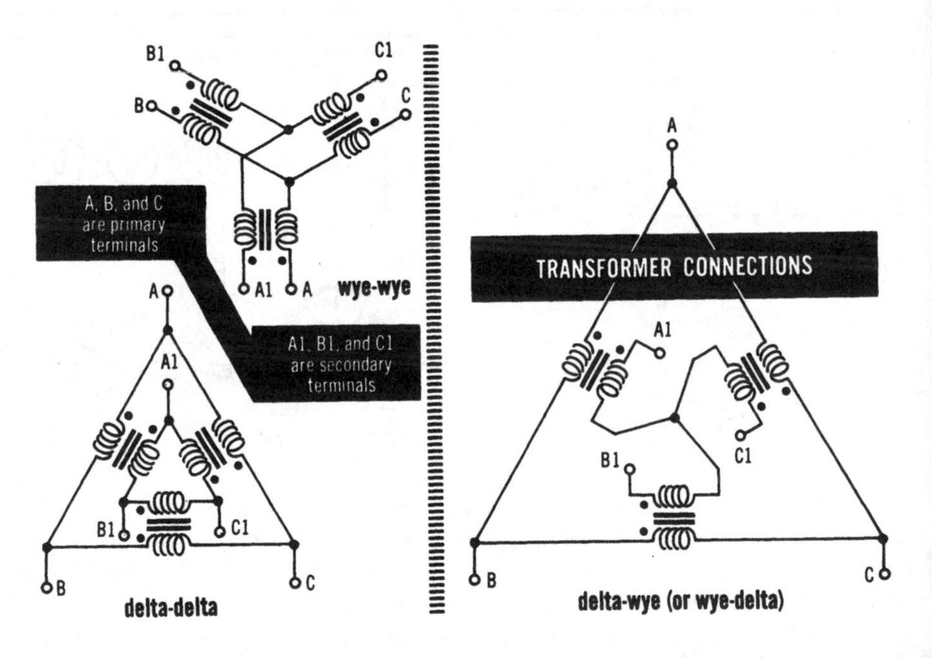

It is important that polarity marking be observed in the delta connection; otherwise a large current will flow in the loop when the delta is closed, either opening up circuit breakers or burning out transformers. If you get the polarity backward on one of the windings for a wye connection, you will reverse the direction of phase rotation.

If you are connecting transformers in a delta configuration, you can check for the right polarity by putting a voltmeter between any two legs (before connecting them). If the voltmeter reads zero or only a small value, the polarity is correct.

Multiphase Transformers (continued)

Transformers in delta-delta and wye-wye configurations behave like single-phase transformers, and the line voltages *out* are equal to the line voltage *in*, times the turns ratio.

With a delta-wye (or wye-delta) configuration the voltages are related not only by the turns ratio but also by the fact that the line-to-line voltages for a wye connection are from two windings and, as you know, are equal to $\sqrt{3}$ times the phase (transformer winding) voltage.

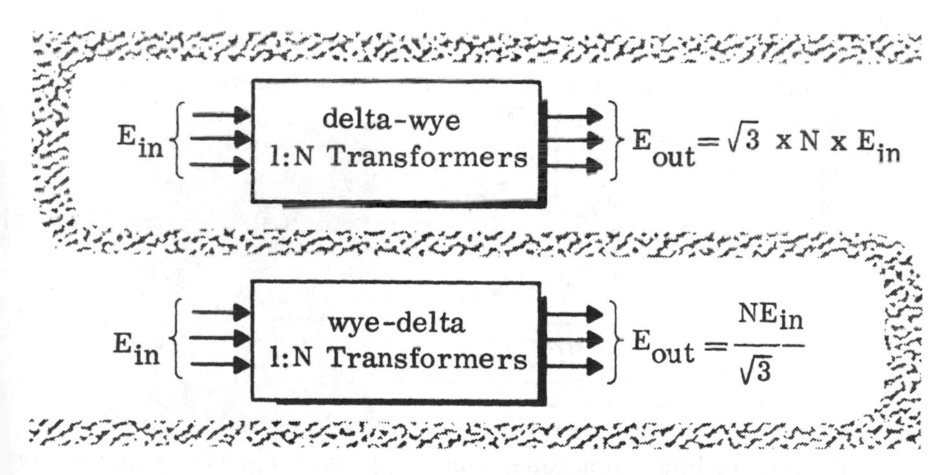

As you can see, delta-wye transformer configuration can be used to obtain voltages other than specified by the transformer turns ratio. The delta-to-wye connection is often used to step down transmission line voltages to lower voltages used in homes and industry. This is particularly convenient if a number of single-phase loads are to be supplied. For example, if the line-to-line voltages are 208 volts, the phase voltages (line to neutral) will be 120 volts.

Experiment/Application—Wye and Delta Transformer Connections

You can use three equal transformers to show the connection of wye and delta circuits and how to transform from one to the other. The first thing you must do is determine the phasing on each transformer. To do this, connect one transformer to the line as a *reference*. Mark one side of this transformer primary and secondary with a dot. Connect each of the other transformers (one at a time) across the same primary voltage and mark the same primary lead also with a dot. Connect the secondary of the reference transformer and the transformer under test in series and measure the total voltage. If the voltage is twice the single transformer voltage, then the windings are in series aiding and should be so marked.

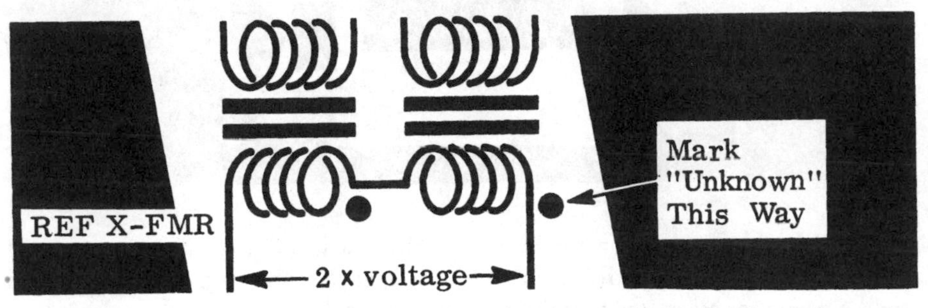

If the voltage is near zero, reverse the secondary leads of the transformer under test and try again. *Do not reverse any other connections.* Repeat the above with the third transformer. You now have a set of properly phased transformers.

Connect the transformers in a *delta to wye* as shown below:

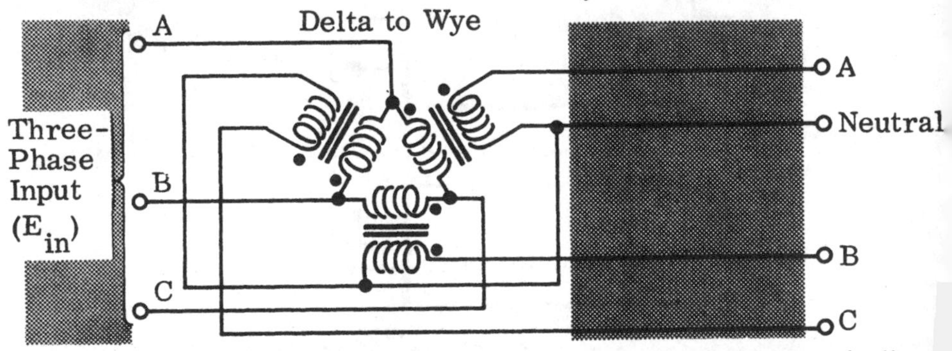

Measure the line to line voltage on the primary. Then measure the line to line voltage on the secondary. If the voltage transformation ratio for each transformer is N, you will note that the secondary (wye) voltages are $E_{in} \times N$ for line to neutral and $E_{in} \times \sqrt{3} \times N$ from line to line, showing that the results are as predicted earlier.

You can also reverse the configuration for a wye to delta connection and show that the secondary voltage is $E_{in} \times N/\sqrt{3}$. You can also try wye to wye and delta to delta connections and show that the secondary voltage are $N \times E_{in}$ for either connection.

Review of AC Generators—Alternators

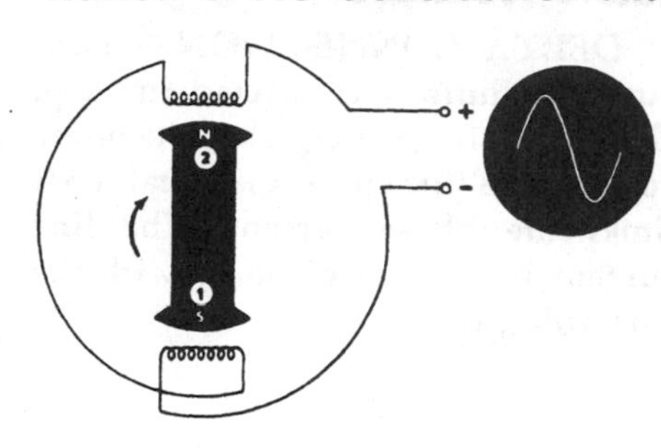

1. SINGLE-PHASE ALTERNATOR—A single-phase alternator has an armature that consists of a number of windings placed symmetrically around the stator and connected in series. The voltages generated in each winding add to produce the total voltage across the two output terminals.

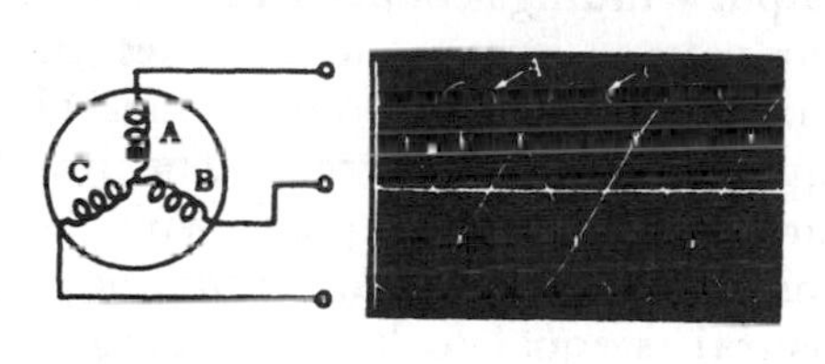

2. THREE-PHASE ALTERNATOR—In the three-phase alternator, the windings have voltages generated in them that are 120° out of phase. Three-phase alternators are most often used to generate ac power.

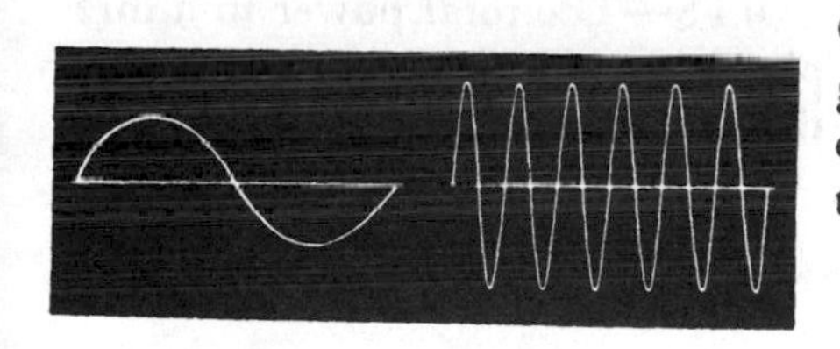

3. ALTERNATOR FREQUENCY—The frequency (f) of the ac generated by an alternator depends on the speed of rotation (N) and on the number of rotor poles (P).

$$f = \frac{NP}{120}$$

$$E_{out} = E_{gen} - IR - IX_L - IX_L^1$$

4. VOLTAGE REGULATION—The voltage regulation of an alternator is poorer than that of a dc generator because of the IX_L drop in the armature winding, in addition to the IR drop and armature reaction (X_L).

Review of AC Generators—Alternators (continued)

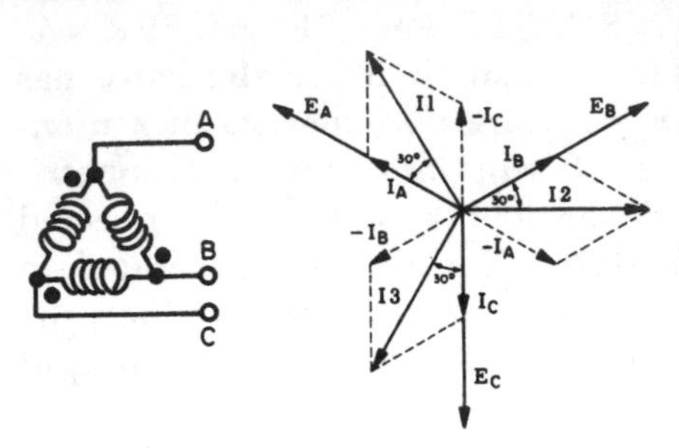

5. DELTA CONNECTION—Stator output windings connected in loop. The line voltage is equal to the phase voltage; the line current is equal to $\sqrt{3}$ times the phase current. The line current is 30° out of phase with the line voltage.

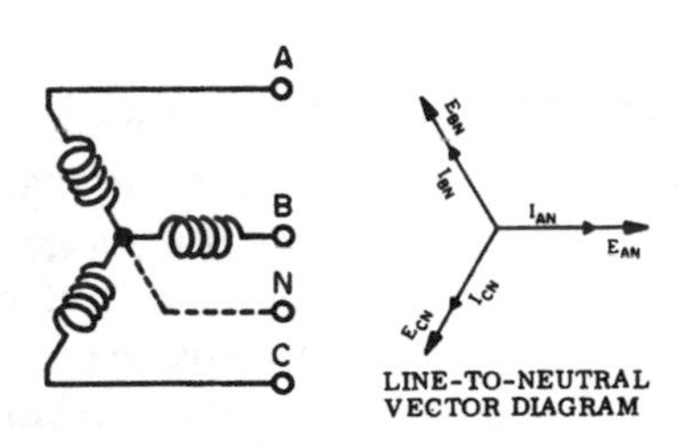

LINE-TO-NEUTRAL VECTOR DIAGRAM

6. WYE CONNECTION—Stator output windings connected as Y. The line voltage is equal to $\sqrt{3}$ times the phase voltage. The line current equals the phase current. The line current is 30° out of phase with the line voltage. The voltage from line to neutral is equal to line voltage/$\sqrt{3}$.

Power $= \sqrt{3}\ E_L I_L \cos\theta$

7. POWER IN THREE-PHASE CIRCUITS—The total power in a three-phase circuit, either wye or delta, is the same.

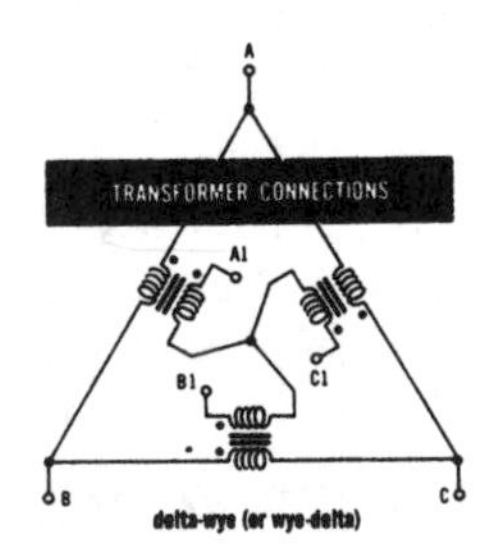

8. TRANSFORMERS—In three-phase circuits transformers can be used to modify the line voltages and to convert from wye to delta or vice versa.

Self-Test—Review Questions

1. List the essential elements of the ac generator. What is the difference between a high-speed and a low-speed machine?
2. Sketch single-phase, two-phase, and three-phase generator configurations. Show two-phase three-wire and three-phase delta and wye wiring arrangements.
3. An alternator has 6 pairs of poles and is to generate 120 volts at 60 Hz. How fast must it be driven? If the number of pole pairs were tripled, what would happen to the motor speed?
4. How do the regulation characteristics of ac generators differ from dc generators? If the dc resistance of a generator is 0.05 ohm, with an inductance of 0.5 mH and an armature reaction equivalent to 0.2 mH, what is the voltage regulation if the no load-voltage is 130 volts and the full-load current is 50 amperes? (Hint: Use vectors to calculate total impedance.)
5. What is the power factor of the output from question three into a resistive load?
6. Sketch the circuit and draw a vector diagram for a wye connected generator, both for line to neutral and line to line.
7. Sketch the circuits and draw a vector diagram for a delta-connected generator.
8. List the relationships between the phase and line voltage and currents for the delta- and the wye-connected generators. If you had a three-phase delta-connected generator that delivered 120 volts ac at 20 amperes/line and you reconnected it as a wye, what would be the voltage and current output available?
9. Draw multiphase transformers with delta-delta, wye-wye, delta-wye, and wye-delta configurations. If the input line voltage is 4,400 volts, and the transformers are all 10:1 step down, what are the output line and phase voltages?
10. If the input line currents are 100 amperes in question 9, and the load has a power factor of 80%, what are the powers (apparent and true) being delivered?

Learning Objectives—Next Section

Overview—Now that you know about ac generators, you are ready to learn about ac motors. Ac motors are probably the most common electrical devices, aside from lamps, that you will encounter. You will find that there are many types of ac motors for many different purposes.

Types of AC Motors

Since the major part of all electrical power generated is ac, most motors are designed for ac operation. Ac motors can, in many cases, duplicate the operation of dc motors and are less troublesome to operate. This is because dc machines encounter difficulties from the action of commutation, which involves brushes, brush holders, neutral planes, etc. Most ac motors, in contrast, do not even use slip rings, with the result that they give trouble-free operation over long periods of time. However, ac motors usually only operate well over a very narrow speed range.

Ac motors are particularly well suited for *constant-speed* applications, since the speed is determined by the frequency of the ac applied to the motor terminals. However, ac motors are also manufactured with variable-speed characteristics within certain limits.

Ac motors can be designed to operate from a single-phase or a multiphase ac supply. Whether the motor is single-phase or multiphase, it operates on the same principle: that the ac applied to the motor generates a rotating magnetic field, and this rotating magnetic field causes the rotor of the motor to turn.

Ac motors are generally classified into two basic types: (1) the *synchronous motor* and (2) the *induction motor*. The synchronous motor is an ac generator operated as a motor. In it, ac is applied to the stator and dc is applied to the rotor. The induction motor differs from the synchronous motor in that its rotor is not connected to any source of power, but is powered by magnetic induction.

The ac series motor, which is widely used for some appliances and small tools, is a modified version of the dc series motor. It has the advantage of readily adjustable speed, and can also be used in applications where the dc series motor is used.

Of the two basic types of ac motors mentioned, the induction motor is by far the more commonly used.

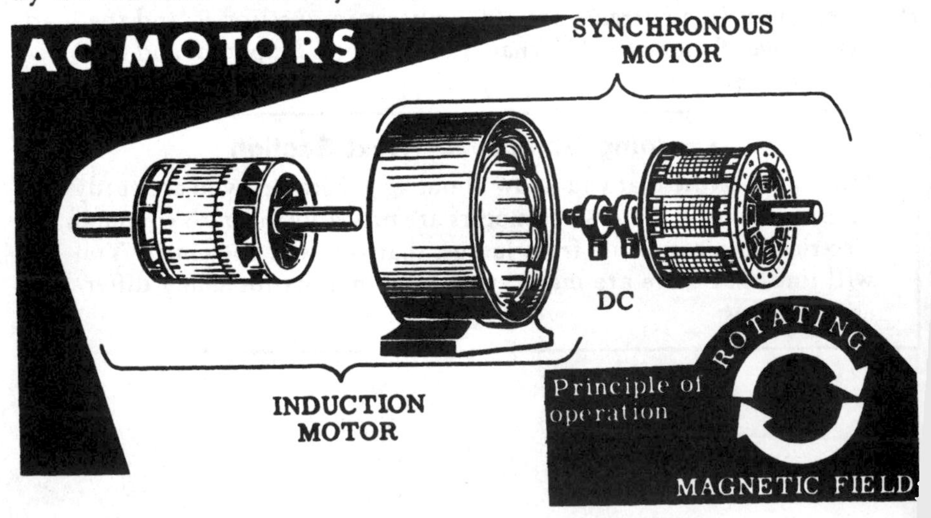

Rotating Field

Before learning how a rotating magnetic field will cause an energized rotor to turn, you must first find out how a rotating magnetic field can be produced. This is easiest to understand with a three-phase system, so we will start with that. The diagram below illustrates a three-phase stator to which three-phase ac is applied. The windings are physically spaced 120° apart and are connected in delta as shown. The two windings in each phase are wound in the same direction.

At any instant the magnetic field generated by one particular phase depends on the current through that phase. If the current is zero, the magnetic field is zero. If the current is a maximum, the magnetic field is a maximum. Since the currents in the three windings are 120° out of phase, the magnetic fields generated will also be 120° out of phase. Now the three magnetic fields that exist at any instant will combine to produce *one field*, which acts on the rotor. You will see on the following page that from one instant to the next, the magnetic fields combine to produce a magnetic field whose position is shifting. At the end of one complete cycle of ac the magnetic field will have shifted through 360°, or one revolution.

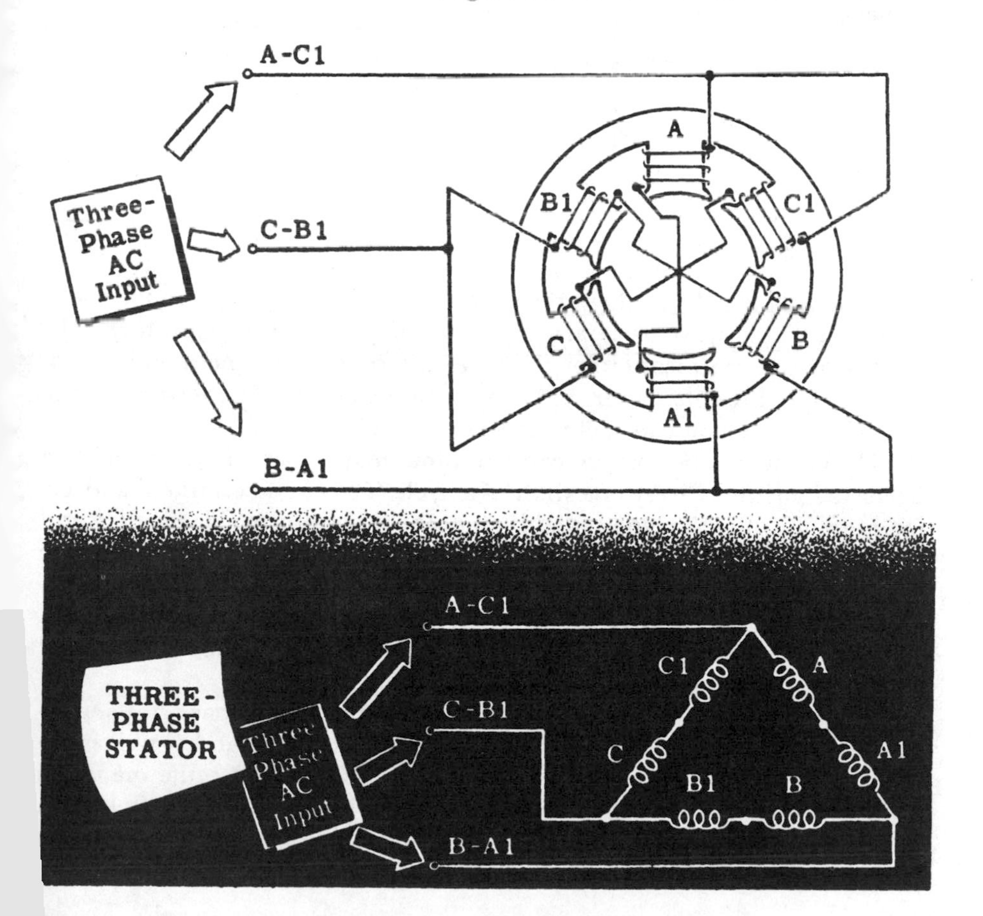

Rotating Field (continued)

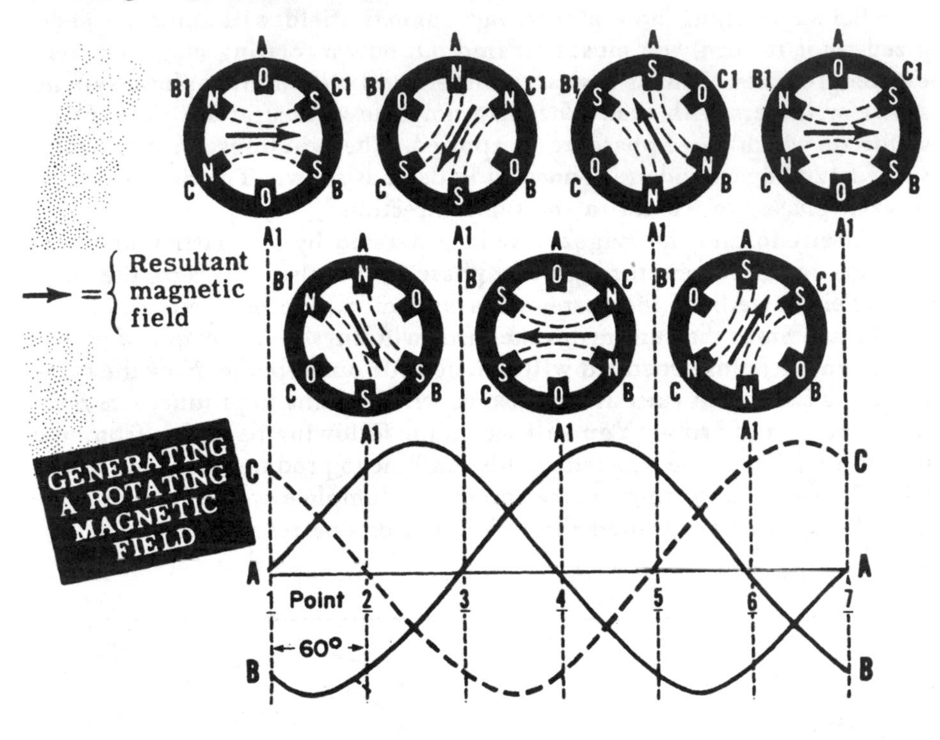

The drawing shows the three current waveforms applied to the stator windings. These waveforms are 120° out of phase with each other. The waveforms can represent either the three alternating magnetic fields generated by the three phases, or the currents in the phases. The waveforms are lettered to correspond to their associated phase.

Using the waveforms, we can combine the magnetic fields generated at selected positions every one sixth of a cycle (60°) to determine the direction of the resultant magnetic field. At point 1, waveform C is positive and waveform B is negative. This means that the current flows in opposite directions through phases B and C, and so establishes the magnetic polarity of phases B and C. The polarity is shown on the simplified diagram above point 1. Observe that B1 is a north pole and B is a south pole, and that C is a north pole and C1 is a south pole.

Since at point 1 there is no current flowing through phase A, its magnetic field is zero. The magnetic fields leaving poles B1 and C will move toward the nearest south poles C1 and B as shown. Since the magnetic fields of B and C are equal in amplitude, the resultant magnetic field will lie between two fields, and will have the direction shown.

Rotating Field (continued)

At point 2, 60° later, the input current waveforms to phases A and B are equal and opposite, and waveform C is zero. The resultant magnetic field has rotated through 60°. At point 3, waveform B is zero, and the resultant magnetic field has rotated through another 60°. From points 1 to 7 (corresponding to 1 cycle of ac) you can see that the resultant magnetic field rotates through one revolution for every cycle of ac supplied to the stator.

The conclusion is that the application of three-phase ac to three windings symmetrically spaced around a stator causes a rotating magnetic field to be generated.

You can apply similar reasoning and show that a two-phase system will also generate a rotating magnetic field. In fact, any number of phases will generate a rotating field. A single-phase system will not start, however, as you will learn later in this section. Therefore, special arrangements are necessary in single-phase ac motors to make them operate properly.

As you noticed on the previous page, the field rotated 1 cycle for 1 current cycle. Then, if the current were supplied from a 60-Hz source, the field would rotate 60 times per second or 3,600 times per minute. However, if the number of stator coils were doubled (a 4-pole machine), the field would rotate half as fast. This is exactly analogous to what you learned about ac generators where, for a given frequency output, the prime mover speed went down in proportion to the number of poles. Thus, the speed of the field can be easily calculated as

$$N = \frac{120F}{P}$$

Field Rotation Rate Is Proportional to Frequency and Inversely Proportional to the Number of Poles

You will notice that this is the same equation you used to calculate the frequency of the output from a generator, except that it is solved for N (motor field speed in rpm) rather than f (frequency). As before, P represents the number of poles. You will notice that the number of poles referred to is the number of poles per phase. Thus, a 2-pole 2-phase motor will actually have 4 poles, and a 2-pole 3-phase motor will actually have 6 poles, but they will both rotate at the same speed.

The speed formed by the equation above is called the *synchronous speed* because it is synchronized to the power line frequency. Thus, a 2-pole machine operating at 60 Hz has a synchronous speed of 3,600 rpm, and a 4-pole machine has a synchronous speed of 1,800 rpm, etc.

As you can see, the speed of the rotating field depends only on the line frequency and is independent of load. The revolving field provides the driving force for most ac motors; hence, within narrow limits most ac motors are constant-speed devices.

The Synchronous Motor

Although the synchronous motor is not as commonly used as the induction motor, we will study it first because it is very similar in construction to the ac generator.

The synchronous motor is so called because its rotor is synchronized with the rotating field set up by the stator. Its construction is essentially the same as that of the salient-pole alternator. You know that the application of three-phase ac to the stator causes a rotating magnetic field to be set up around the rotor. If the rotor is energized with dc, it will act like a bar magnet. If a bar magnet is suspended in a magnetic field, it will turn until it lines up with the magnetic field. In the same way, the dc-energized rotor of a synchronous motor is a magnet that will line up with the magnetic field caused by the application of three-phase ac to the stator. When the magnetic field turns, the rotor will turn synchronously with the field. If the rotating magnetic field is strong, it will exert a strong turning force on the rotor, which will therefore be able to turn a load as it rotates.

IN A SYNCHRONOUS MOTOR THE ROTOR TURNS SYNCHRONOUSLY WITH THE MAGNETIC FIELD

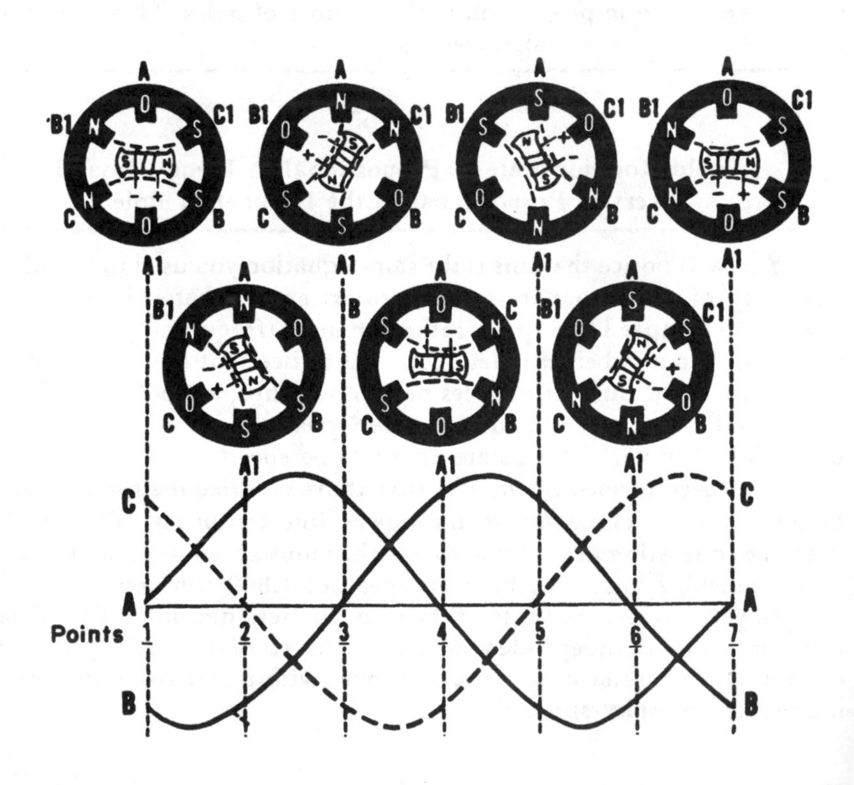

The Synchronous Motor (continued)

As you know, the speed of rotation of the magnetic field depends on the frequency of the ac supply. And since the supply frequency is fixed, synchronous motors are, in practice, single-speed motors. They are used for loads that require strictly constant speed from no-load through to full-load conditions. Note that in a synchronous motor, operating at a synchronous speed, there is no ac induced into the rotor, since there is no relative motion between the rotating field and the rotor. Some small synchronous motors use permanent magnet rotors and do not then need a dc supply.

One of the disadvantages of a purely synchronous motor is that it cannot be started from a standstill by applying ac to the stator.

The instant ac is applied to the stator, a high-speed rotating field appears. This rotating field rushes past the rotor poles so quickly that the rotor does not have a chance to get started; it is repelled first in one direction and then in the other. In other words, a synchronous motor in its pure form has no starting torque. It is usually started, therefore, with the help of a small induction motor, or with windings equivalent to this (called a *squirrel cage*) incorporated into the synchronous motor. When the rotor has been brought close to synchronous speed by the starting device, it is energized by connecting it to the dc voltage source. The rotor then falls into step with the rotating field. Such a motor is called an *induction-start synchronous-run motor.*

You will learn about the induction motor later in this section.

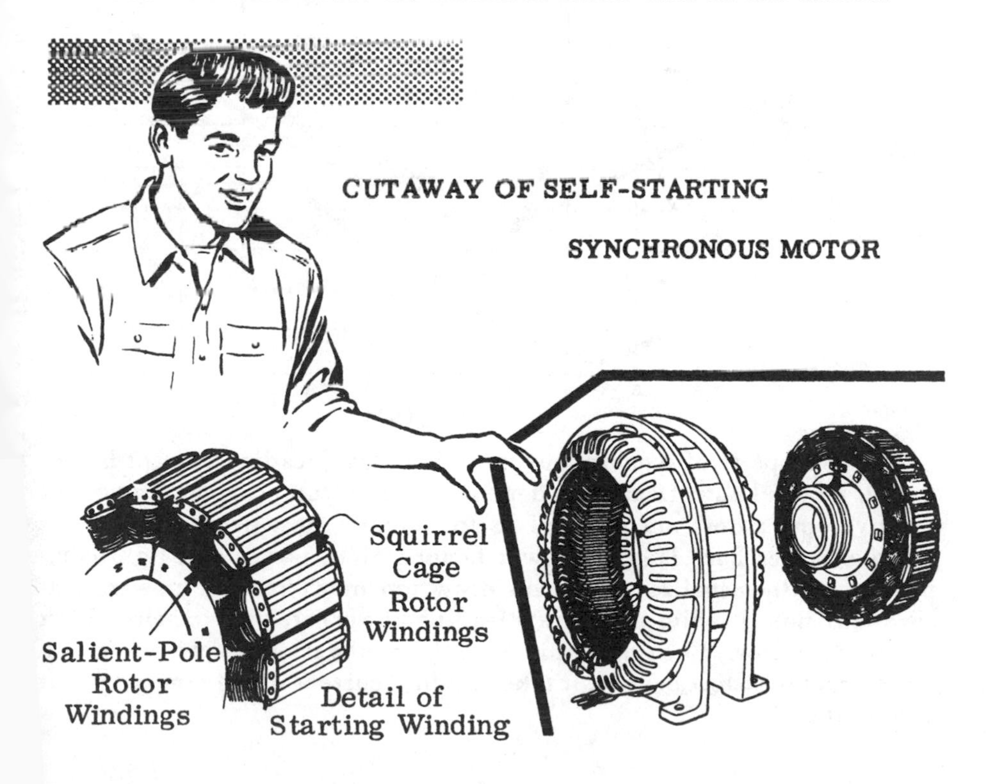

The Synchronous Motor (continued)

When a synchronous motor is starting, there is relative motion between the field and the rotor so an ac voltage is set up in the salient-pole rotor winding. This can aid the squirrel cage rotor by providing additional starting torque for starting synchronous motors.

The power factor of most loads, for example, of induction motors, is inductive (lagging); the power factor of a synchronous motor can be unity or either leading or lagging, depending on the dc excitation. Under no-load conditions, the current drawn by a synchronous motor is small. When a load is applied, the rotor responds by pulling back by a phase angle with respect to the field. Note that the speed is still synchronous but that the phase between the rotor and the field varies. Under these conditions, the power factor is usually lagging. As the dc excitation current is increased, the power factor goes to unity and then becomes leading. Thus, for a given load, the synchronous motor power factor is determined by the dc excitation. This is shown graphically in the synchronous motor E-curves below, which shows the variations in field current under load for various power factor conditions.

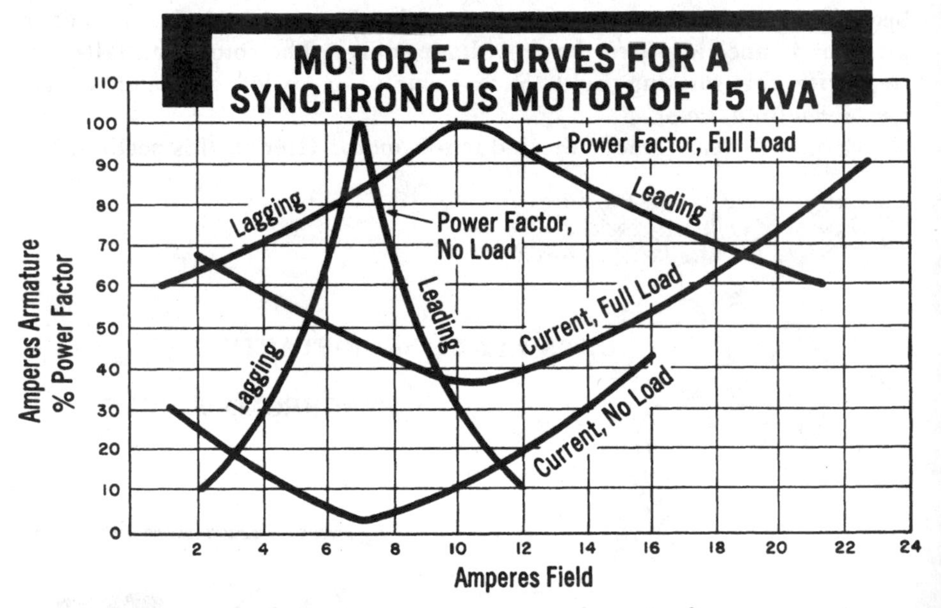

The property of synchronous motors to draw leading current is very useful. As you know, IR drop is dependent on the total current drawn, but reactive power does not produce useful output. Most devices that draw reactive power have lagging power factors. Since power and light companies usually penalize users who draw too much reactive power, synchronous motors (adjusted to draw leading power) are used to help correct the power factor. When used this way, the motor is often referred to as a *synchronous condenser* because it takes leading current in the same way that a capacitor does.

Induction Motors

The induction motor is the most commonly used ac motor because of its simplicity, its robust construction, and its low manufacturing cost. These characteristics of the induction motor are due to the fact that the rotor is self-contained and usually not connected physically to the external source of voltage. The induction motor derives its name from the fact that ac currents are *induced* in the rotor circuit by the rotating magnetic field in the stator.

The stator construction of the induction motor and of the synchronous motor are almost identical, but their rotors are different. The rotor of the induction motor is a laminated cylinder with slots in its surface. The windings in these slots are of two types. The most common is called a *squirrel cage winding*, which is made up of heavy copper bars connected together at either end by a metal ring made of copper or brass. No insulation is required between the core and the bars because of the very low voltages generated in these rotor bars. The air gap between the rotor and stator is kept very small so as to obtain maximum field strength. The other type of winding contains coils placed in the rotor slots. The rotor is then called a *wound rotor*.

Regardless of the type of rotor used, the basic principle of operation is the same. The rotating magnetic field generated in the stator induces an emf in the rotor. The current in the rotor circuit caused by this induced emf sets up a magnetic field. The two fields interact and cause the rotor to turn.

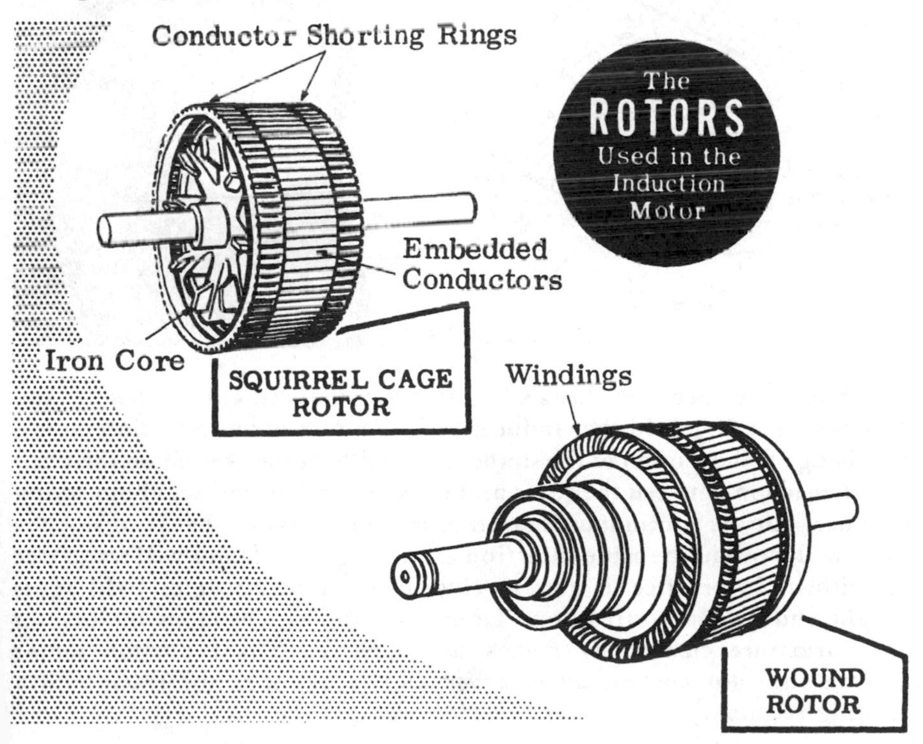

Induction Motors (continued)

Wound rotor motors often have slip rings connecting the windings to external resistances. The variable resistances provide a means for increasing rotor resistance during starting to give better starting characteristics. When the motor is up to speed, the windings are shorted and the operation is like a squirrel cage rotor.

When ac is applied to the stator windings, a rotating magnetic field is generated. This rotating field cuts the bars of the rotor and induces a current in them. As you know from your study of meter movements and transformers, this induced current will generate a magnetic field around the conductors of the rotor, which will try to line up with the stator field. However, since the stator field is rotating continuously, the rotor must always follow along behind it.

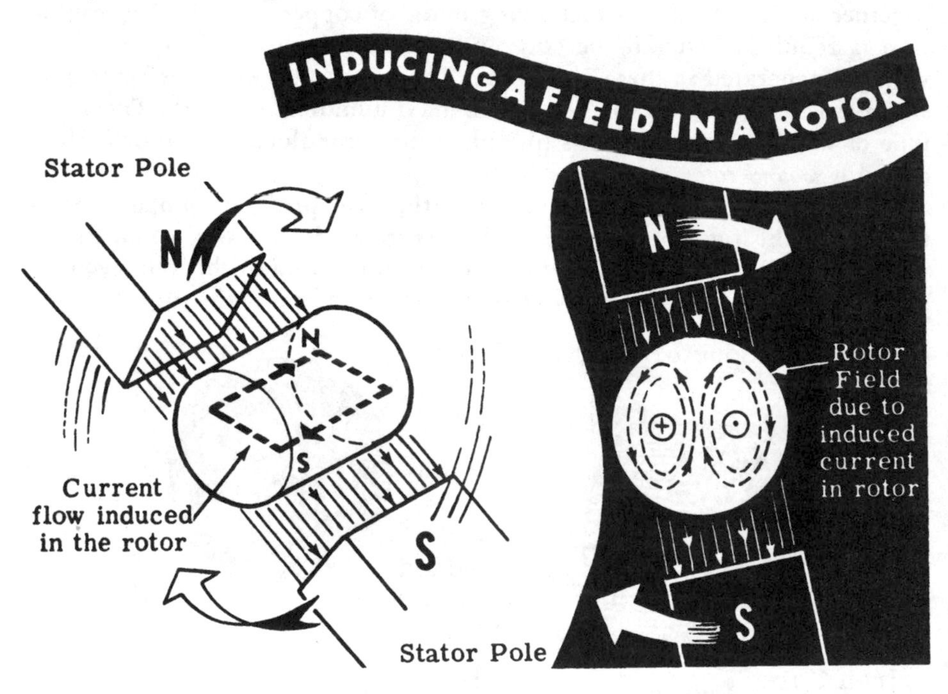

You remember from Lenz's law that any induced current tries to oppose the changing field that induces it. In the case of an induction motor, the change is the motion of the stator field. So the force exerted on the rotor by the reaction between the rotor and the stator fields will set about trying to cancel out the continuous motion of the stator field. That is to say, the rotor will move in the same direction as the stator field, and will try to line up with it. In practice, it gets as close to the moving stator field as its weight and its load will allow. Dc motors and synchronous motors get their armature current by means of *conduction.* The induction motor receives its rotor current by *induction* and is like a transformer with a rotating secondary.

Induction Motors—Slip

It is impossible for the rotor of an induction motor to turn as fast as the rotating magnetic field. If the speeds were the same, no relative motion would exist between the two, and no induced emf would result in the rotor. Without induced emf no turning force would be exerted on the rotor. The rotor must rotate at a speed lower than that of the rotating magnetic field if *relative* motion is to exist between the two. (Remember, it can't go faster than synchronous speed.)

This percentage difference between the speed of the rotating stator field and the speed of the rotor is called *slip*. The smaller the slip, the closer the rotor speed will approach the speed of the stator field.

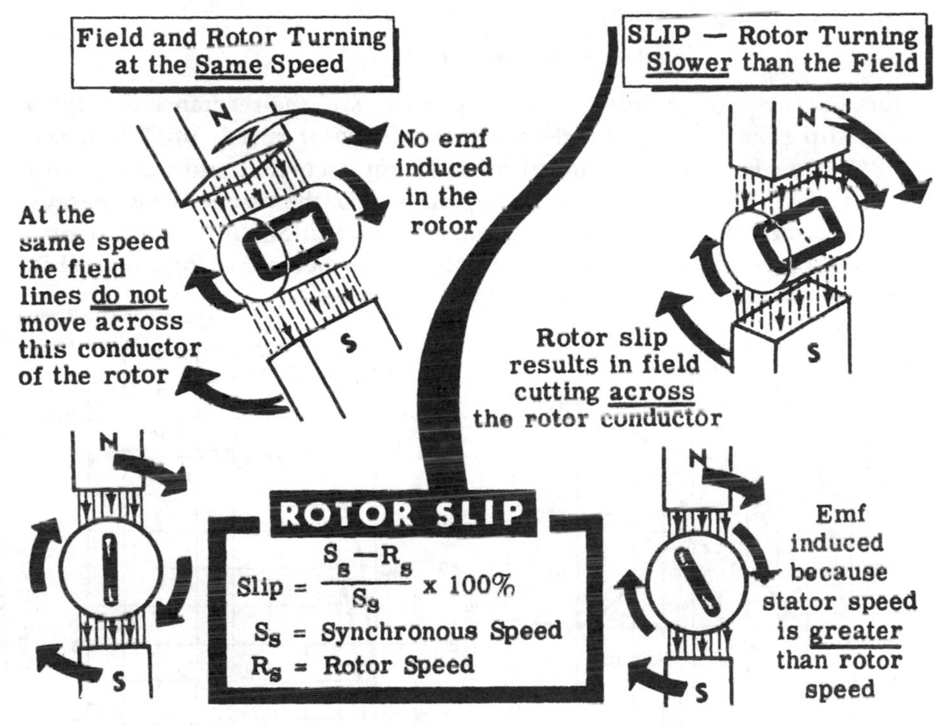
Field and Rotor Turning at the Same Speed
No emf induced in the rotor
At the same speed the field lines do not move across this conductor of the rotor
SLIP — Rotor Turning Slower than the Field
Rotor slip results in field cutting across the rotor conductor
ROTOR SLIP
Slip = (Ss—Rs)/Ss x 100%
Ss = Synchronous Speed
Rs = Rotor Speed
Emf induced because stator speed is greater than rotor speed
N
S

The speed of the rotor depends on the requirements of the load. The greater the load, the stronger the turning force needed to turn the rotor.

But the turning force can only increase if the rotor-induced emf increases. And this emf can only increase if the magnetic field cuts through the rotor at a faster rate.

Now, to increase the relative speed between the field and the rotor, the rotor must *slow down*.

For heavier loads, therefore, the induction motor will turn more slowly than it will for lighter loads. Actually, only a slight change in speed is necessary to produce the current changes required for normal changes in load. This is because the rotor windings have such a very low resistance. Thus induction motors are, for all practical purposes, *constant-speed motors*.

Induction Motors—Torque and Efficiency

At start, the frequency of the ac induced in the rotor is the same as the line frequency, but as the rotor speeds up, this frequency is reduced—at synchronous speed, it would be zero. The rotor frequency (f_r) is proportional to the percentage of slip (S) and the supply frequency (f_s) in Hz or

$$f_r = \frac{Sf_s}{100}$$

For example, if the frequency is 60 Hz and the motor has 2.78% slip (4-pole induction motor with speed of 1,750 rpm), then the rotor ac frequency is 2.78 × 60/100 = 1.67 Hz. The reactance of the rotor is also proportional to slip so that

$$\text{(Rotor) } X_L = 2\pi f_s SL/100$$

where L is the rotor inductance. As you can see, the reactance is large at high slip (low speeds) and decreases as the rotor speeds up. Thus, near 100% slip, the rotor reactance limits the rotor current and the torque is low; and at the other extreme, if the slip is zero, the torque is low because of low rotor current. Because of rotor reactance, induction motors show a lagging power factor. The characteristic curves below are for a typical industrial three-phase induction motor.

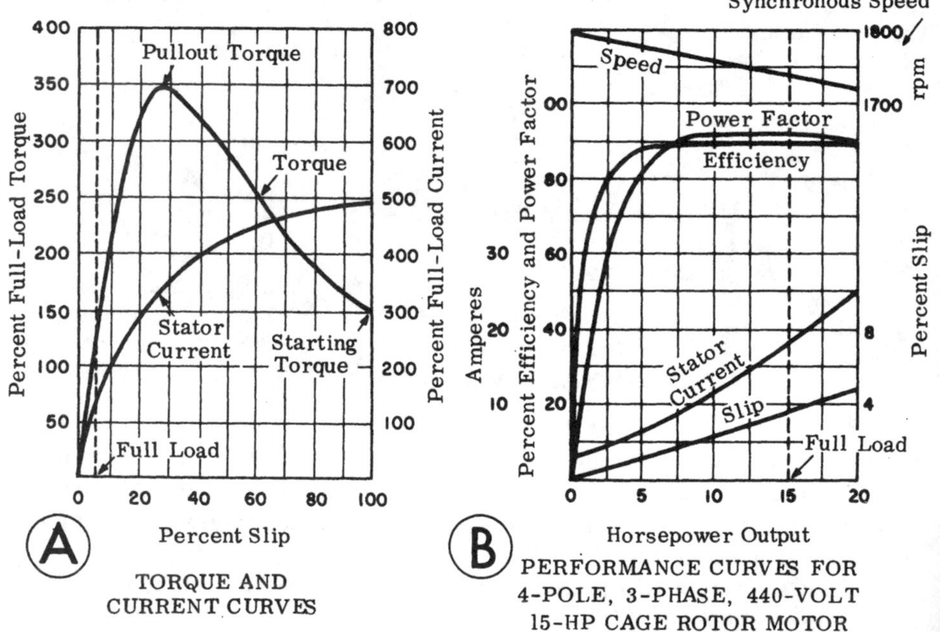

(A) TORQUE AND CURRENT CURVES

(B) PERFORMANCE CURVES FOR 4-POLE, 3-PHASE, 440-VOLT 15-HP CAGE ROTOR MOTOR

The *pullout* torque point shown on graph A is the point where rotor reactance and resistance are equal, and therefore the rotor power factor is 70%. Beyond this point, the motor speed drops off rapidly and the motor will stall.

Two-Phase Induction Motors

Induction motors are designed for three-phase, two-phase, or single-phase operation. But in every case, the ac applied to the stator must generate a rotating field that will pull the rotor with it.

You have already seen how three-phase ac applied to a three-phase symmetrically distributed winding will generate a rotating magnetic field. A two-phase induction motor has its stator made up of two windings placed at right angles to each other around the stator. Although you will not find many two-phase motors in use, they are important to study because many single-phase motors are started as two-phase motors, as you will learn a little later in this section. The simplified drawing below illustrates a two-phase stator. The other drawing is a schematic of a two-phase induction motor. The dotted circle represents the rotor winding.

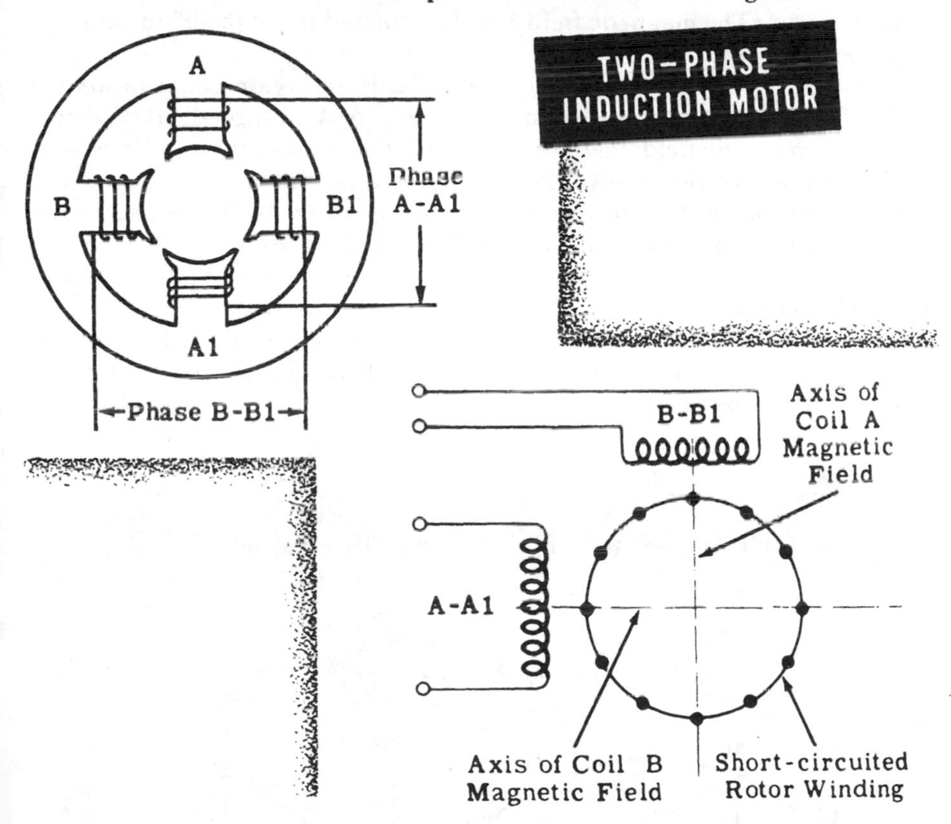

If the voltages applied to phases A-A1 and B-B1 are 90° out of phase, the currents which will flow in the phases will be displaced by 90°. Since the magnetic fields generated in the coils will be in phase with their respective currents, the magnetic fields will also be 90° out of phase with each other. These two out-of-phase magnetic fields, whose coil axes are at right angles to each other, will add together at every instant during their cycle to produce a field that will rotate one revolution for each cycle of ac.

Two-Phase Induction Motors (continued)

The diagram below shows a graph of the two alternating magnetic fields which are displaced 90° in phase. The waveforms are lettered to correspond to their associated phase. The similarity to the three-phase generated rotating field is evident.

At position 1, the current flow and magnetic field in winding A-A1 is a maximum, and the current flow and magnetic field in winding B-B1 is zero. The resultant magnetic field will therefore be in the direction of the winding A-A1 axis. At the 45° point (position 2), the magnetic field will lie midway between windings A-A1 and B-B1, since the coil currents and magnetic fields are equal in strength. At 90° (position 3), the magnetic field in winding A-A1 is zero, and the magnetic field in winding B-B1 is a maximum. Now the magnetic field lies along the axis of the B-B1 winding, as shown. The magnetic field has thus rotated through 90° to get from position 1 to position 3.

At 135° (position 4), the magnetic fields are again equal in amplitude. However, the magnetic field in winding A-A1 has reversed its direction. The magnetic field, therefore, lies midway between the windings, and points in the direction shown. At 180° (position 5), the magnetic field is zero in winding B-B1 and a maximum in winding A-A1. The magnetic field will, therefore, lie along the axis of winding A-A1.

From 180° to 360° (position 5 to 9), the magnetic field rotates through another half cycle and completes a revolution.

Thus, by placing two windings at right angles to each other, and by exciting these windings with voltages 90° out of phase, a rotating magnetic field can be created.

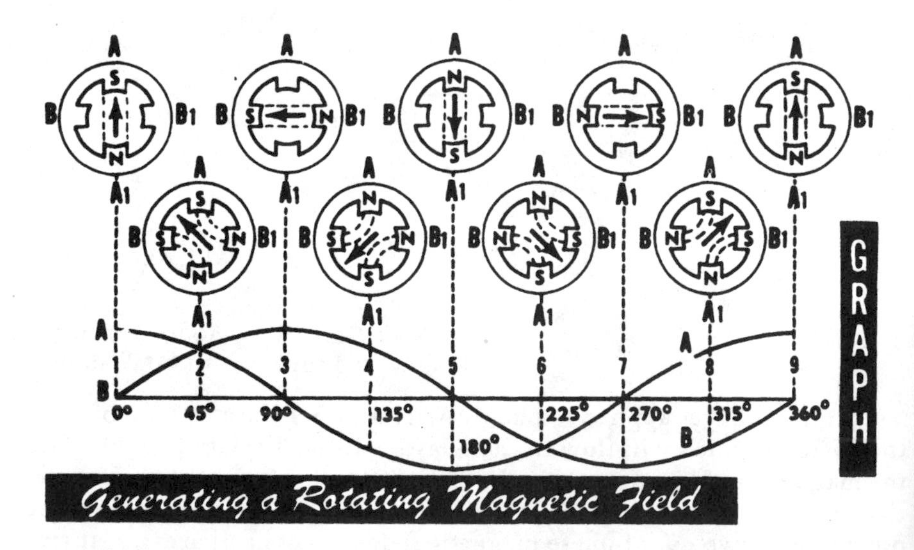

Generating a Rotating Magnetic Field

Other characteristics of two-phase motors are similar to those of three-phase motors.

Single-Phase Motors

A single-phase induction motor has only one phase and runs on single-phase ac. This motor is extensively used in applications that require small low-output motors. The advantage gained by using single-phase motors is that in small sizes they are less expensive to manufacture than other types. Also, they eliminate the need for three-phase ac lines. Single-phase motors are used in communication equipment, fans, refrigerators, portable drills, grinders, etc.

Single-phase motors are divided into two kinds: (1) *induction motors* and (2) *series motors.* Induction motors use the squirrel cage rotor and a suitable starting device. Series motors resemble dc motors in that they have commutators and brushes.

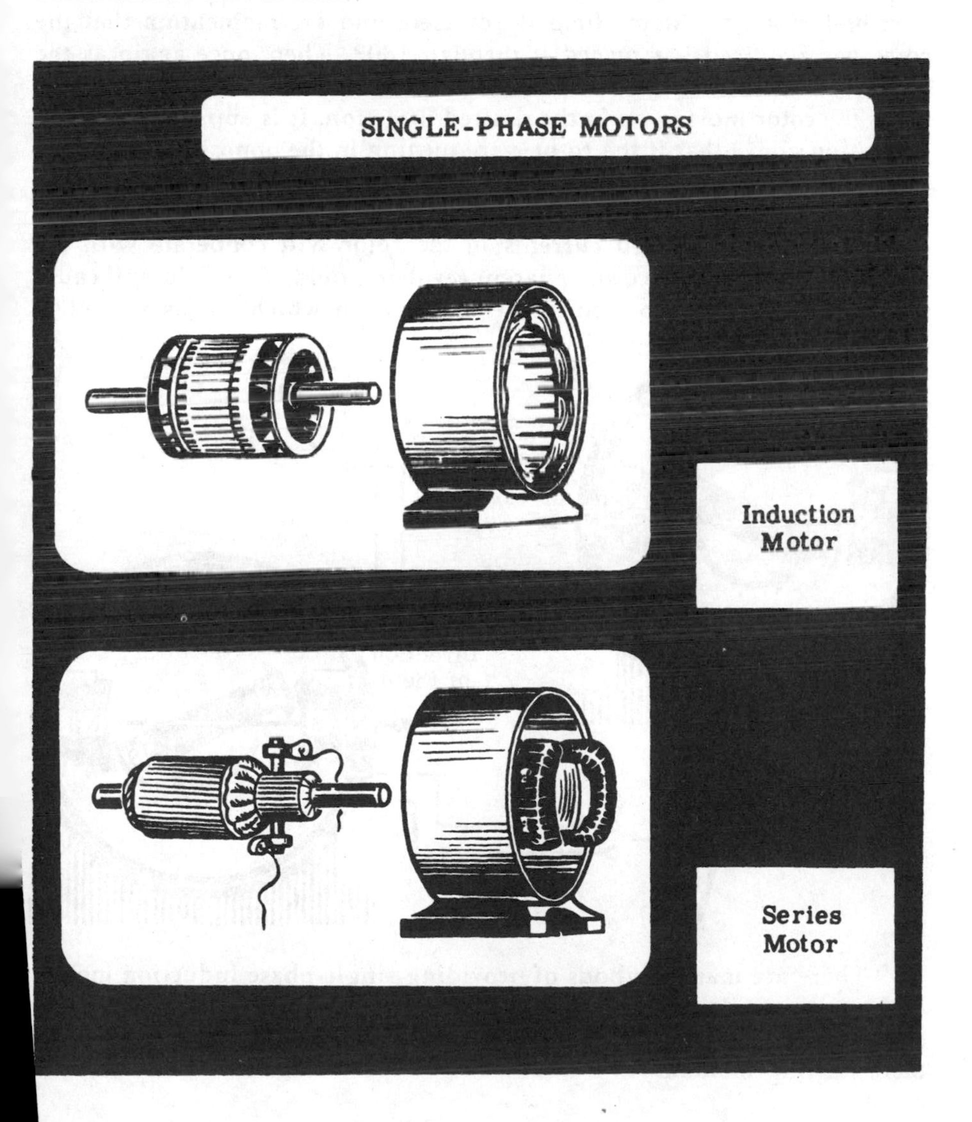

Single-Phase Induction—Stator Fields

Now you will see how single-phase ac applied to a stator winding will produce a reversing magnetic field, which can also drive a motor.

If the rotor is turning, every time a rotor pole approaches a stator winding, the direction of the stator field must be such that it can attract the pole and give it a torque in the direction of its motion. So the field current in a stator winding must pass through half a cycle in the interval between the approach of rotor poles of opposite polarity.

The diagrams below indicate the direction of the stator field created by the application of single-phase ac. Diagram 1 at the top shows one half cycle; and you will see that, as the rotor lines itself up with the stator field, its poles are being attracted by the windings of the stator. In diagram 2, the next half cycle, the stator field is reversed, and the momentum that the rotor has acquired has turned it through 180°. Then, once again as the rotor poles approach the stator winding, there is a force of attraction that keeps the rotor moving on in the desired direction. It is apparent from the discussion above that if the rotor were turning in the opposite direction it would continue to turn. Thus, with a single phase, there is no rotating magnetic field and no starting torque. But if the rotor can be started by other means, the induced currents in the rotor will cooperate with the stator currents to produce an apparent revolving field. This field will cause the rotor to continue to turn in the direction in which it was started.

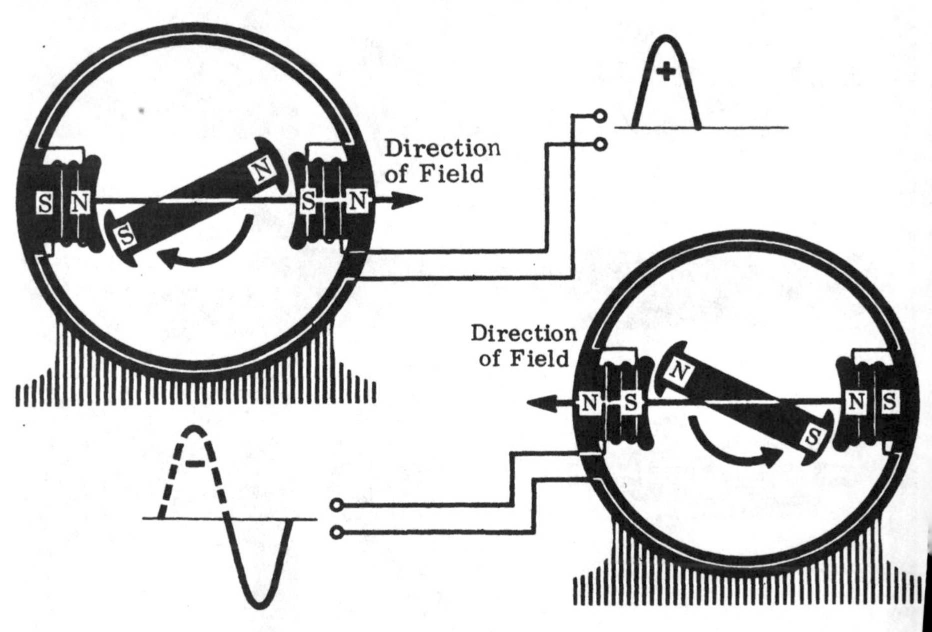

There are many methods of providing single-phase induction motors with starting torque—the most common of these are the split-phase capacitor and the shaded pole. Single-phase induction motors are classified by their starting method.

Single-Phase Induction Motors—Rotor Fields

A single-phase induction motor has only one stator winding while running. This winding generates a field that alternates along the axis of the single winding.

When the rotor is stationary, the expanding and collapsing stator field induces currents in the rotor that generate a rotor field. The opposition of these fields exerts a force on the rotor, the force trying to turn it 180° from its position. Since this force is exerted through the center of the rotor, the rotor does not turn.

However, if something starts the rotor turning, the turning force in that direction is aided by the momentum of the rotor. The rotor will increase speed until it turns nearly 180° for each alternation of the stator field. Remember that, since slip is necessary to cause an induced rotor current, at maximum speed the rotor must turn slightly less than 180° every time the stator field reverses polarity.

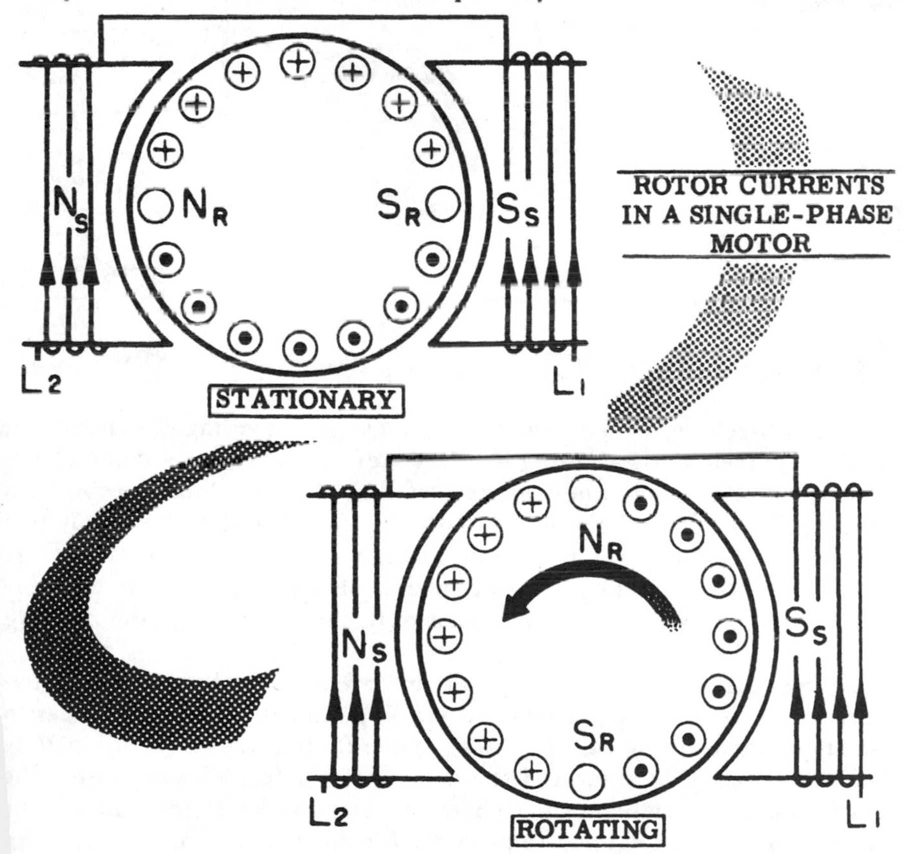

ROTOR CURRENTS IN A SINGLE-PHASE MOTOR

Since the field created by the single-phase ac voltage applied to the stator winding is pulsating, single-phase ac induction motors develop a *pulsating* torque. They are, therefore, less efficient than three-phase or two-phase motors whose torque acts more uniformly.

Single-Phase Induction Motors—Split Phase

You have seen that once the single-phase motor has been started turning by some means, it will continue to rotate by itself. It is impractical to start a motor by turning it over by hand, so an electric device must be incorporated into the stator circuit that will cause a rotating field to be generated on starting. Once the motor has started, this device can be switched out of the stator, since the rotor and stator together will generate their own rotating field to keep the motor turning.

The split-phase motor has a stator with an auxiliary winding (starting winding), as well as the main (running) winding. The axes of these two windings are displaced physically by 90°.

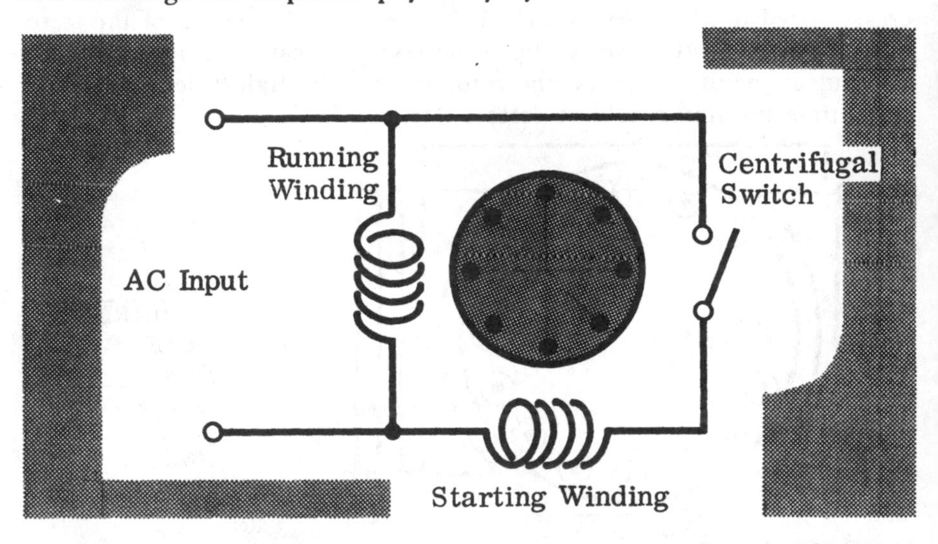

The current in the starting winding lags the line voltage by about 30°, while the current in the running winding lags the voltage by about 45°. On starting, these two windings, because of their differing phase, produce a rotating field. The rotor current lags the rotor voltage by about 90° because of the high rotor reactance (high frequency of rotor voltage because the slip is large). The interaction of the rotor currents and the stator field causes the rotor to accelerate in the direction of the rotating stator field.

When the rotor has come up to about 75% of normal speed, the centrifugal switch disconnects the starter winding, and the motor continues to operate on the running winding alone. The rotating field is maintained by the interaction of the rotor and stator magnetic fields. This motor has the constant-speed, variable-torque characteristics of the shunt motor. The starting torque can attain up to twice the full-load torque, and the starting current is six to eight times full-load current. These motors are used in home appliances—washers, dryers, oil burners, ventilating fans, etc. The rotation direction is reversed by interchanging the starter winding leads.

Single-Phase Induction Motors—Capacitor Motors

The capacitor motor is a modified version of the split-phase motor. The diagram below shows a simplified schematic of a typical capacitor-start motor. The stator consists of the main winding and a starting winding that is connected in parallel with it and spaced at right angles to it. The 90° electrical phase difference between the two windings is obtained by connecting the auxiliary winding in series with a capacitor and starting switch.

On starting, the switch is closed, placing the capacitor in series with the auxiliary winding. The value of the capacitor is such that the auxiliary winding is effectively a resistive-capacitive circuit in which the current leads the line voltage by approximately 45°. The main winding has enough inductance to cause the current to lag the line voltage by approximately 45°. The two currents are therefore 90° out of phase, and so are the magnetic fields that they generate. The effect is that the two windings act like a two-phase stator and produce the revolving field required to start the motor.

When about 75% of full speed has been attained, a centrifugal switch cuts out the starting winding, and the motor runs as a single-phase induction motor.

Because a two-phase induction motor is more efficient than a single-phase motor, it is often desirable to keep the auxiliary winding permanently in the circuit so that the motor will run as a two-phase induction motor. The starting capacitor is usually made quite large, so as to allow a large current to flow through the auxiliary winding. The motor can thus build up a large starting torque.

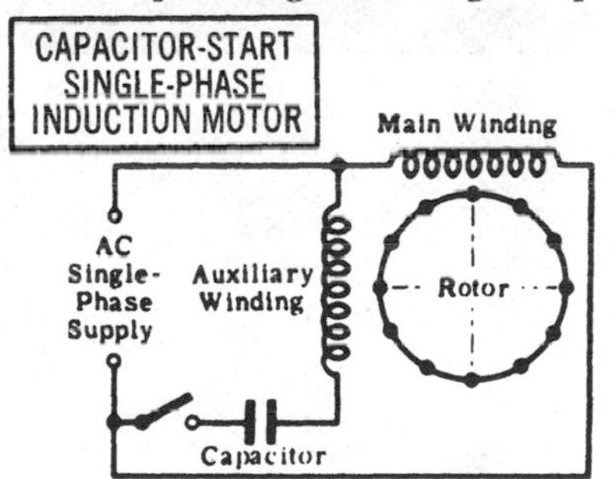

When the motor comes up to speed, it is not necessary that the auxiliary winding continue to draw the full starting current, and the capacitor can be reduced. Therefore, two capacitors are used in parallel for starting, and one is cut out when the motor comes up to speed. Such a motor is called a *capacitor-start, capacitor-run induction motor.*

The capacitor motor produces a much greater phase difference between the windings and, as a result, has a higher starting torque than the split-phase motor (as high as 400% of the full-load torque). Special non-polar electrolytic capacitors are used in sizes from 80 μF for a ⅛-horsepower motor, to 400 μF for 1-horsepower motors. These motors typically are used to drive grinders, drill presses, refrigerator compressors, and other loads needing *high* starting torque. As with the split-phase motor, the rotation direction is reversed by reversing the starting winding leads.

Single-Phase Induction Motors—Centrifugal Switch

The split-phase and capacitor motors use a centrifugal switch to switch out the starting winding when a sufficient speed has been reached (usually about 75%). Centrifugal switches usually have a pair of normally closed contacts that are opened when the rotor speed exceeds a given value. One of the most common centrifugal switches uses a pair of weights to overcome spring tension holding the switch closed.

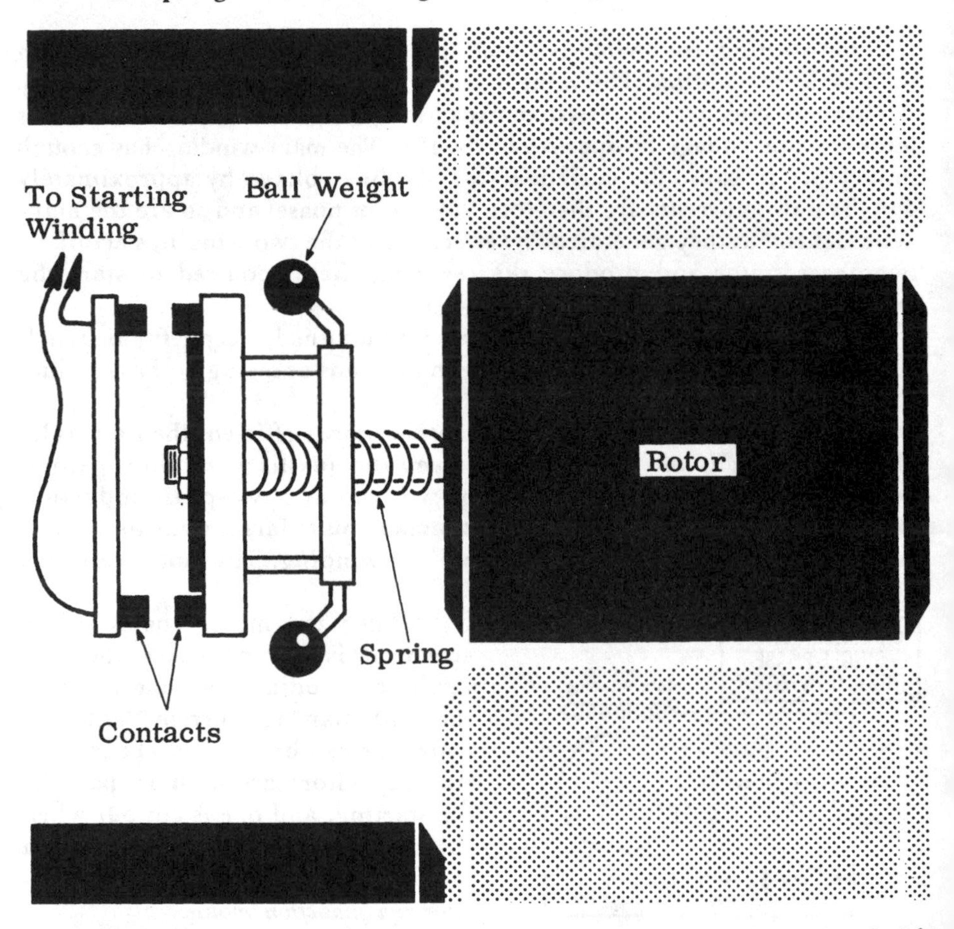

In the arrangement shown above (a ball-type switch), the spring holds the contact block in touch with the contacts to the starter winding so that the circuit is closed for starting. As the motor speeds up, the ball weights are thrown outward, and when they are going fast enough, they overcome the spring tension and pull back the contact block—opening the starting winding. It remains open as long as the motor is operating. When the motor is stopped, the weights move in, and the spring tension closes the contacts again.

There are many variations on the simple centrifugal switch. Each ha the same basic elements: a set of contacts actuated by centrifugal force

Single-Phase Induction Motors—Shaded-Pole Motors

The shaded-pole induction motor is a single-phase induction motor that uses a different method to start the rotor turning.

In this motor, a moving magnetic field is produced by constructing the stator in a special way. The motor has projecting pole pieces and portions of the pole piece surfaces are surrounded by a copper strap called a *shading coil.* The pole piece with the strap in place is shown below.

The strap moves the field back and forth across the face of the pole piece in the following manner. As the alternating stator field starts increasing from 0°, the lines of force expand across the face of the pole piece and cut through the strap. A current is induced in the strap, which generates a field to oppose the cutting action of the main field. Therefore, as the field increases to a maximum at 90°, a large portion of the magnetic lines of force is concentrated in the unshaded portion of the pole (diagram 1 below). At 90° the field reaches its maximum value. Since the lines of force have stopped expanding, no emf is induced in the strap, and no opposing magnetic field is generated. As a result, the main field is uniformly distributed across the pole as shown in diagram 2.

From 90° to 180°, the main field starts decreasing or collapsing inward. The field generated in the strap opposes the collapsing field, which concentrates the lines of force in the shaded portion of the pole face as shown in diagram 3. Thus, from 0° to 180°, the main field has shifted across the pole face. From 180° to 360°, the main field undergoes the same change that it did from 0° to 180°, but in the *opposite* direction. Its motion will be the same during the second half cycle as it was during the first half cycle.

The motion of the field produces a weak torque to start the motor. It is because of this weakness of the starting torque that shaded-pole motors are only built in small sizes for driving small devices such as fans, switches, and turntable motors. A variation of the shaded-pole motor uses a hardened magnet steel rotor, and the rotor turns synchronously with the field. This motor is called a *warren synchronous motor* and is used to operate clocks and timing devices.

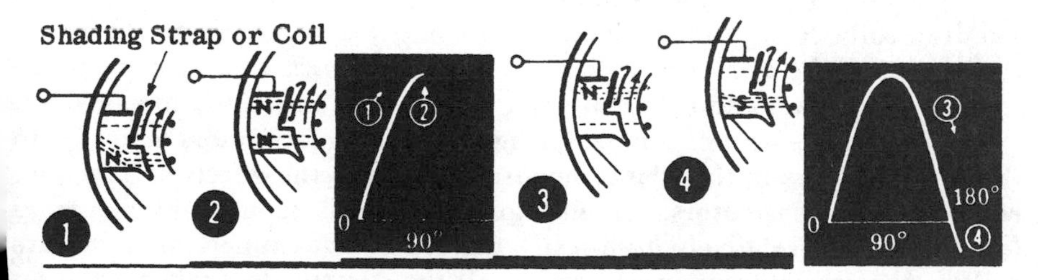

Since the direction of rotation of the field is determined by the position of the shading pole, these motors are not usually reversible. The shaded-pole motor is simple and inexpensive but has very low starting torque.

AC Series Motors—Universal Motors

You have learned that if the dc current through a series motor field and armature is reversed, the direction of rotation remains unchanged. When ac is applied to a series motor, the current through the armature and field change simultaneously, and therefore, the motor will rotate in one direction.

The number of field turns in the ac series motor is smaller than in the dc series motor. This is necessary in order to decrease the reactance of the field so that the required amount of current will flow. In addition, laminated construction is used throughout the magnetic circuit to minimize hysteresis and eddy current losses.

The characteristics of the ac series motor are similar to those of the dc series motor. It is a varying speed machine, with low speeds for large loads and high speeds for light loads. The starting torque is also very high. Very large ac series motors are used in locomotives, subways, and similar applications. In such cases, the motor performance improves as the frequency is reduced, since the reactances are correspondingly reduced. For many large series motor applications, the line frequency is 25 Hz.

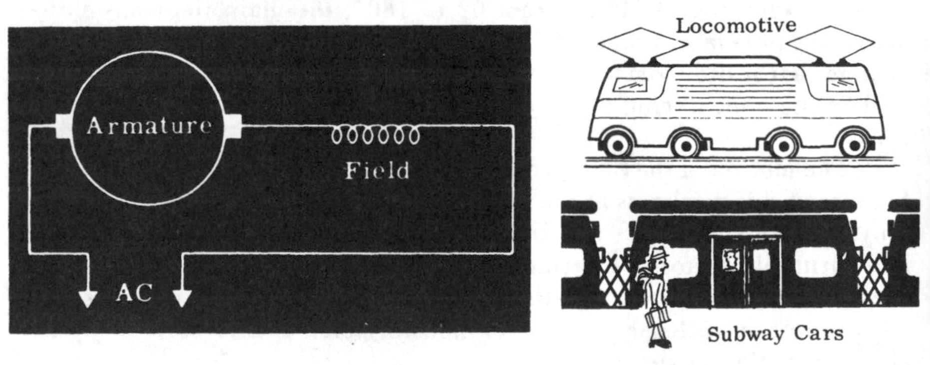

The reactance of the armature is reduced by using a compensating winding embedded in the pole pieces. A compensating winding connected in series with the armature is *conductively compensated*; while a compensating winding that is short-circuited upon itself is *inductively compensated*.

The arrangement of compensating windings in ac motors is similar to that in dc motors in that the compensators reduce the effects of armature reaction. In large motors, commutator sparking is reduced by *preventive leads*. These are relatively high-resistance leads that connect the armature coils to the commutator in order to limit the current flow when adjacent segments are shorted by the brushes.

Low-power (fractional horsepower) series motors can operate on ac *and* dc. These motors, called *universal motors*, are commonly used in small household appliances where high torque or high speed are required. Fo

AC Series Motors—Universal Motors (continued)

this reason they are used in electric drills, sanders, and similar power tools because of their high torque. They are used in centrifugal pumps, vacuum cleaners, and similar devices because of their high running speed. These motors do not use compensating windings.

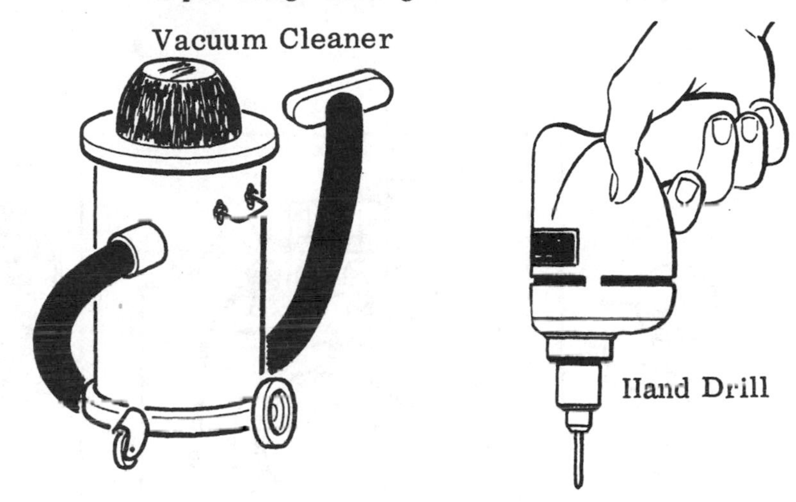

Universal motors are also very useful because (unlike induction motors) their speed can be readily controlled, since it depends on the input voltage and the load. You will learn about this in the next section of this volume.

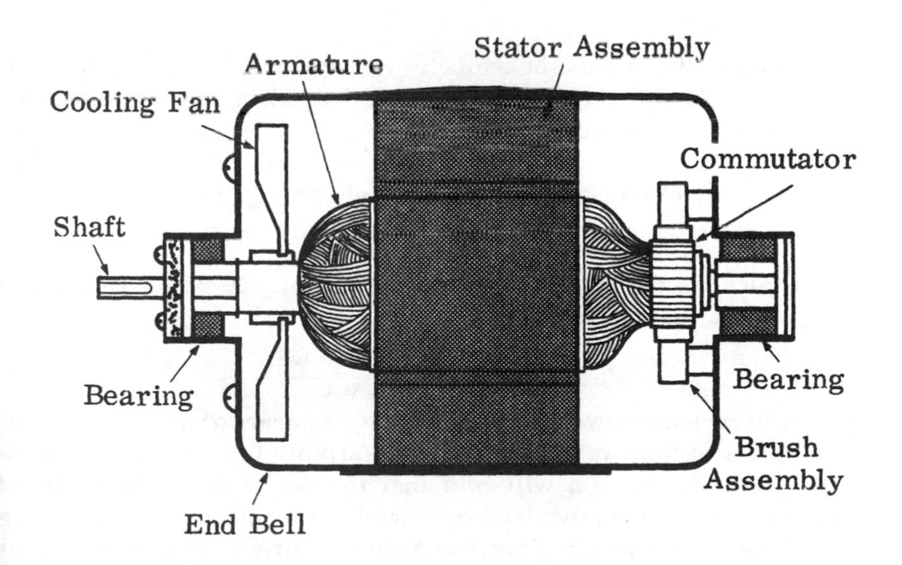

Experiment/Application—AC Motor Power Factor

Most small single-phase ac motors have a lagging power factor between 0.6 and 0.8 (60% to 80%). You can see this readily by measuring the volt-ampere input and comparing it to the reading of a wattmeter properly connected into the circuits. Suppose you had the circuit shown below:

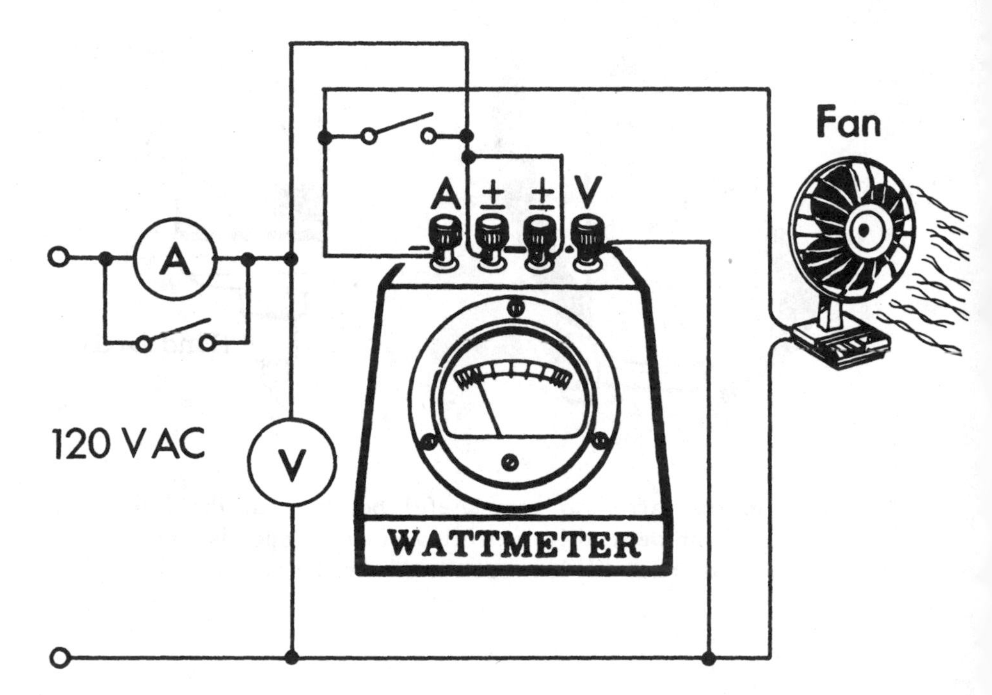

The switches shown across the ammeter and the current coil of the wattmeter must be closed—shorting out these devices when the motor is started to avoid damage to them. After the motor is running, open the switches and read the voltage, current, and wattmeter reading. You can calculate the volt-amperes (VA) as the product of voltage and current.

$$VA = E \times I$$

You can calculate the power factor as the ratio of true power (wattmeter reading) to the volt-amperes.

$$\text{Power factor} = \frac{\text{True power}}{\text{VA}}$$

You can improve the power factor by placing a capacitor across the motor terminals. Try various values of capacitor (around 0.5-2 μF) and recalculate the power factor; you will find that the power factor is improved because the capacitor draws leading current to compensate the lagging current drawn by the motor. Thus, the power factor will approach unity as the right capacitor is chosen.

Review of AC Motors

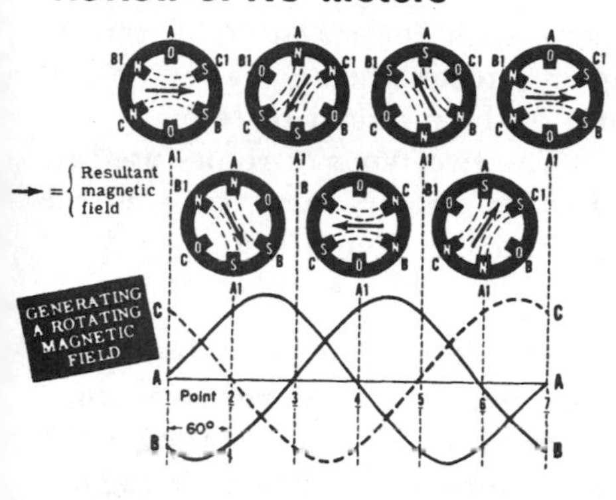

1. ROTATING MAGNETIC FIELD—If three windings are placed 120° apart around a stator frame, and three-phase ac is applied to the windings, the magnetic fields generated in each of the three windings will combine into a rotating magnetic field.

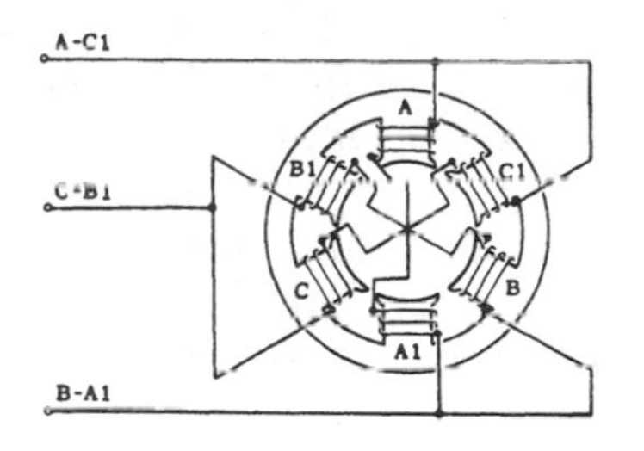

2. THREE-PHASE STATOR The magnetic fields generated in three-phase ac motors are 120° apart. At any given instance, these fields are combining to produce one resultant field, which acts on the rotor. The rotor turns because the magnetic field rotates. The speed of rotation (N) of the field is given by

$$N = 120F/P$$

This is called the synchronous speed.

Squirrel Cage Winding

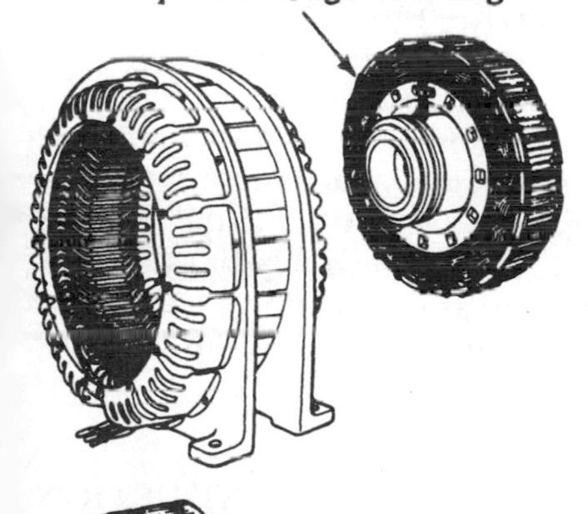

3. SYNCHRONOUS MOTOR—A synchronous motor uses a stator to generate a rotating magnetic field and an electromagnetic rotor that is supplied with dc. The rotor is a magnet and is attracted by the rotating stator field. This attraction will exert a torque on the rotor and cause it to rotate synchronously with the field.

Synchronous motors are not self-starting and must be brought up to near synchronous speed before they can continue rotating by themselves.

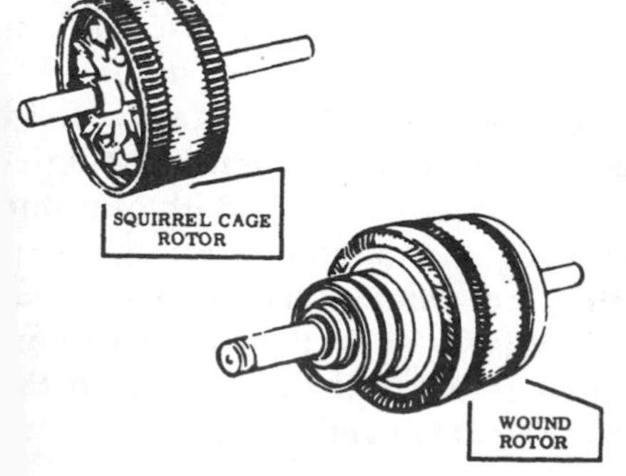

4. INDUCTION MOTOR—The induction motor has the same stator as the synchronous motor. The rotor is different in that it does not require an external source of power. Current is induced in the rotor by the action of the rotating field cutting through the rotor conductors. This rotor current generates a magnetic field which in-

Review of AC Motors (continued)

teracts with the stator field, resulting in a torque being exerted on the rotor and causing it to rotate.

The two types of rotors used in induction motors are the squirrel cage and the wound rotor.

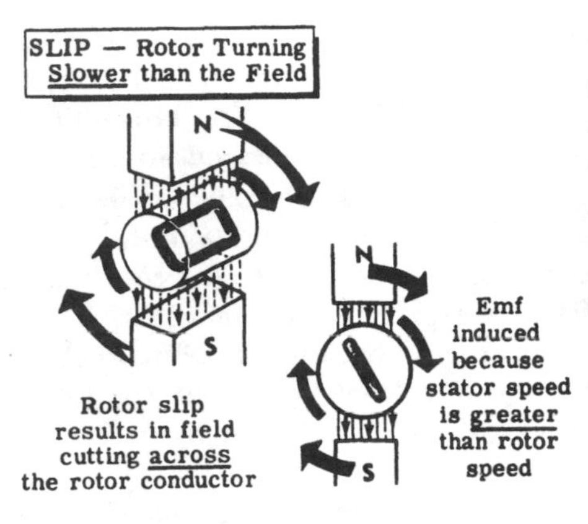

5. SLIP—The rotor of an induction motor rotates at less than synchronous speed so that the rotating field can cut through the rotor conductors and induce a current flow in them. This percentage difference between the synchronous speed (N_s) and the rotor speed (N_r) is known as slip (S). Slip varies very little with normal load changes, and the induction motor is therefore considered a constant-speed motor.

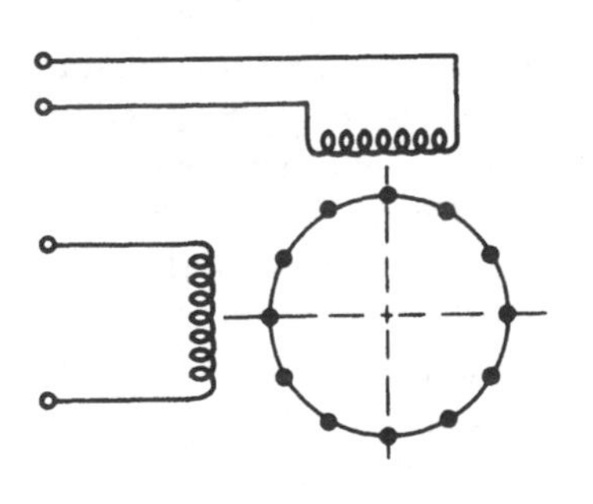

6. TWO-PHASE INDUCTION MOTOR—Induction motors are designed for three-phase, two-phase, and single-phase operation. The two-phase stator generates a rotating field by positioning two windings at right angles to each other. If the voltages applied to the two windings are 90° out of phase, a rotating field will be generated the same as for a three-phase system.

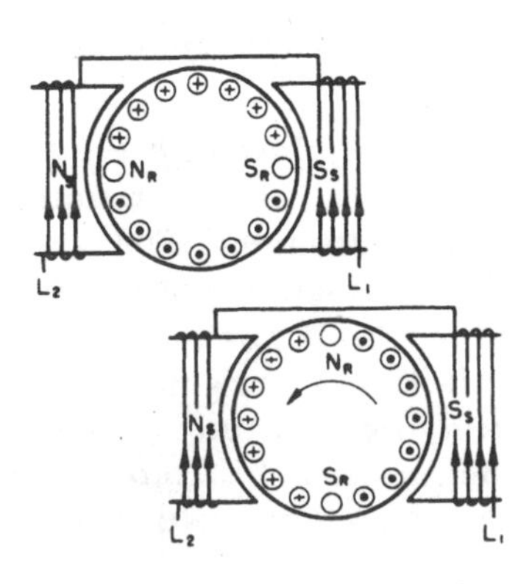

7. SINGLE-PHASE INDUCTION MOTOR—A single-phase induction motor has only one stator winding; therefore the magnetic field generated does not rotate. A single-phase induction motor with only one winding cannot start rotating by itself. Once rotation has been effected a field is set up in the rotating rotor which is 90° out of phase with the stator field. Together these two field

Review of AC Motors (continued)

produce a rotating field that keeps the rotor in motion. Single-phase induction motors are named by their method of starting.

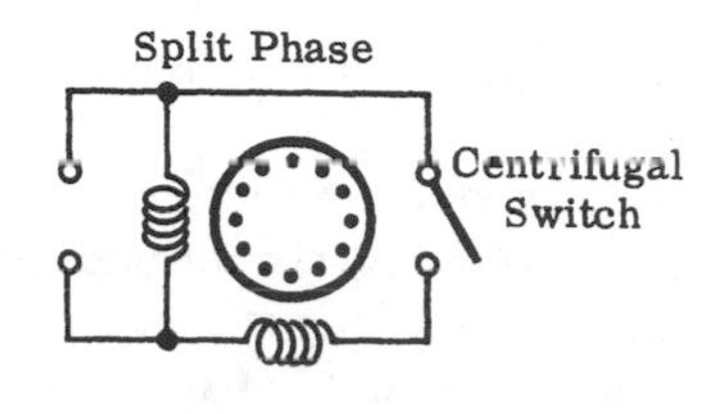

8. STARTING SINGLE-PHASE INDUCTION MOTORS—To make a single-phase motor self-starting, a starting winding is often added to the stator. If this starting winding is used with an arrangement so that the current in the starting winding will be out of phase with the current in the running winding, a rotating magnetic field will therefore be generated and the rotor will rotate. Once the rotor comes up to speed, the starting winding circuit can be switched out, and the motor will continue running as a single-phase motor.

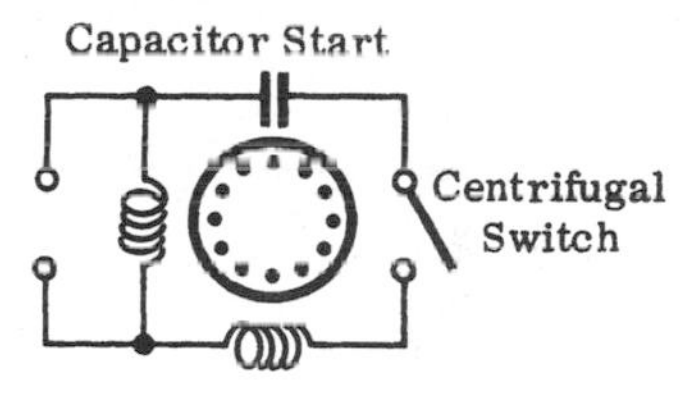

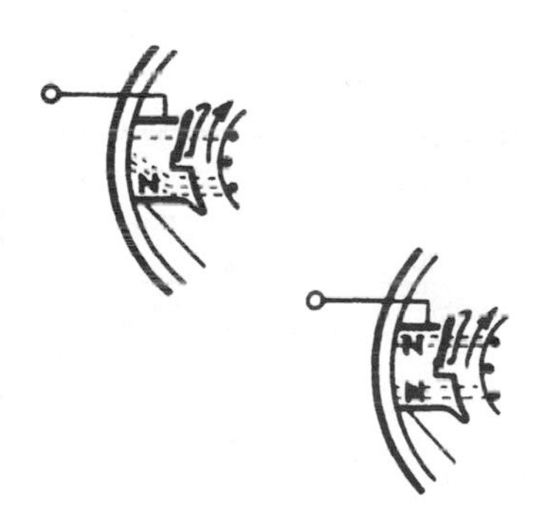

9. SHADED-POLE INDUCTION MOTOR—In this motor, a section of each pole face in the stator is shorted out by a metal strap. This has the effect of moving the magnetic field back and forth across the pole face. The moving magnetic field has the same effect as a rotating field, so the motor is self-starting when switched on.

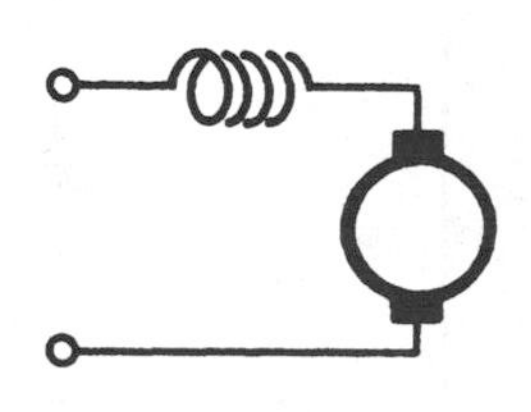

10. SERIES MOTOR—UNIVERSAL MOTOR—The series ac motor has the same properties as the dc series motor—high torque at low speeds and high-speed operation under light loads. It is used in every size motor, from electric railway locomotives to small home appliances. A version of the series motor called the universal motor can be used with ac or dc.

Self-Test—Review Questions

1. Describe (with sketches) how a rotating magnetic field is generated using a three-phase stator and power source.
2. Repeat the above for a two-phase stator and power source.
3. What are the essential parts of a synchronous motor? How does it operate (three phase)?
4. What are the essential parts of an induction motor? How does it operate (three phase)?
5. Define synchronous speed and slip. If a two-pole motor is connected to a 60-Hz source and rotates at 1,800 rpm, is it an induction motor or a synchronous motor?
6. Why will a single-phase induction motor run once started?
7. Sketch and describe the various methods of starting single-phase induction motors. Explain the operation of the centrifugal switch.
8. Sketch and show the principle of operation of the shaded-pole induction motor. Why is it self-starting?
9. How does the ac series motor differ from the dc series motor?
10. Choose a single-phase motor for each application listed below:
    a. Medium-torque start, constant-speed running
    b. High-torque start, constant-speed running
    c. High starting torque, variable-speed operation
    d. Light duty, self-starting

Learning Objectives—Next Section

Overview—Now that you have learned about ac generators and motors, you are ready to learn about control circuits for these devices and how to troubleshoot them.

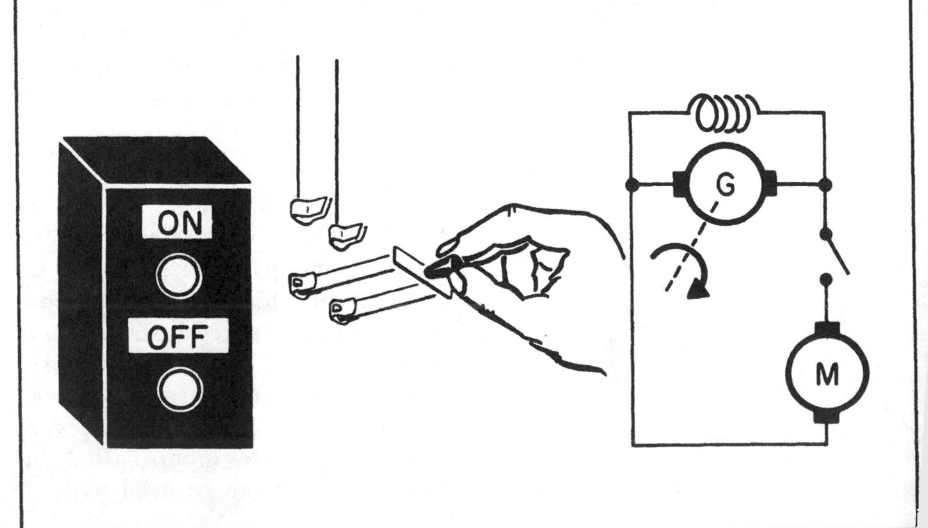

Introduction to AC Controls—The Controller

The basic circuit arrangement for ac system control is the same as for dc systems. The only significant difference is that ac relays and contactors use a laminated core in the electromagnet, and the current flow in the relay winding depends on its ac impedance (resistance and inductance) rather than on dc resistance alone. The dc control circuit shown on page 5-67 could readily be used to control a single-phase ac circuit. To extend it to a three-phase circuit, additional contacts are added to the contactor, as shown below. Note also that a commonly used symbol in industrial schematic diagrams (a circle with the relay designation in it) is also used.

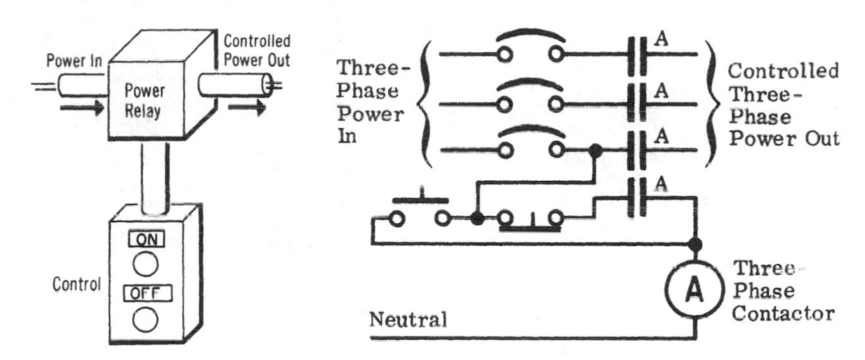

Notice that the neutral is *never* switched and *never* fused. If the neutral *should* become open, the voltages for devices connected across the line and neutral would then be in series from line to line. The voltage across them would be in inverse proportion to their impedance, and excessive voltage could appear across some devices, causing them to burn out.

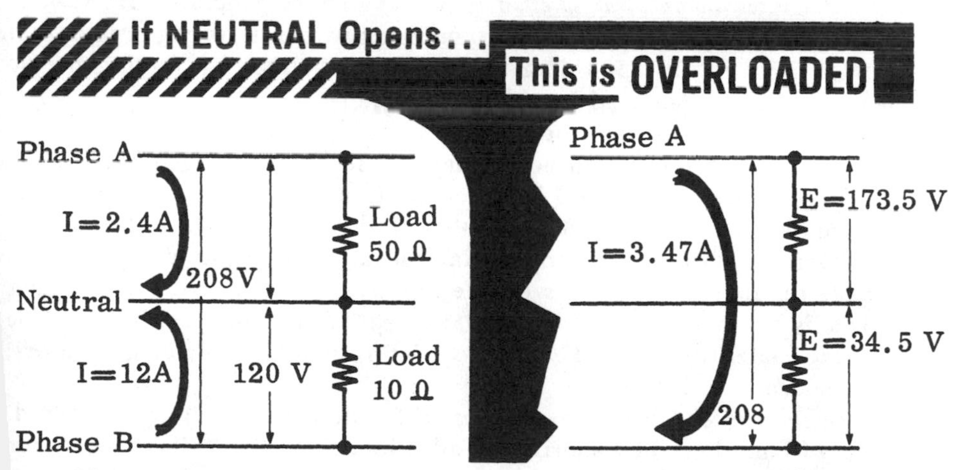

If you were to examine an actual installation, you would see that three-phase circuits are usually arranged so that all circuits open and close simultaneously. The purpose of this arrangement is to avoid overloading the other circuits if one phase should be shorted or become open.

Alternators—Voltage Regulators

The magnitude of the voltage generated by an alternator is varied by varying the field strength (field current). The speed is usually held constant for constant frequency.

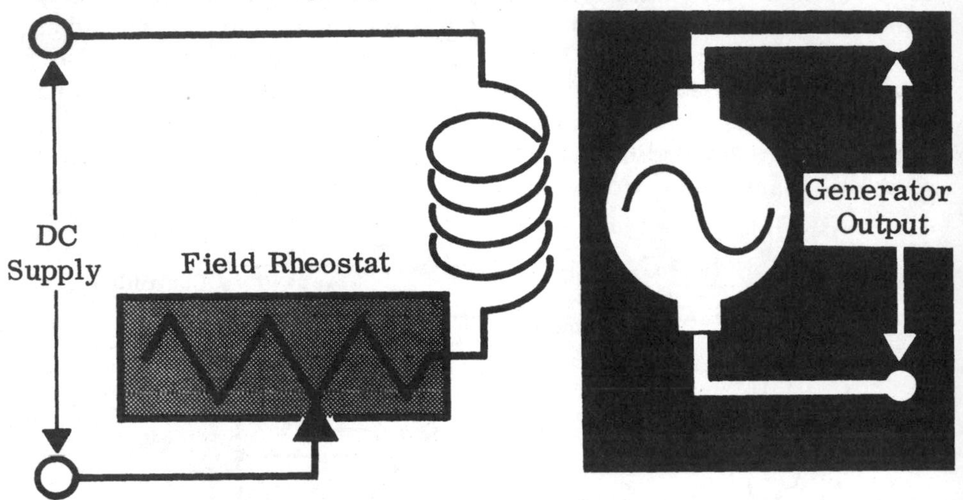

In an alternator, just as in a dc generator, the output voltage varies with the load. As you know from your earlier study, there is another voltage drop in the windings, called the IX_L drop, in addition to the IR drop. This IX_L drop is due to the inductive reactance of the windings. Both the IR drop and the IX_L drop decrease the output voltage as the load increases.

The change in voltage from a no-load to a full-load condition is called the *voltage regulation of an alternator,* just as in a dc generator. Ac voltage regulation is calculated using the same formula for dc voltage regulation on page 5-37. A constant voltage output is maintained by varying the dc field strength as required by changes in load.

A manually operated rheostat in the field circuit can be used to vary the field strength for this purpose. But as the load on many generators fluctuates continuously, it would be necessary to make too many adjustments. Automatic regulating devices have therefore evolved that use the varying load current to bring about appropriate variations in the field current. The voltage regulator associated with the alternator in a car battery charging system is an example of an automatic regulator. This regulator, to be described later, operates like the dc vibrating regulator for dc generators, described on page 5-70.

Carbon piles (similar to those used for dc generators) are also used. They will be described shortly. In addition, there are voltage regulator schemes based on the variable transformer that you studied earlier Saturable transformers are used, as well. Thus, ac voltage regulators can b divided into those that regulate the dc field and those that directly vary th ac between the generator and the load.

The Semiconductor Rectifier Diode

Before proceeding further in your study of ac systems and controls, you must learn about an important semiconductor device—the *rectifier diode.* As you remember, you learned something about semiconductor rectifier diodes and circuits in your study of ac meters in Volume 3 of this series. The rectifier diode consists of a junction of semiconductor materials that conduct electricity very well in one direction but very poorly in the other. They are usually made of silicon, selenium, or germanium material.

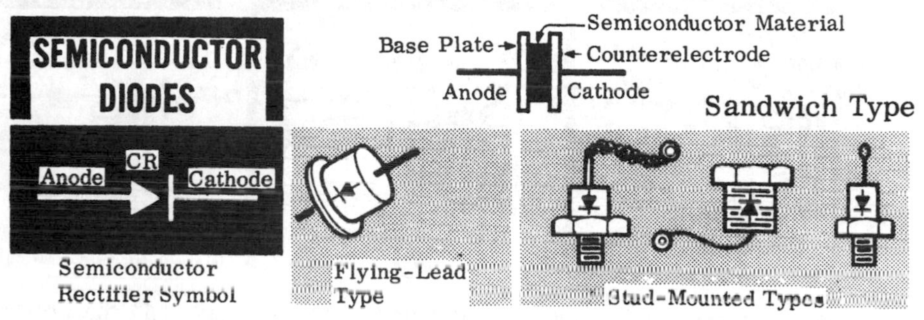

Most diodes are marked with the semiconductor symbol to indicate the polarity, so that when the arrowhead side, or anode, is relatively *positive* with respect to the cathode, the diode conducts well. And when the arrowhead side is *negative* with respect to the cathode (cathode is less negative than the anode), the diode looks like an open circuit.

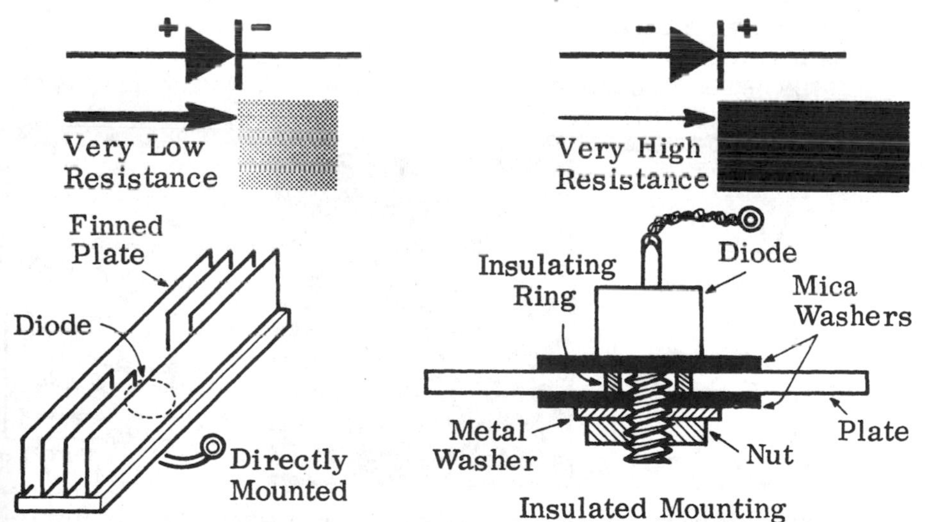

In many designs, the diodes are arranged to screw directly into metal blocks (often finned) so as to radiate heat. In other designs, the diode case is insulated from the metal frame by thin mica washers. Rectifier diodes handling power get quite hot because there is some voltage drop in them when they are conducting. Semiconductor diodes are relatively frail, and even a momentary short circuit will often burn them out.

Rectifier Diode Circuits

Semiconductor rectifier diodes are usually used in either the *full-wave* or *full-wave bridge* configuration. They are called *full-wave circuits* because they use both halves of the sine wave. In most power applications, the bridge circuit is used because a center-tapped input is not required. The full-wave rectifier is shown schematically below.

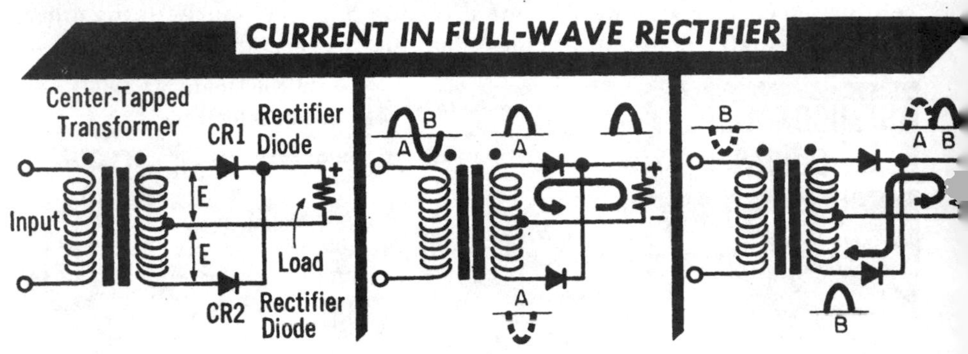

As indicated during the first half of the ac sine wave (A) the upper diode (CR1) conducts. CR2 does not conduct because the voltage across it is of the wrong polarity. During the second half of the ac wave, CR2 conducts and CR1 does not. The result is the pulsating dc voltage shown, with the peak of the pulsating voltage proportional to the transformer input voltage.

The full-wave bridge circuit can be used without a balanced line, but has the disadvantage of needing four diodes.

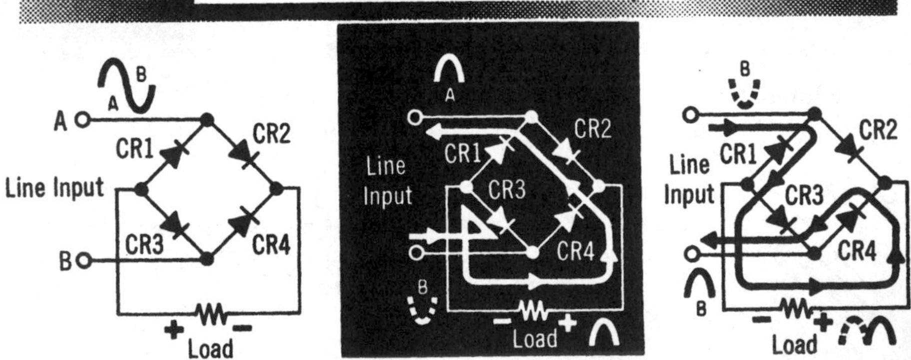

In the full-wave bridge circuit (shown above) when line A is positive with respect to line B, current flows through CR3 to the load through CR2 to the other side of the line. On the other half of the ac cycle (line B is positive with respect to line A), current flows through CR1 to the load, through CR4 back to the line. Again as you can see, a pulsating dc waveform results, since the direction of the current through the load is the same for both halves of the ac cycle.

Automatic Voltage Regulators—Field Controlled

A typical automatic ac voltage regulator that controls the field of an alternator is shown below. It uses the carbon pile, which you studied with dc generators, as the control device. Other dc voltage control devices can also be used in the same circuit.

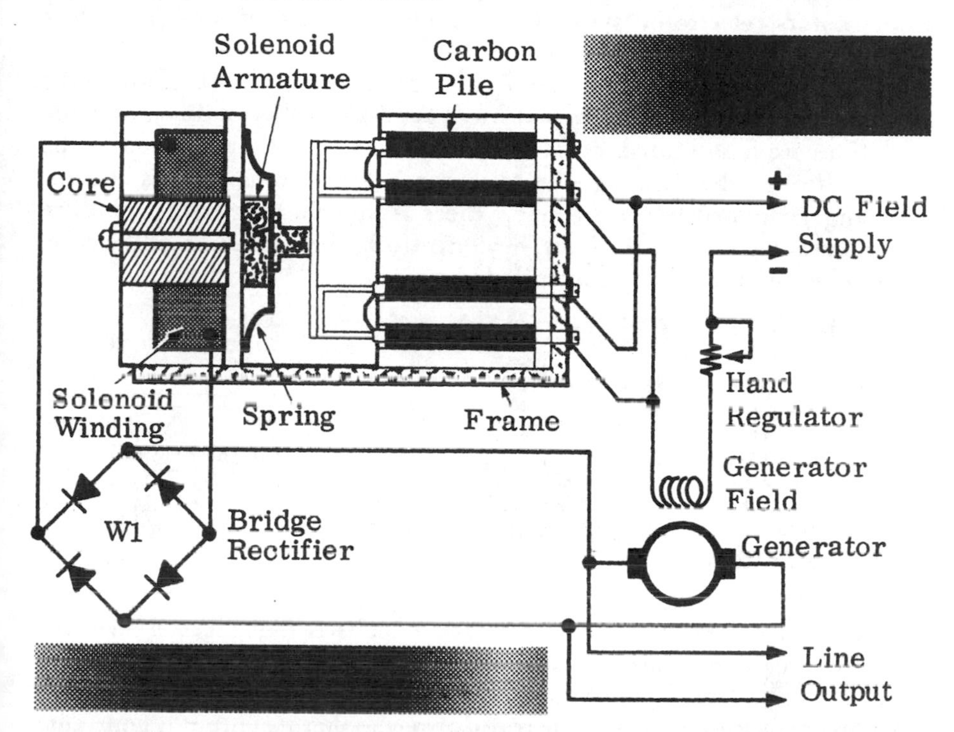

The ac output voltage of the generator is applied to a bridge rectifier circuit (W1, in the diagram above). The dc output of the rectifier bridge is connected across the solenoid winding. The armature of this solenoid is attached to a ring which is held in contact with carbon piles by a spring. The carbon piles are connected in series with the generator field supply.

Assume that the output voltage of the generator increases. There will be a higher voltage applied to the bridge rectifier, more current will flow through the solenoid winding, and the solenoid armature will be attracted toward the core of the solenoid. This will relieve the pressure on the carbon piles, increasing their electrical resistance and reducing the generator field current. The output voltage will then fall toward its original value. Conversely, when the generator voltage decreases, the voltage applied to the solenoid will decrease, allowing the spring to compress the carbon pile. Its electrical resistance is thereby decreased and its field current and the voltage output increased.

A hand-operated rheostat is connected into the field circuit for the purpose of adjusting the output voltage to the value at which it is required to be maintained by the regulator.

Line Voltage Regulators

Line voltage regulators work to keep the line voltage constant as the load fluctuates. Since, in general, ac is supplied via transmission lines, often remote from the generator, some means is often necessary to regulate line voltage for sensitive applications. This is usually only required for small motors, electronic systems, or other apparatus where small line voltage changes are detrimental.

The most common of the line regulators for small installations are the *constant voltage transformers.* These devices are transformers with a saturable core. If a core is saturated, no more lines of force can be generated as the current through the winding tries to increase. Since no more lines of force are being generated by the primary, there is no changing field to induce voltage with the secondary. Thus, by controlling the saturation of the core, the voltage output can be controlled.

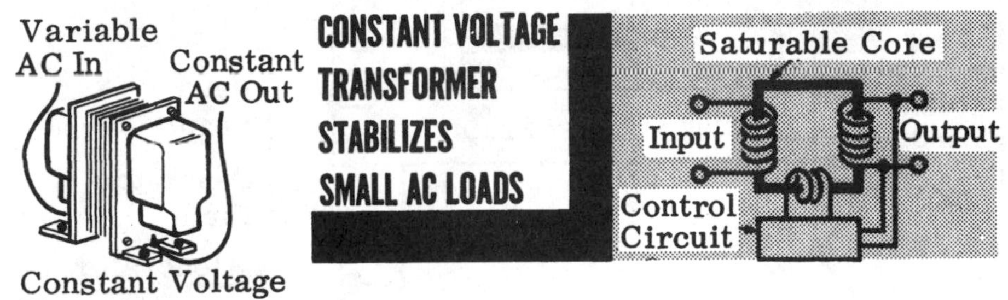

Another common regulator for larger loads is the *induction regulator.* Induction regulators use the variable transformers that you studied in Volume 4 of this series. They are motor driven by a suitable control circuit with the purpose of adjusting the transformer so that its output is constant.

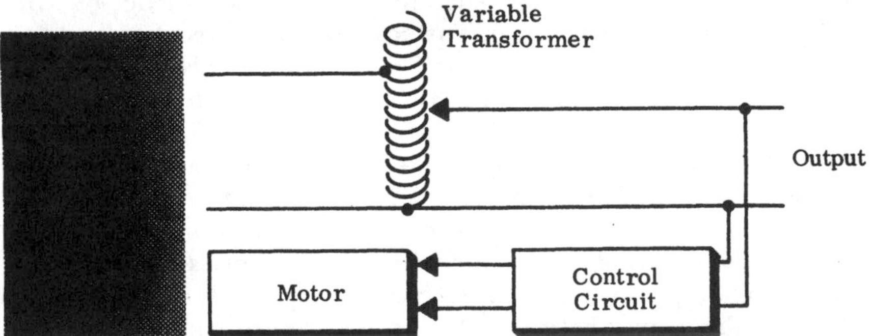

When the voltage is low, the control circuit activates the motor to move the arm on the variable transformer so as to increase the voltage. When the voltage is high, the arm is moved to decrease it. In very large regulators, the windings of the variable transformer are arranged so that they can be moved relative to each other, thus avoiding moving contacts.

The Automobile Alternator System

Most modern automobiles use an alternator and rectifier combination to supply the dc necessary for battery charging and operating the other electrical parts of an automobile. This is done because an alternator can be built to give useful output at low engine speeds that, with the elimination of a commutator, results in a more reliable unit.

Many variations on the alternator design exist. There are small units with permanent magnet rotors such as those used in motorcycles and marine outboard engines and much larger units, such as the three-phase alternator with a separately excited field that is used today in most automobiles, trucks, buses, etc. The two major components of these systems are the *alternator-rectifier* and the *voltage regulator.*

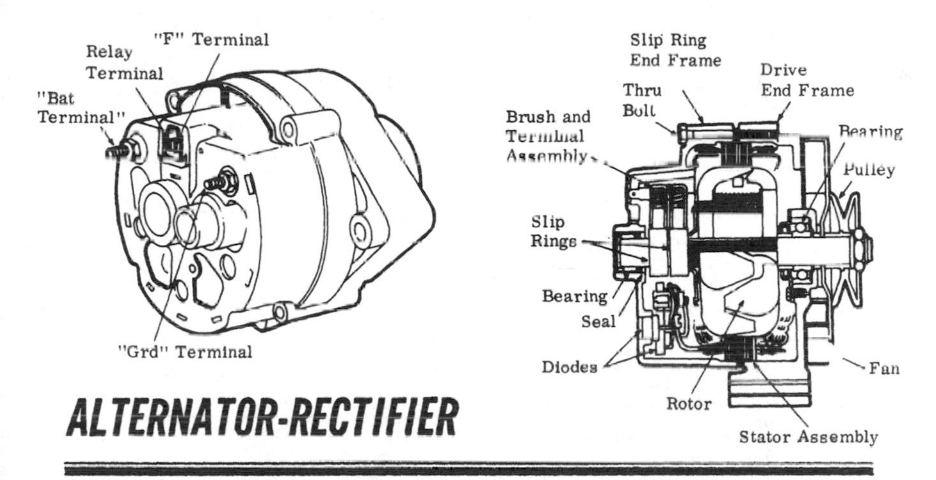

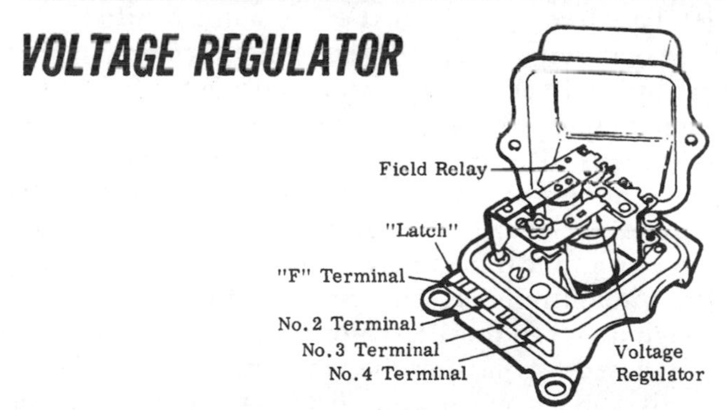

As you can see in the drawings above, the alternator-rectifier has the rotor and stator elements that you learned about earlier in your study of ac generators. The voltage regulator is of the vibrating reed type that you studied in connection with the voltage control of dc generators.

A schematic diagram of a typical automobile alternator system is shown at the top of the next page.

The Automobile Alternator System (continued)

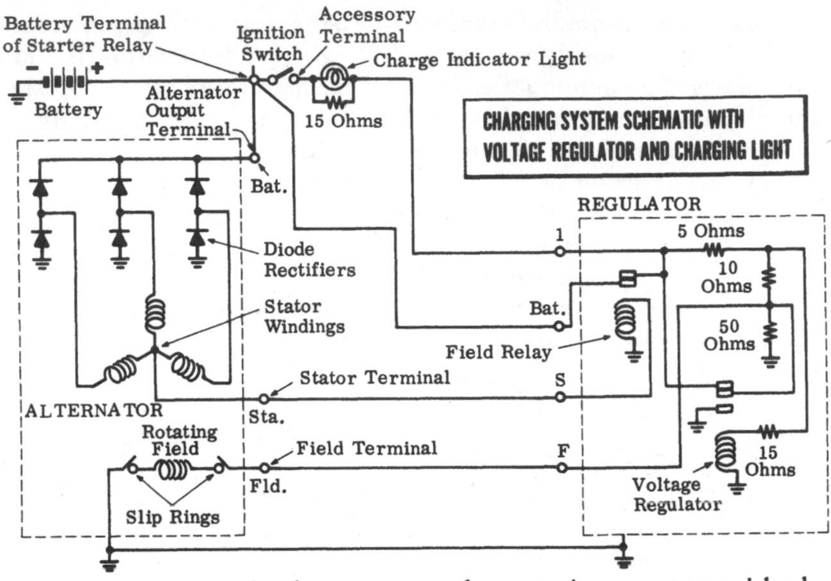

As you can see, it shares many elements in common with the dc automobile electrical system that you studied earlier. The alternator stator is three-phase, wye-connected and is fed to the three-phase full-wave bridge rectifier to produce pulsating dc. The battery acts as a large capacitor and stabilizes the pulsations, so that almost ripple-free output is obtained. Ripple reduction is also helped by the three-phase circuit. As you learned in the discussion of the single-phase bridge, there are two pulsations of dc per cycle. In a three-phase full-wave system, there are six pulsations per cycle and they overlap. Thus, the ripple is reduced just as more commutator segments reduce ripple. The dc output from the alternator-rectifier is controlled by varying the excitation to the field wound on the rotor and fed via slip rings. The vibrating reed voltage regulator controls the field current and thus the alternator output. The regulator switches the resistances in and out of the field circuits. The field relay closes when the alternator starts to deliver power and applies the full battery voltage to the voltage regulator. It also bypasses the charge indicator light when closed. Before the alternator delivers power, the field current flows through the indicator light and the parallel 15-ohm resistor. The indicator lights as a result of the current flow through it. When the alternator is delivering power, however, the field relay is closed—bypassing the indicator light—and it goes out. Since this can only happen when these field relay contacts close, it provides an indication that the electrical system is operating properly. As you will note, the rectifier diodes prevent the battery from discharging through the alternator, so no reverse current cutout is necessary.

Partially (or completely) solid-state regulator systems are now in use but are beyond the scope of a study of basic electricity as they require some knowledge of basic electronics.

Induction Motor Starters—Direct

Most small ac induction motors (1 horsepower or less) are started by connecting them directly across the line. This is possible because the inductive reactance of the windings helps to limit current to a safe value. In addition, the starting current does not go through a commutator or other relatively delicate circuits. The starting current for small induction motors is usually between five and ten times the full-load current, so the protective devices on the line have to be able to handle these peak currents of short duration called *transients.* Lines with motors on them are protected by slow-blowing fuses (designed to open only if an overload is sustained), or by circuit breakers that incorporate a thermal delay, so that they do not trip on a short-term overload, such as starting the motor. Some of these overload relays have manual reset so that the circuit must be reactivated manually before the motor can be started.

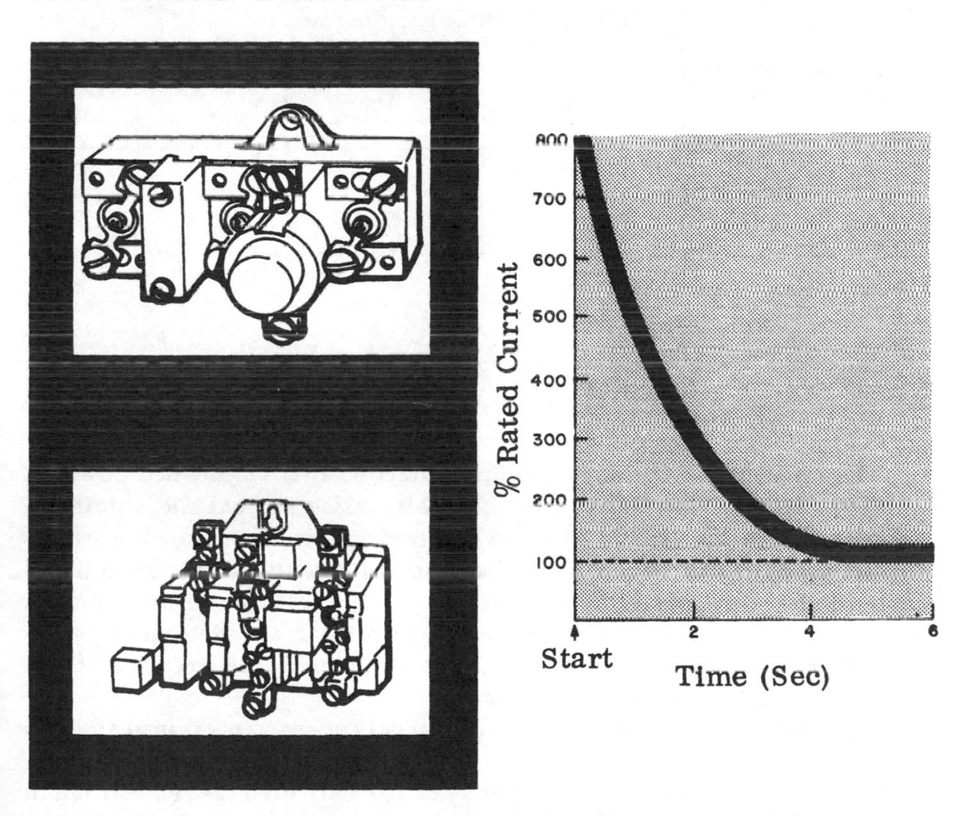

A motor control starter circuit where the full voltage is applied would have components and a circuit as shown on page 5-137 for a three-phase motor or a single-phase motor of moderate size (up to 5 horsepower). Small single-phase (or three-phase) motors, designed to handle the full line voltage at start, may have only a manual switch in the line to turn the motor on or off.

Induction Motor Starters—Reduced Voltage

Larger induction motors use reduced voltage starting just as you learned in your study of dc motor starting. One major difference is that the reduced voltage can be obtained from an autotransformer as well as from resistors. As with dc starters, time delay relays can be used to switch out the voltage-reducing elements when the motor speed is high enough.

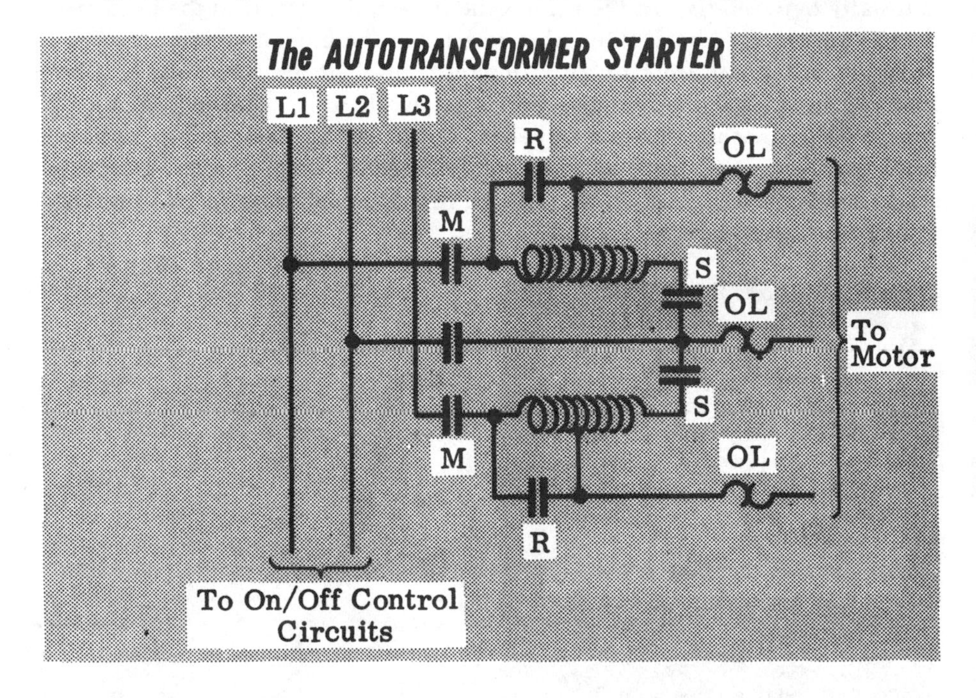

Contacts M are the main circuit contactors that close when power is applied. At that time relay contacts S also close so that the autotransformers are in the circuit, and the reduced voltage from the autotransformer is applied to the motor via the overload cutouts (OL). After a timed interval, the relay contacts R are closed, applying full voltage to the motor. At the same time, contacts S are opened to remove power from the autotransformers. Note that only two autotransformers are necessary for a three-phase delta-connected system.

A variation of this circuit uses resistances (or reactances) instead of an autotransformer.

In some cases, multistep time delay circuits are used (as you learned in your study of dc motor starting). These have the advantage of reducing the starting current even further than for a single-step starter.

Another starter that you may see in three-phase systems uses the motor windings themselves for starting by bringing out both ends of each of the three windings. For starting, the three windings are wye connected, while for running, the three windings are delta connected. In this way, the starting voltage (per phase) is $1/\sqrt{3}$ of the running voltage.

AC Motor Speed Controls

As you learned, synchronous and induction motors are essentially constant-speed devices, although some motors have extra sets of poles that can be activated to get two or sometimes even three speeds. This method is often used in devices like multispeed electric fans. To get variable speed, however, particularly in smaller motors, a series or universal motor is used with an appropriate voltage controller.

The simplest way to do this is with a rheostat or variable transformer.

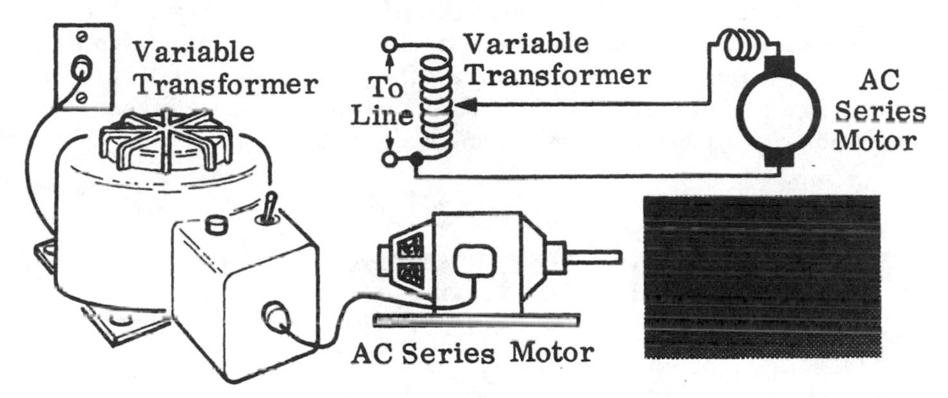

A less expensive way to control small motors is to use a solid-state voltage controller. This is essentially the same device used in lamp dimmers. Many small power tools, such as drills, sanders, routers, etc., use these devices in conjunction with a universal motor to attain speed control. We shall study these devices and how they are used on the next page.

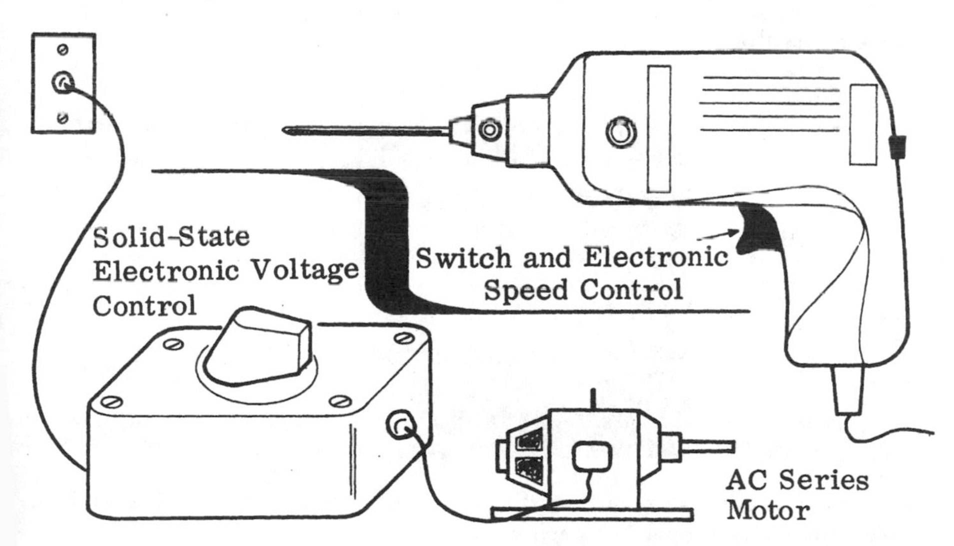

Another method sometimes used involves the conversion of ac to dc via rectifiers and the use of a dc motor with an appropriate field rheostat or automatic controller.

Electronic Motor Speed Controllers

Solid-state devices have become very readily available for ac voltage control. While there are many different configurations, we shall study some simple ones to see how these devices work. To do this, we shall need to know about some semiconductor devices in addition to the diode.

The *silicon-controlled rectifier* (SCR) is like a diode with an extra terminal called *the gate.* Unlike the rectifier diode, it will not conduct when the proper polarity is present unless it is triggered (turned on) by means of an appropriate voltage applied momentarily to the gate input. Once the SCR has started to conduct, however, it will continue to do so (even if the gate signal is removed) until the current flowing through it falls to a low value. (In a practical ac circuit this occurs when the sine wave goes to the other polarity). This is shown schematically below.

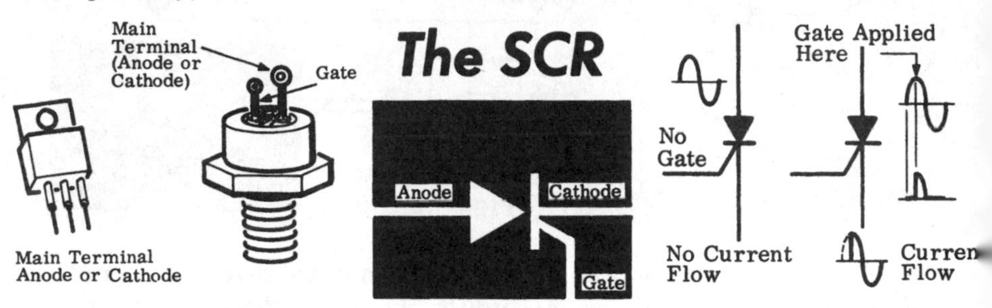

As you can see, the SCR only conducts when its anode is positive and when it is triggered, so that at best only a half wave of current can flow. While this problem of getting a full wave of current can be solved with a circuit containing an additional SCR and some diodes, etc., it has been solved essentially by putting all of the hardware into a single package called a *thyristor* or *triac.*

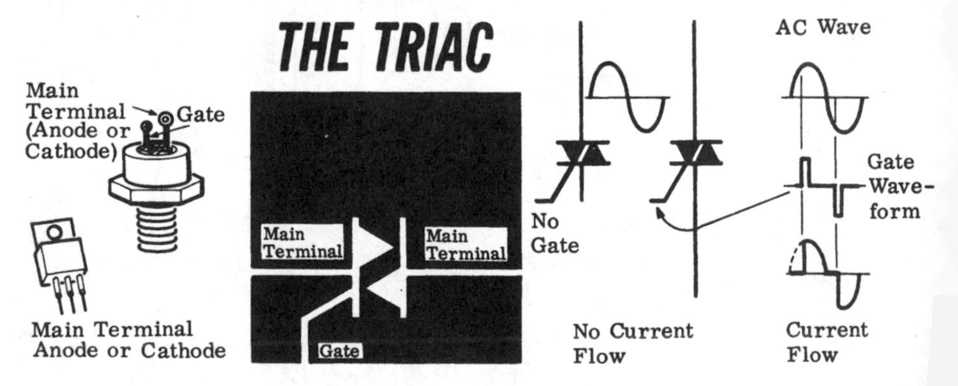

As you can see from the diagrams, the triac will conduct in *either* direction upon proper triggering.

As the diagram indicates, the average power delivered to the load is determined by where the gate occurs with respect to the ac wave. This is done merely by changing the triggering point.

Electronic Motor Speed Controllers (continued)

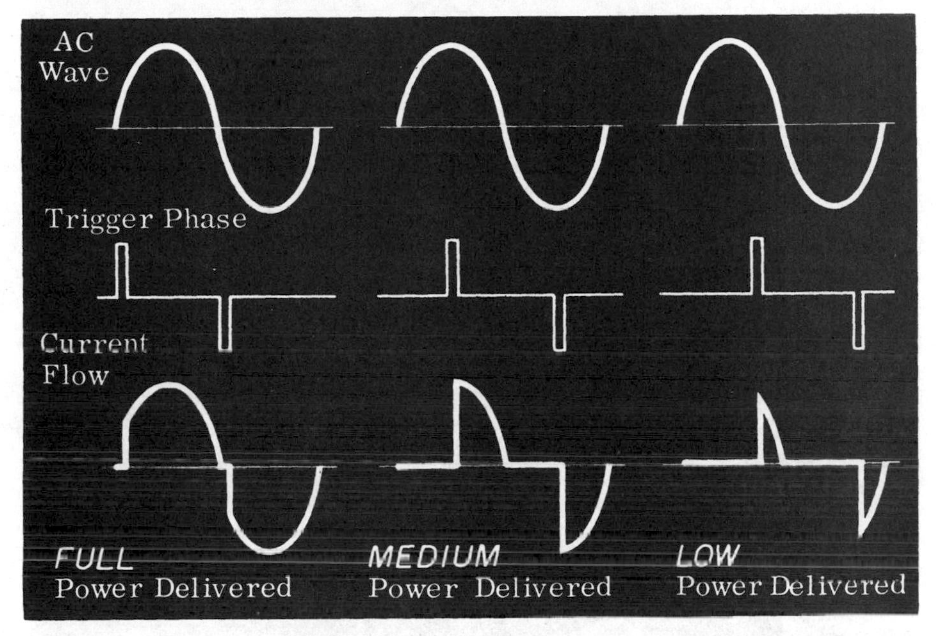

The triggering point can be varied by a simple RC circuit coupled to a device called a *diac*. The diac consists of a pair of special back-to-back diodes that conduct very suddenly when the voltage exceeds a threshold value. They are designed specifically to trigger triacs. Since they handle almost no power, they are physically small.

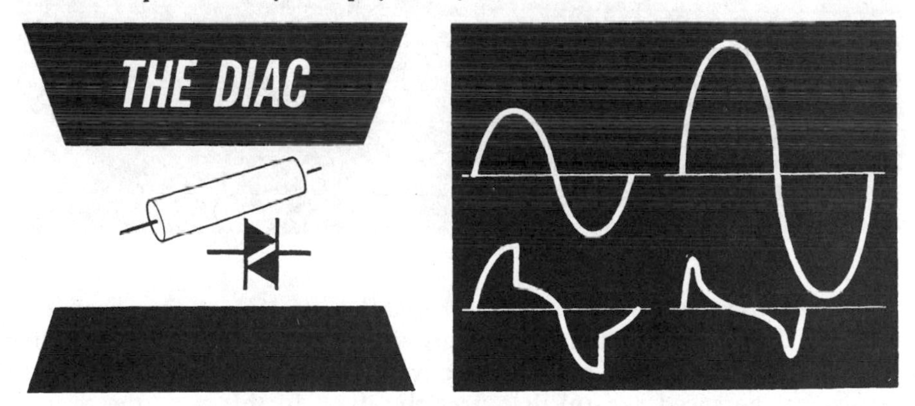

As you can see, the symbol for a diac is like a triac without the gate lead. The waveforms also show that the breakover point occurs at the same voltage. Therefore, if the voltage is varied, the breakover point moves closer to the beginning of a cycle at high voltage input and moves more quickly into the waveform as the voltage decreases. By using an RC network, a combination of voltage change and phase shift allows for a wide range of control.

Electronic Motor Speed Controllers and Lamp Dimmers

The circuit for a universal electronic motor speed controller is shown below.

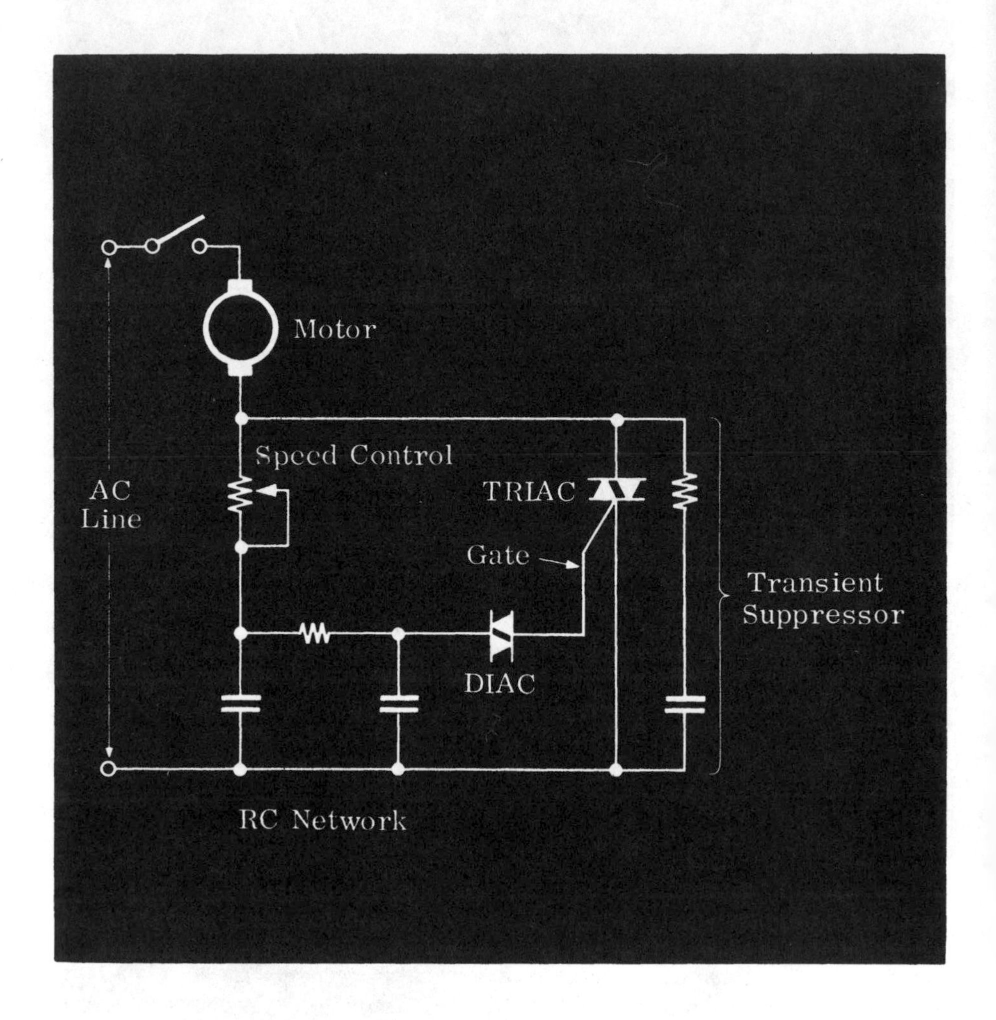

The circuit shown is typical of small universal motor speed controllers. The RC circuit and the diac provide trigger pulses to the triac that let power into the motor on each half cycle, the amount depending on the setting of the speed control that fires the diac. In this way, the average voltage to the motor, and therefore the motor's speed, is controlled. The transient suppressor is used to eliminate interference in radio and TV caused by the rapid switching.

A lamp dimmer has an essentially identical circuit except that the range of control is more limited. Since a lamp does not usually glow at all until about 50% of the rated voltage is applied, the controls are arranged to start functioning, not at 0 volts, but at about 50 to 60 volts.

Experiment/Application—Lamp Dimmers

The operation of lamp dimmers and ac series motor speed controls can be readily observed by the following experiment. Suppose you connected a lamp and lamp dimmer as shown below. An autotransformer is necessary in this experiment to avoid the hazard of electric shock since the ground side of the oscilloscope must be connected to one side of the line.

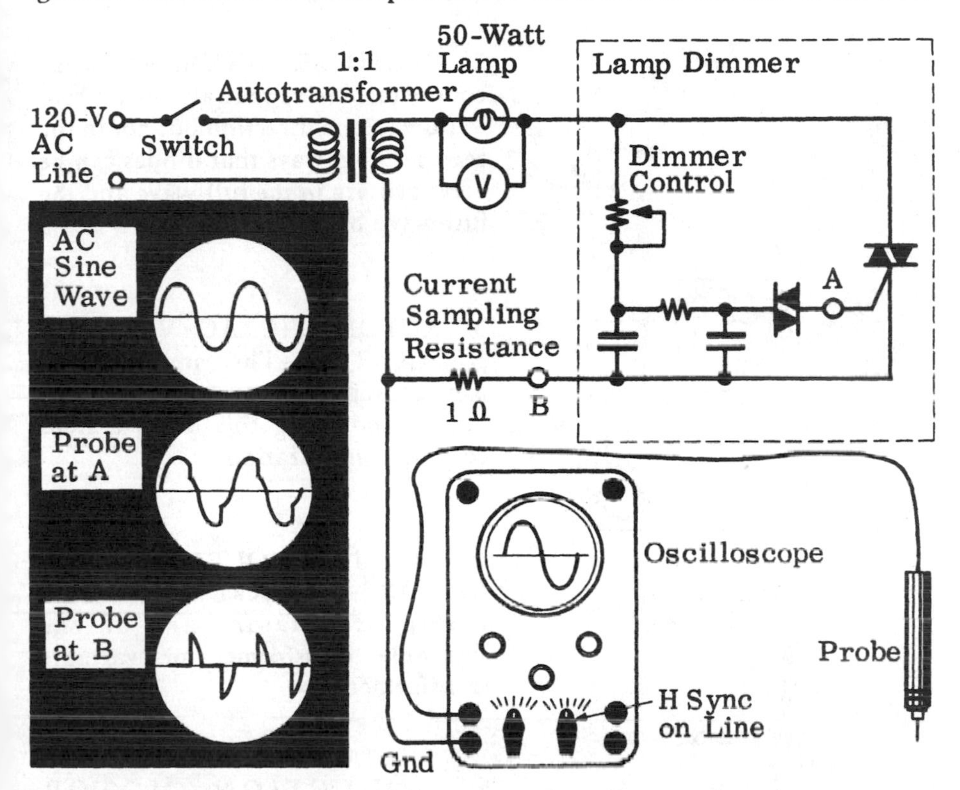

Observe that the lamp brightness changes as the control is varied. Note also that the voltage changes in proportion to the control setting. Connect the oscilloscope as shown and set the horizontal sweep on line synch and adjust the sweep length for two ac cycles. Connect the scope probe to point A on the lamp dimmer and observe the firing point (abrupt voltage change) and how it varies as the dimmer control is varied. Note these points on the oscilloscope screen for several known positions of the dimmer control.

Connect the scope probe to point B to observe the current through the sampling resistance. You will see that the current begins to flow at the point of firing for the known lamp dimmer control positions and continues till the ac sine wave crosses the zero axis. Although the current flows in pulses and the voltage is intermittent (circuit is closed twice each cycle), the meters cannot respond to these pulses and integrate or average to show the average value of current or voltage.

Review of AC Systems and Controls

1. AC CONTROLLER—A circuit containing switches and contactors for controlling an ac line input or output.

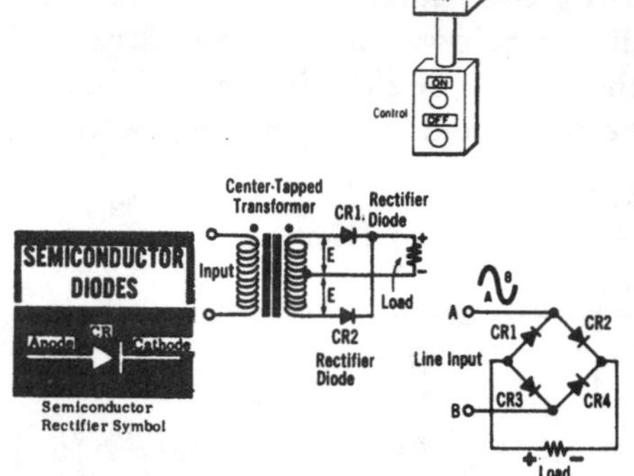

2. RECTIFIER DIODE—A semiconductor device that allows current to flow in one direction but not in the reverse. Two ways that diodes can be arranged are in the full-wave and the full-wave bridge circuits.

3. AUTOMATIC AC VOLTAGE REGULATOR—The ac generator can be regulated by rectifying the output and using this dc in a field control configuration.

4. AC LINE VOLTAGE REGULATORS—Typical small line voltage regulators are of the saturable transformer or variable transformer type.

5. INDUCTION MOTOR STARTING—Small induction motors are put directly on line. Large motors use autotransformers or some other scheme to lower the voltage on starting.

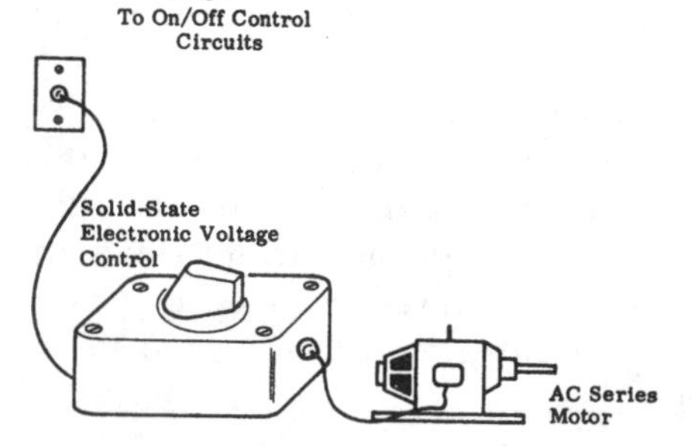

6. AC MOTOR SPEED CONTROLS—Ac motors in controlled speed applications are usually universal or ac series motors. Speed is controlled by varying the motor voltage with a variable transformer or a solid-state controller.

Self-Test—Review Questions

1. Draw a schematic diagram of a simple three-phase ac controller. Describe its operation.
2. Why should you not fuse or switch the neutral in a three-phase wye system?
3. How are ac generators regulated? Draw a schematic diagram of a simple regulator and describe its operation.
4. What is a semiconductor diode? How does it function in a circuit? Draw a single-phase full-wave bridge and describe its operation.
5. Why are line voltage regulators commonly used in ac circuits?
6. Draw a simple schematic diagram of an automobile alternator system. Describe how it works. Why are these generally three-phase?
7. How are large ac motors started? What special characteristics do fuses and circuit breakers have when used in motor starting circuits?
8. Why are series or universal motors used when variable speed is needed?
9. Draw a schematic diagram of a simple series motor speed controller. Describe how it works.
10. Draw a schematic diagram of an electronic speed controller. Describe how it works to control motor speed and how each component works.

Learning Objectives—Next Section

Overview—You have now essentially completed your study of basic electricity. As a final application of your knowledge, you will study the troubleshooting of ac circuits in this, the last, section.

Troubleshooting AC Controls

Aside from simple switches, ac control systems generally use a contactor to supply power. There are several interlocks on these contactors to prevent application of power whenever there is a problem. These interlocks are usually connected in series with the OFF button (normally closed). The simplest way to check them is with a trouble light. A trouble light is nothing more than a neon bulb and a resistor in series in a holder.

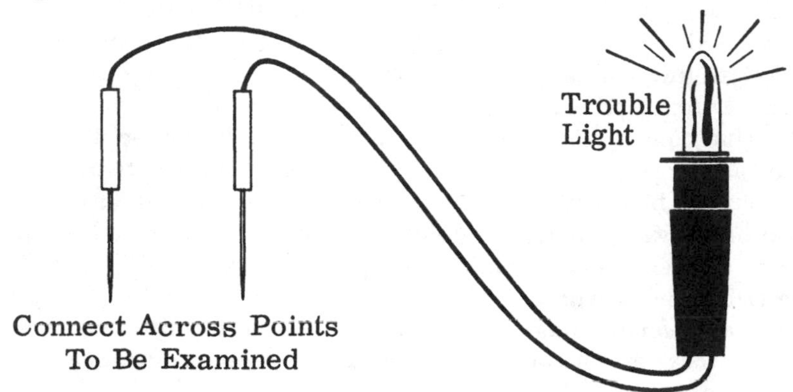

The trouble light can quickly tell you whether voltage is present at each point in the circuit. If an interlock or other circuit is open, the trouble light will glow.

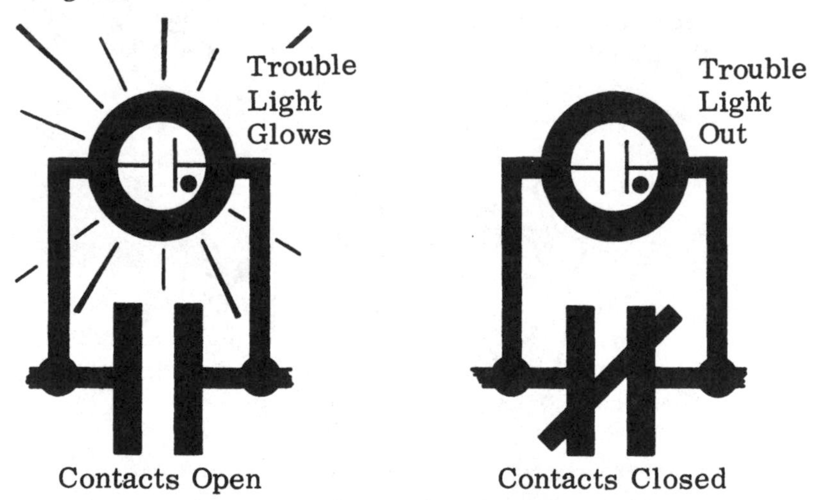

Another very useful device for troubleshooting is the clamp-on ammeter. It allows the measurement of current in an ac line without opening the line. Most of these meters have a voltmeter built in as well, and they can be extremely useful in tracing power through a circuit. Tracing voltages and currents through an ac circuit by means of the trouble light and the clamp-on ammeter will allow you to localize the difficulty.

Troubleshooting AC Circuits—Semiconductors

In addition to circuit tracing, you may encounter semiconductor devices that you suspect are defective. Testing a diode is easy with an ohmmeter. If a diode shows a low resistance with the ohmmeter connected across it and a high resistance with the diode reversed, the diode is probably all right. On the other hand, if the diode shows either high or low resistance in *both* directions, it is open or shorted, respectively, and must be replaced.

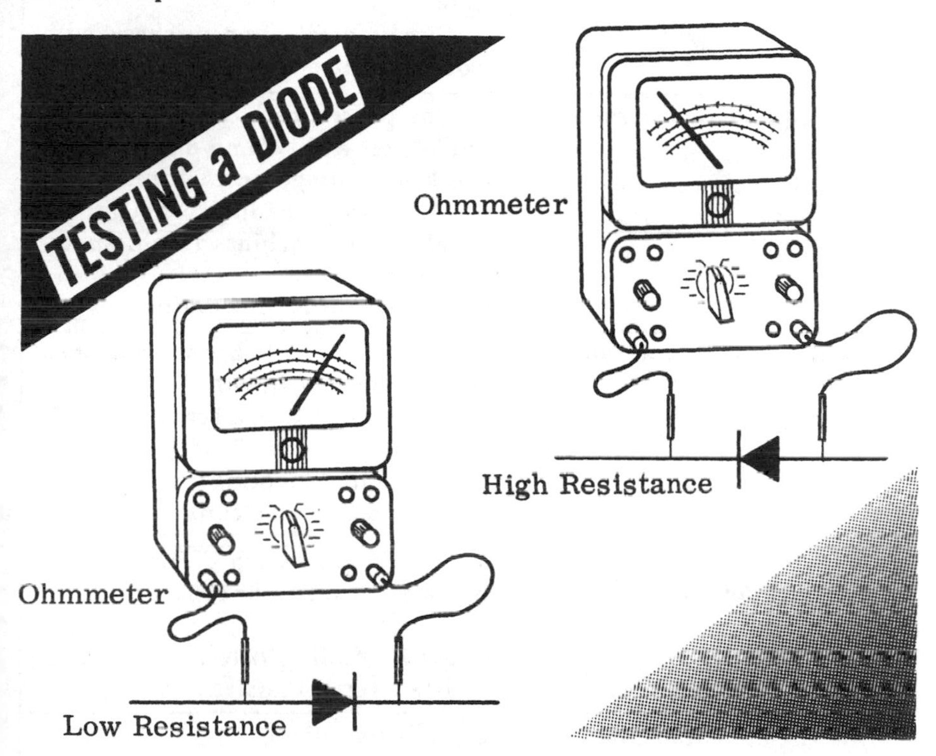

Triacs and diacs are much more difficult to test. Normally, with an ohmmeter both devices should show an open circuit in *both* directions. If they do not, they are defective. On the other hand, there is no simple way to determine whether the open circuit observed indicates that the device is good or that it is blown out.

As discussed earlier, there is no substitute for careful observation. Look for hot devices, lack of voltage at a point, broken or frayed wiring, unintentional short circuits, etc. Being systematic is very important. It is often helpful to prepare a simple schematic or distribution diagram with the pertinent voltages and currents noted. As mentioned earlier, there is no substitute for experience and care. Use what you know about circuits to evaluate whether the voltage and current measurements are *realistic*. By thinking about the problem and making appropriate observations, all system problems can be solved.

Troubleshooting Chart—AC Electrical Machines

For mechanical problems and problems with universal motors, you should refer to the dc troubleshooting section.

Symptom	*Probable Causes*
1. No power	a. Interlock open; check with trouble light b. Circuit breaker open; check with trouble light
2. Three-phase motor runs hot	a. One phase open; check power input/phase with clamp-on ammeter; check for voltage on each phase b. Mechanical loading problem; see dc electrical machines troubleshooting chart c. Shorted turns or windings
3. Single-phase motor runs hot	a. Centrifugal switch does not open starter winding in run b. Load is excessive; mechanical difficulty with motor or load (see dc machines troubleshooting chart) c. Shorted turns or coils
4. Single-phase motor hums, but will not start	a. Centrifugal switch defective b. Starter winding or capacitor defective c. Main winding defective d. Load very excessive
5. Alternator output low	a. Defective diodes b. Field current too low; defective regulator c. Excessive load or short circuit
6. Universal motor with speed control not running	a. Open triac b. No triggering of triac; defective diac, pot, or central circuit c. Defective motor (try substituting a lamp for motor to see whether voltage control works)
7. Universal motor with speed control—full speed only	a. Shorted triac b. No control of triac triggering; check components

Overview—Basic Electronics

You now have completed your course in basic electricity and should have a good working knowledge of the principles of operation of dc and ac circuits and equipment.

For those of you who will be going on to the study of basic electronics (*Basic Solid-State Electronics*) this will provide you with needed background. Whereas in electricity you were concerned primarily with the transfer of power, in electronics you will be concerned primarily with transfer/communication of information—voice, music, video, data, etc. You'll find that your understanding of electricity will provide you with the foundation that will help you master electronics.

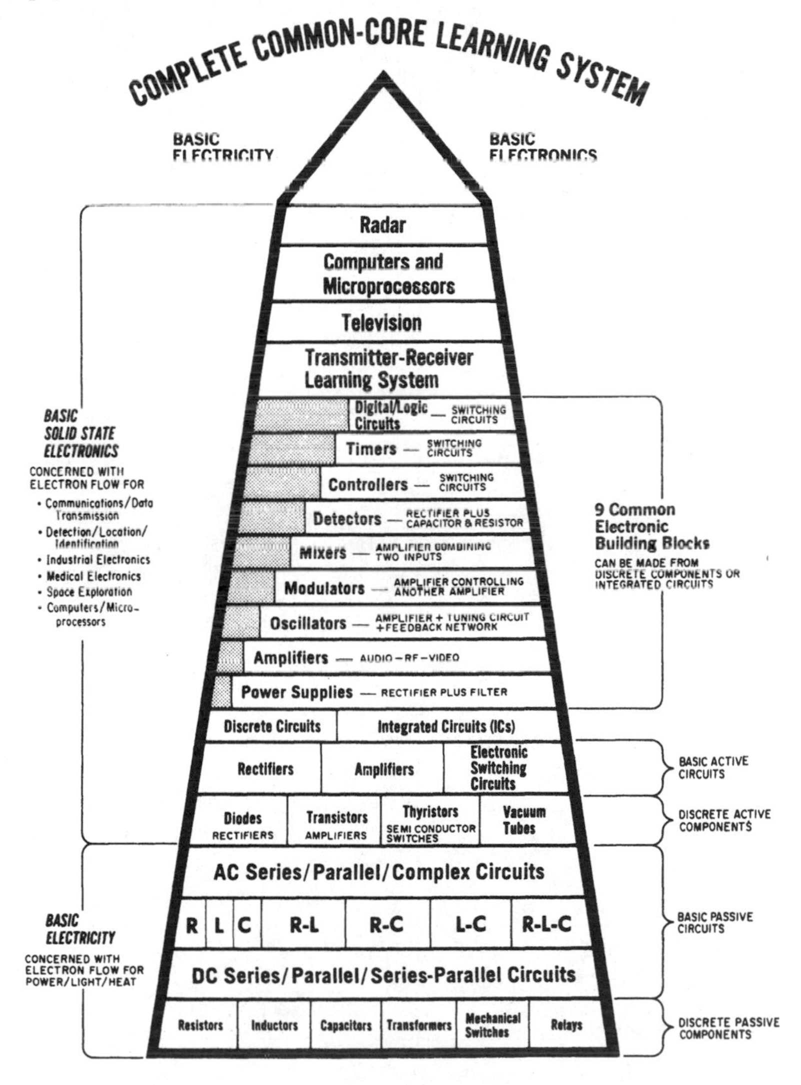

Overview—Basic Electronics

CUMULATIVE INDEX

(Note: The first number in each entry identifies the *volume* in which the information is to be found; the second number identifies the *page*.)